NUMERICAL METHODS

(FOR B.SC./M.SC./B.TECH/M.TECH STUDENTS)

NUMERICAL METHODS

(FOR B.SC./M.SC./B.TECH/M.TECH STUDENTS)

MATHEMATICS MADE EASY BY

PROF. K. SAMBAIAH

Formerly Professor of Mathematics

Kakatiya University, Warangal,

Telangana, India

White Falcon Publishing

www.whitefalconpublishing.com

Numerical Methods
Prof. K. Sambaiah

www.whitefalconpublishing.com

ISBN - 978-93-89932-25-6

Preface

The book *Numerical Methods* has been written to meet the requirements of graduate and postgraduate students of engineering, mathematics and computer science. One of the important features of this book lies in introducing the procedures like algorithms to implement each of the numerical method which are given in the book. Also some shortcut methods have been given to solve the boundary value problems. Many examples have been given in the chapters to inculcate the concepts of numerical methods in the students.

The book is divided into thirteen chapters and every chapter is divided into some sections based on the subject matter. Chapter 1 comprises discussion regarding various types of errors. The various operators and finite differences are defined in Chapter 2. Interpolation, numerical differentiation and integration are thoroughly discussed in the Chapters 3, 4 and 5, respectively. The various methods of obtaining the numerical solution of first order differential equation have been discussed in Chapter 6. The curve fitting and approximations of functions have been discussed in Chapters 7 and 8. Methods of solving algebraic and transcendental equations along with the convergence of some important methods have been given in Chapter 9. In Chapter 10, the numerical methods of solving the system of linear equations are presented. The different methods of finding the eigen values and eigen vectors of a matrix have been discussed in Chapter 11. The solutions of difference equations have been discussed in Chapter 12. Finally, the solutions of boundary value problems have been discussed in Chapter 13.

Various shortcut methods have been introduced in this book to assists the students with the solutions of boundary value problems. Numerous examples, illustrations and exercises have been given to facilitate the understanding of theoretical concepts. The proofs of all the theorems have been presented in detail. Important theories of various mathematicians have been presented in a clear and lucid manner. In a nutshell, the approach of this book has been kept student friendly and easy to comprehend.

Acknowledgements

I would like to express my gratitude to all those authors whose works have inspired me to write this book. I am thankful to the reviewers who provided detailed reviews of one or more chapters. I am also thankful to my students R. Srinivas, T. Sree Laxmi and E. Rama. A special mention must be made of wife Radha and son K. Srinivas for their support. I am also thankful to management of White Falcon Publishing Solutions LLP for publishing the book so nicely and elegantly.

The readers are requested to give their feedback regarding any error or typos they might have spotted. It would help in making the book more enrich and better.

K. Sambaiah

0
Preliminaries

In this chapter, for the convenience of reader, some of the results and concepts pertaining to mathematics are presented, which are useful to understand the contents of subsequent chapters of this text book and also helpful in solving some typical problems.

0.1 POLYNOMIAL EQUATION

If n is a non-negative integer, $a_0, a_1, \ldots, a_n$ are real numbers and $a_0 \neq 0$, then $f(x) = a_0 x^n + a_1 x^{n-1} + \ldots + a_{n-1} x + a_n$ is called a polynomial in x of degree n and $f(x) = 0$ is called a polynomial equation. The numbers $a_0, a_1, a_2, \ldots, a_n$ are called coefficients. We consider polynomial equations with real coefficients in our discussions in subsequent chapters. A polynomial with zero degree is called a constant polynomial. For example, 5 is a constant polynomial and $3x^4 - 4x^3 + 7x + 8$ is a polynomial of degree 4.

0.2 ROOT OR ZERO OF AN EQUATION

A number a is said to be a root of an equation $f(x) = 0$ if $f(a) = 0$.

For example, 2 is a root of $x^2 - 5x + 6 = 0$ and $x = \pi/6$ is a root of $2 \sin x - 1 = 0$.

0.3 REMAINDER THEOREM

If $f(x)$ is a polynomial of degree $n \geq 1$ and $a \in C$ (set of complex numbers), then there exists a polynomial of degree $(n - 1)$ such that

$$f(x) = (x - a) \, q(x) + f(a)$$

Here $q(x)$ is called quotient and $f(a)$ as remainder when $f(x)$ is divided by $(x - a)$.

0.4   FUNDAMENTAL THEOREM OF ALGEBRA

Every non-constant polynomial equation has at least one root.

Theorem:

Every polynomial equation of nth degree where n is a positive integer has n roots and no more.

Note that the n roots need not be distinct, i.e., some roots may be equal for some polynomial equations.

For example, $x^2 - 4x + 4 = 0$ has two roots and they are 2, 2, which are not distinct. The equation $x^2 - 5x + 6 = 0$ has roots 2 and 3, which are distinct.

Theorem:

If $f(x) = 0$ is a polynomial equation of degree $n \geq 1$, then complex roots of it occurs in pairs. This is the case when $f(x)$ has complex roots only. In other words, if $a + i\beta$ is a complex root of $f(x) = 0$, then $a - i\beta$ is also a root of $f(x) = 0$.

Note that this result is valid only when the coefficients of polynomial are real.

Theorem:

If $x = a$ is a root of a polynomial equation $f(x) = 0$, then $(x - a)$ is a factor of $f(x)$ and $f(a) = 0$.

0.5   CONTINUOUS FUNCTION

A function $f(x)$ is said to be continuous at $x = a$ if for given any $\in\, > 0$ $\exists\ a\ \delta > 0$ such that $|f(x) - f(a)| < \in$ for $|x - a| < \delta$.

If $f(x)$ is continuous at each and every point of $[a, b]$, then $f(x)$ is said to be continuous on $[a, b]$.

For example, x^2 and $\sin x$ are continuous functions on IR (set of real numbers). The function $f(x) = \tan x$ is not continuous at $x = \pi/2$.

Note: Every polynomial function $f(x)$ is continuous on IR.

0.6   GEOMETRICAL INTERPRETATION OF A ROOT

Let $f(x) = 0$ be an equation and C be the curve $y = f(x)$. If the curve C and x-axis intersect at $(a, 0)$, then $f(a) = 0$. Thus, a root of $f(x) = 0$ is the abscissa of the point of intersection of the curve $y = f(x)$ and x-axis.

For example, the graph $y = \sin x$ (see Fig. 0.1) intersects the x-axis at $x = 0, \pm\pi, \pm 2\pi, \pm 3\pi, \ldots$ etc.

Thus, the roots of $\sin x = 0$ are $x = 0, \pm\pi, \pm 2\pi, \pm 3\pi, \ldots$ etc.

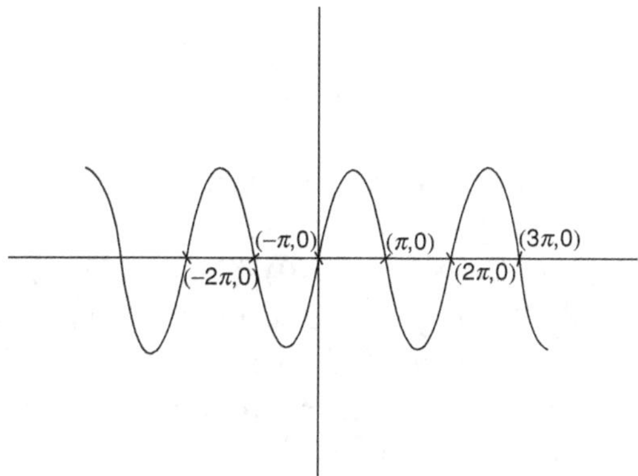

Figure 0.1

Theorem:

If $f(x)$ is a real-valued continuous function on $[a, b]$ and $f(a)f(b) < 0$ (i. e., $f(a)$ and $f(b)$ have opposite signs), then there is at least one or an odd number of real roots of $f(x) = 0$ in $[a, b]$ (see Figs. 0.2, 0.3).

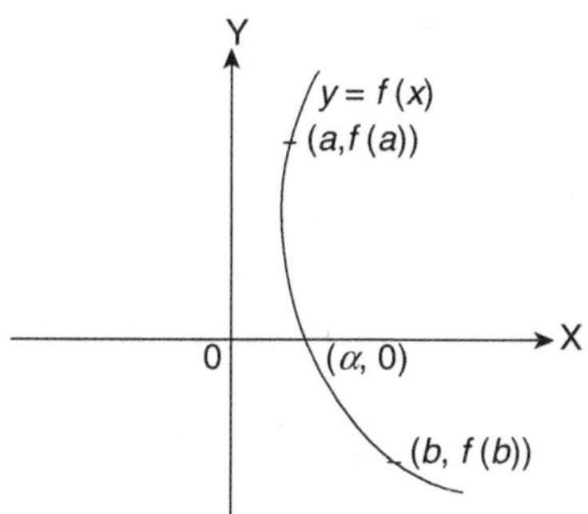

Figure 0.2

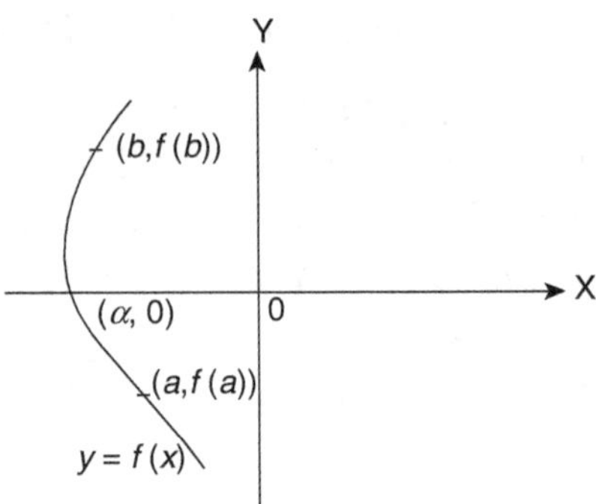

Figure 0.3

This theorem is useful in tracing an approximate root of an equation. For example, if $f(x) = x^3 - 5x + 1 = 0$, then it is a continuous function on IR, $f(0) = 1$ and $f(1) = -3$. Thus, $f(0) f(1) < 0$. Hence, there is at least one real root of $f(x) = 0$ lying between 0 and 1. A number lying between 0 and 1 can be taken as an approximate root of $f(x) = x^3 - 5x + 1 = 0$. Using the numerical methods, this approximate root can be improved up to a desired degree of accuracy.

0.7   CHANGE OF SIGN IN A POLYNOMIAL

When the terms of a polynomial $f(x)$ are in increasing (or decreasing) powers of x if a + sign follows a − sign or − sign follows a + sign, then we say that a change of sign occurred in $f(x)$.

For example, for the polynomial equation $f(x) = 16x^5 + 14x^4 - 19x^3 + 2x^2 - 9x + 5 = 0$, the number of changes of signs is 4 (see Fig. 0.4).

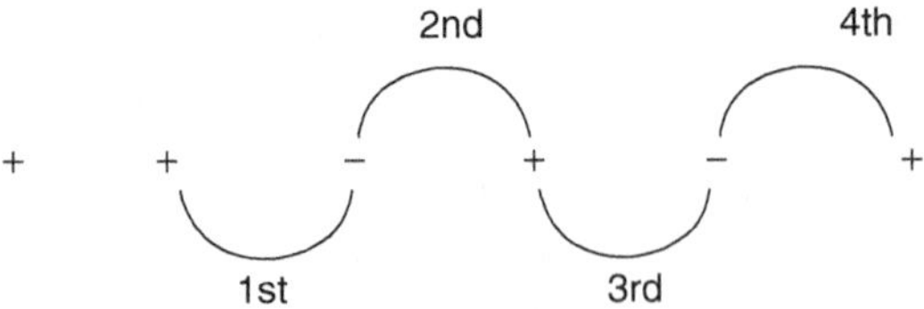

Figure 0.4

0.8   DESCARTES' RULE OF SIGNS

(i) The number of positive roots of a polynomial equation $f(x) = 0$ cannot exceed the number of changes in sign in $f(x)$.

(ii) The number of negative roots of a polynomial equation $f(x) = 0$ cannot exceed the number of changes in sign in $f(-x)$.

This rule does not give an exact number of real roots but it gives an upper bound of real roots.

For example, the polynomial equation $f(x) = x^4 - 1 = 0$ has one change in sign. Therefore, the maximum number of positive roots is 1. Now, $f(-x) = x^4 - 1 = 0$. Therefore, number of changes in sign in $f(-x)$ is one. Hence, $f(x)$ has maximum of one negative root. Note that the roots of $x^4 - 1 = 0$ are $1, -1, i$ and $-i$.

Note: Let $f(x) = 0$ be a polynomial equation of degree n.

(i) the number of changes in sign in $f(x)$ is k (k is a positive integer).

(ii) the number of changes in sign in $f(-x)$ is l (l is a positive integer). If $k + l < n$, then number of complex roots is $n - (k + l)$ or more. Suppose $f(x) = 0$ is a polynomial equation of degree 7, the maximum number of positive roots of it is 3 and the minimum number of negative roots is 2 and then the number complex roots of it are 2 or more.

0.9   SOME USEFUL RESULTS

(i) If $f(x) = 0$ is a polynomial equation of degree n and n is odd, then it has at least one real root.

(ii) The approximate largest root of $a_0 x^n + a_1 x^{n-1} + a_2 x^{n-2} + \ldots + a_{n-1} x + a_n = 0$ is a root of $a_0 x + a_1 = 0$ or the numerically the largest root of $a_0 x^2 + a_1 x + a_2 = 0$.

(iii) The approximate smallest root of $f(x) = a_0 x^n + a_1 x^{n-1} + a_2 x^{n-2} + \ldots + a_{n-1} x + a_n = 0$ is a root of $a_{n-1} x + a_n = 0$ or the numerically smallest root of $a_{n-2} x^2 + a_{n-1} x + a_n = 0$.

0.10   MULTIPLE ROOT

A root a is said to be of multiplicity k of $f(x) = a_0 x^n + a_1 x^{n-1} + \ldots + a_n = 0$ $(a_0 \neq 0)$ if $f(x) = (x - a)^k \phi(x)$ and $\phi(a) \neq 0$.

Note that, if $f(a) = 0, f'(a) = 0, \ldots, f^{(k)}(a) = 0$ and $f^{(k+1)}(a) \neq 0$, then a is a root of $f(x) = 0$ with multiplicity k.

For example, suppose $f(x) = x^3 - 4x^2 + 5x - 2 = 0$, then $f(1) = 0$, $f'(1) = 0$ and $f''(1) \neq 0$. Hence, 1 is a double root of $f(x) = 0$.

0.11 STURM SEQUENCE

The exact number of real roots of a polynomial equation can be obtained by Sturm's theorem. This theorem uses Sturm sequence, which is defined as follows.

Let $f(x)$ be a polynomial of degree n. A sequence $f(x), f_1(x), f_2(x), \ldots, f_n(x)$ is called Sturm sequence or Sturm functions of $f(x)$ where

(i) $f_1(x)$ is $\dfrac{df}{dx}$

(ii) $f_i(x) = -\{\text{Remainder of } f_{(i-2)}(x) \text{ divided by } f_{(i-1)}(x)\}$, where $i = 2$, 3, ..., n with $f_0(x) = f(x)$.

Note that $f_i(x)$ or $f_i(x)/k$ (k is positive) can be taken as $f_i(x)$.

Example

Suppose if $f(x) = x^3 - 6x + 4$, we find the Sturm sequence of it.
Here $f_0(x) = x^3 - 6x + 4$ and $f_1(x) = 3x^2 - 6$.
Now,

$$f_2(x) = -\{\text{Remainder of } f_{(0)}(x) \text{ divided by } f_{(1)}(x)\}$$
$$= -\{\text{Remainder of } x^3 - 6x + 4 \text{ divided by } (3x^2 - 6)\}$$
$$= -(-4x + 4) = 4(x - 1)$$

Since,

$$x^3 - 6x + 4 = (3x^2 - 6)\left(\frac{x}{3}\right) + (-4x + 4).$$

We take,

$$f_2(x) = (x - 1)$$

Now

$$f_3(x) = -[\text{Remainder of } f_{(1)}(x) \text{ divided by } f_{(2)}(x)]$$
$$= -[\text{Remainder of } 3x^2 - 6 \text{ divided by } (x - 1)]$$
$$= 3.$$

Thus, the Sturm sequence of $x^3 - 6x + 4$ is $x^3 - 6x + 4$, $3x^2 - 6$, $x - 1$, 3.

0.12 STURM THEOREM

The number of real roots of the polynomial equation $f(x) = 0$ on $[a, b]$ equals the difference between the number of changes in sign in the Sturm sequence at $x = a$ and $x = b$, provided that $f(a) \neq 0$ and $f(b) \neq 0$.

Example

We determine the number of real roots and complex roots of the polynomial equation $f(x) = 4x^4 + 2x^2 - 1 = 0$.

Here, $f_0(x) = 4x^4 + 2x^2 - 1$, $f_1(x) = 16x^3 + 4x$,

i. e., $f_1(x) = 4(4x^3 + x)$ since 4 is positive we can take $f_1(x) = 4x^3 + x$.

Now,

$$f_2(x) = - \text{[Remainder of } f_0(x) \text{ divided by } f_1(x)]$$

$$= - [x^2 - 1]$$

$$= - x^2 + 1$$

Thus, $\quad f_2(x) = - x^2 + 1$.

$$f_3(x) = - \text{[Remainder of } f_1(x) \text{ divided by } f_2(x)]$$

$$= - [5x]$$

$$= - 5x$$

$$= 5(- x).$$

We take, $\quad f_3(x) = - x$.

$$f_4(x) = - \text{[Remainder of } f_2(x) \text{ divided by } f_3(x)]$$

$$= - [1] = - 1.$$

Thus, Sturm sequence is $f(x) = 4x^4 + 2x^2 - 1$; $f_1(x) = 4x^3 + x$; $f_2(x) = - x^2 + 1$; $f_3(x) = - x$ and $f_4(x) = - 1$.

Let $V(a)$ denote the number of sign changes in Sturm sequence at $x = a$. Now, we construct the following table to compute $V(a)$ for different values of a:

a	$f(x)$	$f_1(x)$	$f_2(x)$	$f_3(x)$	$f_4(x)$	$V(x)$
$-\infty$	$+$	$-$	$-$	$+$	$-$	3
-1	$+$	$-$	$-*$	$+$	$-$	3
0	$-$	$-*$	$+$	$+*$	$-$	2
1	$+$	$+$	$+*$	$-$	$-$	1
∞	$+$	$+$	$-$	$-$	$-$	1

The signs shown at * places are the signs of the immediate preceding places. The values obtained at * places are zero.

Since $V(-1) = 3$ and $V(0) = 2$, there will be one root in $(-1, 0)$ and $V(0) = 2$, $V(1) = 1$ implies that there will be one root in $(0, 1)$. Therefore, the given polynomial has 2 real roots. As the degree of given polynomial equation is 4, it will have two complex roots.

0.13   TAYLOR'S THEOREM

(i) Let $f(x)$ be a real valued function in $[a, a + h]$ such that $f^{(n+1)}(x)$ exists for every $x \in [a, a + h]$ and $f^{(n+1)}(x)$ is continuous on $[a, a + h]$ then $\exists$ a θ such that $0 \le \theta \le 1$ and

$$f(a + h) = f(a) + \frac{f'(a)}{\lfloor 1} h + \frac{f''(a)}{\lfloor 2} h^2 + \ldots + \frac{f^{(n)}(a)}{\lfloor n} h^n$$
$$+ \frac{f^{(n+1)}(a + \theta h)}{\lfloor n+1} h^{n+1}$$

The following series is known as Taylor's series:

$$f(a + h) = f(a) + \frac{f'(a)}{\lfloor 1} h + \frac{f''(a)}{\lfloor 2} h^2 + \ldots$$

Similarly, the Taylor's series in two variables is given by

$$f(a + h, b + k) = f(a, b) + \left(h\frac{\partial}{\partial x} + k\frac{\partial}{\partial y} \right) f(a,b)$$
$$+ \frac{1}{\lfloor 2} \left(h\frac{\partial}{\partial x} + k\frac{\partial}{\partial y} \right)^2 f(a,b) + \frac{1}{\lfloor 3} \left(h\frac{\partial}{\partial x} + k\frac{\partial}{\partial y} \right)^3 f(a,b) + \ldots$$

where $\dfrac{\partial f(a,b)}{\partial x} = \left(\dfrac{\partial f}{\partial x} \right)_{(a,b)}$, the value of $\dfrac{\partial f}{\partial x}$ at (a, b),

$$\frac{\partial f(a,b)}{\partial y} = \left(\frac{\partial f}{\partial y} \right)_{(a,b)}, \text{ etc.}$$

0.14 WEIERSTRASS APPROXIMATION THEOREM

Let $f(x)$ be any function on $[a, b]$ and also continuous on $[a, b]$. Then given $\in > 0$ there exists a polynomial $P(x)$ such that

$$\left|P(x) - f(x)\right| < \in \ \forall \ x \in [a,b].$$

0.15 CONVERGENT SEQUENCE

A sequence $\{x_n\}$ of real numbers is said to be convergent to L if for every $\in > 0$ $\exists$ a positive integer N such that

$$\left|x_n - L\right| < \in \ \forall \ n \geq N.$$

In this case, we say the $\{x_n\}$ is a convergent sequence.

1

Errors

We have one or more numerical methods to solve a particular problem. For example, Gauss–Seidel method is one of the numerical method to solve a system of linear equations and other methods to solve them are Gauss-elimination method, Jacobi method, etc. Further, we have several numerical methods to solve different types of problems of applied mathematics, engineering and physical sciences.

Due to the development of high-speed computers, these methods can be implemented on computers to get the solutions of problems with help of a programming language like FORTRAN, C and C++. Due to certain limitations of computer, there will be two sources of errors namely round-off errors and truncation errors when these numerical methods are implemented to get the solutions of the problems.

A number in the memory of the computer is stored as an integer or floating-point (real) number. An integer is a sequence of digits whereas a floating-point number consists of integral part and a fraction part, e.g., 23.532 or 67.00. A variable x is considered an integer or floating-point number in a programming language; the computer allots certain amount of space in the memory of the computer for the variable x. The amount of space be in certain number of words and it is machine dependent (a word consists of 16 bits). If $\dfrac{375}{99}$ is assigned to a variable x, then the rounded number 3.787878……. stored in the address of the variable x in binary form (machine code) in a fixed word length. This may cause an error in computations involving x. The expressions $3 \times \dfrac{1}{3}$ and $\dfrac{(3 \times 1)}{3}$ are evaluated differently by the computers as per the precedence of operators.

To compute $f(x) = 1 + 2x + \dfrac{3x^2}{2^2} + \ldots\ldots$ for any given x, we consider only a finite number of terms in its calculations. This also causes an error in the computations. Thus, we have two sources of errors in numerical computations and they are round-off errors and truncation errors. In this section, we learn in detail about these errors and also define relative and percentage errors. Before proceeding to learn these errors, we learn the meaning of certain terms that are used in numerical methods like significant digits, approximations, etc.

1.1   SIGNIFICANT DIGITS

The digits that are used to express a number are called significant digits (or figures). If a number is greater than or equal to 1 then the number of significant digits in it are counted from the left to right of that-number otherwise counting of significant digits of the number starts from its non-zero digit. For example, the numbers of significant digits in the numbers 3.1416, 0.666667, 2.10653 and 0.0005632 are 5, 6, 6 and 4 respectively.

1.2   APPROXIMATE VALUE

Let x and y be two numbers. Then, x is said to be the exact value of y if $x = y$. In case, if x represents a value of a number y to certain degree of accuracy, then x is called the approximated value of y or approximate number of y. For example, 3.141 is an approximate number of π upto four significant digits and 0.3333 is an aapproximate number of 1/3 correct to four decimal places.

1.3   ERROR

The error of an approximation is defined as (true value - approximate number). In other words, if $x^{(a)}$ is the approximation of $x^{(t)}$ then $x^{(t)} - x^{(a)}$ will be error of the approximation of $x^{(a)}$.

If 1.4142 is an approximate value of $\sqrt{2}$, then the error in the approximation is $\sqrt{2} - 1.4142$.

The absolute value of an error is defined as an absolute error. The absolute error in the approximation $x^{(a)}$ is $\left| x^{(t)} - x^{(a)} \right|$ where $x^{(a)}$ is the approximate number of true value $x^{(t)}$.

1.4   ROUNDING NUMBERS

If a number has a large number of digits, then we cut them to a usable number of digits following certain rules for convenience. This process is called rounding-off a number. The number so obtained by rounding-off is called rounding number or rounding-off number.

The following are convention rules for rounding-off a number to n significant digits and discard the digits after nth significant digits.

To round-off a number x to n significant digits:

(a) If $(n + 1)$th significant digit of x is less than 5, then discard the digits present after the nth significant digit.

(b) If $(n + 1)$th significant digit of x is greater than 5, then add 1 to the nth significant digit and discard the digits after nth significant digit.

(c) If $(n + 1)$th digit of x is exactly 5, then:

(i) nth significant digit of x is unaltered if $(n + 2)$th digit of x is even.

(ii) nth significant digit of x is increase by 1 if $(n + 2)$th digit of x is odd.

Example 1:   Rounding-off number of 3.05623527 to four significant digits is 3.056.

Example 2:   Rounding-off number of 52.06383215 to five significant digits is 52.064.

Example 3:   Rounding-off number of 52.06532168 to four significant digits is 52.07.

Example 4:   Rounding-off number of 52.06522168 to four significant digits is 52.06.

1.5 TRUNCATION NUMBER

A number obtained by ignoring on and after certain digit of a number x is called a truncation number or truncated number of x. For example, 1.3 is a truncated number of 1.33333….

Some of real numbers are stored in the computer's memory with a limited number of bits. In the computer it is not possible to store the value of π as it is. However, the truncated number of π can be stored in the computer's memory. This causes an error and it is known as truncation error. Further, a finite number of terms in the infinite series expansion of e^x, sin x, cos x, etc. causes errors, such errors are also called truncation errors.

For example: $e^x = 1 + x + \dfrac{x^2}{\lfloor 2} + \dfrac{x^3}{\lfloor 3} + \ldots$ values for $x = 1.4$ for various number of terms of this infinite series is shown in the following table. These values converge to 4.055199967 (approximately).

n (no. of terms)	$e^{1.4}$
1	1
2	2.4
3	3.38
4	3.3837
5	3.5437667

Note that we can obtain the truncation error of such series from the remainder term of Taylor's theorem.

1.6 ROUND-OFF ERROR

If $x^{(r)}$ is the rounding-off number of x then $x - x^{(r)}$ is defined as round-off error in the approximation $x^{(r)}$.

If 1.3768 is the rounding-off number of 1.376814236 then the rounding error is $1.376814236 - 1.3768 = 0.000014236$.

1.7 TRUNCATION ERROR

If x^{tr} is the truncated number of x then $x - x^{\mathrm{tr}}$ is called truncation error of the approximation x^{tr}.

If 1.3333 is the truncated number of 1.33334444, then the truncation error of the approximation 1.3333 is $1.33334444 - 1.3333 = 0.00004444$.

1.8 RELATIVE ERROR

If $x^{(a)}$ is an approximation of an exact value $x^{(e)}$, then $\left| \dfrac{x^{(e)} - x^{(a)}}{x^{(e)}} \right|$ is defined as relative error in $x^{(a)}$ For example, if 0.99×10^4 is an approximation of 1×10^4 then the relative error in the approximation is $\left| \dfrac{10^4 - 0.99 \times 10^4}{10^4} \right| = 0.01$.

In numerical methods, usually we find successive approximation x_1, x_2, x_3, x_4, ... of an exact value which may be known or unknown value. In such case, the relative error in its approximation is considered as $\left| \dfrac{x_i - x_{i+1}}{x_i} \right|$.

1.9 PERCENTAGE OF ERROR

If $x^{(a)}$ is the approximation of an exact number $x^{(e)}$, then $\left| \dfrac{x^{(e)} - x^{(a)}}{x^{(e)}} \right| \times 100$ or relative error of $x^{(a)} \times 100$ is defined as percentage of error in the approximation of $x^{(a)}$. For example, if 0.99×10^4 is an approximation of 10^4, then the percentage of error in the approximation 0.99×10^4 is

$$\left| \frac{10^4 - 0.99 \times 10^4}{10^4} \right| \times 100 = 1.$$

1.10 INHERENT ERROR

The inherent error is an error, which is already present in the statement of the problem before obtaining its solution. Usually the inherent error arises either due to the presence of errors in physical measurement of the parameter of the problem or due to simplified assumption in the formulation of problem.

1.11 AN APPROXIMATION OF A NUMBER TO *K* SIGNIFICANT DIGITS

Suppose $x^{(a)}$ is an approximation to the exact number $x^{(e)}$. We define $x^{(a)}$ as an approximation of x to k significant digits if $\left| x^{(e)} - x^{(a)} \right| < \dfrac{1}{2} 10^{1-k}$.

For example, $x^{(a)} = 7.1634$ is correct to five significant digits of x means $7.1634 - 0.00005 \leq x \leq 7.1634 + 0.00005$. The relative error in $x^{(a)}$ is $\dfrac{0.00001}{7.1634}$.

Note that to guarantee k significant digits accuracy of a number, we should ensure the relative error less than or equal to 10^{-k}.

SOLVED PROBLEMS

Problem 1: Find round-off number of 69.21786 to two decimal places.

Solution: As the third decimal place number is 7 which is greater than 5, the round-off number of given number is 69.22.

Problem 2: If 6.1432 is the round-off number of 6.1432162, then find its relative and percentage of error.

Solution: The relative error is $\left| \dfrac{6.1432162 - 6.1432}{6.1432162} \right| = 3.5192 \times 10^{-5}$

and the percentage of error is $3.5192 \times 10^{-5} \times 100 = 0.0035$.

Problem 3: Find the value of $e^x = 1 + x + \dfrac{x^2}{\lfloor 2} + \dfrac{x^3}{\lfloor 3} + \ldots$ with an absolute error less than 0.005 for $x = 0.35$. Assume that the value of $e = 2.7183$.

Solution: The sum of different number terms of the series

$e^x = 1 + x + \dfrac{x^2}{\lfloor 2} + \dfrac{x^3}{\lfloor 3} + \ldots$ for $x = 0.35$ and the absolute error are shown in the following table:

No. of terms	Exact values	Approximate value	Absolute error
1	1.419071	1	0.419071
2	1.419071	1.35	0.069071
3	1.419071	1.411250	0.007821

The required value is 1.411250.

Problem 4: How many terms of $e^x = 1 + x + \dfrac{x^2}{\lfloor 2} + \dfrac{x^3}{\lfloor 3} + \ldots$ are to be added for $x = 1$, so that e^x is correct to five decimal places.

Solution: By Taylor's theorem, the remainder term is $\dfrac{x^n}{\lfloor n} . e^\xi$. The maximum absolute error is at $\xi = x$. Hence the maximum relative error is

$\dfrac{x^n}{\lfloor n}$. For five-decimal accuracy at $x = 1$, we must have

$$\frac{1}{\lfloor n} < \frac{1}{2}\left(10^{-5}\right)$$

i.e. $\qquad\qquad\qquad \lfloor n > 2 \times 10^5$

The least value of n for which the above inequality true is 9. Therefore, we need to take 9 number of terms of the given exponentiation series so that the sum is correct to five decimal places for $x = 1$.

EXERCISE 1

1. Find the truncated numbers of the following to five significant digits 6.3215689, 2.16532, 1.000563 and 0.00067893.

2. Find the rounding-off numbers of the following numbers to three significant digits 64.637842, 0.00379248, 1002.620034 and 165.300896.

3. If 1.6328 is an approximate value of 1.63279432, find its relative and percentage of error.

4. Given that $\pi = 3.1415926$. If 3.1412 is an approximate value of π, then find its absolute error, relative error and percentage of error.

5. Given that $\cos x = 1 - \dfrac{x^2}{\lfloor 2} + \dfrac{x^4}{\lfloor 4} - \dfrac{x^6}{\lfloor 6} \ldots.$, where x is in radians. Use the series to compute $\cos 60°$ so that its absolute error is less than 0.01.

6. How many terms of $e^x = 1 + x + \dfrac{x^2}{\lfloor 2} + \dfrac{x^3}{\lfloor 3} + \ldots.$ are to be added for $x = 2$ so that e^x is correct to three decimal places?

ANSWERS

1. 6.32152, 1653, 1.005, 0.00067893   2. 64.6, 0.004, 100, 165

3. 3.4787×10^{-6}, 3.4787×10^{-4} 4. 3.926×10^{-4}, 1.2497×10^{-4}, 0.0125

5. 0.501796   6. 12.

2

Operators and Finite Differences

An expression of the form $f(x+l) - f(x+m)$ where l and m are constants is called a finite difference. It has an application in a numerical solution of ordinary and partial differential equations. In this chapter, we define forward, backward and central differences, which are particular cases of finite differences. These differences have applications in interpolation, numerical differentiation and numerical integration. Further, we define few operators such as forward-difference operator, backward-difference operator, etc. and various difference tables. We also discuss the relations between the operators, the properties of each operator and dividend differences.

2.1 DEFINITIONS

2.1.1 Arguments and entries

A value of the independent variable of a function $y = f(x)$ is called an argument whereas the value of the dependent variable at an argument is called the entry of the function corresponding to the argument.

For example, Table 2.1 shows the values of x and a function $y = f(x)$.

Table 2.1 Arguments and entries

x	2	5	7	9	10
y	4	25	36	75	103

The numbers 4, 25, 36, 75 and 103 are the entries corresponding to the arguments 2, 5, 7, 9 and 10 respectively.

2.1.2 Interval of differencing

If the arguments are in an arithmetic progression, then the common difference of it is called an interval of differencing or step size. Usually it is denoted by h.

For example, if 2, 5, 8, 11, 14 and 17 are the arguments, then the interval of differencing is 3, i.e., $h = 3$.

2.1.3 Forward-difference operator, forward difference

The forward-difference operator Δ is defined by the equation

$$\Delta f(x) = f(x + h) - f(x),$$

and $\Delta f(x)$ is called first–order forward difference. For example $\Delta f(a) = f(a + h) - f(a)$, $\Delta f(a + h) = f(a + 2h) - f(a + h)$ etc.

2.1.4 *n*th-order forward difference

The nth–order forward difference is defined by

$$\Delta^n f(x) = \Delta^{n-1} (\Delta f(x))$$

where n is a positive integer, we define $\Delta^0 = 1$.
For $n = 2$, $\Delta^2 f(x) = \Delta (\Delta f(x))$
$$= \Delta\{f(x + h) - f(x)\} = \Delta f(x + h) - \Delta f(x)$$
$$= f(x + 2h) - 2 f(x+h) + f(x);$$

There, note that $\Delta^2 f(x) = \Delta f(x + h) - \Delta f(x)$. Hence, the second–order forward differences of $f(x)$ are the differences of first-order forward differences. In general, nth-order differences of $f(x)$ are the differences of $(n - 1)th$-order forward differences.

2.1.5 Forward-difference table

Let $x_0 = a, x_1 = a + h, x_2 = a + 2h, \ldots\ldots, x_n = a + nh$ be the values of independent variable of a function $y = f(x)$ and $y_0, y_1, y_2 \ldots\ldots, y_n$ be entries corresponding to these arguments, where $y_k = f(x_k)$.

$y_1 - y_0, y_2 - y_1, y_3 - y_2, \ldots\ldots\ldots, y_n - y_{n-1}$ are the first-order forward differences and they are equal to $\Delta y_0, \Delta y_1, .., \Delta y_{n-1}$ respectively; $\Delta y_1 - \Delta y_0$, $\Delta y_2 - \Delta y_1, \ldots., \Delta y_{n-1} - y_{n-2}$ are the second–order forward differences and they are equal to $\Delta^2 y_0$, $\Delta^2 y_1$,...... $\Delta^2 y_{n-2}$. Similar argument can be given to higher order forward differences.

A table that shows the arguments, entries and forward differences of first order, second order, ... etc. is called a forward-difference table, and it is shown below for 5 entries (Table 2.2).

Table 2.2 Forward difference table

x	$f(x)$	Δy	$\Delta^2 y$	$\Delta^3 y$	$\Delta^4 y$	$\Delta^5 y$
x_0	y_0					
		$\to\Delta y_0$				
x_1	y_1		$\to\Delta^2 y_0$			
		$\to\Delta y_1$		$\to\Delta^3 y_0$		
x_2	y_2		$\to\Delta^2 y_1$		$\to\Delta^4 y_0$	
		$\to\Delta y_2$		$\to\Delta^3 y_1$		$\to\Delta^5 y_0$
x_3	y_3		$\to\Delta^2 y_2$		$\to\Delta^4 y_1$	
		$\to\Delta y_3$		$\to\Delta^3 y_2$		
x_4	y_4		$\to\Delta^2 y_3$			
		$\to\Delta y_4$				
x_5	y_5					

The entry y_0 is called leading term, and Δy_0, $\Delta^2 y_0$, $\Delta^3 y_0$,......, etc. are called the leading differences.

2.2 PROPERTIES OF FORWARD-DIFFERENCE OPERATOR Δ

Property 1: If c in any constant, then $\Delta c = 0$.

Proof: Let $f(x) = c$; Then, $\Delta c = \Delta f(x) = \Delta f(x + h) - f(x) = c - c = 0$.

Property 2: If k is any constant, then $\Delta\{kf(x)\} = k\,\Delta f(x)$.

Proof: Let $F(x) = k f(x)$. Then,

$$\Delta\{k f(x)\} = \Delta F(x)$$
$$= F(x + h) - F(x)$$
$$= k f(x + h) - k f(x)$$
$$= k\{f(x + h) - f(x)\}$$
$$= k \Delta f(x).$$

Property 3: For any two functions $f(x)$ and $g(x)$,

$$\Delta\{f(x) + g(x)\} = \Delta f(x) + \Delta g(x).$$

Proof: Let $F(x) = f(x) + g(x)$. Then,

$$\Delta\{f(x) + g(x)\} = \Delta F(x)$$
$$= F(x + h) - F(x)$$
$$= \{f(x + h) + g(x + h)\} - \{f(x) + g(x)\}$$
$$= \{f(x + h) - f(x)\} + \{g(x + h) - g(x)\}$$
$$= \Delta f(x) + \Delta g(x).$$

Property 4: For any two functions $f(x)$ and $g(x)$,
$$\Delta\{f(x) - g(x)\} = \Delta f(x) - \Delta g(x).$$

Proof: Let $F(x) = f(x) - g(x)$. Then

$$\Delta\{f(x) - g(x)\} = \Delta F(x)$$
$$= F(x + h) - F(x)$$
$$= \{f(x + h) - g(x + h)\} - \{f(x) - g(x)\}$$
$$= \{f(x + h) - f(x)\} - \{g(x + h) - g(x)\}$$
$$= \Delta f(x) - \Delta g(x).$$

Property 5: The operator Δ is linear, i.e., $\Delta\{a f(x) + b g(x)\} = a \Delta f(x) + b \Delta g(x)$, where $f(x)$ and $g(x)$ are any two functions and a and b are any constants. This follows from property 2 and 3.

Property 6: If $f(x)$ and $g(x)$ are any two functions, then
$$\Delta\{f(x) g(x)\} = f(x + h) \Delta g(x) + g(x) \Delta f(x)$$
$$= f(x) \Delta g(x) + g(x + h) \Delta f(x)$$
$$= f(x) \Delta g(x) + g(x) \Delta f(x) + \Delta f(x) \Delta g(x).$$

Proof: Consider $\Delta\{f(x)\,g(x)\}$

$$\Delta\{f(x)\,g(x)\} = f(x+h)\,g(x+h) - f(x)\,g(x)$$
$$= f(x+h)\,g(x+h) - f(x+h)\,g(x) + f(x+h)\,g(x)$$
$$- f(x)g(x)$$
$$= f(x+h)\,\{g(x+h) - g(x)\} + g(x)\{f(x+h) - f(x)\}$$
$$= f(x+h)\,\Delta g(x) + g(x)\,\Delta f(x)$$

Similarly, it can be shown $\Delta\{f(x)\,g(x)\} = f(x)\,\Delta g(x) + g(x+h)\,\Delta f(x)$
Now, we proceed to prove a third identity as

$$\Delta f(x)\,g(x) = f(x+h)\,g(x+h) - f(x)\,g(x)$$
$$= \{\Delta f(x) + f(x)\}\,\{\Delta g(x) + g(x)\} - f(x)\,g(x)$$
$$= \Delta f(x)\,\Delta g(x) + g(x)\,\Delta f(x) + f(x)\,\Delta g(x) + f(x)\,g(x)$$
$$- f(x)\,g(x)$$
$$= f(x)\Delta g(x) + g(x)\,\Delta f(x) + \Delta f(x)\,\Delta g(x).$$

Property 7: For any two functins $f(x)$ and $g(x)$,

$$\Delta\left\{\frac{f(x)}{g(x)}\right\} = \frac{g(x)\Delta f(x) - f(x)\Delta g(x)}{g(x)g(x+h)}.$$

Proof: By the definition of forward-difference operator

$$\Delta\left\{\frac{f(x)}{g(x)}\right\} = \frac{f(x+h)}{g(x+h)} - \frac{f(x)}{g(x)}$$

$$= \frac{f(x+h)g(x) - f(x)g(x+h)}{g(x+h)g(x)}$$

$$= \frac{f(x+h)g(x) - f(x)g(x) + f(x)g(x) - f(x)g(x+h)}{g(x+h)g(x)}$$

$$= \frac{g(x)\Delta f(x) - f(x)\Delta g(x)}{g(x+h)g(x)}.$$

SOLVED PROBLEMS

Problem 1: Construct the forward difference table for the data given overleaf.

x	2	4	6	8	10
$f(x)$	25	36	70	84	105

Solution: The following is the forward difference table (Table 2.3) for the given data.

Table 2.3   Forward difference table

x	y	Δy	$\Delta^2 y$	$\Delta^3 y$	$\Delta^4 y$
2	$25 = y_0$				
		$\rightarrow 11 = \Delta y_0$			
4	36		$\rightarrow 23 = \Delta^2 y_0$		
		$\rightarrow 34$		$\rightarrow -43 = \Delta^3 y_0$	
6	70		$\rightarrow -20$		$\rightarrow 70 = \Delta^4 y_0$
		$\rightarrow 14$		$\rightarrow 27$	
8	84		$\rightarrow 7$		
		$\rightarrow 21$			
10	105				

Problem 2: Express $\Delta^3 f(a)$ in term of entries.

Solution: By definition, $\Delta^3 f(a) = \Delta\left(\Delta^2 f(a)\right)$

$$= \Delta\left[\Delta(\Delta f(a)\right] = \Delta\left[\Delta\{f(a+h) - f(a)\}\right]$$

$$= \Delta[\Delta f(a+h) - \Delta f(a)]$$

$$= \Delta[\{f(a+2h) - f(a+h)\} - \{f(a+h) - f(a)\}]$$

$$= \Delta\{f(a+2h) - 2f(a+h) + f(a)\}$$

$$= \Delta f(a+2h) - 2\Delta f(a+h) + \Delta f(a)$$

$$= \{f(a+3h) - f(a+2h)\} - 2\{f(a+2h) - f(a+h)\}$$

$$+ \{f(a+h) - f(a)\}$$

$$= f(a+3h) - 3f(a+2h) + 3f(a+h) - f(a)$$

$$= y_3 - 3y_2 + 3y_1 - y_0$$

Problem 3: If $f(x) = x^2 + 5x + 7$, then find $\Delta f(x)$, $\Delta^2 f(x)$ and $\Delta^3 f(x)$ with $h = 3$.

Solution: Since $h = 3$, we have

$$\Delta f(x) = f(x+3) - f(x)$$
$$= \{(x+3)^2 + 5(x+3) + 7\} - \{x^2 + 5x + 7\}$$
$$= 6x + 24 = 6(x+4);$$

Now, $\Delta^2 f(x) = \Delta(\Delta f(x))$
$$= 6\,\Delta\,(x+4)$$
$$= 6.\,\{(x+4+3) - (x+4)\}$$
$$= 6(3)$$
$$= 18$$

Therefore, $\Delta^3 f(x) = \Delta(18) = 0$.

Problem 4: From the following data, find $\Delta f(3)$, $\Delta^2 f(3)$ and $\Delta f(6)$

x	0	3	6	9	12	15
$f(x)$	16	25	42	59	65	78

Solution: From the given data, we have $h = 3$. Therefore,

$$\Delta f(3) = f(3+3) - f(3) = f(6) - f(3) = 42 - 25 = 17.$$

Similarly,

$$\Delta f(6) = f(9) - f(6) = 59 - 42 = 17.$$

Now, $\Delta^2 f(3) = f(3 + 2 \times 3) - 2f(3+3) + f(3)$
$$= f(9) - 2f(6) + f(3)$$
$$= 59 - 2(42) + 25$$
$$= 0.$$

Other method:

We can obtain the forward differences from the forward-difference table.

x	$y = f(x)$	Δy	$\Delta^2 y$	$\Delta^3 y$	$\Delta^4 y$	$\Delta^5 y$
0	16					
		$\to 9 = \Delta f(0)$				
3	25		$\to 8 = \Delta^2 f(0)$			
		$\to 17 = \Delta f(3)$		$\to -8 = \Delta^3 f(0)$		
6	42		$\to 0 = \Delta^2 f(3)$		$\to 7 = \Delta^4 f(0)$	
		$\to 17 = \Delta f(6)$		$\to -1 = \Delta^3 f(3)$		$\to -8 = \Delta^5 f(0)$
9	59		$\to -1 = \Delta^2 f(6)$		$\to -1 = \Delta^4 f(3)$	
		$\to 16 = \Delta f(9)$		$\to -2 = \Delta^3 f(6)$		
12	65		$\to -3 = \Delta^2 f(9)$			
		$\to 13 = \Delta f(12)$				
15	78					

From the above table, $\Delta f(3) = 17$, $\Delta f(6) = 17$ and $\Delta^2 f(3) = 0$.

2.3   FUNDAMENTAL THEOREM OF DIFFERENCE CALCULUS

The nth-order forward difference of nth-degree polynomial in x is constant, and its $(n + 1)$th forward difference is zero.

Proof: First, we prove this result for $f(x) = x^n$, where n is a positive integer or zero. We prove it by the principle of mathematical induction. Clearly, the result is true for $n = 0$, and suppose $n = 1$, then $f(x) = x$.

Hence, $\Delta f(x) = (x + h) - x = h$ and $\Delta^2 f(x) = 0$;

Therefore, the result is true for $n = 1$.

It can be shown that $\Delta^2 x^2 = \lfloor 2 \; h^2$, $\Delta^3 x^3 = \lfloor 3 \; h^3$, etc

Suppose the result is true for $n = k$, then

$$\Delta^k x^k = \lfloor k \; h^k \text{ and } \Delta^{k+1} x^k = 0.$$

Now for $\quad n = k + 1,$

$$\Delta^{k+1} x^{k+1} = \Delta^k(\Delta x^{k+1})$$

$$= \Delta^k\{(x+h)^{k+1} - x^{k+1}\}$$

$$= \Delta^k\{{}^{(k+1)}C_1 \, x^k h + {}^{(k+1)}C_2 \, x^{k-1}.h^2 + \ldots\ldots + h^{k+1}\}$$

$$= {}^{(k+1)}C_1.h.\ \Delta^k\ x^k + {}^{(k+1)}C_2.h^2.\ \Delta^k\ x^{(k-1)} + \ldots.. + \Delta^k\ h^{k+1}$$

$$= (k+1)\ h.\ \lfloor k\ \ h^k + 0 + \ldots..+ 0\ \text{(by induction hypothesis)}$$

$$= \lfloor k+1.\ h^{k+1}$$

which is a constant, and hence, $\Delta^{k+2}\ x^{k+1} = 0$.

Thus, the result is true for $n = (k + 1)$. Hence, the result is true for every positive integer, i.e., the nth forward difference of x^n is constant and it is equal to $\lfloor n\ .\ h^n$, and its $(n + 1)$th-forward difference is zero. Now, we prove the fundamental theorem of calculus.

Let $p_n(x) = a_0 + a_1\ x + a_2\ x^2 + \ldots\ldots + a_{n-1}\ x^{n-1} + a_n\ x^n$ be an nth-degree polynomial, where $a_0, a_1, \ldots.. a_n$ are constants and $a_n \neq 0$ and now

$$\Delta^n\ p_n(x) = \Delta^n\ a_0 + a_1\ \Delta^n\ x + a_2\ \Delta^n\ x^2 + \ldots\ldots.. + a_{n-1}\ \Delta^n\ x^{n-1} + a_n\ \Delta^n\ x^n$$

$$= 0 + 0 + 0 + \ldots\ldots + 0 + a_n\ \lfloor n\ h^n$$

$$= a_n\ \lfloor n\ h^n,$$

which is constant and $\Delta^{n+1}\ p_n(x) = 0$ and hence the theorem.

2.4   FACTORIAL FUNCTION

If $f(x)$ is a polynomial, then the evaluation of the forward differences of various orders of $f(x)$ are tedious. To overcome this problem, the factorial functions are introduced. It is possible to express every polynomial in factorial functions. Computations of forward differences become easy once we express the polynomial in terms of factorial functions.

A function of the form $x(x - h)\ (x - 2h)\ldots\ldots\ (x- \overline{r-1}\ h)$ is called a factorial function of x and is denoted by $x^{(r)}$. Here, h is the interval of differencing. For example $x^{(3)} = x(x - h)\ (x - 2h)$. If $h = 1$, then $x^{(3)} = x(x - 1)(x - 2)$. We define $x^{(0)} = 1$.

Theorem:

$$\Delta x^{(r)} = rh\ x^{(r-1)}$$

Proof: By the definition of Δ operator, we have

$$\Delta x^{(r)} = (x + h)^{(r)} - x^{(r)}$$

$$= (x + h)\ (x + h - h)\ (x + h - 2h)\ldots\ldots(x + h - \overline{r-1}\ h)$$

$$-x\ (x - h)\ (x - 2h)\ldots\ldots\ldots(x - \overline{r-1}\ h)$$

$$= (x + h) \, x \, (x - h) \ldots\ldots\ldots (x - \overline{r-2} \ h)$$
$$- x \, (x - h) \ldots\ldots\ldots (x - \overline{r-1} \ h)$$
$$= x(x - h)\ldots\ldots\ldots(x - \overline{r-2} \ h) \, \{(x + h) - (x - \overline{r-1} \ h)\}$$
$$= x(x - h)\ldots\ldots\ldots(x - \overline{r-2} \ h) \, \{x + h - x + rh - h\}$$
$$= rh. \, x^{(r-1)}.$$

Theorem:

$$\Delta^n \, x^{(n)} = \lfloor n \ . h^n.$$

Proof: We prove it by the principle of mathematical induction.

For $n = 1$,

$$\Delta^1 \, x^{(1)} = \Delta x = (x + h) - (x) = h. = \lfloor 1 \, . \, h^1;$$

Therefore, the result is true for $n = 1$; suppose there the result is true for $n = k$, i.e.,

$$\Delta^k \, x^{(k)} = \lfloor k \ h^k,$$

Now for $n = k + 1$,

$$\Delta^{k+1} \, x^{(k+1)} = \Delta^k(\Delta x^{(k+1)})$$

$$\Delta^k \{(x + h)^{(k+1)} - x^{(k+1)}\}$$

$$= \Delta^k [\{(x + h)(x + h - h)(x + h - 2h)\ldots\ldots(x + h - \overline{k+1-1}h)\}$$
$$- \{x(x - h)(x - 2h)\ldots(x - \overline{k+1-1}h)\}]$$

$$= \Delta^k [\{(x + h)(x)(x - h)\ldots\ldots(x - \overline{k-1}h)\}$$
$$- \{x(x - h)(x - 2h)\ldots\ldots(x - kh)\}]$$

$$= \Delta^k [x(x - h)\ldots\ldots\ldots(x - \overline{k-1}h)\{(x + h) - (x - kh)\}]$$

$$= \Delta^k x^{(k)}.\{(k + 1)h\}$$

$$= (k + 1)h.\Delta^k x^{(1k)}$$

$$= (k + 1)h.\lfloor kh^k \text{ (by induction hypothesis)}$$

$$= \lfloor k + 1.h^{k+1}$$

Therefore, the result is true for $n = k + 1$, and hence by the principle of mathematical induction, the result is true for every positive integer n.

Theorem:

If $f(x) = a_0 x^n + a_1 x^{n-1} + ... + a_{n-1} x + a_n$,

then, $f(x) = f(0) + \dfrac{\Delta f(0)}{\lfloor 1} x^{(1)} + \dfrac{\Delta^2 f(0)}{\lfloor 2} x^{(2)} + ... + \dfrac{\Delta^n f(0)}{\lfloor n} x^{(n)}$,

where $x^{(r)} = x(x-1)\,(x-2) .... (x - \overline{r-1})$.

Proof: Without loss of generality let

$$f(x) = A_0 + A_1 x^{(1)} + A_2 x^{(2)} + ...A_n x^{(n)}$$

i.e. $f(x) = A_0 + A_1 x + A_2 x(x-1) + ...$

$$+ A_n x(x-1)........(x+n-1). \tag{1}$$

Put $x = 0$ in (1) we get

$$f(0) = A_0 \tag{2}$$

Next, put $x = 1$ in (1), we have

$$f(1) = A_0 + A_1 + 0$$

$$= f(0) + A_1$$

$\therefore \qquad\qquad A_1 = f(1) - f(0) = \Delta f(0) \tag{3}$

Put $x = 2$ in equation (1), we get

$$f(2) = A_0 + 2A_1 + A_2 \lfloor 2 + 0$$

$$f(2) = f(0) + 2\Delta f(0) + A_2 \lfloor 2$$

$$\therefore A_2 \lfloor 2 = f(2) - f(0) - 2\Delta f(0)$$

$$= f(2) - f(0) - 2\{f(1) - f(0)\}$$

$$= f(2) - f(0) - 2f(1) + 2f(0)$$

$$= f(2) - 2f(1) + f(0)$$

$$= \Delta^2 f(0)$$

Hence, $\qquad\qquad A_2 = \dfrac{\Delta^2 f(0)}{\lfloor 2}.$

Similarly, it can be shown that $A_3 = \dfrac{\Delta^3 f(0)}{\lfloor 3}, \ldots\ldots\ldots, A_n = \dfrac{1}{\lfloor n} \Delta^n f(0)$

Thus, $f(x) = f(0) + \dfrac{\Delta f(0)}{\lfloor 1} x^{(1)} + \dfrac{\Delta^2 f(0)}{\lfloor 2} x^{(2)} + \ldots\ldots + \dfrac{\Delta^n f(0)}{\lfloor n} x^{(n)}.$

SOLVED PROBLEMS

Problem 1: Express $4x^3 + 2x^2 - 7x + 12$, in terms of factorial functions.

Solution: We present the solution of this problem in two different methods:

Method 1:

Let $f(x) = 4x^3 + 2x^2 - 7x + 12$ by theorem of section 2.4.

we have

$$f(x) = 4x^3 + 2x^2 - 7x + 12$$

$$= f(0) + \dfrac{\Delta f(0)}{\lfloor 1} x^{(1)} + \dfrac{\Delta^2 f(0)}{\lfloor 2} x^{(2)} + \dfrac{\Delta^3 f(0)}{\lfloor 3} x^{(3)} \tag{1}$$

To get the solution of the problem, we have to find $f(0)$, $\Delta f(0)$, $\Delta^2 f(0)$ and $\Delta^3 f(0)$. To know these values, we construct a forward-difference table with arguments 0, 1, 2, 3 and the entries corresponding to these arguments.

x	$f(x)$	$\Delta f(x)$	$\Delta^2 f(x)$	$\Delta^3 f(x)$
0	$12 = f(0)$			
		$\rightarrow -1 = \Delta f(0)$		
1	11		$\rightarrow 28 = \Delta^2 f(0)$	
		$\rightarrow 27$		$\rightarrow 24 = \Delta^3 f(0)$
2	38		$\rightarrow 52$	
		$\rightarrow 79$		
3	117			

Substitute the values $f(0) = 12$, $\Delta f(0) = -1$, $\Delta^2 f(0) = 28$ and $\Delta^3 f(0) = 24$ in the Eq. (1), we get $f(x) = 12 - x^{(1)} + 14\,x^{(2)} + 4\,x^{(3)}$;
Hence, the given polynomial in factorial function is

$$4x^{(3)} + 14x^{(2)} - x^{(1)} + 12.$$

Method 2:

Let $4x^3 + 2x^2 - 7x + 12 = Ax^{(3)} + Bx^{(2)} + Cx^{(1)} + D$,

where A, B, C and D are constants,

i.e., $4x^3 + 2x^2 - 7x + 12$

$$= A\,x(x-1)\,(x-2) + Bx\,(x-1) + Cx + D \tag{2}$$

Put $x = 0$ in (2); we get

$$D = 12 \tag{3}$$

Next put $x = 1$ in (2); we get

$$C + D = 11.$$

Using (3), we have

$$C = -1 \tag{4}$$

Put $x = 2$ in (2); we get

$$32 + 8 - 14 + 12 = 0 + 2B + 2C + D.$$

Using (3) and (4), we have

$$B = 14 \tag{5}$$

Comparing the coefficient of x^3 on both sides of (2), we get $A = 4$.

Thus, $\qquad f(x) = 4x^{(3)} + 14x^{(2)} - x^{(1)} + 12.$

We can also compute the constants A, B, C, and D by the method of detached coefficient of the given polynomial

```
1 | 4          2          -7    |12 = D
  | 0     +(1)(4)       +(1)(6)
2 | 4          6              |    -1 = C
  | 0     +(2)(4)
3 | 4              |14 = B
  | 0
  | 4 = A
```

Problem 2: Express $x^4 - 12x^3 + 24x^2 - 30x + 9$ in factorial function and find its differences of all orders.

Solution: Let $f(x) = x^4 - 12x^3 + 24x^2 - 30x + 9$

$$= Ax^{(4)} + Bx^{(3)} + Cx^{(2)} + Dx^{(1)} + E,$$

where A, B, C, D and E are the constants.

$$
\begin{array}{r|lllll}
1 & 1 & -12 & 24 & -30 & \underline{9 = E} \\
  & 0 & +(1)(1) & +(1)(-11) & +(1)(13) & \\ \hline
2 & 1 & -11 & 13 & \underline{-17 = D} & \\
  & 0 & +(2)(1) & +(2)(-9) & & \\ \hline
3 & 1 & -9 & \underline{-5 = C} & & \\
  & 0 & +(3)(1) & & & \\ \hline
4 & 1 & \underline{-6 = B} & & & \\
  & 0 & & & & \\ \hline
  & \underline{1 = A} & & & &
\end{array}
$$

Thus, $\quad f(x) = x^{(4)} - 6x^{(3)} - 5x^{(2)} - 17x^{(1)} + 9$

Hence, $\Delta f(x) = 4x^{(3)} - 18x^{(2)} - 10x^{(1)} - 17$,

$$\Delta^2 f(x) = 12x^2 - 36x^{(1)} - 10x, \ \Delta^3 f(x) = 24x^{(1)} - 36$$

$$\Delta^4 f(x) = 24 \text{ and } \Delta^k f(x) = 0 \quad \text{ for } k \geq 5.$$

Problem 3: Show that $x^{(-1)} = \dfrac{1}{x+h}$.

Solution: We know that

$$x^{(n)} = (x - \overline{n-1}\ h)\, x^{(n-1)}.$$

Put $\qquad\qquad\qquad n = 0;$

$$x^{(0)} = (x + h)\, x^{(-1)}$$

$\therefore \qquad\qquad (x + h)\, x^{(-1)} = 1;$

Hence, $\qquad\qquad x^{(-1)} = \dfrac{1}{x+h}.$

Problem 4: Show that $x^{(-n)} = \dfrac{1}{(x + nh)^{(n)}}$, where n is a positive integer.

Solution: We will prove it by the principle of mathematical induction, for $n = 1$.

$$x^{(-1)} = \frac{1}{(x+h)^{(1)}}$$

Clearly, it is true for $n = 1$ (see problem 3).
Suppose, the result is true for $n = k$, then

$$x^{(-k)} = \frac{1}{(x+kh)^{(k)}} \cdot \qquad (1)$$

We know that

$$x^{(n)} = (x - \overline{n-1}\ h)\, x^{(n-1)};$$

Therefore,

$$x^{(-k)} = \{x - \overline{-k-1}\ h\}\, x^{(-k-1)};$$

Thus,

$$x^{(-k-1)} = \frac{1}{(x + \overline{k+1}h)} \cdot x^{(-k)}$$

In view of (1),

$$x^{(-k-1)} = \frac{1}{(x + \overline{k+1}h)(x+kh)^{(k)}}$$

$$= \frac{1}{(x + \overline{k+1}h)^{(k+1)}} \cdot$$

Hence, the result is true for $n = k + 1$.

Problem 5: Show that $\Delta x^{(-n)} = (-n)h\, x^{(-n-1)}$.

Solution: By the definition of $\Delta x^{(-n)} = (x+h)^{(-n)} - x^{(-n)}$. Using the result of problem 4 we have

$$\Delta x^{(-n)} = \frac{1}{(x+h+nh)^{(n)}} - \frac{1}{(x+nh)^{(n)}} = \frac{1}{(x+\overline{n+1}h)^{(n)}} - \frac{1}{(x+nh)^{(n)}}$$

$$= \frac{1}{(x+\overline{n+1}h)(x+nh)(x+\overline{n-1}h).......(x+2h)}$$

$$- \frac{1}{(x+nh)(x+\overline{n-1}h).....(x+2h)(x+h)}$$

$$= \frac{1}{(x+nh)(x+\overline{n-1}h).....(x+2h)} \left\{ \frac{1}{x+(n+1)h} - \frac{1}{x+h} \right\}$$

$$= \frac{x+h-x-(n+1)h}{(x+\overline{n+1}h)(x+nh).....(x+2h)(x+h)}$$

$$= \frac{-nh}{(x+\overline{n+1}h)^{(n+1)}} = -nh.x^{(-n-1)}.$$

2.5   BACKWARD DIFFERENCES

Let y_0, y_1, ... y_n be $(n+1)$ values of $y = f(x)$ at $x = x_0$, x_1, x_2, ... x_n, where $x_k = a + kh(k = 0, 1, 2,... n)$ and h is the interval of differencing.

The backward-difference operator ∇ is defined by the equation $\nabla f(x) = f(x) - f(x-h)$ and $\nabla f(x)$ is called first-order backward difference; for example, $\nabla f(a+h) = f(a+h) - f(a)$, $\nabla f(a+2h) = f(a+2h) - f(a+h)$, etc.

The higher order backward difference is defined by

$$\nabla^n f(x) = \nabla(\nabla^{n-1} f(x)) \tag{1}$$

For example, $\nabla^2 f(x) = \nabla(\nabla f(x) = \nabla\{f(x) - f(x-h)\}$
$$= \nabla f(x) - \nabla f(x-h) = \{f(x) - f(x-h)\}$$
$$- \{f(x-h) - f(x-2h)\}$$
$$= f(x) - 2f(x-h) + f(x-2h).$$

Note that $\nabla^2 f(x) = \nabla f(x) - \nabla f(x-h)$. Hence, the second-order backward differences are the differences of first-order backward differences.

A table showing the arguments, entries and backward differences is called the backward difference table. The backward difference table with five entries is shown below.

Table 2.4

x	y	∇y	$\nabla^2 y$	$\nabla^3 y$	$\nabla^4 y$
x_0	y_0				
x_1		$\rightarrow \nabla y_1$	$\rightarrow \nabla^2 y_2$		
	y_1	$\rightarrow \nabla y_2$		$\rightarrow \nabla^3 y_3$	

$$
\begin{array}{llll}
x_2 & y_2 & \rightarrow \nabla^2 y_3 & \rightarrow \nabla^4 y_4 \\
 & & \rightarrow \nabla y_3 & \rightarrow \nabla^3 y_4 \\
x_3 & y_3 & \rightarrow \nabla^2 y_4 & \\
 & & \rightarrow \nabla y_4 & \\
x_4 & y_4 & &
\end{array}
$$

where $y_k = f(x_0 + kh)$ or $f(a + kh)$.

2.6 SHIFT OPERATOR, CENTRAL-DIFFERENCE OPERATOR, AVERAGE OPERATOR AND DIFFERENTIAL OPERATOR

The shift operator E is defined by the following equation

$$E f(x) = f(x + h),$$

and in general

$$E^n f(x) = f(x + nh).$$

The central-difference operator is defined by

$$\delta f(x) = f\left(x + \frac{h}{2}\right) - f\left(x - \frac{h}{2}\right),$$

and in general,

$$\delta^n f(x) = \delta(\delta^{n-1} f(x)).$$

The average difference operator is defined by

$$\mu f(x) = \frac{f\left(x + \dfrac{h}{2}\right) + f\left(x - \dfrac{h}{2}\right)}{2}$$

In general,

$$\mu^n f(x) = \mu(\mu^{n-1} f(x))$$

The differential operator D is defined by

$$D f(x) = f'(x)$$

and

$$D^n f(x) = \frac{d^n}{dx^n} f(x);$$

we consider 1 as a unit operator since $1 \cdot f(x) = f(x)$.

2.7   EQUALITY OF OPERATORS

If M and N are expressions in terms of one more operators $\Delta, \nabla, E, 1,$ D, δ and μ, then we say that M and N are equal if

$$M f(x) = N f(x),$$

and in this case, we write $M = N$.

2.8   RELATIONS BETWEEN THE OPERATORS

1. $\Delta - \nabla = \Delta \nabla$.

Proof: Let $f(x)$ be any arbitrary function.
Consider

$$\begin{aligned}
(\Delta \nabla) f(x) &= \Delta (\nabla f(x)) \\
&= \Delta \{ f(x) - f(x - h) \} \\
&= \Delta f(x) - \{ f(x - h + h) - f(x - h) \} \\
&= \Delta f(x) - \nabla f(x) \\
&= (\Delta - \nabla) f(x)
\end{aligned}$$

$$\therefore \qquad \Delta \nabla = \Delta - \nabla.$$

2. $E = 1 + \Delta$.

Proof:
$$\begin{aligned}
E f(x) &= f(x + h) \\
&= f(x + h) - f(x) + f(x) \\
&= \Delta f(x) + 1 . f(x) \\
&= (\Delta + 1) f(x)
\end{aligned}$$

$$\therefore \qquad E = 1 + \Delta.$$

3. $E \nabla = \nabla E = \Delta$.

Proof: For any arbitrary function $f(x)$

$$\begin{aligned}
E \nabla f(x) &= E \{ f(x) - f(x - h) \} \\
&= E f(x) - E f(x - h) \\
&= f(x + h) - f(x) \\
&= \Delta f(x);
\end{aligned}$$

Similarly,

$$\nabla E f(x) = \Delta f(x)$$

$$\therefore E \Delta = \Delta E = \Delta.$$

4. $(1 + \Delta)(1 - \nabla) = 1.$

Proof: For any function $f(x)$

$$(1 + \Delta)(1 - \nabla) f(x)$$
$$= (1 + \Delta)\{f(x) - \nabla f(x)\}$$
$$= (1 + \Delta)\{f(x) - (f(x) - f(x - h))\}$$
$$= (1 + \Delta) f(x - h)$$
$$= f(x - h) + \Delta f(x - h)$$
$$= f(x - h) + f(x) - f(x - h)$$
$$= f(x)$$
$$= 1 . f(x)$$

$\therefore (1 + \Delta)(1 - \Delta) = 1.$

5. $E = e^{hD} = 1 + \Delta.$

Proof: By definition of E, we have

$$Ef(x) = f(x + h)$$

$$= f(x) + hf'(x) + \frac{h^2}{\lfloor 2} f''(x) + \ldots\ldots\text{(By Taylor's theorem)}$$

$$= 1.f(x) + h.Df(x) + \frac{h^2}{\lfloor 2} D^2 f(x)\ldots\ldots$$

$$= \left(1 + hD + \frac{h^2 D^2}{\lfloor 2} + \ldots\ldots\right) f(x)$$

$$= e^{hD} f(x)$$

$\therefore E = e^{hD}$

By 2 of 2.8, and $E = e^{hD}$, we have

$$E = e^{hD} = 1 + \Delta.$$

6. $\delta = E^{\frac{1}{2}} - E^{-\frac{1}{2}}.$

Proof: Since $\delta f(x) = f\left(x + \frac{h}{2}\right) - f\left(x - \frac{h}{2}\right)$, we have

$$\delta f(x) = E^{\frac{1}{2}} f(x) - E^{-\frac{1}{2}} f(x)$$

$$= \left(E^{\frac{1}{2}} - E^{-\frac{1}{2}}\right) f(x)$$

$$\therefore \qquad \delta = E^{\frac{1}{2}} + E^{-\frac{1}{2}}.$$

7. $\mu = \dfrac{E^{\frac{1}{2}} + E^{-\frac{1}{2}}}{2}.$

Proof: As $\quad \mu f(x) = \dfrac{f\left(x + \dfrac{h}{2}\right) + f\left(x - \dfrac{h}{2}\right)}{2}$

$$= \frac{1}{2}\left(E^{\frac{1}{2}} + E^{-\frac{1}{2}}\right) f(x)$$

$$\therefore \qquad \mu = \frac{1}{2}\left(E^{\frac{1}{2}} + E^{-\frac{1}{2}}\right).$$

8. $\mu^2 = 1 + \dfrac{\delta^2}{4}.$

Proof: Consider $\mu^2 = \left(\dfrac{E^{\frac{1}{2}} + E^{-\frac{1}{2}}}{2}\right)^2$

$$= \frac{\left(E^{\frac{1}{2}} + E^{-\frac{1}{2}}\right)^2}{4}$$

$$= \frac{\left(E^{\frac{1}{2}} - E^{-\frac{1}{2}}\right)^2 + 4}{4} = \frac{\delta^2 + 4}{4}$$

$$= \frac{\delta^2}{4} + 1$$

Thus $\mu^2 = 1 + \dfrac{\delta^2}{4}.$

9. $E^{\frac{1}{2}} = \mu + \dfrac{\delta}{2}.$

Proof: We know that

$$\mu = \frac{1}{2}\left(E^{\frac{1}{2}} + E^{-\frac{1}{2}}\right),$$

or

$$\delta = \left(E^{\frac{1}{2}} - E^{-\frac{1}{2}}\right)$$

$$\therefore \qquad \mu + \frac{\delta}{2} = E^{\frac{1}{2}}.$$

10. $hD = -\log(1 - \nabla)$.

Proof: By Taylor's expansion,

$$f(x - h) = f(x) - \frac{h}{\lfloor 1} f'(x) + \frac{h^2}{\lfloor 2} f''(x) - \ldots$$

$$E^{-1}f(x) = [1 - \frac{hD}{\lfloor 1} + \frac{h^2}{\lfloor 2} D^2 \ldots\ldots]f(x)$$

$$= e^{-hD} f(x)$$

$$\therefore \qquad E^{-1} = e^{-1D} \qquad\qquad (1)$$

Now,

$$\nabla f(x) = f(x) - f(x - h)$$

$$= (1 - E^{-1})f(x)$$

$$\therefore \qquad \nabla = 1 - E^{-1}$$

$$E^{-1} = 1 - \nabla \qquad\qquad (2)$$

From (1) and (2),

$$1 - \nabla = e^{-hD}$$

$$-hD = \log(1 - \nabla)$$

$$hD = -\log(1 - \nabla);$$

Using the relation $\Delta = E - 1$, the solution of problem 2 of section 2.3 can be obtained in less number of computations in the following way.
$\Delta^3 f(a) = (E - 1)^3 f(a) = E^3 f(a) - 3 E^3 f(a) + 3 E^2 f(a) + 3 E f(a) + f(a) = f(a + 3h) - 3f(a + 2h) + 3f(a + h) + f(a)$.

SOLVED PROBLEMS

Problem 1: Construct a backward difference table for the data given below.

x	2	4	6	8
$f(x)$	15	25	37	49

Solution: The backward difference table for the given data is as follows:

x	$f(x)$	$\nabla f(x)$	$\nabla^2 f(x)$	$\nabla^3 f(x)$
2	$y_0 = 15$			
		$\to \nabla y_1 = 10$		
4	$y_1 = 25$		$\to \nabla^2 y_2 = 2$	
		$\to \nabla y_2 = 12$		$\to \nabla^3 y_3 = -2$
6	$y_2 = 37$		$\to \nabla^2 y_3 = 0$	
		$\to \nabla y_3 = 12$		
8	$y_3 = 49$			

Note that the forward-difference table may be similar to it as per the numerical values concerned but first backward differences $\nabla y_1 = 10$, $\nabla y_2 = 12$ and $\nabla y_3 = 12$ represent Δy_0, Δy_1 and Δy_2, respectively.

Problem 2: Obtain the missing term in the following table:

x	0	1	2	3	4
$f(x)$	1	3	9	–	81

Solution: We were given the values of $f(x)$ at $x = 0, 1, 2, 4$; since $f(x)$ is known at 4 points of x, we can find $f(x)$ as a polynomial of third degree that assumes 1, 3, 9, 81 at $x = 0, 1, 2, 4$ respectively. Therefore, we assume $f(x)$ is polynomial of a third degree. Thus, by theorem 2.4, we have

$$\Delta^4 f(x) = 0$$

i.e., $(E - 1)^4 f(x) = 0$

i.e., $f(x + 4h) - 4f(x + 3h) + 6f(x + 2h) - 4f(x + h) + f(x) = 0 \quad (1)$

For $h = 1$ and $x = 0$, the above equation becomes

$$f(4) - 4f(3) + 6f(2) - 4f(1) + f(0) = 0$$

Substituting $f(0), f(1), f(2)$ and $f(4)$ values, we have
$$81 - 4f(3) + 6(9) - 4(3) + 1 = 0;$$
Thus, $f(3) = 31.$

Problem 3: Obtain two missing terms in the following table:

x	2	2.1	2.2	2.3	2.4	2.5	2.6
$f(x)$	5.132	–	5.236	5.346	–	5.632	5.842

Solution: As $f(x)$ is known at 5 points of x, we assume that $f(x)$ is a polynomial of fourth degree. Therefore, by theorem 2.4,
$$\Delta^5 f(x) = 0,$$
i.e., $f(x + 5h) - 5 f(x + 4h) + 10 f(x + 3h) - 10 f(x + 2h)$
$$+ 5 f(x + h) - f(x) = 0. \tag{1}$$

Put $x = 2.0$ and $h = 0.1$ in (1); we get
$$f(2.5) - 5 f(2.4) + 10 f(2.3) - 10 f(2.2) + 5 f(2.1) - f(2.0) = 0. \tag{2}$$
Now, put $x = 2.1$ and $h = 0.1$, in Eq. (1); we obtain
$$f(2.6) - 5 f(2.5) + 10 f(2.4) - 10 f(2.3) + 5 f(2.2) - f(2.0) = 0. \tag{3}$$

Now, substituting the values of $f(2), f(2.2), f(2.3), f(2.5)$ and $f(2.6)$ in Eqs. (2) and (3) we obtain simultaneous equations involving $f(2.1)$ and $f(2.4)$ solving them, we get $f(2.1) = 5.4253$ and $f(2.4) = 5.7453$.

Problem 4: Show that $f(a + nh) = \displaystyle\sum_{r=0}^{n} {}^nC_r \, \Delta^r f(a).$

Solution: Consider $f(a + nh) = E^n f(a)$. We know that $E = 1 + \Delta$. Therefore,
$$f(a + nh) = (1 + \Delta)^n f(a)$$
$$= \left(\sum_{r=0}^{n} {}^nC_r \Delta^n \right) f(a) = \sum_{r=0}^{n} {}^nC_r \, \Delta^n f(a).$$

Problem 5: Show that
$$\Delta^n \cos(ax + b) = \left(2 \sin\frac{ah}{2} \right)^n \cos\left\{ ax + b + n\left(\frac{ah + \pi}{2} \right) \right\} \tag{1}$$
where n is a positive integer.

Solution: For $n = 1$,

$$\Delta \cos(ax + b) = \cos\{a(x + h) + b\} - \cos(ax + b)$$

$$= \cos(ax + ah + b) - \cos(ax + b)$$

$$= -2\sin \frac{ax + ah + b + ax + b}{2} \, \sin \frac{ax + ah + b - ax - b}{2}$$

$$= -2\sin \left(ax + b + \frac{ah}{2} \right) \sin \left(\frac{ah}{2} \right)$$

$$= \left(2\sin \frac{ah}{2} \right) \cdot \cos \left(ax + b + \frac{ah}{2} + \frac{\pi}{2} \right)$$

$$= 2\sin \frac{ah}{2} \, \cos \left\{ ax + b + \left(\frac{ah + \pi}{2} \right) \right\} ; \tag{2}$$

$\therefore$ The result (1) is true for $n = 1$; Suppose the result (1) is true for $n = k$, then

$$\Delta^k \cos(ax + b) = \left(2\sin \frac{ah}{2} \right)^k \cdot \cos \left\{ ax + b + k \left(\frac{ah + \pi}{2} \right) \right\} . \tag{2}$$

Now, for $n = k + 1$,

$$\Delta^{k+1} \cos(ax + b) = \Delta\{\Delta^k \cos(ax + b)\}$$

$$= \Delta \left[\left(2\sin \frac{ah}{2} \right)^k \cdot \cos \left\{ ax + b + k \left(\frac{ah + \pi}{2} \right) \right\} \right] \quad \text{(by eqution 2)}$$

$$= \left(2\sin \frac{ah}{2} \right)^{k.} \Delta \cos \left\{ ax + b + k \left(\frac{ah + \pi}{2} \right) \right\} \tag{3}$$

Since $\Delta \cos(ax + b) = 2\sin \frac{ah}{2} \cdot \cos \left(ax + b + \frac{ah}{2} + \frac{\pi}{2} \right)$,

we have $\Delta \cos \left\{ ax + b + k \left(\frac{ah + \pi}{2} \right) \right\}$

$$= 2\sin \frac{ah}{2} \cdot \cos \left\{ ax + b + k \left(\frac{ah + \pi}{2} \right) + \left(\frac{ah + \pi}{2} \right) \right\}$$

$$= 2\sin \frac{ah}{2} \, \cos \left\{ ax + b + (k + 1) \left(\frac{ah + \pi}{2} \right) \right\} . \tag{4}$$

From Eqs. (3) and (4), we get

$$\Delta^{k+1}\cos(ax + h) = \left(2\sin\frac{ah}{2}\right)^{k+1} \cdot \cos\left\{ax + b + (k+1)\left(\frac{ah + \pi}{2}\right)\right\}.$$

$\therefore$ The result (1) is true for $n = k + 1$. Hence, by the principle of mathematical induction, the result is true for every positive integer n.

Problem 6: Express $f(x) = \dfrac{x-3}{(x+1)(x+3)}$ in terms of reciprocal factorial functions.

Solution: $f(x) = \dfrac{x-3}{(x+1)(x+3)}$ can be expressed as

$$f(x) = \frac{(x-3)(x+2)}{(x+1)(x+2)(x+3)}$$

$$= \frac{x^2 - x - 6}{(x+1)(x+2)(x+3)} = \frac{x(x+3) - 4(x+3) + 6}{(x+1)(x+2)(x+3)}$$

$$= \frac{(x+3)(x-4) + 6}{(x+1)(x+2)(x+3)} = \frac{x-4}{(x+1)(x+2)} + \frac{6}{(x+1)(x+2)(x+3)}$$

$$= \frac{(x+2-6)}{(x+1)(x+2)} + 6x^{(-3)} = \frac{1}{x+1} - \frac{6}{(x+1)(x+2)} + 6x^{(-3)}$$

$$= x^{(-1)} - 6x^{(-2)} + 6x^{(-3)}.$$

Problem 7: Prove that

$$u_0 + \frac{u_1 x}{\lfloor 1} + \frac{u_2 x^2}{\lfloor 2} + \frac{u_3 x^3}{\lfloor 3} + \ldots = e^x\left[u_0 + x\Delta u_0 + \frac{x^2}{\lfloor 2}\Delta^2 u_0 \ldots\ldots\right]. \quad (1)$$

Solution: Observe that the suffixes of u in the L.H.S. of Eq. (1) are 0, 1, 2, ..., whereas the suffixes of u on R.H.S of (1) are zero. So, we consider the L.H.S. of (1) and try to make each suffix of u as 0 using the operator E. Now consider

$$u_0 + \frac{u_1 x}{\lfloor 1} + \frac{u_2 x^2}{\lfloor 2} + \frac{u_3 x^3}{\lfloor 3} + ...$$

$$= u_0 + \frac{x}{\lfloor 1} Eu_0 + \frac{x^2}{\lfloor 2} E^2 u_0 + \frac{x^3}{\lfloor 3} E^3 u_0 ......$$

$$= \left(1 + \frac{x}{\lfloor 1} E + \frac{x^2}{\lfloor 2} E^2 + \frac{x^3}{\lfloor 3} E^3 + ... \right) u_0$$

$$= e^{xE} u_0$$

$$= e^{x(1+\Delta)} u_0$$

$$= e^x . e^{x\Delta} u_0$$

$$= e^x \left[1 + x\Delta + \frac{x^2 \Delta^2}{\lfloor 2} + \frac{x^3 \Delta^3}{\lfloor 3} + .. \right] u_0$$

$$= e^x \left[u_0 + x\Delta u_0 + \frac{x^2}{\lfloor 2} \Delta^2 u_0 + ... \right],$$

which is the R.H.S. of (1).

Problem 8: Show that

$$u_x = u_{x-1} + \Delta u_{x-2} + \Delta^2 u_{x-3} + ...+ \Delta^{n-1} u_{x-n} + \Delta^n u_{x-n}. \tag{1}$$

Solution: Observe that the term $\Delta^n u_{x-n}$ of R.H.S. of (1) is odd as per its earlier/previous terms, therefore, there is a possibility of obtaining it by showing

$$u_x - \Delta^n u_{x-n} = u_{x-1} + \Delta u_{x-2} + \Delta^2 u_{x-3} + ... + \Delta^{n-1} u_{x-n}. \tag{2}$$

Consider

$$u_x - \Delta^n u_{x-n}$$

$$= u_x - \Delta^n E^{-n} u_x$$

$$= (1 - \Delta^n E^{-x}) u_x$$

$$= \frac{E^n - \Delta^n}{E^n} u_x$$

$$= \frac{(E-\Delta)(E^{n-1} + E^{n-2}\Delta + E^{n-3}\Delta^2 + \ldots\ldots\Delta^{n-1})}{E^n} u_x$$

$$= (E^{-1} + E^{-2}\Delta + E^{-3}\Delta^2 \ldots\ldots\ldots + E^{-n}\Delta^{n+1})u_x \ (\because E - \Delta = 1)$$

$$= u_{x-1} + \Delta u_{x-2} + \Delta^2 u_{x-3} + \ldots\ldots + \Delta^{n-1} u_{x-n}.$$

Hence,

$$u_x = u_{x-1} + \Delta u_{x-2} + \Delta^2 u_{x-3} + \ldots\ldots + \Delta^{n-1} u_{x-n} + \Delta^n u_{x-n}.$$

Problem 9: Prove that $u_1 x + u_2 x^2 + u_3 x^3 + \ldots\ldots$

$$= \frac{x}{1-x}u_1 + \frac{x^2}{(1-x)^2}\Delta u_1 + \frac{x^3}{(1-x)^3}\Delta^2 u_1 + \frac{x^4}{(1-x)^4}\Delta^3 u_1 + \ldots\ldots(1).$$

Solution: R.H.S of (1)

$$= \frac{x}{1-x}u_1 + \frac{x^2}{(1-x)^2}(E-1)u_1 + \frac{x^3}{(1-x)^3}(E-1)^2 u_1$$

$$+ \frac{x^4}{(1-x)^4}(E-1)^3 u_1 + \ldots.$$

$$= \frac{x}{1-x}u_1 + \frac{x^2}{(1-x)^2}Eu_1 - \frac{x^2}{(1-x)^2}u_1 + \frac{x^3}{(1-x)^3}E^2 u_1 - \frac{2x^3}{(1-x)^3}Eu_1$$

$$+ \frac{x^3}{(1-x)^3}u_1 + \frac{x^4}{(1-x)^4}E^3 u_1 - \frac{3x^4}{(1-x)^4}E^2 u_1$$

$$+ \frac{3x^4}{(1-x)^4}E u_1 - \frac{x^4}{(1-x)^4}u_1 + \ldots\ldots$$

$$= \frac{x}{1-x} - \frac{x^2}{(1-x)^2} + \frac{x^3}{(1-x)^3} - \frac{x^4}{(1-x)^4} + \ldots.\}u_1$$

$$+ \left\{\frac{x^2}{(1-x)^2} - \frac{2x^3}{(1-x)^3} + \frac{3x^4}{(1-x)^4} \ldots\ldots\right\}Eu_1$$

$$+ \left\{\frac{x^3}{(1-x)^4} - \frac{3x^4}{(1-x)^4} + \ldots\ldots\right\}E^2 u_1$$

$$+ \left\{\frac{x^4}{(1-x)^4} - \frac{4.x^5}{(1-x)^5} \ldots\ldots\right\}E^3 u_1 + \ldots.$$

$$= \frac{x}{1-x}\left\{1 - \frac{x}{1-x} + \frac{x^2}{(1-x)^2} + \frac{x^3}{(1-x)^3} - \frac{x^4}{(1-x)^4} + ...\right\}u_1$$

$$+ \frac{x^2}{(1-x)^2}\left\{1 - 2\frac{x}{1-x} + \frac{3x^2}{(1-x)^2} ....\right\}u_2$$

$$+ \frac{x^3}{(1-x)^3}\left\{1 - 3\frac{x}{1-x} + ...\right\}u_3$$

$$+ \frac{x^4}{(1-x)^4}\left\{1 - 4\frac{x}{1-x} ....\right\}u_4 + ...$$

$$= \frac{x}{1-x}\left\{1 + \frac{x}{1-x}\right\}^{-1} u_1 + \frac{x^2}{(1-x)^2}\left\{1 + \frac{x}{1-x}\right\}^{-2} u_2$$

$$+ \frac{x^3}{(1-x)^3}\left\{1 + \frac{x}{1-x}\right\}^{-3} u_3 + \frac{x^4}{(1-x)^4}\left\{1 + \frac{x}{1-x}\right\}^{-4} u_4 + ...$$

$$= x\,u_1 + x^2\,u_2 + x^3\,u_3 + x^4\,u_4 + ...$$

Problem 10: Prove that

$$u_0 - u_1 + u_2 + ... = \frac{1}{2}u_0 - \frac{1}{4}\Delta u_0 + \frac{1}{8}\Delta^2 u_0 - \frac{1}{16}\Delta^3 u_0 ....(1)$$

Solution: L.H.S of $(1) = u_0 - Eu_0 + E^2u_0 + E^3u_0 + ...$

$$= (1 - E + E^2 - E^3 + ...)u_0$$
$$= (1 + E)^{-1} u_0$$
$$= (1 + 1 + \Delta)^{-1} u_0$$
$$= (2 + \Delta)^{-1} u_0$$
$$= \frac{1}{2}\left(1 + \frac{\Delta}{2}\right)^{-1} u_0$$
$$= \frac{1}{2}\left\{1 - \frac{\Delta}{2} + \frac{\Delta^2}{4} - \frac{\Delta^3}{8} + ...\right\}u_0$$
$$= \frac{1}{2}u_0 - \frac{\Delta u_0}{4} + \frac{\Delta^2 u_0}{8} - \frac{\Delta^3 u_0}{16} + ...$$

Problem 11: Find $\Delta^n e^{ax}$, where n is a positive integer by definition of forward-difference operator.

Solution: $\Delta e^{ax} = e^{a(x+h)} - e^{ax} = e^{ax}(e^{ah} - 1).$ $\qquad\qquad$ (1)

$$\Delta^2 e^{ax} = \Delta(\Delta e^{ax}) = \Delta\{e^{ax}(e^{ah} - 1)\}$$
$$= (e^{ah} - 1)\,\Delta\,e^{ax}$$
$$= (e^{ah} - 1)\,(e^{ah} - 1)\,e^{ax} \text{ by (1)}$$
$$= (e^{ah} - 1)^2\,e^{ax}.$$

Similarly, it can be shown that

$$\Delta^3 e^{ax} = (e^{ah} - 1)^3.e$$

In general,

$$\Delta^n e^{ax} = (e^{ah} - 1)^n.e^{ah}.$$

Problem 12: Prove that

$$u_x - \frac{1}{8}\Delta^2 u_{x-1} + \frac{1.3}{8.16}\Delta^4 u_{x-2} - \frac{1.3.5}{8.16.24}.\Delta^6 u_{x-3} + \ldots\ldots$$

$$= u_{x+\frac{1}{2}} - \frac{1}{2}\Delta u_{x+\frac{1}{2}} + \frac{1}{4}\Delta^2 u_{x+\frac{1}{2}}\ldots\ldots \qquad\qquad (1)$$

Solution: The L.H.S. of (1)

$$= u_x - \frac{\Delta^2 E^{-1}}{8}u_x + \frac{1.3}{8.16}\Delta^4 E^{-2}u_x - \frac{1.3.5}{8.16.24}\Delta^6 E^{-3}u_x + \ldots\ldots$$

$$= \left[1 - \frac{\Delta^2 E^{-1}}{4.2} + \frac{1.3}{2.4}\frac{\Delta^4 E^{-2}}{4^2} - \frac{1.3.5}{2.4.6}\frac{\Delta^6 E^{-3}}{4^3} + \ldots\ldots\right]u_x$$

$$= \left[1 - \frac{1}{2}\frac{\Delta^2 E^{-1}}{4} + \frac{\frac{1}{2}\frac{3}{2}}{\lfloor 2}\left(\frac{\Delta^2 E^{-1}}{4}\right)^2 - \frac{\frac{1}{2}\cdot\frac{3}{2}\cdot\frac{5}{2}}{\lfloor 3}\left(\frac{\Delta^2 E^{-1}}{4}\right)^3 + \ldots\right]u_x$$

$$= \left[1 + \frac{\Delta^2 E^{-1}}{4}\right]^{-\frac{1}{2}}u_x$$

$$= \left[E^{-1} \cdot \left(E + \frac{\Delta^2}{4} \right) \right]^{-\frac{1}{2}} u_x = E^{+\frac{1}{2}} \left(E + \frac{\Delta^2}{4} \right)^{-\frac{1}{2}} u_x$$

$$= \left[E + \frac{(E-1)^2}{4} \right]^{-\frac{1}{2}} u_{x+\frac{1}{2}} = \left[\frac{(E+1)^2}{4} \right]^{-\frac{1}{2}} u_{x+\frac{1}{2}}$$

$$= \left[\frac{(2+\Delta)^2}{4} \right]^{-\frac{1}{2}} u_{x+\frac{1}{2}} = \left[\left(1 + \frac{\Delta}{2} \right)^2 \right]^{-\frac{1}{2}} u_{x+\frac{1}{2}}$$

$$= \left(1 + \frac{\Delta}{2} \right)^{-1} u_{x+\frac{1}{2}} = \left(1 - \frac{\Delta}{2} + \frac{\Delta^2}{4} - \frac{\Delta^3}{8} + \dots \right) u_{x+\frac{1}{2}}$$

$$= u_{x+\frac{1}{2}} - \frac{1}{2} \, \Delta u_{x+\frac{1}{2}} + \frac{1}{4} \Delta^2 u_{x+\frac{1}{2}} - \frac{1}{8} \Delta^3 u_{x+\frac{1}{2}}.$$

2.9   ANTI-DIFFERENCE OPERATOR Δ^{-1}

We know there that if $\dfrac{d}{dx} f(x) = F(x)$, then $\int F(x)dx = f(x) + c$, where c is an arbitrary constant; $f(x) + c$ is called anti-derivatives of $F(x)$. Likewise, if $\Delta f(x) = g(x)$, then $f(x) + c$ is defined as anti-difference of $g(x)$ and is denoted by

$$\frac{1}{\Delta} g(x) = f(x) + c,$$

or

$$\Delta^{-1} g(x) = f(x) + c;$$

For example, $\Delta e^{ax} = e^{ax}(e^a - 1)$; therefore

$$\frac{1}{\Delta} e^{ax} = \frac{e^{ax}}{(e^a - 1)} + c.$$

The operator Δ^{-1} is called anti-difference operator.

Theorem:

$$\frac{1}{\Delta} x^{(r)} = \frac{x^{(r+1)}}{(r+1)} + c, \text{ where } r \text{ is an integer and } r \neq -1.$$

Proof: Consider $\Delta \left\{ \dfrac{x^{(r+1)}}{r+1} + c \right\}$

$$= x^{(r)}$$

Therefore, by the definition of Δ^{-1} i.e., $\left(\dfrac{1}{\Delta} \right)$, we have

$$\frac{1}{\Delta} x^{(r)} = \frac{x^{(r+1)}}{(r+1)} + c.$$

SOLVED PROBLEMS

Problem 1: Find a polynomial function whose first difference is $6x^2 + 2$.

Solution: Let $f(x)$ be the required polynomial; then
$$\Delta f(x) = 6x^2 + 2 \tag{1}$$
Now, we express $6x^2 + 2$ in factorial functions
Let $\qquad\qquad 6x^2 + 2 = Ax^{(2)} + Bx^{(1)} + c$

$$
\begin{array}{r|lll}
1 & 6 & 0 & \underline{\,2 = C} \\
  & 0 & +(1)(6) & \\
\cline{2-3}
2 & 6 & \underline{\,6 = B} & \\
  & 0 & & \\
\cline{2-2}
  & \underline{\,6 = A} & &
\end{array}
$$

Thus,
$$6x^2 + 2 = 6x^{(2)} + 6x^{(1)} + 2 \tag{2}$$
From equations (1) and (2), we have
$$\Delta f(x) = 6x^{(2)} + 6x^{(1)} + 2x^{(0)} \; (\because\; x^{(0)} = 1).$$
Hence,
$$f(x) = \frac{1}{\Delta} \{ 6x^{(2)} + 6x^{(1)} + 2x^{(0)} \}$$

$$= \frac{6x^{(3)}}{3} + \frac{6x^{(2)}}{2} + \frac{2x^{(1)}}{1} + c$$

$$= 2x^3 - 3x^2 + 3x + c.$$

Problem 2: Find $\dfrac{1}{\Delta} e^{5x}$.

Solution: $\Delta e^{5x} = e^{5x}(e^{5h} - 1)$,

i.e.,
$$\Delta \left\{ \frac{e^{5x}}{e^{5h} - 1} \right\} = e^{5x}$$

$\therefore \qquad \dfrac{1}{\Delta} e^{5x} = \dfrac{e^{5x}}{(e^{5h} - 1)} + c.$

2.10   DIFFERENCES OF ZERO

If n and m are positive integers, then $[\Delta^n x^m]_{x = 0}$ are defined as zero differences and they are denoted by $\Delta^n 0^m$. Usually h is taken as a unity.

Theorem:

$\Delta^n 0^m = n^m - {}^nC_1(n - 1)^m + {}^nC_2(n - 2)^m + \ldots + (-1)^{n-1}\, {}^nC_{n-1}.1^m$,
where n and m are positive integers.

Proof: By definition of zero differences

$$\Delta^n 0^m = [\Delta^n x^m]_{x=0}$$
$$= [(E - 1)^n x^m]_{x=0}$$
$$= [\{E^n - {}^nC_1 E^{n-1} + {}^nC_2 E^{n-2} + \ldots\ldots\ldots\ldots$$
$$(-1)^{n-1}\, {}^nC_{n-1}E + (-1)^{n-1}\, {}^nC_n \}x^m]_{x=0}$$
$$= [(x + n)^m - {}^nC_1(x + n - 1)^m + {}^nC_2(x + x - 2)^m + \ldots\ldots$$
$$+ (-1)^{n-1}\, {}^nC_{n-1}(x + 1)^m + (-1)^n\, {}^nC_n x^m]_{x=0}$$
$$= n^m - {}^nC_1(n - 1)^m + {}^nC_2(n - 2)^m \ldots\ldots\ldots + (-1)^{n-1}.1^m.$$

SOLVED PROBLEMS

Problem 1: Compute $\Delta^2 0^3$.

Solution: Here, $n = 2$ and $m = 3$

$\therefore \qquad\qquad \Delta^2 0^3 = 2^3 - {}^2C_1\, 1^3 = 8 - 2 = 6.$

Problem 2: Compute $\Delta^3 0^3$.

Solution: Here, $n = 3$ and $m = 3$

$$\therefore \qquad \Delta^3 0^3 = 3^3 - {}^3C_1\, 2^3 + {}^3C_2\, 1^3$$
$$= 27 - 24 + 3 = 6.$$

Problem 3: Compute $\Delta 0^4$, $\Delta^2 0^4$, $\Delta^3 0^4$ and $\Delta^4 0^4$.

Solution: $\Delta 0^4 = 1^4$,

$$\Delta^2 0^4 = 2^4 - {}^2C_1.1^4 = 1^4$$
$$\Delta^3 0^4 = 3^4 - {}^3C_1\, 2^4 + {}^3C_2\, 1^4$$
$$= 81 - 48 + 3 = 36$$
$$\Delta^4 0^4 = 4^4 - {}^4C_1\, 3^4 + {}^4C_2\, 2^4 - {}^4C_3\, 1^4$$
$$= 256 - 324 + 64 - 4 = -8.$$

Problem 4: Show that $\Delta^n 0^m = 0$ if $n > m$.

Solution: We know that

$$\Delta^n x^m = 0 \text{ If } n > m;$$

Therefore, for $n > m$, $\Delta^n 0^m = [\Delta^n\, x^m]_{n=0} = [0]_{n=0} = 0$.

Problem 5: Show that $\Delta^n 0^n = \lfloor n$ where n is a positive integer.

Solution: We know that $\Delta^n x^n = \lfloor n$ (with $h = 1$)

$$\therefore\ \Delta^n 0^n = [\Delta^n x^n]_{x=0} = [\lfloor n\,]_{x=0} = \lfloor n\,.$$

Problem 6: Show that the number of onto functions from A having m distinct elements into B having n distinct elements is $\Delta^n 0^m$.

Solution: We use the notation $|X| =$ number of distinct elements of a set X. Suppose the elements of B be $b_1, b_2 \ldots \ldots b_n$

and $U = \{f : A \to B \,|\, f$ is a mapping$\}$; Then $|U| = n^m$.

Let $A_i = \{f : A \to B \,|\, b_i \in B$ and b_i has no pre-image in $A\}$
$(i = 1, 2, 3, \ldots n)$. Then

$$\overline{A_i} = \{f : A \to B \,|\, b_i \subset B, \text{ and } b_i \text{ has pre-image in } A\}$$

and

$$\overline{A_1} \cap \overline{A_2} \cap .... \cap \overline{A_n} = \{f : A \to B \mid b_1, b_2, ..., b_n \in B \text{ and each } b_i \text{ has a pre-image in } A\}$$

Thus, the number of onto mapping from

$$A \text{ to } B = \left|\; \overline{A_1} \cap \overline{A_2} \cap .\overline{A_3} \cap ... \cap \overline{A_n}\;\right| = |U| - |A_1 \cup A_2 \cup ... \cup A_n|$$

Now from the principle of inclusion and exclusion,

$$|A_1 \cup A_2 \cup ... \cup A_n| = \sum_{1=1}^{n}|A_i| - \sum_{1 \le i \le j \le n}|A_i \cap A_j| + \sum_{1 \le i < j < k \le n}$$

$$|A_c \cap A_j \cap A_k| + ... + (-1)^{n-2}$$

$$\sum_{i \le i_1 < i_2 < ..... < i_{n-1} \le n} \left|A_{i_1} \cap A_{i_2} \cap .... \cap A_{i_{n-1}}\right| + (-1)^{n-1}\left|A_1 \cap A_2 \cap .... \cap A_n\right|$$

$\left|A_1\right|$ = The number of mapping from $A \to B$ such that b_1 has no pre-image in A

$$= (n-1)^m$$

$$\therefore \sum_{i=1}^{n}|A_i| = n(n-1)^m = {}^nC_1(n-1)^m;$$

Similarly

$$\sum_{1 \le i < j \le n}\left|A_i \cap A_j\right| = {}^nC_2\,(n-2)^m$$

$$............$$

$$\sum\left|A_{i_1} \cap A_{i_2} \cap .... \cap A_{i_{n-1}}\right| = {}^nC_{n-1}1^m,$$

and

$$\left|A_1 \cap A_2 \cap .... \cap A_n\right| = 0.$$

Hence,

$$\left|\; \overline{A_1} \cap \overline{A_2} \cap ..... \cap \overline{A_n}\;\right| = n^m - \{{}^nC_1(n-1)^m - {}^nC_2(n-2)^m + ...$$

$$+ (-1)^{n-1}\,{}^nC_{n-1}\,1^m + (-1)^n\,0\}$$

$$= n^m - {}^nC_1(n-1)^m + {}^nC_2(n-2)^m + ...$$

$$+ (-1)^{n-1}\,{}^nC_{n-1}\,1^m$$

$$= \Delta^n 0^m \text{ (By theorem of 2.10).}$$

Problem 7: Prove that $\Delta^n\, 0^m = n[\Delta^{n-1}\, 0^{m-1} + \Delta^n\, 0^{m-1}]$.

Solution: We know that

$$\Delta^n 0^m = n^m - {}^nC_1(n-1)^m + {}^nC_2(n-2)^m + \ldots\ldots + (-1)^{n-1}\, {}^nC_{n-1}\, 1^m$$

$$= n^m - n(n-1)^{m+n} + \frac{(n-1)}{\underline{|2}}(n-2)^m + \ldots\ldots\ldots + (-1)^{n-1}\, n.1^m$$

$$= n\{n^{m-1} - (n-1)^m + \frac{(n-1)}{\underline{|2}}(n-2)^m + \ldots\ldots + (-1)^{n-1}\, 1^m\}$$

$$= n[(1 + n - 1)^{m-1} - (n-1)(n-1)^{m-1} + \frac{(n-1)(n-2)}{\underline{|2}}.(n-2)^{m-1}$$

$$+ \ldots\ldots + (-1)^{n-1}\,(1)^{m-1}]\ (\because 1^m = 1^{m-1})$$

$$= n[(1 + n - 1)^{m-1} - (n-1)C_1\,(1 + n - 2)^{m-1} - {}^{(n-1)}C_2$$
$$(1 + n - 3)^{m-1} + \ldots\ldots + (-1)^{n-1}\, {}^{n-1}C_{n-1}\, 1^{m-1}]$$

$$= [E^{n-1}.1^{m-1} - {}^{(n-1)}C_1\, E^{n-2}.1^{m-1} + {}^{(n-1)}C_2\, E^{n-3}.1^{m-1}$$
$$+ \ldots\ldots\ldots + (-1)^{n-1}\, {}^nC_{n-1}\, 1^{m-1}]$$

$$= n[(E-1)^{n-1}.1^{m-1}]$$

$$= n\,\Delta^{n-1}.\, 1^{m-1}$$

$$= n\,\Delta^{n-1}\, E0^{m-1}$$

$$= n\{\Delta^{n-1}\,(1+\Delta)\, 0^{m-1}\}$$

$$= n(\Delta^{n-1} + \Delta^n)\, 0^{m-1}$$

$$= n[\Delta^{n-1}\, 0^{m-1} + \Delta^n\, 0^{m-1}].$$

Problem 8: Show that $\Delta^n\, 0^{n+1} = \dfrac{n(n+1)}{2}\,\Delta^n\, 0^n$.

Solution: We will prove this result using following recurrence relation:

$$\Delta^n\, 0^n = n[\Delta^{n-1}\, 0^{m-1} + \Delta^n\, 0^{m-1}]$$

Replacing m by $n+1$ in (1), we obtain

$$\Delta^n\, 0^{n+1} = n[\Delta^{n-1}\, 0^n + \Delta^n\, 0^n] \tag{A1}$$

Similarly,

$$\Delta^{n-1}\, 0^n = (n-1)[\Delta^{n-2}\, 0^{n-1} + \Delta^{n-1}\, 0^{n-1}] \tag{A2}$$

$$\Delta^{n-2}\, 0^{n-1} = (n-2)[\Delta^{n-3}\, 0^{n-2} + \Delta^{n-2}\, 0^{n-2}] \tag{A3}$$

$$\Delta^{n-3}\, 0^{n-2} = (n-3)[\Delta^{n-4}\, 0^{n-3} + \Delta^{n-3}\, 0^{n-3}] \tag{A4}$$

$$\ldots\ldots\ldots\ldots\ldots\ldots$$

$$\Delta^2 \, 0^3 = 2[\Delta \, 0^2 + \Delta^2 \, 0^2] \tag{A_{n-1}}$$

$$\Delta^1 \, 0^2 = 1[\Delta^0 \, 0^1 + \Delta^1 \, 0^1] \tag{A_n}$$

Substituting (A2) in (A1), we get

$$\Delta^n \, 0^{n+1} = n \, \Delta^n \, 0^n + n(n-1)\Delta^{n-2} \, 0^{n-1} + n(n-1) \, \Delta^{n-1} \, 0^{n-1} \tag{2}$$

Substituting (A2) in (2), we get

$$\Delta^n \, 0^{n+1} = n \, \Delta^n \, 0^n + n(n-1)\Delta^{n-1} \, 0^{n-1} + n(n-1)(n-2) \, \Delta^{n-3} \, 0^{n-2}$$

$$+ \, n(n-1)(n-2) \, \Delta^{n-2} \, 0^{n-2}$$

Continuing this process, we get

$$\Delta^n \, 0^{n+1} = n\lfloor n + n(n-1)\lfloor n-1 + n(n-1)(n-2)\lfloor n-2 + \ldots\ldots$$

$$+ \, n(n-1)(n-2)..3.\,2.\,1.$$

$$= n\lfloor n + (n-1)\lfloor n + (n-2)\lfloor n + \ldots\ldots + \lfloor n$$

$$= \lfloor n \, [n + (n-1) + (n-2) + \ldots.. + 1]$$

$$= \lfloor n \, n(n+1)$$

$$= \frac{n(n+1)}{2} \, \Delta^n \, 0^n.$$

2.11   DIVIDED DIFFERENCES

Let $f(x_0), f(x_1),\ldots\ldots f(x_n)$ be $(n+1)$ entries corresponding to the distinct argument $x_0, x_1\ldots, x_{n-1}, x_n$, which are need not be equally spaced, i.e, $x_1 - x_0, x_2 - x_1,\ldots, x_n - x_{n-1}$ are equal, or some of them equal or all distinct.

The first-order divided difference of $f(x)$ for the argument x_0 and x_1 is defined by $\dfrac{f(x_1) - f(x_0)}{x_1 - x_0}$ and is denoted by $f[x_0, x_1]$, i.e., $f[x_0, x_1]$

$$= \frac{f(x_1) - f(x_0)}{x_1 - x_0}.$$

Divided-difference table

x	$f(x)$	$f[x_r, x_{r+1}]$	$f[x_r, x_{r+1}, x_{r+2}]$	$f[x_r, x_{r+1}, x_{r+2}, x_{r+3}]$	$f(x_r, x_{r+1}, x_{r+2}, x_{r+3}, x_{r+4})$
x_0	$f(x_0)$				
		$\rightarrow \dfrac{f(x_1) - f(x_0)}{x_1 - x_0} = f[x_0, x_1]$			
x_1	$f(x_1)$		$\rightarrow \dfrac{f(x_1, x_2) - f(x_0, x_1)}{x_2 - x_0} = f[x_0, x_1, x_2]$		
		$\rightarrow \dfrac{f(x_2) - f(x_1)}{x_2 - x_1} = f[x_1, x_2]$		$\rightarrow \dfrac{f(x_1, x_2, x_3) - f(x_0, x_1, x_2)}{x_3 - x_0} = f[x_0, x_1, x_2, x_3]$	
x_2	$f(x_2)$		$\rightarrow \dfrac{f(x_2, x_3) - f(x_1, x_2)}{x_3 - x_1} = f[x_1, x_2, x_3]$		$\rightarrow \dfrac{f(x_1, x_2, x_3, x_4) - f(x_0, x_1, x_2, x_3)}{x_4 - x_0} = f[x_0, x_1, x_2, x_3, x_4]$
		$\rightarrow \dfrac{f(x_3) - f(x_2)}{x_3 - x_2} = f[x_2, x_3]$		$\rightarrow \dfrac{f(x_2, x_3, x_4) - f(x_1, x_2, x_3)}{x_4 - x_1} = f[x_1, x_2, x_3, x_4]$	
x_3	$f(x_3)$		$\rightarrow \dfrac{f(x_3, x_4) - f(x_2, x_3)}{x_4 - x_2} = f[x_2, x_3, x_4]$		
		$\rightarrow \dfrac{f(x_4) - f(x_3)}{x_4 - x_3} = f[x_3, x_4]$			
x_4	$f(x_4)$				

It is also denoted by $x_1 f(x_0)$. Note that $f[x_0, x_1] = f[x_1, x_0]$. In general, the first-order divided difference $f[a, b] = \dfrac{f(b) - f(a)}{b - a}$,

Similarly, the second-order divided difference of $f(x)$ for the arguments x_r, x_{r+1} and x_{r+2} is defined by $\dfrac{f[x_{r+1}, x_{r+2}] - f[x_r, x_{r+1}]}{x_{r+2} - x_r}$ and is denoted by $f[x_r, x_{r+1}, x_{r+2}]$, i.e.,

$$f[x_r, x_{r+1}, x_{r+2}] = \frac{f[x_{r+1}, x_{r+2}] - f[x_r, x_{r+1}]}{x_{r+2} - x_r};$$

For example, $f[a, b, c] = \dfrac{f[b, c] - f[a, b]}{c - a}$; The nth-order divided difference for the arguments $x_0, x_1, x_2, \ldots \ldots x_n$ is denoted by $f[x_0, x_1, x_2, \ldots \ldots x_{n-1}, x_n] =$

$$\ldots \ldots x_{n-1}, x_n] = \frac{f[x_0, x_1, x_2, \ldots \ldots x_{n-1}, x_n] - f[x_0, x_1, \ldots \ldots x_{n-1}]}{x_n - x_0}.$$

Note that there are $(r + 1)$ arguments in the rth–order divided difference where $r = 1, 2, 3\ldots\ldots$

A table showing the values of various orders of divided difference is known as a divided-difference table. The divided-difference table for the data (x_0, y_0), (x_1, y_1), (x_2, y_2), (x_3, y_3) and (x_4, y_4) where $y_i = f(x_i)(i = 1, 2, 3, 4$ is shown in page 2-37).

SOLVED PROBLEMS

Problem 1: Construct the divided–difference table for the data given below:

x	1	3	4	6	8
$f(x)$	1	27	64	216	512

Solution: The following is the divided–difference table for the given data.

x	$f(x)$	1st d.d	2nd d.d	3rd d.d	4th d.d
1	1				
		$\rightarrow \dfrac{27-1}{3-1}$ $=13$			
3	27		$\rightarrow \dfrac{37-13}{4-1}=8$		
		$\rightarrow \dfrac{64-27}{4-3}$ $=37$		$\rightarrow \dfrac{13-8}{6-1}=1$	
4	64		$\rightarrow \dfrac{76-37}{6-3}=13$		$\rightarrow \dfrac{1-1}{8-1}$ $=0$
		$\rightarrow \dfrac{216-64}{6-4}$ $=76$		$\rightarrow \dfrac{18-13}{8-3}=1$	
6	216		$\rightarrow \dfrac{148-76}{8-4}=18$		
		$\rightarrow \dfrac{512-216}{8-6}$ $=148$			
8	512				

In the above table, $f[1, 3] = 13, f[3, 4] = 37, f[4, 6] = 76, f[6, 8] = 148, f[1, 3, 4] = 8, f[3, 4, 6] = 13, f[4, 6, 8] = 18, f[1, 3, 4, 6] = 1, f[3, 4, 6, 8] = 1$ and $f[1, 3, 4, 6, 8] = 0$.

Problem 2: Find the divided-difference table for the data given below:

x	0	2	4	6	8
$f(x)$	1	9	65	217	513

Solution: Note that the arguments are equally speed points. Still it possible to construct the divided-difference table

x	$f(x)$	1st d.d	2nd d.d	3rd d.d	4th d.d
0	1				
		$\rightarrow \dfrac{9-1}{2-0}=4$			
2	9		$\rightarrow \dfrac{28-4}{4-0}=6$		
		$\rightarrow \dfrac{65-9}{4-2}=28$		$\rightarrow \dfrac{12-6}{6-0}=1$	
4	65		$\rightarrow \dfrac{76-28}{6-2}=12$		$\rightarrow \dfrac{1-1}{8-0}=0$
		$\rightarrow \dfrac{217-65}{6-4}=76$		$\rightarrow \dfrac{18-12}{8-2}=1$	
6	217		$\rightarrow \dfrac{148-76}{8-4}=18$		
		$\rightarrow \dfrac{513-217}{8-6}=148$			
8	513				

Problem 3: Construct the divided-difference table for the data given below.

x	2	6	8	9	10	12	16
$f(x)$	76	98	121	156	178	190	230

Solution: The required divided-difference table as shown:

x	$f(x)$	1st d.d	2nd d.d	3rd d.d	4th d.d	5th d.d	6th d.d
2	76						
		$\rightarrow 5.5$					

6	98		$\to 1$		
		$\to 11.5$		$\to 0.976$	
8	121		$\to 7.833$		$\to -0.57$
		$\to 35$		$\to -3.583$	$\to 0.122$
9	156		$\to -6.5$		$\to 0.646$ $\to -0.013$
		$\to 22$		$\to 0.292$	$\to -0.058$
10	178		$\to -5.333$		$\to 0.071$
		$\to 6$		$\to 0.857$	
12	190		$\to 0.667$		
		$\to 10$			
16	230				

Problem 4: If $f(x) = x^2$ find $f[a, b, c]$.

Solution: Here, the arguments are a, b and c and the function $f(x) = x^2$. Therefore, $f(a) = a^2$, $f(b) = b^2$ and $f(c) = c^2$

x	$f(x)$	1st d.d	2nd d.d
a	a^2		
		$\to b + a = f(a, b)$	
b	b^2		$\to 1 = f[a, b, c]$
		$\to c + b = f[b, c]$	
c	c^2		

$\therefore f[a, b, c] = 1$.

Problem 5: If $f(x) = \dfrac{1}{x}$, then find $f[a, b, c, d]$.

Solution: We present the solution of this problem in two different ways:

Method 1:

By the definition of divided differences,

$$f[a, b] = \frac{f(b) - f(a)}{b - a} = \frac{\dfrac{1}{b} - \dfrac{1}{a}}{b - a} = \frac{-1}{ab}$$

Similarly, $f[b, c] = \dfrac{-1}{bc}, f[c, a] = \dfrac{-1}{cd}$.

Now,

$$f[a, b, c] = \frac{f[b,c]-f[a,b]}{c-a} = \frac{\left(\frac{-1}{bc}\right)-\left(\frac{-1}{ab}\right)}{c-a} = \frac{1}{abc},$$

and $f[b, c, d] = \dfrac{1}{bcd}$.

Now consider

$$f[a, b, c, d] = \frac{f[b,c,d]-f[a,b,c]}{d-a}$$

$$= \frac{\dfrac{1}{bcd}-\dfrac{1}{abc}}{d-a} = \frac{-1}{abcd}$$

Thus, $f[a, b, c, d] = \dfrac{-1}{abcd}$.

Method 2:

x	$f(x)$	1st d.d	2nd d.d	3rd d.d
a	$\dfrac{1}{a}$			
		$\rightarrow \dfrac{-1}{ba} = f[a, b]$		
b	$\dfrac{1}{b}$		$\dfrac{1}{abc} = f[a,b,c]$	
		$\rightarrow -\dfrac{1}{cb}$		$\rightarrow \dfrac{-1}{abcd} = f[a,b,c,d]$
c	$\dfrac{1}{c}$		$\rightarrow \dfrac{1}{bcd}$	
		$\rightarrow \dfrac{-1}{dc}$		
d	$\dfrac{1}{d}$			

$$\therefore f[a, b, c, d] = \frac{-1}{abcd}.$$

Problem 6: If $f(x) = \dfrac{1}{x^3}$, then find $f[a, b, c]$.

Solution: Consider $f[a, b] = \dfrac{f(b) - f(a)}{b - a} = \dfrac{\dfrac{1}{b^3} - \dfrac{1}{a^3}}{b - a}$

$$= \frac{a^3 - b^3}{a^3 b^3 (b - a)} = \frac{-(a^2 + ab + b^2)}{a^3 b^3}$$

Similarly, $f[b, c] = \dfrac{-(b^2 + bc + c^2)}{b^3 c^3}.$

Now, $f[a, b, c] = \dfrac{f[b,c] - f[a,b]}{c - a}$

$$= \frac{\left\{ -\dfrac{(b^2 + bc + c)}{b^3 c^3} \right\} - \left\{ -\dfrac{(a^2 + ab + b^2)}{a^3 b^3} \right\}}{c - a}$$

$$= \frac{b^4 + b^2 ca + b^2 a^2 + abc^2 + a^2 bc + c^2 a^2}{a^3 b^3 c^3}.$$

2.12 PROPERTIES OF DIVIDED DIFFERENCES

Property 1: If $f(x) = c$, then $f[x_r, x_{r+1}] = 0$, where c is a constant and $x_r \neq x_{r+1}$, the divided difference of constant function is zero for distinct arguments.

Proof: Since $f(x) = c$, we have

$$f[x_r, x_{r+1}] = \frac{f(x_{r+1}) - f(x_r)}{x_{r+1} - x_r} = \frac{c - c}{x_{r+1} - x_r} = 0$$

Property 2: If $f(x) = k\, g(x)$, then

$$f[x_r, x_{r+1}] = k\, g[x_r, x_{r+1}].$$

Proof: By the definition of divided difference,

$$f[x_r, x_{r+1}] = \frac{f(x_{r+1}) - f(x_r)}{x_{r+1} - x_r}$$

$$= \frac{kg(x_{r+1}) - kg(x_r)}{x_{r+1} - x_r} = k\left\{\frac{g(x_{r+1}) - g(x_r)}{x_{r+1} - x_r}\right\}$$

$$= k \; g[x_r, x_{r+1}].$$

Problem 3: If $F(x) = f(x) \pm g(x)$, then

$$F[x_r, x_{r+1}] = f[x_r, x_{r+1}] \pm g[x_r, x_{r+1}],$$

i.e., divided difference of sum or difference of two functions is the sum or difference of their divided differences.

Proof: Consider

$$F[x_r, x_{n+1}] = \frac{F(x_{r+1}) - F(x_r)}{x_{r+1} - x_r}$$

$$= \frac{\{f(x_{r+1}) \pm g(x_{r+1})\} - \{f(x_r) \pm g(x_r)\}}{x_{r+1} - x_r}$$

$$= \left\{\frac{f(x_{r+1}) - f(x_r)}{x_{r+1} - x_r}\right\} \pm \left\{\frac{g(x_{r+1}) - g(x_r)}{x_{r+1} - x_r}\right\}$$

$$= f[x_{r+1}, x_r] \pm g[x_{r+1}, x_r].$$

Property 4: If $F(x) = f(x) \, g(x)$, then

$$F[x_r, x_{r+1}] = f(x_{r+1}) \, g[x_r, x_{r+1}] + g(x_r) f[x_r, x_{r+1}].$$

Proof: The divided difference

$$F[x_r, x_{r+1}] = \frac{F(x_{r+1}) - F(x_r)}{x_{r+1} - x_r}$$

$$= \frac{f(x_{r+1}) g(x_{r+1}) - f(x_r) g(x_r)}{x_{r+1} - x_r}$$

$$= \frac{f(x_{r+1}) g(x_{r+1}) - f(x_{r+1}) g(x_r) + f(x_{r+1}) g(x_r) - f(x_r) g(x_r)}{x_{r+1} - x_r}$$

$$= \frac{f(x_{r+1})\left\{g(x_{r+1}) - g(x_r)\right\} + g(x_r)\left\{f(x_{r+1}) - f(x_r)\right\}}{(x_{r+1} - x_r)}$$

$$= f(x_{n+1})g[x_n, x_{n+1}] + g(x_n)f[x_n, x_{n+1}].$$

Property 5: If $F(x) = \dfrac{f(x)}{g(x)}$, then

$$F[x_r, x_{r+1}] = \frac{g(x_r)f[x_r, x_{r+1}] - f(x_r)g[x_r, x_{r+1}]}{g(x_r)g(x_{r+1})},$$

provided $g(x_r)$ and $g(x_{n+1}) \neq 0$.

Proof: Consider

$$F[x_r, x_{n+1}] = \frac{F(x_{r+1}) - F(x_r)}{x_{r+1} - x_r}$$

$$= \frac{\dfrac{f(x_{r+1})}{g(x_{r+1})} - \dfrac{f(x_r)}{g(x_r)}}{x_{r+1} - x_r}$$

$$= \frac{f(x_{r+1})g(x_r) - f(x_r)g(x_{r+1})}{(x_{r+1} - x_r)g(x_r)g(x_{r+1})}$$

$$= \frac{f(x_{r+1})g(x_r) - f(x_r)g(x_r) + f(x_r)g(x_r) - f(x_r)g(x_{r+1})}{(x_{r+1} - x_r)g(x_r)g(x_{r+1})}$$

$$= \frac{g(x_r)\left\{\dfrac{f(x_{r+1}) - f(x_r)}{x_{r+1} - x_r}\right\} - f(x_r)\left\{\dfrac{g(x_{r+1}) - g(x_r)}{x_{r+1} - x_r}\right\}}{g(x_r)g(x_{r+1})}$$

$$= \frac{g(x_r)f[x_r, x_{r+1}] - f(x_r)g[x_r, x_{r+1}]}{g(x_r)g(x_{r+1})}.$$

Property 6: Divided differences of all order are symmetric functions of their arguments,

i.e., $f[x_0, x_1, x_2, \ldots\ldots x_n] = f[\text{Any permutation of } x_0, x_1, \ldots\ldots x_n].$

Proof: We will prove it by the principle of mathematical induction. Let n denote the order of divided difference, for $n = 1$,

$$f[x_0, x_1] = \frac{f(x_1) - f(x_0)}{x_1 - x_0}$$

$$= \frac{f(x_1)}{x_1 - x_0} + \frac{f(x_0)}{x_0 - x_1} \tag{1}$$

Clearly, first-order divided differences are symmetric over their arguments. For $n = 2$,

$$f[x_0, x_1, x_2] = \frac{f[x_1, x_2] - f[x_0, x_1]}{x_2 - x_0}$$

$$= \frac{1}{(x_2 - x_0)}\left[\left\{ \frac{f(x_1)}{x_1 - x_2} + \frac{f(x_2)}{x_2 - x_1} \right\} - \left\{ \frac{f(x_0)}{x_0 - x_1} + \frac{f(x_1)}{x_1 - x_0} \right\} \right]$$

$$= \frac{1}{(x_2 - x_0)}\left[-\frac{f(x_0)}{x_0 - x_1} + f(x_1)\left\{ \frac{1}{x_1 - x_2} - \frac{1}{x_1 - x_0} \right\} + \frac{f(x_2)}{(x_2 - x_1)} \right]$$

$$= \frac{f(x_0)}{(x_0 - x_1)(x_0 - x_2)} + \frac{f(x_1)}{(x_1 - x_0)(x_1 - x_2)} + \frac{f(x_2)}{(x_2 - x_0)(x_2 - x_0)} \tag{2}$$

Here, $n = 2$ is considered to know the general form like (1) of the divided differences of kth-order divided differences.

Using (1) and (2), for $n = k$,

$$f[x_0, x_1, x_2, \ldots\ldots x_k] = \frac{f(x_0)}{(x_0 - x_1)(x_0 - x_2)\ldots(x_0 - x_k)}$$

$$+ \frac{f(x_1)}{(x_1 - x_0)(x_1 - x_2)\ldots(x_1 - x_k)} + \ldots\ldots$$

$$+ \frac{f(x_k)}{(x_k - x_0)(x_k - x_1)\ldots(x_k - x_{k-1})} \tag{3}$$

Now, for $n = k + 1$,

$$f[x_0, x_1, x_2, \ldots\ldots x_k, x_{k+1}] = \frac{f[x_1, x_2, \ldots\ldots, x_k, x_{k+1}] - f[x_0, x_1, \ldots\ldots, x_k]}{x_{k+1} - x_0}$$

$$= \frac{1}{x_{k+1} - x_0} \left[\left\{ \frac{f(x_1)}{(x_1 - x_2)........(x_1 - x_k)(x_1 - x_{k+1})} \right. \right.$$

$$+ \frac{f(x_2)}{(x_2 - x_1)(x_2 - x_3)........(x_2 - x_4)(x_2 - x_{k+1})} + ...$$

$$+ \frac{f(x_k)}{(x_k - x_1)(x_k - x_2)....(x_k - x_{k-1})(x_k - x_{k+1})}$$

$$+ \left. \frac{f(x_{k+1})}{(x_{k+1} - x_1)(x_{k+1} - x_2).....(x_{k+1} - x_k)} \right\}$$

$$- \left\{ \frac{f(x_0)}{(x_0 - x_1)(x_0 - x_2)........(x_0 - x_k)} \right.$$

$$+ \frac{f(x_1)}{(x_1 - x_0)(x_1 - x_2)........(x_1 - x_k)} + ...$$

$$+ \left. \left. \frac{f(x_k)}{(x_k - x_0)(x_k - x_1)........(x_k - x_{k-1})} \right\} \right]$$

$$= \frac{1}{(x_{k+!} - x_0)} \left[\frac{-f(x_0)}{(x_0 - x_1)(x_1 - x_2).......(x_0 - x_{k-1})(x_0 - x_k)} \right.$$

$$+ \frac{f(x_1)}{(x_1 - x_2)(x_1 - x_3)....(x_1 - x_k)} \left\{ \frac{1}{x_1 - x_{k+1}} - \frac{1}{x_1 - x_0} \right\}$$

$$+ \frac{f(x_2)}{(x_2 - x_1)(x_2 - x_3)....(x_2 - x_k)} \left\{ \frac{1}{x_2 - x_{k+1}} - \frac{1}{x_2 - x_0} \right\}$$

$$+.......$$

$$+ \frac{f(x_k)}{(x_k - x_1)(x_k - x_2)....(x_k - x_{k+1})} \left\{ \frac{1}{x_k - x_{k+1}} - \frac{1}{x_k - x_0} \right\}$$

$$+ \left. \frac{f(x_{k+1})}{(x_{k+1} - x_0)(x_{k+1} - x_1)....(x_{k+1} - x_k)} \right]$$

$$= \frac{f(x_0)}{(x_0 - x_1)(x_0 - x_2)\text{.......}(x_0 - x_k)(x_0 - x_{k+1})}$$

$$+ \frac{f(x_1)}{(x_1 - x_0)(x_1 - x_2)\text{....}(x_1 - x_k)(x_1 - x_{k+1})} + \text{.......}$$

$$+ \frac{f(x_k)}{(x_k - x_0)(x_k - x_1)\text{....}(x_k - x_{k-1})(x_k - x_{k+1})}$$

$$+ \frac{f(x_{k+1})}{(x_{k+1} - x_0)(x_{k+1} - x_1)\text{....}(x_{k+1} - x_{k-1})(x_{k+1} - x_k)},$$

which is a symmetric function over the arguments $x_0, x_1, x_2 \text{.....} x_n$ and x_{k+1}.

Since the R.H.S. of Eq. (3) have terms such that for each argument of $(k + 1)$ arguments, we have a term having an entry corresponding to the argument in the numerator and the product of differences of the remaining arguments from that argument in the denominator. Hence by the principle of mathematical induction, the result is in true for every positive integer.

Property 7: If $f(x) = x^n$, then $f[x_r, x_{r+1}, \text{....}x_{r+n}] = 1$, i.e., nth d.d of x^n is 1.

Proof: First, we compute the first-order d.d. of $f(x)$ for the argument x_r, x_{n+1}.

$$f[x_r, x_{r+1}] = \frac{x_{r+1}^n - x_r^n}{x_{r+1} - x_2} = x_{r+1}^{n-1} + x_{r+1}^{n-2} x_r + \text{.....} + x_r^{n-1} \tag{1}$$

Clearly, it is a symmetric function of $(n - 1)^{th}$ degree in the arguments x_r and x_{r+1}.

Now consider $f[x_r, x_{r+1}, x_{r+2}] = \dfrac{f[x_{r+1}, x_{r+2}] - f[x_r, x_{r+1}]}{x_{r+2} - x_2}$

$$= \frac{1}{x_{r+2} - x_r}\left[\left(x_{r+2}^{n-1} + x_{r+2}^{n-2} x_{r+1} + \text{.....} + x_{r+1}^{n-1}\right) - \left(x_{r+1}^{n-1} + x_{r+1}^{n-2} x_r + \text{.....} + x_r^{n-1}\right)\right]$$

$$= \frac{1}{x_{r+2} - x_r}\left[\left(x_{r+2}^{n-1} - x_r^{n-1}\right) + x_{r+1}\left(x_{r+2}^{n-2} - x_r^{n-2}\right) + \text{.........} + x_{r+1}^{n-2}\left(x_{r+2} - x_r\right)\right]$$

$$= \left(x_{r+2}^{n-2} + x_{r+2}^{n-3} + \ldots\ldots x_r^{n-2}\right) + x_{r+1}\left(x_{r+2}^{n-3} + x_{r+2}^{n-4}.x_r + \ldots\ldots\ldots x_r^{n-3}\right) + \ldots\ldots +$$

$$x_{r+1}^{n-2}.$$

Clearly, it is a symmetric function of $(n-2)$ degree in x_r, x_{r+1} and x_{r+2}.

Continuing this process, it can be shown that $f[x_r, x_{r+1}, x_{r+2}, \ldots, x_{r+n}]$ is a symmetric function of $(n-n)^{\text{th}}$ degree in x_r, x_{r+1}, x_{r+2}, $\ldots$, x_{r+n}. Hence it is a constant. Let this constant be k. For $f(x) = x$, $f[x_r, x_{r+1}] = 1$, therefore $k = 1$. Hence, $f[x_r, x_{r+1}, x_{r+2}, \ldots, x_{r+n}] = 1$.

Property 8: If $f(x) = x^n$ then $f[x_r, x_{r+1}, x_{r+2}, \ldots, x_{r+n}, x_{r+n+1}] = 0$.

Proof: Since, $f[x_r, x_{r+1}, x_{r+2}, \ldots, x_{r+n}, x_{r+n+1}]$

$$= \frac{f[x_{r+1}, x_{r+2}, \ldots, x_{r+n}, x_{r+n+1}] - f[x_r, x_{r+1}, x_{r+2}, \ldots, x_{r+n}]}{x_{r+n+1} - x_r}$$

the result follows by the property 7.

Property 9: If $f(x) = a_0 x^n + a_1 x^{n-1} + \ldots + a_{n-1}x + a_n$ then $f[x_r, x_{r+1}, \ldots, x_{r+n}] = a_0$.

Proof: Let $f_i(x) = x^i$ $(i = 0, 1, 2, \ldots, n)$. Then

$$f(x) = a_0 f_n(x) + a_1 f_{n-1}(x) + \ldots + a_{n-1} f_1(x) + a_n f_0(x)$$

we know that the divided difference of sum of functions is equal to sum of their differences. Therefore,

$$f[x_r, x_{r+1}, \ldots, x_{r+n}] = a_0 f_n[x_r, x_{r+1}, \ldots, x_{r+n}] + a_1 f_{n-1}[x_r, x_{r+1}, \ldots, x_{r+n}] + \ldots +$$
$$a_{n-1}f_1[x_r, x_{r+1}, \ldots, x_{r+n}] + a_n f_0[x_r, x_{r+1}, \ldots, x_{r+n}] \tag{1}$$

By the property 7, we have

$$f_n[x_r, x_{r+1}, \ldots, x_{r+n}] = 1 \qquad\qquad (\therefore f_n(x) = x^n)$$

and property 8,

$$f_{n-1}[x_r, x_{r+1}, \ldots, x_{r+n}] = 0,$$
$$f_{n-2}[x_r, x_{r+1}, \ldots, x_{r+n}] = 0$$
$$\ldots\ldots$$
$$f_1[x_r, x_{r+1}, \ldots, x_{r+n}] = 0$$
$$f_0[x_r, x_{r+1}, \ldots, x_{r+n}] = 0$$

Since, $f_{n-1} = x^{n-1}, f_{n-2} = x^{n-2}, \ldots, f_1(x) = x$ and $f_0(x) = 1$ and, their order of divided difference is n. Therefore, eqn. (1) becomes

$$f[x_r, x_{r+1}, \ldots, x_{r+n}] = a_0.$$

Note: The nth order divided difference of nth degree polynomial is the coefficient of x^n and $(n+1)^{\text{th}}$ divided difference of nth degree polynomial is zero.

2.13 RELATION BETWEEN DIVIDED DIFFERENCES AND FORWARD DIFFERENCES WHEN THE ARGUMENTS ARE EQUALLY SPACED

If $x_0, x_1, \ldots x_n$ are equally spaced points with interval of differencing h, then

$$f[x_0, x_1, x_2 \ldots x_n] = \frac{\Delta^n f(x_0)}{\lfloor nh^n},$$

where $x_i = x_0 + ih$ $(i = 1, 2, \ldots n)$.
We will prove it by the principle of mathematical induction.

For $n = 1, f[x_0, x_1] = \dfrac{f(x_1) - f(x_2)}{x_1 - x_0} = \dfrac{f(x_0 + h) - f(x_0)}{h}$

$$= \frac{\Delta f(x_0)}{h}$$

$\therefore$ The result is true for $n = 1$.

Suppose the result is true for $n = k$, then

$$f[x_0, x_1, \ldots x_k] = \frac{\Delta^k f(x_0)}{\lfloor k\, h^k}. \tag{1}$$

Now, for $n = k + 1$,

$f[x_0, x_1, \ldots x_{k,x\ k+1}]$

$$= \frac{f[x_1, \ldots x_k, x_{k+1}] - f[x_0, x_1, \ldots x_k]}{x_{k+1} - x_0}$$

$$= \frac{\dfrac{\Delta^k f(x_1)}{\lfloor k\, h^k} - \dfrac{\Delta^k f(x_0)}{\lfloor k\, h^k}}{(k+1)h}$$

$$= \frac{1}{\lfloor k+1.h^{k+1}} \, \Delta^k \{f(x_0 + h) - f(x_0)\}$$

$$= \frac{1}{\lfloor k+1.h^{k+1}} \, \Delta^{k+1} f(x_0).$$

Hence, the result is true for $n = k + 1$, and by the principle of induction, the result is true for every positive integer.

EXERCISE 2

1. Construct forward-difference table for the data given below:

x	2	4	6	8	10
$f(x)$	25	36	47	54	65

2. From the data given below, find $\Delta^2 f(2)$ and $\Delta^3 f(2)$.

x	2	5	8	11
$f(x)$	16	20.52	40.63	54.93

3. Find $\Delta^4(1 - x^2)(3 - 4x)(16 - 3x)$.
4. Find $\Delta^2(e^{ax} + \sin bx)$.
5. Find $\Delta(x^3 + 5x^2 + 6)$ with $h = 1$.
6. Find first-, second-, third- and fourth-order forward differences of $x^3 - 5x^2 + 7x + 3$ when $h = 1$.
7. Find $\dfrac{1}{\Delta}(x^4 + 6x^3 + 12x^2 + 7x + 9)$.
8. On what function Δ has to be operated to get a function $x^3 + 7x + 2$?
9. Find a function whose first difference is $4x^2 + 7x + 3$.
10. Find a function whose second difference is $6x^3 - 8x + 9$.
11. Find the missing term in the follow data

x	1	3	5	7	9
$f(x)$	3	11	?	51	83

12. Find the two missing term of the following data:

X	0	2	4	6	8	10
$f(x)$	8	?	12	?	16	18

13. If $u_x = x^4$, find $(\Delta u_x)^2$ and $\Delta^2 u_x$ when the interval of differencing is h.
14. Show that $(n + 1)\, \Delta^n 0^n = 2[\Delta^{n-1} 0^n + \Delta^n 0^n]$.
15. Find $\Delta^4 0^4$, $\Delta^2 0^4$, $\Delta 0^4$, $\Delta^4 0^5$ and $\Delta^5 0^5$.
16. Express $5x^4 - 13x^2 + 7x + 8$ in factorial function.

17. Express $13x^3 - 8x^2 + 7x + 3$ and its first-, second-, third- and fourth-order forward differences in factorial functions when the interval of differencing is 1, i.e., $h = 1$.

18. Show that $\Delta^n \cos (ax + b) = \left(2\sin\dfrac{ah}{2} \right)^n \cos\left\{ ax + b + n\left(\dfrac{ah + \pi}{2} \right) \right\}$, where n is a positive integer.

ANSWERS

1.

x	$f(x)$	$\Delta f(x)$	$\Delta^2 f(x)$	$\Delta^3 f(x)$
2	25			
		$\rightarrow 11$		
4	36		$\rightarrow 0$	
		$\rightarrow 11$		$\rightarrow -4$
6	47		$\rightarrow -4$	$\rightarrow 12$
		$\rightarrow 7$		$\rightarrow 8$
8	54		$\rightarrow 4$	
		$\rightarrow 11$		
10	65			

2. $\Delta^2 f(2) = 15.59$, $\Delta^3 f(2) = -21.40$

3. -288

4. $(e^{ah} - 1)^2 e^{ax} + \left\{ 2\sin\left(\dfrac{bh}{2} \right) \right\}^2 \sin\{ b(x + h) + \pi \}$

5. $3x^2 + 13x + 6$

6. $\Delta f(x) = 3x^2 - 4$, $\Delta^2 f(x) = 6x + 3$, $\Delta^3 f(x) = 6$, $\Delta^4 f(x) = 0$

7. $\dfrac{x^{(5)}}{5} + 3x^{(4)} + \dfrac{37}{3}x^{(3)} + 13x^{(2)} + 9x^{(1)} + c$ (c is an arbitrary constant)

8. $\dfrac{x^{(4)}}{4} + \dfrac{x^{(3)}}{3} + 4x^{(2)} + 2x^{(1)} + c$ 9. $\dfrac{4x^{(3)}}{3} + \dfrac{11x^{(2)}}{2} + 3x^{(1)} + c$

10. $\dfrac{3x^{(5)}}{10} + \dfrac{3x^{(4)}}{2} - \dfrac{x^{(3)}}{3} + \dfrac{9x^{(2)}}{2} + c_1 x^{(1)} + c_2$ 11. 27   12. $f(2) = 14, f(6) = 10$

13. $\Delta u_x = 4x^3 + 6x^2 + 4x + 1$, $(\Delta u_x)^2 = (4x^3 + 6x^2 + 4x + 1)^2$, $\Delta^2 u_x = 12x^2 + 28x + 10$

15. 24, 14, 1, 243, 120   16. $5x^{(4)} + 15x^{(3)} + 22x^{(2)} - x^{(1)} + 8$

17. $13x^{(3)} + 31x^{(2)} + 12x^{(1)} + 3$, $39x^{(2)} + 62x^{(1)} + 12$, $78x^{(1)} + 62$.

3

Interpolation

Interpolation is the process of computing values of $y = f(x)$ at $x = c$, $a < c < b$ from known data (x_i, y_i). $(i = 0, 1, 2, \ldots n)$ where the arguments x_i are such that $a \le x_i \le b$ with $x_0 = a$ and $x_n = b$. The general problem of interpolation is reconstructing a known or unknown function from a set of tabulated values (x_i, y_i). $(i = 0, 1, 2, \ldots n)$ so that reconstructed function has the values y_i at $x = x_i$ $(i = 0, 1, 2, \ldots n)$. The interpolation is the basis for numerical differentiation, numerical integration and numerical solution of ordinary and partial differential equations. Interpolation methods are used in various disciplines like applied mathematics, economics, business, population studies and many branches of engineering. The extrapolation is the process of computing values of $y = f(x)$, at $x = d$ where $d < a$ or $d > b$ from a given set of values of (x_i, y_i). $(i = 0, 1, 2, \ldots n)$ with $x_0 = a$ and $x_n = b$. Now I wish to explain you these concepts with an example. Supposing the following information is known for the unknown function $y = f(x)$,

x	0	2	4	6	8
$f(x)$	13	17	21	29	34

then the process of finding y at $x = 3$ or 3.5 or 5 or 6.2 or 7.9 using the given data constitutes an interpolation; on the other hand, the process of finding y at $x = -0.5$ or -1 or 8.5 or 8.7 or any value less than 0 and greater than 8 will be an extrapolation.

3.1   DEFINITION

Given a set of values of a function $y = f(x)$ known or unknown if the function $f(x)$ is approximated by a function $\phi(x)$ such that $f(x)$ and $\phi(x)$ agree with the set of tabulated points then the function $\phi(x)$ is called formula of interpolation. In case, $\phi(x)$ is a polynomial, then it is called an interpolating polynomial of $y = f(x)$. In this case, we write $f(x) \approx \phi(x)$. The polynomials are relatively simple in a situation where a function has to replace by a formula of interpolation. These situations arise in numerical differentiation and numerical integration.

3.2   NEWTON'S FORWARD INTERPOLATION FORMULA

Now, we proceed to derive the Newton's forward interpolation formula, which approximates a function by an interpolating polynomial.

Theorem:

Let $y_0, y_1, \ldots\ldots y_n$ be the values of $y = f(x)$ at $x = x_0, x_1 \ldots\ldots x_n$, where $x_i = x_0 + ih$ ($i = 0, 1, 2, \ldots.. n$) and h is the interval of differencing. The interpolating polynomial $p_n(x)$ is given by

$$p_n(x) = y_0 + u\Delta y_0 + \frac{u(u-1)}{\underline{|2}}\Delta^2 y_0 + \ldots\ldots$$

$$+ \frac{u(u-1)(u-2)\ldots..(u-\overline{n-1})}{\underline{|n}}\Delta^n y_0, \qquad (3.1)$$

where

$$u = \frac{x - x_0}{h} \qquad (3.2)$$

Proof: Let $p_n(x)$ be an interpolating polynomial of degree n of the function $y = f(x)$; then

$$y_i = p_n(x_i) \ (i = 0, 1, 2, \ldots.. n) \qquad (3.3)$$

Note that a polynomial of nth degree $p_n(x)$ can be supposed in many forms.

For example $p_n(x) = a_0 + a_1 x + \ldots. + a_n x^n$ or $p_n(x) = a_0 + a_1(x - 1) + a_2(x - 1)^2 + \ldots\ldots +a_n (x - 1)^n$ or $p_n(x) = a_0 + a_1(x - 1) + a_2 (x - 1)(x - 2) + \ldots\ldots + a_n (x - 1)(x - 2) \ldots\ldots (x - n)$. The coefficient $a_0, a_1,$

... a_n can be obtained by substituting $(n + 1)$ values of x_i and $p_n(x_i)$ $(i = 0, 1, 2, ….. n)$. To get the coefficients of $p_n(x)$ in terms of a forward differences, we suppose $p_n(x)$ as follows:

$$p_n(x) = A_0 + A_1(x - x_0) + A_2(x - x_0)(x - x_0 - h)$$
$$+ A_3(x - x_0)(x - x_0 - h)(x - x_0 - 2h) + ………..$$
$$+ A_n(x - x_0)(x - x_0 - h)…….\{x - x_0 - (n - 1)h\}, \qquad (3.4)$$

where $A_0, A_1, …….. A_n$ are to be determined constants. Substituting $x = x_0$ in (3.4), we get

$$p_n(x_0) = A_0$$

From (3.3), we have $p_n(x_0) = y_0$. Therefore,

$$\therefore A_0 = y_0 \qquad (3.5)_0$$

Putting $x = x_0 + h$ in (3.4) and using $p_n(x_1) = y_1$ from (3.3), we have

$$y_1 = A_0 + A_1(x_0 + h - x_0).$$

Using Eq. $(3.5)_0$, the above equation becomes

$$y_1 = y_0 + A_1(h),$$

i.e.,
$$A_1 = \frac{\Delta y_0}{h} \qquad (3.5)_1$$

Now, substituting $x = x_0 + 2h$ in (3.4) and using $p_n(x_2) = y_2$, we get

$$y_2 = A_0 + (x_0 + 2h - x_0) A_1 + (x_0 + 2h - x_0)(x_0 + 2h - x_0 - h)A_2.$$

Using $A_0 = y_0$ and $A_1 = \dfrac{\Delta y_0}{h}$, the above equation reduces to

$$y_2 = y_0 + 2h.(\Delta y_0) + (2h)(h)A_2,$$

i.e.,
$$y_2 = y_0 + 2h.\frac{(y_1 - y_0)}{h} + \lfloor 2.h^2 A_2,$$

i.e.,
$$A_2 \lfloor 2\, h^2 = y_2 - 2y_1 + y_0$$

Hence,
$$A_2 = \frac{\Delta^2 y_0}{\lfloor 2 h^2} \qquad (3.5)_2$$

Similarly, substituting the values of $x = x_0 + 3h, x_0 + 4h, …., x_0 + nh$ in (3.4) and using (3.3) it can be shown that

$$A_3 = \frac{\Delta^3 y_0}{\lfloor\underline{3}\, h^3}$$

(3.5)$_3$

$$\cdots\cdots\cdots$$

$$A_n = \frac{\Delta^n y_0}{\lfloor\underline{n}\, h^n}$$

(3.5)$_n$

Now substituting the values of $A_0, A_1, A_2 \ldots A_n$ in (3.4), we get

$$p_n(x) = y_0 + (x - x_0).\frac{\Delta y_0}{\lfloor\underline{1}.h} + (x - x_0)(x - x_0 - h).\frac{\Delta^2 y_0}{\lfloor\underline{2}\, h^2}$$

$$+ (x - x_0)(x - x_0 - h)(x - x_0 - 2h)\frac{\Delta^3 y_0}{\lfloor\underline{3}\, h^3} + \ldots$$

$$+ (x - x_0)(x - x_0 - h)(x - x_0 - 2h)\ldots\{x - x_0 - (n-1)h\}\frac{\Delta^n y_0}{\lfloor\underline{n}\, h^n}$$

(3.6)

Let $u = \dfrac{x - x_0}{h}$; then Eq. (3.6) can be reduced to

$$p_n(x) = p_n(x_0 + hu)$$

$$= y_0 + u\Delta y_0 + \frac{u(u-1)}{\lfloor\underline{2}}\Delta^2 y_0 + \ldots\ldots\ldots +$$

$$+ \frac{u(u-1)(u-2)\ldots(u-\overline{n-1})}{\lfloor\underline{n}}\Delta^n y_0$$

(3.7)

which can be written as $p_n(x) = \sum_{k=0}^{n} u^{(k)} \dfrac{\Delta^k y_0}{\lfloor\underline{k}}$,

with $\Delta^0 y_0 = y_0$ and $u^{(0)} = 1$. Here, $u^{(k)}$ is the factorial function with $h = 1$.

Since $p_n(x)$ is the interpolating polynomial of the function $f(x)$, we have $p_n(x) \approx f(x)$. Therefore,

$$f(x) = f(x_0 + hu) = \sum_{k=0}^{n} u^{(k)} \frac{\Delta^k y_0}{\lfloor\underline{k}},$$

(3.8)

which is known as Newton's forward interpolation formula or Newton formula for forward interpolation. The other forms of it are given by (3.6) and (3.7).

3.3   PROCEDURE TO FIND $y = f(x)$ VALUE AT $x = c$ USING NEWTON'S FORWARD INTERPOLATION FORMULA

Step 1: Construct the forward difference table and note down the values of $y_0, \Delta y_0, \Delta^2 y_0$ ..... etc.

Step 2: Compute $u = \dfrac{x - x_0}{h}$ corresponding to the point $x = c$. i.e.,

$$u = \dfrac{c - x_0}{h}.$$

Step 3: Write down the Newton's forward interpolation formula upto k terms, where k is the number of paired observations or data (x_i, y_i).

Step 4: Substitute the value of u (obtained in step 2) and $y_0, \Delta y_0, \Delta^2 y_0$ ..... etc (obtained in step 1) in Newton's forward interpolation formula (written in Step 3).

Step 5: Simplify the right-hand side of the equation obtained in Step 4 and it will be the value of $f(x)$ at $x = c$.

SOLVED PROBLEMS

Problem 1: From the data given below find $f(3)$ by Newton's forward interpolation formula.

x	0	2	4	6
$f(x)$	1	9	25	49

Solution: The forward difference table for the given data is shown below:

x	$f(x)$	$\Delta f(x)$	$\Delta^2 f(x)$	$\Delta^3 f(x)$
0	1			
		$\to 8$		
2	9		$\to 8$	
		$\to 16$		$\to 0$
4	25		$\to 8$	
		$\to 24$		
6	49			

From the table, we have $y_0 = 1$, $\Delta y_0 = 8$, $\Delta^2 y_0$ and $\Delta^3 y_0 = 0$. We have to find $f(x)$ value at $x = 3$. Here, $c = 3$. Hence, $u = \dfrac{3-0}{2}\left(u = \dfrac{x - x_0}{h}\right) = \dfrac{3}{2}$ $(x_0 = 0, h = 2)$.

Now, the Newton's forward interpolation formula is

$$f(x) = f(x_0 + hu) = y_0 + u\Delta y_0 + \frac{u(u-1)}{\lfloor 2} \Delta^2 y_0$$

$$+ \frac{u(u-1)(u-2)}{\lfloor 3} \Delta^3 y_0 + \ldots.$$

$$\therefore f(3) = f\left(0 + \frac{3}{2}.2\right) = 1 + \frac{3}{2}(8) + \frac{\frac{3}{2}\left(\frac{3}{2}-1\right)}{\lfloor 2}(8) + \frac{\frac{3}{2}\left(\frac{3}{2}-1\right)\left(\frac{3}{2}-2\right)}{\lfloor 3}(0)$$

$$= 16.$$

Problem 2: Given $\sin 45° = 0.7071$, $\sin 50° = 0.7660$, $\sin 55° = 0.8192$, and $\sin 60° = 0.8660$. Find $\sin 52°$ using Newton's forward interpolation formula.

Solution: Let $y = f(x) = \sin x$. The forward difference table for the given data is shown below:

x	$y = \sin x$	Δy	$\Delta^2 y$	$\Delta^3 y$
45°	0.7071			
		→ 0.0589		
50°	0.7660		→−0.0057	
		→ 0.0532		→−0.0007
55°	0.8192		→−0.0064	
		→ 0.0468		
60°	0.8660			

From the table, we have $y_0 = 0.7071$, $\Delta y_0 = 0.0589$, $\Delta^2 y_0 = -0.0057$, $\Delta^3 y_0 = -0.0007$, $x_0 = 45°$ and $h = 5°$.

Let $c = 52°$; Then $u = \dfrac{52° - 45°}{5} = 1.4$.

Newton's forward interpolation formula is

$$f(x) = f(x_0 + hu) = y_0 + u\Delta y_0 + \frac{u(u-1)}{\underline{2}}\Delta^2 y_0 + \frac{u(u-1)(u-2)}{\underline{3}}\Delta^3 y_0 + ...$$

$$f(52°) = f\left(45° + \frac{7}{5}\times 5\right) = 0.7071 + (1.4)(0.0589)$$

$$+ \frac{(1.4)(1.4-1)}{\underline{2}}(-0.0057) + \frac{(1.4)(1.4-1)(1.4-2)}{\underline{3}}(-0.0007)$$

$$= 0.7880$$

$$\therefore \qquad \sin 52° = 0.7880.$$

Problem 3: An alloy is formed with two types of materials A and B. The following table gives the melting point of the alloy against the percentage of the material A present in the alloy.

Percentage of material A in the alloy	40	45	50	55	60
Melting point °C	210	229	248	260	274

Find the melting point of the alloy when it contains 42% of material A.

Solution: Let x be the percentage of material A in the alloy and $y = f(x)$ be the melting point of the alloy when the percentage of material A in the alloy is x units. The difference table for the given data is shown in the following table:

x	$y = f(x)$	Δy	$\Delta^2 y$	$\Delta^3 y$	$\Delta^4 y$
40	210				
		$\to 19$			
45	229		$\to 0$		
		$\to 19$		$\to -7$	
50	248		$\to -7$		$\to 16$
		$\to 12$		$\to 9$	
55	260		$\to 2$		
		$\to 14$			
60	274				

From the above table, we have $x_0 = 40$, $h = 5$, $y_0 = 210$; $\Delta y_0 = 19$; $\Delta^2 y_0 = 0$; $\Delta^3 y_0 = -7$ and $\Delta^4 y_0 = 16$. We have to find melting point the alloy when it contains 42% of the material A. Therefore,

$$u = \frac{42 - 40}{5} = \frac{2}{5} = 0.4.$$

Now, the Newton's forward interpolation formula is,

$$f(x) = f(x_0 + hu) = y_0 + u\Delta y_0 + \frac{u(u-1)}{\lfloor 2} \Delta^2 y_0 + \frac{u(u-1)(u-2)}{\lfloor 3} \Delta^3 y_0$$

$$+ \frac{u(u-1)(u-2)(u-3)}{\lfloor 4} \Delta^4 y_0 + \ldots\ldots$$

$$\therefore f(42) = 210 + 0.4(19) + \frac{(0.4)(0.4-1)}{\lfloor 2}(0)$$

$$+ \frac{(0.4)(0.4-1)(0.4-2)}{\lfloor 3} \times (-7)$$

$$+ \frac{(0.4)(0.4-1)(0.4-2)(0.4-3)}{\lfloor 4} \times 16$$

$$= 210 + 7.6 + 0 - 0.448 - 0.6656$$

$$= 216.4864.$$

Thus, the melting point of the alloy with a presence of 42% of material A is 216.4864.

Problem 4: Find the Newton's forward-difference interpolating polynomial for the following data:

x	0	1	2	3
$f(x)$	1	0	1	10

Solution: The forward difference table for the given data is as follows:

x	$y = f(x)$	Δy	$\Delta^2 y$	$\Delta^3 y$
0	1			
		$\rightarrow -1$		
1	0		$\rightarrow 2$	
		$\rightarrow 1$		$\rightarrow 6$
2	1		$\rightarrow 8$	
		$\rightarrow 9$		
3	10			

From the above table, we have $y_0 = 1$, $\Delta y_0 = -1$; $\Delta^2 y_0 = 2$; $\Delta^3 y_0 = 6$; $x_0 = 0$ and $h = 1$, and further $u = \dfrac{x - x_0}{h} = \dfrac{x - 0}{1} = x$.

The Newton's forward interpolation formula is

$$f(x) = f(x_0 + hu) = y_0 + u\Delta y_0 + \frac{u(u-1)}{\underline{|2}}\Delta^2 y_0 + \frac{u(u-1)(u-2)}{\underline{|3}}\Delta^3 y_0 + \ldots$$

Substituting y_0, Δy_0, $\Delta^2 y_0$, $\Delta^3 y_0$ and u in the right-hand side of above equation, we get

$$f(x) = 1 + x(-1) + \frac{(x)(x-1)}{\underline{|2}}(2) + \frac{x(x-1)(x-2)}{\underline{|3}}(6)$$

$$= x^3 - 2x^2 + 1,$$

which is the required interpolating polynomial.

Problem 5: The populations of a town in the decinial census are given below. Estimate the population for the year 1925. Also find the increase in the population from 1925 to 1931.

Year	1921	1931	1941	1951	1961
Population (in thousands)	46	66	84	93	104

Solution: Let $y = f(x)$ represent the population of the town corresponding to the year x. The forward difference table for the given data is shown below:

x	$y = f(x)$	Δy	$\Delta^2 y$	$\Delta^3 y$	$\Delta^4 y$
1921	46				
		$\rightarrow 20$			
1931	66		$\rightarrow -2$		
		$\rightarrow 18$		$\rightarrow -7$	
1941	84		$\rightarrow -9$		$\rightarrow 18$
		$\rightarrow 9$		$\rightarrow 11$	
1951	93		$\rightarrow 2$		
		$\rightarrow 11$			
1961	104				

From the above table, we have $x_0 = 1921$, $h = 10$, $y_0 = 46$, $\Delta y_0 = 20$, $\Delta^2 y_0 = -2$, $\Delta^3 y_0 = -7$ and $\Delta^4 y_0 = 18$. We have to find the population for the year 1925. Hence, $c = 1925$ and $u = \dfrac{c - x_0}{h} = \dfrac{1925 - 1921}{10} = 0.4$.

Now, the Newton's forward interpolation formula is

$$f(x) = f(x_0 + hu) = y_0 + u\Delta y_0$$
$$+ \frac{u(u-1)}{\lfloor 2} \Delta^2 y_0 + \frac{u(u-1)(u-2)}{\lfloor 3} \Delta^3 y_0$$
$$+ \frac{u(u-1)(u-2)(u-3)}{\lfloor 4} \Delta^4 y_0 + ...$$

Now, substituting y_0, Δy_0, $\Delta^2 y_0$, $\Delta^3 y_0$, $\Delta^4 y_0$ and u in the right-hand side of the Newton's forward interpolation formula, we get

$$f(1925) = 46 + 0.4(20) + \frac{(0.4)(0.4 - 1)}{\lfloor 2}(-2)$$
$$+ \frac{(0.4)(0.4 - 1)(0.4 - 2)}{\lfloor 3} \times (-7)$$
$$+ \frac{(0.4)(0.4 - 1)(0.4 - 2)(0.4 - 3)}{\lfloor 4}(18)$$

$$= 53.0432.$$

Thus, the population of the town for the year 1925 is 53,043.

The population increase from the year 1925 to 1931 = population of the town for the year 1931 − population of the town for the year 1925

$$= 66,000 - 53,043$$
$$= 12957$$

Problem 6: From the following data, find the number of students who obtained less than 52 marks

Marks	0–20	20–40	40–60	60–80	80–100
Number of students	8	12	28	16	20

Solution: We have to reformulate the given data so that the Newton's forward interpolation formula can be applied to find the number of students who obtained less than 52 marks.

Let x represent marks and $y = f(x)$ is the number of students who obtained less than x marks. Now, the given data in terms of x and $f(x)$ are as follows:

x	20	40	60	80	100
$f(x)$	8	20	48	64	90

and the forward difference table for this data in shown below.

x	$y = f(x)$	Δy	$\Delta^2 y$	$\Delta^3 y$	$\Delta^4 y$
20	8				
		$\to 12$			
40	20		$\to 16$		
		$\to 28$		$\to -28$	
60	48		$\to -12$		$\to 50$
		$\to 16$		$\to 22$	
80	64		$\to 10$		
		$\to 26$			
100	90				

From the above table, we have $x_0 = 20$, $h = 20$, $y_0 = 8$, $\Delta y_0 = 12$, $\Delta^2 y_0 = 16$, $\Delta^3 y_0 = -28$ and $\Delta^4 y_0 = 50$.

We have to find $f(52)$; therefore, $c = 52$. Hence,

$$u = \frac{c - x_0}{h} = \frac{52 - 20}{20} = \frac{32}{20} = 1.6.$$

Newton's forward interpolation formula is

$$f(x) = f(x_0 + hu) = y_0 + u\Delta y_0 + \frac{u(u-1)}{\lfloor 2} \Delta^2 y_0 + \frac{u(u-1)(u-2)}{\lfloor 3} \Delta^3 y_0$$

$$+ \frac{u(u-1)(u-2)(u-3)}{\lfloor 4} \Delta^4 y_0 + \ldots\ldots$$

Now, substituting y_0, Δy_0, $\Delta^2 y_0$, $\Delta^3 y_0$, $\Delta^4 y_0$ and u in the right-hand side of the Newton's forward interpolation formula, we obtain

$$f(52) = 8 + 1.6(12) + \frac{(1.6)(1.6-1)}{\lfloor 2}(16) + \frac{(1.6)(1.6-1)(1.6-2)}{\lfloor 3} \times (-28)$$

$$+ \frac{(1.6)(1.6-1)(1.6-2)(1.6-3)}{\lfloor 4}(50).$$

$$= 8 + 19.2 + 7.68 + 1.792 + 1.12$$
$$= 37.792$$
$$\approx 38.$$

Problem 7: From the following data, find $f(0.6)$.

x	0.1	0.3	0.5	0.7	0.9	1.1
$f(x)$	0.003	0.067	0.148	0.248	0.370	0.518

Solution: The forward difference table for the given data is shown below:

x	$y = f(x)$	Δy	$\Delta^2 y$	$\Delta^3 y$	$\Delta^4 y$	$\Delta^5 y_0$	$\Delta^6 y_0$
0.1	0.003						
		$\to$ 0.064					
0.3	0.067		$\to$ 0.017				
		$\to$ 0.081		$\to$ 0.002			
0.5	0.148		$\to$ 0.019		$\to$ 0.001		
		$\to$ 0.100		$\to$ 0.003		$\to$ 0	
0.7	0.248		$\to$ 0.022		$\to$ 0.001		$\to$ 0
		$\to$ 0.122		$\to$ 0.004		$\to$ 0	
0.9	0.370		$\to$ 0.026		$\to$ 0.001		
		$\to$ 0.148		$\to$ 0.005			
1.1	0.518		$\to$ 0.031				
		$\to$ 0.179					
1.3	0.697						

Note that we have $x_0 = 0.1$, $h = 0.2$, $y_0 = 0.003$, $\Delta y_0 = 0.064$, $\Delta^2 y_0 = 0.017$, $\Delta^3 y_0 = 0.002$, $\Delta^4 y_0 = 0.001$, $\Delta^5 y_0 = \Delta^6 y_0 = 0$. We have $c = 0.6$ and find $f(0.6)$, therefore $c = 0.6$ and $u = \dfrac{c - x_0}{h} = \dfrac{0.6 - 0.1}{0.2} = 2.5$.

Substituting the above values in Newton's forward-difference interpolation formula, we obtain

$$f(0.6) = 0.003 + 2.5(0.064) + \frac{(2.5)(2.5-1)}{\lfloor 2} (0.017)$$

$$+ \frac{(2.5)(2.5-1)(2.5-2)}{\lfloor 3} \times (0.02)$$

$$+ \frac{(2.5)(2.5-1)(2.5-2)(2.5-3)}{\lfloor 4} (0.001)$$

$$= 0.1955.$$

Theorem: (Newton's backward–difference interpolation formula)

Let $y_0, y_1 \ldots\ldots y_n$ be $(n + 1)$ values of $y = f(x)$ at $x = x_0, x_1 \ldots\ldots x_n$, where $x_i = x_0 + nh$ $(i = 0, 1, 2, 3, \ldots n)$ and h be the interval of differencing. Then

$$f(x) = y_n + u\nabla y_n + \frac{u(u+1)}{\lfloor 2}\nabla^2 y_n + \frac{u(u+1)(u+2)}{\lfloor 3}\nabla^3 y_n + \ldots\ldots$$

$$+ \frac{u(u+1)\ldots(u+\overline{n-1})}{\lfloor n}\nabla^n y_n \tag{3.9}$$

or

$$f(x) = \sum_{k=0}^{n} \frac{(u+k-1)^{(k)}}{\lfloor k}\nabla^k y_n, \tag{3.10}$$

where

$$u = \frac{x - x_n}{h}. \tag{3.11}$$

Proof: Suppose $p_n(x)$ be the required nth degree interpolating polynomial of $y = f(x)$ for the given values of $y = f(x)$ at $x = x_0, x_1 \ldots\ldots, x_n$ Then

$$p_n(x_0) = y_0, \, p_n(x_1) = y_1, \ldots\ldots, p_n(x_n) = y_n \tag{3.12}$$

Now we supose

$$p_n(x) = B_0 + B_1(x - x_n) + B_2(x - x_n)(x - x_{n-1}) + B_3(x - x_n)(x - x_{n-1})(x - x_{n-2})$$
$$+ \ldots\ldots + B_n(x - x_n)(x - x_{n-1})\ldots\ldots (x - x_1)$$

where $B_0, B_1, \ldots\ldots, B_n$ are to be determined constants. Since $x_i = x_0 + ih$, the above equation reduces to

$$p_n(x) = B_0 + B_1\left\{x - (x_0 + nh)\right\} + B_2\left\{x - (x_0 + nh)\right\}\left\{x - (x_0 + \overline{n-1}h)\right\}$$

$$+ B_3\left\{x - (x_0 + nh)\right\}\left\{x - (x_0 + \overline{n-1}h)\right\}\left\{x - (x_0 + \overline{n-2}h)\right\} + \ldots\ldots$$

$$+ B_n\left\{x - (x_0 + nh)\right\}\left\{x - (x_0 + \overline{n-1}h)\right\} \times$$

$$\left\{x - (x_0 + \overline{n-2}h)\right\}\ldots\ldots\left\{x - (x_0 + h)\right\}. \tag{3.13}$$

Now, we find the co nstants $B_0, B_1 \ldots\ldots B_n$ using (3.12). Substituting $x = x_n$ (i.e. $x = x_0 + nh$) in (3.13), we get

$$p_n(x_n) = B_0.$$

From (3.12), $p_n(x_n) = y_n$. Therefore,

$$B_0 = y_n. \tag{3.14$_0$}$$

Next, Substituting $x = x_{n-1}$ in (3.13), we obtain

$$p_n(x_{n-1}) = B_0 + \left\{(x_0 + \overline{n-1}h) - (x_0 + nh)\right\}B_1 + 0,$$

i.e., $\qquad p_n(x_{n-1}) = B_0 + (-h)B_1$

In view of (3.12) and (3.14)$_0$, we have

$$y_{n-1} = y_n - hB_1,$$

i.e., $\qquad B_1 = \dfrac{y_n - y_{n-1}}{h} = \dfrac{\nabla y_n}{\underline{|1}\,h}. \tag{3.14$_1$}$

Similarly, by substituting $x = x_{n-2}, x_{n-3} \ldots\ldots\ldots x_1$, it can be shown that

$$B_2 = \frac{\nabla^2 y_n}{\underline{|2}\,h^2} \tag{3.14$_2$}$$

$$B_3 = \frac{\nabla^3 y_n}{\underline{|3}\,h^3} \tag{3.14$_3$}$$

$$\ldots\ldots\ldots\ldots\ldots$$

$$B_n = \frac{\nabla^n y_n}{\underline{|n}\,h^n} \tag{3.14$_n$}$$

Substituting $B_0, B_1, \ldots\ldots B_n$ values and $u = \dfrac{x - x_n}{h}$ in Eq. (3.13), we obtain

$$p_n(x) = p_n(x_n + hu)$$

$$= y_n + u\nabla y_n + \frac{u(u+1)}{\lfloor 2}\nabla^2 y_n$$

$$+ \frac{u(u+1)(u+2)}{\lfloor 3}\nabla^3 y_n + \ldots\ldots + \frac{u(u+1)\ldots\ldots(u+\overline{n-1})}{\lfloor n}\nabla^n y_n.$$

$$(3.15)$$

Since $p_n(x)$ is the interpolating polynomial of $f(x)$, we have $p_n(x) \approx f(x)$. Thus, Eq. (3.15) reduces to

$$f(x) = f(x_n + hu) = y_n + u\nabla y_n + \frac{u(u+1)}{\lfloor 2}\nabla^2 y_n$$

$$+ \frac{u(u+1)(u+2)}{\lfloor 3}\nabla^3 y_n + \ldots\ldots + \frac{u(u+1)\ldots\ldots(u+\overline{n-1})}{\lfloor n}\nabla^n y_n$$

$$(3.16)$$

or

$$f(x) = \sum_{k=0}^{n} \frac{(u+k-1)^{(k)}}{\lfloor k}\nabla^k y_n. \qquad (3.17)$$

with $(u-1)^0 = 1$ and $\nabla^0 y_n = y_n$.

The formula given by (3.16) or 3.17 is known as Newton's backward-difference interpolation formula.

Note: Newton's forward formula is used to interpolate the values of y near the beginning of the set of tabulated arguments or for extrapolating a value of y to the left of initial argument (i.e., a value less than the initial argument); on the other hand, Newton's backward–interpolation formula is used to interpolate the values of y near the end of the set of tabulated arguments or for extrapolating the values of y to the right of the last argument. Note that both formulae can be used to find the values of $y = f(x)$ for any x that may lie inside or outside the set of arguments of the given data. To incorporate the variations of $f(x)$ at the beginning, we use Newton's forward difference formula to find y value near the initial argument, and to incorporate the variation of $f(x)$ at the end, we use Newton's backward difference formula to find $f(x)$ at a value near the end of the argument of the given data.

3.4   PROCEDURE TO FIND A VALUE OF $y = f(x)$ AT $x = c$ USING NEWTON'S BACKWARD-DIFFERENCE INTERPOLATION FORMULA

Step 1: Construct backward difference table for the given data and note down y_n, ∇y_n, $\nabla^2 y_n$, ... values.

Step 2: Compute $u = \dfrac{x - x_n}{h}$ with $x = c$; c is the value of x at which y is required. Here, x_n is the last argument and h is the interval of differencing.

Step 3: Write down the Newton's backward difference formula (3.16).

Step 4: Substitute the values of u obtained in Step 2 and y_n, ∇y_n, $\nabla^2 y_n$, ... etc. obtained in Step 1 in the formula written in Step 3. Simplifying it, we get the interpolating value of $y = f(x)$ at $x = c$.

SOLVED PROBLEMS

Problem 1: Using the following tabulated values, find $f(3)$ by Newton's backward interpolation formula.

x	0	2	4	6
$f(x)$	1	9	25	49

Solution: The backward difference table for the given data is shown below:

x	$f(x)$	∇y	$\nabla^2 y$	$\nabla^3 y$
0	1			
		$\rightarrow 8$		
2	9		$\rightarrow 8$	
		$\rightarrow 16$		$\rightarrow 0$
4	25		$\rightarrow 8$	
		$\rightarrow 24$		
6	49			

From the above table, we have $x_n = 6, h = 2, \nabla y_n = 24, \nabla^2 y_n = 8$ and $\nabla^3 y_n = 0$.

We have to find $f(3)$. Hence, $c = 3$ and $u = \dfrac{x - x_n}{h} = \dfrac{3 - 6}{2} = -1.5$.

The Newton's backward-difference interpolation formula is

$$f(x) = f(x_n + uh) = y_n + u\nabla y_n + \frac{u(u+1)}{\lfloor 2}\nabla^2 y_n$$

$$+ \frac{u(u+1)(u+2)}{\lfloor 3}\nabla^3 y_n + \ldots\ldots \qquad (1)$$

Substituting the values of y_n, ∇y_n, $\nabla^2 y_n$, $\nabla^3 y_n$, $x = c = 3$ and u in the right-hand side of Eq. (1), we get

$$f(3) = 49 + (-1.5)(24) + \frac{(-1.5)(-1.5+1)}{\lfloor 2} \times 8$$

$$+ \frac{(-1.5)(-1.5+1)(-1.5+2)}{\lfloor 3}(0).$$

$$= 16$$

Hence, the value of $f(x)$ at $x = 3$ is 16.

Problem 2: The population of a town in the decinial census was given below. Estimate the population for the year 1930.

Year	1891	1901	1911	1921	1931
Population (in lakhs)	25	32	41	49	58

Solution: The backward difference table for the given data is shown below:

x	$f(x)$	∇y	$\nabla^2 y$	$\nabla^3 y$	$\nabla^4 y$
1891	25				
		$\to 7$			
1901	32		$\to 2$		
		$\to 9$		$\to -3$	
1911	41		$\to -1$		$\to 5$
		$\to 8$		$\to 2$	
1921	49		$\to 1$		
		$\to 9$			
1931	58				

From the above table, we have $x_n = 1931$, $y_n = 58$, $\nabla y_n = 9$, $\nabla^2 y_n = 1$ and $\nabla^3 y_n = 2$, $\nabla^4 y_n = 5$ and $h = 10$.

Now, we have to find the population for the year 1930. Therefore $c = 1930$ and

$$u = \frac{c - x_n}{h} = \frac{1930 - 1931}{10} = -0.1.$$

The Newton's backward-difference interpolation formula is,

$$f(x) = f(x_n + uh) = y_n + u\nabla y_n + \frac{u(u+1)}{\lfloor 2} \nabla^2 y_n$$

$$+ \frac{u(u+1)(u+2)}{\lfloor 3} \nabla^3 y_n + \frac{u(u+1)(u+2)(u+3)}{\lfloor 4} \nabla^4 y_n + \ldots\ldots(1)$$

$$f(1930) = 58 + (-0.1)9 + \frac{(-0.1)(-0.1+1)}{\lfloor 2}(1)$$

$$+ \frac{(-0.1)(-0.1+1)(-0.1+2)}{\lfloor 3}(2)$$

$$+ \frac{(-0.1)(-0.1+1)(-0.1+2)(-0.1+3)}{\lfloor 4}(5)$$

$$= 58 - 0.9 - 0.045 - 0.057 - 0.10331$$
$$= 56.89469$$

Problem 3: Given $\sin 45° = 0.7071$, $\sin 50° = 0.7660$, $\sin 55° = 0.8192$ and $\sin 60° = 0.8660$, find $\sin 58°$ using Newton's backward-difference interpolation formula.

Solution: Let $y = \sin x$. The backward difference table for the data is shown below:

x	$y = \sin x$	∇y	$\nabla^2 y$	$\nabla^3 y$
45°	0.7071			
		$\rightarrow 0.0589$		
50°	0.7660		$\rightarrow -0.0057$	
		$\rightarrow 0.0532$		$\rightarrow -0.0007$
55°	0.8192		$\rightarrow -0.0064$	
		$\rightarrow 0.0468$		
60°	0.8660			

From the above table, observe that $x_n = 60°$, $h = 5°$, $y_n = 0.8660$, $\nabla y_n = 0.0468$, $\nabla^2 y_n = -0.0064$ and $\nabla^3 y_n = -0.0007$.

Since $y = \sin x$ value is required at $x = 58°$, $c = 58°$ and

$$u = \frac{c - x_n}{h} = \frac{58 - 60}{5} = -0.4.$$

Newton's backward-difference interpolation formula is

$$f(x) = f(x_n + uh) = y_n + u\nabla y_n + \frac{u(u+1)}{\lfloor 2}\nabla^2 y_n$$

$$+ \frac{u(u+1)(u+2)}{\lfloor 3}\nabla^3 y_n + \ldots\ldots \tag{1}$$

Substituting the values of $u, y_n, \nabla y_n, \nabla^2 y_n, \ldots$ etc in (1), we get

$$\mathrm{Sin}\,58° = 0.8660 + (-0.4)(0.0468) + \frac{(-0.4)(-0.4+1)}{\lfloor 2}(-0.0064)$$

$$+ \frac{(-0.4)(-0.4+1)(0.4+2)}{\lfloor 3}(-0.0007)$$

$$= 0.8660 - 0.01872 + 0.000768 + 0.0000448$$

$$= 0.8480928.$$

EXERCISE 3(a)

1. Use Newton's forward interpolation formula to find $f(1)$, given that

x	0	2	4	6
y	4	12	68	220

2. Find the interpolating polynomial for the data given below, using Newton's forward interpolation formula.

x	0	1	2	3
$y = f(x)$	3	6	11	18

3. Using Newton's forward interpolation formula, find $f(1.5)$ from the data given below:

x	1.4	1.6	1.8	2.0	2.2
$f(x)$	0.21	0.69	1.25	1.89	2.61

4. The following table gives the specific heat of a material at different temperatures. Estimate the specific heat corresponding to 5°c.

Temperature (0°c)	0	10	20	30	40	50
Specific heat	0.54	0.58	0.6	0.62	0.65	0.70

5. Find the polynomial, which fits the data given below:

x	3	5	7	9	11
y	6	25	58	109	175

6. The sales of a company in lakhs of rupees for the years 2002 to 2012 are given below.

Year	2002	2004	2006	2008	2010	2012
Sales	30	44	60	82	94	104

Find the sales in the year 2003

7. Given that $\sqrt{120} = 10.9544$, $\sqrt{130} = 11.4017$, $\sqrt{140} = 11.8321$, $\sqrt{150} = 12.2474$, $\sqrt{160} = 12.64491$, find $\sqrt{125}$ using Newton's forward-difference interpolation formula.

8. From the following data, find the number of persons earning wages between 65 and 70 rupees.

Wages in ₹	Number of persons
Below 50	200
50–60	140
60–70	110
70–80	80
80–90	45
90–100	24

9. Applying Newton's forward interpolation formula, compute the value $\sqrt{5.5}$ given that $\sqrt{5} = 2.2360$, $\sqrt{6} = 2.4494$, $\sqrt{7} = 2.6457$, $\sqrt{8} = 2.8284$ and $\sqrt{9} = 3$.

10. Given that $\cos 45° = 0.7071$, $\cos 50° = 0.6427$, $\cos 55° = 0.5735$, $\cos 60° = 0.5$, $\cos 65° = 0.4226$ and $\cos 70° = 0.3420$, find $\cos 47°$, using Newton's forward interpolation formula.

11. The following table gives the values of the function

$$P(x) = \frac{1}{\sqrt{2\pi}} \int_{-\infty}^{x} e^{-\left(\frac{t^2}{2}\right)} dt$$

x	1.0	1.05	1.10	1.15	1.20
$P(x)$	0.6827	0.7062	0.7287	0.7198	0.7699

Evaluate $P(x)$ for $x = 1.04$.

12. The following table gives the values of the function $f(x) = e^{-x}$ for $x = 1, 2, 3, 4$ and 5.

x	1	2	3	4	5
$f(x)$	0.3678	0.1353	0.0497	0.0186	0.0067

Find $f(1.5)$ using Newton's forward difference interpolation formula.

13. From the given data, find $f(44)$ using Newton's backward interpolation formula,

x	20	25	30	35	40	45
$f(x)$	354	332	291	260	231	204

14. Using the Newton's backward-difference interpolation formula, find the interpolating polynomial of $f(x)$ for the data given below.

x	2	4	6	8
$f(x)$	7	21	43	73

15. Below are given the production of a fertilizer factory. In the years 2002, 2004, 2006, 2008 and 2010.

Year	2002	2004	2006	2008	2010
Production: (in tonnes)	40	45	52	60	66

Find the production of the factory for the year 2009.

16. Given that $\text{Sin}30° = 0.5$, $\sin35° = 0.5735$, $\sin40° = 0.6428$, $\sin45° = 0.7071$ and $\sin50° = 0.7660$, find $\sin47°$ using Newton's backward-difference interpolation formula.

17. Find the curve $y = f(x)$ passing through the points (0, 1), (1, 0), (2, 1) and (3, 10). Also, find the ordinate of a point on the curve when its abscissa is 2.5.

18. The amounts of carbon dioxide emanated by a car engine against the age-old of the car are given in the following table.

Age of car: (in years)	2	4	6	8	10
Amount of carbon dioxide	0.5	0.7	1.0	1.2	1.6

Find the amount of carbon dioxide emanated by the car engine if the age of the car is 9 years.

19. Find $f(3.5)$ from the data given below:

x	1	2	3	4
$f(x)$	1	16	81	256

20. Find y_{30}, given that $y_{20} = 254$, $y_{24} = 296$, $y_{28} = 342$ and $y_{32} = 396$.

ANSWERS

1. $f(1) = 5$

2. $x^2 + 2x + 3$

3. 0.44

4. 0.564413

5. $-\dfrac{7}{384}x^4 + \dfrac{25}{48}x^3 + \dfrac{1297}{192}x^2 + \dfrac{719}{48}x - \dfrac{2845}{128}$

6. 0.38984375

7. 11.18028

8. 112

9. 2.34509

10. 0.68196

11. 0.7015532

12. 0.3297414

13. 207.82208

14. $x^2 + x + 1$

15. 63.515625

16. 0.73131696

17. $x^3 - 22x^2 + 1$, 4.125

18. 1.336

19. 150.875

20. 367.75.

3.5   INTERPOLATION WITH UNEQUAL INTERVALS

The Newton's forward and backward difference formulae are applicable only when the values of a function given at the arguments that are equally spaced. These formulae are not applicable in the case of the arguments are not equally spaced. In such cases, either we can use the Lagrange's or Newton's divided-difference interpolation formula. Now, we proceed to learn interpolation formulae for unequal intervals (i.e., the arguments are not equally spaced points).

Theorem: (Lagrange's interpolation formula)

If $y_0, y_1 \ldots\ldots y_n$ are $(n + 1)$ values of $y = f(x)$ at $x = x_0, x_1, x_2 \ldots\ldots x_n$ (not necessarily at equal intervals), then the interpolating polynomial of $f(x)$ is given by

$$f(x) = \frac{(x - x_1)(x - x_2)\ldots\ldots\ldots\ldots(x - x_n)}{(x_0 - x_1)(x_0 - x_2)\ldots\ldots\ldots(x_0 - x_n)}\, y_0$$

$$+ \frac{(x - x_0)(x - x_2)\ldots\ldots\ldots\ldots(x - x_n)}{(x_1 - x_0)(x_1 - x_2)\ldots\ldots\ldots(x_1 - x_n)}\, y_1 + \ldots\ldots$$

$$+ \frac{(x - x_0)(x - x_2)\ldots\ldots\ldots\ldots(x - x_{n-1})}{(x_n - x_0)(x_n - x_1)\ldots\ldots\ldots(x_n - x_{n-1})}\, y_n \qquad (3.18)$$

Proof: Let the interpolating polynomial of $y = f(x)$ be $P_n(x)$ and

$$P_n(x) = A_0(x - x_1)(x - x_2) \ldots\ldots\ldots\ldots\ldots (x - x_n) +$$

$$A_1(x - x_0)(x - x_2)\ldots\ldots\ldots(x - x_n) + A_2(x - x_0)(x - x_1)(x - x_3)\ldots\ldots\ldots(x - x_n) +$$

$$\ldots\ldots\ldots + A_n(x - x_0)(x - x_1)\ldots\ldots\ldots(x - x_{n-1}), \qquad (3.19)$$

where $A_0, A_1, \ldots A_n$ are constants to be determined. As $P_n(x)$ is an interpolating polynomial of $y = f(x)$, we have

$$P_n(x_0) = y_0, \ P_n(x_1) = y_1, \ldots\ldots P_n(x_n) = y_n, \qquad (3.20)$$

and

$$P_n(x) \approx f(x). \qquad (3.21)$$

Now, substituting $x = x_0$ in (3.19) and using $P_n(x_0) = y_0$, We have

$$y_0 = A_0(x_0 - x_1)(x_0 - x_2)\ldots\ldots(x_0 - x_n)$$

$$\therefore \qquad A_0 = \frac{y_0}{(x_0 - x_1)(x_0 - x_2)\ldots\ldots\ldots(x_0 - x_n)}. \qquad (3.22)_0$$

Similarly, by substituting $x = x_1, x_2, \ldots\ldots, x_n$ and using $P_n(x_i) = y_i$ $(i = 1, 2, \ldots, n)$, we get

$$A_1 = \frac{y_1}{(x_1 - x_0)(x_1 - x_2)\ldots\ldots\ldots(x_1 - x_n)} \qquad (3.22)_1$$

$$A_2 = \frac{y_2}{(x_2 - x_0)(x_2 - x_1)(x_2 - x_3)\ldots\ldots\ldots(x_2 - x_n)} \qquad (3.22)_2$$

$$A_n = \frac{y_n}{(x_n - x_0)(x_n - x_1)(x_n - x_2)\ldots\ldots\ldots(x_n - x_{n-1})}. \quad (3.22)_n$$

Substituting $A_0, A_1, A_2, \ldots\ldots, A_n$ and using (3.21), we get

$$f(x) = \frac{(x - x_1)(x - x_2)\ldots\ldots\ldots\ldots(x - x_n)}{(x_0 - x_1)(x_0 - x_2)\ldots\ldots\ldots(x_0 - x_n)} y_0$$

$$+ \frac{(x - x_0)(x - x_2)\ldots\ldots\ldots\ldots(x - x_n)}{(x_1 - x_0)(x_1 - x_2)\ldots\ldots\ldots(x_1 - x_n)} y_1$$

$$+ \frac{(x - x_0)(x - x_1)(x - x_3)\ldots\ldots\ldots\ldots(x - x_n)}{(x_2 - x_0)(x_2 - x_1)(x_2 - x_3)\ldots\ldots\ldots(x_2 - x_n)} y_2 + \ldots\ldots$$

$$+ \frac{(x - x_0)(x - x_1)\ldots\ldots\ldots\ldots(x - x_{n-1})}{(x_n - x_0)(x_n - x_1)\ldots\ldots\ldots(x_n - x_{n-1})} y_n,$$

which is known as Lagrange's interpolation formula.

Note:

(i) The Lagrange's formula can also be used for equally spaced arguments.

(ii) The Lagrange's formula can also be written as $f(x) = \sum_{k=0}^{n} L_k(x) y_k$,

where $L_k(x) = \dfrac{(x - x_0)(x - x_1)\ldots(x - x_{k-1})(x - x_{k+1})\ldots(x - x_n)}{(x_k - x_0)(x_k - x_1)\ldots(x_k - x_{k-1})(x_k - x_{k+1})\ldots(x_k - x_{k+1})}$

(iii) Lagrange's formula for 3 tabulated points (x_0, y_0), (x_1, y_1) and (x_2, y_2) is

$$f(x) = \frac{(x - x_1)(x - x_2)}{(x_0 - x_1)(x_0 - x_2)} y_0 + \frac{(x - x_0)(x - x_2)}{(x_1 - x_0)(x_1 - x_2)} y_1$$

$$+ \frac{(x - x_0)(x - x_1)}{(x_2 - x_0)(x_2 - x_1)} y_2.$$

(iv) Lagrange's formula for 4 tabulated points (x_0, y_0), (x_1, y_1), (x_2, y_2) and (x_3, y_3) is,

$$f(x) = \frac{(x-x_1)(x-x_2)(x-x_3)}{(x_0-x_1)(x_0-x_2)(x_0-x_3)} y_0 + \frac{(x-x_0)(x-x_2)(x-x_3)}{(x_1-x_0)(x_1-x_2)(x_1-x_3)} y_1$$

$$+ \frac{(x-x_0)(x-x_1)(x-x_3)}{(x_2-x_0)(x_2-x_1)(x_2-x_3)} y_2 + \frac{(x-x_0)(x-x_1)(x-x_2)}{(x_3-x_0)(x_3-x_1)(x_3-x_2)} y_3$$

SOLVED PROBLEMS

Problem 1: Use Lagrange's formula to find y at $x = 2$ for the data given below:

x	0	1	3	4
$f(x)$	1	4	16	25

Solution: Here $x_0 = 0$, $x_1 = 1$, $x_2 = 3$, $x_3 = 4$, $y_0 = 1$, $y_1 = 4$, $y_2 = 16$ and $y_3 = 25$. Using the Lagrange's formula for 4 data points (iv of Note 3.5) we get

$$f(x) = \frac{(x-1)(x-3)(x-4)}{(0-1)(0-3)(0-4)}(1) + \frac{(x-0)(x-3)(x-4)}{(1-0)(1-3)(1-4)}(4)$$

$$+ \frac{(x-0)(x-1)(x-4)}{(3-0)(3-1)(3-4)}(16) + \frac{(x-0)(x-1)(x-3)}{(4-0)(4-1)(4-3)}(25)$$

Substituting $x = 2$ in the above equation, we obtain $f(2) = 9$.

Problem 2: Find the form of the function of $y = f(x)$ when $f(0) = -12$, $f(1) = 0$, $f(3) = 6$ and $f(4) = 12$.

Solution: Here $x_0 = 0$, $x_1 = 1$, $x_2 = 3$, $x_3 = 4$, $y_0 = -12$, $y_1 = 0$, $y_2 = 6$ and $y_3 = 12$. Therefore, by the Lagrange's formula the form of $f(x)$ is

$$f(x) = \frac{(x-1)(x-3)(x-4)}{(0-1)(0-3)(0-4)}(-12) + \frac{(x-0)(x-3)(x-4)}{(1-0)(1-3)(1-4)}(0)$$

$$+ \frac{(x-0)(x-1)(x-4)}{(3-0)(3-1)(3-4)}(6) + \frac{(x-0)(x-1)(x-3)}{(4-0)(4-1)(4-3)}(12)$$

Upon simplifying the right-hand side of the above equation, we get

$$f(x) = x^3 - 7x^2 + 18x - 12.$$

Problem 3: Applying the Lagrange's formula, find the partial fractions of $\dfrac{x^2 - 3x + 6}{x(x-1)(x-2)}$.

Solution: Let $f(x) = x^2 - 3x + 6$. The roots of $x(x-1)(x-2) = 0$ are 0, 1 and 2. At these points, we compute $f(x) = x^2 - 3x + 6$ values and they are as follows:

x	0	1	2
$f(x)$	6	4	4

Applying the Lagrange's formula to the above data, we get

$$f(x) = \frac{(x-1)(x-2)}{(0-1)(0-2)} \times 6 + \frac{(x-0)(x-2)}{(1-0)(1-2)} \times 4 + \frac{(x-0)(x-1)}{(2-0)(2-1)} \times 4,$$

i.e.,
$$x^2 - 3x + 6 = 3(x-1)(x-2) - 4x(x-2) + 2x(x-1)$$

$\therefore$
$$\frac{x^2 - 3x + 6}{x(x-1)(x-2)} = \frac{3(x-1)(x-2) - 4x(x-2) + 2x(x-1)}{x(x-1)(x-2)}$$

$$= \frac{3}{x} - \frac{4}{x-1} + \frac{2}{x-2}.$$

Problem 4: The following table gives the normal weights of babies for various months of life.

Age in months (x)	0	2	4	7	9
Weights in kg (y)	3.2	5.1	6	7.6	8.4

Estimate the normal weight of babies at the age of 6 months.

Solution: Let $y = f(x)$ be the weight of babies at the age of x months. Then $x_0 = 0$, $x_1 = 2$, $x_2 = 4$, $x_3 = 7$, $x_4 = 9$, $y_0 = 3.2$, $y_1 = 5.1$, $y_2 = 6$ and $y_3 = 7.6$, $y_4 = 8.4$ for the given data. By the Lagrange's formula, with $x = 6$, we have

$$f(6) = \frac{(6-2)(6-4)(6-7)(6-9)}{(0-2)(0-4)(0-7)(0-9)}(3.2) + \frac{(6-0)(6-4)(6-7)(6-9)}{(2-0)(2-4)(2-7)(2-9)}(5.1)$$

$$+\frac{(6-0)(6-2)(6-7)(6-9)}{(4-0)(4-2)(4-7)(4-9)}(6)+\frac{(6-0)(6-2)(6-4)(6-9)}{(7-0)(7-2)(7-4)(7-9)}(7.6)$$

$$+\frac{(6-0)(6-2)(6-4)(6-7)}{(9-0)(9-2)(9-4)(9-7)}(8.4)$$

$$= 7.01238.$$

Problem 5: Find a polynomial satisfied by (0, 2), (1, 3), (2, 12) and (5, 147).

Solution: Suppose $(x_0, y_0) = (0, 2)$, $(x_1, y_1) = (1, 3)$, $(x_2, y_2) = (2, 12)$ and $(x_3, y_3) = 147$, Hence, by the Lagrange's interpolation formula,

$$f(x) = \frac{(x-1)(x-2)(x-5)}{(0-1)(0-2)(0-5)}(2)+\frac{(x-0)(x-2)(x-5)}{(1-0)(1-2)(1-5)}(3)$$

$$+\frac{(x-0)(x-1)(x-5)}{(2-0)(2-1)(2-5)}(12)+\frac{(x-0)(x-1)(x-2)}{(5-0)(5-1)(5-2)}(147)$$

$$= x^3 + x^2 - x - 12.$$

Problem 6: Find the partial fraction of $\dfrac{f(x)}{(x-3)(x-2)(x-1)(x+1)}$ using the following data and the Lagrange's interpolation formula

x	-1	1	2	3
$f(x)$	-21	15	12	3

Solution: Let $x_0 = -1$, $y_0 = -21$, $x_1 = 1$, $y_1 = 15$, $x_2 = 2$, $y_2 = 12$, $x_3 = 3$ and $y_3 = 3$. For this data, the Lagrange's interpolation gives

$$f(x) = \frac{(x-1)(x-2)(x-3)}{(-1-1)(-1-2)(-1-3)}(-21)+\frac{(x+1)(x-2)(x-3)}{(1+1)(1-2)(1-3)}(15)$$

$$+\frac{(x+1)(x-1)(x-3)}{(2+1)(2-1)(2-3)}(12)+\frac{(x+1)(x-1)(x-2)}{(3+1)(3-1)(3-2)}(3)$$

$$= \frac{7}{8}(x+1)(x-2)(x-3)+\frac{15}{4}(x+1)(x-2)(x-3)$$

$$-4(x+1)(x-1)(x-3)+\frac{3}{8}(x+1)(x-1)(x-2)$$

$$\therefore \quad \frac{f(x)}{(x+1)(x-1)(x-2)(x-3)} = \frac{7}{8(x+1)} + \frac{15}{4(x-1)} - \frac{4}{(x-2)} + \frac{3}{8(x-3)},$$

which are the required partial functions.

Problem 7: Let $f(a), f(b)$ and $f(c)$ be values of $y = f(x)$ at $x = a, b$ and c respectively. Show that $f(x)$ attains maximum or minimum at

$$x = \frac{\sum (b^2 - c^2) f(a)}{\sum (b - c) f(a)}$$

Solution: For the problem at our disposal, the data in tabular form are

x	a	b	c
$f(x)$	$f(a)$	$f(b)$	$f(c)$

For this data, applying the Lagrange's interpolation formula, we obtain

$$f(x) = \frac{(x-b)(x-c)}{(a-b)(a-c)} f(a) + \frac{(x-a)(x-c)}{(b-a)(b-c)} f(b) + \frac{(x-a)(x-b)}{(c-a)(c-b)} f(c),$$

i.e., $$f(x) = \sum \frac{\{x^2 - (b+c)x + bc\}}{(a-b)(a-c)} f(a). \tag{1}$$

For the existence of maximum or minimum $f'(x) = 0$; therefore, differentiating (1) w. r. t. x and equating to zero, we obtain

$$\sum \frac{\{2x - (b+c)\}}{(a-b)(a-c)} f(a) = 0,$$

i.e., $$\frac{\{2x - (b+c)\}}{(a-b)(a-c)} f(a) + \frac{\{2x - (c+a)\}}{(b-a)(b-c)} f(b) + \frac{\{2x - (a+b)\}}{(c-a)(c-b)} f(c) = 0.$$

Multiplying the above equation by $(b - c)(c - a)(a - b)$, we get

$$- (b - c)\{2x - (b + c)\} f(a) - (c - a)\{2x - (c + a)\}f(b) - (a - b)\{2x - (a + b)\} f(c) = 0$$

i.e. $2x(b - c) f(a) + 2x(c - a) f(b) + 2x(a - b) f(c)$
$$= (b^2 - c^2) f(a) + (c^2 - a^2) f(b) + (a^2 - b^2) f(c),$$

i. e, $2x \sum (b - c) f(a) = \sum (b^2 - c^2) f(a);$

$$\therefore x = \frac{1}{2} \cdot \frac{\sum \{(b^2 - c^2) f(a)\}}{\sum \{(b-c)f(a)\}}.$$

Problem 8: Using the Lagrange's formula, prove that

$$y_1 = y_3 - 0.3(y_5 - y_{-3}) + 0.2(y_{-3} - y_{-5}).$$

Solution: Let y_x be the value of y at x; we need to find y_1 in terms of y_{-5}, y_{-3}, y_3 and y_5. The tabular form of arguments and entries is as follows:

x	-5	-3	3	5
y_x	y_{-5}	y_{-3}	y_3	y_5

For the above data, the Lagrange's formula at $x = 1$ gives

$$y_1 = \frac{(1+3)(1-3)(1-5)}{(-5+3)(-5-3)(-5-5)}(y_{-5}) + \frac{(1+5)(1-3)(1-5)}{(-3+5)(-3-3)(-3-5)}(y_{-3})$$

$$+ \frac{(1+5)(1+3)(1-5)}{(3+5)(3+3)(3-5)}(y_3) + \frac{(1+5)(1+3)(1-3)}{(5+5)(5+3)(5-3)}(y_5)$$

$$= -0.2(y_{-5}) + 0.5\, y_{-3} + y_3 - 0.3\, y_5$$

$$= y_3 - 0.3(y_5 - y_{-3}) + 0.2(y_{-3} - y_{-5}).$$

Theorem: (Newton's divided-difference interpolation formula)

Let $y_0,\ y_1,\ y_2,\ \dots\ y_n$ be $(n + 1)$ values of $y = f(x)$ at $x_0,\ x_1,\ \dots\ x_n$ (not necessarily equally spaced points); then

$$f(x) = f(x_0) + (x - x_0)\, f[x_0, x_1] + (x - x_0)(x - x_1)\, f[x_0, x_1, x_2] +$$
$$(x - x_0)(x - x_1)\dots\dots(x - x_{n-1}) f[x_0, x_1, x_2, \dots, x_n],$$

where $f[x_0, x_1, \dots, x_r]$ denote the rth divided difference over the arguments $x_0, x_1, \dots, x_r$.

Proof: Since we were given $(n + 1)$ values of y at $(n + 1)$ arguments, we can interpolate a polynomial $p_n(x)$ of nth degree such that $f(x) = p_n(x)$.

Hence, $f(x)$ is assumed to be a polynomial of nth degree.

Now, by the definition of divided difference of first order for x and x_0

$$f[x, x_0] = \frac{f(x) - f(x_0)}{x - x_0}$$

$$f(x) = f(x_0) + (x - x_0)f[x, x_0]. \tag{3.23}$$

Consider the second-order dividend difference

$$f[x, x_0, x_1] = \frac{f\left[x, x_0\right] - f\left[x_0, x_1\right]}{x - x_1}.$$

From this equation, we obtain

$$f[x, x_0] = f[x_0, x_1] + (x - x_1)f[x, x_0, x_1]. \tag{3.24}$$

From (3.23) and (3.24), we have

$$f(x) = f(x_0) + (x - x_0)f[x_0, x_1] + (x - x_0)(x - x_1)f[x, x_0, x_1]. \tag{3.25}$$

Similarly, by considering third, fourth, ... , $(n + 1)$th order divided differences and repeating the above process, it can be shown that

$$f(x) = f(x_0) + (x - x_0)f[x_0, x_1] + (x - x_0)(x - x_1)f[x_0, x_1, x_2] +$$
$$............ + (x - x_0)(x - x_1).....(x - x_{n-1})f[x_0, x_1, x_2, ..., x_n]$$
$$+ (x - x_0)(x - x_1)......(x - x_{n-1})(x - x_n)f[x, x_0, x_1,..., x_{n-1}, x_n]. \tag{3.26}$$

Note that the divided difference in the last term on the right–hand side of (3.26) is of $(n + 1)$th order divided difference, namely $f[x, x_0, x_1, ... x_{n-1}, x_n]$. Since $f(x)$ is assumed to be a polynomial of nth degree, we have $f[x, x_0, x_1, ... x_{n-1}, x_n] = 0$. Thus, Eq. (3.26) reduces to

$$f(x) = f(x_0) + (x - x_0)f[x_0, x_1] + (x - x_0)(x - x_1)f[x_0, x_1, x_2] + ...$$
$$+ (x - x_0)(x - x_1).....(x - x_{n-1})f[x_0, x_1, x_2,..., x_n] \tag{3.27}$$

which is called as Newton's divided difference interpolation formula.

3.6   PROCEDURE TO FIND y VALUE AT $x = c$ USING NEWTON'S DIVIDED DIFFERENCE INTERPOLATION FORMULA

Step 1:   For the given data construct the divided difference table.

Step 2:   Note down $x_0, x_1, x_2, ..., y_0, y_1, y_2.........$ values and divided differences $f[x_0, x_1], f[x_0, x_1, x_2], ...$ we also note down $x = c$ at which $f(x)$ is required

Step 3:   Write down Newton's divided difference interpolation formula

Step 4: Substitute the values obtained in Step 2 in the formula written in Step 3, simplifying it we obtain $f(c)$ value.

SOLVED PROBLEMS

Problem 1: For the data given below, find $f(14)$ using Newton's divided-difference interpolation formula

x	0	1	2	5
$f(x)$	2	3	12	147

Solution: The divided difference table for the given data is shown below:

x	$y = f(x)$	$f[x_r, x_{r+1}]$	$f[x_r, x_{r+1}, x_{r+2}]$	$f[x_r, x_{r+1}, x_{r+2}, x_{r+3}]$
0	2			
		$\rightarrow \dfrac{3-2}{1-0} = 1$		
1	3		$\rightarrow \dfrac{9-1}{2-0} = 4$	
		$\rightarrow \dfrac{12-3}{2-1} = 9$		$\rightarrow \dfrac{9-4}{5-0} = 1$
2	12		$\rightarrow \dfrac{45-9}{5-1} = 9$	
		$\rightarrow \dfrac{147-12}{5-2} = 45$		
5	147			

From the above table, we have $x_0 = 0, x_1 = 1, x_2 = 2, x_3 = 5, y_0 = 2, y_1 = 3,$
$y_2 = 12, y_3 = 147, f[x_0, x_1] = 1, f[x_0, x_1, x_2] = 4$ and $f[x_0, x_1, x_2, x_3] = 1$
Since, we have to find $f(4)$, the value of $(x =) c = 4$
The Newton's divided-difference interpolation formula is

$$f(x) = f(x_0) + (x - x_0)f[x_0, x_1] + (x - x_0)(x - x_1)f[x_0, x_1, x_2]$$
$$+ (x - x_0)(x - x_1)(x - x_2)f[x_0, x_1, x_2, x_3] + \ldots\ldots\ldots \tag{1}$$

Substituting the values of $y_0 = f(x_0)$, $(x =)c = 4$ and divided differences in (1), we obtain

$$f(4) = 2 + (4 - 0)(1) + (4 - 0)(4 - 1)(4) + (4 - 0)(4 - 1)(4 - 2)(1)$$
$$= 2 + 4 + 48 + 24$$
$$= 78.$$

Thus, the value of $y = f(x)$ at $x = 4$ is 78.

Problem 2: Using the Newton's divided-difference interpolation formula, find the polynomial satisfying the following data:

x	-4	-1	0	2	5
$y = f(x)$	1245	33	5	9	1335

Solution: The divide difference table for the given data is shown below:

x	$y = f(x)$	$f[x_r, x_{r+1}]$	$f[x_r, x_{r+1}, x_{r+2}]$	$f[x_r, x_{r+1}, x_{r+2}, x_{r+3}]$	$f[x_r, x_{r+1}, x_{r+2}, x_{r+3}, x_{r+4}]$
-4	1245				
		$\rightarrow -404$			
-1	33		$\rightarrow 94$		
		$\rightarrow -28$		$\rightarrow -14$	
0	5		$\rightarrow 10$		$\rightarrow 3$
		$\rightarrow 2$		$\rightarrow 13$	
2	9		$\rightarrow 88$		
		$\rightarrow 442$			
5	1335				

From the above table, we have $x_0 = -4$, $x_1 = -1$, $x_2 = 0$, $x_3 = 2$, $x_4 = 5$, $f(x_0) = 1245$, $f[x_0, x_1] = -404$, $f[x_0, x_1, x_2] = 94$ and $f[x_0, x_1, x_2, x_3] = -14$ and $f[x_0, x_1, x_2, x_3, x_4] = 3$.

The Newton's divided difference interpolation formula is

$$f(x) = f(x_0) + (x - x_0)f[x_0, x_1] + (x - x_0)(x - x_1)f[x_0, x_1, x_2] + \ldots \ldots$$

$$\therefore f(x) = 1245 + (x + 4)(-404) + (x + 4)(x + 1)\,(94) + (x + 4)(x + 1)$$
$$(x - 0)(-14) + (x + 4)(x + 1)(x - 0)(x - 2)\,(3)$$
$$= 3x^4 - 5x^3 + 6x^2 - 14x + 5,$$

which is the required polynomial satisfying the given data.

Problem 3: Using the Newton's divided-difference interpolation formula, find $f(9)$ and $f(12)$ from the following data:

x	4	5	7	10	11	13
$f(x)$	48	100	294	900	1210	2028

Solution: The divided difference formula for the given data is shown below:

x	$y = f(x)$	1ˢᵗ d.d	2ⁿᵈ d.d	3ʳᵈ d.d	4ᵗʰ d.d	5ᵗʰ d.d
4	48					
		$\to 52$				
5	100		$\to 15$			
		$\to 97$		$\to 1$		
7	294		$\to 21$		$\to 0$	
		$\to 202$		$\to 1$		$\to 0$
10	900		$\to 27$		$\to 0$	
		$\to 310$		$\to 1$		
11	1210		$\to 33$			
		$\to 409$				
13	2028					

From the above table, $x_0 = 4,\ x_1 = 5,\ x_2 = 7,\ x_3 = 10,\ x_4 = 11,\ x_5 = 13,$
$y_0 = 48,\ f[x_0, x_1] = 52,\ f[x_0, x_1, x_2] = 15,$ and $f[x_0, x_1, x_2, x_3] = 1$ and
$f(x_0, x_1, x_2, x_3, x_4) = 0,\ f[x_0, x_1, x_2, x_3, x_4, x_5] = 0$
The Newton's divided-difference interpolation formula is
$$f(x) = f(x_0) + (x - x_0)f[x_0, x_1] + (x - x_0)(x - x_1)f[x_0, x_1, x_2] + \ldots \quad (1)$$
To find $f(x)$ at $x = 9$
Put $x = 9$ in (1) and use the values of $y_0 (= f(x_0)), f[x_0, x_1], f[x_0, x_1, x_2]\ldots$
etc., in the equation (1) we get

$$f(9) = 48 + 5(52) + (5)(4)15 + (5)\,(4)\,(2)\,(1) + 0$$
$$= 48 + 260 + 300 + 40 = 648.$$

Similarly, substituting $x = 12$ in Eq. (1) we obtain

$$f(12) = 48 + (12 - 4)52 + (12 - 4)(12 - 5)(15)$$
$$+ (12 - 4)(12 - 5)(12 - 7)(1)$$
$$= 48 + 416 + 840 + 280$$
$$= 1584.$$

Problem 4: Given the following data, find $f(x)$ as a polynomial is powers of $(x-5)$

x	0	2	3	4	7	9
$f(x)$	4	26	58	112	466	922

Solution: First, we find the interpolating polynomial of $f(x)$ for the data given, using the Newton's divided-difference interpolation formula. Later, we express $f(x)$ = interpolating polynomial of $f(x)$ in power of $(x - 5)$.

The divided difference table for the given data is as follows:

x	$f(x)$	1st d.d	2nd d.d	3rd d.d	4th d.d	5th d.d
0	4					
		$\to 11$				
2	26		$\to 7$			
		$\to 32$		$\to 1$		
3	58		$\to 11$		$\to 0$	
		$\to 54$		$\to 1$		$\to 0$
4	112		$\to 26$		$\to 0$	
		$\to 118$		$\to 1$		
7	466		$\to 22$			
		$\to 228$				
9	922					

The Newton's divided-difference interpolation formula is

$$f(x) = f(x_0) + (x - x_0)f[x_0, x_1] + (x - x_0)(x - x_1)f[x_0, x_1, x_2] + (x - x_0)$$
$$(x - x_1)(x - x_2)f[x_0, x_1, x_2, x_3] + \ldots\ldots\ldots \tag{1}$$

For the given problem, $x_0 = 0$, $x_1 = 2$, $x_2 = 3$, $x_3 = 4$, $x_4 = 7$, $x_5 = 9$, $f(x_0) = 4$, $f[x_0, x_1] = 11$, $f[x_0, x_1, x_2] = 7$, $f[x_0, x_1, x_2, x_3] = 1$, $f[x_0, x_1, x_2, x_3, x_4] = f[x_0, x_1, x_2, x_3, x_4, x_5] = 0$. Substituting these values in Eq. (1), we get

$$f(x) = 4 + (x - 0)(11) + (x - 0)(x - 2)(7) + (x - 0)(x - 2)(x - 3)(1) + 0$$

i.e., $f(x) = x^3 + 2x^2 + 3x + 4$ \hfill (2)

Now, we express the right–hand side of Eq. (2) as a polynomial in powers of $(x - 5)$.

Suppose

$$f(x) = x^3 + 2x^2 + 3x + 4 = A + B(x - 5) + C(x - 5)^2 + D(x - 5)^3 \quad (3)$$

Then

$$x^3 + 2x^2 + 3x + 4 = x^3 (D) + x^2 (-15D + C) + x (75D - 10C + B) + (A + 125D + 25C - 5B).$$

Comparing the coefficient of like powers of x, we obtain

$$D = 1$$
$$-15D + C = 2$$
$$75D - 10C + B = 3 \hspace{3em} (4)$$
$$A + 125D + 25C - 5B = 4$$

Solving the simultaneous equations of (4) for A, B, C and D, we get $A = 794$, $B = 98$, $C = 17$, $D = 1$. Thus,

$$f(x) = 794 + 98(x - 5) + 17(x - 5)^2 + (x - 5)^3.$$

EXERCISE 3(b)

1. Use the Newton's dividend difference formula to find the value of y at $x = 4$, given that

x	0	1	2	5
y	3	6	8	9

2. Find the form of the function y for the following data using Newton's divided difference formula:

x	5	6	9	11
y	12	13	14	17

3. Given that

x	5	11	27	34	42
$f(x)$	23	899	17315	35606	68510

Find $f(x)$ in powers of $(x - 2)$.

4. Given that

x	650	654	658	661
$f(x) = \log_{10} x$	2.8129	2.8156	2.8182	2.8202

Find $\log_{10} 657$ using (i) Lagrange's interpolation formula (ii) Newton's divided difference formula.

5. Use the Lagrange's interpolation formula to find the curve passing through the points (0, 1), (1, 3) and (3, 55).

6. The population of a town in the decenial census was as given below:

year	1891	1901	1911	1921	1931
Population in lakhs	48	64	79	K	93

Find K.

7. Given that

x	10	14	16	20	23
$f(x)$	19.97	21.51	22.47	23.52	24.65

Find $f(18)$.

8. Given that $\sqrt[3]{12} = 2.2894$, $\sqrt[3]{15} = 2.4602$, $\sqrt[3]{20} = 2.7144$ and $\sqrt[3]{22} = 2.8020$; Find $\sqrt[3]{17}$, Using (i) Lagrange's formula of interpolation (ii) Newton's divided difference formula of interpolation.

9. Estimate the production of cotton in the year 1960 from the data given below:

Year	Production (in tonnes)
1940	17.23
1946	18.46
1950	20.24
1956	30.16
1962	38.24

10. Suppose $f(x)$ = The number of students who obtained less than or equals to x marks in an examination. The values of $f(x)$ are given in the following table for $x = 25, 40, 50, 70, 85$ and 100:

x	25	40	50	70	85	100
$f(x)$	4	12	16	34	43	60

Find the number of students who obtained less than 60 marks. Also, find the number of students who obtained the marks between 60 and 70.

11. For the data given below, find the approximate function of $f(x)$.

x	5	6	9	11
$f(x)$	12	14	18	20

12. For the data given below, find $f(7)$ and $f(10)$.

x	5	6	9	11
$f(x)$	13	14	16	19

13. The following table gives the average weights of babies during the first 12 months of life:

Age in months	0	2	5	7	11	12
Weight in kgs	2.5	3.0	3.5	4.0	4.9	5

Estimate the average weight of baby at the age of 10 months.

14. From the following data, estimate the number of employees having income between ₹11,000 and 14,000:

Income (Rs)	No. of employees
Below 8000	250
8000 to 9000	60
9000 to 10000	42
10001 to 12000	28
12001 to 14000	14

15. Given that, $\tan 15° = 0.2679$ $\tan 18° = 0.3249$ $\tan 20° = 0.3639$ $\tan 24° = 0.4452$ $\tan 30° = 0.5773$. Find $\tan 25°$.

ANSWERS

1. 9.38984

2. $(0.0666)x^3 - (1.49987)x^2 + 11.432151x - 15.9583$

3. $(x-2)^3 + 3(x-2)^2 - 7(x-2) - 10$ 4. 2.817462, 2.817554

5. $8x^2 - 6x + 1$ 6. 89.75 7. 23.114551

8. $2.571468, 2.57149$ 9. 37.256973 10. $31, 3$

11. $0.0166x^3 - 0.4999x^2 + 5.98833x - 7.5$

12. $f(7) = 14.6667, f(10) = 17.1666$

13. 4.417027 14. 18 15. $0.46632.$

3.7   REMARK

Suppose $y_x = f(x)$ and the variable x is transformed in the variable u by the relation $u = \dfrac{x - x_0}{h}$, substituting $u = \dfrac{x - x_0}{h}$ in $y_x = f(x)$ we obtain a function in terms of the variable u. Let this function be $y_u = F(u)$.

Then $[y_x]_{x = c} = [y_u]_{u = \frac{c - x_0}{h}}$ i.e, the value of y_x at $x = c$ is same as the value of y_u at $u = \dfrac{c - x_0}{h}$. We illustrate with an example.

Suppose, $f(x) = x^2 + 4$. The values of it at $x = 1, 3, 5, 7$ and 9 are respectively $5, 13, 29, 53$ and 85. Let $u = \dfrac{x - 5}{2}$; then the function $f(x) = x^2 + 4$ in the variable u is

$$y_u = F(u) = (2u + 5)^2 + 4 = 4u^2 + 20u + 29.$$

The values of u at $x = 1, 3, 5, 7, 9$ are $-2, -1, 0, 1, 2$, respectively. The values of $F(u)$ at $u = -2, -1, 0, 1$ and 2 are $5, 13, 29, 53$ and 85, respectively.

Thus, $[f(x)]_{x = 1} = [F(u)]_{u = -2}$, $[f(x)]_{x = 3} = [F(u)]_{u = -1}$, $[f(x)]_{x = 5} = [F(u)]_{u = 0}$, $[f(x)]_{x = 7} = [F(u)]_{u = 1}$ and $[f(x)]_{x = 9} = [F(x)]_{u = 2}$.

3.8   NOTATION

In the subsequent sections, the following notations are used. If the values of $y = f(x)$ at $x_0, x_1, x_2, \ldots, x_n$ are $y_0, y_1, \ldots \ldots \ldots y_n$ with $x_i = x_0 + ih(i = 0, 1, 2, \ldots., n)$ and

$$u = \frac{x - (middle \; or \; near \; middle \; of \; the \; arguments)}{h}$$

then the values of u at $x_0, x_1, x_2, \ldots, x_n$ are ...., $-3, -2, -1, 0, 1, 2,$ .... The values of y for the arguments $u = \ldots, -3, -2, -1, 0, 1, 2,$ $3, \ldots\ldots$ are denoted by........ $y_{-3}, y_{-2}, y_{-1}, y_0, y_1, y_2, y_{3\ldots}$. Note that the entries $y_{-3}, y_{-2}, y_{-1}, y_0, y_1, y_2, y_3$ are corresponding to arguments of the variable $u = \ldots, -3, 2, -1, 0, 1, 2, 3$, and we refer middle entry or near-middle entry used in u as a change of origin or origin.

3.9 CENTRAL DIFFERENCE

Suppose, we need to find a variable of y at the middle or near the middle of the arguments from the given data where the arguments are equidistant, in such cases to involve the differences of the middle of the difference table, we employ the central difference formula for interpolation as the variation of $y = f(x)$ at the middle of the arguments may differ from variation at the beginning and end of the arguments. Now, we proceed to derive such formulae.

3.10 PATH OF THE INTERPOLATION FORMULA

In a difference table, the line segments joining from an entry (or entries) involved in an interpolation formula to the lowest order central difference (or forward or backward) involved in that interpolation formula and from this lower order central (or forward or backward) differnce to the next lowest order central (or forward or backward) difference involved in the interpolation and so on yield a path. This path is known as the path of the interpolation.

For example, the Newton's forward difference formula is

$$f(x) = y_0 + u\Delta y_0 + \frac{u(u-1)}{\lfloor 2} \Delta^2 y_0 + \frac{u(u-1)(u-2)}{\lfloor 3} \Delta^3 y_0$$

$$+ \frac{u(u-1)(u-2)(u-3)}{\lfloor 4} \Delta^4 y_0 + \ldots\ldots$$

The formula involved $y_0, \Delta y_0, \Delta^2 y_0, \Delta^3 y_0, \Delta^4 y_0, \ldots\ldots$ etc. In forward differnce table the line segment joining from y_0 to Δy_0, Δy_0 to $\Delta^2 y_0$ $\Delta^2 y_0$ to $\Delta^3 y_0$, $\Delta^3 y_0$ to $\Delta^4 y_0$, ... etc yield a path and is shown in the following table.

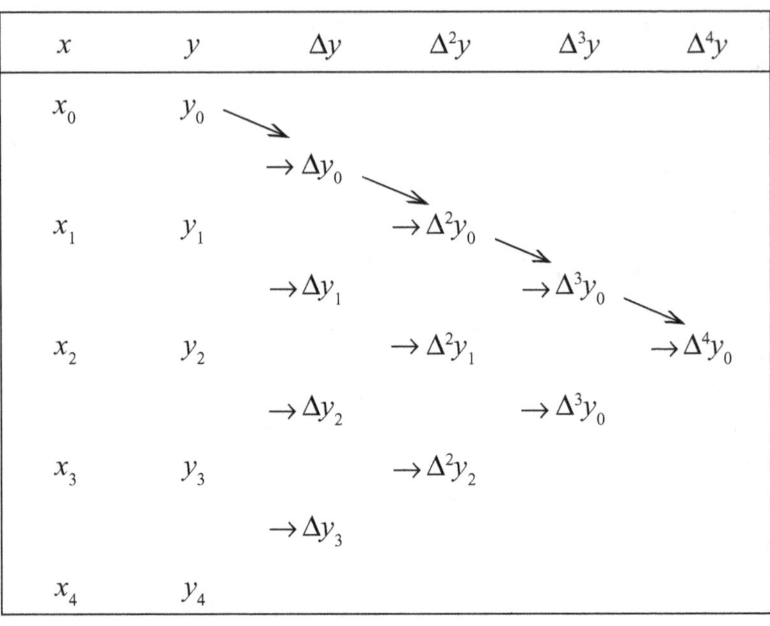

$y_0 \rightarrow \Delta y_0 \rightarrow \Delta^2 y_0 \rightarrow \Delta^3 y_0 \rightarrow \Delta^4 y_0$ is the path of forward-difference interpolation formula.

3.11 GAUSS'S FORWARD INTERPOLATION FORMULA

Let $.....y_{-3}, y_{-2}, y_{-1}, y_0, y_1, y_2, ...$ be the values of $y = f(x)$ at the arguments $x_{-3}, x_{-2}, x_{-1}, x_0, x_1, x_2, x_3 .....$

Suppose we wish to have an interpolation formula involving the differences along the path shown in the following table:

x	y	Δy	$\Delta^2 y$	$\Delta^3 y$	$\Delta^4 y$	$\Delta^5 y$	$\Delta^6 y$
x_{-3}	y_{-3}						
		$\rightarrow \Delta y_{-3}$					
x_{-2}	y_{-2}		$\rightarrow \Delta^2 y_{-3}$				
		$\rightarrow \Delta y_{-2}$		$\rightarrow \Delta^3 y_{-3}$			
x_{-1}	y_{-1}		$\rightarrow \Delta^2 y_{-2}$		$\rightarrow \Delta^4 y_{-3}$		
		$\rightarrow \Delta y_{-1}$		$\rightarrow \Delta^3 y_{-2}$		$\rightarrow \Delta^5 y_{-3}$	

$$
\begin{array}{llllllll}
x_0 & y_0 & & \to \Delta^2 y_{-1} & & \to \Delta^4 y_{-2} & & \to \Delta^6 y_{-3} \\
 & & \to \Delta y_0 & & \to \Delta^3 y_{-1} & & \to \Delta^5 y_{-2} & \\
x_1 & y_1 & & \to \Delta^2 y_0 & & \to \Delta^4 y_{-1} & & \\
 & & \to \Delta y_1 & & \to \Delta^3 y_0 & & & \\
x_2 & y_2 & & \to \Delta^2 y_1 & & & & \\
 & & \to \Delta y_2 & & & & & \\
x_3 & y_3 & & & & & &
\end{array}
$$

Let the required interpolating polynomial involving the differences as shown in the table be

$$y_p = y_0 + G_1\,\Delta y_0 + G_2\,\Delta^2 y_{-1} + G_3\,\Delta^3 y_{-1} + G_4\,\Delta^4 y_{-2} + G_5\,\Delta^5 y_{-2} + \ldots \quad (3.28)$$

where $G_1, G_2, G_3, \ldots$, are to be determined coefficients and $p = \dfrac{x - x_0}{h}$.

Note that x_0 is not an initial argument, and it is either middle or near-the middle argument of the given arguments. Now,

$$y_p = E^p\, y_0 = (1 + \Delta)^p\, y_0$$

$$= y_0 + {}^pC_1\,\Delta y_0 + {}^pC_2\Delta^2 y_0 + {}^pC_3\Delta^3 y_0 + \ldots \quad (3.29)$$

Consider

$$y_{-1} = E^{-1}y_0 = (1 + \Delta)^{-1}y_0 = y_0 - \Delta y_0 + \Delta^2 y_0 - \Delta^3 y_0 + \ldots$$

$$\therefore \qquad \Delta^2 y_{-1} = \Delta^2 y_0 - \Delta^3 y_0 + \Delta^4 y_0 - \Delta^5 y_0 + \ldots \quad (3.30)$$

$$\Delta^3 y_{-1} = \Delta^3 y_0 - \Delta^4 y_0 + \Delta^5 y_0 - \Delta^6 y_0 + \ldots \quad (3.31)$$

Similarly,

$$\Delta^4 y_{-2} = \Delta^4 y_0 - 2\Delta^5 y_0 + 3\Delta^6 y_0 - \ldots \quad (3.32)$$

$$\Delta^5 y_{-2} = \Delta^5 y_0 - 2\Delta^6 y_0 + 3\Delta^7 y_0 - \ldots \quad (3.33)$$

Substituting (3.30), (3.31), (3.32) and (3.33) in (3.28), we get

$$y_p = y_0 + G_1\,\Delta y_0 + G_2\{\Delta^2 y_0 - \Delta^3 y_0 + \Delta^4 y_0 - \Delta^5 y_0 + \ldots\}$$

$$+ G_3(\Delta^3 y_0 - \Delta^4 y_0 + \Delta^5 y_0 - \Delta^6 y_0 - \ldots)$$

$$+ G_4(\Delta^4 y_0 - 2\Delta^5 y_0 + 3\Delta^6 y_0 - \ldots)$$

$$+ G_5(\Delta^5 y_0 - 2\Delta^6 y_0 + 3\Delta^7 y_0 - \ldots) + \ldots$$

i.e., $y_p = y_0 + G_1\Delta y_0 + G_2\Delta^2 y_0 + \{-G_2 + G_3\}\,\Delta^3 y_0 + \{G_2 - G_3 + G_4\}\Delta^4 y_0$

$$+ \{-G_2 + G_3 - 2G_4 + G_5\}\Delta^5 y_0 + \ldots\ldots \tag{3.34}$$

From (3.29) and (3.34), we have

$$y_0 + {}^pC_1\,\Delta y_0 + {}^pC_2\,\Delta^2 y_0 + {}^pC_3\,\Delta^3 y_0 + \ldots\ldots\ldots$$

$$= y_0 + G_1\,\Delta y_0 + G_2\Delta^2 y_0 + \{-G_2 + G_3\}\,\Delta^3 y_0$$

$$+ \{G_2 - G_3 + G_4\}\,\Delta^4 y_0 + \{-G_2 + G_3 - 2G_4 + G_5\}\,\Delta^5 y_0 + \ldots.$$

Comparing the coefficients of $\Delta^k y_0 (k = 0, 1, 2, 3, \ldots\ldots..)$, we get

$$G_1 = p,\ G_2 = {}^pC_2,\ -G_2 + G_3 = {}^pC_3$$

$$G_2 - G_3 + G_4 = {}^pC_4,\ -G_2 + G_1 - 2G_4 + G_5 = {}^pC_5 \ldots\ldots.$$

Using ${}^nC_{r-1} + {}^nC_r = {}^{n+1}C_r$, it can be shown that

$G_1 = {}^pC_1,\ G_2 = {}^pC_2,\ G_3 = {}^{(p+1)}C_3,\ G_4 = {}^{(p+1)}C_4,\ G_1 = {}^{(p+2)}C_5$, and so on

Thus,

$$y_p = y_0 + {}^pC_1\,\Delta y_0 + {}^pC_2\,\Delta^2 y_{-1} + {}^{(p+1)}C_3\,\Delta^3 y_{-1} + {}^{(p+1)}C_4\,\Delta^4 y_{-2}$$

$$+ {}^{(p+2)}C_5\,\Delta^5 y_{-2} + \ldots\ldots\ldots \tag{3.35a}$$

or

$$y_p = y_0 + p\Delta y_0 + \frac{p(p-1)}{\lfloor 2} \Delta^2 y_{-1} + \frac{(p+1)p(p-1)}{\lfloor 3} \Delta^3 y_{-1}$$

$$+ \frac{(p+1)p(p-1)(p-2)}{\lfloor 4} \Delta^4 y_{-2} + \frac{(p+2)(p+1)p(p-1)(p-2)}{\lfloor 5} \Delta^5 y_{-2} + \ldots.$$

$$\tag{3.35b}$$

The equations given by (3.35a) or (3.35b) are known as Gauss's forward interpolation formula.

3.12 THE PATH OF GAUSS'S FORWARD INTERPOLATION FORMULA

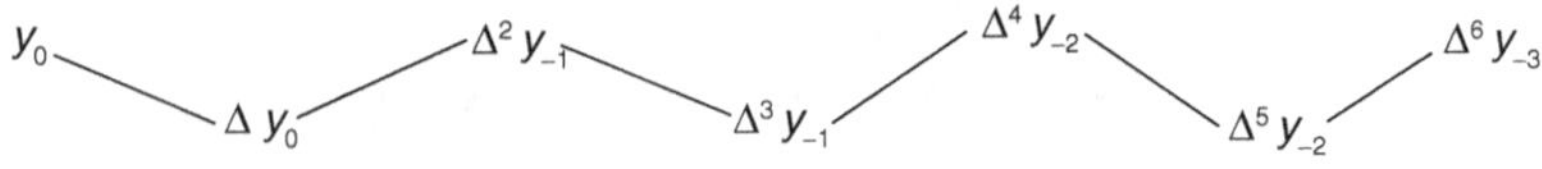

The path shown above is the path of Gauss's forward interpolation formula. Note that it involves the odd differences below the central

line drawn through x_0 and y_0, and the even differences on the central line.

Note: Since $\delta y_{\frac{1}{2}} = \Delta y_0$, $\delta^2 y_0 = \Delta^2 y_{-1}$, $\delta^3 y_{\frac{1}{2}} = \Delta^3 y_{-1}$, $\delta^4 y_0 = \Delta^4 y_{-2}$ etc,

The Gauss's forward interpolation formula can be expressed as

$$y_p = y_0 + p\,\delta y_{\frac{1}{2}} + \frac{p(p-1)}{\lfloor 2} \delta^2 y_0 + \frac{(p+1)p(p-1)}{\lfloor 3} \delta^3 y_{\frac{1}{2}}$$

$$+ \frac{(p+1)p(p-1)(p-2)}{\lfloor 4} \delta^4 y_0 + \frac{(p+2)(p+1)p(p-1)(p-2)}{\lfloor 5} \delta^5 y_{\frac{1}{2}} + \ldots$$

3.13 PROCEDURE TO FIND A VALUE OF $f(x)$ AT $x = c$ USING GAUSS'S FORWARD INTERPOLATION FORMULA

Step 1: Choose x_0 as the middle or near the middle of the given arguments. For example, if 2, 5, 8, 11, 14 are the given arguments, then choose $x_0 = 8$. If the arguments are 2, 5, 8, 11, 14, 17 are the given argments then choose x_0 as either 8 or 11.

Step 2: Transform the variable y in x to y in p by the relation

$$p = \frac{x - x_0}{h}$$

Step 3: Construct the difference table: (i) having x values in first column; (ii) having p values in second column; and (iii) having y values in third column. Next compute the forward differences of the entries of column 3.

Step 4: Draw the path of Gauss's forward formula in the table constructed in Step 3 and note down the values of y_0, Δy_0, $\Delta^2 y_{-1}$, $\Delta^3 y_{-1}$, $\Delta^4 y_{-2}$, $\Delta^5 y_{-2}$, ... etc. appeared along the path of Gauss's forward formula

Step 5: Note down the value of x at which $f(x)$ is required by Gauss's forward formula. Let this value be c.

Step 6: Compute p value corresponding to $x = c$.

Step 7: Write down the Gauss's forward interpolation formula.

Step 8: Substitute p value obtained in Step 6, y_0 and the forward differences obtained in Step 4 in the formula written in Step 7.

Step 9: Simplifying the right-hand side of the equation obtained in Step 8 yields the required result, i.e., $f(c)$.

SOLVED PROBLEMS

Problem 1: Find $f(26)$ by Gauss's forward interpolation formula from the data given below:

x	20	24	28	32
$f(x)$	26	34	37	42

Solution: Here number of arguments in the given data is 4. Now, we can choose the origin an either 24 and 28. Therefore, we choose $x_0 = 24$. The interval of differencing is $h = 4$. Hence, $p = \dfrac{x - 24}{4}$ is the relation used to transform the function $f(x)$ in u.

$\therefore f(x) = f(24 + 4p)$. Now we construct the central difference table

x	$p = \dfrac{x-24}{4}$	y_p	Δy_p	$\Delta^2 y_p$	$\Delta^3 y_p$
20	-1	26			
			$\to 8$		
24	0	$34\ (y_0)$		$\to -5(\Delta^2 y_{-1})$	
			$\to 3(\Delta y_0)$		$\to 7(\Delta^3 y_{-1})$
28	1	37		$\to 2$	
			$\to 5$		
32	2	42			

At $x = 26$ (i.e., $c = 26$) the value of $p = \dfrac{26 - 24}{4} = \dfrac{1}{2}$: Gauss's forward interpolation formula is

$$f(x) = f(x_0 + ph) = y_0 + p\,\Delta y_0 + \frac{p(p-1)}{\lfloor 2}\Delta^2 y_{-1}$$

$$+ \frac{(p+1)p(p-1)}{\lfloor 3}\Delta^3 y_{-1} + \quad \dots\dots\dots\dots (1)$$

Substituting the values of p, y_0, Δy_0, $\Delta^2 y_{-1}$ and, $\Delta^3 y_{-1}$ in (1), we get

$$f(26) = f\left(24 + 4\left(\frac{1}{2}\right)\right) = 34 + \frac{1}{2}(3) + \frac{\frac{1}{2}\left(\frac{1}{2}-1\right)}{\underline{|2}}(-5)$$

$$+ \frac{\left(\frac{1}{2}+1\right)\left(\frac{1}{2}\right)\left(\frac{1}{2}-1\right)}{\underline{|3}}(7)$$

$$= 35.6875.$$

Problem 2: From the following data, find $f(42)$ using the Gauss's forward formula.

x	30	35	40	45	50
$f(x)$	15.8	14.7	14.0	13.2	12.4

Solution: Note that the number of arguments given in the data is 5. Therefore, we choose the middle argument namely 40 as origin, Here, $h = 5$, and hence, $p = \dfrac{x-40}{5}$. The central difference table for the function y_p (which is transformed function of $f(x)$) is given below.

x	$p = \dfrac{x-40}{5}$	y_p	Δy_p	$\Delta^2 y_p$	$\Delta^3 y_p$	$\Delta^4 y_p$
30	-2	15.8				
			$\to -1.1$			
35	-1	14.7		$\to 0.4$		
			$\to -0.7$		$\to -0.5$	
40	0	14.0		$\to -0.1$		$\to 0.6$
			$\to -0.8$		$\to 0.1$	
45	1	13.2		$\to 0$		
			$\to -0.8$			
50	2	12.4				

The path of Gauss's forward formula is shown in the table. Hence, $y_0 = 14.0$, $\Delta y_0 = -0.8$, $\Delta^2 y_{-1} = -0.1$, $\Delta^3 y_{-1} = 0.1$ and $\Delta^4 y_{-2} = 0.6$.

We have to find $f(42)$, and therefore $c = 42$ and $p = \dfrac{42 - 40}{5} = 0.4$

Gauss's forward formula is given by

$$f(x) = f(x_0 + ph) = y_0 + p\,\Delta y_0 + \frac{p(p-1)}{\lfloor 2} \Delta^2 y_{-1} + \frac{(p+1)p(p-1)}{\lfloor 3} \Delta^3 y_{-1}$$

$$+ \frac{(p+1)p(p-1)(p-2)}{\lfloor 4} \Delta^4 y_{-2} + \dots \tag{1}$$

Substituting the value of p, y_0, Δy_0, $\Delta^2 y_{-1}$, $\Delta^3 y_{-1}$ and $\Delta^4 y_{-2}$ in (1), we get

$$f(42) = f(40 + 5 \times 0.4)$$

$$= 14.0 + 0.4(-0.8) + \frac{(0.4)(0.4-1)}{\lfloor 2}(-0.1) + \frac{(0.4)(0.4-1)(04+1)}{\lfloor 3}(0.1)$$

$$+ \frac{(0.4+1)(0.4)(0.4-1)(0.4-2)}{\lfloor 4}(0.6)$$

$$= 14.0 - 0.32 + 0.012 - 0.0056 + 0.01344$$
$$= 13.69984.$$

3.14 GAUSS'S BACKWARD INTERPOLATION FORMULA

Suppose $\dots y_{-3},\ y_{-2},\ y_{-1},\ y_0,\ y_1,\ y_2 \dots$ be the values of $y = f(x)$ at the arguments$\dots\dots$ $x_{-3},\ x_{-2},\ x_{-1},\ x_0,\ x_1,\ x_2,\ \dots\dots$ and we wish to have an interpolation formula involving the differences along the path shown in the following table, the path is known as the path of Gauss's backward formula.

x	y_p	Δy_p	$\Delta^2 y_p$	$\Delta^3 y_p$	$\Delta^4 y_p$	$\Delta^5 y_p$	$\Delta^6 y_p$
x_{-3}	y_{-3}						
		$\rightarrow\Delta y_{-3}$					
x_{-2}	y_{-2}		$\rightarrow\Delta^2 y_{-3}$				
		$\rightarrow\Delta y_{-2}$		$\rightarrow\Delta^3 y_{-3}$			
x_{-1}	y_{-1}		$\rightarrow\Delta^2 y_{-2}$		$\rightarrow\Delta^4 y_{-3}$		

$$
\begin{array}{ccccccc}
 & & \nearrow \Delta y_{-1} & \searrow & \nearrow \Delta^3 y_{-2} & \searrow & \nearrow \Delta^5 y_{-3} & \searrow \\
x_0 & y_0 & & \to \Delta^2 y_{-1} & & \to \Delta^4 y_{-2} & & \to \Delta^6 y_{-3} \\
 & & \to \Delta y_0 & & \to \Delta^3 y_{-1} & & \to \Delta^5 y_{-2} \\
x_1 & y_1 & & \to \Delta^2 y_0 & & \to \Delta^4 y_{-1} \\
 & & \to \Delta y_1 & & \to \Delta^3 y_0 \\
x_2 & y_2 & & \to \Delta^2 y_1 \\
 & & \to \Delta y_2 \\
x_3 & y_3
\end{array}
$$

Let the interpolating polynomial involving the differences along the path shown in the table be

$$
y_p = y_0 + G_1\,\Delta y_{-1} + G_2\,\Delta^2 y_{-1} + G_3\,\Delta^3 y_{-2} + G_4\,\Delta^4 y_{-2} + \ldots\ldots\ldots\ldots, \tag{3.36}
$$

where G_0, G_1, G_2, ……. are to be determined coefficients of the forward differences. Now,

$$
\begin{aligned}
y_p = E^p\,y_0 &= (1 + \Delta)^p y_0 \\
&= y_0 + {}^pC_1\,\Delta y_0 + {}^pC_2\,\Delta^2 y_0 + {}^pC_3\Delta^3 y_0 + \ldots\ldots
\end{aligned} \tag{3.37}
$$

Consider

$$
\begin{aligned}
\Delta y_{-1} = \Delta E^{-1}\,y_0 &= \Delta(1 + \Delta)^{-1}\,y_0 \\
&= \Delta\{1 - \Delta + \Delta^2 - \Delta^3 + \Delta^4 \ldots\ldots\ldots\}y_0 \\
&= \Delta y_0 - \Delta^2 y_0 + \Delta^3 y_0 - \Delta^4 y_0 + \Delta^5 y_0 + \ldots\ldots
\end{aligned} \tag{3.38}
$$

Similarly,

$$
\Delta^2 y_{-1} = \Delta^2 y_0 - \Delta^3 y_0 + \Delta^4 y_0 - \Delta^5 y_0 \ldots\ldots\ldots \tag{3.39}
$$

$$
\begin{aligned}
\Delta^3 y_{-2} = \Delta^3 E^{-2} y_0 &= \Delta^3(1 + \Delta)^{-2} y_0 \\
&= \Delta^3 y_0 - 2\Delta^4 y_0 + 3\Delta^5 y_0 - 4\Delta^6 y_0 + \ldots\ldots
\end{aligned} \tag{3.40}
$$

and

$$
\Delta^4 y_{-2} = \Delta^4 y_0 - 2\Delta^5 y_0 + 3\Delta^6 y_0 - 4\Delta^7 y_0 + \ldots\ldots.. \tag{3.41}
$$

so on…

Substituting (3.38) to (3.41) in 3.36, we obtain

$$y_p = y_0 + G_1\{\Delta y_0 - \Delta^2 y_0 + \Delta^3 y_0 - \Delta^4 y_0 + \Delta^5 y_0 + \ldots\ldots\}$$

$$+ G_2\{\Delta^2 y_0 - \Delta^3 y_0 + \Delta^4 y_0 - \Delta^5 y_0 + \ldots\ldots\} + G_3\{\Delta^3 y_0 - 2\Delta^4 y_0 + 3\Delta^5 y_0$$

$$- 4\Delta^6 y_0 \ldots\ldots\} + G_4\{\Delta^4 y_0 - 2\Delta^5 y_0 + 3\Delta^6 y_0 - 4\Delta^7 y_0 + \ldots\ldots\}$$

$$\text{i.e., } y_p = y_0 + G_1\Delta y_0 + (G_2 - G_1)\,\Delta^2 y_0 + (G_1 - G_2 + G_3)\,\Delta^3 y_0$$

$$+ (- G_1 + G_2 - 2G_3 + G_4)\,\Delta^4 y_0 + \ldots\ldots \tag{3.42}$$

Comparing the coefficients of $\Delta^k y_0$ on the right-hand sides of (3.37) and (3.42), we get

$$G_1 = {}^pC_1; \; -G_1 + G_2 = {}^pC_2; \; G_1 - G_2 + G_3 = {}^pC_3; \; -G_1 + G_2 - 2G_3 + G_4$$

$$= {}^pC_4\ldots \text{ etc.}$$

Solving the above equations, we obtain

$$G_1 = {}^pC_1, \; G_2 = {}^{(p+1)}C_2, \; G_3 = {}^{(p+1)}C_3, \; G_4 = {}^{(p+2)}C_4\ldots \text{ etc.}$$

Substituting G_1, G_2, G_3, G_4, …. etc in (3.36), we get

$$y_p = y_0 + {}^pC_1\,\Delta y_{-1} + {}^{(p+1)}C_2\,\Delta^2 y_{-1} + {}^{(p+1)}C_3\,\Delta^3 y_{-2}$$

$$+ {}^{(p+2)}C_4\,\Delta^4 y_{-2} + \ldots\ldots\ldots\ldots \tag{3.43a}$$

or

$$y_p = y_0 + p\,\Delta y_{-1} + \frac{(p+1)p}{\lfloor 2}\,\Delta^2 y_{-1} + \frac{(p+1)p(p-1)}{\lfloor 3}\,\Delta^3 y_{-2}$$

$$+ \frac{(p+2)(p+1)p(p-1)}{\lfloor 4}\,\Delta^4 y_{-2} + \ldots\ldots \tag{3.43b}$$

The formula given by (3.43a) or (3.43b) is known as Gauss's backward formula.

3.15   THE PATH OF GAUSS'S BACKWARD FORMULA

The following is the path of Gauss's backward interpolation as formula:

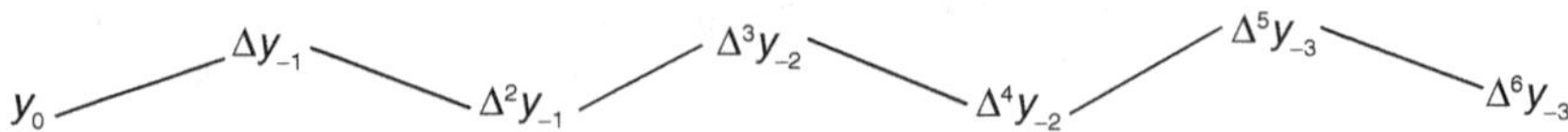

Note that the odd-order forward difference of the path of Gauss's backward formula appears just above the central line and even-order forward differences of it are along the central line.

3.16 PROCEDURE TO FIND A VALUE OF $f(x)$ AT $x = c$ USING GAUSS BACKWARD INTERPOLATION FORMULA

Step 1: Choose x_0 as the middle or near middle of the given arguments (See step1 of 3.13)

Step 2: Transfer the variable y in x to y in p by the relation $p = \dfrac{x - x_0}{h}$.

Step 3: Construct the difference table: (i) having x values in first column; (ii) having p values in second column; and (iii) having of values in the third column; next compute the forward differences of entries of column 3.

Step 4: Draw the path of Gauss's backward formula in the table constructed in Step 3 and note down the values of y_0, Δy_{-1}, $\Delta^2 y_{-1}$, $\Delta^3 y_{-2}$, $\Delta^4 y_{-2}$, $\Delta^5 y_{-3}$, $\Delta^6 y_{-3}$, ...

Step 5: Note down the values of x at which $f(x)$ is required. Let this value be c.

Step 6: Compute p value corresponding to $x = c.$, i.e., $p = \dfrac{c - x_0}{h}$.

Step 7: Write down Gauss's backward formula.

Step 8: Substitute p value obtained in Step 6, entry(y_0), differences obtained in Step 4 in the formula written in Step 7.

Step 9: Simplifying the right–hand side of the equation obtained in Step 8 yields the required result, i.e., $f(c)$.

SOLVED PROBLEMS

Problem 1: Use Gauss's backward-interpolation formula to find $f(34)$ using the following data:

x	25	30	35	40
$f(x)$	0.2707	0.3027	0.3386	0.3794

Solution: The number of arguments of the given data is 4, Therefore, we choose $x_0 = 35$. Since $h = 5$, we have $p = \dfrac{x - x_0}{h} = \dfrac{x - 35}{5}$

x	$p = \dfrac{x-35}{5}$	y_p	Δy_p	$\Delta^2 y_p$	$\Delta^3 y_p$
25	-2	0.2707			
			→ 0.032		
30	-1	0.3027		→ 0.0039	
			→0.0359		→ 0.0010
35	0	0.3386		→ 0.0049	
			→0.0408		
40	1	0.3794			

From the table, we have $y_0 = 0.3386$, $\Delta y_{-1} = 0.0359$, $\Delta^2 y_{-1} = 0.0049$,

$\Delta^3 y_{-2} = 0.0010$; At $x = 34$, i.e., $c = 34$, $p = \dfrac{34-35}{5} = \dfrac{-1}{5} = -0.2$.

Gauss's backward-difference interpolation formula is

$$f(x) = f(x_0 + pu) = y_0 + p\Delta y_{-1} + \frac{(p+1)p}{\lfloor 2} \Delta^2 y_{-1}$$

$$+ \frac{(p+1)p(p-1)}{\lfloor 3} \Delta^3 y_{-2} + \dots \qquad (1)$$

Substituting $p = -0.2$, $y_0 = 0.3386$, $\Delta y_{-1} = 0.0359$, $\Delta^2 y_{-1} = 0.0049$, and $\Delta^3 y_{-2} = 0.0010$ values in (1), we get

$$f(34) = 0.3386 + (-0.2)(0.0359) + \frac{(-0.2+1)(-0.2)}{\lfloor 2}.0.0049$$

$$+ \frac{(-0.2+1)(-0.2)(-0.2-1)}{\lfloor 3} \times 0.0010$$

$$= 0.327532.$$

Problem 2: Use Gauss's backward–interpolation formula to find y_{28}, from the data given $y_{20} = 49225$, $y_{25} = 48316$, $y_{30} = 47236$, $y_{35} = 45926$ and $y_{40} = 44316$.

Solution: Here the arguments are 20, 25, 30, 35 and 40. Therefore, the number of arguments is 5. We choose the middle argument 30 as the origin, i.e., $x_0 = 30$. As the interval of differencing is 5, we have

$$p = \frac{x-30}{5}.$$

Now, we construct the difference table.

x	$p = \dfrac{x-30}{5}$	y_p	Δy_p	$\Delta^2 y_p$	$\Delta^3 y_p$	$\Delta^4 y_p$
20	-2	49225				
			$\to -909$			
25	-1	48316		$\to -171$		
			$\to -1080$		$\to -59$	
30	0	47236		$\to -230$		$\to -21$
			$\to -1310$		$\to -80$	
35	1	45926		$\to -310$		
			$\to -1620$			
40	2	44306				

From the above difference table, $y_0 = 47236$, $\Delta y_{-1} = -1080$, $\Delta^2 y_{-1} = -230$, $\Delta^3 y_{-2} = -59$, $\Delta^4 y_{-2} = -21$

At $x = 28$ (i.e., $c = 28$), $p = \dfrac{28-30}{5} = -0.4$.

Gauss's backward-difference interpolation formula is

$$f(x) = f(x_0 + hp)$$

$$= y_0 + p\Delta y_{-1} + \frac{(p+1)p}{\underline{2}}\Delta^2 y_{-1}$$

$$+ \frac{(p+1)p(p-1)}{\underline{3}}\Delta^3 y_{-2} + \frac{(p+2)(p+2)p(p-1)}{\underline{4}}\Delta^4 y_{-2} + \dots \dots (1)$$

Substituting p, y_0, Δy_{-1}, $\Delta^2 y_{-1}$, $\Delta^3 y_{-2}$, $\Delta^4 y_{-2}$ in (1), we get

$$f(28) = f(30 + 5(-0.4))$$

$$= 47236 + (-0.4)(-1080) + \frac{(-0.4+1)(-0.4)}{\underline{2}} \times (-230)$$

$$+ \frac{(-0.4+1)(-0.4)(-0.4-1)}{\underline{3}} \times (-59) +$$

$$+ \frac{(-0.4+2)(-0.4+1)(-0.4)(-0.4-1)}{\underline{4}} \times (-21)$$

$$f(28) = 47940.2256.$$

3.17 STIRLING'S INTERPOLATION FORMULA

Stirling's interpolation formula is the mean of Gauss's forward and Gauss's backward–difference interpolation formula.

Thus, the mean of (3.35b) and (3.43b) is

$$y_p = y_0 + p\left\{\frac{\Delta y_0 + \Delta y_{-1}}{2}\right\} + \frac{p^2}{\lfloor 2}\Delta^2 y_{-1} + \frac{p(p^2 - 1^2)}{\lfloor 3}\left\{\frac{\Delta^3 y_{-1} + \Delta^3 y_{-2}}{2}\right\}$$

$$+ \frac{p^2(p^2 - 1^2)}{\lfloor 4}\left\{\Delta^4 y_{-2}\right\} + \dots \tag{3.44}$$

which is the Stirling's interpolation formula.

3.18 THE PATH OF STIRLING'S INTERPOLATION FORMULA

The following is the path of Stirling's interpolation formula.

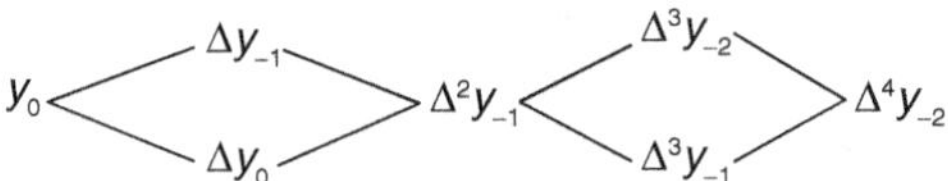

The differences along the central line and the differences above and below the central line appear in the Stirling's interpolation formula.

SOLVED PROBLEMS

Problem 1: Use the Stirling's formula, find $f(5)$ from the data given below:

x	0	2	4	6	8
$f(x)$	1	19	261	1303	4105

Solution: For the given data the interval of differencing $h = 2$

Let $p = \dfrac{x - 4}{2}$. The central difference table for the given data with this transformation is given below:

x	$p = \dfrac{x-4}{2}$	y_p	Δy_p	$\Delta^2 y_p$	$\Delta^3 y_p$	$\Delta^4 y_p$
0	-2	1				
			$\to 18$			
2	-1	19		$\to 224$		
			$\to 242$		$\to 576$	
4	0	261		$\to 800$		$\to 384$
			$\to 1042$		$\to 960$	
6	0	1303		$\to 1760$		
			$\to 2802$			
8	1	4105				

From the above table, $y_0 = 261$, $\Delta y_{-1} = 242$, $\Delta y_0 = 1042$, $\Delta^2 y_{-1} = 800$, $\Delta^3 y_{-2} = 576$, $\Delta^3 y_{-1} = 960$ and $\Delta^4 y_{-2} = 384$

We have to find $f(x)$ at $x = 5$; Therefore, $p = \dfrac{5-4}{2} = 0.5$. Note that

$f(5) = [y_p]_{p=0.5}$

The Stirling's formula is

$$y_p = y_0 + p\left\{\frac{\Delta y_0 + \Delta y_{-1}}{2}\right\} + \frac{p^2}{\lfloor 2} \Delta^2 y_{-1} + \frac{p(p^2-1^2)}{\lfloor 3}\left\{\frac{\Delta^3 y_{-1} + \Delta^3 y_{-2}}{2}\right\}$$

$$+ \frac{p^2(p^2-1^2)}{\lfloor 4}\left\{\Delta^4 y_{-2}\right\} + \dots \tag{1}$$

Substituting the values of p, y_0 and the forward differences in (1), we get

$$y_{0.5} = 261 + (0.5)\frac{242+1042}{2} + \frac{(0.5)^2}{\lfloor 2} \times 800$$

$$+ \frac{0.5\{(0.5)^2-1\}}{\lfloor 3} \times \frac{576+960}{2} + \frac{(0.5)^2\{(0.5)^2-1\}}{\lfloor 4} \times 384$$

$$= 261 + 321 + 100 - 48 - 3$$
$$= 631.$$

Hence, $f(5) = 631$.

Problem 2: The following table gives the specific heat of ethyl alcohol at different temperatures. Estimate the specific heat of ethyl alcohol at the temperature 20 using the Stirling's formula.

Temperature (x) (in °c)	0°	10°	20°	30°	40°	50°
Specific heat $y = f(x)$	0.51	0.55	0.57	0.59	0.62	0.67

Solution: Here, the interval of differencing $h = 10°$. Let $p = \dfrac{x-20}{10}$.

The central difference table for the transformed data is given below:

x	$p = \dfrac{x-20}{10}$	y_p	Δy_p	$\Delta^2 y_p$	$\Delta^3 y_p$	$\Delta^4 y_p$	$\Delta^5 y_p$
0	-2	0.51					
			$\to 0.04$				
10	-1	0.55		$\to -0.02$			
			$\to 0.02$		$\to 0.02$		
20	0	0.57		$\to 0$		$\to -0.01$	
			$\to 0.02$		$\to 0.01$		$\to 0.01$
30	1	0.59		$\to 0.01$		$\to 0$	
			$\to 0.03$		$\to 0.01$		
40	2	0.62		$\to 0.02$			
			$\to 0.05$				
50	3	0.67					

From the above table, we have $y_0 = 0.57$, $\Delta y_{-1} = 0.02$, $\Delta y_0 = 0.02$, $\Delta^2 y_{-1} = 0$, $\Delta^3 y_{-2} = 0.02$, $\Delta^3 y_{-1} = 0.01$ and $\Delta^4 y_{-2} = -0.01$. For $x = 25°$,

$$p = \frac{25-20}{10} = \frac{5}{10} = 0.5$$

Stirling's formula is

$$y_p = y_0 + p\left\{\frac{\Delta y_0 + \Delta y_{-1}}{2}\right\} + \frac{p^2}{\lfloor 2} \Delta^2 y_{-1}$$

$$+ \frac{p(p^2-1^2)}{\lfloor 3}\left\{\frac{\Delta^3 y_{-1} + \Delta^3 y_{-2}}{2}\right\} + \frac{p^2(p^2-1^2)}{\lfloor 4}\left\{\Delta^4 y_{-2}\right\} + ...$$

Now, substituting p, y_0 and forward differences in (1), we get

$$y_{0.5} = 0.57 + (0.5)\left\{\frac{0.02 + 0.02}{2}\right\} + \frac{(0.5)^2}{\lfloor 2}\times 0$$

$$+ \frac{0.5\{(0.5)^2 - 1\}}{\lfloor 3}\times\frac{0.02 + 0.01}{2} + \frac{(0.5)^2\{(0.5)^2 - 1\}}{\lfloor 4}(-0.01)$$

$$= 0.57 + 0.01 + 0 - 0.0009375 + 0.000098125$$
$$= 0.57914065.$$

3.19 GAUSS'S THIRD FORMULA FOR INTERPOLATION

If we start with y_1 and the differences along the path parallel to Gauss's backward-difference interpolation formula, then the indices of all forward differences are increased to 1 and it reduces to

$$y_v = y_1 + {}^{v}C_1\Delta y_0 + {}^{(v+1)}C_2\Delta^2 y_0 + {}^{(v+1)}C_3\Delta^3 y_1 + {}^{(v+2)}C_4\Delta^4 y_1 + \ldots\ldots$$

$$(3.44)$$

where $v = \dfrac{x - x_1}{h}$.

If $p = \dfrac{x - x_0}{h}$, then $v = \dfrac{x - x_0 - h}{h} = p - 1$. Now (3.44) reduces to

$$y_p = y_1 + {}^{(p-1)}C_1\Delta y_0 + {}^{p}C_2\Delta^2 y_0 + {}^{p}C_3\Delta^3 y_{-1} + {}^{(p+1)}C_4\Delta^4 y_{-1} + \ldots \quad (3.45)$$

It is known as Gauss's third interpolation formula. And it is used in the derivation of Bessel's formula of interpolation.

3.20 BESSEL'S INTERPOLATION FORMULA

It is the mean of Gauss's forward interpolation formula and Gauss's third formula of interpolation. Now, the mean of (3.35b) and (3.45) is

$$y_p = \frac{1}{2}(y_0 + y_1) + \left(p - \frac{1}{2}\right)\Delta y_0 + \frac{p(p-1)}{\lfloor 2}\left\{\frac{\Delta^2 y_0 + \Delta^2 y_{-1}}{2}\right\}$$

$$+ \frac{p\left(p - \dfrac{1}{2}\right)(p-1)}{\lfloor 3}\{\Delta^3 y_{-1}\} + \frac{(p+1)p(p-1)(p-2)}{\lfloor 4}\left\{\frac{\Delta^4 y_{-1} + \Delta^4 y_{-2}}{2}\right\}$$

$$(3.46)$$

which is known as Bessel's interpolation formula.

3.21   THE PATHS OF BESSEL'S FORMULA

The path of Bessel's formula is shown below:

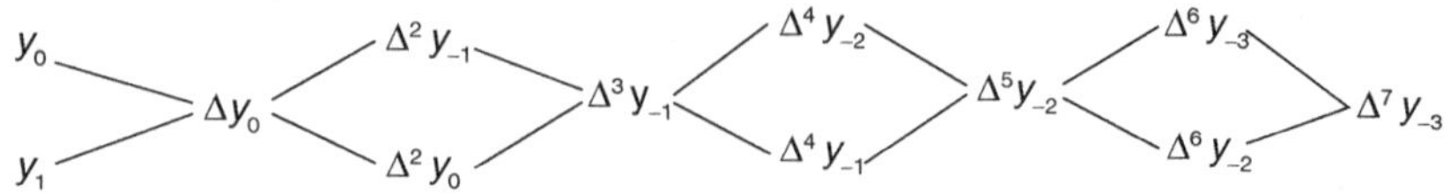

SOLVED PROBLEMS

Problem 1: For the data given below, find $f(5)$ using Bessel's formula

x	0	2	4	6	8
$f(x)$	2	20	262	1304	4106

Solution: Here $h = 2$, suppose $p = \dfrac{x-4}{2}$

The central difference table for the transformed data is given below:

x	$p = \dfrac{x-4}{2}$	y_p	Δy_p	$\Delta^2 y_p$	$\Delta^3 y_p$	$\Delta^4 y_p$
0	−2	2				
			→ 18			
2	−1	20		→ 224		
			→ 242		→ 576	
4	0	262		→800		→ 384
			→1042		→960	
6	1	1304		→1760		
			→2802			
8	2	4106				

From the above table, we have $y_0 = 262$, $y_1 = 1304$, $\Delta y_0 = 1042$, $\Delta^2 y_{-1} = 800$, $\Delta^2 y_0 = 1760$, $\Delta^3 y_{-1} = 960$.

At $x = 5$, $p = \dfrac{5-4}{2} = \dfrac{1}{2}$.

The Bessel's formula is

$$y_p = \frac{1}{2}(y_0 + y_1) + \left(p - \frac{1}{2}\right)\Delta y_0 + \frac{p(p-1)}{\lfloor 2}\left\{\frac{\Delta^2 y_0 + \Delta^2 y_{-1}}{2}\right\}$$

$$+ \frac{p\left(p - \frac{1}{2}\right)(p-1)}{\lfloor 3}\left\{\Delta^3 y_{-1}\right\} + \dots \tag{1}$$

Note that fifth term of Bessel's formula is not included as $\Delta^4 y_{-1}$ difference is not available in the central difference.

Now, substituting p, y_0, y_1 and forward differences in (1), we get

$$y_{\frac{1}{2}} = \frac{1}{2}\,(262 + 1304) + \left(\frac{1}{2} - \frac{1}{2}\right)(1042) + \frac{\left(\frac{1}{2}\right)\left(\frac{1}{2} - 1\right)}{\lfloor 2}\left(\frac{800 + 1760}{2}\right)$$

$$+ \frac{\frac{1}{2}\left(\frac{1}{2} - \frac{1}{2}\right)\left(\frac{1}{2} - 1\right)}{\lfloor 3}(960).$$

$$= 783 + 0 - 160 + 0$$

$$= 623.$$

Problem 2: From the data given below, find y_{25}:
$y_{20} = 2854,\ y_{24} = 3162,\ y_{28} = 3544$ and $y_{32} = 3992$.

Solution: Here $h = 4$. Suppose $p = \dfrac{x - 24}{4}$ The central difference table for the transformed data of y is shown below:

x	$p = \dfrac{x-24}{4}$	y_p	Δy_p	$\Delta^2 y_p$	$\Delta^3 y_p$
20	−1	2854			
			→ 308		
24	0	3162		→ 74	
			→ 382		→ −8
28	1	3544		→ 66	
			→ 448		
32	2	3992			

From the above table, we have $y_0 = 3162$, $y_1 = 3544$, $\Delta y_0 = 382$, $\Delta^2 y_{-1} = 74$, $\Delta^2 y_0 = 66$ and $\Delta^3 y_{-1} = -8$.

We have to find $y_{25}(x = 25)$. Therefore, at $x = 25$, the value of $p = \dfrac{25-24}{4} = \dfrac{1}{4}$. Note that $[y]_{x\,=\,25} = [y]_{p=\frac{1}{4}}$.

Now, the Bessel's formula is

$$y_p = \frac{y_0 + y_1}{2} + \left(p - \frac{1}{2}\right)\Delta y_0 + \frac{p(p-1)}{\lfloor 2}\left\{\frac{\Delta^2 y_0 + \Delta^2 y_{-1}}{2}\right\}$$

$$+ \frac{p(p-\frac{1}{2})(p-1)}{\lfloor 3}\left\{\Delta^3 y_{-1}\right\} + \cdots\cdots \tag{1}$$

Substituting the values of y_0, y_1, p and the forward differences in (1), we get

$$y_{\frac{1}{4}} = \frac{1}{2}(3162 + 3544) + \left(\frac{1}{4} - \frac{1}{2}\right)(382) +$$

$$\frac{\left(\frac{1}{4}\right)\left(\frac{1}{4} - 1\right)}{\lfloor 2}\left(\frac{74 + 66}{2}\right) + \frac{\frac{1}{4}\left(\frac{1}{4} - \frac{1}{2}\right)\left(\frac{1}{4} - 1\right)}{\lfloor 3}(-8)$$

$$= 3353 - 95.5 - 6.5625 - 0.0625$$
$$= 3250.875$$

3.22   LAPLACE–EVERETT'S FORMULA

This formula can be obtained from Gauss's forward formula by expressing the odd order differences in even order differences. Recall the Gauss's forward interpolation formula given by(3.35a)

$$y_p = y_0 + {}^pC_1\,\Delta y_0 + {}^pC_2\,\Delta^2 y_{-1} + {}^{(p+1)}C_3\,\Delta^3 y_{-1} + {}^{(p+1)}C_4\,\Delta^4 y_{-2}$$
$$+ {}^{(p+2)}C_5\,\Delta^5 y_{-2} + \cdots\cdots\cdots \tag{3.47}$$

Now for any non-negative integer k, we have

$$\Delta^{2k+1}y_{-k} = \Delta^{2k}(\Delta y_{-k}) = \Delta^{2k}\{y_{-k+1} - y_{-k}\}$$

i.e., $$\Delta^{2k+1}y_{-k} = \Delta^{2k}y_{-k+1} - \Delta^{2k}y_{-k} \tag{3.48}$$

Replacing the odd-order differences involved in (3.47) by (3.48), Eq. (3.47) reduces to

$$y_p = y_0 + {}^pC_1\,(y_1 - y_0) + {}^pC_2\,\Delta^2 y_{-1} + {}^{(p+1)}C_3\{\Delta^2 y_0 - \Delta^2 y_{-1}\}$$
$$+ {}^{(p+1)}C_4\,\Delta^4 y_{-2} + {}^{(p+2)}C_5\{\Delta^4 y_{-1} - \Delta^4 y_{-2}\} + \ldots\ldots + \ldots..$$

i.e.,
$$y_p = (1 - p)\,y_0 + py_1 + \{\,{}^pC_2 - {}^{(p+1)}C_3\Delta^2 y_{-1} + {}^{(p+1)}C_3\,\Delta^2 y_0$$
$$+ \{{}^{(p+1)}C_4 - {}^{(p+1)}C_5\}\Delta^4 y_{-2} + {}^{(p+2)}C_5\Delta^4 y_{-1} + \ldots\ldots$$

Using the relation ${}^nC_{r-1} - {}^{(n+1)}C_r = -({}^nC_r)$, the above equation can be expressed as

$$y_p = (1 - p)y_0 + py_1 - {}^pC_3\,\Delta^2 y_{-1} + {}^{(p+1)}C_3\Delta^2 y_0 - {}^{(p+1)}C_5\,\Delta^4 y_{-2}$$
$$+ {}^{(p+2)}C_5\,\Delta^4 y_{-1} \ldots\ldots$$
$$= \{py_1 + {}^{(p+1)}C_3\Delta^2 y_0 + {}^{(p+2)}C_5\,\Delta^4 y_{-1} + \ldots.\} + \{(1 - p)\,y_0$$
$$- {}^pC_3\,\Delta^2 y_{-1} - {}^{(p+1)}C_5\,\Delta^4 y_{-1}\}$$

For convenience, we suppouse $q = 1 - p$ in the terms written in the second bracket of the above equation and it reduces to

$$y_p = \{py_1 + {}^{(p+1)}C_3\,\Delta^2 y_0 + {}^{(p+2)}C_5\,\Delta^4 y_{-1} + \ldots.\}$$
$$+ \{qy_0 + {}^{(q+1)}C_3\,\Delta^2 y_{-1} + {}^{(q+1)}C_5\,\Delta^4 y_{-2} + \ldots.\} \qquad (3.49(a))$$

or

$$y_p = \left[\,py_1 + \frac{p(p^2 - 1^2)}{\underline{|3}}\Delta^2 y_0 + \frac{p(p^2 - 1^2)(p^2 - 2^2)}{\underline{|5}}\Delta^4 y_{-4} + \ldots..\right]$$

$$+ \left[\,qy_0 + \frac{q(q^2 - 1^2)}{\underline{|3}}\Delta^2 y_{-1} + \frac{q(q^2 - 1^2)(q^2 - 2^2)}{\underline{|5}}\Delta^4 y_{-2} + \ldots..\right]\,3.49(b)$$

The equations given by (3.49a) or (3.49b) are known as Laplace–Everett formula

3.23　THE PATH OF LAPLACE–EVERETT'S FORMULA

The following is the path of Laplace–Everett's formula:

$$
\begin{array}{llll}
y_0 \!\!-\!\!-\!\!-\!\!- & \Delta^2 y_{-1} \!\!-\!\!-\!\!-\!\!- & \Delta^4 y_{-2} \!\!-\!\!-\!\!-\!\!- \\[4pt]
\mid \\[4pt]
y_1 \!\!-\!\!-\!\!-\!\!- & \Delta^2 y_0 \!\!-\!\!-\!\!-\!\!- & \Delta^4 y_{-1} \!\!-\!\!-\!\!-\!\!-
\end{array}
$$

The differences in first bracket of (3.49a) are obtained along the horizontal line drawn from y_1 and the differences in second bracket of it are obtained along the horizontal line drawn from y_0 of the difference table.

SOLVED PROBLEMS

Problem 1: Apply Laplace–Everett's formula to find $f(3)$ from the data given below:

x	0	2	4	6
$f(x)$	2	10	66	218

Solution: Here $h = 2$. Suppose $p = \dfrac{x-2}{2}$,

The central difference table for the transformed function y_p is shown below:

x	$p = \dfrac{x-2}{2}$	y_p	Δy_p	$\Delta^2 y_p$	$\Delta^3 y_p$
0	−1	2			
			→ 8		
2	0	10	...... → 48		.....
			→ 56		→48
4	1	66	...... → 96		.....
			→152		
6	2	218			

From the above differences table, we have $y_0 = 10$, $y_1 = 66$, $\Delta^2 y_{-1} = 48$ and $\Delta^2 y_0 = 96$ and we have to find $f(x)$ at 3. Therefore

$$p = \frac{3-2}{2} = \frac{1}{2} \text{ and } q = 1 - p - \frac{1}{2}.$$

The Laplace–Everett's formula is

$$y_p = [py_1 + p\frac{(P^2 - 1^2)}{\lfloor 3} \Delta^2 y_0] + \dots + [qy_0 + \frac{(q^2 - 1^2)}{\lfloor 3} .\Delta^2 y_{-1} + \dots] \quad (1)$$

Now, substituting p, q, y_0, y_1 and the forward differences in (1), we get

$$y_{1/2} = \left[\frac{1}{2}(66) + \frac{\frac{1}{2}\left\{\left(\frac{1}{2}\right)^2 - 1\right\}}{\lfloor 3} (96) + \ldots\ldots \right]$$

$$+ \left[\frac{1}{2}(10) + \frac{\frac{1}{2}\left\{\left(\frac{1}{2}\right)^2 - 1\right\}}{\lfloor 3} (48) + \ldots\ldots \right]$$

$$= [33 - 6] + [5 - 3]$$
$$= 27 + (2)$$
$$= 29.$$

Problem 2: For the data given below, find $\log_{10}(337.5)$ using Laplace–Evevett's formula.

x	310	320	330	340	350	360
$\log_{10}x$	2.4913	2.5057	2.5185	2.5312	2.5441	2.5563

Solution: For the given problem, the interval of differencing $h = 10$. Suppose $p = \dfrac{x-330}{10}$. The central difference table for the transformed function y_p is shown below:

x	$p = \dfrac{x-330}{10}$	y_p	Δy_p	$\Delta^2 y_p$	$\Delta^3 y_p$	$\Delta^4 y_p$	$\Delta^5 y_p$
310	−2	2.4913					
			→ 0.0144				
320	−1	2.5057		→ −0.0016			
			→ 0.0128		→ 0.0015		
330	0	2.5185		→ −0.0001		→ −0.0012	
			→ 0.0127		→ 0.0003		→ 0
340	1	2.5312		→ +0.0002		→ −0.0012	
			→ 0.0129		→ −0.0009		
350	2	2.5441		→ −0.0007			
			→ 0.0122				
360	3	2.5563					

We have to find $f(x)$ at $x = 337.5$; therefore,

$$p = \frac{337.5 - 330}{10} = \frac{7.5}{10} = 0.75.$$

Hence, $q = 1 - p = 0.25$.

Laplace–Everett's Formula is

$$y_p = \left[Py_1 + \frac{p(p^2 - 1^2)}{\lfloor 3} \Delta^2 y_0 + \frac{p(p^2 - 1^2)}{\lfloor 5} \Delta^4 y_{-1} + \ldots \ldots \right]$$

$$+ \left[qy_0 + \frac{q(q^2 - 1^2)}{\lfloor 3} \Delta^2 y_{-1} + \frac{q(q^2 - 1)(q^2 - 2^2)}{\lfloor 5} \Delta^4 y_{-2} + \ldots \ldots \right]$$

Substituting the values of q, p, y_0, y_1 and the forward differences, we obtain

$$y_{0.75} = \left[0.75(2.5312) + \frac{\{(0.75)^2 - 1^2\}}{\lfloor 3}(0.0002) \right.$$

$$\left. + \frac{(0.75)\{(0.75)^2 - 1\}\{(0.75)^2 - 2^2\}}{\lfloor 5}(-0.0012) \right] +$$

$$\left[0025(2.5185) + \frac{(0.23)\{(0.25)^2 - 1\}}{\lfloor 3}(-0.0001) \right.$$

$$\left. + \frac{(0.25)\{(0.25)^2 - 1^2\}\{(0.25)^2 - 2^2\}}{\lfloor 5}(-0.0012) \right]$$

$$= [1.8984 - 0.0000109375 - 0.0000112793]$$
$$+ [0.629625 + 0.0000390625 - 0.00000922852]$$
$$= 2.527997461.$$

ADDITIONAL SOLVED PROBLEMS

Problem 1: Given that

x	0	2	4	6	8
$f(x)$	4	20	260	1300	4100

Estimate $f(5)$ using (i) Gauss's forward (ii) Stirling's (iii) Bessel's (iv) Laplace–Everett's formulae of interpolation.

Solution: The interval of differencing for the problem at our disposal is 2, i.e., $h = 2$,

Suppose $p = \dfrac{x-4}{2}$; with this transformation, the central difference table is shown below:

x	$p = \dfrac{x-4}{2}$	y_p	Δy_p	$\Delta^2 y_p$	$\Delta^3 y_p$	$\Delta^4 y_p$
0	−2	4				
			→16			
2	−1	20		→224		
			→240		→576	
4	0	260		→800		→ 384
			→1040		→960	
6	1	1300		→1760		
			→2800			
8	2	4100				

(i) To find $f(5)$ using Gauss's forward formula:

The entry and forward differences of y_p of the path of Gauss's forward interpolation formula is shown below:

$$260(y_0) \underset{1040(\Delta\, y_0)}{\overset{240(\Delta\, y_{-1})}{<\qquad>}} 800(\Delta^2\, y_{-1}) \underset{960(\Delta^3\, y_{-1})}{\overset{576(\Delta^3\, y_{-2})}{<\qquad>}} 384(\Delta^4\, y_{-2})$$

At $x = 5$, the value of $p = \dfrac{5-4}{2} = 0.5$.

Gauss's forward formula for interpolation is

$$y_p = y_0 + p\,\Delta y_0 + \frac{p(p-1)}{\lfloor 2} \Delta^2 y_{-1} + \frac{(p+1)p(p-1)}{\lfloor 3} \Delta^3 y_{-1}$$

$$+ \frac{(p+1)p(p-1)(p-2)}{\lfloor 4} \Delta^4 y_{-1} + \ldots\ldots \tag{1}$$

Substituting the value of p, y_0 and the forward differences in (1), we get

$$y_{0.5} = 260 + (0.5)(1040) + \frac{(0.5)(0.5-1)}{\lfloor 2}(800)$$

$$+ \frac{(0.5+1)(0.5)(0.5-1)}{\lfloor 3}(960)$$

$$+ \frac{(0.5+1)(0.5)(0.5-1)(0.5-2)}{\lfloor 4}(384)$$

$$= 260 + 520 - 100 - 60 + 9$$
$$= 629.$$

(ii) To find $f(5)$ using Stirling's formula:

The entry and forward differences along the path of Stirling's formula is shown below:

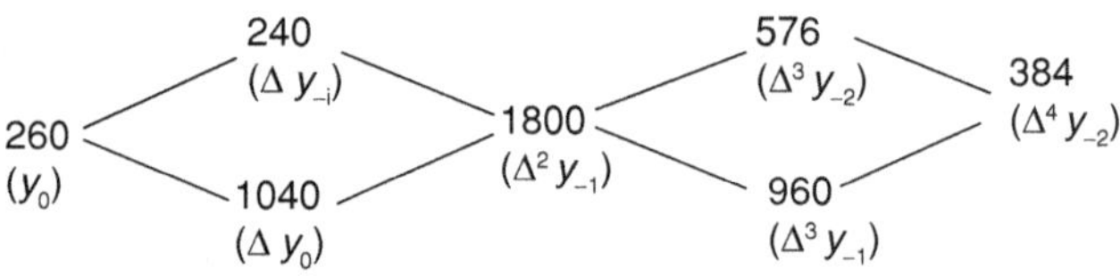

At $x = 5$, $p = 0.5$. The Stirling's formula is

$$y_p = y_0 + p\, \frac{\Delta y_0 + \Delta y_{-1}}{2} + \frac{p^2}{\lfloor 2}\, \Delta^2 y_{-1} +$$

$$\frac{p(p^2-1)}{\lfloor 3}\, \frac{\Delta^3 y_{-1} + \Delta^3 y_{-2}}{2} + \frac{p^2(p^2-1^2)}{\lfloor 4}\,\Delta^4 y_{-2} + \ldots\ldots \qquad (2)$$

Substituting p and entry, forward differences along the path of Stirling's formula, we obtain

$$y_{0.5} = 260 + (0.5)\,\frac{240 + 1040}{2} + \frac{(0.5)^2}{12}(800) + \frac{(0.5)\{(0.5)^2 - 1\}}{\lfloor 3}$$

$$\left(\frac{576 + 960}{2}\right) + \frac{(0.5)^2\{(0.5)^2 - 1^2\}}{\lfloor 4}(384)$$

$$= 260 + 320 + 100 - 48 - 3 = 629$$

$$\therefore f(5) = 629.$$

(iii) To find $f(5)$ using Bessel's formula:

The entries and forward differences along the path of Bessel's formula are shown below:

$$260(y_0) \qquad\qquad 800(\Delta^2 y_{-1})$$
$$\searrow 1040(\Delta\, y_0) \qquad\qquad \searrow 960(\Delta^3 y_{-1})$$
$$1300(y_1) \qquad\qquad 1760(\Delta^2 y_0)$$

Note that at $x = 5$, $p = 0.5$, The Bessel's formula is

$$y_p = \frac{y_0 + y_1}{2} + \left(p - \frac{1}{2}\right)\Delta y_0 + \frac{p(p-1)}{\lfloor 2} \cdot \frac{\Delta^2 y_{-1} + \Delta^2 y_0}{2}$$

$$+ \frac{\left(p - \dfrac{1}{2}\right)p(p-1)}{\lfloor 3} \Delta^2 y_{-1} + \dots\dots \tag{3}$$

Substituting p, y_0, y_1 and the forward differences in Eq. (3), we get

$$y_{0.5} = \frac{260 + 1300}{2} + \left(0.5 - \frac{1}{2}\right)(1040) + (0.5)\frac{(0.5-1)}{\lfloor 2}\left(\frac{800 + 1760}{2}\right)$$

$$+ \frac{(0.5 - \dfrac{1}{2})(0.5)(0.5 - 2)}{\lfloor 3}(960)$$

$$= 780 + 0 - 160 + 0 = 620.$$

Thus, $f(5) = 620$, by the Bessel's formula

(iv) To find f (5) using Laplace–Everette's formula:

The entries and forward differences along the path of Laplace–Everett's formula are shown below:

$$260(y_0) \;\dots\dots\; 800(\Delta^2 y_{-1}) \;\dots\dots\dots\; 384(\Delta^4 y_{-2})$$

.

.

.

$$1300(y_1) \;\dots\dots\; 1760(\Delta^2 y_0)$$

The Laplace–Everett's formula is

$$y_p = \left[py_1 + \frac{p(p^2 - 1)}{\lfloor 3}\Delta^2 y_0 + \frac{p(p^2 - 1^2)(p^2 - 4)}{\lfloor 5}\Delta^4 y_{-1} + \dots\dots \right]$$

$$+ \left[qy_0 + \frac{q(q^2 - 1)}{\lfloor 3}\Delta^2 y_{-1} + \frac{q(q^2 - 1)(q^2 - 4)}{\lfloor 5}\Delta^4 y_{-2} + \dots \right] \tag{4}$$

where $q = 1 - p$

Since $p = 0.5$, we have $q = 0.5$.

Now, substituting p, q, y_0, y_1 and the forward differences in (4), we obtain

$$y_{0.5} = \left[0.5\,(1300) + \frac{(0.5)\{(0.5)^2 - 1\}}{\underline{3}}(1760) \right]$$

$$+ \left[0.5\,(260) + \frac{(0.5)\{(0.5)^2 - 1\}}{\underline{3}}(800) \right.$$

$$\left. + \frac{(0.5)\{(0.5)^2 - 1\}\{(0.5)^2 - 4\}}{\underline{5}}(384) \right]$$

$$= [650 - 110] + [130 - 50 + 4.5]$$

$$= 624.5.$$

Problem 2: For the data given below, find the value of log 338 by (i) Gauss's forward (ii) Stirling's formula (iii) Bessel's and (iv) Laplace–Everett's interpolation formulae.

x	310	320	330	340	350	360
$f(x)$	2.4914	2.5051	2.5185	2.5315	2.5441	2.5563

Solution: Here, the intervals of differencing is $h = 10$. Suppose $p = \dfrac{x - 330}{10}$. The central difference table in the transformed data is shown below:

X	$p = \dfrac{x-330}{10}$	y_p	Δy_p	$\Delta^2 y_p$	$\Delta^3 y_p$	$\Delta^4 y_p$	$\Delta^5 y_p$
310	−2	2.4914					
			→ 0.0137				
320	−1	2.5051		→ −0.0003			
			→ 0.0134		→ −0.0001		
330	0	2.5185		→ −0.0004		→ 0.0001	
			→ 0.013		→ 0		→ −0.0001
340	1	2.5315		→ −0.0004		→ 0	
			→ 0.0126		→ 0		

$$
\left|
\begin{array}{lll}
350 & 2 & 2.5441 \qquad\qquad \rightarrow -0.0004 \\
 & & \qquad\qquad \rightarrow 0.0122 \\
360 & 3 & 2.5563
\end{array}
\right.
$$

At $x = 338$, the value of $p = \dfrac{338 - 330}{10} = 0.8$

(i) To find $f(338)$ using Gauss's forward formula:

The entry and forward differences along the path of Gauss's forward formula are shown below:

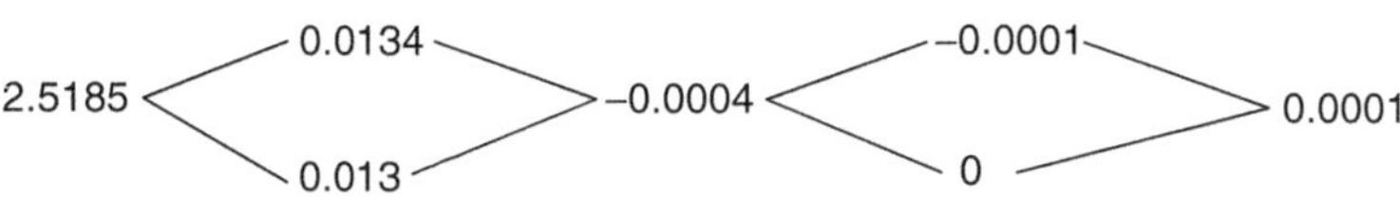

Gauss's forward formula is

$$
y_p = y_0 + p\Delta y_0 + \frac{p(p-1)}{\lfloor 2}\Delta^2 y_{-1} + \frac{(p+1)p(p-1)}{\lfloor 3}\Delta^3 y_{-1}
$$

$$
+ \frac{(p+1)p(p-1)(p-2)}{\lfloor 4}\Delta^4 y_{-1}
$$

$$
+ \frac{(p+2)(p+1)p(p-1)(p-2)}{\lfloor 5}\Delta^5 y_{-2} + \ldots\ldots\ldots \qquad (1)
$$

Substituting p, y_0 and forward differences in (2), we obtain

$$
y_{0.8} = 2.5185 + (0.8)(0.013) + \frac{(0.8)(0.8-1)}{\lfloor 2}(-0.0004)
$$

$$
+ \frac{(0.8+1)(0.8)(0.8-1)}{\lfloor 3}(0) + \frac{(0.8+1)(0.8)(0.8-1)(0.8-2)}{\lfloor 4}(0.0001)
$$

$$
+ \frac{(0.8+2)(0.8+1)(0.8)(0.8-1)(0.8-2)}{\lfloor 5}(-0.0001)
$$

$$
= 2.5185 + 0.0104 + 0.000032 + 0 + 0.0000144 - 0.0000008064
$$

$$
= 2.528932634.
$$

Thus, $f(338) = 2.528932634$, by the Gauss's forward formula

(ii) To find $f(338)$ using Stirling's formula:

The entry and forward differences along the path of Stirling's formula are shown below:

The Stirling's formula is

$$y_p = y_0 + p \cdot \frac{\Delta y_0 + \Delta y_1}{2} + \frac{p^2}{2} \Delta^2 y_{-1}$$

$$+ \frac{p(p^2 - 1)}{\lfloor 3} \cdot \frac{\Delta^3 y_{-1} + \Delta^3 y_{-2}}{2} + \frac{p^2(p^2 - 1)}{\lfloor 4} \Delta^4 y_{-2} + \ldots\ldots(2)$$

Substituting p, y_0 and the forward differences in Eq. (2), we get

$$y_{0.8} = 2.5185 + 0.8 \left(\frac{0.0134 + 0.013}{2} \right)$$

$$+ \frac{(0.8)^2}{2}(-0.0004) + \frac{(0.8)\{(0.8)^2 - 1\}}{\lfloor 3} \cdot \frac{(-0.0001 + 0)}{2}$$

$$+ \frac{(0.8)^2\{(0.8)^2 - 1\}}{\lfloor 4}(0.0001)$$

$$= 2.5185 + 0.01056 - 0.000128 - 0.0000024 - 0.0000096$$

$$= 2.52893344.$$

(iii) To find $f(338)$ using Bessel's formula:
The entries and forward differences along the path of Bessel's formula are shown below:

$$
\begin{array}{ccccccc}
2.5185 & & -0.0004 & & 0.0001 & \\
 & 0.013 & & 0 & & -0.0001 \\
2.5315 & & -0.0004 & & 0 & \\
\end{array}
$$

The Bessel's formula is

$$y_p = \frac{y_0 + y_1}{2} + \left(p - \frac{1}{2} \right) \Delta y_0 + \frac{p(p-1)}{\lfloor 2} \frac{\Delta^2 y_{-1} + \Delta^2 y_0}{2}$$

$$+ \frac{\left(p - \frac{1}{2} \right) p(p-1)}{\lfloor 3} \Delta^3 y_{-1} + \frac{(p+1)p(p-1)(p-2)}{\lfloor 4} \times$$

$$\left(\frac{\Delta^4 y_{-2} + \Delta^4 y_{-2}}{2} \right) + \frac{\left(p - \frac{1}{2} \right)(p+1)(p)(p-1)(p-2)}{\lfloor 5} \Delta^5 y_{-2} + \ldots(3)$$

Substituting y_0, y_1, p and forward differences in (3), we get

$$y_{0.8} = \frac{2.5185 + 2.5315}{2} + \left(0.8\frac{1}{2}\right)(0.013)$$

$$+ \frac{(0.8)(0.8-1)}{\underline{|2}}\left(\frac{-0.0004 - 0.0004}{2}\right) + \frac{\left(0.8 - \frac{1}{2}\right)(0.8)(0.8-1)}{\underline{|3}}(0)$$

$$+ \frac{(0.8+1)(0.8)(0.8-1)(0.8-2)}{\underline{|4}}\left(\frac{0.0001 + 0}{2}\right)$$

$$+ \frac{\left(0.8 - \frac{1}{2}\right)(0.8+1)(0.8)(0.8-1)(0.8-2)}{\underline{|5}}(-0.0001)$$

$$= 2.525 + 0.0039 + 0.000032 + 0 + 0.00000072 - 0.000000864$$

$$= 2.528932.$$

(iv) To find $f(328)$ using Laplace–Everett's formula:

The entries and forward differences along the path of Laplace–Everett's are shown below:

$$2.5185 \underline{\hspace{3cm}} -0.0004 \underline{\hspace{3cm}} +0.0001$$

$$\vdots$$

$$2.5315 \underline{\hspace{3cm}} -0.0004$$

Laplace–Everett's formula is

$$y_p = \left[py_1 + \frac{p(p^2-1)}{\underline{|3}}\Delta^2 y_0 + \frac{p(p^2-1)(p^2-4)}{\underline{|5}}\Delta^4 y_{-1} + \ldots \right]$$

$$+ \left[qy_0 + \frac{q(q^2-1)}{\underline{|3}}\Delta^2 y_{-1} + \frac{q(q^2-1)(q^2-4)}{\underline{|5}}\Delta^4 y_{-2} + \ldots\ldots \right](4)$$

where $q = 1 - p$.

Since $p = 0.8$, we have $q = 0.2$. Now substituting p, q, y_0, y_1 and the forward differences in (4),

we get

$$y_{0\cdot8} = \left[\, 0.8\,(2.5315) + \frac{(0.8)\{(0.8)^2 - 1\}}{\lfloor 3} (-0.0004) \right]$$

$$+ \left[\, 0.2\,(2.5185) + \frac{(0.2)\{(0.2)^2 - 1\}}{\lfloor 3} (-0.0004) \right.$$

$$\left. + \frac{(0.2)\{(0.2)^2 - 1\}\{(0.2)^2 - 4\}}{\lfloor 5} (0.0001) \right]$$

$$= [2.0252 + 0.0000192] + [0.5037 + 0.0000128 + 0.0000066336]$$

$$= 2.528932634.$$

Problem 3: Find $f(1.2)$ using Gauss's forward formula for the data given below:

x	0	0.5	1.0	1.5	2
y	0.5	1.5	2	2.3	2.5

Solution: Here $h = 0.5$. Suppose $p = \dfrac{x-1}{0.5}$.

The difference table for the function y_p is shown below:

x	$p = \dfrac{x-1}{0.5}$	y_p	Δy_p	$\Delta^2 y_p$	$\Delta^3 y_p$	$\Delta^4 y_p$
0.0	−2	0.5				
			→ 1			
0.5	−1	1.5		→−0.5		
			→ 0.5		→ 0.3	
1.0	0	2		→−0.2		→−0.2
			→ 0.3		→ 0.1	
1.5	1	2.3		→−0.1		
			→ 0.2			
2	2	2.5				

The path of Gauss's farword is as follows:

$$\begin{array}{ccccccc}
2 & & -0.2 & & -0.2 \\
(y_0) & & (\Delta^2 y_{-1}) & & (\Delta^4 y_{-2}) \\
& 0.3 & & 0.1 & \\
& (\Delta y_0) & & (\Delta^3 y_{-1}) &
\end{array}$$

At $x = 1.2$, the value of $p = \dfrac{12-1}{0.5} = 0.4$

Gauss's forward formula is

$$y_p = y_0 + p\Delta y_0 \ \frac{p(p-1)}{\underline{|2}}\Delta^2 y_1 + \frac{(p+1)p(p-1)}{\underline{|3}}\Delta^3 y_{-1}$$

$$+ \frac{(p+1)p(p-1)(p-2)}{\underline{|4}}\Delta^4 y_{-2} + \dots \tag{1}$$

Substituting y_0, p and forward differences in (1), we obtain

$$y_{0.4} = 2 + (0.4)(0.3) + \frac{(0.4)(0.4-1)}{\underline{|2}}(-0.2)$$

$$+ \frac{(0.4+1)(0.4)(0.4-1)}{\underline{|3}}(0.1)$$

$$+ \frac{(0.4+1)(0.4)(0.4-1)(0.4-2)}{\underline{|4}}(-0.2)$$

$$= 2 + 0.12 + 0.024 - 0.0036 - 0.00448$$
$$= 2.13392.$$

Problem 4: The numbers of deaths in different age groups are given below. Using it, find the number of deaths in the age group 45 to 50.

Age group (in *years*)	Below 35	35–45	45–55	55–65
No.of deaths	2000	4400	5200	6400

Solution: Let $f(x)$ represent the number of deaths of persons whose ages are less than x *years*. Now, the given data can be expressed as follows:

x	35	45	55	65
$f(x)$	2000	6400	11,600	18,000

Suppose $p = \dfrac{x-45}{10}$; The difference table of y_p is shown below:

x	$p = \dfrac{x-45}{10}$	y_p	Δy_p	$\Delta^2 y_p$	$\Delta^3 y_p$
35	−1	2000			
			→ 4400		
45	0	6400		→800	
			→ 5200		→400
55	1	11600		→1200	
			→ 6400		
65	2	18000			

At $x = 50$, the value of $p = \dfrac{50-45}{10} = \dfrac{1}{2}$.

We use the Gauss's forward formula to find $f(50)$. The path of Gauss's forward formula is shown below:

$$
\begin{array}{ccccc}
6400 & & 800 & & \\
(y_0) & 5200 & (\Delta^2 y_{-1}) & 400 & \\
& (\Delta y_0) & & (\Delta^3 y_{-1}) &
\end{array}
$$

Gauss's forward formula is

$$y_p = y_0 + p\Delta y_0 + \frac{p(p-1)}{\lfloor 2} \Delta^2 y_{-1} + \frac{(p+1)p(p-1)}{\lfloor 3} \Delta^3 y_{-1} + \ldots\ldots\ldots(1)$$

Now, substituting y_0, p and forward difference in (1), we get

$$y_{0.5} = 6400 + (0.5)5200 +$$

$$\frac{(0.5)(0.5-1)}{\lfloor 2}(800) + \frac{(0.5+1)(0.5)(0.5-1)}{\lfloor 3}(400)$$

$$= 6400 + 2600 - 100 + 25 = 8875;$$

Thus, $f(50) = 8875$.

Hence, the number of deaths in the age group 45 to 50

$$= f(50) - f(45)$$
$$= 8875 - 6400$$
$$= 2475.$$

Problem 5: The populations of a certain city for different years are given below:

year (x)	2000	2004	2008	2012
population (y): in lakhs	4	8	11	16

Find the population of the town for the year 2006 using Gauss's backward formula.

Solution: Here $h = 4$; Let $p = \dfrac{x - 2008}{4}$.

The central differences table with this transformation is shown below:

x	p	y_p	Δy_p	$\Delta^2 y_p$	$\Delta^3 y_p$
2000	-2	4			
			$\to 4$		
2004	-1	8		$\to -1$	
			$\to 3$		$\to 3$
2008	0	11		$\to 2$	
			$\to 5$		
2012	1	16			

At $x = 2006$, the value of $p = \dfrac{2006 - 2008}{4} = -0.5$

The path of Gauss's backward formula and the entry and differences along this path are shown below.

$$11(y_0) \quad \nearrow \quad 3(\Delta y_{-1}) \quad \searrow \quad 2(\Delta^2 y_{-1}) \quad \nearrow \quad 3(\Delta^3 y_{-2})$$

Gauss's backward formula is given by

$$y_p = y_0 + p\Delta_{-1} + \frac{(p+1)(p)}{\lfloor 2} \Delta^2 y_{-1} + \frac{(p+1)p(p-1)}{\lfloor 3} \Delta^3 y_{-1} + \ldots\ldots\ldots(1)$$

Substituting y_0, p and forward differences, we get

$$y_{-0.5} = 11 + (0.5)3 + \frac{(-0.5+1)(0.5)}{2}(2) + \frac{(-0.5+1)(-0.5)(-0.5-1)}{6}(3)$$

$$= 11 - 1.5 - 0.25 + 0.1875$$
$$= 9.4375;$$

$\therefore$ The population of the town for the *year* 2006 is 9.4375 lakhs.

Problem 6: For the data given below, find $f(26)$ using Stirling's formula:

x	8	12	16	20	24	28	32	36
$f(x)$	17.35	21.62	24.73	28.64	33.65	37.75	42.54	45.64

Solution: The interval of differencing for the given data is $h = 4$. Let
$$p = \frac{x - 24}{4}.$$

At $x = 26$, the value of $p = 0.5$.

The central differences table is shown in page 75*

The entry and forward differences along the path of Stirlings formula is shown below:

$$
\begin{array}{ccccccccccc}
& & 5.01 & & & & -1.98 & & & & 5.77 & & \\
33.65 & & & & -0.88 & & & & 3.49 & & & & -13.06 \\
& & 4.13 & & & & 1.51 & & & & -7.29 & &
\end{array}
$$

The Stirling's formula is given by

$$
y_p = y_0 + p\frac{\Delta y_0 + \Delta y_1}{2} + \frac{p^2}{\lfloor 2} \Delta^2 y_{-1} + \frac{p(p^2 - 1^2)}{\lfloor 3} \frac{\Delta^3 y_{-1} + \Delta^3 y_{-2}}{2}
$$

$$
+ \frac{p^2(p^2 - 1^2)}{\lfloor 4} \Delta^4 y_{-2} + \frac{p(p^2 - 1^2)(p^2 - 2^2)}{\lfloor 5} \frac{\Delta^5 y_{-3} + \Delta^5 y_{-2}}{2}
$$

$$
+ \frac{p^2(p^2 - 1^2)(p^2 - 2^2)}{\lfloor 6} \Delta^6 y_{-2} + \ldots\ldots\ldots \tag{1}
$$

$$
= 33.65 + 0.5\left(\frac{5.01 + 4.13}{2}\right) + \frac{(0.5)^2}{\lfloor 2}(-0.88)
$$

$$
+ \frac{(0.5)\{(0.5)^2 - 1^2\}}{\lfloor 3} \frac{(-1.98 + 1.51)}{2} + \frac{(0.5)^2\{(0.5)^2 - 1^2\}}{\lfloor 4}(3.49)
$$

x	$p=\dfrac{x-24}{4}$	y_p	Δy_p	$\Delta^2 y_p$	$\Delta^3 y_p$	$\Delta^4 y_p$	$\Delta^5 y_p$	$\Delta^6 y_p$	$\Delta^7 y_p$
8	−4	17.35							
			→ 4.27						
12	−3	21.62		→ −1.16					
			→ 3.11		→ 1.96				
16	−2	24.73		→ 0.8		→ −1.66			
			→ 3.91		→ 0.3		→ −0.62		
20	−1	28.64		→ 1.1		→ −2.28		→ 6.39	
			→ 5.01		→ −1.98		→ 5.77		→ −19.45
24	0	33.65		→ −0.88		→ 3.49		→ −13.06	
			→ 4.13		→ 1.51		→ −7.29		
28	1	37.78		→ 0.63		→ −3.8			
			→ 4.76		→ −2.29				
32	2	42.54		→ −1.66					
			→ 3.1						
36	3	45.64							

$$+\frac{(0.5)\{(0.5)^2-1^2\}\{(0.5)^2-2^2\}}{\lfloor 5}\left(\frac{5.77-7.29}{2}\right)$$

$$+\frac{(0.5)^2\{(0.5)^2-1^2\}\{(0.5)^2-2^2\}}{\lfloor 6}(-13.06)$$

$$\therefore f(26)=35.7908$$

Problem 7: Below are given the production of fertilizer of an item in different *years*:

Year (x)	2006	2008	2010	2012
Production (y) (in tonnes)	40	45	52	66

Find the production of the item produced by the fertilizer factory in the *year* 2009 using Bessel's formula

Solution: Here $h=2$; Let $p=\dfrac{x-2508}{2}$.

At $x=2009$, the value of $p=\dfrac{1}{2}$. The function difference table for the transformed function y_p is shown below:

x	$p=\dfrac{x-2508}{2}$	y_p	Δy_p	$\Delta^2 y_p$	$\Delta^3 y_p$
2006	−1	40			
			→5		
2008	0	45		→2	
			→7		→5
2010	1	52		→7	
			→14		
2012	2	66			

The entries and forward differences along the path of Bessel's formula are shown below

$$45(y_0) \searrow$$
$$\qquad \nearrow 7(\Delta y_0) \quad \swarrow 2(\Delta^2 y_{-1}) \searrow$$
$$52(y_1) \nearrow \qquad \nwarrow 7(\Delta^2 y_0) \nearrow 5(\Delta^3 y_{-1})$$

The Bessel's formula is

$$y_p = \frac{y_0 + y_1}{2} + \left(p - \frac{1}{2}\right)\Delta y_0 + \frac{p(p-1)}{\lfloor 2} \frac{\Delta^2 y_{-1} + \Delta^2 y_0}{2}$$

$$+ \frac{\left(p - \frac{1}{2}\right)p(p-1)}{\lfloor 3}\Delta^3 y_{-1} + \ldots\ldots \tag{1}$$

Substituting p, y_0, y_1 and forward differences in Eq. (1), we obtain

$$y_{0.5} = \frac{45+52}{2} + \left(0.5 - \frac{1}{2}\right)(7) + \frac{(0.5)(0.5-1)}{\lfloor 2}\left(\frac{2+7}{2}\right)$$

$$+ \frac{\left(0.5 - \frac{1}{2}\right)(0.5)(0.5-1)}{\lfloor 3} \tag{5}$$

$$= 48.5 + 0 - 0.5625 + 0$$

$$= 47.9375;$$

Hence, the production of the item in the year 2009 is 47.9375 (tonnes).

Problem 8: If $y_{20} = 49225$, $y_{25} = 48316$, $y_{30} = 47236$ $y_{35} = 45926$ find $y_{40} = 44306$, find y_{29} using
(i) Gauss's forward (ii) Stirling's (iii) Bessel's (iv) Laplace–Everett's formulae of interpolation

Solution: Here $h = 5$; Let $p = \dfrac{x-30}{5}$. At $x = 29$, the value of $p = \dfrac{29-30}{5} = -0.2$.

The difference table for y_p is shown below:

x	$p = \dfrac{x-30}{5}$	y_p	Δy_p	$\Delta^2 y_p$	$\Delta^3 y_p$	$\Delta^4 y_p$
20	-2	49225				
			$\to -909$			
25	-1	48316		$\to -171$		
			$\to -1080$		-59	

30	0	47236		$\rightarrow -230$	$\rightarrow -21$
			$\rightarrow -1310$		-80
35	1	45926		$\rightarrow -310$	
			-1620		
40	2	44306			

(i) To find y_{29} using Gauss forward formula.

The entry and forward differences along the path of Gauss forward formula are given below.

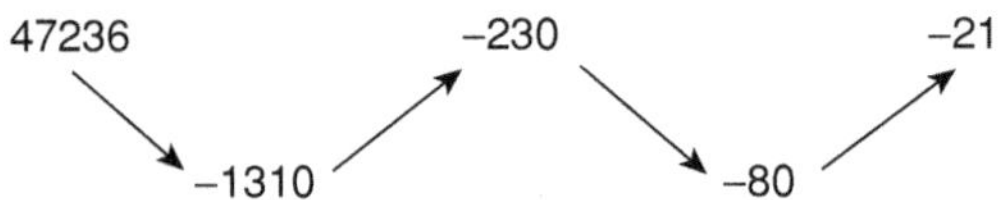

Gauss forward formula is

$$y_p = y_0 + p\Delta y_0 + \frac{p(p-1)}{2!}\Delta^2 y_{-1} + \frac{(p+1)p(p-1)}{3!}\Delta^3 y_{-1}$$

$$+ \frac{(p+1)p(p)(p-1)(p-2)}{4!}\Delta^4 y_{-2} + \dots\dots \tag{1}$$

Now substituting p, y_0 and forward difference of gauss forward formula, we get

$$y_{-0.2} = 47236 + (-0.2)(-1310) + \frac{(-0.2)(-0.2-1)}{2!}(-230)$$

$$+ \frac{(-0.2+1)(-0.2)(-0.2-1)}{3!}(-80)$$

$$+ \frac{(-0.2+1)(-0.2)(-0.2-1)(-0.2-2)}{4!}(-21)$$

$$= 47236 + 262 - 27.6 - 2.56 + 0.3696$$

$$y_{-0.2} = 47468.2096$$

(ii) To find y_{29} using Stirling's formula

The entry and forward difference along the path of Stirling's formula are shown below

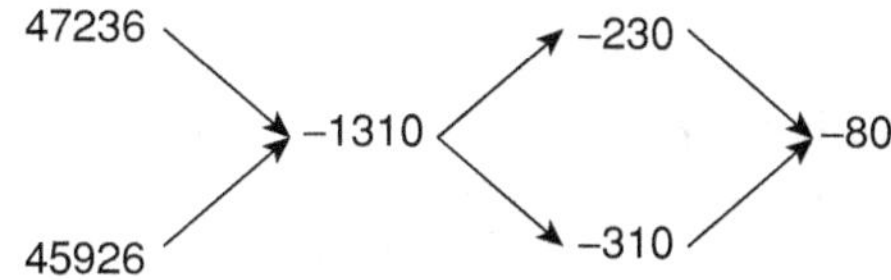

The Stirling's formula is given by

$$y_p = y_0 + p\frac{\Delta y_0 + \Delta y_{-1}}{2!} + \frac{p^2}{2}\Delta^2 y_{-1} + \frac{p(p^2-1)}{3!}\left(\frac{\Delta^3 y_{-1} + \Delta^3 y_{-2}}{2}\right)$$

$$+ \frac{p^2(p^2-1)}{4!}\Delta^4 y_{-2} + \ldots\ldots \tag{2}$$

Substituting p, y_0 and forward differences in (2), we get

$$y_{-0.2} = 47236 + (-0.2)\frac{(-1080-1310)}{2} + \frac{(-0.21)^2}{2!}(-230)$$

$$+ \frac{(-0.2)\left[(-0.2)^2-1\right]}{3!}\left(\frac{-59-80}{2}\right)$$

$$+ \frac{(-0.2)^2\left[(-0.2)^2-1\right]}{4!}(-21)$$

$$= 47236 + 239 - 4.6 - 2.224 + 0.0336$$

$$y_{-0.2} = 47468.2096$$

(iii) to find y_{29} using Bessel's formula.
The entries and the forward differences along the path of Bessel's formula are shown below.

The Bessel's formula is

$$y_p = \frac{y_0 + y_1}{2} + \left(p - \frac{1}{2}\right)\Delta y_0 + \frac{p(p-1)}{2!}\frac{\Delta^2 y_{-1} + \Delta^2 y_0}{2}$$

$$+ \frac{\left(p - \frac{1}{2}\right)p(p-1)}{3!}\Delta^3 y_{-1} + \ldots\ldots\ldots \tag{3}$$

Substituting p, y_0, y_1 and forward difference in (3), we get

$$y_{-0.2} = \frac{47236 + 45926}{2} + \left(-0.2 - \frac{1}{2}\right)(-1310)$$

$$+ \frac{(-0.2)(-0.2-1)}{2!}\left(\frac{-230-310}{2}\right)$$

$$+ \frac{\left(-0.2 - \dfrac{1}{2}\right)(-0.2)(-0.2-1)}{3!}(-80)$$

$$= 46581 + 917 - 324 + 2.24$$

$$y_{-0.2} = 47467.84$$

(iv) To find y_{2q} by Laplace–Everett's formula:
The entry and forward differences of the path of Laplace–Everett's formula are shown below:

$$47236 \quad\text{------}\quad 230 \text{------} (-21)$$

$$45926 \quad\text{------}\quad -310$$

The Laplace–Everett's formula is

$$y_p = \left[py_1 + \frac{p(p^2-1^2)}{\lfloor 3}\Delta^2 y_0 + \frac{p(p^2-1^2)(p^2-2^2)}{\lfloor 5}\Delta^4 y_{-1} + ... \right]$$

$$+ \left[qy_o + \frac{q(q^2-1^2)}{\lfloor 3}\Delta^2 y_{-1} + \frac{q(q^2-1^2)(q^2-2^2)}{\lfloor 5}\Delta^4 y_{-2} + ... \right], \quad (4)$$

where $q = 1-p$
Since $p = -0.2$, we have $q = 1.2$

Thus, substituting y_0, y_1, p, q and forward differences in (4), we get

$$y_{-0.2} = [(-0.2)(45926) + \frac{(-0.2)\{(-0.2)^2-1\}}{\lfloor 3}(-310)]$$

$$+ [(1.2)(47236) + \frac{(1.2)\{(1.2)^2-1\}}{\lfloor 3}(-230) +$$

$$\frac{1.2\{(1.2)^2-1\}\{(1.2)^2-4\}}{\lfloor 5}(-21)]$$

$$= -9185.2 - 9.92 + 56683.2 - 20.24 + 0.236544$$
$$= 47468.07654$$

$\therefore$ The value of y is $x = 29$ is given as 47468.07654 by Laplace–Everett's formula for the given data.

EXERCISE 3(c)

1. If $f(x)$ is defined by the integral $f(x) = \dfrac{2}{\sqrt{\pi}} \int\limits_{0}^{x} e^{-t^2}\, dt$, then find $f(0.08)$ using the following data:

x	0.05	0.10	0.15	0.20
$f(x)$	0.0563	0.1124	0.1680	0.2227

2. If $\sqrt{40} = 6.3245$ $\sqrt{45} = 6.7082$ $\sqrt{50} = 7.0710$ $\sqrt{55} = 7.4162$ and $\sqrt{60} = 7.7460$, then find $\sqrt{52}$ using Gauss's forward formula.

3. Apply Stirling's formula to find $f(26)$ for the data given below:

x	20	24	28	32
$f(x)$	2800	3100	3500	3900

4. Find $\log_{10} 339$ using Bessel's formula for the data given below:

x	310	320	330	340	350
$f(x)$	2.4914	2.5052	2.5185	2.5315	2.5441

5. Find the sin $53°$ by using Laplace–Everett's formula from the table given below:

x	48°	50°	52°	54°	56°
Sin x	0.7431	0.7660	0.7880	0.8090	0.8290

6. The following table gives the specific heats of ethyl alcohol at different temperatures:

Temperature ($x°$c)	0°	8°	16°	24°	32°	40°
specific heat (y)	0.51	0.52	0.51	0.56	0.61	0.64

Find the specific heat corresponding to the temperature at 17°C using
(i) Stirling's formula (ii) Bessel's formula and (iii) Laplace–
Everett's formula.

7. Use Stirling's formula to find y_{35}, given that $y_{20} = 516$, $y_{30} = 350$ and $y_{50} = 247$.

8. If $\sqrt{12500} = 111.8034$, $\sqrt{12510} = 111.8481$, $\sqrt{12520} = 111.8928$ and $\sqrt{12530} = 111.9375$, find $\sqrt{12517}$ using Gauss's forward formula and Gauss's backward formula.

9. From the following table, find $e^{-1.64}$ using (i) Bessel's formula and (ii) Laplace–Everett's formula:

x	e^{-x}
+ 1.2	0.3012
+ 0.4	0.2466
+ 1.6	0.2019
+ 1.8	0.1653
+ 2.0	0.1353

10. The values of the elliptic integral $E(m) = \int_0^{\frac{\pi}{2}} (1 - m\sin^2 \theta)^{\frac{-1}{2}} d\theta$ for some equidistant values of m are given below. Use Bessel's formula to find $E(0.254)$.

m	$E(m)$
0.20	1.6596
0.22	1.6698
0.24	1.6803
0.26	1.6912
0.28	1.7024
0.30	1.7139

11. From the data given below, find $\sqrt{156.5}$ using Stirling's formula and Bessel's formula:

x	$f(x) = \sqrt{x}$
152	12.3288
154	12.4096

156	12.4899
158	12.5698
160	12.6491

12. Use Gauss's backward-interpolation formula to find $f(33)$ given that

x	25	30	35	40
$f(x)$	0.2707	0.3027	0.3386	0.3794

13. Given that $y_{20} = 24$, $y_{24} = 32$, $y_{28} = 37$ and $y_{32} = 46$. Find y_{26} using Laplace–Everett's formula.

14. Express the Stirling's formula given by (3.44) in the following form:

$$y_p = y_0 + \mu\,y_0 + \frac{p^2}{\lfloor 2}\,{}^2y_0 + \frac{(p+1)p(p-1)}{\lfloor 3}\,\mu\delta^3\,y_0 + \frac{p^2(p^2-1^2)}{\lfloor 4}\,\delta^4 y_0$$

$$+ \frac{(p+2)(p+1)p(p-1)(p-2)}{\lfloor 5}\,\mu\delta^5\,y_0 + \dots$$

where δ and μ are central difference operators.

15. Express the Bessel's formula given by (3.46) in the following form:

$$y_p = \frac{y_0 + y_1}{2} + \left(p - \frac{1}{2}\right)\delta y_{1/2} + \frac{p(p-1)}{\lfloor 2}\cdot\frac{\delta^2 y_0 + \delta^2 y_1}{2}$$

$$+ \frac{(p-\frac{1}{2})p(p-1)}{\lfloor 3}\cdot\delta^3 y_{1/2} + \frac{(p+1)p(p-1)(p-2)}{\lfloor 4}\cdot\frac{\delta^4 y_0 + \delta^4 y_1}{2} + \dots$$

where δ is central difference operator.

16. Given that

x	0.4	0.5	0.6	0.7	0.8
$f(x)$	1.5836	1.7974	2.0442	2.3275	2.6511

Find $f(0.65)$ using Bessel's formula.

17. Given that

$\theta\,^\circ$	10°	15°	20°	25°	30°
$\tan\theta$	0.1763	0.2679	0.3639	0.4663	0.5773

Find $\tan 22°$ using (i) Gauss's forward formula and (ii) Bessel's formula.

18. The gamma function $\Gamma(x)$ values are given for various values of x in the following table:

x	2	3	4	5	6
$f(x) = \Gamma(x)$	1	2	6	24	120

Find $\Gamma(x)$ value at $x = 4.25$ using Stirling's formula.

ANSWERS

1. 0.09002   2. 7.21104416   3. 3287.5

4. 2.53024   5. 0.798626171

6. 0.542226928, 0.541779323, 0.541779022

7. 399   8. 11.87939, 11.87937   9. 0.19398, 0.193982419

10. 1.6878985   11. 12.50991519, 12.50992031   12. 0.323708

13. 34.4375   16. 2.18105

17. 17. 0.40397328, 0.4039688   18. 7.598717.

3.24   HERMITE INTERPOLATION FORMULA

The formulae derived for interpolation in earlier sections are applicable only when the function's values are given at a certain number of arguments. If we were given the values of a function and its first-order derivate at some points of independent-variable, then we have to apply the Hermite interpolation formula to find a value of a dependent variable. Further, we can also obtain an interpolating polynomial using Hermite interpolation formula. Such a polynomial is called Hermite interpolating polynomial. Now, we proceed to derive the Hermite interpolation formula.

Assume that the following data are given:

x	x_0	x_1	$x_2 \dots\dots\dots\dots x_n$
$f(x)$	y_0	y_1	$y_2 \dots\dots\dots\dots y_n$
$f'(x)$	$f'(x_0)$	$f'(x_1)$	$f'(x_2)\dots\dots\dots\dots f'(x_n)$;

From the table, we have $(2n + 2)$ conditions, namely $y_i = f(x_i)(i = 0, 1, 2, \ldots\, n)$ and $f'(x_i) = \left(\dfrac{df}{dx}\right)_{(x=x_i)}$ $(i = 0, 1, 2, \ldots\ldots n)$ for the function $y = f(x)$, which may be known or unknown. Therefore, it is possible to find an interpolating polynomial $p(x)$ of degree less than or equal to $(2n + 1)$ for the given data of $y = f(x)$. Let

$$p(x) = \sum_{i=0}^{n} A_i(x)y_i + \sum_{i=0}^{n} B_i(x)f'(x_i) \tag{3.50}$$

be such a polynomial, where $A_i(x)$ and $B_i(x)$ are polynomials of a degree less than or equal to $(2n + 1)$. Since it is an interpolating polynomial of $f(x)$, we have

$$p(x_i) = y_i \tag{3.51}$$

and

$$p'(x_i) = f'(x_i) \tag{3.52}$$

From (3.50) and (3.51), we have

$$A_i(x_j) = \begin{cases} 0 \text{ if } i \neq j \\ 1 \text{ if } i = j \end{cases} \tag{3.53}$$

$$B_i(x_j) = 0 \ \forall \ i, j \tag{3.54}$$

Similarly, from (3.50) and (3.52), we obtain

$$A_i'(x_j) = 0 \ \forall \ i, j \tag{3.55}$$

$$B_i'(x_j) = \begin{cases} 0 \text{ if } i \neq j \\ 1 \text{ if } i = j \end{cases} \tag{3.56}$$

Let

$$A_i(x) = a_i(x)\, l_i^2(x) \tag{3.57}$$

$$B_i(x) = \beta_i(x)\, l_i^2(x), \tag{3.58}$$

where

$$l_i(x) = \frac{(x - x_0)(x - x_1)\ldots\ldots(x - x_{i-1})(x - x_{i+1})\ldots\ldots(x - x_n)}{(x_i - x_0)(x_i - x_1)\ldots(x_i - x_{i-1})(x_i - x_{i+1})\ldots\ldots(x_i - x_n)} \tag{3.59}$$

$$a_i(x) = a_i x + b_i \tag{3.60}$$

$$\beta_i(x) = c_i x + d_i. \tag{3.61}$$

and a_i, b_i, c_i, d_i are constants.

Clearly, each $l_i^2(x)$ $(i = 0, 1, 2, \ldots, n)$ is a polynomial of degree $2n$. Hence, $p(x)$ is a polynomial of degree less than or equal to $(2n + 1)$ as $a_i(x)$ and $\beta_i(x)$ are linear in x. The interpolating polynomial $p(x)$ is completely known once we found the values of a_i, b_i, c_i and d_i $(i = 0, 1, 2, \ldots, n)$. Now, our task is to find these constants.

Differentiating (3.57) with respect to x, we get

$$A_i'(x) = 2\,(a_i x + b_i)\, l_i(x)\, l_i'(x) + a_i\, l_i^2\,(x)$$

$$\therefore \quad A_i'(x_i) = 2\,(a_i x_i + b_i)\, l_i(x_i)\, l_i'(x_i) + a_i\, l_i^2\,(x_i)$$

In view of (3.55), we get

$$(a_i x_i + b_i)\, 2\, l_i(x_i)\, l_i'(x_i) + a_i\, l_i^2\,(x_i) = 0$$

As $l_i(x_i) = 1$, the above equation reduces to

$$2(a_i x_i + b_i)\,(l_i'(x_i)) + a_i = 0 \tag{3.62}$$

Now, substituting $x = x_i$ in (3.57), we have

$$A_i(x_i) = a_i\,(x_i)\, l_i^2\,(x_i),$$

$$\text{i.e., } A_i(x_i) = (a_i x_i + b_i)\, l_i^2\,(x_i)\ (\because a_i(x) = a_i x + b_i).$$

Using the fact $l_i(x_i) = 1$ and (3.53), we obtain

$$(a_i x_i + b_i) = 1. \tag{3.63}$$

Solving (3.62) and (3.63) for a_i and b_i, we get

$$a_i = -2\, l_i'\,(x_i) \tag{3.64}$$

and

$$b_i = 1 + 2\, x_i\, l_i'\,(x_i) \tag{3.65}$$

Now, differentiating (3.58) with respect to x and using the fact $\beta_i(x) = c_i x + d_i$, we have

$$B_i'(x) = 2(c_i x + d_i)\, l_i\,(x)\, l_i'\,(x) + c_i\, l_i^2\,(x).$$

Hence,

$$B_i'(x) = 2(c_i x_i + d_i)\, l_i(x_i)\, l_i'\,(x_i) + c_i\, l_i^2\,(x_i) \tag{3.66}$$

Using $B_i'(x_i) = 1$ (by 3.56) and $l_i(x_i) = 1$, the Eqn (3.66) reduce to

$$2(c_i x_i + d_i)\ l_i'\ (x_i) + c_i = 1. \tag{3.67}$$

Substituting $x = x_i$ in (3.58), we have

$$B_i(x_i) = (c_i x_i + d_i)\ l_i^2\ (x_i),$$

which reduces to the following equation using the fact $B_i(x_i) = 0$ (3.54) and $l_i(x_i) = 1$

$$c_i x_i + d_i = 0. \tag{3.68}$$

Solving (3.67) and (3.68) for c_i and d_i, we get

$$c_i = 1 \tag{3.69}$$

and

$$d_i = -x_i. \tag{3.70}$$

Thus,

$$A_i(x) = [1 - 2(x - x_i)\ l_i'\ (x_i)]\ l_i^2\ (x)$$

and

$$B_i(x) = (x - x_i)\ l_i^2\ (x).$$

Now, the interpolating polynomial of $f(x)$ is

$$p(x) = \sum_{i=0}^{n} [1 - 2(x - x_i)l_i'(x)]\ l_i^2(x) y_i + \sum_{i=0}^{n} (x - x_i)l_i^2(x) f'(x_i),$$

which is known as Hermite interpolation formula or Hermite interpolating polynomial of $f(x)$. Thus, the approximate form of $f(x)$ is

$$p(x) = \sum_{i=0}^{n} [1 - 2(x - x_0)l_i'(x_i)]\ l_i^2(x) y_i + \sum_{i=0}^{n} (x - x_i)l_i^2(x) f'(x_i) \tag{3.71}$$

3.25　PROCEDURE TO FIND $F(X)$ AT $X = C$ USING HERMITE INTERPOLATION FORMULA

Step 1: Note down the values of $x_0, x_1, \ldots\ldots, x_n, y_0, y_1, \ldots\ldots\ldots y_n$ and $f'(x_0), f'(x_1) \ldots, f'(x_n)$.

Step 2: For the given arguments $x_0, x_1, \ldots\ldots, x_n$, write $l_0(x), l_1(x)$ $l_2(x), \ldots l_n(x)$.

$$\text{using } l_i(x) = \frac{(x - x_0)\ldots\ldots(x - x_{i-1})(x - x_{i+1})\ldots\ldots(x_i - x_n)}{(x_i - x_0)\ldots(x_i - x_{i-1})(x_i - x_{i+1})\ldots(x_i - x_n)}.$$

Step 3: Find $l_0'(x)$, $l_1'(x)$, $l_2'(x)$.......... $l_n'(x)$.

Step 4: Compute $l_0'(x_0)$, $l_1'(x_1)$, $l_2'(x_2)$.......... $l_n'(x_n)$.

Step 5: Write down Hermite interpolation formula

Step 6: Substitute $l_i'(x_i)$ and $l_i(x)$, y_i and $f'(x_i)$ values in the formula written in Step 5. next, put $x = c$.

Step 7: Simplifying the equation obtained in Step 6, we get $f(c)$ by the Hermite interpolation formula.

SOLVED PROBLEMS

Problem 1: Find the Hermite polynomial in the data given below:

x	0	2
$f(x)$	2	12
$f'(x)$	5	5

Solution: Here $x_0 = 0$, $x_1 = 2$, $y_0 = 2$, $y_1 = 12$, $f'(x_0) = 5$ and $f'(x_1) = 5$. Now

$$l_0(x) = \frac{x-2}{0-2}, \quad l_1(x) = \frac{x-0}{2-0},$$

i.e., $l_0(x) = \frac{-1}{2}(x-2)$, $l_1(x) = \frac{x}{2}$.

Hence,

$$l_0'(x) = \frac{-1}{2} \text{ and } l'_1(x) = \frac{1}{2}.$$

Now, we find $\qquad l_0'(x_0) = l_0'(0) = \frac{-1}{2}$

$$l_0'(x_1) = l_0'(2) = \frac{-1}{2},$$

$$l_1'(x_0) = l_1'(0) = \frac{1}{2} \text{ and } l_1'(x_1) = l_1'(1) = \frac{1}{2}.$$

The Hermite interpolation formula for data points x_0 and x_1 is

$$f(x) = \sum_{i=0}^{1} [1-2(x-x_i)\, l'_i(x_i)]\, l_i^2\,(x_i)y_i + \sum_{i=0}^{1} [x-x_i]\, l_i^2\,(x)f'(x_i)$$

$$= [1 - 2(x - x_0)\, l'_0\,(x_0)]\, l_o^2\,(x)\, y_0 + [1 - 2(x - x_1)]\, l'_1\,(x_1)]\, l_1^2\,(x)y_1 +$$

$$(x - x_1)\, l_o^2\,(x)f'(x_0) + (x - x_1)\, l_1^2\,(x)f'(x_1)$$

$$= \left[1- 2(x-0)\cdot\left(\frac{-1}{2}\right)\right]\left\{\frac{-1}{2}(x-2)\right\}^2 (2) + \left[1-2(x-2)\frac{1}{2}\right]\left(\frac{x}{2}\right)^2 (12)$$

$$+x\left\{\frac{-1}{2}(x-2)\right\}^2 (5) + (x-2)\left(\frac{x}{2}\right)^2 .5$$

$$= 5x + 2.$$

Note that though there are four conditions, we have a linear polynomial that is less than a third degree.

Problem 2: Given the following values of $f(x)$ and $f'(x)$:

x	$f(x)$	$f'(x)$
−1	1	−5
0	1	1
1	3	7

Find the approximate form of $f(x)$ using Hermite interpolation formula. Also find $f(2)$.

Solution: Here, $x_0 = -1$, $x_1 = 0$, $x_2 = 1$, $y_0 = 1$, $y_1 = 1$, $y_2 = 3$, $f'(x_0) = -5$, $f'(x_1) = 1$ and $f'(x_2) = 7$. Now,

$$l_0(x) = \frac{(x-0)(x-1)}{(-1-0)(-1-1)} = \frac{x^2 - x}{2}$$

$$l_1(x) = \frac{(x+1)(x-1)}{(0+1)(0-1)} = -(x^2-1)$$

$$l_2(x) = \frac{(x+1)(x-0)}{(1+1)(1-0)} = \frac{x^2+x}{2}.$$

Hence,
$$l_0'(x) = \frac{2x-1}{2}, \ l_1'(x) = -2x, \ l_2'(x) = \frac{2x+1}{2}.$$

Further,
$$l_0'(x_0) = l_0'(-1) = \frac{-3}{2},$$

$$l_1'(x_1) = l_1'(0) = 0 \text{ and } l_2'(x_2) = l_2'(1) = \frac{3}{2}$$

The Hermite interpolation formula for 3 data points is

$$f(x) = \sum_{i=0}^{2} [1 - 2(x-x_i) \ l_i^1(x_i)] \ l_i^2(x)y_i + \sum_{i=0}^{2}(x-x_i) \ l_i^2(x).f'(x_i)$$

$$= [1 - 2(x-x_0) \ l_0'(x_0)] \ l_o^2(x)y_0 + [1 - 2(x-x_1)l_1'(x_1)] \ l_1^2(x)y_1$$

$$+ [1 - 2(x-x_2) \ l_2'(x_2)]l_2^2(x)y_2 + (x-x_0) \ l_o^2(x)f'(x_0)$$

$$+ (x-x_1) \ l_1^2(x)f'(x_1) + (x-x_2) \ l_2^2(x)f'(x_2)$$

$$\therefore f(x) = [1 - 2(x+1).\left(\frac{-3}{2}\right)] \frac{x^2(x-1)^2}{4} \ (1) + [1 - 2(x-0)(0)] \times$$

$$(x^2-1)^2(1) + [1 - 2(x-1).\frac{3}{2}] \frac{x^2(x+1)^2}{4} \ (3) + (x+1) \times$$

$$\frac{x^2(x-1)^2}{4}.(-5) + (x-0)(x^2-1)^2.(1) + (x-1) \frac{x^2(x+1)^2}{4} \ (7)$$

$$= 2x^4 - x^2 + x + 1.$$

Putting $x = 2$ in the above equation, we get

$$f(2) = 2(2)^4 - 2^2 + 2 + 1$$

$$= 32 - 4 + 2 + 1 = 31$$

$\therefore$ The value of f (2) at $x = 2$ is 31.

3.26 SPLINE INTERPOLATION

The word spline is borrowed from the drafting instrument of the same name, and it is a flexible strip used in drawing a curve. In olden days, the draftsmen had used this device to draw a smooth curve through a given set of points such that the slope and curvature are also continuous along the curve.

Given any two points in a plane, they can join by a straight line or curve. For example, the two points are joined by a line representing $y = ax + b$ or joined by a curve representing $y = ax^2 + bx + c$ or $y = ax^3 + bx^2 + cx + d$ etc. for the given data (x_i, y_i) $(i = 0, 1, 2, ..., n)$, where $x_0 < x_1 < x_2 < ...... < x_n$, the spline interpolation consists of an interpolating polynomial of kth degree (k is a positive integer) in each subinterval $[x_i, x_{i+1}](i = 0, 1, 2...,(n - 1))$ of the interval $[x_0, x_n]$, i.e., we have n polynomials of kth degree whose domains are respectively $[x_0, x_1]$, $[x_1, x_2]$, ..., and $[x_{n-1}, x_n]$.

3.27   DEFINITIONS

1. Nodes (*knots*)

   If $a = x_0 < x_1 < x_2 < ... < x_{n-1} < x_n = b$, then $[x_0, x_1]$, $[x_1, x_2]....[x_{n-1}, x_n]$ are called the subintervals of $[a, b]$ and, $x_1, x_2....... $ and x_{n-1} are called nodes (*knots*) of the spline.

2. Spline function of degree k

   Let $y_0, y_1, ... y_n$ be $(n + 1)$ values of $y = f(x)$ at $x = x_0, x_1, x_2, ..., x_n$. A spline function of degree k with nodes $x_1, x_2, ..., x_{n-1}$ is a function $S(x)$ such that

   (i) $S(x_i) = y_i$ $(i = 0, 1, 2, ..., n)$

   (ii) $S(x)$ is a polynomial of degree k on each $[x_i, x_{i+1}]$ $(i = 0.1.2, ..., (n - 1))$

   (iii) $S(x)$ and its first $(k - 1)$ derivatives are continuous on (a, b).

Note:

   1. The spline function of a degree 1 is known as linear spline, and that of degree 2 and 3 are known as quadratic and cubic splines, respectively.

   2. A linear polynomial $Ax + B$ can be expressed as $a(x - c) + b$.

   3. $S(x; [a, b])$ is used to denote the spline function on the interval $[a, b]$

3.28   LINEAR SPLINE

Let $y_0, y_1, ... y_n$ be $(n + 1)$ values of $y = f(x)$ at the arguments $a = x_0 < x_1 < x_2 < ... < x_{n-1} < x_n = b$. Further, let $S(x; [x_i, x_{i+1}])$ be the spline function of degree one (i.e., linear spline) on the interval $[x_i, x_{i+1}]$ $(i = 0, 1, 2, ..., \overline{n-1})$, clearly $S(x; [x_i, x_{i+1}])$ represents a straight line joining the points (x_i, y_i) and (x_{i+1}, y_{i+1}).

Therefore,

$$S(x; [x_i, x_{i+1}]) - y_i = \frac{y_{i+1} - y_i}{x_{i+1} - x_i}(x - x_i)$$

$$\text{i.e., } S(x; [x_i, x_{i+1}]) = y_i + \frac{y_{i+1} - y_i}{x_{i+1} - x_i}(x - x_i). \qquad (3.72)$$

Thus, the linear spline $S(x; [x_i, x_{i+1}])$ is given by

$$S(x; [x_0, x_n]) = \begin{cases} S(x; [x_0, x_1]) & \text{for } x_0 \leq x \leq x_1 \\ S(x; [x_1, x_2]) & \text{for } x_1 \leq x \leq x_2 \\ \cdots\cdots\cdots\cdots\cdots\cdots\cdots\cdots\cdots \\ S(x; [x_{n-1}, x_n]) & \text{for } x_{n-1} \leq x \leq x_n \end{cases},$$

where each $S(x; [x_i, x_{i+1}])$ is given by (3.72) and they are called linear splines.

SOLVED PROBLEMS

Problem 1: Find the linear splines for the data given below:

x	1	2	3
$y = f(x)$	6	10	14

Also find $f(1.5)$ and $f'(1.5)$.

Solution: Here $x_0 = 1$, $x_1 = 2$, $x_2 = 3$ and $y_0 = 6$, $y_1 = 10$, $y_2 = 14$. We need to find two linear splines: $S(x; [1, 2])$ and $S(x; [2, 3])$.

The linear spline on $[x_i, x_{i+1}]$ is

$$S(x; [x_i, x_{i+1}]) = y_i + \frac{y_{i+1} - y_i}{x_{i+1} - x_i}(x - x_i) \ (i = 0, 1) \qquad (1)$$

in view of (3.72).

The linear spline on [1, 2]: Putting $i = 0$ in (1), we get

$$S(x; [\,x_0, x_i]) = y_0 + \frac{y_1 - y_0}{x_1 - x_0}(x - x_0) \tag{2}$$

Substituting the values of x_0, x_1, y_0 and y_1 in (2), we get

$$\therefore \qquad S(x; [1, 2]) = 6 + \frac{10 - 6}{2 - 1}(x - 1),$$

i. e., $\qquad S(x; [1, 2]) = 4x + 2.$

Now, putting $i = 1$ in (1), we obtain

$$S(x; [x_1, x_2]) = y_1 + \frac{y_2 - y_1}{x_2 - x_1}(x - x_1). \tag{3}$$

Substituting the values of x_1, y_1, x_2 and y_2 values in (3), we get

$$S(x; [2, 3]) = 10 + \frac{14 - 10}{3 - 2}(x - 2)$$

i.e., $S(x; [2, 3]) = 4x + 2.$

Thus, the linear spline function $S(x; [1, 3])$ on [1, 3] is

$$S(x; [1, 3]) = \begin{cases} 4x + 2 \text{ if } 1 \le x \le 2 \\ 4x + 2 \text{ if } 2 \le x \le 3 \end{cases}.$$

Now, $\qquad f(1.5) = S(1.5; [1, 2]) = 4.(15)12 = 8,$

and $\qquad f'(1.5) = S'(1.5; [1.2]) = 4.$

Problem 2: Find the linear splines for the data given below:

x	1	2	3
y	-6	1	20

Solution: Here, $x_0 = 1$, $x_1 = 2$, $x_2 = 3$, $y_0 = -6$, $y_1 = 1$, $y_2 = 20$.
Proceeding as in the solution of Problem 1, we have

$$S(x, [1, 2]) = -6 + \frac{1 + 6}{2 - 1}(x - 1) = 7x - 13$$

and

$$S(x, [2, 3]) = 1 + \frac{20-1}{3-2}(x-2) = 19x - 37.$$

Thus, the linear spline $S(x; [1, 3])$ is given by

$$S(x; [1, 3]) = \begin{cases} 7x - 13 \text{ if } 1 \le x \le 2 \\ 19x - 37 \text{ if } 2 \le x \le 3 \end{cases}$$

3.29   QUADRATIC SPLINE

Let $y_0, y_1, \ldots y_n$ be $(n + 1)$ values of $y = f(x)$ at the arguments $x_0, x_1, x_2, \ldots x_n$.

Let $S(x; [x_i, x_{i+1}])$ be the quadratic spline function on the interval $[x_i, x_{i+1}]$ $(i = 0, 1, 2, \ldots \overline{n-1})$.

We need to find n quadratic spline functions to have $S(x; [x_0, x_n])$. By the definition of spline function, we have

$$S(x_i; [x_i, x_{i+1}]) = y_i, \tag{3.73}$$

and $S(x; [x_i, x_{i+1}])$ and $S'(x; [x_i, x_{i+1}])$ are continuous for $i = 0, 1, 2, \ldots (n - 1)$. Since $S(x_i, [x_i, x_{i+1}])$ is connected by a quadratic expression, and the derivative of it is linear, i.e., $S'(x; [x_i, x_{i+1}])$ is linear. Suppose that

$$S'(x; [x_i, x_{i+1}]) = A(x - x_i) + B(x - x_{i+1}). \tag{3.74}$$

Substituting $x = x_i$ in (3.74), we obtain

$$S'(x_i; [x_i, x_{i+1}]) = B(x_i - x_{i+1})$$

$$\therefore \qquad B = \frac{S'\left(x_i; [x_i, x_{i+1}]\right)}{x_i - x_{i+1}} \tag{3.75}$$

Introducing

$$m_i = S'(x_i, [x_i, x_{i+1}]) \ (i = 0, 1, 2, \ldots (n-1)), \tag{3.76}$$

We have

$$B = \frac{m_i}{x_i - x_{i+1}}.$$

Now, substituting $x = x_{i+1}$ in (3.74), we get

$$S'(x_{i+1}; [x_i, x_{i+1}]) = A(x_{i+1} - x_i)$$

$$\therefore \qquad A = \frac{m_{i+1}}{x_{i+1} - x_i} \qquad\qquad (3.77)$$

in view of (3.76).

Let $x_{i+1} - x_i = h_i (i = 0, 1, 2, \ldots\ldots, \overline{n-1})$. Then

$$S'(x; [x_i, x_{i+1}]) = \frac{m_{i+1}}{h_i}(x - x_i) - \frac{m_i}{h_i}(x - x_{i+1})$$

$$= \frac{1}{h_i}[(x_{i+1} - x)m_i + m_{i+1}(x - x_i)]$$

Integrating the above equations w. r. t x, we get

$$S(x; [x_i, x_{i+1}]) = \frac{1}{h_i}\left[\frac{-(x_{i+1} - x)^2}{2} m_i + \frac{(x - x_i)^2}{2} m_{i+1}\right] + c_i, \qquad (3.78)$$

where c_i $(i = 0, 1, 2, \ldots\ldots, \overline{n-1})$ are to be determined constants. Putting $x = x_i$ is (3.78), then we have

$$S(x_i; [x_i, x_{i+1}]) = \frac{1}{h_i}\left[\frac{-h_i^2}{2} m_i\right] + c_i$$

In view of (3.73), the above equation reduces to

$$y_i = \frac{-h_i}{2} m_i + c_i.$$

Thus,

$$c_i = y_i + \frac{h_i}{2} m_i$$

Hence, Eq. (3.78) reduces to

$$S(x; [x_i, x_{i+1}]) = \frac{1}{h_i}\left[\frac{-(x_{i+1} - x)^2}{2} m_i + \frac{(x - x_i)^2}{2} m_{i+1}\right]$$

$$+ y_i + \frac{h_i}{2} m_i \qquad\qquad (3.79)$$

Note that $S(x; [x_i, x_{i+1}])$ will be known if we find m_i. To find m_i, we use the condition of the spline function.

$$\underset{x \to x_{i+1}^-}{Lt}\ S(x;\ [x_i,\ x_{i+1}]) = \underset{x \to x_{i+1}^+}{Lt}\ S(x;\ [x_{i+1},\ x_{i+2}])$$

$$\underset{x \to x_{i+1}^-}{Lt}\left\{ \frac{1}{h_i}\left[\frac{-(x_{i+1} - x)^2}{2}\ m_i + \frac{(x - x_i)^2}{2}\ m_{i+1} \right] + y_i + \frac{hi}{2}\ m_i \right\}$$

$$= \underset{x \to x_{i+1}^+}{Lt}\left\{ \frac{1}{h_{i+1}}\left[\frac{-(x_{i+2} - x)^2}{2}\ m_{i+1} + \frac{(x - x_{i+1})^2}{2}\ m_{i+2} \right] + y_{i+1} + \frac{h_{i+1}}{2}\ m_{i+1} \right\}$$

$$\therefore\ \frac{1}{h_i}\left[\frac{h_i^2}{2}\ m_{i+1} \right] + y_i + \frac{h_i}{2}\ m_i$$

$$= \frac{1}{h_{i+1}}\left[\frac{-h_{i+1}^2}{2}\ m_{i+1} \right] + y_{i+1} + \frac{h_{i+1}}{2}\ m_{i+1}$$

$$\text{i.e. } m_i + m_{i+1} = \frac{2}{h_i}\ (y_{i+1} - y_i),\ i = 0, 1, 2, \ldots, \overline{n-1} \tag{3.80}$$

Equation (3.80) represents n equations in $(n + 1)$ unknowns $(m_0, m_1, m_2, \ldots, m_n)$, so we require one more condition to determine all m'_i s. We choose the adition condition as $m_0 = m_1$. Solving (3.80) and $m_0 = m_1$, we obtain $m'_i s$ values. Substituting m_i values in (3.79), we get the quadratic splines.

SOLVED PROBLEMS

Problem 1: For the following data,

x	1	2	3
$f(x)$	-8	-1	18

find the quadratic splines.

Solution: Here, $x_0 = 1$, $x_1 = 2$, $x_2 = 3$, $y_0 = -8$, $y_1 = -1$ and $y_2 = 18$. Further $h_0 = x_1 - x_0 = 1$ and $h_1 = x_2 - x_1 = 1$ for quadratic splines;

$$m_i + m_{i+1} = \frac{2}{h_i}\ (y_{i+1} - y_i)\ (i = 0, 1) \tag{1}$$

$$\therefore \qquad m_0 + m_1 = \frac{2}{h_0}(y_1 - y_0) \qquad (2)$$

and

$$m_1 + m_2 = \frac{2}{h_1}(y_2 - y_1) \qquad (3)$$

Substituting the values of y_0, y_1, y_2, h_0 and h_1 in (2) and (3), we obtain

$$m_0 + m_1 = 14 \qquad (4)$$

and

$$m_1 + m_2 = 38 \qquad (5)$$

We choose an additional condition $m_0 = m_1$; solving this and (4), (5) for m_0, m_1, m_2, we get $m_0 = 7$, $m_1 = 7$ and $m_2 = 31$. The quadratic spline is given by (3.79)

$$S(x;[x_i, x_{i+1}]) = \frac{1}{h_i}\left[\frac{-(x_{i+1} - x)^2}{2}m_i + \frac{(x - x_i)^2}{2}m_{i+1}\right] + y_i + \frac{h_i}{2}m_i$$

Hence, the quadratic spline on $[1, 2]$ is $S(x; [1, 2]) = 7x - 15$ and the quadratic spline on $[2, 3]$ is $S(x; [2, 3]) = 12x^2 - 41x + 33$.

3.30 CUBIC SPLINES

Let y_0, y_1, y_2, ..., y_n be $(n + 1)$ values of $y = f(x)$ at $a = x_0 < x_1 < x_2 <$ ...... $< x_{n-1} < x_n = b$. Further, let $S(x; [x_i, x_{i+1}])$ denote the cubic spline function on the interval $[x_i, x_{i+1}]$ $(i = 0, 1, 2, ..., \overline{n-1})$.

Then, $S''(x; [x_i, x_{i+1}])$ is linear, considering the following table for the functions $S''(x;[x_i, x_{i+1}])$.

x	x_i	x_{i+1}
$S''(x; x_i, x_{i+1})$	$S''(x_i; [x_i, x_{i+1}])$	$S''(x_{i+1}; [x_i, x_{i+1}])$

By Lagrange's interpolation formula, we have

$$S''(x; [x_i, x_{i+1}]) = \frac{x - x_{i+1}}{x_i - x_{i+1}} \cdot S''(x_i; [x_i, x_{i+1}])$$
$$+ \frac{(x - x_i)}{(x_{i+1} - x_i)} S''(x_i; [x_i, x_{i+1}]).$$

Let $h_i = x_{i+1} - x_i$ ($i = 0, 1, 2, \ldots, \overline{n-1}$). Then

$$S''(x_i; [x_i, x_{i+1}]) = \frac{x - x_{i+1}}{-h_i} S'' (x_i; [x_i, x_{i+1}]) + \frac{(x - x_i)}{h_i} S'' (x_i; [x_i, x_{i+1}])$$

$$= \frac{1}{h_i} [(x_{i+1} - x) S''(x_i; [x_i, x_{i+1}]$$

$$+ (x - x_i) S'' (x_{i+1}; [x_i, x_{i+1}])] \tag{3.81}$$

Integrating (3.81) twice, we get

$$S''(x; [x_i, x_{i+1}]) = \frac{1}{h_i} \times$$

$$\left[\frac{(x_{i+1} - x)^3}{\underline{3}} S\left(x_i; [x_i, x_{i+1}]\right) + \frac{(x - x_i)^3}{\underline{3}} S''(x_{i+1}; [x_i, x_{i+1}]) \right]$$

$$+ C'x + D', \tag{3.82}$$

where C' and D' are arbitrary constants.
Expressing $C'x + D' = C(x_{i+1} - x) + D(x - x_i)$, Eq. (3.82) becomes

$$S(x; [x_i, x_{i+1}]) = \frac{1}{h_i} \times$$

$$\left[\frac{(x_{i+1} - x)^3}{\underline{3}} S''\left(x_i; [x_i, x_{i+1}]\right) + \frac{(x - x_i)^3}{\underline{3}} S''(x_{i+1}; [x_i, x_{i+1}]) \right]$$

$$+ C(x_{i+1} - x) + D(x - x_i). \tag{3.83}$$

Substituting $x = x_i$ in (3.83), we get

$$S(x_i; [x_i, x_{i+1}])$$

$$= \frac{1}{h_i} \left[\frac{(x_{i+1} - x_i)^3}{\underline{3}} S''\left(x_i; [x_i, x_{i+1}]\right) + C(x_{i+1} - x_i) \right]$$

i.e. $S\left(x_i; [x_i, x_{i+1}]\right) = \frac{1}{h_i} \left[\frac{h_i^3}{\underline{3}} S''\left(x_i; [x_i, x_{i+1}]\right) \right] + Ch_i$

Thus,

$$C = \frac{1}{h_i} \left[S(x_i; [x_i, x_{i+1}]) \right] - \frac{h_i^2}{\underline{3}} S''\left(x_i; [x_i, x_{i+1}]\right). \tag{3.84}$$

Now, substituting $x = x_{i+1}$ in (3.83), we get

$$D = \frac{1}{h_i}[S(x_{i+1}; [x_i, x_{i+1}]) - \frac{h_i^2}{\lfloor 3} S''(x_{i+1}; [x_i, x_{i+1}])]$$ (3.85)

Since $S(x_i; [x_i, x_{i+1}]) = y_i$ and $S(x_{i+1}; [x_i, x_{i+1}]) = y_{i+1}$,
Eqs. (3.84) and (3.85) are reduced to

$$C = \frac{1}{h_1}\left[y_i - \frac{h_i^2}{\lfloor 3} S''\left(x_i; [x_i, x_{i+1}]\right) \right]$$ (3.86)

and

$$D = \frac{1}{h_1}\left[y_{i+1} - \frac{h_i^2}{\lfloor 3} S''\left(x_{i+1}; [x_i, x_{i+1}]\right) \right]$$ (3.87)

Substituting (3.86) and (3.87) in (3.83), we get

$$S(x; [x_i, x_{i+1}]) = \frac{1}{h_i} \times$$

$$\left[\frac{(x_{i+1} - x)^3}{\lfloor 3} S''\left(x_i; [x_i, x_{i+1}]\right) + \frac{(x - x_i)^3}{\lfloor 3} .S''\left(x_{i+1}; [x_i, x_{i+1}]\right) \right]$$

$$+ \frac{1}{h_i}(x_{i+1} - x)\left[y_i - \frac{h_i^2}{\lfloor 3} S''\left(x_i; [x_i, x_{i+1}]\right) \right]$$

$$+ \frac{1}{h_i}(x - x_i)\left[y_{i+1} - \frac{h_i^2}{\lfloor 3} S''\left(x_{i+1}; [x_i, x_{i+1}]\right) \right]$$ (3.88)

Let $S''(x_i; [x_i, x_{i+1}]) = M_i$ and $S''(x_{i+1}; [x_i, x_{i+1}]) = M_{i+1}$. Then, Eq. (3.88)
reduces to
$$S(x; [x_i, x_{i+1}]) =$$

$$\frac{1}{6h_i}\left[(x_{i+1} - x)^3 M_i + (x - x_i)^3 M_{i+1} \right] + \frac{1}{h_i}(x_{i+1} - x)\left[y_i - \frac{h_i^2}{6} M_i \right]$$

$$+ \frac{1}{h_i}(x - x_i)\left[y_{i+1} - \frac{h_i^2}{6} M_{i+1} \right]$$ (3.89)

Now, we use the condition of continuity of
$S'(x; [x_i, x_{i+1}])$ and $S'(x; [x_{i-1}, x_i])$ at $x = x_i$.

$$\underset{x \to x_i^+}{Lt} S(x; [x_i, x_{i+1}]) = \underset{x \to x_i^-}{Lt} S'(x; [x_{i-1}, x_i]) \ (i = 1, 2, \ldots, \overline{n-1})$$ (3.90)

Differentiating (3.89) with respect to x, we have

$$S'(x; [x_i, x_{i+1}]) = \frac{1}{6h_i} [-3(x_{i+1} - x)^2 M_i + 3(x - x_i)^2 M_{i+1}]$$

$$- \frac{1}{h_i}\left(y_i - \frac{hi^2}{6} M_i\right) + \frac{1}{h_i}\left(y_{i+1} - \frac{hi^2}{6} M_{i+1}\right) \qquad (3.91)$$

Now,

$$\underset{x \to x_i^+}{Lt}\ S'(x; [x_i, x_{i+1}]) = \frac{-h_i}{6} [2M_i + M_{i+1}] + \frac{1}{h_i}(y_{i+1} - y_i) \qquad (3.92)$$

From (3.89), $S(x; [x_{i-1}, x_i])$ is given by

$$S(x; [x_{i-1}, x_i]) = \frac{1}{6h_{i-1}} [(x_i - x)^3 M_{i-1} + (x - x_{i-1})^3 M_i]$$

$$+ \frac{1}{h_{i-1}} (x_i - x)[y_{i-1} - \frac{h_{i-1}^2}{6} M_{i-1}] + \frac{1}{h_{i-1}} (x - x_{i-1})[y_i - \frac{h_{i-1}^2}{6} M_i]$$

Differentiating it with respect to x, we obtain

$$S'(x; [x_{i-1}, x_i]) = \frac{1}{6h_{i-1}} \left[-3(x_i - x)^2 M_{i-1} + 3(x - x_{i-1})^2 M_i\right]$$

$$- \frac{1}{h_{i-1}}\left[y_{i-1} - \frac{h_{i-1}^2}{6} M_{i-1}\right] + \frac{1}{h_{i-1}}\left[y_i - \frac{h_{i-1}^2}{6} M_i\right]$$

$$\therefore \qquad \underset{x \to x_i^-}{Lt}\ S(x; [x_{i-1}, x_i]) = \frac{h_{i-1}}{6}(2M_i + M_{i-1}) + \frac{1}{h_{i-1}}(y_i - y_{i-1}). \qquad (3.93)$$

From equation (3.92) and (3.93) and the condition of continuity (3.90), we have

$$h_{i-1} M_{i-1} + 2(h_{i-1} + h_i)M_i + h_i M_{i+1} = 6\left[\frac{y_{i+1} - y_i}{h_i} - \frac{y_i - y_{i-1}}{h_{i-1}}\right] \qquad (3.94)$$

$i = 1, 2, 3, \dots \overline{(n-1)}$

Eq. (3.94) gives us a system of $(n - 1)$ linear equation in $(n + 1)$ unknowns $M_0, M_1, M_2, \dots M_n$. In order to have a unique solution for the system of equations, we must add two more condition to it. Usually the two conditions are taken as

$$M_o = 0 \tag{3.95}$$

and

$$M_n = 0 \tag{3.96}$$

The cubic spline functions obtained by using the conditions (3.95) and (3.96) are known as natural cubic splines. Solving the system of Eqs. (3.94), (3.95) and (3.96), we will have the values of M_0, M_1, ..., M_n. With the known values of M_0, M_1, M_2, ..., M_n, one can find

$$S(x; [x_i, x_{i+1}])(i = 0, 1, 2,.. \overline{(n-1)}) \text{ using the Eq. (3.89)}.$$

Note: The cubic spline functions on the interval $[x_0, x_1]$ and $[x_1, x_2]$ are as follows (from 3.89):

$$S(x; [x_0, x_1]) = \frac{1}{6h_0} [(x_1 - x)^3 M_0 + (x - x_0)^3 M_1]$$

$$+ \frac{1}{h_0}(x_1 - x)\left[y_0 - \frac{h_o^2}{6} M_0 \right] + \frac{1}{h_0}(x - x_0)\left[y_1 - \frac{h_o^2}{6} M_1 \right]$$

$$S(x; [x_1, x_2]) = \frac{1}{6h_1} [(x_2 - x)^3 M_1 + (x - x_1)^3 M_2]$$

$$+ \frac{1}{h_1}(x_2 - x)\left[y_1 - \frac{h_1^2}{6} M_1 \right] + \frac{1}{h_1}(x - x_1)\left[y_2 - \frac{h_1^2}{6} M_2 \right],$$

where $h_0 = x_1 - x_0$ and $h_1 = x_2 - x_1$. Similarly, one can write the natural cubic spline functions on the intervals $[x_2, x_3]$, $[x_3, x_4]$, ..., $[x_{n-1}, x_n]$.

Note: If the argument x_0, x_1, ... x_n are equidistant points, i.e., $x_{i+1} - x_i = h$ $(i = 0, 1, 2, .., \overline{n-1})$, then Eq. (3.89) reduces to

$$M_{i-1} + 4M_i + M_{i+1} = \frac{6}{h^2} (y_{i-1} - 2y_i + y_{i+1})(i = 1, 2, 3, ..., \overline{n-1}).$$

We take two more equations $M_0 = 0$ and $M_n = 0$ in order to solve the equations for M_0, M_1, M_2,...,M_n. Solving these equations, we may get M_0, M_1, M_2, ..., M_n values.

If the arguments are equidistant with interval of differencing h, then

(a) The cubic spline on $[x_0, x_1]$ is
$$S(x; [x_0, x_1]) =$$
$$\frac{1}{h}\left[\frac{(x_i - x)^3}{6} M_0 + \frac{(x - x_0)^3}{6} M_1 + (x_1 - x)\left\{y_0 - \frac{h^2}{6} M_0\right\}\right.$$
$$\left. + (x - x_0)\left\{y_1 - \frac{h^2}{6} M_1\right\}\right]$$

(b) The Cubic spline on $[x_1, x_2]$ is
$$S(x; [x_1, x_2]) =$$
$$\frac{1}{h}\left[\frac{(x_2 - x)^3}{6} M_1 + \frac{(x - x_1)^3}{6} M_2 + (x_2 - x)\left\{y_1 - \frac{h^2}{6} M_1\right\}\right.$$
$$\left. + (x - x_1)\left\{y_2 - \frac{h^2}{6} M_2\right\}\right]$$

(c) In general, the cubic spline on $[x_i, x_{i+1}]$ is
$$S(x; [x_i, x_{i+1}]) = \frac{1}{h}\left[\frac{(x_{i+1} - x)^3}{6} M_i + \frac{(x - x_i)^3}{6} M_{i+1}\right.$$
$$\left. + (x_{i+1} - x)\{y_i - M_i\} + (x - x_i)\left(y_{i+1} - \frac{h^2}{6} M_{i+1}\right)\right]$$

3.31   PROCEDURE TO FIND NATURAL CUBIC SPLINE FUNCTIONS ON THE SUBINTERVALS $[x_0, x_1]$, $[x_1, x_2]$ ..., $[x_{n-1}, x_n]$

Step 1: Note down the number of data points and equate it to $n + 1$. It gives the value of n.

Step 2: Suppose the given arguments as $x_0, x_1, x_2, \ldots x_n$ and function values at these arguments as $y_0, y_1, y_2, \ldots y_n$.

Step 3: Write down the $(n - 1)$ equations involving $M_0, M_1, \ldots M_n$ using the following formula for $i = 1, 2, 3, \ldots, n - 1$.

$$h_{i-1} M_{i-1} + 2(h_{i-1} + h_i)M_i + h_i M_{i+1} = 6\left[\frac{y_{i+1} - y_i}{h_i} - \frac{y_i - y_{i-1}}{h_i - 1}\right]$$

$$(i = 1, 2, \ldots, \overline{n-1})$$

Step 4: For the system of linear equations obtained in Step 3, add two more equations

$$M_0 = 0$$

$$M_n = 0$$

Step 5: Solve the equation obtained in Step 3 and Step 4. At this stage, we know the values of M_0, M_1, M_2, …. and M_n.

Step 6: The cubic spline in each subinterval is obtained by substituting M_0, M_1, ….M_n values in the following formula for $i = 0, 1, 2, …, (n-1)$.

$$S(x; [\, x_i, x_{i+1}]) = \frac{1}{6h_i}\left[(x_{i+1}-x)^3 M_1 + (x-x_i)^3 M_{i-1}\right] + \frac{1}{h_i}[x_{i+1}-x]$$

$$\left[y_i - \frac{h_i^2}{6}M_i\right] + \frac{1}{h_i}(x-x_i)\left(y_{i+1} - \frac{h_i^2}{6}M_{i+1}\right)$$

where $h_i = x_{i+1} - x_i$.

Step 7: Simplify the expressions on the right–hand side of the equations obtained in Step 6 for $i = 0, 1, 2, …, \overline{n-1}$ and suppose these expressions are $f_0(x), f_1(x), f_2(x), ….f_{n-1}(x)$.

Step 8: State the spline function as $S(x; [x_0, x_n])$ in the following form:

$$S(x; [x_0, x_n]) = \begin{cases} f_0(x)_i \text{ if } x_0 \le x \le x_1 \\ f_1(x)_i \text{ if } x_1 \le x \le x_2 \\ . \\ . \\ . \\ f_{n-1}(x)_i \text{ if } x_{n-1} \le x \le x_n. \end{cases}$$

SOLVED PROBLEMS

Problem 1: Obtain the cubic spline functions for the data given below:

x	0	1	2	3
$f(x)$	1	2	33	244

Solution: Note that the number of data points is 4. Thus, $(n + 1) = 4$, and hence, $n = 3$. Here, $x_0 = 0$, $x_1 = 1$, $x_2 = 2$, $x_3 = 3$, $y_0 = 1$, $y_1 = 2$, $y_2 = 33$ and $y_3 = 244$. Further, $h_1 = x_1 - x_0 = 1$, $h_2 = x_2 - x_1 = 1$, and $h_3 = x_3 - x_2 = 1$.

Here, the arguments are equidistant with interval of differencing $h = 1$. The relation involving $M_0, M_1, \ldots M_n$ is given by

$$M_{i-1} + 4M_i + M_{i+1} = \frac{6}{h^2}(y_{i+1} - 2y_i + y_{i-1})(i = 1, 2, \ldots, \overline{n-1}). \tag{1}$$

For $i = 1$ and $i = 2$, (since $n = 3$), the above equation gives

$$M_0 + 4\,M_1 + M_2 = 180. \tag{2}$$

$$M_1 + 4M_2 + M_3 = 1080. \tag{3}$$

We suppose two additional conditions as

$$M_0 = 0 \tag{4}$$

$$M_3 = 0 \tag{5}$$

Solving (2), (3), (4) and (5) for M_0, M_1, M_2, and M_3, we obtain $M_0 = 0$, $M_1 = 24$, $M_2 = 276$ and $M_3 = 0$.

The cubic splines on $[x_i, x_{i+1}]$ are computed from

$$S\left(x;[x_i, x_{i+1}]\right) = \frac{1}{h_i}\left[\begin{array}{l} \dfrac{(x_{i+1} - x)^3}{6}M_i + \dfrac{(x - x_i)^3}{6}M_{i-1} \\[2mm] + \left(x_{i+1} - x\right)\left(y_i - \dfrac{h_i^2}{6}M_i\right) \\[2mm] + \left(x - x_i\right)\left(y_{i+1} - \dfrac{h_i^2}{6}i+1\right) \end{array}\right] \tag{6}$$

for $i = 0, 1$, and 2.

(i) **To find the cubic spline function on the subinterval [0, 1].**
   Substituting $i = 0$ and the values of x_0, x_1, M_0, M_1 and h in (6), we get

$$S(x;\,[0,\,1]) = \frac{1}{1}\left[\frac{(1-x)^3(0)}{6} + \frac{(x-0)^2(-24)}{6}\right] + 1(1-x)\left(1 - \frac{1^2}{6}(0)\right)$$

$$+ \frac{1}{1}(x-0)\left\{2 - \frac{1^2}{6}(-24)\right\}$$

$$= -4x^3 + 5x + 1.$$

(ii) To find the cubic spline function at the interval [1, 2].

Substituting $i = 1$ and the value of $x_1, x_2, y_1, y_2, M_1,$ and M_2 in (6), we get

$$S(x; [1, 2]) = \frac{1}{1}\Big[(2-x)^2(-24) + (x-1)^3 276 + (2-x)$$

$$\left\{2 - \frac{1^2}{6}(-24)\right\} + (x-1)\left\{33 - \frac{1^2}{6}(276)\right\}\Big]$$

$$= 50x^3 - 162x^2 + 167x - 53.$$

(iii) To find cubic spline function on the interval [2, 3]

Substituting $i = 2$ and the values of x_2, x_3, y_2, y_3, M_2 and M_3 in Eq. (6), we get

$$S(x; [2, 3]) = \frac{1}{1}\Big[(3-x)^2(276) + (x-2)^2(0) + (3-x)$$

$$\left\{33 - \frac{1^2}{6}(276)\right\} + \frac{1}{1}(x-1)\left\{244 - \frac{1^2}{6}*0\right\}\Big]$$

$$= -46x^3 + 414x^2 - 985x + 715.$$

Thus, the required cubic spline on [0,3] is given below:

$$S(x; [0, 3]) = \begin{cases} -4x^3 + 5x + 1 & \text{if } 0 \le x \le 1 \\ 50x^3 - 162x^2 + 167x - 53 & \text{if } 1 \le x \le 2 \\ -46x^3 + 414x^2 - 985x + 715 & \text{if } 2 \le x \le 3. \end{cases}$$

Problem 2: Obtain the cubic spline function for the given data:

x	0	1	3
$f(x)$	1	2	244

Solution: Here the number of data points is 3.

Hence, $n + 1 = 3$. Therefore $n = 2$. Here $x_0 = 0, x_1 = 1, x_2 = 3, y_0 = 1, y_1 = 2, y_2 = 244, h_0 = x_1 - x_0 = 1 - 0 = 1$ and $h_1 = x_2 - x_1 = 3 - 1 = 2$. The relations in M_i are given by

$$h_{i-1} M_{i-1} + 2(h_{i-1} + h_i)M_i + h_i M_{i+1} = 6\left[\frac{y_{i+1} - y_i}{h_i} - \frac{y_i - y_{i-1}}{h_{i-1}}\right]$$

When $i = 1$, it reduce to

$$M_0 + (2)3M_1 + 2M_2 = 6\left[\frac{244-2}{2} - \frac{2-1}{1}\right],$$

i.e., $M_0 + 6M_1 + 2M_2 = 720.$

The two additional conditions are $M_0 = 0$ and $M_2 = 0$. Thus, the value of $M_0 = 0$, $M_1 = 120$, $M_2 = 0$

(i) **To find the cubic spline function on [0, 1].**
   The cubic spline on $[x_0, x_1]$ is given by note of 3.30

$$S(x; [x_0, x_1]) = \frac{1}{6h_0}[(x_1 - x)^3 M_0 + (x - x_0)^3 M_1] + \frac{1}{h_0}(x_1 - x)$$

$$\left(y_0 - \frac{h_o^2}{6} M_0\right) + \frac{1}{h_0}(x - x_0)\left(y_1 - \frac{h_o^2}{6} M_1\right)$$

Substituting the values of x_0, y_0, x_1, y_1, M_0 and M_1, we get

$$S(x; [0, 1]) = 20x^3 - 19x + 1.$$

(ii) **To find the cubic spline function on [1, 3].**
   The cubic spline on $[x_1, x_2]$ is given by

$$S(x; [x_1, x_2]) = \frac{1}{6h_1}[(x_2 - x)^3 M_1 + (x - x_1)^3 M_2] + \frac{1}{h_1}(x - x_1)$$

$$\left(y_1 - \frac{h^2}{6}.M_1\right) + \frac{1}{h_1}(x - x_1)\left(y_2 - \frac{h^2}{6} M_2\right).$$

Now substituting the values of x_1, x_2, y_1, y_2, M_1 and M_2 in the above equation, we get

$$S(x; [x_1, x_2]) = \frac{1}{6h_2}[3 - x^3(120) + (x - 1)(0)] + \frac{1}{2}(3 - x)$$

$$\left(2 - \frac{2^2}{6}(120)\right) + \frac{1}{2}(x - 1)\left(244 - \frac{1}{6}(0)\right)$$

$$= -10x^3 + 90x^2 - 109x + 31.$$

Thus, the spline function as [0, 3] is

$$S(x;\,[0,\,3]) = \begin{cases} 20x^3 - 19x + 1 & \text{if } 0 \le x \le 1 \\ -10x^3 + 90x^2 - 109x + 31 & \text{if } 1 \le x \le 3. \end{cases}$$

3.32 ERROR IN INTERPOLATION

If the side of length 4 cm of a square is measured as 3.92 cm, then we say that there is an error of 0.08 cm is measuring the side of square. Likewise, if $p_n(x)$ is the interpolating polynomial of a function $f(x)$ for a give data $(x_i,\, y_i)(i = 0, 1, 2, \ldots n)$ with $a = x_0 < x_1 < x_2 < \ldots.. < x_n = b$, then we define $f(c) - p_n(c)$ as an error in the interpolation at $x = c$, where $c \in [a, b]$. For example suppose a function $f(x) = 1 + \sin x$ has the values 1, 1.8415, 1.9093 at $x = 0$, 1 and 2 respectively, Then, we can find an interpolating polynomial of second degree using the data. Let this polynomial be $p_2(x)$. It can be shown that

$p_2(x) = 1 + 0.8415x + 0.3869x\,(x - 1)$ using Newton's forward difference formula.

Thus, the error in interpolation at x is

$$(1 + \sin x) - \{1 + 0.845x - 0.3869x\,(x - 1)\}.$$

The error at $x = 0.5$ is

$$\{1 + \sin(0.5)\} - \{1 - 0.8415\,(0.5) - 0.3896\,(0.5)\,(0.5 - 1)\}$$

$$= 1.4794 - 1.5181 = -0.0387$$

Theorem: (Error in interpolation)

Let $p_n(x)$ be the interpolating polynomial of $f(x)$ that interpolates at $(n + 1)$ distinct arguments $x_0, x_1, \ldots.. x_n$ in $[a, b]$. Then, the error in polynomial interpolation at $x = \bar{x}$ is

$$f(\bar{x}) - p_n(\bar{x}) = \frac{(\bar{x} - x_0)(\bar{x} - x_1)....(\bar{x} - x_n)}{\lfloor n+1}\, f^{(n+1)}\,(z),$$

where $z \in [a, b]$.

Proof: Since $p_n(x)$ is the interpolating polynomial, we have $p_n(x_k) = f(x_k)(k = 0, 1, 2, \ldots, n)$. Thus, $f(x) - p_n(x)$ vanishes at $x = x_0,\ x_1, \ldots, x_n$.

Let $x \ne x_0, x_1, x_2, \ldots, x_n$ and

$$k(\bar{x}) = \frac{f(\bar{x}) - p_n(\bar{x})}{(\bar{x} - x_0)(\bar{x} - x_1).....(\bar{x} - x_n)} \tag{3.97}$$

Suppose

$$W(t) = f(t) - p_n(t) - (t - x_0)(t - x_1) \ ..... \ (t - x_n)k(\bar{x}). \tag{3.98}$$

Clearly, $W(t)$ vanishes at $t = x_0, x_1, x_2 ..., x_n$ and also it vanishes at $t = \bar{x}$, and hence, $W(t)$ vanishes at $(n + 2)$ points. Therefore, by Rolle's theorem there exists a point $z \in [a, b]$ such that

$$W^{(n+1)}(z) = 0. \tag{3.99}$$

Differentiating (3.98) $(n + 1)$ times with respect to t and substituting $t = z$, we get

$$W^{(n+1)}(z) = f^{(n+1)}(z) - 0 - \lfloor n+1 \ .k(\bar{x}).$$

In view of (3.99), the above equations gives

$$k\overline{(x)} = \frac{1}{\lfloor n+1} .f^{(n+1)}(z),$$

i.e., $$f(\bar{x}) - p_n(\bar{x}) = \frac{(\bar{x} - x_0)(\bar{x} - x_1).....(\bar{x} - x_n)}{\lfloor n+1} f^{(n+1)}(z) \tag{3.100}$$

It is denoted by $E(f; \bar{x})$.

Note: The error in interpolation at x is

$$E(f : x) = \frac{(x - x_0)(x - x_1)....(x - x_n)}{\lfloor n+1} f^{(n+1)}(z), \tag{3.101}$$

where $z \in [a, b]$.

Remark:

The truncation error of f is given by

$$|E(f;x)| \leq \frac{M_{n+1}}{\lfloor n+1} .Max \left|(x - x_0)(x - x_1)....(x - x_n)\right|,$$

where $$M_{n+1} = Max \left| f^{(n+1)}(z) \right|.$$

SOLVED PROBLEMS

Problem 1: If $p_2(x)$ is an interpolating polynomial of the data given below, then find $p_2(2.5)$ and an error in interpolation at $x = 2.5$.

x	1	3	5
$f(x) = \dfrac{1}{x}$	1	$\dfrac{1}{3}$	$\dfrac{1}{5}$

Solution: The forward differences table for the given data is shown below:

x	y	Δy	$\Delta^2 y$
1	1		
		$\rightarrow \dfrac{-2}{3}$	
3	$\dfrac{1}{3}$		$\rightarrow \dfrac{4}{5}$
		$\rightarrow -\dfrac{2}{15}$	
5	$\dfrac{1}{5}$		

$$\therefore p_2(x) = 1 + \frac{(x-1)}{2}\left(\frac{-2}{3}\right) + \frac{(x-1)}{2}\frac{(x-3)}{2}\left(\frac{1}{2}\right)\left(\frac{4}{5}\right)$$

$$= 1 - \frac{1}{3}(x-1) + \frac{1}{10}(x-1)(x-3)$$

Hence, $\qquad\qquad p_2(2.5) = 0.425.$

Since, $\qquad\qquad f(x) = \dfrac{1}{x}, f(2.5) = 0.4$

$\therefore$ The error at $x = 2.5$ is $0.4 - 0.425 = -0.025.$

Problem 2: For the given data, find an interpolating $p_2(x)$ and find the upper bound of error in interpolation

x	$\dfrac{-1}{\sqrt{3}}$	0	$\dfrac{1}{\sqrt{3}}$
$f(x) = \cos x$	$\cos\left(\dfrac{-1}{\sqrt{3}}\right)$	1	$\cos\left(\dfrac{1}{\sqrt{3}}\right)$

Solution: From Eq. (3.101), we have

$$f(x) - p_2(x) = \frac{\left(x + \dfrac{1}{\sqrt{3}}\right)(x - 0)\left(x - \dfrac{1}{\sqrt{3}}\right)}{\lfloor 3} f'''(z)$$

where $z \in \left[-\dfrac{1}{\sqrt{3}}, \dfrac{1}{\sqrt{3}}\right]$

Now,
$$\left|f(x) - p_2(x)\right| \leq \frac{1}{\lfloor 3}\left|x\left(x^2 - \frac{1}{3}\right)\right|\left|f'''(z)\right|$$

Clearly,
$$\left|f'''(z)\right| = \left|\sin z\right| \leq 1$$

$\therefore$
$$\left|f(x) - p_2(x)\right| \leq \frac{1}{\lfloor 3}\left|x\left(x^2 - \frac{1}{3}\right)\right|. \tag{1}$$

Let
$$g(x) = x\left(x^2 - \frac{1}{3}\right). \tag{2}$$

Then, $g'(x) = 3x^2 - \dfrac{1}{3} = \dfrac{9x^2 - 1}{3}$

and $g''(x) = 6x.$

For maximum or minimum $g'(x) = 0$

i.e.
$$\frac{9x^2 - 1}{3} = 0$$

$$x = \pm\frac{1}{3}$$

Since $g''\left(\dfrac{1}{3}\right) > 0$ and $g''\left(\dfrac{-1}{3}\right) < 0$, the function $g(x)$ is maximum at

$$x = \left(\frac{-1}{3}\right)$$

The maximum of $g(x)$ is $g\left(\dfrac{-1}{3}\right) = \left(\dfrac{2}{27}\right)$

$\therefore \qquad |f(x) - p_2(x)| \le \dfrac{1}{6}\left(\dfrac{2}{27}\right) = \dfrac{1}{81}.$

Thus, the upper bound of error in interpolation is $\dfrac{1}{81}$.

Problem 3: Determine the interval of differencing h to be used in the tabulation of $f(x) = \cos x$ in the interval $\left[0, \dfrac{\pi}{4}\right]$ at equally speed points so that the truncation error of quadratic interpolating polynomial is less than 5×10^{-5}.

Solution: Let x_{i-1}, x_i, x_{i+1} be three equally speed arguments with interval of differencing h, i.e., $x_k - x_{k-1} = h(k = i, i + 1)$, Let $E(f; x)$ be the error in quadratic interpolating polynomial Then, by the Remark of section 3.32,

$$|E(f;x)| \le \dfrac{M_3}{6} \, \text{Max} \, |(x - x_{i-1})(x - x_i)(x - x_{i+1})|, \qquad (1)$$

where $M_3 = \text{Max}|f'''(z)|$ and $Z \in [x_{i-1}, x_{i+1}]$.

Clearly, $M_3 \le 1$, since $f(x) = \sin x$.

Now, we find the maximum of $|(x - x_{i-1})(x - x_i)(x - x_{i+1})|$

Suppose $\dfrac{x - x_i}{h} = u$, then $\dfrac{x - x_i - 1}{h} = u - 1$ and $\dfrac{x - x_i + 1}{h} = u + 1.$

Further,

$$(x - x_{i-1})(x - x_i)(x - x_{i+1}) = (u - 1)h(u)h(u + 1)h$$

$$= h^3 u(u^2 - 1). \qquad (2)$$

Let $\qquad\qquad\qquad g(u) = u^3 - u.$

Then $g'(u) = 3u^2 - 1$ and $g''(u) = 64$ $\qquad\qquad\qquad (3)$

for maxima or minima $g'(u) = 0$,

i.e., $\quad 3u^2 - 1 = 0,$

i.e., $\qquad\qquad u = \pm\dfrac{1}{\sqrt{3}}.$

$\therefore$ The maximum of $g(x)$ is $= \dfrac{-2}{3\sqrt{3}}$, since $g''\left(\dfrac{-1}{\sqrt{3}}\right) < 0$

Thus, the max $\left|(x-x_{i-1})(x-x_i)(x-x_{i+1})\right| \le h^3 \cdot \dfrac{2}{3\sqrt{3}}$ \qquad (by (2))

Hence,

$$|E(f;x)| \le \frac{1}{6}h^3 \cdot \frac{2}{3\sqrt{3}} = \frac{1}{9\sqrt{3}}h^3$$

$\therefore$ The interval of differencing h to be chosen so that the truncation error is less than 5×10^{-5} is

$$\frac{h^3}{9\sqrt{3}} < 5 \times 10^{-5}$$

$$\text{i.e., } h < \left(9 \times \sqrt{3} \times 5 \times 10^{-5}\right)^{1/3}$$

$$\text{i.e, } h < 0.0920.$$

EXERCISE 3(d)

1. Find the Hermite's interpolating polynomial of $f(x)$ for the following data:

x	2	3
$f(x)$	6	11
$f'(x)$	4	6

Also find $f(2.5)$.

2. Using the Hermite interpolation, find $f(3^\circ)$ for the data given below:

x	2°	4°	6°
$f(x)$	0.0348	0.0697	0.1045
$f'(x)$	0.9999	0.9875	0.9945

3. Find the Hermite interpolating polynomial of $f(x)$ for the data given below:

x	0	0.5	1.0
$f(x)$	0	0.4794	0.8415
$f'(x)$	1	0.8776	0.5403

4. Find $f(2.5)$, given that

x	2	3
$f(x)$	29	105
$f'(x)$	50	105

5. Using Hermite interpolation formula, find $\log_e 4.2$ from the following data:

x	4	5	6
$f(x){:}\log x$	1.3862	1.6094	1.7917
$f'(x){:}\ \dfrac{1}{x}$	0.25	0.2	0.1666

6. Obtain the linear splines for the function $f(x)$ satisfying the following data:

x	1	3	5
$f(x)$	5	9	13

7. Obtain the quadratic splines for the function $f(x)$ satisfying the following data:

x	0	1	2	3
$f(x)$	1	3	7	13

8. Obtain the natural cubic splines for the function $f(x)$ satisfying the following data:

x	0	1	2	3
$f(x)$	3	5	7	39

9. Find the values of a and b such that the function

$$f(x) = \begin{cases} x^2 + ax + 1, & 1 \le x \le 2 \\ 3x + b, & 2 \le x \le 3 \end{cases}$$

is a quadratic spline.

10. Obtain linear spline functions for the function $f(x)$ satisfying the data

x	0	1	2	3
$f(x)$	1	2	5	10

Also find $f(1.5), f(2.5)$ and $f(2.8)$

11. Obtain the cubic spines for the function $f(x)$ satisfying the following data:

x	0	1	2	3
$f(x)$	1	4	10	8

with end conditions $f''(0) = f''(3) = 0$.

12. Obtain the cubic spines for the function $f(x)$ satisfying the following data:

x	−1	0	1
$f(x)$	1	0	1

under the conditions $s'(-1) = -4$ and $s'(1) = 4$.

13. If $p_2(x)$ is an interpolating polynomial of the data given below, then find $p_2(5)$ and an error in interpolation at $x = 5$.

x	2	4	6
$f(x) = e^x$	73890	54.5981	403.4288

14. Determine that the interval of differencing h can be used in the tabulation of $f(x) = \sin x$ in the interval $\left[0, \dfrac{\pi}{2}\right]$ at equally spaced points so that the truncation error of quadratic interpolation is less than 5×10^{-6}.

15. Determine the interval of differencing that can be used in the tabulation of $f(x) = \cos 2x$ on $\left[0, \dfrac{\pi}{4}\right]$ at equally spaced points so that the absolute value of truncation error of interpolating polynomial of cubic degree is less than 10^{-6}.

ANSWERS

1. $x^2 + 2$, 8.25

2. 0.13409141

3. $0.0068 x^5 + 0.002x^4 - 0.1671x^3 - 0.002 x^2 + x$

4. 60.125

5. 1.4350077

6. $S(x;[1, 3]) = 2x + 3$, $S(x;[3, 5]) = 2x + 3$,

7. $S(x;[0, 1]) = 2x + 1$, $S(x;[1, 2]) = 2x^2 - 2x + 3$, $S(x;[2, 3]) = 6x - 5$

8. $S(x;[0, 1]) = \dfrac{1}{5} (6x^3 + 4x + 15)$, $S(x;[1, 2]) = \dfrac{1}{5} (10x^3 - 12 x^2 + 16x - 11)$,

   $S(x;[2, 3]) = \dfrac{1}{5} (-16x^3 + 144x^2 - 296x + 219)$,

9. $a = -1, b = -3$

10. $S(x;[0, 1]) = x + 1$, $S(x;[1, 2]) = 3x - 1$, $S(x;[2, 3]) = 5x - 5$

11. $S(x;[0, 1]) = \dfrac{1}{3} (4x^3 + 5x + 3)$, $S(x;[1, 2]) = \dfrac{1}{3} (-11x^3 + 45 x^2 - 40x + 18)$,

   $S(x;[2, 3]) = \dfrac{1}{3} (7x^3 - 63x^2 + 176x - 126)$,

12. $S(x;[-1, 0]) = -(2x^3 - x^2)$, $S(x;[0, 1]) = 2x^3 - x^2$,

13. 47.8975909

14. $h < 0.0427$

15. $h < 0.0808$.

4

Numerical Differentiation

4.1 INTRODUCTION

Engineering professionals usually deal with changes in systems and processes. In certain cases, it may be required to know the rate of change of one entity with respect to another entity. In such situations, the derivatives act as a tool, since the derivative is the rate of change of the dependent variable with respect to the independent variable. If we know a function $f(x, y) = 0$, then by applying the principles of derivative we can find its derivative $\dfrac{dy}{dx}$. For example, if a particle is moving along a straight line according to $s = t^3 - 3t^2 + st + 6$, where s and t represent distance (in meters) and time (in seconds), respectively, then $\dfrac{ds}{dt} = 3t^2 - 6t + 5$ represents the particle velocity at time t. The velocity of the particle at $t = 2$ sec is 5 m/s. This method fails to find

the velocity of a particle if we know only the positions of the particle at different times. In such cases, we can employ numerical methods to find their rate of change by numerical differentiation.

Numerical differentiation is the process of computing the derivative of a function (need not to be known) at a particular value of the independent variable when the values of the function corresponding to certain values of the independent variable are known. Note that we have to compute the derivative of interpolating polynomial of $f(x)$ to find its derivative. As the computer fails to find the derivative of a function on its own, we need to have numerical methods to find the derivatives of a function at a point so that they can be implemented by a software program.

In Chapter 3, we learnt different interpolation formulas such as Newton's forward interpolation and Lagrange's interpolation formula. The technique involved in solving the problem of differentiation numerically is first we have to find the interpolating polynomial of $f(x)$ for the given data using an appropriate interpolation formula. Next, we differentiate it with respect to the independent variable and substitute the relevant values (like forward or divided differences, etc.) in it to have the numerical value of the derivative at the required point of the independent variable. To find the derivative of a point near the beginning, the middle and the end of arguments, we use the Newton's forward formula, any central difference formula and Newton's backward interpolation formula, respectively, when the arguments are equally spaced. If the arguments are not equally spaced, we use Newton's divided difference interpolation formula in order to obtain a derivative of a function at any point of the independent variable; of course, we can also use the Lagrange's interpolation formula in case the arguments are not equidistant. But it should be avoided as it is laborious to find the derivative using Lagrange's formula since the Lagrange's formula has n factors in each term involving x.

In this chapter, we derive some formulas for the derivatives of a function from the corresponding interpolation formulas. The method of derivation of the derivatives from the remaining interpolation formula is similar to the ones mentioned above, and hence, they are left as an exercise to the reader. Note that the derivatives obtained from these formulas are approximations because the functions are approximated by polynomials, and we are using these interpolating polynomials to compute the derivative at a point. We also find the maxima and minima values of a function from a set of tabulated values of it for different values of independent variable.

4.2 DERIVATIVES USING NEWTON'S FORWARD DIFFERENCE FORMULA

Suppose $y_0, y_1, y_2, \ldots, y_n$ are $(n+1)$ values of $y = f(x)$ at equally spaced arguments $x_0, x_1 = x_0 + h, x_2 = x_0 + 2h, \ldots, x_n = x_0 + nh$, where h is the interval of differencing. Then, by Newton's forward interpolation, we have

$$y = y_0 + u\Delta y_0 + \frac{u(u-1)}{\lfloor 2} \Delta^2 y_0 + \frac{u(u-1)(u-2)}{\lfloor 3} \Delta^3 y_0$$

$$+ \frac{u(u-1)(u-2)(u-3)}{\lfloor 4} \Delta^4 y_0 + \cdots \tag{4.1}$$

where
$$u = \frac{x - x_0}{n} \tag{4.2}$$

Differentiating (4.1) with respect to x, we get

$$\frac{dy}{dx} = \left[\Delta y_0 + \frac{(2u-1)}{\lfloor 2} \Delta^2 y_0 + \frac{(3u^2 - 6u + 2)}{\lfloor 3} \Delta^3 y_0 \right.$$

$$+ \frac{(4u^3 - 18u^2 + 22u - 6)}{\lfloor 4} \Delta^4 y_0$$

$$\left. + \frac{(5u^4 - 40u^3 + 105u^2 - 100u + 24)}{\lfloor 5} \Delta^5 y_0 + \cdots \right] \cdot \frac{du}{dx} \tag{4.3}$$

From (4.2), we have

$$\frac{du}{dx} = \frac{1}{h}$$

Hence, Eq. (4.3) becomes

$$\frac{dy}{dx} = \frac{1}{h} \left[\Delta y_0 + \frac{(2u-1)}{\lfloor 2} \Delta^2 y_0 + \frac{(3u^2 - 6u + 2)}{\lfloor 3} \Delta^3 y_0 \right.$$

$$+ \frac{(4u^3 - 18u^2 + 22u - 6)}{\lfloor 4} \Delta^4 y_0$$

$$\left. + \frac{(5u^4 - 40u^3 + 105u^2 - 100u + 24)}{\lfloor 5} \Delta^5 y_0 + \cdots \right] \tag{4.4}$$

To obtain $\dfrac{d^2 y}{dx^2}$, differentiating (4.4) with respect to x, we get

$$\frac{d^2 y}{dx^2} = \frac{1}{h^2} \left[\Delta^2 y_0 + (u-1)\Delta^3 y_0 + \frac{(12u^2 - 36u + 22)}{4!} \Delta^4 y_0 \right.$$

$$\left. + \frac{(20u^3 - 120u^2 + 210u - 100)}{5!} \Delta^5 y_0 + \cdots \right] \tag{4.5}$$

Further differentiating (4.5) with respect to x, we get

$$\frac{d^3y}{dx^3} = \frac{1}{h^3}\left[\Delta^3 y_0 + \frac{(24u-36)}{\lfloor 4}\Delta^4 y_0 + \frac{(60u^2-240u+210)}{\lfloor 5}\Delta^5 y_0 + \cdots\right]$$

$$(4.6)$$

Similarly, we can derive the formula for higher order derivatives.

4.3 DERIVATIVE OF A FUNCTION AT AN INITIAL ARGUMENT

At the initial argument, u value is $u = \dfrac{x_0 - x_0}{h} = 0$. Hence, $\dfrac{dy}{dx}$, $\dfrac{d^2y}{dx^2}$ and $\dfrac{d^3y}{dx^3}$ at $x = x_0$ are obtained from Eqs. (4.4) – (4.6) by substituting $u = 0$ and they are, respectively,

$$\left(\frac{dy}{dx}\right)_{x=x_0} = \frac{1}{h}\left[\Delta y_0 - \frac{1}{2}\Delta^2 y_0 + \frac{1}{3}\Delta^3 y_0 - \frac{1}{4}\Delta^4 y_0 + \frac{1}{5}\Delta^5 y_0 - \cdots\right] \quad (4.7)$$

$$\left(\frac{d^2y}{dx^2}\right)_{x=x_0} = \frac{1}{h^2}\left[\Delta^2 y_0 - \Delta^3 y_0 + \frac{11}{12}\Delta^4 y_0 - \frac{5}{6}\Delta^5 y_0 + \cdots\right] \quad (4.8)$$

and

$$\left(\frac{d^3y}{dx^3}\right)_{x=x_0} = \frac{1}{h^3}\left[\Delta^3 y_0 - \frac{3}{2}\Delta^4 y_0 + \frac{7}{4}\Delta^5 y_0 + \cdots\right] \quad (4.9)$$

Note that formulas given by (4.7), (4.8) and (4.9) are to be used only to find the derivative at the initial argument, with the arguments being equally spaced. These formulas can also be obtained by using the relation between operators. The relation between the differential and forward differential operators (from: 5 of 2.8) is

$$hD = \log(1 + \Delta) \qquad (4.10)$$

i.e., $$hD = \Delta - \frac{\Delta^2}{2} + \frac{\Delta^3}{2} - \frac{\Delta^4}{4} + \frac{\Delta^5}{5} + \cdots \qquad (4.11)$$

Hence,

$$f'(x) = \frac{1}{h}\left[\Delta f(x) - \frac{\Delta^2}{2}f(x) + \frac{\Delta^3}{3}f(x) - \frac{\Delta^4}{4}f(x) + \frac{\Delta^5}{5}f(x) + \cdots\right]$$

$$\therefore \left(\frac{dy}{dx}\right)_{x=x_0} = \frac{1}{h}\left[\Delta y_0 - \frac{\Delta^2 y_0}{2} + \frac{\Delta^3 y_0}{3} - \frac{\Delta^4 y_0}{4} + \frac{\Delta^5 y_0}{5} + \cdots\right] \quad (4.12)$$

which is the same as Eq. (4.7). For higher order derivatives, the following expression can be used

$$\frac{\left[\log(1+x)\right]^k}{\lfloor k} = \sum_{n=k}^{\infty} \frac{a_k^{(n)} x^n}{\lfloor k} \quad (4.13)$$

where $a_k^{(n)}$ are Stirling's numbers of the first kind, which are given below.

n \ k	1	2	3	4	5	6	7	8
1	1							
2	−1	1						
3	2	−3	1					
4	−6	11	−6	1				
5	24	−50	35	−10	1			
6	−120	274	−225	85	−15	1		
7	720	−1764	1624	−735	175	−21	1	
8	−5040	13,068	−13,132	6769	−1960	322	−28	1

From (4.10), we have

$$h^r D^r = \left[\log(1+D)\right]^r \quad (4.14)$$

where r is a positive integer.

In view of (4.13), Eq. (4.14) becomes

$$D^r = \frac{1}{h^r}\left[a_r^{(r)}\Delta^r + \frac{a_r^{(r+1)}}{r+1}\Delta^{r+1} + \frac{a_r^{(r+2)}}{(r+1)(r+2)}\Delta^{r+2}\right.$$

$$\left. + \frac{a_r^{(r+3)}}{(r+1)(r+2)(r+3)}\Delta^{r+3} + \cdots\right] \quad (4.15)$$

Substituting $r = 2$ in Eq. (4.15), we get

$$D^2 = \frac{1}{h^2}\left[a_2^{(2)}\Delta^2 + \frac{a_2^{(3)}}{3}\Delta^3 + \frac{a_2^{(4)}}{12}\Delta^4 + \frac{a_2^{(5)}}{60}\Delta^5 + \cdots\right]$$

Now substituting the values of Stirling numbers of first kind, we get

$$D^2 = \frac{1}{h^2}\left[\Delta^2 - \Delta^3 + \frac{11}{12}\Delta^4 - \frac{5}{6}\Delta^5 + \cdots\right] \qquad (4.16)$$

Similarly,

$$D^3 = \frac{1}{h^3}\left[\Delta^3 - \frac{3}{2}\Delta^4 + \frac{7}{4}\Delta^5 - \frac{15}{8}\Delta^6 + \ldots\right] \qquad (4.17)$$

Hence,

$$\left(\frac{d^2y}{dx^2}\right)_{x=x_0} = \frac{1}{h^2}\left[\Delta^2 y_0 - \frac{\Delta^3 y_0}{1} + \frac{11}{12}\Delta^4 y_0 - \frac{5}{6}\Delta^5 y_0 + \cdots\right] \qquad (4.18)$$

Similarly, it can be shown that

$$\left(\frac{d^3y}{dx^3}\right)_{x=x_0} = \frac{1}{h^3}\left[\Delta^3 y_0 - \frac{3}{2}\Delta^4 y_0 + \frac{7}{4}\Delta^5 y_0 - \frac{15}{8}\Delta^6 y_0 + \cdots\right] \qquad (4.19)$$

Note: The formulas for computation of $\dfrac{dy}{dx}, \dfrac{d^2y}{dx^2}$ and $\dfrac{d^3y}{dx^3}$ at any argument x_r, when the arguments are equally spaced, are

$$\left(\frac{dy}{dx}\right)_{x=x_r} = \frac{1}{h}\left[\Delta y_r - \frac{\Delta^2 y_r}{2} + \frac{\Delta^3 y_r}{3} - \frac{\Delta^4 y_r}{4} + \cdots\right] \qquad (4.20)$$

$$\left(\frac{d^2y}{dx^2}\right)_{x=x_r} = \frac{1}{h^2}\left[\Delta^2 y_r - \Delta^3 y_r + \frac{11}{12}\Delta^4 y_r - \frac{5}{6}\Delta^5 y_r + \cdots\right] \qquad (4.21)$$

and

$$\left(\frac{d^3y}{dx^3}\right)_{x=x_r} = \frac{1}{h^3}\left[\Delta^3 y_r - \frac{3}{2}\Delta^4 y_r + \frac{7}{4}\Delta^5 y_r - \frac{15}{8}\Delta^6 y_r + \cdots\right] \qquad (4.22)$$

which can be obtained from (4.11), (4.16) and (4.17).

4.4   PROCEDURE TO FIND THE DERIVATIVE OF A FUNCTION AT A POINT WHEN THE ARGUMENTS ARE EQUALLY SPACED

Step 1:   For the given set of values, construct the forward difference table and note down the forward differences such as Δy_0, $\Delta^2 y_0$, ...

Step 2: Note down the value of x at which the derivative is required.

Step 3: Also note down the value of h.

Step 4: Compute $u = \dfrac{x - x_0}{h}$, where x_0 is the initial argument.

Step 5: Write down the relevant formula to compute the derivative.

Step 6: Substitute the values obtained in Steps 1–4 in the formula given in Step 5.

Step 7: Simplify the terms involved in the expression obtained in Step 6, which gives the required result.

SOLVED PROBLEMS

Problem 1: Find the derivative of a function tabulated below at the point $x = 2.2$.

x	2	2.2	2.4	2.6	2.8	3.0
$f(x)$	5	7.448	10.424	13.976	18.152	2.37

Solution: The forward difference table of the tabulated function is given below.

x	y	Δy	$\Delta^2 y$	$\Delta^3 y$	$\Delta^4 y$	$\Delta^5 y$
2.0	5					
		$\to 2.448 = \Delta y_0$				
2.2	7.448		$\to 0.528 = \Delta^2 y_0$			
		$\to 2.976$		$\to 0.048 = \Delta^3 y_0$		
2.4	10.424		$\to 0.576$		$\to 0 = \Delta^4 y_0$	
		$\to 3.552$		$\to 0.048$		$\to 0 = \Delta^5 y_0$
2.6	13.976		$\to 0.624$		$\to 0$	
		$\to 4.176$		$\to 0.048$		
2.8	18.152		$\to 0.672$			
		$\to 4.848$				
3.0	23					

The derivative of y is required at 2.2; hence, $x = 2.2$. The interval of differencing h is 0.2. Therefore, $u = \dfrac{x - x_0}{h} = \dfrac{2.2 - 2}{0.2} = 1$.

By Eq. (4.4), we have

$$\left(\frac{dy}{dx}\right) = \frac{1}{h}\left[\Delta y_0 - \frac{2u-1}{\lfloor 2}\Delta^2 y_0 + \frac{(3u^2 - 6u + 2)}{6}\Delta^3 y_0 + \cdots\right]$$

Substituting the values of h, u and forward differences in the above equation, we obtain

$$\left(\frac{dy}{dx}\right)_{x=2.2} = \frac{1}{0.2}\left[2.448 - \frac{2(1)-1}{2}(0.528) + \frac{(3(1)^2 - 6(1) + 2)}{6}0.048\right]$$

$$= 13.52$$

Thus, the derivative of the function at $x = 2.2$ is 13.52.

Problem 2: From the following tabulated values of x and $y = f(x)$, find $f'(1)$.

x	0	1	2	3	4
$f(x)$	4	5	14	37	80

Solution: The forward difference table for the given data is given below.

x	y	Δy	$\Delta^2 y$	$\Delta^3 y$	$\Delta^4 y$
0	4				
		$\to 1 = \Delta y_0$			
1	5		$\to 8 = \Delta^2 y_0$		
		$\to 9 = \Delta y_1$		$\to 6 = \Delta^3 y_0$	
2	14		$\to 14 = \Delta^2 y_1$		$\to 0 = \Delta^4 y_0$
		$\to 23$		$\to 6 = \Delta^3 y_1$	
3	37		$\to 20$		
		$\to 43$			
4	80				

We have to compute $f'(x)$ at one of the argument. By Eq. (4.20) of note appearing in section 4.3, we have,

$$\left(\frac{dy}{dx}\right)_{x=x_1} = \frac{1}{h}\left[\Delta y_1 - \frac{\Delta^2 y_1}{2} + \frac{\Delta^3 y_1}{3} - \frac{\Delta^4 y_1}{4} + \cdots\right]$$

$$\therefore f'(1) = \frac{1}{1}\left[9 - \frac{14}{2} + \frac{6}{3}\right] = 9 - 7 + 2 = 4$$

since $h = 1$ and $\Delta y_1 = 9$, $\Delta^2 y_1 = 14$, $\Delta^3 y_1 = 6$.

Hence, the derivative of $f(x)$ at $x = 1$ in $f'(1) = 4$.

Other Method:

Forward Difference table

x	y	Δy	$\Delta^2 y$	$\Delta^3 y$	$\Delta^4 y$
0	4				
		$\to 1 = \Delta y_0$			
1	5		$\to 8 = \Delta^2 y_0$		
		$\to\ 9$		$\to 6 = \Delta^3 y_0$	
2	14		$\to 14$		$\to 0 = \Delta^4 y_0$
		$\to\ 23$		$\to\ 6$	
3	37		$\to 20$		
		$\to\ 43$			
4	80				

The derivative of y is required at 1; hence, $x = 1$. The value of $h = 1$.

Therefore, $u = \dfrac{x - x_0}{h} = \dfrac{1 - 0}{1} = 1.$

By Eq. (4.4), we have,

$$\left(\frac{dy}{dx}\right) = \frac{1}{h}\left[\Delta y_0 + \frac{(2u-1)}{\lfloor 2} \Delta^2 y_0 + \frac{(3u^2 - 6u + 2)}{\lfloor 3} \Delta^3 y_0 + \cdots\right]$$

Substituting the values of h, u, Δy_0, $\Delta^2 y_0$ and $\Delta^3 y_0$, we have

$$\left(\frac{dy}{dx}\right)_{x=1} = \frac{1}{1}\left[1 - \frac{2(1)-1}{2}(0.8) + \frac{(3(1)^2 - 6(1) + 2)}{6}(0.6)\right] = 1[1 + 4 - 1] = 4$$

Hence, $f'(1) = 4$.

Problem 3: Find $f'(0.25)$, given

x	0	1	2	3	4
$f(x)$	1	1	15	40	85

Solution: The forward difference table for the tabulated values is given below.

x	y	Δy	$\Delta^2 y$	$\Delta^3 y$	$\Delta^4 y$
0	1				
		$\to 0 = \Delta y_0$			
1	1		$\to 14 = \Delta^2 y_0$		
		$\to \ 14$		$\to -3 = \Delta^3 y_0$	
2	15		$\to \ 11$		$\to 12 = \Delta^4 y_0$
		$\to \ 25$		$\to \ 9$	
3	40		$\to \ 20$		
		$\to \ 45$			
4	85				

The derivative of y is required at 0.25; hence, $x = 0.25$. The interval of differencing $h = 1$ and $x_0 = 0$. Therefore,

$$u = \frac{x - x_0}{h} = \frac{0.25 - 0}{1} = 0.25$$

The derivative formula, given by (4.4), is

$$\left(\frac{dy}{dx}\right) = \frac{1}{h}\left[\Delta y_0 + \frac{(2u-1)}{2}\Delta^2 y_0 + \frac{(3u^2 - 6u + 2)}{6}\Delta^3 y_0\right.$$
$$\left. + \frac{(4u^3 - 18u^2 + 22u - 6)}{24}\Delta^4 y_0 + \cdots\right]$$

Substituting the values of $u = 0.25$, $h = 1$, Δy_0, $\Delta^2 y_0$, $\Delta^3 y_0$, and $\Delta^4 y_0$ in the above equation, we get

$$\left(\frac{dy}{dx}\right)_{x=0.25} = \frac{1}{1}\left[0 + \frac{(2(0.25)-1)}{2}(14) + \frac{(3(0.25)^2 - 6(0.25)+2)}{6}(-3)\right.$$

$$\left. + \frac{(4(0.25)^3 - 18(0.25)^2 + 22(0.25)-6)}{24}(12)\right] = -4.625$$

Hence $f'(0.25) = -4.625$.

Problem 4: Find $\dfrac{dy}{dx}, \dfrac{d^2 y}{dx^2}$ and $\dfrac{d^3 y}{dx^3}$ of the following tabulated function at $x = 2$.

x	2	3	4	5	6
$f(x)$	12	31	68	129	220

Solution: The forward difference table for the given data is shown below.

x	y	Δy	$\Delta^2 y$	$\Delta^3 y$	$\Delta^4 y$
2	12				
		$\to 19 = \Delta y_0$			
3	31		$\to 18 = \Delta^2 y_0$		
		$\to$ 37		$\to 6 = \Delta^3 y_0$	
4	68		$\to$ 24		$\to 0 = \Delta^4 y_0$
		$\to$ 61		$\to$ 6	
5	129		$\to$ 30		
		$\to$ 91			
6	220				

The first three derivatives are required at 2, which is an initial argument. Hence, we use the formulas given by Eqs. (4.12), (4.18) and (4.19) to compute these derivatives. The formula given by Eq. (4.12) is

$$\left(\frac{dy}{dx}\right)_{x=x_0} = \frac{1}{h}\left[\Delta y_0 - \frac{\Delta^2 y_0}{2} + \frac{\Delta^3 y_0}{3} - \frac{\Delta^4 y_0}{4} + \cdots\right]$$

$$\therefore \qquad \left(\frac{dy}{dx}\right)_{x=2} = \frac{1}{1}\left[19 - \frac{18}{2} + \frac{6}{3} - 0\right] = 12$$

Similarly, the formulas to compute the second and third order derivatives are

$$\left(\frac{d^2 y}{dx^2}\right)_{x=x_0} = \frac{1}{h^2}\left[\Delta^2 y_0 - \Delta^3 y_0 + \frac{11}{12}\Delta^4 y_0 + \cdots\right]$$

$$\left(\frac{d^3 y}{dx^3}\right)_{x=x_0} = \frac{1}{h^3}\left[\Delta^3 y_0 + \frac{3}{2}\Delta^4 y_0 + \cdots\right]$$

Therefore,

$$\left(\frac{d^2 y}{dx^2}\right)_{x=2} = \frac{1}{1^2}[18 - 6] = 12$$

and

$$\left(\frac{d^3 y}{dx^3}\right)_{x=2} = \frac{1}{1^3}[6] = 6$$

Thus, $\dfrac{dy}{dx}, \dfrac{d^2 y}{dx^2}$ and $\dfrac{d^3 y}{dx^3}$ at $x = 2$ are 12, 12 and 6 respectively.

Problem 5: If $f(0) = 3, f(1) = 5, f(2) = 19, f'(0) = 2$ and $f'(1) = 4$ for a function $y = f(x)$, then find $\Delta^3 f(0)$ and $\Delta^4 f(0)$.

Solution: The forward difference table for the given data is shown below.

x	y	Δy	$\Delta^2 y$
0	3		
		$\rightarrow 2 = \Delta y_0$	
1	5		$\rightarrow 12 = \Delta^2 y_0$
		$\rightarrow \ 14$	
2	19		

Let $\Delta^3 f(0) = l$, that is, $\Delta^3 y_0 = l$ and $\Delta^4 f(0) = m$, that is, $\Delta^4 y_0 = m$. For the problem at our disposal, the initial argument $x_0 = 0$, interval of differencing $h = 1$ and $u = \dfrac{x - x_0}{h} = \dfrac{x - 0}{1} = x$, we know that

$$\left(\frac{dy}{dx}\right)_{x=x_0} = \frac{1}{h}\left[\Delta y_0 - \frac{1}{2}\Delta^2 y_0 + \frac{1}{3}\Delta^3 y_0 - \frac{1}{4}\Delta^4 y_0\right]$$

$$\therefore \qquad \left(\frac{dy}{dx}\right)_{x=0} = \frac{1}{1}\left[2 - \frac{1}{2}(12) + \frac{l}{3} - \frac{m}{4}\right]$$

Since $f'(0) = 2$, we have

$$2 = -4 + \frac{l}{3} - \frac{m}{4}$$

that is, $4l - 3m = 72$ \hfill (1)

Now we compute $\dfrac{dy}{dx}$ at $x = 1$ involving l and m. The formula for computing $\dfrac{dy}{dx}$ at any x is

$$\left(\frac{dy}{dx}\right) = \frac{1}{h}\left[\Delta y_0 + \frac{(2u-1)}{\lfloor 2}\Delta^2 y_0 + \frac{(3u^2 - 6u + 2)}{\lfloor 3}\Delta^3 y_0\right.$$

$$\left. + \frac{(4u^3 - 18u^2 + 22u - 6)}{\lfloor 4}\Delta^4 y_0 + \cdots\right]$$

where $u = \dfrac{x - x_0}{h} = \dfrac{x - 0}{1} = x$

$$\therefore \qquad \left(\frac{dy}{dx}\right)_{x=1} = \frac{1}{1}\left[2 + \frac{1}{2}(12) - \frac{1}{6}(l) + \frac{1}{12}(m)\right] = 8 - \frac{l}{6} + \frac{m}{12}$$

Since $f'(1) = 4$, the above equation becomes

$$4 = 8 - \frac{l}{6} + \frac{m}{12}$$

that is, $2l - m = 48$ \hfill (2)

Solving (1) and (2) for l and m, we obtain $l = 36$ and $m = 24$. Hence, $\Delta^3 f(0) = 36$ and $\Delta^4 f(0) = 24$.

Problem 6: A rod is rotating in a plane. The following table gives the angle θ (in radians) through which the rod has turned about an axis for various values of time t in seconds. Find the angular velocity of the rod when $t = 0.2$ s.

t	0	0.4	0.8	1.2	1.6	2.0
θ	0	0.14	0.26	0.38	0.49	0.62

Solution: The forward difference table for the given data is shown below.

t	θ	$\Delta\theta$	$\Delta^2\theta$	$\Delta^3\theta$	$\Delta^4\theta$	$\Delta^5\theta$
0	0					
		$\to 0.14 = \Delta\theta_0$				
0.4	0.14		$\to -0.02 = \Delta^2\theta_0$			
		$\to 0.12$		$\to 0.02 = \Delta^3\theta_0$		
0.8	0.26		$\to 0$		$\to -0.03 = \Delta^4\theta_0$	
		$\to 0.12$		$\to -0.01$		$\to 0.07 = \Delta^5\theta_0$
1.2	0.38		$\to -0.01$		$\to 0.04$	
		$\to 0.11$		$\to 0.03$		
1.6	0.49		$\to 0.02$			
		$\to 0.13$				
2.0	0.62					

Replacing the dependent variable y by θ and the independent variable x by t in Eq. (4.4), we have

$$\frac{d\theta}{dt} = \frac{1}{h}\left[\Delta\theta_0 + \frac{(2u-1)}{\lfloor 2}\Delta^2\theta_0 + \frac{(3u^2-6u+2)}{\lfloor 3}\Delta^3\theta_0 \right.$$

$$+ \frac{(4u^3-18u^2+22u-6)}{\lfloor 4}\Delta^4\theta_0$$

$$\left. + \frac{(5u^4-40u^3+105u^2-100u+24)}{\lfloor 5}\Delta^5\theta_0 + \cdots \right]$$

where $u = \dfrac{t - t_0}{h}$. Here $t_0 = 0$, $t = 0.2$ and $h = 0.4$. Hence, $u = 0.5$.

Substituting the values of u, h and forward differences in Eq.(1), we get

$$\left(\frac{d\theta}{dt}\right)_{t=0.2} = \frac{1}{0.4}\left[0.14 + \frac{(2(0.5)-1)}{\lfloor 2}(-0.02) + \frac{(3(0.5)^2 - 6(0.5)+2)}{\lfloor 3}(0.02)\right.$$

$$+ \frac{(4(0.5)^3 - 18(0.5)^2 + 22(0.5) - 6)}{\lfloor 4}(-0.03)$$

$$\left. + \frac{(5(0.5)^4 - 40(0.5)^3 + 105(0.5)^2 - 100(0.5) + 24)}{\lfloor 5}(0.07) + \cdots\right]$$

$$= 0.3383125.$$

4.5 DERIVATIVES USING NEWTON'S BACKWARD DIFFERENCE INTERPOLATION FORMULA

Suppose that $y_0, y_1, y_2, \ldots, y_n$ are $(n+1)$ values of a function $y = f(x)$ at $(n+1)$ equally spaced arguments $x_0, x_0 + h, x_0 + 2h, \ldots, x_0 + nh$. Then by Newton's backward difference interpolation formula, we have

$$y = y_n + u\nabla y_n + \frac{u(u+1)}{\lfloor 2}\nabla^2 y_n + \frac{u(u+1)(u+2)}{\lfloor 3}\nabla^3 y_n + \cdots$$

$$+ \frac{u(u+1)(u+2)\ldots(u+n-1)}{\lfloor n}\nabla^n y_n \qquad (4.23)$$

where

$$u = \frac{x - x_n}{h} \qquad (4.24)$$

Differentiating (4.23) with respect to x, we get

$$\frac{dy}{dx} = \left[\nabla y_n + \frac{(2u+1)}{\lfloor 2}\nabla^2 y_n + \frac{(3u^2 + 6u + 2)}{\lfloor 3}\nabla^3 y_n\right.$$

$$+ \frac{(4u^3 + 18u^2 + 22u + 6)}{\lfloor 4}\nabla^4 y_n$$

$$\left. + \frac{(5u^4 - 40u^3 + 105u^2 - 100u + 24)}{\lfloor 5}\nabla^5 y_n + \cdots\right]\cdot\frac{du}{dx}$$

From (4.24), we have $\dfrac{du}{dx} = \dfrac{1}{h}$. Hence,

$$\frac{dy}{dx} = \frac{1}{h}\left[\nabla y_n + \frac{(2u+1)}{\lfloor 2} \nabla^2 y_n + \frac{(3u^2+6u+2)}{\lfloor 3} \nabla^3 y_n \right.$$

$$+ \frac{(4u^3+18u^2+22u+6)}{\lfloor 4} \nabla^4 y_n$$

$$\left. + \frac{(5u^4-40u^3+105u^2-100u+24)}{\lfloor 5} \nabla^5 y_n + \cdots \right] \qquad (4.25)$$

Equation (4.25) is usually used to find the derivative of y at a point near to the end of the argument or end of the argument of tabulated arguments to incorporate the variations at the end of the argument, because the variation of function at the beginning may differ from the variation at the end of values of independent variable. Of course, it can also be used to find the derivative at any point of the independent variable.

Differentiating (4.25) successively two times, we get

$$\frac{d^2 y}{dx^2} = \frac{1}{h^2}\left[\nabla^2 y_n + (u+1)\nabla^3 y_n + \frac{(12u^2+36u+22)}{\lfloor 4} \nabla^4 y_n \right.$$

$$\left. + \frac{(204u^3+120u^2+210u+100)}{\lfloor 5} \nabla^5 y_n + \cdots \right] \qquad (4.26)$$

and

$$\frac{d^3 y}{dx^3} = \frac{1}{h^3}\left[\nabla^3 y_n + \frac{(24u+36)}{\lfloor 4} \nabla^4 y_n + \frac{(60u^2-240u+210)}{\lfloor 5} \nabla^5 y_n + \cdots \right]$$

$$(4.27)$$

4.6   DERIVATIVE OF A FUNCTION AT AN END OF THE TABULATED ARGUMENT

Substituting $u = \dfrac{x_n - x_n}{h} = 0$, in Eqs. (4.25), (4.26) and (4.27), we have the following formulas to find $\dfrac{dy}{dx}, \dfrac{d^2 y}{dx^2}$ and $\dfrac{d^3 y}{dx^3}$ at $x = x_n$.

$$\left(\frac{dy}{dx}\right)_{x=x_n} = \frac{1}{h}\left[\nabla y_n + \frac{1}{2}\nabla^2 y_n + \frac{1}{3}\nabla^3 y_n + \frac{1}{4}\nabla^4 y_n + \ldots\right] \quad (4.28)$$

$$\left(\frac{d^2 y}{dx^2}\right)_{x=x_n} = \frac{1}{h^2}\left[\nabla^2 y_n + \nabla^3 y_n + \frac{11}{12}\nabla^4 y_n + \ldots\right] \quad (4.29)$$

$$\left(\frac{d^3 y}{dx^3}\right)_{x=x_n} = \frac{1}{h^3}\left[\nabla^3 y_n + \frac{3}{2}\nabla^4 y_n + \frac{7}{4}\nabla^5 y_n + \cdots\right] \quad (4.30)$$

Note that the above formulas are to be used only when y', y'' and y''' are required at the end of the argument of tabulated values.

4.7 PROCEDURE TO FIND THE DERIVATIVE USING NEWTON'S BACKWARD INTERPOLATION

Step 1: For the given set of values, construct the backward difference table and note down the backward differences such as $\nabla y_n, \nabla^2 y_n, \nabla^3 y_n$ etc.

Step 2: Note down the value of x at which the derivative is required.

Step 3: Also note down the value of h.

Step 4: Compute $u = \dfrac{x - x_n}{h}$, where x_n is the end argument in the table.

Step 5: Write down the formula to find the derivative using backward interpolation.

Step 6: Substitute the values obtained in Steps 1–4 in the formula given in Step 5.

Step 7: Simplify the terms in the expression obtained in Step 6, which gives the required result.

SOLVED PROBLEMS

Problem 1: Given that

x	2	4	6	8
$f(x)$	7	19	39	67

Find $f'(8)$.

Solution: As the derivative is required at the end of the argument of tabulated values, we use Eq. (4.28), that is,

$$\left(\frac{dy}{dx}\right)_{x=x_n} = \frac{1}{h}\left[\nabla y_n + \frac{\nabla^2 y_n}{2} + \frac{\nabla^3 y_n}{3} + \frac{\nabla^4 y_n}{4} + \cdots\right] \quad (1)$$

To know the backward differences, we construct the backward difference table, which is given below.

x	y	∇y	$\nabla^2 y$	$\nabla^3 y$
2	7			
		$\rightarrow$ 12		
4	19		$\rightarrow$ 8	
		$\rightarrow$ 20		$\rightarrow 0 = \nabla^3 y_n$
6	39		$\rightarrow 8 = \nabla^2 y_n$	
		$\rightarrow 28 = \nabla y_n$		
8	$67 = y_n$			

Substituting $h = 2$ and backward differences ∇y_n, $\nabla^2 y_n$,... in (1), we get

$$\left(\frac{dy}{dx}\right)_{x=8} = \frac{1}{2}\left[28 + \frac{8}{2} + \frac{0}{3}\right] = 16$$

Hence, $f'(8) = 16$.

Problem 2: A slider is moving along a fixed straight line. The distance s (in cm) moved by the slider at different times t (in s) is given below. Find the velocity and acceleration of the slider when $t = 4.5$ s.

t	0	1	2	3	4	5
s	0	4.5	6.2	8.3	9.4	10.6

Solution: We know that the velocity is the rate of change of displacement in unit time, and acceleration is the rate of change of velocity.

The velocity and acceleration of the slider are $\dfrac{ds}{dt}$ and $\dfrac{d^2 s}{dt^2}$, respectively, at any time t. Hence, we need to compute $\left(\dfrac{ds}{dt}\right)_{t=4.5}$ and $\left(\dfrac{d^2 s}{dt^2}\right)_{t=4.5}$.

Since 4.5 is near to the end of the argument of tabulated values, we use the formula given by Eq. (4.25) or Eq. (4.26) to compute velocity and acceleration of the slider at $t = 4.5$. The following table is the backward difference table for the given data.

t	x	∇s	$\nabla^2 s$	$\nabla^3 s$	$\nabla^4 s$	$\nabla^5 s$
0	0					
		$\to$ 4.5				
1	4.5		$\to$ -2.8			
		$\to$ 1.7		$\to$ 3.2		
2	6.2		$\to$ 0.4		$\to$ -4.6	
		$\to$ 2.1		$\to$ -1.4		$\to$ $7.1 = \nabla^5 s_n$
3	8.3		$\to$ -1		$\to$ $2.5 = \nabla^4 s_n$	
		$\to$ 1.1		$\to$ $1.1 = \nabla^3 s_n$		
4	9.4		$\to$ $0.1 = \nabla^2 s_n$			
		$\to$ $1.2 = \nabla s_n$				
5	$10.6 = s_n$					

Here $t_n = 5$ and $t = 4.5$; hence, $u = \dfrac{t - t_n}{h} = \dfrac{4.5 - 5}{1} = -0.5$.

Further, $\nabla s_n = 1.2$, $\nabla^2 s_n = 0.1$, $\nabla^3 s_n = 1.1$, $\nabla^4 s_n = 2.5$, $\nabla^5 s_n = 7.1$

Having s as the dependent and t as the independent variables, Eqs. (4.25) and (4.26) become

$$\frac{ds}{dt} = \frac{1}{h}\left[\nabla s_n + \frac{(2u+1)}{\lfloor 2}\nabla^2 s_n + \frac{(3u^2 + 6u + 2)}{\lfloor 3}\nabla^3 s_n \right.$$

$$+ \frac{(4u^3 + 18u^2 + 22u + 6)}{\lfloor 4}\nabla^4 s_n$$

$$\left. + \frac{(5u^4 - 40u^3 + 105u^2 - 100u + 24)}{\lfloor 5}\nabla^5 s_n + \cdots \right]$$

$$\frac{d^2 s}{dt^2} = \frac{1}{h^2}\left[\nabla^2 s_n + (u+1)\nabla^3 s_n + \frac{(12u^2 + 36u + 22)}{\lfloor 4}\nabla^4 s_n \right.$$

$$\left. + \frac{(204u^3 + 120u^2 + 210u + 100)}{\lfloor 5}\nabla^5 s_n + \cdots \right]$$

Hence, the velocity and acceleration of the slider at $t = 4.5$ s are

$$\left(\frac{ds}{dt}\right)_{t=4.5} = \frac{1}{1}\left[1.2 + \{2(-0.5)+1\}\frac{0.1}{\lfloor 2} + \frac{\{3(-0.5)^2 + 6(-0.5)+2\}}{\lfloor 3}(1.1)\right.$$

$$+ \frac{\{4(-0.5)^3 + 18(-0.5)^2 + 22(-0.5)+6\}}{\lfloor 4}(2.5)$$

$$\left. + \frac{\{5(-0.5)^4 - 40(-0.5)^3 + 105(-0.5)^2 - 100(-0.5)+24\}}{\lfloor 5}(.1) + \cdots\right]$$

$$= 7.29578$$

$$\left(\frac{d^2 s}{dt^2}\right)_{t=4.5} = \frac{1}{1^2}\left[0.1 + \{(-0.5)+1\}(1.1) + \frac{(12(-0.5)^2 + 36(-0.5)+22)}{\lfloor 4}(2.5)\right.$$

$$\left. + \frac{(20(-0.5)^3 + 120(-0.5)^2 + 210(-0.5)+100)}{\lfloor 5}(7.1)\right]$$

$$= 2.79166.$$

Problem 3: Population of a certain city for different years is as follows:

Year	1931	1941	1951	1961	1971
Population (in thousands)	20.63	25.72	28.13	30.62	32.72

Find the growth rate of the population of the city for the year 1966.

Solution: Let the population y (in thousands) and year x be connected by a relation $y = f(x)$. Then the growth rate of the population is $\dfrac{dy}{dx}$. To know the growth rate of the population for the year 1966, we have to find $\dfrac{dy}{dx}$ at $x = 1966$. As the year 1966 is close to the end of the argument of tabulated values, we use the formula obtained from Newton's backward interpolation formula.

The backward difference table of the given data is as follows:

x	$y = f(x)$	∇y	$\nabla^2 y$	$\nabla^3 y$	$\nabla^4 y$
1931	20.63				
		→ 5.09			
1941	25.72		→ −2.68		
		→ 2.41		→ 2.76	

1951	28.13	$\to 0.08$		$\to$ $\begin{array}{c} -3.23 \\ = \nabla^4 y_n \end{array}$
		$\to 2.49$	$\to -0.47 = \nabla^3 y_n$	
1961	30.62	$\to -0.39 = \nabla^2 y_n$		
		$\to 2.1 = \nabla y_n$		
1971	$32.72 = y_n$			

Here $h = 10$, $x_n = 1971$ (end of argument) and $x = 1966$; hence,

$$u = \frac{1966 - 1971}{10} = -0.25.$$ The different order backward differences are given in the difference table. Substituting the values of u, h and backward differences in Eq. (4.25), we get

$$\left(\frac{dy}{dx}\right)_{x=1966} = \frac{1}{10}\left[2.1 + \frac{\{2(-0.5)+1\}}{\lfloor 2}(-0.39) + \frac{\{3(-0.5)^2 + 6(-0.5) + 2\}}{\lfloor 3} \times\right.$$

$$\left. (-0.47) + \frac{\{4(-0.5)^3 + 18(-0.5)^2 + 22(-0.5) + 6\}}{\lfloor 4}(-3.23)\right]$$

$$= 0.2254.$$

Problem 4: A rocket is launched from the ground. Its average velocities recorded are as follows:

t (in s)	0	10	20	30	40	50
V (km/s)	0	200	600	1200	2000	3000

Find the acceleration at $t = 50$ s.

Solution: Acceleration is the rate of change of velocity in unit time. Therefore, we need to find $\dfrac{dV}{dt}$ at $t = 50$ s to find the acceleration of the rocket at $t = 50$ s. As $t = 50$ is the end of the argument of tabulated values, we use formula (4.28) with appropriate change of variables to compute the acceleration of rocket at $t = 50$ s. Formula (4.28) with appropriate change of variables is

$$\frac{dV}{dt} = \frac{1}{h}\left[\nabla V_n + \frac{1}{2}\nabla^2 V_n + \frac{1}{3}\nabla^3 V_n + \frac{1}{4}\nabla^4 V_n + \frac{1}{5}\nabla^5 V_n + \cdots\right] \quad (1)$$

Here, $h = 10$. To know the backward differences, we compute the backward difference table which is given below.

t	V	∇V	$\nabla^2 V$	$\nabla^3 V$	$\nabla^4 V$	$\nabla^4 V$
0	0					
		$\rightarrow$ 200				
10	200		$\rightarrow$ 200			
		$\rightarrow$ 400		$\rightarrow$ 0		
20	600		$\rightarrow$ 200		$\rightarrow$ 0	
		$\rightarrow$ 600		$\rightarrow$ 0		$\rightarrow 0 = \nabla^5 V_n$
30	1200		$\rightarrow$ 200		$\rightarrow 0 = \nabla^4 V_n$	
		$\rightarrow$ 800		$\rightarrow 0 = \nabla^3 V_n$		
40	2000		$\rightarrow 200 = \nabla^2 V_n$			
	$\rightarrow 1000 = \nabla V_n$					
50	$3000 = V_n$					

$$\therefore \left(\frac{dV}{dt}\right)_{t=50} = \frac{1}{10}\left[1000 + \frac{200}{2} + 0 + 0 + 0\right] = 110 \text{ m/s}^2.$$

4.8 DERIVATIVES USING CENTRAL DIFFERENCE FORMULA

We know that the central difference interpolation formulas are used to find the value of a function at a particular point of independent variable, which is near the middle value of the given arguments (when the arguments are in ascending order). Similarly, if we need to find $\frac{dy}{dx}$ or higher order derivatives near the middle of the tabulated arguments, then we use the central difference interpolation formula. We discuss the derivative using the Stirling's formula and obtain a formula to find the derivative of a function using Stirling central difference interpolation formula.

Suppose $y_0, y_1, y_2, \ldots, y_n$ be $(n + 1)$ values of a function $y = f(x)$ at x_0, $x_1 = x_0 + h, x_2 = x_1 + 2h, \ldots, x_n = x_0 + nh$, such that $x_0 < x_1 < x_2 < \cdots < x_n$.

If n is even, then we consider $x_0 + \dfrac{n}{2}h$ as the middle of the argument; otherwise, we consider $x_0 + \dfrac{n-1}{4}h$ or $x_0 + \dfrac{n+1}{2}h$ as the middle of the argument.

We change the scale of independent variable by $u = \dfrac{x - \text{middle of the argument}}{h}$. The value of u at the arguments x_0, $x_0 + h, x_0 + 2h, \ldots, x_0 + nh$ are $\ldots -3, -2, -1, 0, 1, 2, 3\ldots$. Note that 0 corresponds to the middle argument. We rename the middle argument as x_0 and its successive preceding arguments as $x_0 - h, x_0 - 2h, \ldots$, and succeeding arguments as $x_0 + h, x_0 + 2h, \ldots$. The values of $y = f(x)$ at these points is denoted by $\ldots, y_{-3}, y_{-2}, y_{-1}, y_0, y_1, y_2, \ldots$ Now the Stirling's formula is

$$y = y_0 + u\frac{\Delta y_0 + \Delta y_{-1}}{2} + \frac{u^2}{2}\Delta^2 y_{-1} + \frac{u^3 - u}{12}(\Delta^3 y_{-1} + \Delta^3 y_{-2})$$

$$+ \frac{u^4 - u^2}{24}\Delta^4 y_{-2} + \frac{u^5 - 5u^3 + 4u}{120}\left(\frac{\Delta^5 y_{-3} + \Delta^5 y_{-2}}{2}\right) + \cdots \quad (4.31)$$

where

$$u = \frac{x - x_0}{h} \qquad (4.32)$$

Note that x_0 is the middle of the argument.
Differentiating (1) w.r.t. x, we get

$$\frac{dy}{dx} = \frac{1}{h}\left[\frac{1}{2}(\Delta y_0 + \Delta y_{-1}) + u\Delta^2 y_{-1} + \frac{3u^2 - 1}{12}(\Delta^3 y_{-1} + \Delta^3 y_{-2})\right.$$

$$\left. + \frac{1}{12}(2u^3 - u)\Delta^4 y_{-2} + \frac{5u^4 - 15u^2 + 4}{240}(\Delta^5 y_{-2} + \Delta^5 y_{-3}) + \cdots\right]$$

$$(4.33)$$

Again differentiating (3) successively two times, we obtain

$$\frac{d^2 y}{dx^2} = \frac{1}{h^2}\left[\Delta^2 y_{-1} + \frac{u}{2}(\Delta^3 y_{-1} + \Delta^3 y_{-2}) + \frac{1}{12}(6u^2 - 1)\Delta^4 y_{-2}\right.$$

$$\left. + \frac{2u^3 - 3u}{24}(\Delta^5 y_{-2} + \Delta^5 y_{-3}) + \cdots\right] \tag{4.34}$$

and

$$\frac{d^3 y}{dx^3} = \frac{1}{h^3}\left[\frac{1}{2}(\Delta^3 y_{-1} + \Delta^3 y_{-2}) + u\Delta^4 y_{-2}\right.$$

$$\left. + \frac{1}{8}(2u^2 - 1)(\Delta^5 y_{-2} + \Delta^5 y_{-3}) + \cdots\right] \tag{4.35}$$

The formulas for computing the derivatives $\dfrac{dy}{dx}, \dfrac{d^2 y}{dx^2}$ and $\dfrac{d^3 y}{dx^3}$ at the middle of the argument of tabulated values are

$$\left(\frac{dy}{dx}\right)_{x=x_0} = \frac{1}{h}\left[\frac{1}{2}(\Delta y_0 + \Delta y_{-1}) - \frac{1}{12}(\Delta^3 y_{-1} + \Delta^3 y_{-2})\right.$$

$$\left. + \frac{1}{60}(\Delta^5 y_{-2} + \Delta^5 y_{-3}) + \cdots\right] \tag{4.36}$$

$$\left(\frac{d^2 y}{dx^2}\right)_{x=x_0} = \frac{1}{h^2}\left[\Delta^2 y_{-1} - \frac{1}{12}\Delta^4 y_{-2} + \frac{1}{90}\Delta^6 y_{-3} + \cdots\right] \tag{4.37}$$

$$\left(\frac{d^3 y}{dx^3}\right)_{x=x_0} = \frac{1}{h^3}\left[\frac{1}{2}(\Delta^3 y_{-1} + \Delta^3 y_{-2}) - \frac{1}{8}(\Delta^5 y_{-2} + \Delta^5 y_{-3}) + \cdots\right] \tag{4.38}$$

We can also find the formulas to find the derivative using Bessel's and other central difference formulas following the method described above.

4.9 PROCEDURE TO FIND THE DERIVATIVE USING STIRLING'S FORMULA

Step 1: First find the middle value of the arguments given in the table and it is supposed to be x_0.

Step 2: Note down the value of h.

Step 3: Transform the independent variable x (or the symbol of independent variable) by the relation

$$u = \frac{x - x_0}{h}$$

Step 4: Construct the central difference table having the first column as given arguments, second column having u values corresponding to these arguments, third column having y values and successive columns having forward differences of different order (which are equivalent to central differences).

Step 5: Next draw the path of Stirling's formula along which we have the differences involved in the Stirling's formula. Note down the values of $\Delta y_0, \Delta y_{-1}, \Delta^2 y_{-1}, \Delta^3 y_{-1}, \Delta^3 y_{-2}, \ldots$

Step 6: Note down the x value at which the derivative of y is required.

Step 7: Compute $u = \dfrac{x - x_0}{h}$.

Step 8: Write the formula to compute the derivative using the Stirling's formula.

Step 9: Substitute the values of u, h and forward differences in the formula given in Step 8.

Step 10: Simplify the terms obtained in Step 9. The required derivative is obtained numerically.

SOLVED PROBLEMS

Problem 1: From the following data, obtain the first derivative $\dfrac{dy}{dx}$ at $x = 5$.

x	0	2	4	6	8
y	3	7	51	183	451

Solution: Here the arguments are equally spaced with interval of differences $h = 2$. We use the following change of scale for the independent variable.

$$u = \frac{x - 4}{2} \tag{1}$$

The central difference table is shown below.

x	u	y_u	Δy_u	$\Delta^2 y_u$	$\Delta^3 y_u$	$\Delta^4 y_u$
0	−2	$3 = y_{-2}$				
			$\rightarrow 4 = \Delta y_{-2}$			
2	−1	$7 = y_{-2}$		$\rightarrow 40 = \Delta^2 y_{-2}$		
			$\rightarrow 44 = \Delta y_{-1}$		$\rightarrow 48 = \Delta^3 y_{-2}$	
4	0	$51 = y_0$		$\rightarrow 88 = \Delta^2 y_{-1}$		$\rightarrow 0 = \Delta^4 y_{-2}$
			$\rightarrow 132 = \Delta y_0$		$\rightarrow 48 = \Delta^3 y_{-1}$	
6	1	$183 = y_1$		$\rightarrow 136 = \Delta^2 y_{-0}$		
			$\rightarrow 268 = \Delta y_1$			
8	1	$451 = y_2$				

At $x = 5$, $u = \dfrac{5-4}{2} = 0.5$.

The derivative using Stirling's formula is

$$\left(\frac{dy}{dx}\right) = \frac{1}{h}\left[\frac{1}{2}(\Delta y_0 + \Delta y_{-1}) - u\Delta^2 y_{-1} + \frac{3u^2 - 1}{12}(\Delta^3 y_{-1} + \Delta^3 y_{-2}) + \cdots\right]$$

$$\therefore \left(\frac{dy}{dx}\right)_{x=5} = \frac{1}{2}\left[\frac{1}{2}(132 + 44) + 0.5(88) + \frac{1}{12}\{3(0.5)^2 - 1\}(48 + 48)\right]$$

$$= \frac{1}{2}[88 + 44 + 2] = 67.$$

Problem 2: From the following data, obtain the first derivative $y = \log_e x + x$ at $x = 520$.

x	500	510	520	530	540	550
$y = \log_e x$	506.2146	516.2344	526.2538	536.2729	546.2916	556.3099

Solution: Here the arguments are equally spaced and the interval differencing $h = 10$. Let

$$u = \frac{x - 520}{10} \tag{1}$$

The difference table for the given data is shown below.

x	$u = \dfrac{x-520}{10}$	y_u	Δy_u	$\Delta^2 y_u$	$\Delta^3 y_u$	$\Delta^4 y_u$	$\Delta^5 y_u$
500	-2	506.2146					
			$\to 10.0198$				
510	-1	516.2344		$\to -0.0004$			
			$\to 10.0194$		$\to 0.0001$		
520	0	526.2538		$\to -0.0003$		$\to -0.0002$	
			$\to 10.0191$		$\to -0.0001$		$\to 0.0003$
530	1	536.2729		$\to -0.0004$		$\to 0.0001$	
			$\to 10.0187$		$\to 0$		
540	2	546.2916		$\to -0.0004$			
			$\to 10.0183$				
550	3	556.3099					

Using the Stirling's formula, the formula for computing the derivative at the middle of the argument can be given as

$$\left(\frac{dy}{dx}\right)_{x=\text{middle}} = \frac{1}{h}\left[\frac{(\Delta y_0 + \Delta y_{-1})}{2} - \frac{1}{12}(\Delta^3 y_{-1} + \Delta^3 y_{-2}) \right.$$

$$\left. + \frac{1}{60}(\Delta^5 y_{-2} + \Delta^5 y_{-3}) + \ldots\ldots \right]$$

$$\therefore \left(\frac{dy}{dx}\right)_{x=520} = \frac{1}{10}\left[\frac{(10.0191 + 10.0194)}{2} - \frac{1}{12}(-0.0001 + 0.0001) \right.$$

$$\left. + \frac{1}{60}(0.0003 + 0.0003) \right] \quad \text{We assumed the fifth difference as constant.}$$

$$= \frac{1}{10}\left[10.01925 - 0 + 0.000005\right] = 1.001925.$$

4.10 DERIVATIVE OF A FUNCTION USING THE DIVIDED DIFFERENCE INTERPOLATION FORMULA

So far we have seen how to find the derivative of a function from a given set of values of x and y, when the arguments are equally spaced. Furthermore, the formulas derived so far to find the derivatives are not applicable for the given data when the arguments are not equally spaced. Hence, there is a need for deriving a formula to find the derivative in such cases. Now in this section, we learn the process of finding the derivative of a function when a set of values x and y are given and the arguments are not equally spaced.

Let $y_0, y_1, y_2, \ldots, y_n$ be $(n+1)$ values of $y = f(x)$ at $x = x_0, x_1, x_2, \ldots, x_n$ (not necessarily equally spaced and need not be in ascending or descending order). Then, by the Newton divided difference interpolation formula, the interpolating polynomial of $f(x)$ is given by

$$f(x) = f(x_0) + (x - x_0)f[x_0, x_1] + (x - x_0)(x - x_1)f[x_0, x_1, x_2]$$

$$+ (x - x_0)(x - x_1)(x - x_2)f[x_0, x_1, x_2, x_3]$$

$$+ (x - x_0)(x - x_1)(x - x_2)(x - x_3)f[x_0, x_1, x_2, x_3, x_4]$$

$$+ (x - x_0)(x - x_1)(x - x_2)(x - x_3)(x - x_4)f[x_0, x_1, x_2, x_3, x_4, x_5] + \ldots$$

$$+ (x - x_0)(x - x_1)\ldots(x - x_{n-1})f[x_0, x_1, x_2, \ldots, x_n]$$

$$(4.39)$$

Differentiating the above equation, we get

$$f'(x) = f[x_0, x_1] + [(x - x_0) + (x - x_1)]f[x_0, x_1, x_2]$$

$$+ [(x - x_0)(x - x_1) + (x - x_0)(x - x_2) + (x - x_1)(x - x_2)] \times f(x_0, x_1, x_2, x_3)$$

$$+ \big[(x - x_0)(x - x_1)(x - x_2) + (x - x_0)(x - x_1)(x - x_3) + (x - x_0)(x - x_2)$$

$$(x - x_3) + (x - x_1)(x - x_2)(x - x_3)\big] \times f[x_0, x_1, x_2, x_3, x_4] + \cdots$$

$$(4.40)$$

Again differentiating the equation twice, we get

$$f''(x) = 2f[x_0, x_1, x_2] + 2[(x - x_0) + (x - x_1) + (x - x_2)] \times$$
$$f[x_0, x_1, x_2, x_3] + 2[(x - x_1)(x - x_3) + (x - x_1)(x - x_2)$$
$$+ (x - x_2)(x - x_3) + (x - x_0)(x - x_1)$$
$$+ (x - x_1)(x - x_2) + (x - x_0)(x - x_3)] \, f[x_0, x_1, x_2, x_3, x_4] + \text{.......} \quad (4.41)$$

and

$$f'''(x) = 6f[x_0, x_1, x_2, x_3]$$
$$+ 6[(x - x_0) + (x - x_2) + (x - x_3)] f[x_0, x_1, x_2, x_3, x_4] + \cdots \quad (4.42)$$

Similarly, we can find higher order derivatives. Note that the above formulas can also be used to find the derivative of a function from a set of values when the arguments are equally spaced.

4.11 PROCEDURE TO FIND THE DERIVATIVE WHEN THE ARGUMENTS ARE NOT EQUALLY SPACED

(This method can also be applied when the arguments are equally spaced.)

Step 1: Construct the divided difference table for the given data and note down the first, second, third, divided differences and x_0, x_1, x_2, ... values.

Step 2: Note down the x value at which the derivative is required.

Step 3: Write down the formula to find the derivative using Newton's divided difference formula.

Step 4: Substitute the value of x and the divided difference in the formula given in Step 3 and simplify to get the required result.

SOLVED PROBLEMS

Problem 1: Given that

x	0	1	3	4
$f(x)$	1	4	16	25

Find $f'(2)$ and $f''(2)$.

Solution: Since the arguments are not equidistant, we use the formula of derivative obtained from Newton's divided difference formula. The divided difference table is given below.

x	y	1st d.d	2nd d.d	3rd d.d
0	1			
		$\rightarrow 3 = f[x_0, x_1]$		
1	4		$\rightarrow 1 = f[x_0, x_1, x_2]$	
		$\rightarrow 6$		$\rightarrow\ 0 = f[x_0, x_1, x_2, x_3]$
3	16		$\rightarrow 1$	
		$\rightarrow 9$		
4	25			

Here $x_0 = 0$, $x_1 = 1$, $x_2 = 3$, $x_3 = 4$.
By Eq. (4.38), we have

$$f'(x) = f[x_0, x_1] + [(x - x_0)(x - x_1)] f(x_0, x_1, x_2)$$

$\therefore \qquad f'(2) = 3 + [(2 - 0) + (2 - 1)](1) + 0 = 6$

By Eq. (4.39), we have

$$f''(x) = 2f[x_0, x_1, x_2] + 2[(x - x_0) + (x - x_1) + (x - x_2)] \times$$
$$f[x_0, x_1, x_2, x_3] + \cdots$$

$\therefore \qquad f''(2) = 2 \times 1 + 0 = 2.$

Problem 2: From the following data, obtain the first derivative of $y = \log_e x$ at $x = 2$.

x	1	1.2	1.6	2	2.3
$f(x)$	0	0.1823	0.4700	0.6931	0.8329

Solution: The divided difference table for the given data is shown below.

x	y	1ˢᵗ d.d	2ⁿᵈ d.d	3ʳᵈ d.d	4ᵗʰ d.d
1	0				
		$\to 0.9115$			
1.2	0.1823		$\to -0.3204$		
		$\to 0.71925$		$\to 0.1185$	
					$\to -0.0416$
1.6	0.470		$\to -0.2019$		
		$\to 0.55775$		$\to 0.0644$	
2	0.6931		$\to -0.1311$		
		$\to 0.466$			
2.3	0.8329				

$$f'(x) = f[x_0, x_1] + [(x - x_0) + (x - x_1)]f[x_0, x_1, x_2] +$$
$$[(x - x_0)(x - x_1) + (x - x_0)(x - x_2) + (x - x_1)(x - x_2)]f[x_0, x_1, x_2]$$
$$+ [(x - x_0)(x - x_1)(x - x_2) + (x - x_0)(x - x_1)(x - x_3)$$
$$+ (x - x_0)(x - x_2)(x - x_3) + (x - x_1)(x - x_2)(x - x_3)]$$
$$f[x_0, x_1, x_2 x_3] + \cdots$$

$$f'(2) = 0.9115 + [(2 - 1) + (2 - 1.2)](-0.3204) + [(2 - 1)(2 - 1.2)$$
$$+ (2 - 1)(2 - 1.6) + (2 - 1.2)(2 - 2)](0.1185)$$
$$+ [(2 - 1)(2 - 1.2)(2 - 1.6) + (2 - 1)(2 - 1.2)(0)$$
$$+ (2 - 1)(2 - 1.6)(2 - 2) + 0](-0.0416)$$

$$= 0.9115 + [1 + 0.8](-0.3204) + [(1)(0.8) + (1)(0.4) + (0.8)](0.1185)$$
$$+ [(1)(.8)(.4) + (1)(0.8)(0) + (1)(0.4)(2.2) + 0](-0.0416)$$
$$= 0.9115 - (1.8)(.3204) + [1.2](0.1185) + [.32](-0.0416) = 0.9115$$

Remark: Suppose a function $f(x)$ is continuous on $[a, b]$, and $f(c_1)$, $f(c_2), \ldots, f(c_k)$ are local maxima of $f(x)$ on $[a, b]$ where $c_1, c_2, \ldots, c_k$ are the critical points. Then, the maximum of $f(x)$ on $[a, b]$ is the greatest value of $f(a)$, $f(b)$, $f(c_1)$, $f(c_2)$, $\ldots$, $f(c_k)$. Similarly, if $f(d_1)$, $f(d_2)$, $\ldots$, $f(d_l)$ are local minima of $f(x)$ on $[a, b]$, then the least value of $f(a)$, $f(b)$, $f(d_1)$, $f(d_2)$, $\ldots$ and $f(d_l)$ is the minimum of $f(x)$ on $[a, b]$. Where $d_1, d_2, \ldots, d_l$ are critical points.

The procedure to obtain the maximum and minimum of a function $y = f(x)$ on an interval $[a, b]$ is as follows. We find the interpolating polynomial $p(x)$ from a given set of values of x and y. The unknown function $f(x)$ is approximated by $p(x)$ that is, $f(x) = p(x)$. We find the derivative of it and equate it to zero. Next, we find the roots of this equation, which are the critical points of $y = f(x)$. Further, we find $\dfrac{d^2 y}{dx^2}$ and compute its values at the critical points. If c is any critical point and $f''(c) < 0$, then $f(c)$ is local maximum; on the other hand, if $f''(c) > 0$, then $f(c)$ is local minimum. In case $f''(c) = 0$, it is not possible to decide whether f is local maximum or local minimum. The maximum of $y = f(x)$ on $[a, b]$ is the greatest value of $f(a), f(b)$ and its local maxima, and the minimum of it on $[a, b]$ is the least value of $f(a), f(b)$ and its local minima.

SOLVED PROBLEMS

Problem 1: From the data given below, find the point of maximum and maximum value of y in $[2, 2.8]$.

x	2.0	2.2	2.4	2.6	2.8
$f(x)$	0	0.16	0.24	0.24	0.16

Solution: The arguments are equally spaced points and the interval of difference is $h = 0.2$.

x	y	Δy	$\Delta^2 y$	$\Delta^3 y$
2.0	0			
		→ 0.16		
2.2	0.16		−0.08	
		→ 0.8		→0
2.4	0.24		−0.08	
		→ 0		→0
2.6	0.24		→−0.08	
		→−0.08		
2.8	0.16			

By Newtons forward interpolation formula

$$f(x) = 0 + u(0.16) + \frac{u(u-1)}{\lfloor 2} (-0.08)$$

where
$$u = \frac{x-2}{0.2} = 5(x-2)$$

$$f(x) = 5(x-2)(0.16) + \frac{\{5(x-2)\}\{5(x-2)-1\}}{\lfloor 2}(-0.08) = -x^2 + 5x - 6$$

The root of $f'(x) = 0$ is the critical point and it is 2.5. Further $f''(2.5) < 0$.

$$\frac{d^2 y}{dx^2} = -2 < 0$$

$\therefore$ The function has local maximum at $x = 2.5$.

The maximum of f is the largest of $\{f(2) \, (= 0), f(2.8) \, (= 0.16), f(2.5) \, (= 0.25)\}$.

Hence, the maximum of f is 0.25 and the point of maximum is 2.5.

Problem 2: Find the local maximum and local minimum of y on $[1, 4]$ from the given data.

x	-1	0	1	3	4
y	-16	3	16	48	259

Also find maximum of $f(x)$ on $[0, 4]$.

Solution: The given arguments are not equidistant. So, we use Newton's divided difference formula to interpolate the given function y. The divided difference table of the given data is shown below.

x	y	*First d. d.*	*Second d.d.*	*Third d.d.*	*Fourth d.d.*
-1	-16				
		$\rightarrow 19$			
0	3		-3		
		$\rightarrow 13$		$\rightarrow 1$	

1	16	$\rightarrow 1$	$\rightarrow 3$
	$\rightarrow 16$		$\rightarrow 16$
3	48	$\rightarrow 65$	
	$\rightarrow 211$		
4	259		

The interpolating polynomial of y by the Newton's divided difference formula is given by

$$y = -16 + (x+1)19 + (x+1)(x-0)(-3) + (x+1)x(x-1)1$$
$$+ (x+1)x(x-1)(x-3)3$$
$$= -16 + 19x + 19 - 3x^2 - 3x + x^3 - x + 3x^4 - 9x^3 - 3x^2 + 9x$$
$$y = 3x^4 - 8x^3 - 6x^2 - 24x + 3 \tag{1}$$

Differentiating (1) w.r.t. x, we get

$$\frac{dy}{dx} = 12x^3 - 24x^2 - 12x + 24 = 12(x^3 - 2x^2 - x + 2) \tag{2}$$

The roots of $\dfrac{dy}{dx} = 0$ are critical points, and they are -1, 1 and 2. Now differentiating (2) w.r.t. x, we get

$$\frac{d^2 y}{dx^2} = 12(3x^2 - 6x - 1)$$

The signs of $\dfrac{d^2 y}{dx^2}$ at $x = -1$, 1 and 2 are positive, negative and negative, respectively. Therefore, the function y has local minimum at $x = -1$ and it has local maxima at $x = 1$ and 2.

The values of y at -1, 1, 2 are given in the table. The local minimum at $x = -1$ is -16 and the local maximum at $x = 1$ is 16.

To know the local maximum of y at $x = 2$, we compute $f(2)$ from Eq.(1) and it is 11.

Thus, the required maximum is the largest value of $\{f(-1) (= -16)$, $f(4) (=259)$ and $f(2) (=11)\}$. Hence, the maximum is 259.

Problem 3: Find the maximum and minimum values of y on $[0, 5]$ from the following table.

x	0	1	2	3	4	5
y	1	1.25	1	3.25	17	57.25

Solution: Here the arguments are equally spaced with interval of differencing $h = 1$.

The forward difference table for the given data is as follows:

x	y	Δy	$\Delta^2 y$	$\Delta^3 y$	$\Delta^4 y$	$\Delta^5 y$
0	1					
		→0.25				
1	1.25		→−0.5			
		→−0.25		→3		
2	1		→2.50		→6	
		→2.25		→9		→0
3	3.25		→11.50		→6	
		→13.75		→15		
4	17		→26.50			
		→40.25				
5	57.25					

Since $u = \dfrac{x - 0}{1} = x$, by the Newton's forward formula of interpolation (3.8), we get

$$y = 1 + x(0.25) + \frac{x(x-1)}{\lfloor 2} \times (-0.50) + \frac{x(x-1)(x-2)}{\lfloor 3} \times 3$$

$$+ \frac{x(x-1)(x-2)(x-3)}{\lfloor 4} \times 6$$

$$= 1 + \frac{x}{4} - \frac{(x^2 - x)}{4} + \frac{(x^3 - 3x^2 + 2x)}{2} + \frac{x^4 - 6x^3 + 11x^2 - 6x}{4}$$

$$y = \frac{1}{4}[4 + x - x^2 + x + 2x^3 - 6x^2 + 4x + x^4 - 6x^3 + 11x^2 - 6x]$$

$$= \frac{1}{4}[x^4 - 4x^3 + 4x^2 + 4]$$

$$\frac{dy}{dx} = \frac{1}{4}(4x^3 - 12x^2 + 8x)$$

The roots of $4x^3 - 12x^2 + 8x = 0$ are critical points. Therefore, 0, 1 and 2 are the critical points. Now

$$\frac{d^2y}{dx^2} = \frac{1}{4}(12x^2 - 24x + 8) = 3x^2 - 6x + 2$$

At $x = 0$, $\dfrac{d^2y}{dx^2} = 2 > 0$. Therefore, $f(x)$ has local minimum at $x = 0$.

At $x = 1$, $\dfrac{d^2y}{dx^2} = -1 < 0$. Therefore, $f(x)$ has local maximum at $x = 1$.

At $x = 2$, $\dfrac{d^2y}{dx^2} = 2 > 0$. Therefore, $f(x)$ has local minimum at $x = 2$.

Thus, the maximum of $f(x)$ on [0, 5] is the greatest value of $\{f(0), f(5), f(1)\}$, that is, $\{1, 57.25, 1.25\}$. Hence, the maximum of y is 57.25.

The minimum of y on [0, 5] is the least value of $\{f(0), f(5), f(0), f(2)\}$, i.e., $\{1, 57.25, 1\}$. Hence, the minimum of y on [0.5] is 1. The point of minimum is $x = 0$.

4.12   NUMERICAL DIFFERENTIATION USING FINITE DIFFERENCE

We can reduce the error in the derivative's estimate either by decreasing the interval of differencing or using higher order difference. Of course, to have the higher order differences, we have to add more points to the tabulated data. There is another method of reducing the error in approximating the derivative of a function, which is known as Richardson extrapolation method. This method uses finite differences.

We know that
$$f'(x) = \underset{h \to 0}{\text{Lt}} \frac{f(x+h) - f(x)}{h}$$

For small values of h,

$$f'(x) \approx \frac{f(x+h) - f(x)}{h} \qquad (4.43)$$

which is known as two-point formula to estimate the derivative.
By the Taylor's expansion

$$f(x+h) = f(x) + h f'(x) + \frac{h^2}{\lfloor 2} f''(x) + \frac{h^3}{\lfloor 3} f'''(x) + \cdots \qquad (4.44)$$

which can be expressed as

$$f'(x) = \frac{f(x+h) - f(x)}{h} + E_T \qquad (4.45)$$

where E_T contains terms of h or higher orders of it. This approximation is said to be of the first order.

An alternative two-point formula to (4.43) is

$$f'(x) = \frac{f(x+h) - f(x-h)}{2h} \qquad (4.46)$$

which can be derived as follows. Expanding $f(x - h)$ by Taylor's series, we have

$$f(x-h) = f(x) - hf'(x) + \frac{h^2}{\lfloor 2} f''(x) + \cdots \qquad (4.47)$$

Subtracting (4.47) from (4.44), we get

$$f'(x) = \frac{f(x+h) - f(x-h)}{2h} + O_2(h) \qquad (4.48)$$

where $O_2(h)$ represents the terms of h^2 and higher order. Formula (4.46) follows from (4.48) by neglecting $O_2(h)$. This approximation is of second order. Further, it gives better approximation of $f'(x)$ than the approximation computed from (4.43). Now we proceed to describe the Richardson extrapolation method of estimating the derivative of a function.

Suppose the derivative of a function $y = f(x)$ is computed at $x = x_P$ from the relation

$$f'(x_P) = \frac{f(x_P + h_R) - f(x_P - h_R)}{2h_R} \tag{4.49}$$

where h_R is very small. Then,

$$f'(x_P) = \frac{f(x_P + h_R) - f(x_P - h_R)}{2h_R} + E_T \tag{4.50}$$

where E_T is the error due to truncation of terms. Clearly, the error E_T depends on h_R. Using the Taylor's expansion, Eq. (4.50) can be expressed as

$$f'(x_P) = \frac{\left\{ \begin{array}{c} f(x_P) + h_R f'(x_P) \\ + \dfrac{h_R^2}{\lfloor 2} f''(x_P) + \cdots \end{array} \right\} - \left\{ \begin{array}{c} f(x_P) - h_R f'(x_P) \\ + \dfrac{h_R^2}{\lfloor 2} f''(x_P) + \cdots \end{array} \right\}}{2h_R} + E_T$$

$$= f'(x_P) + h_R^2 f'''(x_P) + h_R^4 f'^v(x_P) + \cdots + E_T$$

$$\therefore \qquad E_T = -[h_R^2 f'''(x_P) + h_R^4 f^v(x_P) + \cdots]$$

Suppose

$$E_T = C_1 h_R^2 + C_2 h_R^4 + C_3 h_R^6 + \cdots \tag{4.51}$$

where $C_i = f^{(2i+1)}(x_P)$ $(i = 1, 2, 3, \ldots)$ are constants and they are independent of h_R. Let

$$F(h_R; x_P) = \frac{f(x_P + h_R) - f(x_P - h_R)}{2h_R} \tag{4.52}$$

In view of (4.51) and (4.52), Eq. (4.50) becomes

$$f'(x_P) = F(h_R; x_P) + C_1 h_R^2 + C_2 h_R^4 + C_3 h_R^6 + \cdots \tag{4.53}$$

Now, suppose the value of h_R is reduced to $\dfrac{h_R}{2}$. Then Eq. (4.53) becomes

$$f'(x_P) = F\left(\frac{h_R}{2}; x_P\right) + C_1 \frac{h_R^2}{4} + C_2 \frac{h_R^4}{16} + C_3 \frac{h_R^6}{64} + \cdots \tag{4.54}$$

Note that both $F(h_R; x_P)$ and $F\left(\dfrac{h_R}{2}; x_P\right)$ are the approximations of the derivative of $f(x)$ at x_P. Now eliminating C_1 from Eqs. (4.53) and (4.54), we get

$$f'(x_p) = \frac{4}{3} F\left(\frac{h_R}{2}; x_P\right) - \frac{1}{3} F(h_R; x_p) + O_4(h_R) \tag{4.55}$$

where $O_4(h_R)$ contains the terms of h^i, $i \geq 4$ and i is an even number. Now, suppose

$$F_1\left(\frac{h_R}{2}; x_p\right) = \frac{4}{3} F\left(\frac{h_R}{2}; x_P\right) - \frac{1}{3} F(h_R; x_p) \tag{4.56}$$

Then Eq. (4.55) becomes

$$f'(x_p) = F_1\left(\frac{h_R}{2}; x_P\right) + d_1 h_R^4 + O_6(h_R) \tag{4.57}$$

where d_1 is a constant.

Repeating the argument for $\dfrac{h_R}{2^2}$ in place of $\dfrac{h_R}{2}$, we have

$$f'(x_p) = F_1\left(\frac{h_R}{2^2}; x_P\right) + d_1 \frac{h_R^4}{16} + O_6(h_R) \tag{4.58}$$

Eliminating the coefficient of h_R^4 from Eqs. (4.57) and (4.58), we get

$$f'(x_p) = F_2\left(\frac{h_R}{2^2}; x_P\right) + O_6(h_R)$$

which is accurate of the sixth order, where

$$F_2\left(\frac{h_R}{2^2}; x_P\right) = \frac{4^2 F_1\left(\dfrac{h_R}{2^2}; x_P\right) - F_1\left(\dfrac{h_R}{2}; x_P\right)}{4^2 - 1} \tag{4.59}$$

Continuing this process and eliminating the coefficient of h^{2k}, we get

$$F_k\left(\frac{h_R}{2^k}; x_P\right) = \frac{4^k F_{k-1}\left(\dfrac{h_R}{2^k}; x_P\right) - F_{k-1}\left(\dfrac{h_R}{2^{k-1}}; x_P\right)}{4^k - 1} \quad (k = 1, 2, 3, \ldots)$$

$$(4.60)$$

where $F_0(h_R; x_P) = F(h_R; x_P)$.

The set of approximation of $f'(x_P)$ calculated by the Richardson extrapolation method is displayed in the following form.

	F	F_1	F_2	F_3
h_R	$F(h_R; x_P)$			
		$F_1\left(\dfrac{h_R}{2}; x_P\right)$		
$\dfrac{h_R}{2}$	$F\left(\dfrac{h_R}{2}; x_P\right)$		$F_2\left(\dfrac{h_R}{2^2}; x_P\right)$	
		$F_1\left(\dfrac{h_R}{2^2}; x_P\right)$		$F_3\left(\dfrac{h_R}{2^3}; x_P\right)$
$\dfrac{h_R}{2^2}$	$F\left(\dfrac{h_R}{2^2}; x_P\right)$		$F_2\left(\dfrac{h_R}{2^3}; x_P\right)$	
		$F_1\left(\dfrac{h_R}{2^3}; x_P\right)$		
$\dfrac{h_R}{2^3}$	$F\left(\dfrac{h_R}{2^3}; x_P\right)$			

Note that each entry in the above table except the first column entries is the approximation of the derivative of $y = f(x)$ at x_P. It can also be noted that h_R is not the interval of differencing of the arguments of the tabulated data. For a particular value of h_R, the calculation $F_i\left(\dfrac{h_R}{2^i}; x_P\right)$ needs the values of $f(x)$ at certain points and they must be given in the tabulated values in order to apply this method to find $f'(x_P)$. This situation does not arise when the function $y = f(x)$ is known as we can find those values by using the known function.

SOLVED PROBLEMS

Problem 1: Find $f'(2)$ from the data given below

x	-2	-1	0	1	3	4	6
$f(x)$	16	1	0	1	81	256	1296

using Richardson extrapolation method with $h = 4, 2, 1$.

Solution: Here $x_P = 2$ and $h_R = 4$.

$$F(h_R; x_P) = \frac{f(x_P + h_R) - f(x_P - h_R)}{2h_R},$$

we have

$$F(4;2) = \frac{f(2+4) - f(2-4)}{8} = \frac{1296 - 16}{8} = 160$$

$$F(2;2) = \frac{f(4) - f(0)}{4} = \frac{256 - 0}{4} = 64$$

$$F(1;2) = \frac{f(3) - f(1)}{2} = \frac{81 - 1}{2} = 40$$

Thus,

h_R	F	F_1	F_2
4	160		
		$\rightarrow \dfrac{4 \times 64 - 160}{4 - 1} = 32$	
2	64		$\rightarrow \dfrac{4^2 \times 32 - 32}{4^2 - 1} = 32$
		$\rightarrow \dfrac{4 \times 40 - 64}{4 - 1} = 32$	
1	40		

$\therefore \qquad\qquad\qquad f'(2) = 32$

Problem 2: Using Richardson's extrapolation method, find $f'(\pi/6)$ where $f(x) = \sin x$ with $h_R = 0.8, 0.4, 0.2, 0.1$.

Solution: Here $x_P = \pi/6$ and $h_R = 0.8$. Therefore,

$$F(0.8; \pi/6) = \frac{f(\pi/6+0.8) - f(\pi/6-0.8)}{0.16} = 0.776466$$

$$F(0.4; \pi/6) = \frac{f(\pi/6+0.4) - f(\pi/6-0.4)}{0.8} = 0.843013$$

$$F(0.2; \pi/6) = \frac{f(\pi/6+0.2) - f(\pi/6-0.2)}{0.4} = 0.860159$$

$$F(0.1; \pi/6) = \frac{f(\pi/6+0.1) - f(\pi/6-0.1)}{0.2} = 0.864478$$

Thus,

Each of the symbol (like a_1, a_2, ...etc) written in the bracket in the following table represent the value adjacent to it. For example 0.776466 (a_1) indicater $a_1 = 0.776466$. This notation is used for convenience.

h	F	F_1	F_2	F_3
0.8	0.776466(a_1)			
		→ 0.865195(b_1)		
0.4	0.843013(a_2)		→ 0.85919(c_1)	
		→ 0.865874(b_2)		→ 0.8659(d_1)
0.2	0.860159(a_3)		→ 0.865920(c_2)	
		→ 0.865917(b_3)		
0.1	0.864478(a_4)			

The computations of b_1, b_2, b_3, c_1, c_2 and d_1 are shown below

$$b_1 = \frac{4a_2 - a_1}{4-1}; b_2 = \frac{4a_3 - a_2}{4-1}; b_3 = \frac{4a_4 - a_3}{4-1}; c_1 = \frac{4^2.b_2 - b_1}{4^2 - 1}$$

$$c_2 = \frac{4^2 b_3 - b_2}{4^2 - 1}; d_1 = \frac{4^3 c_2 - c_1}{4^3 - 1}.$$

Problem 3: Using Richardson's extrapolation method, find $f'(6)$ when $f(x) = x^4 + 4x^3 + 5$ with $h_R = 1.6, 0.8, 0.4, 0.2, 0.1$.

Solution: Here $x_P = 6$, $h_R = 0.8$ and $f(x) = x^4 + 4x^3 + 5$. Therefore,

$$F(1.6;\ 6) = \frac{f(7.6) - f(4.4)}{3.2} = 1367.679932$$

$$F(0.8;\ 6) = \frac{f(6.8) - f(5.2)}{1.6} = 1313.920288$$

$$F(0.4;6) = \frac{f(0.4) - f(5.6)}{0.8} = 1300.480469$$

$$F(0.2;6) = \frac{f(6.2) - f(5.8)}{0.2} = 1297.118774$$

$$F(0.1;6) = \frac{f(6.1) - f(5.9)}{0.2} = 1296.279297$$

We can compute $F_1(0.8;\ 6)$, $F_1(0.4;\ 6)$, $F_1(0.2;\ 6)$, $F_1(0.1;\ 6)$, $F_2(0.4;\ 6)$, … from the relation

$$F_k\!\left(\frac{h_R}{2^k};x_P\right) = \frac{4^k F_{k-1}\!\left(\dfrac{h_R}{2^k};x_P\right) - F_{k-1}\!\left(\dfrac{h_R}{2^{k-1}};x_P\right)}{4^k - 1}$$

and their values are given in the following table.

h	F	F_1	F_2	F_3	F_4
1.6	1367.679932(a_1)				
		$\to b_1$			
0.8	1313.920288(a_2)		$\to c_1$		
		$\to b_2$		$\to d_1$	
0.4	1300.480469(a_3)		$\to c_2$		$\to e_1$
		$\to b_3$		$\to d_2$	
0.2	1297.118774(a_4)		$\to c_3$		
		$\to b_4$			
0.1	1296.279297(a_5)				

$$b_1 = \frac{4a_2 - a_1}{4-1} = 1296.000366; \qquad b_2 = \frac{4a_3 - a_2}{4-1} = 1296.000488$$

$$b_3 = \frac{4a_4 - a_3}{4-1} = 1295.998169; \qquad b_4 = \frac{4a_5 - a_4}{4-1} = 1295.999512$$

$$c_1 = \frac{4^2 b_2 - b_1}{4^2 - 1} = 1296.000488; \qquad c_2 = \frac{4^2 b_3 - b_2}{4^2 - 1} = 1295.998047$$

$$c_3 = \frac{4^2 b_4 - b_3}{4^2 - 1} = 1295.99634; \qquad d_1 = \frac{4^3 c_2 - c_1}{4^3 - 1} = 1295.998047$$

$$d_2 = \frac{4^3 c_3 - c_2}{4^3 - 1} = 1295.999634; \qquad e_1 = \frac{4^4 d_2 - d_1}{4^4 - 1} = 1295.99963.$$

Problem 4: Using Richardson extrapolation method, find $f'(0.05)$ with $h_R = 0.0256, 0.0128, 0.0064, 0.0032, 0.0016$, where $f(x) = \dfrac{-1}{x}$.

Solution: Here $x_P = 0.05$, $h_R = 0.0256$ and $f(x) = \dfrac{-1}{x}$

$$F(0.0256; 0.05) = \frac{f(0.05 + 0.0256) - f(0.05 - 0.0256)}{2 \times 0.0256} = 542.111145$$

$$F(0.0128; 0.05) = \frac{f(0.05 + 0.0128) - f(0.05 - 0.0128)}{2 \times 0.0128} = 428.052826$$

$$F(0.0064; 0.05) = \frac{f(0.05 + 0.0064) - f(0.05 - 0.0064)}{2 \times 0.0064} = 406.662842$$

$$F(0.0032; 0.05) = \frac{f(0.05 + 0.0032) - f(0.05 - 0.0032)}{2 \times 0.0032} = 401.644714$$

$$F(0.0016; 0.05) = \frac{f(0.05 + 0.0016) - f(0.05 - 0.0016)}{2 \times 0.0016} = 400.410309$$

The values of h_R, F, F_1, F_2, F_3 and F_n are given below.

h	F	F_1	F_2	F_3	F_4
0.0256	592.111145(a_1)				
		$\to b_1$			
0.0128	428.052826(a_2)		$\to c_1$		
		$\to b_2$		$\to d_1$	
0.0064	406.662842(a_3)		$\to c_2$		$\to e_1$
		$\to b_3$		$\to d_2$	
0.0032	401.644714(a_4)		$\to c_3$		
		$\to b_4$			
0.0016	400.410309(a_5)				

The values of b_1, b_2, ... e_1 are as follows

$$b_1 = \frac{4a_2 - a_1}{4-1} = 390.033386; \qquad b_2 = \frac{4a_3 - a_2}{4-1} = 399.532837$$

$$b_3 = \frac{4a_4 - a_3}{4-1} = 399.972015; \qquad b_4 = \frac{4a_5 - a_4}{4-1} = 399.998840$$

$$c_1 = \frac{4^2 b_2 - b_1}{4^2 - 1} = 400.166138; \qquad c_2 = \frac{4^2 b_3 - b_2}{4^2 - 1} = 400.001282$$

$$c_3 = \frac{4^2 b_4 - b_3}{4^2 - 1} = 400.000641; \qquad d_1 = \frac{4^3 c_2 - c_1}{4^3 - 1} = 399.998657$$

$$d_2 = \frac{4^3 c_3 - c_2}{4^3 - 1} = 400.000641; \qquad e_1 = \frac{4^4 d_2 - d_1}{4^4 - 1} = 4000.000641$$

Therefore, $f'(0.05) = 400.00064 \approx 400$.

SHORT ANSWER TYPE SOLVED PROBLEMS

Problem 1: From the following data, find $f'(2)$.

x	2	5	8	11
$y = f(x)$	5	26	65	122

Solution: The forward difference table for the above data is given below.

x	$f(x)$	Δy	$\Delta^2 y$	$\Delta^3 y$
2	$5(y_0)$			
		$\to 21(=\Delta y_0)$		
5	26		$\to 18(=\Delta^2 y_0)$	
		$\to 39$		$\to 0(=\Delta^3 y_0)$
8	65		$\to 18$	
		$\to 57$		
11	122			

The formula for the derivative of $f(x)$ at the initial argument is

$$f'(x_0) = \frac{1}{h}\left[\Delta y_0 - \frac{\Delta^2 y_0}{2} + \frac{\Delta^3 y_0}{3}\cdots\right]$$

$$\therefore \qquad f'(2) = \frac{1}{3}\left[21 - \frac{18}{2}\right] = 4$$

Problem 2: Given that

x	0	4	8	12
$y = f(x)$	-1	15	63	143

Find $f'(12)$.

Solution: The backward difference table for the above data is given below:

x	$y = f(x)$	∇y	$\nabla^2 y$	$\nabla^3 y$
0	-1			
		$\rightarrow$ 16		
4	15		$\rightarrow$ 32	
		$\rightarrow$ 48		$\rightarrow$ $0 = \nabla^3 y_n$
8	63		$\rightarrow$ $32 = \nabla^2 y_n$	
		$\rightarrow$ $80 = \nabla y_n$		
12	143			

The formula for estimating the derivative of $f(x)$ at the end argument x_n is

$$f'(x_n) = \frac{1}{h}\left[\nabla y_n + \frac{\nabla^2 y_n}{2} + \frac{\nabla^3 y_n}{3} + \cdots\right]$$

$$\therefore \qquad f'(12) = \frac{1}{4}\left[80 + \frac{32}{2}\right] = 24$$

Problem 3: Given that

x	2	4	6	8	10
$y = f(x)$	6	20	42	72	110

Find $f'(6)$.

Solution: Transforming the independent variable x by the relation $u = \dfrac{x-6}{2}$, we have the following central difference table.

x	$u = \dfrac{x-6}{2}$	y	Δy	$\Delta^2 y$	$\Delta^3 y$	$\Delta^4 y$
2	-2	$6 = y_{-2}$				
			$\rightarrow 14 = \Delta y_{-2}$			
4	-1	$20 = y_{-1}$		$\rightarrow 8 = \Delta^2 y_{-2}$		

$$\begin{array}{lll}
& & \rightarrow 22 = \Delta y_{-2} \\
6 \quad 0 \quad 42 = y_0 & & \rightarrow 8 = \Delta^2 y_{-1} \\
& \rightarrow 30 = \Delta y_0 & \\
8 \quad 1 \quad 72 = y_1 & & \rightarrow 8 = \Delta^2 y_{-0} \\
& \rightarrow 38\ \Delta^2 y_1 & \\
10 \quad 2 \quad 110 = y_2 & &
\end{array}$$

The formula for computing the derivative of $y = f(x)$ at the middle of the argument is

$$\left[f'(x) \right]_{x=\text{middle argument}} = \frac{1}{h}\left[\frac{1}{2}(\Delta y_0 + \Delta y_{-1}) - \frac{1}{12}(\Delta^3 y_{-1} + \Delta^3 y_{-2}) + \cdots \right]$$

$$f'(6) = \frac{1}{2}\left[\frac{1}{2}(22 + 30) \right] = 13$$

Problem 4: If $f(1) = 3, f(2) = 5, f(5) = 23$ and $f(6) = 33$, find $f'(1)$.

Solution: Here the arguments 1, 2, 5 and 6 are not equidistant. So, we use the derivative formula obtained from Newton's divided difference interpolation formula. The divided difference table for the given data is as follows.

x	$y = f(x)$	First d.d.	Second d.d.	Third d.d.
1	3			
		$\rightarrow 2$		
2	5		$\rightarrow 1$	
		$\rightarrow 6$		$\rightarrow 0$
5	23		$\rightarrow 1$	
		$\rightarrow 10$		
6	33			

The derivative formula using divided differences is

$$f'(x) = f\left[x_0, x_1\right] + \left[(x - x_0) + (x - x_1)\right] f\left[x_0, x_1, x_2\right] + \cdots$$

$$\therefore \quad f'(1) = 2 + \left[(1-1) + (1-2)\right](1)$$

$$= 2 + (-1) = 1.$$

Problem 5: Given that

x	2	4	6	8
$f(x)$	12	68	220	516

Find the approximate form of $f'(x)$.

Solution: The forward difference table for the above data is given below.

x	$y = f(x)$	Δy	$\Delta^2 y$	$\Delta^3 y$
2	$12 (= y_0)$			
		$\to 56 (= \Delta y_0)$		
4	68	$\to$	$96 = \Delta^2 y_0$	
		$\to \ 152$		$\to 48 (= \Delta^3 y_0)$
6	220	$\to$	144	
		$\to \ 296$		
8	516			

We know that

$$f'(x) = \frac{1}{h}\left[\Delta y_0 + \frac{(2u-1)}{\lfloor 2} \Delta^2 y_0 + \frac{(u^2 - 6u + 2)}{\lfloor 3} \Delta^3 y_0 + \cdots \right]$$

where $u = \dfrac{x - x_0}{h}$. Therefore, $u = \dfrac{x-2}{2}$ and

$$f'(x) = \frac{1}{2}\left[56 + \frac{(x-2-1)}{\lfloor 2} \times 96 + \frac{\left\{ 3\left(\dfrac{x-2}{2}\right)^2 - 6\left(\dfrac{x-2}{2}\right) + 2 \right\}}{\lfloor 3} 48 \right]$$

$$= 3x^2$$

Thus, the approximate form of $f'(x)$ is $3x^2$.

EXERCISE 4

1. Find $f'(2.5)$ from the following data.

x	2	3	4	5
$f(x)$	5	10	17	26

2. Find $f''(2)$ from the following data.

x	1	2	3	4
$f(x)$	1	8	27	64

3. Construct the divided difference table for the data given below.

x	0	2	3	5
$f(x)$	−3	3	9	27

Hence, or otherwise, find the derivative of $f(x)$ at $x = 2$.

4. Using Bessel's formula of interpolation, find the derivative of $f(x)$ at $x = 3$ from the data given below.

x	1	2	3	4	5
$f(x)$	5	9	15	23	33

5. Find $f'(0)$ from the following data.

x	0	0.2	0.4	0.6	0.8	1.0	1.2
$f(x)$	0	0.12	0.49	1.12	2.02	3.20	4.67

6. From the following tabulated values of x and $f(x)$, find $f'(1.10)$.

x	1.00	1.05	1.10	1.15	1.20
$f(x)$	1.000	1.0249	1.0488	1.0724	1.0956

7. Find $f'(0.6)$, $f''(0.6)$ and $f'''(0.6)$ from the following tabulated values of x and $f(x)$.

x	0.4	0.5	0.6	0.7	0.8
$f(x)$	1.5836	1.7976	2.0542	2.3278	2.6514

8. Find the value of cos(1.74) from the following tabulated values of x and $f(x) = \sin x$.

x	1.7	1.74	1.78	1.82	1.86
$f(x)$	0.9918	0.9856	0.9782	0.9691	0.9585

9. The population of a certain city for different years is given in the following table.

x (year)	1951	1961	1971	1981	1991
y (population in thousands)	26.54	32.64	38.21	42.64	50.21

Find the rate of growth of the population in the year 1971.

10. Find $f'(3)$ from the following data.

x:	3	5	7	9	11	13
$f(x)$:	0.1023	0.1047	0.1071	0.1096	0.1122	0.1148

11. Find the approximate value of $\dfrac{1}{\sqrt{20}}$ from the following tabulated values of x and $f(x) = 2\sqrt{x}$.

x	10	14	18	22	26
$f(x): = 2\sqrt{x}$	6.324	7.483	8.485	9.381	10.198

12. Find $f'(1.8)$ from the following data.

x	1.00	1.20	1.40	1.60	1.80
$f(x)$	0.2420	0.1942	0.1496	0.1107	0.0788

13. Find $f'(1.6)$ from the data given below.

x	1.4	1.7	2.1	2.3	2.4
$f(x)$	−0.256	1.913	6.261	9.167	10.824

14. Find $f'(15)$ from the data given below.

x	0	5	10	15	20	25	30
$f(x)$	0	0.0875	0.1763	0.2679	0.3640	0.4663	0.5772

15. Find an approximate interpolating polynomial of $f'(x)$ and $f''(x)$ from the data given below.

x	0	1	3	4
$f(x)$	1	4	40	85

16. Given that

x	0	2	3	5
$f(x)$	1	5	10	26

Find k and l, such that $k\ f'(2) + l\ f''(2) = f(2)$ and $kf'(2) - 2lf''(2) = f(3)$.

17. Given that $f(0) = 8, f(2) = 8\, f(4) = 16$ and $f(6) = 32$. Find the value of x for which $f(x)$ is minimum. Also find its value.

18. The values of $y = f(x)$ at different values of the independent variable are given in the following table. Find the value of x for which y is maximum. Also, find the maximum value.

x	45	55	65	75	85
$f(x)$	126.2	136.2	142.2	140.8	138.6

19. Find the value of x for which y is maximum from the data given below.

x	0.5	1.5	2.5	3.5
$f(x)$	−3.75	−0.75	0.25	−0.75

20. Find the values of maximum and minimum of the function $y = f(x)$ from the data given below.

x	0	1	2	3	4	5
$f(x)$	5	5.25	5	5.25	21.00	61.25

21. A curve $y = f(x)$ is passing through the points (0.4, 1.58), (0.5, 1.82), (0.6, 2.06) (0.7, 2.43) and (0.8, 2.68). Find the slope of the tangent at $x = 0.3$.

22. The values of temperatures θ of water in a vessel recorded at different values of time t are given below. Find an approximate rate of cooling at $t = 2.5$.

Time (s)	0	1	2	3	4
θ (temperature in °C)	90.3	85.4	72.7	65.4	54.9

23. The observation of a certain experiment pertaining to pressure and volume are recorded as given below.

V (volume)	10.6	8.2	7.1	6.5	5.8
P (pressure)	15	18	21	24	27

Find the rate of change of pressure with respect to volume, when the volume is equal to 8.

24. In a machine, a slider moves along a fixed rod. The distance travelled by the slider x (in m) along the rod is given below for various values of time t (s). Find the acceleration of the slider when $t = 0.2$ s.

t (s)	0	0.2	0.4	0.6	0.8
x (m)	6.14	8.23	9.14	11.02	12.56

25. The values of velocity v of a particle for different values of time t are given in the following table. Find the acceleration of the particle at $t = 2$ s.

t	1	1.5	2	2.5	3	3.5
v	126	132	139	144	152	161

ANSWERS

1. 5

2. $f''(2) = 12$

3. 5

4. 7

5. –0.0083333

6. 0.474

7. 2.644833, 3.7

8. –0.0002593

9. 0.469416

10. 0.001229

11. 0.22371875

12. –0.141333

13. 7.68

14. 0.0187

15. $3x^2 + 2x + 1$, $6x + 2$

16. $k = \dfrac{5}{3}, l = \dfrac{-5}{6}$

17. f is minimum at $x = 1$, minimum value is 7

18. f is maximum at 46.2180

19. f is maximum at $x = 2.5$, maximum value is 0.25

20. f is minimum at $x = 0.3386$, minimum value is 4.0207, f is maximum at $x = 1.4246$, maximum value is 5.7021

21. 12.7

22. –6.9416

23. –1.45326105

24. 4.2667, –22.2917

25. 10.6667.

5

Numerical Integration

5.1 INTRODUCTION

We cannot move in the backward direction as fast as we can move in the forward direction. Likewise, it is not possible to evaluate integrals as fast as we find the derivates of functions in many situations. In some cases, it may not be possible to evaluate an integral in a closed form.

The methods we learnt to evaluate a definite integral $\int_a^b f(x)dx$ in the calculus are inadequate to evaluate many integrals, for example $\int_0^6 \frac{(\sin x + e^x)^{\cos x}}{1 + \log x + x \tan^{-1} x} dx$. These methods also fails to evaluate a definite integral when the values of the integrand $f(x)$ are given at the points $a = x_0, x_1, x_2, \ldots, x_n = b$ only. In such cases, we can employ the numerical methods to evaluate them and numerical integration methods are the tools for evaluation of definite integrals: i) when the values of integrand are known at some points in its independent variable; ii) when fails to obtain $\int_a^b f(x)dx$ by the methods of integration in the calculus.

The definite integrals that are possible to evaluate by the methods of integration can also be evaluated by the numerical integration methods. The methods of numerical integration have applications in differential and integral equations. In this section, we learn some of these methods to evaluate the definite integrals and we also discuss the errors involved in those methods.

5.2 DEFINITION

The process of evaluating a definite integral $I = \int\limits_a^b f(x)dx$ $(a < b)$ from a set of tabulated values of the integrand (known or unknown function) at $a = x_0, x_1, \ldots, x_n = b$ such that $a = x_0 < x_1 < x_2 < \ldots < x_{n-1} < x_n = b$ is known as numerical integration.

Note that if the integrand is known, we can compute the values of the integrand at the points $x = x_0, x_1, \ldots, x_n$ in order to evaluate I using a numerical method.

5.3 GENERAL QUADRATURE FORMULA

The process of evaluation of an integrand of a single variable is known as quadrature. We obtain a general quadrature formula to evaluate $\int\limits_a^b f(x)dx$. Let

$$I = \int\limits_a^b f(x)dx \tag{5.1}$$

and the values of $y = f(x)$ are given at $a = x_0$, $x_1 = x_0 + h$, $x_2 = x_0 + 2h$, $\ldots, x_n = x_0 + nh = b$ and $y_i = f(x_0 + ih)$ $(i = 0, 1, 2, \ldots, n)$.

Since $(n + 1)$ values of $y = f(x)$ are known at $x_0, x_0 + h, \ldots, x_0 + nh$, we can approximate $f(x)$ by nth degree polynomial using the Newton's forward difference interpolation formula. Hence,

$$f(x) = y_0 + u\Delta y_0 + \frac{u(u-1)}{\lfloor 2}\Delta^2 y_0 + \ldots$$

$$+ \frac{u(u-1)(u-2)\ldots(u-\overline{n-1})}{\lfloor n}\Delta^n y_0 \tag{5.2}$$

where

$$u = \frac{x - x_0}{h} \tag{5.3}$$

In view of Eq. (5.2), Eq. (5.1) becomes

$$I = \int\limits_{x_0}^{x_0+nh}\left[y_0 + u\Delta y_0 + \frac{u(u-1)}{\lfloor 2}\Delta^2 y_0 + \cdots + \frac{u(u-1)-(u-\overline{n-1})}{\lfloor n}\Delta^n y_0\right]dx$$

$$\tag{5.4}$$

From (5.3), we have $dx = hdu$, $u = 0$ when $x = x_0$ and $u = n$ when $x = x_n$. Thus, Eq. (5.4) reduces to

$$I = h\int_0^n \left[y_0 + u\Delta y_0 + \frac{u(u-1)}{\underline{|2}}\Delta^2 y_0 + \cdots + \frac{u(u-1)\cdots(u-\overline{n-1})}{\underline{|n}}\Delta^n y_0 \right] du$$

$$\therefore I = h\left[ny_0 + \frac{n^2}{2}\Delta y_0 + \left(\frac{n^3}{3} - \frac{n^2}{2}\right)\frac{\Delta^2 y_0}{\underline{|2}} \right.$$

$$+\left(\frac{n^4}{4} - n^3 + n^2\right)\frac{\Delta^3 y_0}{\underline{|3}} + \left(\frac{n^5}{5} - \frac{3n^4}{2} + \frac{11n^3}{3} - 3n^2\right)\frac{\Delta^4 y_0}{\underline{|4}}$$

$$+\left(\frac{n^6}{6} - 2n^5 + \frac{35n^3}{3} - \frac{50n^3}{3} + 12n^2\right)\frac{\Delta^5 y_0}{\underline{|5}}$$

$$\left. +\left(\frac{n^7}{7} - \frac{5n^6}{2} + \frac{17n^5}{5} - \frac{215n^4}{4} + \frac{217n^3}{3} - 60n^2\right)\frac{\Delta^6 y_0}{\underline{|6}} + \cdots \right] \quad (5.5)$$

which is known as a general quadrature formula. Now, we can derive various special formulas assigning different positive integral values to n.

5.4   TRAPEZOIDAL RULE

It can be derived from general quadrature formula by taking $n = 1$and neglecting the second and higher order differences. So, substituting $n = 1$ in (5.5) and neglecting the second and higher order differences, we obtain

$$\int_{x_0}^{x_0+h} f(x)dx = h\left[y_0 + \frac{1}{2}\Delta y_0 \right]$$

$$\text{i.e.} \quad \int_{x_0}^{x_0+h} f(x)dx = \frac{h}{2}(y_0 + y_1), \quad (5.6)$$

which is known as Trapezoidal rule or Basic Trapezoidal rule. It can also be expressed as

$$\int_a^b f(x)dx = \frac{(b-a)}{2}[f(a) + f(b)].$$

Geometrically, it is the area of trapezium having parallel sides of lengths $f(a), f(b)$ and distance between them is $(b - a)$. It is an approximation to the area under the curve $y = f(x)$, $x = a$, $x = b$ and the x-axis (see fig 5.1)

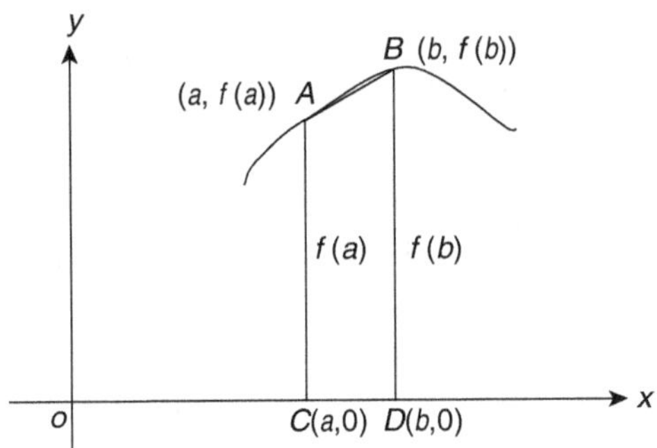

Figure 5.1

Thus, $\int_{a}^{b} f(x)dx \approx ABCD$ trapezium area. Note the curves approximated by the straight line joining A and B. This line is nothing but interpolating linear polynomial of the data $(a, f(a))$ and $(b, f(b))$.

5.5   COMPOSITE TRAPEZOIDAL RULE

The approximation of $I = \int_{a}^{b} f(x)dx$ obtained from Trapezoidal rule is closer to the exact value of I if the step size $(b - a)$ is smaller and smaller. To achieve this, the interval $[a, b]$ is divided into n equal intervals so that the step size of interval of differencing is $h = \dfrac{b-a}{n}$.

Suppose the n intervals be $[x_0, x_1]$, $[x_1, x_2]$, ... $[x_{n-1}, x_n]$, where $x_i = x_0 + ih$ with $x_0 = a$ and $x_n = b$

Now,

$$I = \int_{a}^{b} f(x)dx = \int_{x_0}^{x_0+h} f(x)dx + \int_{x_0+h}^{x_0+2h} f(x)dx + \cdots + \int_{x_0+(n-1)h}^{x_0+nh} f(x)dx$$

$$= \frac{h}{2}(y_0 + y_1) + \frac{h}{2}(y_1 + y_2) + \cdots + \frac{h}{2}(y_{n-1} + y_n)$$

$$= \frac{h}{2}[(y_0 + y_n) + 2(y_1 + y_2 + \cdots + y_{n-1})] \qquad (5.7)$$

which is known as composite Trapezoidal rule. It can be remembered as

$$I = \int_{x_0}^{x_0 + nh} f(x)dx = \frac{h}{2} \ \ [(\text{sum of first and last ordinates}) + 2 \ (\text{remaining ordinates})].$$

5.6   PROCEDURE TO EVALUATE $I = \int_a^b f(x)dx$ BY COMPOSITE TRAPEZOIDAL RULE

Step 1: Note down lower limit a of I and its upper limit b.

Step 2: Note down the number of divisions n of $[a, b]$ desired in the question. [*Note:* The number of divisions is number ordinates minus one.]

Step 3: Compute

$$h = \frac{b-a}{n}$$

Step 4: Compute the value of y at x_0, $x_0 + h$, $x_0 + 2h$, ..., $x_0 + nh$ and formulate the following table:

x	$y = f(x)$
x_0	$f(x_0)$ or y_0
x_1	$f(x_1)$ or y_1
$\vdots$	$\vdots$
x_n	$f(x_n)$ or y_n

Step 5: Write down the trapezoidal rule for n divisions ($n + 1$ ordinates)

$$\int_{x_0}^{x_0 + nh} f(x)dx = \frac{h}{2}[(y_0 + y_n) + 2(y_1 + y_2 + \cdots + y_{n-1})].$$

Step 6: Substitute the values of $y_0, y_1, ..., y_n$ found in Step 4 in the formula given in Step 5. Simplifying it, we get the value of

$$\int_a^b f(x)dx.$$

SOLVED PROBLEMS

Problem 1: Evaluate $\int_1^4 x^2 dx$ using the trapezoidal rule with three divisions (4 ordinates) of $[1, 4]$.

Solution: Clearly, $a = 1$, $b = 4$ and $n = 3$.

Therefore, the interval of differencing $h = \dfrac{b-a}{n} = \dfrac{4-1}{3} = 1$ and $x_0 = 1$, $x_1 = x_0 + h = 2$, $x_2 = x_0 + 2h = 3$ and $x_3 = x_0 + 3h = 4$. Now, we compute the $y = f(x) = x^2$ values at these points, and they are shown in the following table:

x	$y = f(x) = x^2$
1	$1 = y_0$
2	$4 = y_1$
3	$9 = y_2$
4	$16 = y_3$

The trapezoidal rule for three divisions (or four ordinates) is given by

$$\int_{x_0}^{x_0+4h} f(x)dx = \frac{h}{2}[(y_0 + y_3) + 2(y_1 + y_2)]$$

$$\therefore \qquad \int_1^3 x^2 dx = \frac{1}{2}[(1+16) + 2(4+9)] = 21.5$$

Note that the exact value of $\int_0^4 x^2 dx = 21$. Hence the error is 0.5. This error can be reduced by taking more divisions $[1, 4]$.

Problem 2: Evaluate $\int_0^1 \dfrac{dx}{1+x^2}$ using trapezoidal rule with $h = 0.2$. Hence, determine the approximation of π.

Solution: Here, $a = 0$, $b = 1$, $h = 0.2$ and $f(x) = \dfrac{1}{1+x^2}$. Therefore, the number of division of [0, 1] is $n = \dfrac{b-a}{h} = \dfrac{1-0}{0.2} = 5$ and $x_0 = 0$,

$x_1 = x_0 + h = 0.2$, $x_2 = x_0 + 2h = 0.4$, $x_3 = x_0 + 3h = 0.6$, $x_4 = x_0 + 4h = 0.8$

and $x_5 = x_0 + 5h = 1$. The computed values of $y = f(x) = \dfrac{1}{1+x^2}$ at these

points are shown below:

x	$y = f(x) = \dfrac{1}{1+x^2}$
0	$1 = y_0$
0.2	$0.9615 = y_1$
0.4	$0.8621 = y_2$
0.6	$0.7353 = y_3$
0.8	$0.6098 = y_4$
1.0	$0.5 = y_5$

The trapezoidal rule for five divisions (six ordinates) is

$$\int_{x_0}^{x_0+5h} f(x)\,dx = \frac{h}{2}[(y_0 + y_5) + 2(y_1 + y_2 + y_3 + y_4)]$$

$$\therefore \quad \int_0^1 \frac{1}{1+x^2}\,dx = \frac{0.2}{2}[(1+0.5) + 2(0.9615 + 0.8621 + 0.7353 + 0.6098)]$$

$$= 0.7837. \tag{1}$$

From the calculus, we know that

$$\int_0^1 \frac{dx}{1+x^2} = \tan^{-1}(x)\Big]_0^1 = \frac{\pi}{4} \tag{2}$$

From (1) and (2),

$$\frac{\pi}{4} \approx 0.7837$$

i.e., $\qquad\qquad\qquad\qquad \pi \approx 3.1348.$

Problem 3: Evaluate $\int_0^1 (e^{-x^2} + x)dx$ by dividing the range of integration into 4 equal parts and using the trapezoidal rule.

Solution: Here, $a = 0$, $b = 1$, and $n = 4$. Therefore, $h = \dfrac{b-a}{n} = \dfrac{1-0}{4}$

$= 0.25$ and $x_0 = 0$, $x_1 = 0.25$, $x_2 = 0.5$, $x_3 = 0.75$ and $x_4 = 1.0$

x	$y = f(x) = e^{-x^2} + x$
0	$1 = y_0$
0.25	$1.1894 = y_1$
0.5	$1.2788 = y_2$
0.75	$1.3198 = y_3$
1.0	$1.3679 = y_4$

$$\int_{x_0}^{x_0+4h} f(x)dx = \frac{h}{2}\left[(y_0 + y_4) + 2(y_1 + y_2 + y_3)\right]$$

$$\therefore \int_0^1 (e^{-x^2} + x)dx = \frac{0.25}{2}\left[(1 + 1.3679) + 2(1.1894 + 1.2788 + 1.3198)\right]$$

$$= 1.2429875$$

$$\approx 1.243.$$

5.7 SIMPSON'S 1/3 RULE

Substituting $n = 2$ in Eq. (5.5) and neglecting third and higher order differences, we get

$$\int_{x_0}^{x_0+2h} f(x)dx = h\left[2y_0 + \frac{4}{2}\Delta y_0 + \frac{1}{2}\left[\frac{8}{3} - \frac{4}{2}\right]\Delta^2 y_0\right]$$

$$= h[2y_0 + 2\Delta y_0 + \frac{1}{3}\Delta^2 y_0]$$

$$= h\left[2y_0 + 2(y_1 - y_0) + \frac{1}{3}(y_2 - 2y_1 + y_0)\right]$$

$$= \frac{h}{3}(y_0 + 4y_1 + y_2), \tag{5.8}$$

which is known as Basic Simpson's 1/3 rule and it is also called Simpson's 1/3 rule. It can also be expressed as

$$\int_a^b f(x)dx = \frac{\left(\dfrac{b-a}{2}\right)}{3}\left[f(a) + 4f\left(\frac{a+b}{2}\right) + f(b)\right].$$

5.8   COMPOSITE SIMPSON'S 1/3 RULE

Suppose the interval $[a, b]$ is divided into n equal integrals where n is an even number. Then, the subintervals of $[a, b]$ are $[x_0, x_1]$, $[x_1, x_2]$, $[x_2, x_3]$, ..., $[x_{n-1}, x_n]$, where $h = \dfrac{b-a}{n}$ and $x_i = x_0 + ih$ ($i = 0, 1, 2, ...,$ n) with $x_0 = a$, $x_n = b$

Now,

$$I = \int_a^b f(x)dx = \int_{x_0}^{x_0+nh} f(x)dx$$

It can be expressed as

$$I = \int_{x_o}^{x_0+2h} f(x)dx + \int_{x_0+2h}^{x_0+4h} f(x)dx + \cdots + \int_{x_0+n-2h}^{x_0+nh} f(x)dx$$

Using basic Simpson's 1/3 rule, each of interval on the right-hand side of above equation, we get

$$I = \frac{h}{3}(y_0 + 4y_1 + y_2) + \frac{h}{3}(y_2 + 4y_3 + y_4) + \cdots + \frac{h}{3}(y_{n-2} + 4y_{n-1} + y_n)$$

$$= \frac{h}{3}\left[(y_0 + y_n) + 4(y_1 + y_3 + \cdots + y_{n-1}) + 2(y_2 + y_4 + \cdots + y_{n-2})\right] \tag{5.9}$$

which is known as composite Simpson's 1/3 rule.
It can be remembered as

$$\int_{x_0}^{x_0+nh} f(x)dx = \frac{h}{3}\Big[(\text{sum of 1}^{st}\text{ and last ordinates})$$

$$+ 4(\text{sum of the ordinates whose suffixes are odd})$$

$$+ 2(\text{sum of the ordinates whose suffixes are}$$

$$\text{even excluding the last ordinate})\Big]$$

where n is an even number.

5.9   PROCEDURE TO EVALUATE $\int_{a}^{b} f(x)dx$ BY COMPOSITE SIMPSON'S 1/3 RULE

Step 1: Note down lower limit a and upper limit b of definite integral.

Step 2: Select number of divisions of $[a, b]$ n such that n is even.

Step 3: Compute $h = \dfrac{b-a}{n}$.

Step 4: Compute $f(x)$ (integrand) values at $x_0 = a$, $x_1 = x_0 + h$, ..., $x_n = x_0 + nh$. Mark these values as $y_0, y_1, y_2, ..., y_n$.

Step 5: Write composite Simpson's 1/3 rule for the selected n divisions.

Step 6: Substituting the values of h and $y_0, y_1, y_2, ..., y_n$ in the formula written at Step 5,

We get the required result after simplifying it.

SOLVED PROBLEMS

Problem 1: Evaluate $\int_{0}^{1} \dfrac{x}{1+x^2}$ by Simpson's 1/3 rule. Take four divisions of $[0, 1]$.

Solution: Here, $a = 0$, $b = 1$ and $n = 4$. Therefore,

$$h = \frac{b-a}{n} = \frac{1-0}{4} = 0.25. \text{ Further } x_0 = 0;$$

$$x_1 = 0.25, x_2 = 0.5, x_3 = 0.75, x_4 = 1.0.$$

The values of $y = f(x) = \dfrac{1}{1+x^2}$ are shown in the following table:

x	$y = f(x) = \dfrac{1}{1+x^2}$
0	$1 = y_0$
0.25	$0.9412 = y_1$
0.5	$0.8 = y_2$
0.75	$0.64 = y_3$
1.0	$0.5 = y_4$

The composite Simpson's 1/3 rule for 4 divisions is given by

$$\int_{x_0}^{x_0+4h} f(x)dx = \frac{h}{3}\left[(y_0 + y_4) + 4(y_1 + y_3) + 2y_2\right]$$

$$\therefore \quad \int_0^1 \frac{1}{1+x^2} dx = \frac{0.25}{3}\left[(1+0.5) + 4(0.9412 + 0.64) + 2(0.8)\right]$$

$$= 0.78539.$$

Problem 2: Find an approximate value of $\displaystyle\int_0^1 \frac{dx}{1+x}$ using Simpson's 1/3 rule.

Solution: We divide the range of integration in eight equal parts. The interval of differencing $h = \dfrac{1-0}{8} = \dfrac{1}{8}$. The values of $y = f(x) = \dfrac{1}{1+x}$ at $x_0 = 0$, $x_1 = 1/8$, $x_2 = 2/8$, $x_3 = 3/8$, $x_4 = 1/2$, $x_5 = 5/8$, $x_6 = 6/8$, $x_7 = 7/8$, $x_8 = 1$ are shown in the following table:

x	$f(x) = \dfrac{1}{1+x}$
0	$1 = y_0$
$\dfrac{1}{8}$	$0.8888 = y_1$
$\dfrac{1}{4}$	$0.8 = y_2$

$\dfrac{3}{8}$	$0.7272 = y_3$
$\dfrac{1}{2}$	$0.6666 = y_4$
$\dfrac{5}{8}$	$0.6154 = y_5$
$\dfrac{6}{8}$	$0.5714 = y_6$
$\dfrac{7}{8}$	$0.5333 = y_7$
1	$0.5 = y_8$

For eight divisions of the range of integration, the Simpson's 1/3 rule is

$$\int_{x_0}^{x_0+8h} f(x)dx = \frac{h}{3}\left[(y_0 + y_8) + 4(y_1 + y_3 + y_5 + y_7) + 2(y_2 + y_4 + y_6)\right]$$

$$\therefore \ \int_0^1 \frac{1}{1+x}dx = \frac{1}{24}\left[(1+0.5) + 4(0.8888 + 0.7272 + 0.6154\right.$$

$$+ 0.5333) + 2(0.8 + 0.6666 + 0.5714)\Big]$$

$$= \frac{1}{24} \times 16.6348 = 0.6931.$$

Problem 3: Evaluate $\displaystyle\int_0^6 \frac{\sin x}{1+x^2}dx$ by Simpson's 1/3 rule.

Solution: Dividing [0, 6] into six equal parts, we have the interval of differencing $h = \dfrac{b-a}{n} = \dfrac{6-0}{6} = 1$. Thus, we have to find $f(x) = \dfrac{\sin x}{1+x^2}$ at 0, 1, 2, 3, 4, 5 and 6 and they are shown in the following table:

x	$f(x) = \dfrac{\sin x}{1 + x^2}$
0	$0 = y_0$
1	$0.42075 = y_1$
2	$0.18186 = y_2$
3	$0.01411 = y_3$
4	$-0.0445 = y_4$
5	$-0.0369 = y_5$
6	$-0.00754 = y_6$

The Simpson's 1/3 rule for six divisions is

$$\int_{x_0}^{x_0+6h} f(x)dx = \frac{h}{3}\left[(y_0 + y_6) + 4(y_1 + y_3 + y_5) + 2(y_2 + y_4)\right]$$

$$\therefore \int_{0}^{6} \frac{\sin x}{1 + x^2} dx = \frac{1}{3}\left[(0 - 0.00745) + 4(0.42075\right.$$

$$\left. + 0.01411 - 0.0369) + 2(.18186 - 0.00754)\right]$$

$$= \frac{1}{3} \times 1.93294$$

$$= 0.64431.$$

5.10   SIMPSON'S 3/8 RULE

Substituting $n = 3$ in the general quadrature formula (5.5) and neglecting the third and higher differences, we get

$$\int_{x_0}^{x_0+3h} f(x)dx = h\left[3y_0 + \frac{9}{2}\Delta y_0 + \left(9 - \frac{9}{2}\right)\frac{\Delta^2 y_0}{\lfloor 2} + \left(\frac{81}{4} - 27 + 9\right)\frac{\Delta^3 y_0}{\lfloor 3}\right]$$

$$= h\left[3y_0 + \frac{9}{2}\Delta y_0 + \frac{9}{4}\Delta^2 y_0 + \frac{3}{8}\Delta^3 y_0\right]$$

$$= 3h\left[y_0 + \frac{3}{2}(y_1 - y_0) + \frac{3}{4}(y_2 - 2y_1 + y_0) + \frac{1}{8}(y_3 - 3y_2 + 3y_1 + y_0)\right]$$

$$= \frac{3h}{8}\left[y_0 + 3y_1 + 3y_2 + y_3\right], \tag{5.10}$$

which is known as Simpson's 3/8 rule or basic Simpson's 3/8 rule

5.11   COMPOSITE SIMPSON'S 3/8 RULE

Assuming n is a multiple of 3, we have

$$I = \int_{x_0}^{x_0+nh} f(x)dx = \int_{x}^{x_0+3h} f(x)dx + \int_{x_0+3h}^{x_0+6h} f(x)dx + \cdots + \int_{x_0+n-3h}^{x_0+nh} f(x)dx.$$

Using the Simpson's 3/8 rule for each integral on the right-hand side of the above equation, we get

$$I = \int_{x_0}^{x_0+nh} f(x)dx = \frac{3h}{8}\left[(y_0 + 3y_1 + 3y_2 + y_3) + (y_3 + 3y_4 + 4y_5 + y_6)\right.$$

$$\left. + \cdots + (y_{n-3} + 3y_{n-2} + 3y_{n-1} + y_n)\right]$$

$$= \frac{3h}{8}\left[(y_0 + y_n) + 3(y_1 + y_2 + y_4 + y_5 + \cdots + y_{n-2} + y_{n-1})\right.$$

$$\left. + 2(y_3 + y_6 + \cdots + y_{n-3})\right]. \tag{5.11}$$

The equation given by (5.11) is known as composite Simpson's 3/8 rule. It can be remembered as

$$I = \int_{x_0}^{x_0+nh} f(x)dx = \frac{3h}{8}\left[(\text{sum of first and last ordinates}) + 3(\text{sum of}\right.$$

the remaining ordinates whose suffixes are not multiple of three) + 2 (excluding the last ordinate the sum of the ordinates whose suffixes are multiple of 3)].

5.12   PROCEDURE TO EVALUATE $\int_a^b f(x)dx$ BY COMPOSITE SIMPSON'S 3/8 RULE

Step 1: Note down lower limit a and upper limit b of the given definite integral.

Step 2: Select the number of divisions of $[a, b]$ such that n is multiple of 3.

Step 3: Compute $h = \dfrac{b-a}{n}$.

Step 4: Compute $f(x)$ (integrand) values at $x_0 = a$, $x_1 = x_0 + h$, ..., $x_n = x_0 + nh$. Mark these vales as $y_0, y_1, ..., y_n$.

Step 5: Write composite Simpson's 3/8 rule for the selected divisions of $[a, b]$.

Step 6: Substitute the values of h and $y_0, y_1, ..., y_n$ in the formula written at Step 5. Simplifying it, we get the required approximate value of $\displaystyle\int_a^b f(x)dx$.

SOLVED PROBLEMS

Problem 1: Using Simpson's 3/8 rule, evaluate $\displaystyle\int_0^{1.5} \sqrt{\sin x + \cos x}\,dx$.

Solution: To apply the Simpson's 3/8 rule, the number of divisions must be multiple of three. Therefore, we select six equal divisons of $[0, 1.5]$. Thus, the interval of differencing $h = \dfrac{1.5-0}{6} = 0.25$. Now, we compute $f(x) = \sqrt{\sin x + \cos x}$ at 0, 0.25, 0.5, 0.75, 1, 1.25 and 1.5. They are shown in the following table:

x	$f(x)$
0	$1 = y_0$
0.25	$1.1028 = y_1$
0.5	$1.1648 = y_2$
0.75	$1.1887 = y_3$
1.0	$1.1754 = y_4$
1.25	$1.1244 = y_5$
1.5	$1.0335 = y_6$

The composite Simpson's 3/8 rule for six divisions is given by

$$I = \int_{x_0}^{x_0+6h} f(x)dx = \frac{3h}{8}\left[(y_0 + y_6) + 3(y_1 + y_2 + y_4 + y_5) + 2(y_3)\right]$$

$$\therefore \int_0^{1.5} \sqrt{\sin x + \cos x}\ dx = \frac{(3)(0.25)}{8}\left[(1+1.0335) + 3(1.1028 + 1.1648\right.$$

$$+ 1.1754 + 1.1244) + 2(1.1887)]$$

$$= 0.09375\left[2.0335 + 3(4.5674) + 2(1.1887)\right]$$

$$= 0.09375 \times 18.1131 = 1.6981.$$

Problem 2: Given

x	0	1	2	3	4	5	6
$f(x)$	0	1	8	27	64	125	216

Evaluate $\int_0^6 f(x)dx$ by Simpson's 3/8 rule.

Solution: From the given data, we have $a = 0$, $b = 6$, $h = 1$ and $y_0 = 0$, $y_1 = 1$, $y_2 = 8$, $y_3 = 27$, $y_4 = 64$, $y_5 = 125$ and $y_6 = 216$. For six divisions of the range of integration, the Simpson's 3/8 rule is

$$\int_{x_0}^{x_0+6h} f(x)dx = \frac{3h}{8}\left[(y_0 + y_6) + 3(y_1 + y_2 + y_4 + y_5) + 2y_3\right]$$

$$\therefore \int_0^6 f(x)dx = \frac{(3)(1)}{8}\left[(0 + 216) + 3(1 + 8 + 64 + 125) + 2(27)\right]$$

$$= 324.$$

Problem 3: Evaluate $\int_0^9 \frac{1}{1+x}\,dx$ using Simpson's 3/8 rule.

Solution: We select nine equal divisions of [0, 9]. Then, $h = \dfrac{9-0}{9} = 1$. We need to compute the values of $f(x) = \dfrac{1}{1+x}$ at 0, 1, 2, 3, 4, 5, 6, 7, 8 and 9. These values are shown in the following table:

x	$f(x) = \dfrac{1}{1+x}$	x	$f(x) = \dfrac{1}{1+x}$
0	$1 = y_0$	5	$0.1666 = y_5$
1	$0.5 = y_1$	6	$0.1428 = y_6$
2	$0.3333 = y_2$	7	$0.125 = y_7$
3	$0.25 = y_3$	8	$0.1111 = y_8$
4	$0.2 = y_4$	9	$0.1 = y_9$

The Simpson's 3/8 rule for nine equal parts of the range of integration is given by

$$I = \int_{x_0}^{x_0+9h} f(x)\,dx = \frac{3h}{8}\left[(y_0 + y_9) + 3(y_1 + y_2 + y_4 + y_5 + y_7 + y_8) + 2(y_3 + y_6)\right]$$

$$\therefore \quad \int_0^9 \frac{1}{1+x}\,dx = \frac{3(1)}{8}\left[(1+0.1) + 3(0.5 + 0.3333 + 0.2\right.$$
$$\left. + 0.1666 + 0.125 + 0.1111) + 2(0.25 + 0.1428)\right]$$
$$= \frac{3}{8}\left[1.1 + 3(1.2353) + 2(.3928)\right]$$
$$= \frac{3}{8}\left[1.1 + 3.7059 + 0.7856\right]$$
$$= 2.0968125$$

5.13 WEDDLE'S RULE

The general quadrature formula is given by Eq. (5.5). We recall it stating the terms up to sixth–order forward differences.

$$I = \int_{x_0}^{x_0+nh} f(x)dx = h\left[ny_0 + \frac{n^2}{2}\Delta y_0 + \left(\frac{n^3}{3} - \frac{n^2}{2}\right)\frac{\Delta^2 y_0}{\lfloor 2}\right.$$

$$+ \left(\frac{n^4}{4} - n^3 + n^2\right)\frac{\Delta^3 y_0}{\lfloor 3} + \left(\frac{n^5}{5} - \frac{3}{2}n^4 + \frac{11}{3}n^3 - 3n^2\right)\frac{\Delta^4 y_0}{\lfloor 4}$$

$$+ \left(\frac{n^6}{6} - 2n^5 + \frac{35}{4}n^4 - \frac{50}{3}n^3 + 12n^2\right)\frac{\Delta^5 y_0}{\lfloor 5}$$

$$\left. + \left(\frac{n^7}{7} - \frac{5}{2}n^6 + 17n^5 - \frac{225}{4}n^4 + \frac{274}{3}n^3 - 60n^2\right)\frac{\Delta^6 y_0}{\lfloor 6} + \cdots \right] \quad (5.12)$$

Substituting $n = 6$ in Eq. (5.12) and neglecting seventh and higher order forward differences, we get

$$I = \int_{x_0}^{x_0+6h} f(x)dx = h\left[6y_0 + 18\Delta y_0 + 27\Delta^2 y_0 + 24\Delta^3 y_0 \right.$$

$$\left. + \frac{123}{10}\Delta^4 y_0 + \frac{33}{10}\Delta^5 y_0 + \frac{41}{140}\Delta^6 y_0 \right]. \quad (5.13)$$

Using the following relation,

$$\Delta^n y_0 = y_n - {}^n C_1 y_{n-1} + {}^n C_2 y_{n-2} + \cdots + (-1)^n y_0,$$

and replacing the term $\frac{41}{140}\Delta^6 y_0$ by $\frac{42}{140}\Delta^6 y_0$. The Eq. (5.13) reduces to

$$I = \int_{x_0}^{x_0+6h} f(x)dx = \frac{3h}{10}\left[y_0 + 5y_1 + y_2 + 6y_3 + y_4 + 5y_5 + y_6 \right]. \quad (5.14)$$

Eq. (5.14) is known as Weddle's rule or Basic Weddle's rule.
If n is multiple of 6, then the composite Weddle's rule is given by

$$I = \int_{x_0}^{x_0+nh} f(x)dx = \frac{3h}{10}\left[y_0 + 5y_1 + y_2 + 6y_3 + y_4 + 5y_5 + 2y_6 + 5y_7 \right.$$

$$+ y_8 + 6y_9 + y_{10} + 5y_{11} + 2y_{12} + \ldots + 5y_{n-5} + y_{n-4}$$

$$\left. + 6y_{n-3} + y_{n-2} + 5y_{n-1} + y_n \right] \quad (5.15)$$

SOLVED PROBLEMS

Problem 1: Evaluate $\int_0^{12} \sqrt{1+x^2}\, dx$ by Weddle's rule.

Solution: Here, $a = 0$, $b = 12$ and $f(x) = \sqrt{1+x^2}$. Choose number of equal divisions of $[0, 12]$ as 12. Then, $h = \dfrac{12-0}{12} = 1$. The values of $f(x)$ at $x = 0, 1, 2, 3, \ldots, 10, 11, 12$ are shown below:

x	$f(x) = \sqrt{1+x^2}$	x	$f(x) = \sqrt{1+x^2}$	x	$f(x) = \sqrt{1+x^2}$
0	$1 = y_0$	5	$5.0990 = y_5$	10	$10.0498 = y_{10}$
1	$1.4142 = y_1$	6	$6.0827 = y_6$	11	$11.0453 = y_{11}$
2	$2.2360 = y_2$	7	$7.0710 = y_7$	12	$12.0416 = y_{12}$
3	$3.1622 = y_3$	8	$8.0622 = y_8$		
4	$4.1231 = y_4$	9	$9.0533 = y_9$		

For twelve equal divisions of the range of integration, the Weddle's rule is

$$\int_{x_0}^{x_0+12h} f(x)\,dx = \frac{3h}{10}\big[y_0 + 5y_1 + y_2 + 6y_3 + y_4 + 5y_5 + 2y_6$$

$$+ 5y_7 + y_8 + 6y_9 + y_{10} + 5y_{11} + y_{12}\big]$$

$$\therefore \int_0^{12} \sqrt{1+x^2}\,dx = \frac{3(1)}{10}\big[1 + 5(1.4142) + 2.2360 + 6(3.1622) + 4.1231$$

$$+ 5(5.0990) + 2(6.0827) + 5(7.0710) + 8.0622$$

$$+ 6(9.0553) + 10.0498 + 5(11.0453) + 12.0416\big]$$

$$= 73.8392$$

Problem 2: Evaluate $\int_0^1 \dfrac{dx}{1+x}$ by Weddle's rule

Solution: Choose $n = 6$. Then $h = 1/6$ as $a = 0$, $b = 1$ and $n = 6$.

The values of $f(x) = \dfrac{1}{1+x}$ at $x = 0$, 1/6, 2/6, 3/6, 4/6, 5/6 and 6/6 are shown below:

x	$f(x) = \dfrac{1}{1+x}$
0	$1\ (= y_0)$
1/6	$0.8571\ (= y_1)$
2/6	$0.75\ (= y_2)$
3/6	$0.6667\ (= y_3)$
4/6	$0.6\ (= y_4)$
5/6	$0.5455\ (= y_5)$
6/6	$0.5\ (= y_6)$

Now,

$$\int_0^1 \frac{1}{1+x}\,dx = \frac{3\left(\dfrac{1}{6}\right)}{10}\left[1 + 5(0.8571) + 0.75 + 6(0.6667) + (0.6) + 5(0.5455) + 0.5\right]$$

$$= 0.69320.$$

Problem 3: Brakes are applied when a train is moving at 30m/sec. The velocity of the train at t (sec) are given below:

time (t)	0	5	10	15	20	25	30	35	40
Speed (v)	30	24	19.5	16	13.6	11.7	10	8.5	7.0

Determine the distance moved by the train in 40 seconds using Simpson's 1/3 rule

Solution: We know that $\dfrac{ds}{dt} = v$, where s is distance and v is the velocity of the train. Integrating it from 0 to 40, we have

$$\int_0^{40} ds = \int_0^{40} v\,dt$$

i.e., the distance moved in 40 seconds

$$= \int_0^{40} v \, dt \tag{1}$$

For eight divisions of the range of integration, the Simpson's 1/3 rule is

$$\int_{x_0}^{x_0+8h} f(x)dx = \frac{h}{3}\left[(y_0 + y_8) + 4(y_1 + y_3 + y_5 + y_7) + 2(y_2 + y_4 + y_6)\right]$$

$$\therefore \int_0^{40} V \, dt = \frac{5}{3}\left[(30 + 70) + 4(24 + 16 + 11.7 + 8.5) + 2(19.5 + 13.6 + 10)\right]$$

$$= 573.437 \text{ meters} . \tag{2}$$

From (1) and (2), we can conclude that the train moved a distance of 573.437 meters in 40 seconds.

Problem 4: A rocket is launched from the ground. Its acceleration measured every 5 seconds and they are as follows:

t:	0	5	10	15	20	25	30	35	40
$a(t)$:	40	45.25	48.50	51.25	54.35	59.48	61.5	64.3	68.7

Find the velocity and the position of the rocket at $t = 40$ seconds using the trapezoidal rule.

Solution: Let s be the distance travelled by the rocket in time t and v be the velocity of it at time t. Then

$$\text{Acceleration} = a(t) = \frac{dv}{dt}$$

$$\therefore \qquad dv = a(t)dt$$

Integrating the above equation on both sides between the time limits 0 and 40, we have

$$\int_0^{40} dv = \int_0^{40} a dt$$

Velocity at 40 sec. – velocity at 0 sec.

$$= \int_0^{40} a\,dt,$$

i.e., Velocity at 40 sec − 0

$$= \int_0^{40} a\,dt \tag{1}$$

Now, by the trapezoidal rule

$$\int_0^{40} a\,dt = \frac{5}{2}[(40+68.7)+2(45.25+48.50+51.25+54.35$$

$$+\,59.48+61.5+64.3]$$

$$=2194.9.$$

We know that

$$s = u + at.$$

The position of the rocket after 40 seconds is $s = 0 + 2194.9 \times 40 = 87796$ units.

Problem 5: A curve $y = f(x)$ passes through the points as given in the following table:

x	1	1.5	2	2.5	3
$f(x)$	2	3.25	5	7.25	10

Find the area bounded by the curve, x-axis, $x = 1$ and $x = 3$.

Solution: The required area is $\int_{+1}^{3} f(x)dx$. Let $I = \int_{+1}^{3} f(x)dx$. For five ordinates, the Simpson's 1/3 rule is

$$\int_{x_0}^{x_0+4h} f(x)dx = \frac{h}{3}\left[y_0 + y_4 + 4(y_1 + y_3) + 2y_2\right]$$

Here, $x_0 = 1$ and $h = 0.5$

$$\therefore \qquad \int_1^3 f(x)dx = \frac{0.5}{3}\left[2+10+4(3.25+7.25)+2(5)\right]$$

$$= \frac{32}{3} = 10.6666.$$

Problem 6: A curve is passing through $(-2, 0)$ $(-1, \sqrt{3})$, $(0, 2)$, $(1, \sqrt{3})$ and $(2, 0)$. Find the volume of solid generated by revolving the area bounded by the curve, x-axis, $x = -2$ and $x = 2$ about x-axis.

Solution: We know that the volume of surface of revolution $= \pi \int_{-2}^{2} y^2 \, dx$

$= \pi \int_{-2}^{2} [f(x)]^2 \, dx$. Now, we compute y^2 values

x	y	y^2
-2	0	$0 = y_0$
-1	$\sqrt{3}$	$3 = y_1$
0	2	$4 = y_2$
1	$\sqrt{3}$	$3 = y_3$
2	0	$0 = y_4$

Suppose that $y = f(x)$, $F(x) = [f(x)]^2$. Then, required volume of solid generated by the area given in the problem about x-axis is $\pi \int_{-2}^{2} F(x) dx$.

Let
$$I = \pi \int_{-2}^{2} F(x) dx. \tag{1}$$

By Simpson's 1/3 rule,

$$\int_{-2}^{2} F(x) dx = \frac{1}{3}\left[(0^2 + 0^2) + 4(3+3) + 2 \times 4\right] = \frac{1}{3} \times 32 = \frac{32}{3} \tag{2}$$

From (1) and (2),

$$I = \frac{32\pi}{3} \text{ units}.$$

Hence, the volume of the solid generated by the area bounded by the curve, x-axis, $x = -2$ and $x = 2$ about x-axis is

$$\frac{32\pi}{3} \text{ units}.$$

5.14   NEWTON–COTES INTEGRATION METHODS

In general problem of numerical integration, the integral

$I = \int_a^b w(x) f(x) dx$ is approximated by a linear combination of values

$f(x_0), f(x_1), \ldots, f(x_n)$ in the form

$$I = \int_a^b w(x) f(x) dx = \sum_{k=0}^{n} \lambda_k f(x_k), \qquad (5.16)$$

where x_k ($k = 0, 1, 2, \ldots, n$) are either fixed in advance or some points in $[a, b]$. The arguments x_k's are called nodes or abscissas. Further, λ_i ($i = 0, 1, 2, \ldots, n$) and $w(x)$ involved in Eq. (5.16) are called weights and weight function of quadrature formula (5.16) respectively.

If the nodes $x_0, x_1, \ldots, x_n$ are prescribed in advance, and thus to determine a method of integration of type (5.16), we have to find only weights. These weights can be found by replacing $f(x)$ by an interpolating polynomial $P_n(x)$ for the data $(x_i, f(x_i))$ ($i = 0, 1, 2, \ldots, n$). They are also found by another method that we describe in section (5.20).

If $w(x) = 1$ and the nodes x_i's are equally spaced, such that x_0, $x_1 = x_0 + h, x_2 = x_0 + 2h, \ldots, x_n = x_0 + nh$ with $x_0 = a, x_n = b$ and h is the interval difference, then the integration methods of the type (5.16) reduces to

$$\int_a^b f(x) dx = \sum_{k=0}^{n} \lambda_k y_k, \qquad (5.17)$$

which are known as Newton–Cotes type of integration method.
Now, we wish to find the integration methods of type (5.17) using Lagrange's interpolation formula.

5.15   NEWTON–COTES INTEGRATION FORMULA

Let $y_0, y_1, \ldots, y_n$ be ($n + 1$) values of $y = f(x)$ at $x = x_0, x_1, x_2, \ldots, x_n$ with $x_0 = a, x_n = b$ and $x_i = x_0 + ih$ where h is the interval of differencing. Further let

$$I = \int_a^b f(x) dx. \qquad (5.18)$$

By Lagrange's interpolation formula, we have

$$f(x) \approx \sum_{k=0}^{n} L_k(x) y_k, \tag{5.19}$$

where

$$L_k(x) = \frac{(x-x_0)(x-x_1)\cdots(x-x_{k-1})(x-x_{k+1})\cdots(x-x_n)}{(x_k-x_0)(x_k-x_1)\cdots(x_k-x_{k-1})(x_k-x_{k+1})\cdots(x_k-x_n)} \tag{5.20}$$

Let $s = \dfrac{x-x_0}{h}$ \hfill (5.21)

Then, Eq. (5.20) becomes

$$L_k(s) = \frac{s(s-1)\cdots\{s-(k-1)\}\{s-(k+1)\}\cdots(s-n)}{k(k-1)\cdots\{1\}\{-1\}\cdots(k-n)}, \tag{5.22}$$

and we get

$$I = \int_{x_0}^{x_0+nh} f(x)\,dx$$

$$= \int_{x_0}^{x_0+nh} \sum_{k=0}^{n} L_k(x) y_k\,dx$$

$$= \int_0^n \left\{ \sum_{k=0}^{n} L_k(s) y_k \right\} h\,ds \qquad (\because \quad h\,ds = dx)$$

$$= nh \int_0^n \left\{ \sum_{k=0}^{n} \frac{1}{n} L_k(s) y_k \right\} ds$$

$$= nh \sum_{k=0}^{n} \left\{ \frac{1}{n} \int_0^n L_k(s)\,ds \right\} y_k \tag{5.23}$$

$$\therefore \quad I = nh \sum_{k=0}^{n} C_k^n y_k, \tag{5.24}$$

where

$$C_k^n = \frac{1}{n} \int_0^n L_k\,ds. \tag{5.25}$$

The formula given by (5.24) is known as Newton–Cotes formula and C_k^n are called Cotes numbers.

We can also derive the Trapezoidal, Simpson's 1/3 rule and Simpson's 3/8 rule from Newton–Cotes formula.

SOLVED PROBLEMS

Problem 1: Find Newton–Cotes numbers C_0^2, C_1^2, and C_2^2

Solution: From Eq. (5.25), we have

$$C_k^2 = \frac{1}{2}\int_0^2 L_k(s)\,ds\,,$$

where $L_k(s)$ is given by Eq. (5.22). Therefore, $L_0(s)$, $L_1(s)$ and $L_2(s)$ are respectively $\dfrac{(s-1)(s-2)}{(0-1)(0-2)}$, $\dfrac{(s-0)(s-2)}{(1-0)(1-2)}$ and $\dfrac{(s-0)(s-1)}{(2-0)(2-1)}$. Now,

$$C_0^2 = \frac{1}{2}\int_0^2 L_0(s)\,ds$$

$$= \frac{1}{2}\int_0^2 \frac{(s-1)(s-2)}{(-1)(-2)}\,ds$$

$$= \frac{1}{4}\int_0^2 (s^2 - 3s + 2)\,ds$$

$$= \frac{1}{6}.$$

Similarly,

$$C_1^2 = \frac{1}{2}\int_0^2 \frac{s(s-2)}{(1)(-1)}\,ds = \frac{2}{3},$$

and

$$C_2^2 = \frac{1}{2}\int_0^2 \frac{s(s-1)}{(2)(1)}\,ds = \frac{1}{6}.$$

Problem 2: Obtain trapezoidal rule from Newton–Cotes formula.

Solution: For $n = 1$, the Newton–Cotes formula given by 5.24 reduces to

$$I = \int_{x_0}^{x_0+h} f(x)\,dx = 1 \cdot h\left[C_0^1 y_0 + C_1^1 y_1\right] \tag{1}$$

Now,

$$C_0^1 = \frac{1}{1}\int_0^1 \frac{(s-1)}{(0-1)}\,ds = -\left[\frac{s^2}{2} - s\right]_0^1 = -\left[\frac{1}{2} - 1\right] = \frac{1}{2},$$

and

$$C_1^1 = \frac{1}{1}\int_0^1 \frac{(s-0)}{(1-0)}\,ds = \frac{1}{2};$$

$$\therefore \qquad I = \int_{x_0}^{x_0+h} f(x)\,dx = h\left[\frac{1}{2}y_0 + \frac{1}{2}y_1\right] = \frac{h}{2}(y_0 + y_1),$$

which is the required trapezoidal rule.

Problem 3: Obtain Simpson's 1/3 rule from the Newton–Cotes formula.

Solution: For $n = 2$, the Newton–Cotes formula given by 5.24 reduces to

$$I = \int_{x_0}^{x_0+2h} f(x)\,dx = 2\cdot h\left[C_0^2 y_0 + C_1^2 y_1 + C_2^2 y_2\right] \qquad (1)$$

We have shown that $C_0^2 = \dfrac{1}{6}, C_1^2 = \dfrac{2}{3}$ and $C_2^2 = \dfrac{1}{6}$ in the solution of Problem 1. Therefore,

$$I = \int_{x_0}^{x_0+2h} f(x)\,dx = 2h\left[\frac{1}{6}y_0 + \frac{2}{3}y_1 + \frac{1}{6}y_2\right]$$

$$= \frac{h}{3}\left[y_0 + 4y_1 + y_2\right],$$

which is the required Simpson's 1/3 rule.

5.16 GAUSS–LEGENDRE QUADRATURE

In Newton–Cotes integration methods, the nodes are fixed in advance, whereas in Gauss–Legendre quadrature, the nodes are not known in advance. In the former case to determine an integration method of type (5.17), we need only to find weights, whereas in the later case, we have to determine both nodes and weights. As an illustration, we describe the three-point Gauss–Legendre quadrature in this section. Before proceeding to describe it, we mention a point about the conversion of integral $\int_a^b f(x)dx$ into an integral of a function having -1 and 1 as the limits of integration, i.e.,

$$I = \int_a^b f(x)dx = \int_{-1}^1 f\left(\frac{(a+b)+(b-a)t}{2}\right) \cdot \frac{(b-a)}{2} dt. \tag{5.26}$$

It can be shown by considering the substitution $x = p + qt$ and using the fact $t = -1$ when $x = a$ and $t = +1$ when $x = b$. Now, we derive the three-point Gauss–Legendre quadrature formula. Suppose

$$I = \int_{-1}^1 f(x)dx = C_1 f(x_1) + C_2 f(x_2) + C_3 f(x_3), \tag{5.27}$$

where the nodes are x_1, x_2, x_3 and weights are $C_1, C_2,$ and C_3. The method of finding the weights when nodes are known is described in section (5.18). To determine nodes and weights in Eq. (5.27), we assume that the method (5.27) is exact for the polynomials $1, x, x^2, x^3, x^4$ and x^5 to determine six unknowns namely $x_1, x_2, x_3, C_1, C_2,$ and C_3. Now, we proceed to determine these values. Substituting $f(x) = 1, x, x^2, x^3, x^4$, and x^5 in (5.27), we get the following equations:

$$C_1 + C_2 + C_3 = 2 \tag{A1}$$

$$C_1 x_1 + C_2 x_2 + C_3 x_3 = 0 \tag{A2}$$

$$C_1 x_1^2 + C_2 x_2^2 + C_3 x_3^2 = \frac{2}{3} \tag{A3}$$

$$C_1 x_1^3 + C_2 x_2^3 + C_3 x_3^3 = 0 \tag{A4}$$

$$C_1 x_1^4 + C_2 x_2^4 + C_3 x_3^4 = \frac{2}{5} \tag{A5}$$

$$C_1 x_1^5 + C_2 x_2^5 + C_3 x_3^5 = 0 \tag{A6}$$

Multiplying (A1) by x_3 and subtracting from (A2), we get

$$(x_3 - x_1)C_1 + (x_3 - x_2)C_2 = 2x_3 \tag{B1}$$

Similarly, multiplying (A2) by x_3, (A3) by x_3, (A4) by x_3, (A5) by x_3 and subtracting respectively from A3, A4, A5 and A6, we obtain

$$x_1(x_3 - x_1)C_1 + x_2(x_3 - x_2)C_2 = \frac{-2}{3} \tag{B2}$$

$$x_1^2(x_3 - x_1)C_1 + x_2^2(x_3 - x_2)C_2 = \frac{2}{3}x_3 \tag{B3}$$

$$x_1^3(x_3 - x_1)C_1 + x_2^3(x_3 - x_2)C_2 = \frac{-2}{5} \tag{B4}$$

$$x_1^4(x_3 - x_1)C_1 + x_2^4(x_3 - x_2)C_2 = \frac{2}{5}x_5 \tag{B5}$$

Now, multiplying (B1) by x_2, (B2) by x_2, (B3) by x_2, (B4) by x_2 and subtracting respectively from (B2), (B3), (B4) and (B5), we get

$$(x_3 - x_1)(x_2 - x_1)C_1 = 2x_2 x_3 + \frac{2}{3} \tag{D1}$$

$$x_1(x_3 - x_1)(x_2 - x_1)C_1 = \frac{-2}{3}(x_2 + x_3) \tag{D2}$$

$$x_1^2(x_3 - x_1)(x_2 - x_1)C_1 = \frac{2}{5} + \frac{2}{3}x_2 x_3 \tag{D3}$$

$$x_1^3(x_3 - x_1)(x_2 - x_1)C_1 = -\frac{2}{5}(x_2 + x_3) \tag{D4}$$

From (D2) and (D4), we have $x_1 = \pm\sqrt{\frac{3}{5}}$ and from (D3) and (D1) we

have $x_2 x_3 = 0$. By taking $x_1 = \sqrt{\frac{3}{5}}$, $x_2 = 0$, it can be shown that $x_3 = -x_1$,

i.e., $x_3 = -\sqrt{\frac{3}{5}}$. From Eq. (D1), we get $C_1 = \frac{5}{9}$. From Eq. (B1), we get

$C_2 = \frac{8}{9}$ by substituting $x_3 = -x_1$ and $C_1 = \frac{5}{9}$. Finally substituting C_1, C_2

values in Eq. (A1), we get $C_3 = \frac{5}{9}$. Thus,

$$\int_{-1}^{1} f(x)dx = \frac{5}{9}f\left(-\sqrt{\frac{3}{5}}\right) + \frac{8}{9}f(0) + \frac{5}{9}f\left(\sqrt{\frac{3}{5}}\right), \tag{5.28}$$

which is known as three-point Gauss–Legendre formula. Note that to

apply the method given by (5.28), to evaluate $\int_{a}^{b} f(x)dx$, we must be given the integrand. It is also to be noted that it is applicable to interval having -1 and $+1$ only the limits of integration. Further, the nodes are the roots of second-order Legendre polynomial, i.e.,

$$P_3(x) = \frac{1}{2}(5x^3 - 3x).$$

SOLVED PROBLEMS

Problem 1: Evaluate $\displaystyle\int_0^3 x^2 \sin x \, dx$ using three-point Gauss–Legendre quadrature formula.

Solution: Let $I = \displaystyle\int_0^3 x^2 \sin x \, dx$ and $f(x) = x^2 \sin x$. In order to apply Gauss–Legendre formula, we have to convert the limits of integration 0 and 3 to -1 and 1. From the Eq. (5.26), we have

$$I = \int_{-1}^{1} f\left(\frac{(0+3)+(3-0)t}{2}\right)\cdot\left(\frac{3-0}{2}\right) dt$$

$$= \frac{3}{2}\int_{-1}^{1} f\left(\frac{3}{2}(t+1)\right) dt$$

$$\therefore \quad I = \frac{3}{2}\int_{-1}^{1} \left(\frac{3}{2}(t+1)\right)^2 \sin\left(\frac{3}{2}(t+1)\right) dt.$$

Now suppose $F(t) = \left(\dfrac{3}{2}(t+1)\right)^2 \sin\left(\dfrac{3}{2}(t+1)\right)$. Then by three-point Gauss–Legendre formula,

$$I = \frac{3}{2}\left[\frac{5}{9}F\left(-\sqrt{\frac{3}{5}}\right) + \frac{8}{9}F(0) + \frac{5}{9}F\left(\sqrt{\frac{3}{5}}\right)\right]$$

Now $F\left(-\sqrt{\dfrac{3}{5}}\right) = 0.0379$, $F(0) = 2.2444$ and $F\left(\sqrt{\dfrac{3}{5}}\right) = 3.2701.$

Therefore, $I = 5.7835$

Problem 2: Derive two-point Gauss–Legendre quadrature formula.

Solution: Let $I = \int_{-1}^{1} f(x)dx = C_1 f(x_1) + C_2 f(x_2)$ (1)

Since this method is supposed to be exact for $f(x) = 1, x, x^2$ and x^3, we substitute $f(x) = 1, x, x^2$ and x^3 in (1). Then Eq. (1) gives the following equations:

$$C_1 + C_2 = 2 \tag{A1}$$

$$C_1 x_1 + C_2 x_2 = 0 \tag{A2}$$

$$C_1 x_1^2 + C_2 x_2^2 = \frac{2}{3} \tag{A3}$$

$$C_1 x_1^3 + C_2 x_2^3 = 0 \tag{A4}$$

Multiplying (A2) by x_1 and subtracting from (A4), we get

$$C_2 x_2 (x_1 + x_2)(x_1 - x_2) = 0.$$

For the unique solution $C_2 \neq 0$, $x_2 \neq 0$, $x_1 \neq x_2$. Therefore, $x_2 = -x_1$. Substituting $x_2 = -x_1$ in (A2), we have $C_1 - C_2 = 0$. Solving $C_1 + C_2 = 2$ and $C_1 - C_2 = 0$, we have $C_1 = C_2 = 1$; substituting $C_1 = C_2 = 1$ and $x_2 = -x_1$ in (A3), we get $x_1 = -\dfrac{1}{\sqrt{3}}$. Thus, $C_1 = 1, C_2 = 1, x_1 = -\dfrac{1}{\sqrt{3}}$, $x_2 = \dfrac{1}{\sqrt{3}}$. Hence the two-point Gauss–Legendre formula is

$$\int_{-1}^{1} f(x)\, dx = f\left(-\frac{1}{\sqrt{3}}\right) + f\left(+\frac{1}{\sqrt{3}}\right).$$

Problem 3: Evaluate $\displaystyle\int_{3}^{5} \frac{e^x}{1+x}\, dx$ using two-point Gauss–Legendre quadrature formula.

Solution: Let $I = \displaystyle\int_{3}^{5} f(x)\, dx$ and $f(x) = \dfrac{e^x}{1+x}$. The integral I with limits of integration -1 and 1 is given by

$$I = \int_{-1}^{1} f\left(\frac{(3+5)+(5-3)t}{2}\right)\left(\frac{5-3}{2}\right) dt = \int_{-1}^{1} f(4+t)\, dt$$

Suppose $F(t) = f(4 + t)$. Then $F(t) = \dfrac{e^{4+t}}{5+t}$.

Now

$$I = \int_{-1}^{1} F(t)\, dt$$

$$= F\left(-\frac{1}{\sqrt{3}}\right) + F\left(+\frac{1}{\sqrt{3}}\right)$$

$$= \frac{e^{4-\frac{1}{\sqrt{3}}}}{5 - \dfrac{1}{\sqrt{3}}} + \frac{e^{4+\frac{1}{\sqrt{3}}}}{5 + \dfrac{1}{\sqrt{3}}}$$

$$= 24.3681.$$

Problem 4: Evaluate $\displaystyle\int_{-1}^{1} x^4\, dx$ using three-point Gauss–Legendre formula.

Solution: Here, $f(x) = x^4$ and the limits of integration are -1 and 1.
The three-point Gauss–Legendre formula is

$$\int_{-1}^{1} f(x)\, dx = \frac{5}{9} f\left(-\sqrt{\frac{3}{5}}\right) + \frac{8}{9} f(0) + \frac{5}{9} f\left(\frac{\sqrt{3}}{5}\right)$$

$$\therefore \quad \int_{-1}^{1} x^4\, dx = \frac{5}{9}\left(\frac{\sqrt{3}}{5}\right)^4 + \frac{8}{9}(0)^4 + \frac{5}{9}\left(\sqrt{\frac{3}{5}}\right)^4$$

$$= 2 \cdot \frac{5}{9} \cdot \frac{9}{25} = \frac{2}{5} = 0.4.$$

Note that exact value $\left[\dfrac{x^5}{5}\right]_{-1}^{1} = \left(\dfrac{1}{5}\right) - \left(-\dfrac{1}{5}\right) = \dfrac{2}{5} = 0.4.$

Hence, the error $E(I) = 0$.

5.17   ERROR IN QUADRATURE FORMULA

The error in approximating an integration method of the form

$$I = \int_{a}^{b} f(x)\, dx = \sum_{k=0}^{n} \lambda_k f(x_k) \quad \text{is denoted by } E(I), \text{ and it is given by}$$

$$E(I) = \int\limits_{a}^{b} f(x)\,dx - \sum_{k=0}^{n} \lambda_k f(x_k). \tag{5.29}$$

The integration method of the type (5.28) is said to be of order P if it produces exact values for all polynomials of degree less than or equal to P. If $f(x)$ values are given at some fixed points in $[a, b]$, then we can find $E(I)$ using the remainder term of Newton's formula for forward interpolation or Lagrange's interpolation formula. It can also be obtained using Taylor's series expansion. Now, we compute the errors in Trapezoidal rule, Simpson's 1/3 rule and Simpson's 3/8 rule.

5.18(A)   ESTIMATION OF ERROR IN TRAPEZOIDAL RULE

Let
$$I = \int\limits_{x_0}^{x_0+h} f(x)\,dx = \frac{h}{2}\big[f(x_0) + f(x_0 + h))\big] \tag{5.30}$$

be the value of the integral obtained from the Trapezoidal rule. Assume that $f(x)$ is continuous function and have continuous derivatives in $[x_0, x_0 + h]$. Let

$$I_e = \int\limits_{x_0}^{x_0+h} f(x)\,dx = F(x_0 + h) - F(x_0), \tag{5.31}$$

where $\dfrac{d}{dx} F(x) = f(x)$. Then I_e is the exact value of the integral. Hence the error in trapezoidal rule is

$$E(I_1) = \text{Exact vlaue of} \int\limits_{x_0}^{x_0+h} f(x)\,dx - \text{Approximate value of} \int\limits_{x_0}^{x_0+h} f(x)\,dx$$

by the trapezoidal rule

i.e.,
$$E(I_1) = F(x_0 + h) - F(x_0) - \frac{h}{2}\big[f(x_0) + f(x_0 + h)\big]. \tag{5.32}$$

Now, by Taylor's series, we obtain

$$F(x_0 + h) = F(x_0) + hF'(x) + \frac{h^2}{\underline{2}} F''(x_0) + \frac{h^3}{\underline{3}} F'''(x_0) + \cdots$$

Using the fact $F'(x) = f(x)$, $F''(x) = f'(x)$, $F'''(x) = f''(x)$... etc. the above equation reduces to

$$F(x_0 + h) = F(x_0) + hf(x_0) + \frac{h^2}{\lfloor 2} f'(x_0) + \frac{h^3}{\lfloor 3} f''(x_0) + \cdots \tag{5.33}$$

Further

$$f(x_0 + h) = f(x_0) + hf'(x_0) + \frac{h^2}{\lfloor 2} f''(x_0) + \cdots \tag{5.34}$$

Substituting (5.33) and (5.34) in (5.32), the error $E(I_1)$ becomes

$$E(I_1) = \left[F(x_0) + hf(x_0) + \frac{h^2}{\lfloor 2} f'(x_0) + \frac{h^3}{\lfloor 3} f''(x_0) + \cdots \right] - F(x_0)$$

$$- \frac{h}{2} \left[f(x_0) + f(x_0) + h f'(x_0) + \frac{h^2}{\lfloor 2} f''(x) + \frac{h^3}{\lfloor 3} f'''(x) \right]$$

$$= \frac{-h^3}{12} f''(x_0) + O_4(h),$$

where $O_4(h)$ denotes sum of the terms of h^i, $i \geq 4$. Neglecting the $O_4(h)$, the error in trapezoidal rule is

$$E(I_1) \approx -\frac{h^3}{12} f''(x_0) \tag{5.35}$$

The trapezoidal rule is exact for polynomials of degree less than or equal to 1. Therefore, the integration method of trapezoidal rule is of order one. Now, we compute the upper bound for the error in composite trapezoidal rule. Let

$$I_1^c = \int_{x_0}^{x_0+nh} f(x)dx = \int_{x_0}^{x_0+h} f(x)dx + \int_{x_0+h}^{x_0+2h} f(d)dx + \cdots + \int_{x_0+n-1h}^{x_0+nh} f(x)dx$$

$$\tag{5.36}$$

The error in I_1^c is the sum of errors in each the integral on the right-hand side. Hence,

$$E(I_1^c) = -\frac{h^3}{12} f''(x_0) - \frac{h^3}{12} f''(x_1) - \frac{h^3}{12} f''(x_2) \cdots - \frac{h^3}{12} f''(x_{n-1}) \tag{5.37}$$

Suppose $m = \max\{f''(x_0), f''(x_1), \cdots, f''(x_{n-1}))$. Then

$$E(I_1^c) \le -\frac{h^3}{12} nm$$

$$= -\frac{h^2}{12}(nh)m$$

$$= -\frac{h^2}{12}(b-a)m \tag{5.38}$$

The upper bound of the error in the composite trapezoidal rule is $-\frac{h^2}{12}(b-a)m$.

5.18(B)   ERROR IN SIMPSON'S 1/3 RULE

Let $I_2 = \frac{h}{3}\left[f(x_0 - h) + 4f(x_0) + f(x_0 + h)\right]$ be an approximate value

of the integral $\int\limits_{x_0-h}^{x_0+h} f(x)dx$ obtained from Simpson's 1/3 rule. Assume

that $f(x)$ has continuous derivatives in $[x_0 - h, x_0 + h]$. Let

$$I_e = \int\limits_{x_0-h}^{x_0+h} f(x)dx = F(x_0 + h) - F(x_0 - h),$$

where $\frac{d}{dx}F(x) = f(x)$; clearly, I_e is an exact value of the integral.

Since $\frac{d}{dx}F(x) = f(x)$, we have $F''(x) = f'(x)$, $F'''(x) = f''(x)$… etc.

Now the error in Simpson's 1/3 rule is

$$E(I_2) = F(x_0 + h) - F(x_0) - \frac{h}{3}\left[f(x_0 - h) + 4f(x_0) + f(x_0 + h)\right] \tag{5.39}$$

Expanding $F(x_0 + h)$, $f(x_0 - h)$ and $f(x_0 + h)$ in Taylor series and using the fact $F^{(k)}(x) = f^{(k-1)}(x)$ $(k = 1, 2, \ldots)$ Eq. (5.39) becomes

$$E(I_2) = F(x_0) + hf(x_0) + \frac{h^2}{\underline{2}} f'(x_0) + \frac{h^3}{\underline{3}} f''(x_1)$$

$$+ \frac{h^4}{\underline{4}} f'''(x_0) + \frac{h^5}{\underline{5}} f'^v(x_0) + \cdots - F(x_0) +$$

$$\frac{h}{3}\left[\left\{ f(x_0) - hf'(x_0) + \frac{h^2}{\underline{2}} f''(x_0) + \frac{h^3}{\underline{3}} f'''(x_0) + \frac{h^4}{\underline{4}} f'^v(x_0) \right.\right.$$

$$\left. - \frac{h^5}{\underline{5}} f^v(x_0) + \ldots \right\} + 4f(x_0)$$

$$+ \left\{ f(x_0) + hf'(x_0) + \frac{h^2}{\underline{2}} f''(x_0) + \frac{h^3}{\underline{3}} f'''(x_0) + \frac{h^4}{\underline{4}} f'^v(x_0) \cdots \right\}\Big]$$

$$= -\frac{h^5}{90} f'^v(x_0) + 0_5(h).$$

where $O_5(h)$ contains the terms of order h^i, $i \geq 5$.

$\therefore$ The error in estimating $\displaystyle\int_{x_0-h}^{x_0+h} f(x)dx$ by Simpson's 1/3 rule is

$$E(I_2) = -\frac{1}{90} \cdot h^5 f'^v(x_0). \tag{5.40}$$

Note that

the error in $\displaystyle\int_{x_0}^{x_0+2h} f(x)dx$ will be $-\dfrac{h^5}{90} f'^v(x_1)$. $\tag{5.41}$

Now, we estimate the error in composite Simpson's 1/3 rule:
Let

$$I_2^c = \int_{x_0}^{x_0+nh} f(x)dx = \frac{h}{3}\big[(y_0 + y_n) + 4(y_1 + y_3 + \cdots + y_{n-1})$$

$$+ 2(y_2 + y_4 + \cdots + y_{n-2})\big] \tag{5.42}$$

be the value obtained from composite Simpson's 1/3 rule. Then the error in composite Simpson's 1/3 rule is

$$E(I_2^C) = \text{Error in} \int_{x_0}^{x_0+2h} f(x)dx + \text{Error in}$$

$$\int_{x_0+2h}^{x_0+4h} f(x)dx + \cdots + \text{Error in} \int_{x_0+(n-2)h}^{x_0+nh} f(x)dx,$$

where n is an even number. Then

$$E(I_2^C) = -\frac{h^5}{90} f^{\prime\prime v}(x_1) - \frac{h^5}{90} f^{\prime\prime v}(x_3) - \cdots - \frac{h^5}{90} f^{\prime\prime v}(x_{n-1}).$$

The number of terms on the R.H.S. of above equation is $\dfrac{n}{2}$.
Let m be the maximum of $\{ f^{\prime\prime v}(x_1), \ldots f^{\prime\prime v}(x_{n-1}) \}$;

Then, $E(I_2^C) \leq -\dfrac{h^5}{90} \times \dfrac{n}{2} \times m = -\dfrac{h^5}{180} m = -\dfrac{(b-a)h^4}{180} m .$ (5.43)

5.18(C)   ERROR IN SIMPSON'S 3/8 RULE

Let $I_3 = \dfrac{3h}{8}\left[f(x_0 - h) + 3f(x_0) + 3f(x_0 + h) + f(x_0 + 2h) \right]$

be the value of the integral $\displaystyle\int_{x_0-h}^{x_0+2h} f(x)$ obtained by the integration

method Simpson's 3/8 rule. Let

$$I_e = \int_{x_0-h}^{x_0+2h} f(x)dx = F(x_0 + 2h) - F(x_0),$$

where $\dfrac{d}{dx} F(x) = f(x)$. Proceeding as in estimating the error in

Simpson's 1/3 rule, the error estimate in Simpson's 3/8 rule is

$$E(I_3) = F(x_0 + 2h) - F(x_0 - h) - \frac{3h}{8}$$

$$\left[f(x_0 - 2h) + 3f(x_0) + 3f(x_0 + h) + f(x_0 + 2h) \right],$$

and it can be shown that

$$E(I_3) = -\frac{3}{80} h^5 f^{\prime\prime v}(x_0). \tag{5.44}$$

Note that the error in estimating $\displaystyle\int_{x_0}^{x_0+3h} f(x)dx$ using the Simpson's 3/8 rule is $-\dfrac{3}{80}h^5 f''^v(x_1)$

If I_3^c be the integral obtained from the composite Simpson's 3/8 rule, then the error in this integration method is

$$E(I_3^c) = -\frac{3}{8}h^5\left[f''^v(x_1) + f''^v(x_4) + \cdots + f''^v(x_{n-2}) \right].$$

The upper bound of this error is given by

$$E(I_3^c) \le -\frac{3}{80}h^3 \cdot \frac{n}{3} \cdot m,$$

where $m = \max\left\{ f''^v(x_1), f''^v(x_2), ..., f''^v(x_{n-2}) \right\}.$

Since $h = \dfrac{b-a}{n}$,

$$E(I_3^c) \le -\frac{1}{80}h^4(b-a)m. \tag{5.45}$$

SOLVED PROBLEMS

Problem 1: Apply the trapezoidal rule to integrate $\displaystyle\int_{0.3}^{0.35} \sqrt{x}\,dx$ by taking six ordinates. Find the actual error.

Solution: Here, $a = 0.3$, $b = 0.35$, $f(x) = \sqrt{x}$ and

$$h = \frac{b-a}{n} = \frac{0.35-0.3}{5} = 0.01.$$

The six values of independent variable of the integrand are $x_0 = 0.3$, $x_1 = 0.31$, $x_2 = 0.32$, $x_3 = 0.33$, $x_4 = 0.34$ and $x_5 = 0.35$

The values of $f(x) = \sqrt{x}$ at these arguments are shown in the following table:

x	$f(x) = \sqrt{x}$
0.3	$0.5477 = y_0$

0.31	$0.5567 = y_1$
0.32	$0.5657 = y_2$
0.33	$0.5744 = y_3$
0.34	$0.5831 = y_4$
0.35	$0.5916 = y_5$

The trapezoidal rule for six ordinates (five divisions) of the range of integration is

$$\int_{x_0}^{x_0+5h} f(x)dx = \frac{h}{2}\left[(y_0 + y_5) + 2(y_1 + y_2 + y_3 + y_4)\right]$$

$$\therefore \int_{0.3}^{0.35} \sqrt{x}\,dx = \frac{0.01}{2}\left[(.5477 + .5916) + 2(0.567\right.$$

$$\left. + 0.5657 + 0.5744 + 5831)\right]$$

$$= 0.02849$$

From the rules of integral calculus

$$\int_{0.3}^{0.35} \sqrt{x}\,dx = \left[\frac{2}{3}x^{\frac{3}{2}}\right]_{0.3}^{0.35} = \frac{2}{3}\left[(0.35)^{3/2} - (0.3)^{3/2}\right]$$

$$= 0.02846.$$

Thus, the exact value of the integral up to four decimal places is
$$= 0.02846$$
The actual error $= 0.02846 - 0.02849$
$$= -0.00003.$$

Problem 2: Evaluate $\int_{0}^{0.6} x^4 dx$ by Simpson's 1/3 rule by taking seven ordinates. Compare it with exact value.

Solution: Here, $a = 0, b = 0.6, f(x) = x^4$ and $h = \dfrac{b-a}{n} = \dfrac{0.6-0}{6} = 0.1$

x	$f(x)$
0	$0 = y_0$

0.1	$0.0001 = y_1$
0.2	$0.0016 = y_2$
0.3	$0.0081 = y_3$
0.4	$0.0256 = y_4$
0.5	$0.0625 = y_5$
0.6	$0.1296 = y_6$

The Simpsons's 1/3 rule for seven ordinates (equivalently six divisions of the range of integration) is

$$\int_{x_0}^{x_0+6h} f(x)dx = \frac{h}{3}\left[(y_0 + y_6) + 4(y_1 + y_3 + y_5) + 2(y_2 + y_4)\right]$$

$$\therefore \int_{0}^{0.6} x^4 dx = \frac{0.1}{3}\left[(0 + .1296) + 4(0.0001 + 0.0081\right.$$

$$\left. + 0.0625) + 2(0.0016 + 0.0256)\right]$$

$$= \frac{0.1}{3} \times 0.4704 = 0.01568$$

Exact value of the given definite integral

$$= \int_{0}^{0.6} x^4 dx = \left[\frac{x^5}{5}\right]_{0}^{0.6} = \frac{(0.6)^5}{5} = 0.01552.$$

Hence, the actual error $= 0.01552 - 0.01568$
$$= -0.00016.$$

5.19   ROMBERG INTEGRATION

Using the finite differences, we derived many formulae for numerical integration. Each integration method yields approximate value of a definite integral. Therefore, there is a need to improve the approximation. The Romberg's method is a process to improve the approximation of an integral obtained by a finite difference method. This method uses Richardson extrapolation. It provides a way to improve an integral estimate I by the formulae

$$I \approx I(h_2) + \frac{1}{\left(\dfrac{h_1}{h_2}\right)^2 - 1}\left[I(h_2) - I(h_1)\right], \tag{5.46}$$

where $I(h_1)$ and $I(h_2)$ are the integral estimates with interval of differences h_1 and h_2 respectively. For the convenience of computer algorithm h_2 is taken as $\dfrac{h_1}{2}$, and the formula (5.46) can be expressed as

$$I \approx \frac{4}{3} I\left(\frac{h}{2}\right) - \frac{1}{3} I(h). \tag{5.47}$$

This formula is used repeatedly replacing h by $\dfrac{h}{2}, \dfrac{h}{2^2}, \dfrac{h}{2^3}, \cdots$ till we get two successive values are sufficiently close to each other.

SOLVED PROBLEMS

Problem 1: Evaluate $\displaystyle\int_0^4 \frac{dx}{1+x}$ using Romberg's method by taking $h = 2$, 1 and 0.5.

Solution: Here, $a = 0$, $b = 4$ and $f(x) = \dfrac{1}{1+x}$.

Let

$$I = \int_0^4 \frac{dx}{1+x}.$$

Choose $h = 2$

x	0	2	4
$f(x)$	1	0.3333	0.2

$$I_1 = \int_0^4 f(x)\,dx = \frac{2}{2}\left[(1+0.2) + 2(.3333)\right] = 1.8666.$$

Choose $h = 1$

x	0	1	2	3	4
$f(x)$	1	0.5	0.3333	0.25	0.2

$$I_2 = \frac{1}{2}\left[(1+0.2) + 2(0.5 + 0.3333 + 0.25)\right]$$

$$= 1.6833.$$

Now choose $h = 0.5$.

x	0	0.5	1	1.5	2	2.5	3	3.5	4
$f(x)$	1	0.6666	0.5	0.4	0.3333	0.2857	0.25	0.2222	0.2

$$\int_{x_0}^{x_0+8h} f(x)dx = \frac{h}{2}\left[(y_0 + y_8) + 2(y_1 + y_2 + y_3 + y_4 + y_5 + y_6 + y_7)\right]$$

$$\therefore \qquad I_3 = \int_0^4 \frac{1}{1+x}dx = \frac{0.5}{2}\left[(1+0.2) + 2(.6666 + 0.5 + 0.4 \right.$$

$$\left. + .3333 + 0.2857 + 0.25 + 0.2222)\right]$$

$$= 1.6289.$$

The first improved value of the integral using I_1 and I_2:

$$I = \frac{4}{3}I_2 - \frac{1}{3}I_1$$

$$= \frac{4}{3}(1.6833) - \frac{1}{3}(1.8666)$$

$$= 1.6222.$$

The next improved value of the interval I using I_2 and I_3:

$$I = \frac{4}{3}I_3 - \frac{1}{3}I_2$$

$$= \frac{4}{3}(1.6289) - \frac{1}{3}(1.6833)$$

$$= 1.6107667.$$

Problem 2: Find the approximate value of $\int_0^1 \frac{dx}{1+x}$ by trapezoidal rule taking $h = 1$, $1/2$, $1/4$ and $1/8$ and then use Romberg's method to find the improved approximation of the integral.

Solution: Choose $h = 1$. Then the trapezoidal rule is

$$\int_{x_0}^{x_0+h} f(x)dx = \frac{h}{2}(y_0 + y_1).$$

x	0	1
$f(x) = \dfrac{1}{1+x}$	1	0.5

Let I_1 be the value of $\displaystyle\int_0^1 \frac{dx}{1+x}$ obtained by the trapezoidal rule with

$h = 1$. Then

$$I_1 = \frac{1}{2}(1+0.5) = 0.75.$$

Now choose $h = \frac{1}{2}$.

x	0	0.5	1
$f(x)$	1	0.6666	0.5

Let I_2 be the numerical value of $\displaystyle\int_0^1 \frac{1}{1+x} dx$ obtained by the trapezoidal

rule with $h = 0.5$.
Then

$$I_2 = \frac{0.5}{2}\left[(1+0.5) + 2(.6666)\right] = 0.7083.$$

Choose $h = 1/4$.

x	0	1/4	2/4	3/4	1
$f(x)$	0	0.8	0.6666	0.5714	0.5

Let I_3 the value of $\displaystyle\int_0^1 \frac{dx}{1+x}$ by the trapezoidal rule with $h = \frac{1}{4}$. Then

$$I_3 = \frac{1/4}{2}\left[(1+0.5) + 2(0.8+0.6666+0.5714)\right]$$
$$= 0.697.$$

Finally choose $h = 1/8$.

x	0	1/8	2/8	3/8	4/8	5/8	6/8	7/8	8/8
$f(x)$	1	0.8888	0.8	0.7272	0.6666	0.6153	0.5714	0.5333	0.5

$$I_4 = \frac{1}{16}\big[(1+0.5)+2(0.8888+0.8+0.7272$$
$$+0.6666+0.6153+0.5714+0.5333)\big]$$
$$= 0.694075.$$

The first improved value with I_1 and I_2 is

$$I = \frac{4}{3}I_2 - \frac{1}{3}I_1$$
$$= \frac{4}{3}(0.7083) - \frac{1}{3}(0.75)$$
$$= 0.6944 - 0.25$$
$$= 0.6944$$

The second improved value with I_2 and I_3 is

$$I = \frac{4}{3}I_3 - \frac{1}{3}I_2$$
$$= \frac{4}{3}\times 0.697 - \frac{1}{3}\times 0.7083$$
$$= 0.6932.$$

The third improved value with I_3 and I_4 is

$$I = \frac{4}{3}I_4 - \frac{1}{3}I_3$$
$$= \frac{4}{3}0.694 - \frac{1}{3}.697$$
$$= 0.693.$$

Note that the exact value is 0.69314718.

5.20 NUMERICAL DOUBLE INTEGRATION

To find the numerical value of a definite double integral function of two independent variables, we repeat the application of any quadrature formula of one variable treating other variable as a constant. The formulae used in repeated applications need not be the same quadrature formula. Suppose consider a double interval

$$I = \int_a^b \int_c^d f(x, y)\,dy\,dx. \tag{5.48}$$

Let the values of the function $f(x, y)$ be given at the points (x_0, y_0), (x_0, y_1), ... (x_0, y_m), (x_1, y_0), (x_1, y_1), ..., (x_1, y_m), ... (x_n, y_0), (x_n, y_1), ... (x_n, y_m) i.e, (x_i, y_j), where $(i = 0, 1, 2,\dots n; j = 0, 1, 2, \dots m)$ with $x_i = x_0 + ih$ and $y_i = y_0 + jk$. Here, h and k are the interval of differencing for the arguments of x and y respectively. These values can shown in a rectangular form:

$y\backslash x$	x_0	x_1	x_2	...	x_{n-1}	x_n
y_0	$f(x_0, y_0)$	$f(x_1, y_0)$	$f(x_2, y_0)$		$f(x_{n-1}, y_0)$	$f(x_n, y_0)$
y_1	$f(x_0, y_1)$	$f(x_1, y_1)$	$f(x_2, y_1)$		$f(x_{n-1}, y_1)$	$f(x_n, y_1)$
y_2	.	.	.	.	.	.
.	.	.	.	.	.	.
.					.	.
.					.	.
y_m	$f(x_0, y_m)$	$f(x_1, y_m)$			$f(x_{n-1}, y_m)$	$f(x_n, y_m)$

The process of evaluating I from the numerical data such as above table form of values is known as numerical double integration. It is also referred as mechanical cubaline.

5.21 TRAPEZOIDAL RULE FOR DOUBLE INTEGRAL

Suppose we wish to obtain the trapezoidal rule for the definite double integral

$$\int_{x_0}^{x_0+2h} \int_{y_0}^{y_0+2k} f(x, y)\,dx\,dy$$

Let
$$I = \int_{y_0}^{y_0+2k} \left\{ \int_{x_0}^{x_0+2h} f(x, y)dx \right\} dy. \qquad (5.49)$$

Now, for the inner integral, we apply the trapezoidal rule treating y as constant. Then

$$I = \int_{y_0}^{y_0+2k} \frac{h}{2} \{ f(x_0, y) + 2f(x_1, y) + f(x_2, y) \} \, dy$$

$$= \frac{h}{2} \left[\int_{y_0}^{y_0+2k} f(x_0, y)dy + 2 \int_{y_0}^{y_0+2k} f(x_1, y)dy + \int_{y_0}^{y_0+2k} f(x_2, y)dy \right]$$

$$= \frac{h}{2} \left[\frac{k}{2} \{ f(x_0, y_0) + 2f(x_0, y_1) + f(x_0, y_2) \} + 2 \cdot \frac{k}{2} \right.$$

$$\{ f(x_1, y_0) \} + 2f(x_1, y_1) + f(x_1, y_2) \}$$

$$\left. + \frac{k}{2} \{ f(x_2, y_0) + 2f(x_2, y_1) + f(x_2, y_2) \} \right]$$

For convenience, we express it as

$$I = \frac{k}{2} \left[A_0 + 2A_1 + A_2 \right],$$

where

$$A_0 = \frac{h}{2} \left[f(x_0, y_0) + 2f(x_1, y_0) + f(x_2, y_0) \right]$$

$$A_1 = \frac{h}{2} \left[f(x_0, y_1) + 2f(x_1, y_1) + f(x_2, y_1) \right],$$

and

$$A_2 = \frac{h}{2} \left[f(x_0, y_2) + 2f(x_1, y_2) + f(x_2, y_2) \right].$$

Thus,

$$\int_{x_0}^{x_0+2h} \int_{y_0}^{y_0+2k} f(x, y)dydx = \frac{h}{2} \left[A_0 + 2A_1 + A_2 \right] \qquad (5.50)$$

or

$$\int_{x_0}^{x_0+2h} \int_{y_0}^{y_0+2k} f(x, y)dydx = \frac{hk}{4}\big[f(x_0, y_0)+2f(x_1, y_0)+f(x_2, y_0)$$

$$+2f(x_0, y_1)+4f(x_1, y_1)+2f(x_2, y_1)$$

$$+f(x_0, y_2)+2f(x_1, y_2)+f(x_2, y_2)\big]$$

$$(5.51)$$

The values of A_0, A_1 and A_2 can be calculated in the following way:

$y\backslash x$	x_0	x_1	x_2	
y_0	$f(x_0, y_0)$	$f(x_1, y_0)$ ...	$f(x_2, y_0)$	A_0
y_1	$f(x_0, y_1)$	$f(x_1, y_1)$ ...	$f(x_2, y_1)$	A_1
y_2	$f(x_0, y_2)$	$f(x_1, y_2)$ ...	$f(x_2, y_2)$	A_2

Applying the trapezoidal rule for the entries of first row, second row and third row, we have A_0, A_1 and A_2 respectively. In this case, the interval of differencing is h. For example applying the trapezoidal rule for first row entries, we have

$$A_0 = \frac{h}{2}\big[f(x_0, y_0)+2f(x_1, y_0)+f(x_2, y_0)\big]$$

Now, for the values of A_0, A_1 and A_2 apply the trapezoidal rule to have $\int_{x_0}^{x_0+2h} \int_{y_0}^{y_0+2k} f(x, y)dydx$, assuming as if A_0, A_1 and A_2 are the values of certain function. For these values, the interval of differencing is k.

5.22   SIMPSON'S 1/3 RULE FOR DOUBLE INTEGRAL

Suppose $I = \int_{x_0}^{x_0+2h} \left\{ \int_{y_0}^{y_0+2k} f(x,y)dy \right\} dx$

Applying Simpson's 1/3 rule for inner integral, we have

$$I = \int_{x_0}^{x_0+2h} \frac{k}{3}\{f(x, y_0)+4f(x, y_1)+f(x, y_2)\}dx$$

$$= \frac{k}{3}\left[\int_{x_0}^{x_0+2h} f(x, y_0)dx + 4\int_{x_0}^{x_0+2h} f(x, y_1)dx + \int_{x_0}^{x_0+2h} f(x, y_2)dx\right]$$

$$= \frac{k}{3}\left[\frac{h}{3}\{f(x_0, y_0)+4f(x_1, y_0)+f(x_2, y_0)\}\right.$$

$$+ 4\cdot\frac{h}{3}\{f(x_0, y_1)+4f(x_1, y_1)+f(x_2, y_1)\}$$

$$\left.+ \frac{h}{3}\{f(x_0, y_2)+4f(x_1, y_2)+f(x_2, y_2)\}\right]$$

It can be expressed as

$$I = \frac{k}{3}\left[A_0 + 4A_1 + A_2\right],$$

where

$$A_0 = \frac{h}{3}\left[f(x_0, y_0)+4f(x_1, y_0)+f(x_2, y_0)\right]$$

$$A_1 = \frac{h}{3}\left[f(x_0, y_1)+4f(x_1, y_1)+f(x_2, y_1)\right],$$

and

$$A_2 = \frac{h}{3}\left[f(x_0, y_2)+4f(x_1, y_2)+f(x_2, y_2)\right].$$

Thus,

$$\int_{x_0}^{x_0+2h}\int_{y_0}^{y_0+2k} f(x, y)dxdy = \frac{h}{3}\left[A_0 + 4A_1 + A_2\right]$$

or

$$\int_{x_0}^{x_0+2h}\int_{x_0}^{y_0+2h} f(x, y)dy = \frac{hk}{9}\left[f(x_0, y_0)+4f(x_1, y_0)+f(x_2, y_0)\right.$$

$$+4f(x_1, y_1)+16f(x_1, y_1)+$$

$$\left.4f(x_2, y_1)+f(x_0, y_2)+4f(x_1, y_2)+f(x_2, y_2)\right]$$

$$(5.52)$$

5.23 PROCEDURE TO EVALUATE DOUBLE INTEGRAL NUMERICALLY

Suppose $I = \int\limits_{a}^{b} \int\limits_{c}^{d} f(x, y)\, dy\, dx$ when $a \leq x \leq b,\ c \leq y \leq d$.

Step 1: Divide $[a, b]$ in n equal parts and $[c, d]$ in m equal parts. The interval of difference for the argument x is $h = \dfrac{b-a}{n}$ and that of the argument y is $k = \dfrac{d-c}{m}$. Let $x_i = x_0 + ih,\ y_j = y_0 + jk$ $i = 0, 1, \ldots, n;\ j = 0, 1, \ldots m.$

Step 2: Compute the value of $f(x, y)$ at (x_i, y_i) and show them in the following form of table:

$y\backslash x$	x_0	x_1	$\ldots$	x_n	
y_0	$f(x_0, y_0)$	$f(x_1, y_0)$		$f(x_n, y_0)$	A_0
y_1	$f(x_0, y_1)$	$f(x_1, y_1)$		$f(x_n, y_1)$	A_1
$\ldots$					A_2
y_m	$f(x_0, y_m)$	$f(x_1, y_m)$		$f(x_n, y_m)$	A_m

Step 3: To each row values apply the specified quadrature formula (no. of divisions suitable to apply the quadrature formula) and denote them as $A_0, A_1, A_2, \ldots, A_m$

Step 4: For the values $A_0, A_1, A_2, \ldots, A_m$, apply the quadrature formula to have I.

SOLVED PROBLEMS

Problem 1: Using the Trapezoidal rule, evaluate $\int\limits_{0}^{1} \int\limits_{0}^{1} (xy + 1)\, dx\, dy$.

Solution: Here, $f(x, y) = xy + 1$. We choose $h = 0.5$ and $k = 0.5$.

Now, the values of $f(x, y)$ at the nodal points are shown in the following table:

$y \backslash x$	x_0 0	x_1 0.5	x_2 1
$y_0 = 0$	1	1	1
$y_1 = 0.5$	1	1.25	1.5
$y_2 = 1$	1	1.5	2

$A_0 =$ T.R. for the values $1, 1, 1 = \dfrac{0.5}{2}\left[1 + 2(1) + 1\right] = \dfrac{0.5}{2} \times 4 = 1$

$A_1 =$ T.R. for the values $1, 1.25, 1.5$

$A_1 = \dfrac{0.5}{2}\left[1 + 2(1.25) + 1.5\right] = \dfrac{0.5}{2} \times 5 = 1.25$

$A_2 =$ T.R. for the values $1, 1.5, 2$

$A_2 = \dfrac{0.5}{2}\left[1 + 2(1.5) + 2\right] = \dfrac{0.5 \times 6}{2} = 1.5$

Now

$$\int_0^1 \int_0^1 (xy + 1)\,dx\,dy = \text{T.R. for the values } A_0, A_1, A_2$$

$$= \dfrac{0.5}{2}\left[1 + 2(1.25) + 1.5\right]$$

$$= \dfrac{0.5}{2}[5] = \dfrac{2.5}{2} = 1.25$$

Problem 2: Evaluate $\displaystyle\int_1^2 \int_1^2 \dfrac{1}{x+y}\,dx\,dy$ using Trapezoidal rule with $h = k = 0.5$.

Solution: Since $h = 0.5$, we have $x_0 = 1$, $x_1 = 1.5$, $x_2 = 2$ and $y_0 = 1$, $y_1 = 1.5$, $y_2 = 2$. The values of $f(x, y)$ at (x_i, y_j) $(i, j = 1, 2, 3)$ are shown in the following table:

$y\backslash x$	1	1.5	2
1	1/2	1/2.5	1/3
1.5	1/2.5	1/3	1/3.5
2	1/3	1/3.5	1/4

Now

$$A_0 = \text{T.R. for the values } 1/2, 1/2.5, 1/3 \ (1\text{st row values})$$

$$= \frac{0.5}{2}\left[\frac{1}{2} + 2\frac{1}{2.5} + \frac{1}{3}\right] = 0.4083$$

$$A_1 = \text{T.R. for the values of } 1/2.5, 1/3, 1/3.5 \ (2\text{nd row values})$$

$$= \frac{0.5}{2}\left[\frac{1}{2.5} + 2\left(\frac{1}{3}\right) + \frac{1}{3.5}\right] = 0.3381$$

$$A_2 = \text{T.R. for the values of } 1/3, 1/2.5, 1/4$$

$$= \frac{0.5}{2}\left[\frac{1}{3} + 2\cdot\frac{1}{3.5} + \frac{1}{4}\right] = 0.2887.$$

Now

$$I = \int_1^2\int_1^2 \frac{1}{x+y}\,dxdy = \text{T.R. for the values } A_0, A_1, A_2$$

$$= \frac{0.5}{2}\left[.4083 + 2(.3381) + 0.2887\right]$$

$$= 0.3433$$

EXERCISE 5

1. Given that

x	0	2	4	6	8
$f(x)$	0	5	17	37	65

Evaluate $\int_0^8 f(x)dx$ using trapezoidal rule.

2. Evaluate $\int_0^1 \dfrac{e^x}{1+x}\,dx$ taking six equal divisions of the range of integration $[0, 1]$ using the trapezoidal rule.

3. Using Simpson's 1/3 rule, evaluate $\int_0^{1.2} e^{-x^2}\,dx$ by taking $h = 0.2$.

4. Evaluate $\int_0^{10} \dfrac{dx}{1+x^2}$ by Simpson's 1/3 rule.

5. The velocity of a particle at distance from a point on its path is given in following table:

s (in metres)	0	5	10	15	20	25	30
v (in m/sec)	49	54	62	69	72	80	84

Estimate the time taken to travel 30 meters.

6. Evaluate $\int_0^6 \dfrac{dx}{1+x^2}$ using Simpson's 1/3 rule.

7. Evaluate $\int_0^6 (x^2 + 5x + 2)\,dx$ by Weddle's rule.

8. Evaluate $\int_0^1 \dfrac{x^2}{x^3 + 10}\,dx$ using (i) Trapezoidal rule (ii) Simpson's 1/3 rule (iii) Weddle's rule.

9. Evaluate $\int_0^{\pi/2} \sqrt{\cos x}\,dx$ using (i) Simpson's 1/3 rule (ii) Simpson's 3/8 rule.

10. Evaluate $\int_0^{\pi/2} e^{x\sin x}\,dx$ by Weddle's rule.

11. Evaluate $\int_0^1 \dfrac{1}{1+x^2}\,dx$ by two- and three-point Gaussian quadrature formula.

12. Evaluate $I = \int_0^{\pi/2} e^{\sin x} \cos x\,dx$ by three-point Gaussian Quadrature formula. Also find the exact value of I and error in the approximation of I, which is obtained through the Gaussian quadrature formula.

13. Evaluate $\int_0^{\pi} \dfrac{1}{3+4\sin x}\,dx$ by Simpson's 1/3 rule

14. Obtain an approximate value of log2 by evaluating $\int_0^1 \dfrac{1}{1+x}\,dx$ using the trapezoidal rule.

15. Find the approximate value of $I = \int_0^{0.8} \dfrac{1}{1+x^2}\,dx$ by trapezoidal rule taking $h = 0.8, 0.4, 0.2$ and 0.1, and then apply Romberg's integration method to evaluate I.

16. Evaluate $I = \int_1^2 \dfrac{dx}{4+e^x}$ by trapezoidal rule taking $h = 1, 1/2, 1/4$ and $1/8$, and apply Romberg's integration method to evaluate I

17. Evaluate $\left\{ \int_0^2 \dfrac{4xy}{(1+x^2)(1+y^2)}\,dy \right\} dx$, using the trapezoidal rule with $h = 0.25$ and $k = 0.25$.

18. Evaluate $\int_0^1 \int_1^2 (x^2 + y^2)\,dxdy$, using Simpson's 1/3 rule with $h = 0.25$ and $k = 0.25$.

19. The following table shows the velocities of a two-wheeler that starts from rest at various values of time t. Find the distance travelled by the two-wheeler in 30 minutes.

time	0	5	10	15	20	25	30
velocity	0	12	20	25	32	40	48

20. A solid of revolution is formed by rotating about the x-axis the area bounded by x-axis, the lines $x = 0$, $x = 1$ and the curve $y = f(x)$. Given that the curve $y = f(x)$ is passing through the points $(0, 2)$ $(1, 4)$, $(2, 8)$, $(3, 24)$ and $(4, 32)$.

ANSWERS

1. 183
2. 1.126957
3. 0.866745
4. 1.431666
5. 0.425359
6. 1.366173
7. 174
8. 0.032134, 0.031769, 0.031770
9. 1.187281, 1.84930
10. 3.384835
11. 0.76885, 0.904143
12. 1.687366, 2.025632
13. 0.603342
14. 0.693771
15. 0.6747
16. 0.1180
17. 1.090684
18. 2.66667
19. 638.333
20. 3690.324181.

6

Numerical Solution of Differential Equations

A few types of differential equations can be solved through the methods developed to solve them. For example, variables separable, linear, homogeneous and exact differential equations are solved through the corresponding methods developed to get their solutions. It may not be possible to find the solution of every differential equation in a closed form. For example, it is not possible to find the solution of

$$\frac{dy}{dx} = \frac{\sin(\cos(\exp(\tan^{-1} x))) + \log x}{\sin^{10} x + x^{35} + (\log \cos x) + e^{\cos x}} \quad \text{or} \quad \frac{dy}{dx} = x^2 + y^2 - a^2.$$

Many problems in science and engineering yield ordinary differential equations that satisfy certain condition(s). To interpret the results, the solutions of such differential equations are required. We may fail to obtain the solutions of some differential equations that are related to physical problems. In such cases, they can be solved numerically. Of course, the numerical methods of finding the solution of differential equation with initial or boundary conditions can also be applied to a differential equation whose solutions are given in closed form by the methods developed in the theory of differential equations.

In this chapter, we learn some numerical methods to get the numerical solutions upto any desired degree of accuracy of first-order and first–degree ordinary differential equation with an initial condition. Note that every first-order and first-degree ordinary differential equation can be expressed as $\frac{dy}{dx} = f(x, y)$, where $f(x, y)$ is any function of x and y. In this chapter, a differential equation means an ordinary differential equation.

6.1   NOTATION

(i) $y(x_0) = y_0$ is used for the value of y is y_0 at $x = x_0$.

(ii) $x = a(h)b$ is used to denote to represent a set of values a, $a + h$, $a + 2h$, ..., $a + kh = b$ of x, where k is a non-negative integer. For example, $x = 3(2)11$ represent $x = 3, 5, 7, 9, 11$ and $x = 0.1(0.2)0.9$ represent $x = 0.1, 0.3, 0.5, 0.7$ and 0.9.

6.2   INITIAL VALUE PROBLEM

A differential equation $\dfrac{dy}{dx} = f(x, y)$ with an initial condition $y(x_0) = y_0$ is known as initial value problem. We abbreviate initial value problem as IVP.

6.3   EXAMPLES

Example 1: $\dfrac{dy}{dx} = x + y,\ y(0) = 2$ is an IVP.

Example 2: $\dfrac{dy}{dx} = x^2 + y^2,\ y(0) = 3$ is an IVP.

Example 3: $\dfrac{dy}{dx} = 2x + \cos x$ is not an IVP, as the initial condition is not specified.

6.4   NUMERICAL SOLUTION OF AN IVP

Let

$$\frac{dy}{dx} = f(x, y) \tag{6.1}$$

and

$$y(x_0) = y_0 \tag{6.2}$$

be an IVP. The numerical solution of the IVP is $y_0, y_1, \ldots\ldots, y_n$ at $x = x_0$, $x_1, \ldots\ldots, x_n$, where $x_i = x_0 + ih$, $y(x_i) = y_i$, h is the interval of differencing and n is a positive integer.

In the forthcoming sections, we learn how to find y_i $(i = 1, 2, \ldots..,$ $n)$. We use interchangeably the numerical solution of IVP and solution of the differential equation in this chapter.

6.5   EXPLANATION OF NUMERICAL SOLUTION OF IVP THROUGH AN EXAMPLE

Suppose an initial-value problem is as follows:

$$\frac{dy}{dx} = 1 + xy, \; y\,(0) = 2$$

Assume that $y(0.1) = 2.3467$, $y(0.2) = 2.5678$, $y(0.3) = 2.6793$, $y(0.4) = 2.7516$ and $y(0.5) = 2.8149$ are obtained by employing a numerical method of solving an IVP (which we will learn in subsequent sections). The numerical solution of the IVP is given in the following form:

x	0	0.1	0.2	0.3	0.4	0.5
y	2	2.3467	2.5678	2.6793	2.7516	2.8149

There are several numerical methods to solve IVP numerically, out of which we learn the following methods:

1. Euler's method
2. Modified Euler's method
3. Picard's method of successive approximations
4. Taylor's series method
5. Runge–Kutta second-order method
6. Runge–Kutta fourth-order method
7. Milne's method
8. Adam–Moulton method.

6.6   EULER'S METHOD

Consider the IVP given by (6.1) and (6.2), i.e., $\dfrac{dy}{dx} = f\,(x,\,y)$ and $y(x_0) = y_0.$

Suppose we wish to solve it numerically at the points

$$x_r = x_0 + rh \; (r = 1, 2, \ldots\ldots\, n, (n + 1))$$

Integrating (6.1) between the lower limit x_0 and upper limit x_1 with respect to x, we have

$$\int_{x_0}^{x_1} dy = \int_{x_0}^{x_1} f(x, y)dx$$

i.e.,
$$\left[y(x)\right]_{x_0}^{x_1} = \int_{x_0}^{x_1} f(x,y)\,dx$$

i.e.,
$$y(x_1) - y(x_0) = \int_{x_0}^{x_1} f(x,y)\,dx$$

i.e.,
$$y_1 = y_0 + \int_{x_0}^{x_1} f(x,y)\,dx. \tag{6.3}$$

Here, we assume that $f(x, y) = f(x_0, y_0)$, where $x_0 \le x \le x_1$. Now the Eq. (6.3) reduces to

$$y_1 = y_0 + \int_{x_0}^{x_1} f(x_0, y_0)\,dx,$$

i.e.,
$$y_1 = y_0 + h f(x_0, y_0) \tag{6.4}$$

Similarly, integrating 6.1 between the limits x_1 and x_2, we have

$$y_2 = y_1 + h f(x_1, y_1) \tag{6.5}$$

In general,

$$y_{n+1} = y_n + h f(x_n, y_n) \tag{6.6}$$

Eq. (6.6) is known as Euler's algorithm.

The process of finding y_i ($i = 1, 2,...., k$) for an IVP by the Euler's algorithm is called Euler's method of solving the IVP.

6.7   GEOMETRICAL INTERPRETATION OF EULER'S METHOD OF SOLVING AN IVP

Suppose the IVP is given by (6.1) and (6.2), i.e.,

$$\frac{dy}{dx} = f(x, y), \quad y(x_0) = y_0$$

Let $x_0, x_1, x_2, ..., x_n$ be equispaced points with an interval of differencing h. Further, let $F(x)$ be the exact solution of the IVP. Let the curve of $y = F(x)$ be γ (see Fig. 6.1)

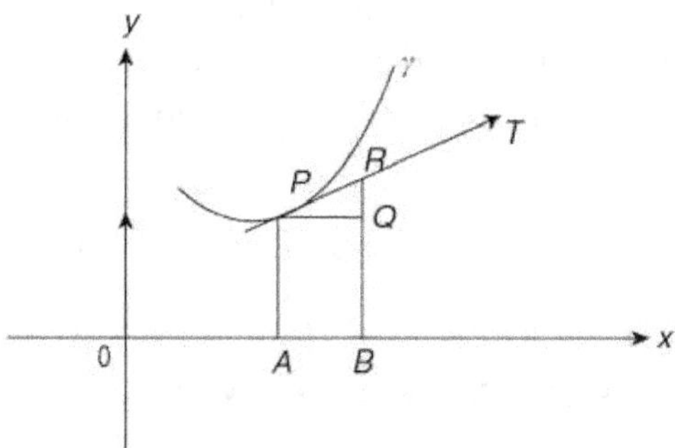

Figure 6.1

Let $P(x_0, y_0)$ be a point lying on the curve γ and PT be tangent to the curve γ at P. Then, the slope of the tangent at P is $\left(\dfrac{dy}{dx}\right)_{(x_0, y_0)}$ or $f(x_0, y_0)$ since $\dfrac{dy}{dx} = f(x, y)$. The equation of tangent at P is

$$y - y_0 = (x - x_0) f(x_0, y_0) \tag{6.7}$$

Suppose $OA = x_0$ and $OB = x_1$.

The ordinate of the point R (see Fig. 6.1), whose abscissa is x_1 and lying on the line (6.7), will be the approximate value of y_1. In Fig. 6.1, $RB = y_1$ and $R(x_1, y_1)$. Note that $RQ = y_1 - y_0$

Thus, y_1 obtained by Euler's algorithm geometrically represent the ordinate of a point with abscissa x_1 lying on a line passing through $P(x_0, y_0)$ and having the slope $f(x_0, y_0)$.

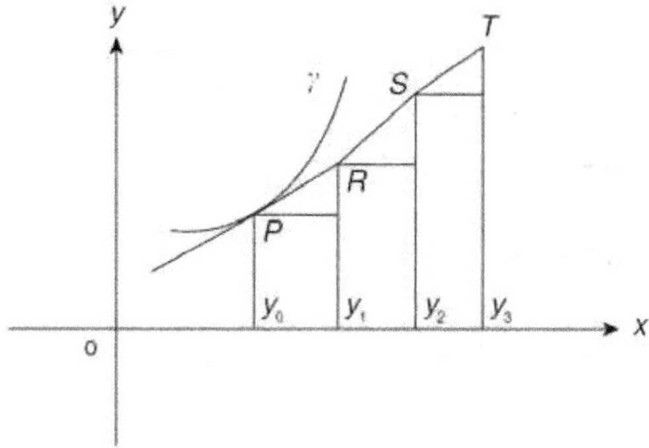

Figure 6.2

Similarly, y_2 obtained from the Euler's algorithm represent geometrically the ordinate of a point with abscissa x_2 lying on a line passing through (x_1, y_1) (it may or may not be a point on γ) and having slope $f(x_1, y_1)$,

i.e., $$y_2 = y_1 + h f(x_1, y_1).$$

The same argument can be extended to $y_3, y_4, \ldots$ and y_n.

6.8 PROCEDURE TO FIND THE NUMERICAL SOLUTION OF IVP FOR $x = x_1(h)x_n$

Step 1: Express the given differential equation as $\dfrac{dy}{dx} = f(x, y)$ and note down the function $f(x, y)$.

Step 2: Note down the values of x_0, y_0 and h.

Step 3: Note down the values of x at which y values are required.

Step 4: Write the Euler's algorithm for the given IVP.

$$\text{i.e., } y_{n+1} = y_n + hf(x_n, y_n), \tag{1}$$

where $f(x, y)$ is the function obtained in Step 1.

Step 5: (i) Put $n = 0$ in (1) and compute y_1.

At this stage, we have y_1 at $x = x_1$,

i.e., $x_1 = x_0 + h$, $y_1 = y_0 + hf(x_0, y_0)$.

(ii) Put $n = 1$ in (1) and obtain y_2 using the values x_1 and y_1 obtained in (*i*).

Continue this process till we get all the desired values of y_i (i.e., $y_1, y_2, \ldots, y_n$).

Step 6: Write the values of $x_0, x_1, \ldots, x_n$ and $y_0, y_1, \ldots, y_n$ in the following tabular form, which will be a numerical solution of the given IVP.

x	x_0	x_1	x_2	$\ldots\ldots$	x_n
y	y_0	y_1	y_2	$\ldots\ldots$	y_n

SOLVED PROBLEMS

Problem 1: Solve $y' = y^2 + x$, $y(0) = 1$ for $0.1(0.1)0.5$ using Euler's method.

Solution: Here, $f(x, y) = y^2 + x$ $\hspace{2em}$ (1)

$x_0 = 0$, $y_0 = 1$ and $h = 0.1$.

We have to find y values at 0.1, 0.2, 0.3, 0.4 and 0.5. Note that, $x_1 = 0.1$, $x_2 = 0.2$, $x_3 = 0.3$, $x_4 = 0.4$, $x_5 = 0.5$.

The Euler's algorithm for the given IVP is

$$y_{n+1} = y_n + h\{y_n^2 + x_n\} \tag{2}$$

To find y at $x_1 = 0.1$:
Putting $n = 0$ in (2), we get

$$y_1 = y_0 + h\{y_0^2 + x_0\} \tag{3}$$

Substituting x_0, y_0 and h values in (3), we have

$$y_1 = 1 + 0.1\{1^2 + 0\} = 1.1$$

Thus, $x_1 = 0.1$ and $y_1 = 1.1$.

To find y at $x_2 = 0.2$:
Substituting $n = 1$ in (2), we get

$$y_2 = y_1 + h\{y_1^2 + x_1\} \tag{4}$$

Now, substituting the values of x_1, y_1 and h, we get

$$y_2 = 1.1 + 0.1\{(1.1)^2 + 0.1\} = 1.231$$

Thus,

$$x_2 = 0.2 \text{ and } y_2 = 1.231.$$

To find y at $x_3 = 0.3$:
Substituting $n = 2$ in (2), we get

$$y_3 = y_2 + h\{y_2^2 + x_2\} \tag{5}$$

Since $x_2 = 0.2$, $y_2 = 1.231$ and $h = 1$, we have

$$y_3 = 1.231 + 0.1\{(1.231)^2 + 0.2\} = 1.4025$$

Thus, $x_3 = 0.3$ and $y_3 = 1.4025$.

To find y at $x_4 = 0.4$:
Putting $n = 3$ in (2), we get

$$y_4 = y_3 + h\{y_3^2 + x_3\} \tag{6}$$

Since, $x_3 = 0.3$ and $y_3 = 1.4025$ and $h = 1$, we have

$$y_4 = 1.4025 + 0.1\{(1.4025)^2 + 0.3\}$$
$$= 1.6292.$$

Thus, $x_4 = 0.4$ and $y_4 = 1.6292$.

To find y at $x_5 = 0.5$:
Substituting $n = 4$ in (2), we get
$$y_5 = y_4 + h\{y_4^2 + x_4\} \tag{7}$$
Since, $x_4 = 0.4$ and $y_4 = 1.6292$ and $h = 1$, we have
$$y_5 = 1.6292 + 0.1\{(1.6292)^2 + 0.4\}$$
$$= 1.9346.$$
Thus, $x_5 = 0.5$ and $y_5 = 1.9346$.

Hence, the numerical solution of given IVP is shown in the following table:

x	0	0.1	0.2	0.3	0.4	0.5
y	1	1.1	1.231	1.4025	16292	1.9346

Problem 2: Solve $\dfrac{dy}{dx} = 1 - y$, $y(0) = 0$ using Euler's method for $0.1(0.1)0.2$. Compare the values found by Eulers's method with exact values.

Solution: Comparing the given differential equation with the initial condition to that of IVP $\dfrac{dy}{dx} = f(x, y)$, $y(x_0) = y_0$. We have $f(x, y) = 1 - y$, $x_0 = 0$ and $y_0 = 0$. The interval of differencing $h = 0.1$.
We have to find y at $x = 0.1$ and 0.2.
The Euler's algorithm for given IVP is
$$y_{n+1} = y_n + h\{1 - y_n\} \tag{1}$$
To find y at $x_1 = 0.1$:
Substituting $n = 0$ in (1), we get
$$y_1 = y_0 + h\{1 - y_0\} \tag{2}$$
$$\therefore \qquad y_1 = 0 + 0.1\{1 - 0\} = 0.1.$$
Thus, $x_1 = 0.1$ and $y_1 = 0.1$.
To find y at $x = 0.2$:
Substituting $n = 1$ in Eq. (1),
We get
$$y_2 = y_1 + h\{1 - y_1\} \tag{3}$$
$$\therefore \qquad y_2 = 0.1 + 0.1\{1 - 0.1\} = 0.19.$$
Thus, $x_2 = 0.2$ and $y_2 = 0.19$
Hence, the numerical solution of given IVP is as follows:

x	0	0.1	0.2
y	0	0.1	0.19

Now, we find the solution (exact) of given differential equation to know the exact values. The given differential equation can be expressed as

$$\frac{dy}{1-y} = dx,$$

which is a variable separable differential equation. Its general solution is obtained by integrating on the both sides, i.e.,

$$\int \frac{dy}{1-y} = \int dx + c,$$

where c is any arbitrary constant. Thus, the solution of given differential equation is

$$-\log(1-y) = x + c.$$

Using the initial condition $y(0) = 0$, we have

$$-\log(1-0) = 0 + c.$$

$$\therefore \qquad c = 0.$$

Thus, the solution of the given initial value problem is

$$y = 1 - e^{-x}.$$

Therefore, the exact solution of given IVP is

$$y = 1 - e^{-x}.$$

The approximate value of y_1, y_2 found by Euler's method and their exact values are shown in the following table:

x	Exact-value	Approximate value
0.1	0.09516	0.1
0.2	0.18126	0.19

Problem 3: Solve $\dfrac{dy}{dx} + y = 0$, $y(0) = 1$, using Euler's method for $x = 0.01(0.01)0.04$.

Solution: Expressing the given differential equation in the form $\dfrac{dy}{dx} = f(x, y)$, we have

$$\frac{dy}{dx} = -y \tag{1}$$

Hence, $f(x, y) = -y$. Further, $y(0) = 1$ gives $x_0 = 0$ and $y_0 = 1$. Note that $h = 0.01$. We have to find y values at $x = 0.01, 0.02\ 0.03$ and 0.04 as per the given problem. Therefore, $x_1 = 0.01$, $x_2 = 0.02$, $x_3 = 0.03$ and $x_4 = 0.04$.

The Euler's algorithm for the given IVP is

$$y_{n+1} = y_n + h\{-y_n\} \tag{2}$$

To find y at $x = 0.01$:

Putting $n = 0$ in (2), we get

$$y_1 = y_0 + h\{-y_0\}$$

$\therefore \qquad y_1 = 1 + 0.01\{-1\} = 0.99$

Thus, $x_1 = 0.01$ and $y_1 = 0.99$.

To find y at $x = 0.02$:

Putting $n = 1$ in (2), we have

$$y_2 = y_1 + h\{-y_1\}$$

$\therefore \qquad y_2 = 0.99 + 0.01\{-0.99\} = 0.9801$

Thus, $x_2 = 0.02$ and $y_2 = 0.9801$.

To find y at $x = 0.03$:

Putting $n = 2$ in (2), we get

$$y_3 = y_2 + h\{-y_2\}$$

$\therefore \qquad y_3 = 0.9801 + 0.01\{-0.9801\} = 0.9703$

Thus, $x_3 = 0.03$ and $y_3 = 0.9703$.

To find y at $x = 0.04$:

Substituting $n = 3$ in (2), we get

$$y_4 = y_3 + h\{-y_3\}$$

$\therefore \qquad y_4 = 0.9703 + 0.01\{-0.9703\}$

$$= 0.9605$$

Thus, $x_4 = 0.04$ and $y_4 = 0.9605$.

Hence, the numerical solution of the given IVP is shown below:

x	0	0.01	0.02	0.03	0.04
y	1	0.99	0.9801	0.9703	0.9605

Problem 4: Solve $y' = (x^3 + xy^2)e^{-x}$, $y(0) = 1$ for $0.1(0.1)0.3$ using Euler's method.

Solution: Here, $f(x, y) = (x^3 + xy^2)e^{-x}$, $x_0 = 0$, $y_0 = 1$ and $h = 0.1$. We have to find y values at $x = 0.1$, 0.2 and 0.3. Thus, $x_1 = 0.1$, $x_2 = 0.2$, $x_3 = 0.3$.

The Euler's algorithm for the given IVP is

$$y_{n+1} = y_n + h\{x_n^3 + x_n y_n^2\}e^{-x_n} \tag{1}$$

To find y at $x = 0.1$:

Substituting $n = 0$ in (1), we have

$$y_1 = y_0 + h\{x_0^3 + x_0 y_0^2\}\, e^{-x_0}$$

$\therefore \qquad y_1 = 1 + 0.1\{(0)^3 + (0)(1)^2\}e^{\,0} = 1$

$\therefore \qquad x_1 = 0.1$ and $y_1 = 1$.

To find y at $x = 0.2$:

Substituting $n = 1$ in Eq. (1), we get

$$y_2 = y_1 + h\{x_1^3 + x_1 y_1^2\}e^{-x_1}$$

$\therefore \qquad y_2 = 1 + 0.1\{(0.1)^3 + (0.1)(1)^2\}e^{-0.1}$

$$= 1.009138.$$

Thus, $x_2 = 0.2$ and $y_2 = 1.009138$.

To find y at $x = 0.3$:

Putting $n = 2$ in (1), we get

$$y_3 = y_2 + h\{x_2^3 + x_2 y_2^2\}e^{-x_2}$$

$\therefore \qquad y_3 = 1.009138 + 0.1\{(0.2)^3 + (0.2)(1.009138)^2\}e^{-0.2}$

$$= 1.02647.$$

Thus, $x_3 = 0.3$ and $y_2 = 1.02647$.

Thus, numerical solution of given IVP is as follows:

x	0	0.1	0.2	0.3
y	1	1	1.009138	1.02647

Problem 5: Use Euler's method to solve the IVP $\dfrac{dy}{dx} = \dfrac{y^2 - x^2}{y^2 + x^2}$, $y(0) = 1$ for $0.2(0.2)0.4$.

Solution: Here, $x_0 = 0$, $y_0 = 1$, $h = 0.2$ and

$$f(x, y) = \frac{y^2 - x^2}{y^2 + x^2} \tag{1}$$

We have to find y at $x = 0.2$ and $x = 0.4$, i.e., $x_1 = 0.2$ and $x_2 = 0.4$. The Euler's algorithm for the given IVP is

$$y_{n+1} = y_n + h\left\{\frac{y_n^2 - x_n^2}{y_n^2 + x_n^2}\right\} \tag{2}$$

To find y at $x = 0.2$:

Substituting $n = 0$ in (2), we get

$$y_1 = y_0 + h\left\{\frac{y_0^2 - x_0^2}{y_0^2 + x_0^2}\right\}$$

Substituting the values of x_0, y_0 and h, we obtain

$$y_1 = 1 + 0.2\left\{\frac{(1)^2 - (0)^2}{(1)^2 + (0)^2}\right\} = 1.2.$$

Thus, $x_1 = 0.2$ and $y_1 = 1.2$.

To find y at $x = 0.4$:

Substituting $n = 1$ in (2), we get

$$y_2 = y_1 + h\left\{\frac{y_1^2 - x_1^2}{y_1^2 + x_1^2}\right\}$$

$$\therefore \qquad y_2 = 1.2 + 0.2\left\{\frac{(1.2)^2 - (0.2)^2}{(1.2)^2 + (0.2)^2}\right\}$$

(since $y_1 = 1.2$, $x_1 = 0.2$ and $h = 0.2$)

$$y_2 = 1.3892.$$

Thus, $x_2 = 0.4$ and $y_2 = 1.3892$. Therefore, the numerical solution of given IVP is as follows:

x	0	0.2	0.4
y	1	1.2	1.3892

EXERCISE 6(a)

1. Given that $(y+x)\dfrac{dy}{dx}+(x-y)=0$ and $y(0)=1$, determine $y(0.1)$ using Euler's method.

2. Using Euler's method solve $\dfrac{dy}{dx}=1+y^2$, $y(0)=0$ for $x=0.2(0.2)0.6$.

3. Determine y at $x=0.1(0.1)0.4$ by the Euler's method, given that $2yy'=\cos x$ and $y(0)=1$.

4. Using Euler's method solve $\dfrac{dy}{dx}-\dfrac{2x}{y}+xy=0$ $y(0)=1$ for $x=0.2(0.2)0.6$.

5. Find the numerical solution of $y'=y+e^x$ $y(0)=0$ for $x=0.2(0.2)0.6$ using the Euler's method.

6. Given the equation $y'+2y=0$, $y(0)=1$, find $y(0.1)$ and $y(0.2)$ by the Euler's method with $h=0.1$.

7. Using the Euler's method, find the numerical solution of $y'=3x^2+y$, $y(0)=4$ for $x=0.1(0.1)0.5$.

8. Determine $y(0.1)$ by the Euler's method given that $y'=1+xy$, $y(0)=2$.

9. Use Euler's method to solve $y'=x^2+y^2$, $y(1)=0$ for $x=1.2(0.2)2$.

10. Solve $y'=\dfrac{x}{y}$, $y(0)=1$ for $x=0.1(0.1)0.3$ by the Euler's method. Compare the results with those obtained from the exact solution.

ANSWERS

1. 1.1
2. $y(0.2)=0.2$, $y(0.4)=0.408$, $y(0.6)=0.641293$
3. $y(0.1)=1.05$, $y(0.2)=1.097381$, $y(0.3)=1.142036$, $y(0.4)=1.183862$
4. $y(0.2)=1$, $y(0.4)=1.04$, $y(0.6)=1.110646$
5. $y(0.2)=0.2$, $y(0.4)=0.484281$, $y(0.6)=0.879502$
6. $y(0.1)=0.8$, $y(0.2)=0.064$
7. $y(0.1)=4.4$, $y(0.2)=4.8431$, $y(0.3)=5.3393$, $y(0.4)=5.90023$, $y(0.5)=6.538253$
8. $y(0.1)=2.1$
9. $y(1.2)=0.2$, $y(1.4)=0.496$, $y(1.6)=0.937203$, $y(1.8)=1.624873$, $y(2.0)=2.800916$,
10. $y(0.1)=1$, $y(0.2)=1.01$, $y(0.3)=1.029802$.

6.9   MODIFIED EULER'S METHOD

In case of h is small, the Euler's method, which we learnt in the earlier section, is too slow to approach the exact value. Even if h is taken very small, the computed values of y by the Euler's method will deviate farther and farther from the exact values of y as long as the curvature of exact solution does not change. Further, this method is inaccurate for large values of h. These drawbacks lead to invention of modified Euler's method to solve an IVP. Now, we describe the modified Euler's method to solve an IVP given by (6.1) and (6.2), i.e.,

$$y' = f(x_0, y_0)$$

and

$$y(x_0) = y_0.$$

We introduce the notation $y_i^{(k)}$, which denotes the kth approximation of y_i.

The first approximation of y_1 (i.e., $y_1^{(1)}$) is computed from

$$y_1^{(1)} = y_0 + h f(x_0, y_0). \tag{6.8}$$

Note that $y_1^{(1)}$ is the value obtained from the Euler method.

Integrating (6.1) from x_0 to x_1, we get

$$\int_{x_0}^{x_1} dy = \int_{x_0}^{x_1} f(x, y)dx,$$

i.e.,
$$\left[y(x)\right]_{x_0}^{x_1} = \int_{x_0}^{x_1} f(x, y)dx$$

$$y(x_1) - y(x_0) = \int_{x_0}^{x_1} f(x, y)dx$$

$$y(x_1) = y(x_0) + \int_{x_0}^{x_1} f(x, y)dx$$

or
$$y(x_1) = y_0 + \int_{x_0}^{x_1} f(x, y)dx \tag{6.9}$$

Evaluating the R.H.S. of (6.11) using the trapezoidal rule with the data (x_0, y_0) and $(x_1, y_1^{(1)})$, we have

$$\int_{x_0}^{x_1} f(x, y)dx = \frac{h}{2}\left[f\left(x_0, y_0\right) + f\left(x_1, y_1^{(1)}\right) \right]. \tag{6.10}$$

In view of (6.10), the Eq. (6.9) reduces to

$$y(x_1) = y_0 + \frac{h}{2}\left[f\left(x_0, y_0\right) + f\left(x_1, y_1^{(1)}\right) \right]. \tag{6.11}$$

The value $y(x_1)$ computed by (6.11) is known as second approximation of y_1 (i.e., $y_1^{(2)}$). Hence,

$$y_1^{(2)} = y_0 + \frac{h}{2}\left[f\left(x_0, y_0\right) + f\left(x_1, y_1^{(1)}\right) \right].$$

By repeating this process, we get

$$y_1^{(k)} = y_0 + \frac{h}{2}\left[f\left(x_0, y_0\right) + f\left(x_1, y_1^{(k-1)}\right) \right], \tag{6.12}$$

where $k = 2, 3, 4\ldots..,$ The iteration is terminated when two successive iterates agree to the desired degree of accuracy. The approximation $y_1^{(k)}$ (for some k) obtained upto the desired degree of accuracy will be taken as y_1. Now, considering $\dfrac{dy}{dx} = f(x, y), y(x_1) = y_1$ as an IVP, the above process can be repeated to get y_2 and so on. The general formula to the obtained mth approximation of y_i is

$$y_i^{(m)} = y_{i-1} + \frac{h}{2}\left[f\left(x_{i-1}, y_{i-1}\right) + f\left(x_i, y_i^{(m-1)}\right) \right], \tag{6.13}$$

where $m = 2, 3, 4, 5\ldots..,$ for fixed $i = 1, 2, 3\ldots.$
and

$$y_i^{(1)} = y_{i-1} + hf(x_{i-1}, y_{i-1}). \tag{6.14}$$

The Eq. (6.14) is used to find the first approximation of each $y_i (i = 1, 2, 3, ..)$, i.e., they are found by Euler's method.

Note: Since $\dfrac{dy}{dx} = f(x, y)$, the Eqs. (6.13) and (6.14) are equivalent to the following equations, respectively:

$$y_i^{(m)} = y_{i-1} + \frac{h}{2}\left[\left(\frac{dy}{dx}\right)_{i-1} + \left(\frac{dy}{dx}\right)_i^{(m-1)}\right]$$ (6.15)

($m = 2, 3, 4\ldots.$ for fixed $i = 1, 2, 3, \ldots..$)
and

$$y_i^{(1)} = y_{i-1} + h\left(\frac{dy}{dx}\right)_i$$ (6.16)

Here, $\left(\dfrac{dy}{dx}\right)_i = \left(\dfrac{dy}{dx}\right)_{(x_i, y_i)}$ and $\left(\dfrac{dy}{dx}\right)_i^{(m-1)} = \left(\dfrac{dy}{dx}\right)_{(x_i, y_i^{m-1})}$.

The representation of (6.13) and (6.14) in (6.15) and (6.16) are helpful in understanding the geometrical interpretation of modified Euler's method.

6.10 GEOMETRICAL REPRESENTATION OF MODIFIED EULER'S METHOD

Let $y' = f(x, y)$, $y(x_0) = y_0$ be an IVP and $x_0, x_1, x_2, \ldots.., x_n$ be equidistant points with an interval of differencing h. Suppose, $y = f(x)$ be the exact solution of this IVP. Let the graph of $y = f(x)$ be as shown in Fig. 6.3.

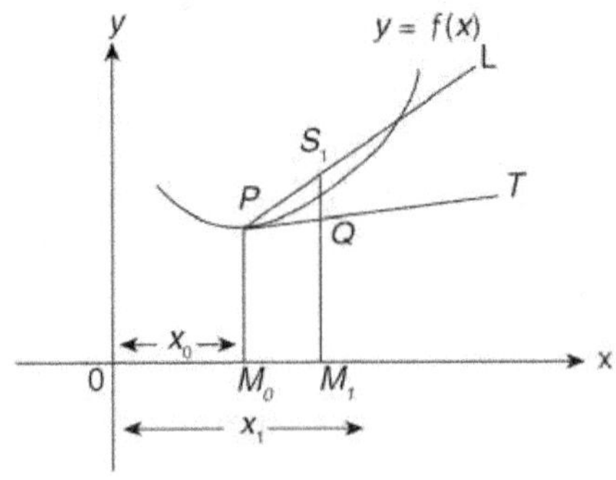

Figure 6.3

The equation of tangent PT at P to the curve $y = f(x)$ is

$$y - y_0 = f(x_0, y_0)(x - x_0)$$ (6.17)

The ordinate of a point Q whose abscissa is x_1 and lying on (6.20) will be the first approximation of y_1, i.e., $y_1^{(1)}$.

In Fig. 6.3, $QM_1 = y_1^{(1)}$.

Suppose $m_0 = f(x_0, y_0)$ and $m_1^{(1)} = f(x_1, y_1^{(1)})$. Let PL be a line through $P(x_0, y_0)$ having the slope $\left(\dfrac{m_0 + m_1^{(1)}}{2} \right)$. Then the equation of PL is

$$y - y_0 = \left(\frac{m_0 + m_1^{(1)}}{2} \right)(x - x_0). \tag{6.18}$$

Now, geometrically $y_1^{(2)}$ represents the ordinate of the point whose abscissa is x_1 and lying on (6.18). In Fig. 6.3, $S_1 M_1 = y_1^{(2)}$. As $(x_1, y_1^{(2)})$ is a point of (6.18), we have

$$y_1^{(2)} - y_0 = \left(\frac{m_0 + m_1^{(1)}}{2} \right)(x_1 - x_0),$$

i.e.,
$$y_1^{(2)} = y_0 + h\{ f(x_0, y_0) + f(x_1, y_1^{(1)})\}.$$

In general, $y_1^{(m)}$ represents the ordinate of the point whose abscissa is x_1 and lying on the line

$$y - y_0 = \frac{\left[f(x_0, y_0) + f(x_1, y_1^{(m-1)}) \right]}{2}(x - x_0)$$

Similar interpretation can be extended to $y_2^{(m)}$, $y_3^{(m)}$, ..., $y_n^{(m)}$.

SOLVED PROBLEMS

Problem 1: Solve $\dfrac{dy}{dx} = 1 + xy$, $y(0) = 2$ for $x = 0.1(0.1)0.4$ using modified Euler's method. Give the approximation correct to 4 decimal places.

Solution: Here, $x_0 = 0$, $y_0 = 2$, $h = 0.1$ and

$$f(x, y) = 1 + xy \tag{1}$$

The first approximation of each y_{n+1} is obtained from

$$y_{n+1}^{(1)} = y_n + h f(x_n, y_n) \tag{2}$$

$$(n = 0, 1, 2,)$$

And mth approximation of y_i is obtained from

$$y_i^{(m)} = y_{i-1} + \frac{h}{2}\left[f\left(x_{i-1}, y_{i-1}\right) + f\left(x_i, y_i^{(m-1)}\right)\right] \qquad (3)$$

We have to find y values at $x = 0.1, 0.2, 0.3$ and 0.4 correct to 4 decimal places. Note that $x_1 = 0.1$, $x_2 = 0.2$, $x_3 = 0.3$, and $x_4 = 0.4$.

To find y at $x = 0.1$ (i.e., $x = x_1$)

The first approximation of y_1 (i.e., $y_1^{(1)}$) by Eq. (2) is

$$y_1^{(1)} = y_0 + hf(x_0, y_0)$$

$\therefore$
$$y_1^{(1)} = 2 + 0.1 f(0.2)$$

i.e.,
$$y_1^{(1)} = 2 + 0.1\{1 + (0)(2)\} = 2.1$$

Thus,
$$y_1^{(1)} = 2.1.$$

Now substituting $m = 2$ and $i = 1$ in Eq. (3), we get

$$y_1^{(2)} = y_0 + \frac{h}{2}\left[f\left(x_0, y_0\right) + f\left(x_1, y_1^{(1)}\right)\right]$$

$\therefore$
$$y_1^{(2)} = 2 + \frac{0.1}{2}\left[f(0,2) + f(0.1, 2.1)\right]$$

$$= 2 + 0.05\left[\{1 + (0)(2)\} + \{1 + (0.1)(2.1)\}\right]$$

$$= 2.1105$$

Thus,
$$y_1^{(2)} = 2.1105.$$

Again substituting $m = 3$ and $i = 1$ in the Eq. (3), we have

$$y_1^{(3)} = y_0 + \frac{h}{2}\left[f\left(x_0, y_0\right) + f\left(x_1, y_1^{(2)}\right)\right]$$

$\therefore$
$$y_1^{(3)} = 2 + \frac{0.1}{2}\left[f(0,2) + f(0.1, 2.1105)\right]$$

i.e.,
$$y_1^{(2)} = 2 + 0.05\left[\{1 + (0)(2)\} + \{1 + (0.1)(2.1105)\}\right]$$

i.e.,
$$y_1^{(2)} = 2.1105.$$

Since $y_1^{(2)} = y_1^{(3)}$ upto four decimal places, we take $y_1 = 2.1105$
Thus, $x_1 = 0.1$ and $y_1 = 2.1105$.

To find y at $x = 0.2$ (i.e., $x = 0.2$)

The first approximation of y_2 i.e., $y_2^{(1)}$ is obtained from Eq. (2) for $n = 1$.

$$y_2^{(1)} = y_1 + hf(x_1, y_1)$$

$\therefore$
$$y_2^{(1)} = y_1 + h\{1 + x_1 y_1\}$$

i.e.,
$$y_2^{(1)} = 2.1105 + 0.1\{1 + (0.1)(2.1105)\}$$
$$= 2.231605.$$

Hence, the first approximation of y_2 is
$$y_2^{(1)} = 2.231605.$$

From the Eq. (3),

$$y_2^{(2)} = y_1 + \frac{h}{2}\left[f(x_1, y_1) + f(x_2, y_2^{(1)})\right]$$

$$\therefore y_2^{(2)} = y_1 + \frac{h}{2}\left[(1 + x_1 y_1) + \{1 + x_2 y_2^{(1)}\}\right]$$

$$\therefore y_2^{(2)} = 2.1105 + \frac{0.1}{2}\left[\{1 + (0.1)(2.1105)\} + \{1 + (0.2)(2.231605\}\right]$$

$$= 2.2433685$$

Let $y_2^{(2)} \approx 2.2434$.
From the Eq. (3),

$$y_2^{(3)} = y_1 + \frac{h}{2}\left[f(x_1, y_1) + f(x_2, y_2^{(2)})\right]$$

$$\therefore y_2^{(3)} = y_1 + \frac{h}{2}\left[(1 + x_1 y_1) + (1 + x_2 y_2^{(2)})\right]$$

$$y_2^{(3)} = 2.1105 + \frac{0.1}{2}\left[\{1 + (0.1)(2.1105)\} + \{1 + (0.2)(2.2434\}\right]$$

$$= 2.2434865$$

$$y_2^{(3)} \approx 2.2434.$$

Since $y_2^{(2)} = y_2^{(3)}$ upto four decimal places, we take $y_2 = 2.2434$
Thus, $x_2 = 0.2$ and $y_2 = 2.2434$.

To find y at $x = 0.3$ (i.e., $x = x_3$):
The first approximation of $y_3^{(1)}$ is
$$y_3^{(1)} = y_2 + hf(x_2, y_2)$$

i.e.,
$$y_3^{(1)} = y_2 + h\{1 + x_2 y_2\}$$

$$\therefore \qquad y_3^{(1)} = 2.2434 + 0.\{1 + 0.2(2.2434)\}$$

i.e., $\qquad y_3^{(1)} \approx 2.3833.$

Now, the second approximation of y_3 i.e., $y_3^{(2)}$ can be obtained from

$$y_3^{(2)} = y_2 + \frac{h}{2}\left[f(x_2, y_2) + f\left(x_3, y_3^{(1)}\right)\right]$$

$$= y_2 + \frac{h}{2}\left[\{1 + x_2 y_2\} + \{1 + x_3 y_3^{(1)}\}\right].$$

Substituting the values of h, x_2, y_2, x_3 and $y_3^{(1)}$,
We have

$$y_3^{(2)} = 2.2434 + \frac{0.1}{2}\left[\{1 + (0.2)(2.2434)\} + \{1 + (0.3)(2.3883)\}\right]$$

$$\approx 2.40166.$$

Similarly,

$$y_3^{(2)} = y_2 + \frac{h}{2}\left[\{1 + x_2 y_2\} + \{1 + x_3 y_3^{(2)}\}\right]$$

$$y_3^{(3)} = 2.2434 + \frac{0.1}{2}\left[\{1 + (0.2)(2.2434)\} + \{1 + (0.3)(2.40166)\}\right]$$

$$y_3^{(3)} \approx 2.40186$$

and

$$y_3^{(4)} = 2.2434 + \frac{0.1}{2}\left[\{1 + (0.2)(2.2434)\} + \{1 + (0.3)(2.40186)\}\right]$$

$$\approx 2.4018$$

$y_3^{(3)} = y_3^{(4)} = 2.4018$ upto 4 decimal places Therefore, we take $y_4 = 2.4018$. Thus,

$$x_3 = 0.3, y_3 = 2.4018.$$

To find y at $x = 0.4$ (i.e., $x_4 = 0.4$):

The first approximation of y_4 i.e., $y_4^{(1)}$ is

$$y_4^{(1)} = y_3 + hf(x_3, y_3)$$

i.e., $\qquad y_4^{(1)} = y_3 + h\{1 + x_3 y_3\}$

$\therefore \qquad y_4^{(1)} = 2.4018 + 0.1\{1 + (0.3)(2.4018)\}$

$$y_4^{(1)} = 2.573854.$$

The second approximation of y_4 i.e., $y_4^{(2)}$ is given by

$$y_4^{(2)} = y_3 + \frac{h}{2}\left[f(x_3, y_3) + f(x_4, y_4^{(1)})\right]$$

i.e., $\quad y_4^{(2)} = y_3 + \frac{h}{2}\left[\{1 + x_3 y_3\} + \{1 + x_4 y_4^{(1)}\}\right]$

$$y_4^{(2)} = 2.4018 + \frac{0.1}{2}\left[\{1 + (0.3)(2.4018)\} + \{1 + (0.4)(2.573854)\}\right]$$

$$y_4^{(2)} = 2.589304.$$

Similarly, it can shown that

$$y_4^{(3)} = 2.589613 \text{ and } y_4^{(4)} = 2.589619.$$

Since, $y_4^{(3)} = y_4^{(4)} = 2.5896$ upto four decimal places, we take
$y_4 = 2.5896$.

Thus, the numerical solution of the IVP by modified Euler's method is as given below:

x	0	0.1	0.2	0.3	0.4
y	2	2.1105	2.2434	2.4018	2.5896

Problem 2: Solve $\dfrac{dy}{dx} = \dfrac{y - x}{y + x}$, $y(0) = 1$ for $0.02(0.02)0.06$ using modified Euler's method. Give the approximation correct to five decimal places.

Solution: Here, $x_0 = 0$, $y_0 = 1$, $h = 0.02$ and

$$f(x, y) = \frac{y - x}{y + x} \tag{1}$$

To find y at $x = 0.02$ (i.e., $x_1 = 0.02$):
The first approximation of y_1 i.e., $y_1^{(1)}$ is given by

$$y_1^{(1)} = y_0 + h f(x_0, y_0)$$

i.e., $\qquad y_1^{(1)} = y_0 + h \cdot \dfrac{y_0 - x_0}{y_0 + x_0}$

Substituting the values of x_0, y_0 and h, we get

$$y_1^{(1)} = 1 + 0.02 \left\{ \frac{1-0}{1+0} \right\}$$

$$= 1.02.$$

The second approximation of y_1 i.e., $y_1^{(2)}$ is given by

$$y_1^{(2)} = y_0 + \frac{h}{2} \left[f(x_0, y_0) + f(x_1, y_1^{(1)}) \right]$$

i.e.,

$$y_1^{(2)} = y_0 + \frac{h}{2} \left[\frac{y_0 - x_0}{y_0 + x_0} + \frac{y_1^{(1)} - x_1}{y_1^{(1)} + x_1} \right]$$

i.e.,

$$y_1^{(2)} = 1 + \frac{0.02}{2} \left[\frac{1-0}{1+0} + \frac{1.02 - 0.02}{1.02 + 0.02} \right]$$

$$y_1^{(2)} = 1.09615.$$

The third approximation of y_1, i.e., $y_1^{(3)}$ is given by

$$y_1^{(3)} = y_0 + \frac{h}{2} \left[\frac{y_0 - x_0}{y_0 + x_0} + \frac{y_1^{(2)} - x_1}{y_1^{(2)} + x_1} \right]$$

$\therefore$

$$y_1^{(3)} = 1 + \frac{0.02}{2} \left[\frac{1-0}{1+0} + \frac{1.019615 - 0.02}{1.019615 + 0.02} \right] = 1.019615.$$

Since $y_1^{(2)} = y_1^{(3)}$, we take $y_1 = 1.01961$ upto five decimal accuracy.
Thus, $x_1 = 0.02$, $y_1 = 1.01961$.
To find y at $x = 0.04$ (i.e $x_2 = 0.02$):
The first approximation of y_2 i.e., $y_2^{(1)}$ is obtained from

$$y_2^{(1)} = y_1 + h f(x_1, y_1)$$

$\therefore$

$$y_2^{(1)} = 1.01961 + 0.02 \left[\frac{1.01961 - 0.02}{1.01961 + 0.02} + \frac{1.01961 + 0.02}{1.01961 + 0.02} \right]$$

i.e., $\quad y_2^{(2)} = 1.038841.$

Now

$$y_2^{(2)} = y_1 + \frac{h}{2} \left[f(x_1, y_1) + f(x_2, y_2^{(1)}) \right]$$

i.e., $\quad y_2^{(2)} = y_1 + \frac{h}{2} \left[\frac{y_1 - x_1}{y_1 + x_1} + \frac{y_2^{(1)} - x_2}{y_2^{(1)} + x_2} \right]$

$$\text{i.e., } y_2^{(2)} = 1.01961 + 0.02\left[\frac{1.0961 - 0.02}{1.0961 + 0.02} + \frac{1.038841 - 0.04}{1.038841 + 0.04}\right]$$

$$y_2^{(2)} = 1.0388484.$$

Since $y_2^{(1)} = y_2^{(1)} = 1.03884$ correct to five decimal places, we take $y_2 = 1.03884$. Thus,

$$x_2 = 0.04 \text{ and } y_2 = 1.03884.$$

To find y at $x = 0.06$ (i.e., $x_3 = 0.06$):
The first approximation of y_3 i.e., $y_3^{(1)}$ is given by

$$y_3^{(1)} = y_2 + h f(x_2, y_2)$$

i.e.,
$$y_3^{(1)} = y_2 + h.\frac{y_2 - x_2}{y_2 + x_2}$$

$$\therefore \qquad y_3^{(1)} = 1.03884 + 0.02\left[\frac{1.03884 - 0.04}{1.03884 + 0.04}\right]$$

$$y_3^{(1)} = 1.057357.$$

Now, the second approximation of y_3 i.e., $y_3^{(2)}$ is obtained from

$$y_3^{(2)} = y_2 + \frac{h}{2}\left[f(x_2, y_2) + f(x_3, y_3^{(1)})\right]$$

i.e.,
$$y_3^{(2)} = y_2 + \frac{h}{2}\left[\frac{y_2 - x_2}{y_2 + x_2} + \frac{y_3^{(1)} - x_3}{y_3^{(1)} + x_3}\right].$$

Hence,

$$y_3^{(2)} = 1.03884 + \frac{0.02}{2}\left[\frac{1.03884 - 0.04}{1.03884 + 0.04} + \frac{1.057357 - 0.06}{1.057357 + 0.06}\right]$$

$$y_3^{(2)} = 1.057025$$

Similarly, $y_3^{(3)} = 1.057024$.
Since, $y_3^{(2)} = y_3^{(3)} = 1.05702$ correct five decimal places, we take $y_3 = 1.05702$. Thus,

$$x_3 = 0.06 \text{ and } y_3 = 1.05702.$$

Hence, the numerical solution of given IVP is

x	0	0.02	0.04	0.06
y	1	1.01961	1.03884	1.05702

Problem 3: Solve $\dfrac{dy}{dx} = x + y$, $y(0) = 1$ for $0.05(0.05)0.15$ using modified Euler's method .

Solution: Here, $x_0 = 0$, $y_0 = 1$, $h = 0.05$ and

$$f(x, y) = x + y \tag{1}$$

We have to find y values at 0.05, 0.10 and 0.15 Note that $x_1 = 0.05$, $x_2 = 0.10$ and $x_3 = 0.15$.

To find y at $x = 0.05$ (i.e $x_1 = 0.05$):

The first approximation of y_1 i.e., $y_1^{(1)}$ is computed from

$$y_1^{(1)} = y_0 + hf(x_0, y_0)$$

i.e., $\qquad\qquad y_1^{(1)} = y_0 + h(x_0 + y_0).$

Substituting x_0, y_0 and h values in the above equation, we get

$$y_1^{(1)} = 1 + 0.05\{0 + 1\} = 1.05$$

The second approximation of y_1 (i.e., $y_1^{(2)}$) is obtained from

$$y_1^{(2)} = y_0 + \frac{h}{2}\left[f(x_0, y_0) + f(x_1, y_1^{(1)})\right]$$

Since $f(x, y) = x + y$, we have

i.e., $\qquad\qquad y_1^{(2)} = y_0 + \dfrac{h}{2}\left[(x_0 + y_0) + (x_1 + y_1^{(1)})\right]$

$\therefore\qquad\qquad y_0^{(2)} = 1 + \dfrac{0.05}{2}\{(0 + 1) + (0.05 + 1.05)\}$

i.e., $\qquad\qquad y_0^{(2)} = 1.0525.$

Similarly, it can be shown that $y_1^{(3)} = 1.052562$ $y_1^{(4)} = 1.052564$. The value of y_1 upto five decimal places of accuracy

We take $y_1 = 1.05256$. Thus,

$$x_1 = 0.05 \text{ and } y_1 = 1.05256.$$

To find y at $x = 0.10$ (i.e., $x_2 = 1.0$):

The first approximation of y_1 i.e., $y_1^{(1)}$ is obtained from the Euler's algorithm

$$y_2^{(1)} = y_1 + hf(x_1, y_1)$$

i.e., $$y_2^{(1)} = y_1 + h\,(x_1 + y_1).$$

Substituting the values of x_1, y_1 and h in the above equation, we get

$$y_2^{(1)} = 1.05256 + 0.05\{0.05 + 1.05256\}.$$
$$y_2^{(1)} = 1.107688.$$

The second approximation of y_2 (i.e., $y_2^{(2)}$) is given by

$$y_2^{(2)} = y_1 + \frac{h}{2}\left[f\left(x_1, y_1\right) + f\left(x_2, y_2^{(1)}\right)\right]$$

i.e., $$y_2^{(2)} = y_1 + \frac{h}{2}\left[\left(x_1 + y_1\right) + \left(x_2 + y_2^{(1)}\right)\right].$$

Substituting x_2, $y_2^{(1)}$, x_1, y_1 and h in the above equation, we have

$$y_2^{(2)} = 1.05256 + \frac{0.05}{2}\left[(0.05 + 1.05256) + (0.10 + 1.07688)\right]$$

i.e., $$y_2^{(2)} = 1.110316.$$

Similarly, it can be shown that $y_2^{(3)} = 1.110382$, $y_2^{(4)} = 1.110384$. Thus,

$$x_2 = 0.10 \text{ and } x_2 = 1.11038.$$

To find y at $x = 0.15$ (i.e., $x_3 = 0.15$):

The first approximation of y_3 (i.e., $y_3^{(1)}$) is given by

$$y_3^{(1)} = y_2 + h f(x_2, y_2)$$

$\therefore$ $$y_3^{(1)} = y_1 + h\,(x_2 + y_2)$$

Substituting the values of h, x_2 and y_2 we get

$$y_3^{(1)} = 1.11038 + 0.05(0.10 + 1.11038)$$

$\therefore$ $$y_3^{(1)} = 1.170899.$$

The second approximation of y_3 (i.e., $y_3^{(2)}$) is

$$y_3^{(2)} = y_2 + \frac{h}{2}\left[f\left(x_2, y_2\right) + f\left(x_3, y_3^{(1)}\right)\right]$$

i.e., $$y_3^{(2)} = y_2 + \frac{h}{2}\left[\left(x_2 + y_2\right) + \left(x_3 + y_3^{(1)}\right)\right]$$

$$\therefore \qquad y_3^{(2)} = 1.11038 + \frac{0.05}{2}\left[(0.10+1.11038)+(0.15+1.170899)\right]$$

i.e., $\qquad y_3^{(2)} = 1.173662.$

Using $\quad y_3^{(3)} = y_2 + \dfrac{h}{2}\left[\left(x_2+y_2\right)+\left(x_3+y_3^{(2)}\right)\right]$ and

$$y_3^{(4)} = y_2 + \frac{h}{2}\left[\left(x_2+y_2\right)+\left(x_3+y_3^{(3)}\right)\right] \text{ we obtain}$$

$y_3^{(3)} = 1.173731$ and $y_3^{(4)} = 1.173733.$ Thus,

$$x_3 = 0.15 \text{ and } y_3 = 1.17373.$$

Therefore, the numerical solution of given IVP is as follows:

x	0	0.05	0.10	0.15
y	1	1.05256	1.11038	1.17373

EXERCISE 6(b)

1. Solve $y' = 4 - 2x, y(0) = 2$ by modified Euler's method for $x = 0.5(0.5)1.5$. Compare the results with those obtained from the exact solution of given IVP.

2. Given that $\dfrac{dy}{dx} + xy^2 = 0$, $y(0) = 2$, Compute $y(0.2)$ using modified Euler's method with $h = 0.1$.

3. Solve $\dfrac{dy}{dx} = \log(x+y)$, $y(0) = 1$, using modified Euler's method for $x = 0.1(0.1)0.3.$

4. Solve $2y\dfrac{dy}{dx} = \cos x$, $y(0) = 1$ for $x = 0.1(0.1)0.3$ by modified Euler's method. Compare the results with those obtained from the exact solution of given IVP.

5. Solve $y' = \dfrac{2y}{x}, y(1) = 2$ by modified Euler's method for $x = 1.5(0.5)2.5.$

6. Given that $y' = x + \sin y$, $y(0) = 1$. Compute $y(0.2)$ and $y(0.4)$ using modified Euler's method with $h = 0.2$.

ANSWERS

1. $y_1 = 3.75, y_2 = 5, y_3 = 5.75$
2. $y_1 = 1.980, y_2 = 1.9238$
3. $y_1 = 1.0021, y_2 = 1.0083, y_3 = 1.0184$
4. $1.048691, 1.096764, 1.138103$
5. $y_1 = 4.499, y_2 = 7.998, y_3 = 12.496$
6. $y_1 = 1.1972, y_2 = 1.3534.$

6.11   PICARD'S METHOD OF SUCCESSIVE APPROXIMATIONS

Now, we describe the Picard's method of successive approximation to solve an IVP given by (6.1) and (6.2). i.e.,

$$y' = f(x_0, y_0)$$

and

$$y(x_0) = y_0$$

Integrating (6.1) with respect to x from the lower limit x_0 to the upper limit x, we have

$$\int_{y_0}^{y} dy = \int_{x_0}^{x} f(x, y)dx$$

i.e.,
$$y(x) = y_0 + \int_{x_0}^{x} f(x, y)dx \qquad (6.19)$$

The initial value problem is equivalent to the integral equation given by Eq. (6.19). The solution of (6.19) is also a solution of IVP given by (6.1) and (6.2). The solution of (6.19) is given by successive approximations.

We introduce $y_{(x)}^{j}$ ($j = 1, 2, \ldots$) to denote the jth approximation of $y(x)$. To have an approximate solution of (6.19), we take $y = y_0$ in the integrand of integral of Eq. (6.19). By this assumption the integrand $f(x, y_0)$ is a function of x alone and the solution of (6.19) is known as first approximation of y. i.e.,

$$y^{(1)}(x) = y_0 + \int_{x_0}^{x} f(x, y_0)dx \qquad (6.20)$$

Similarly, the second approximation of y, i.e., $y^{(2)}(x)$ is given by

$$y^{(2)}(x) = y_0 + \int_{x_0}^{x} f(x, y^{(1)}(x))\,dx$$

and so on. In general, $(n + 1)$th approximation of y by the Picard's method is given by

$$y^{(n+1)}(x) = y_0 + \int_{x_0}^{x} f(x, y^{(n)}(x))\,dx \qquad (6.21)$$

with
$$y^{(0)}(x_0) = y_0.$$

Note: The successive approximations $y^{(n)}(x)$ obtained by Picard's method are functions of x alone. To have a value of y at $x = c$, i.e., $y(c)$ substitute $x = c$ in the results of right-hand side of (6.21), which is obtained after the evaluation of the integral $\int_{x_0}^{x} f(x, y^{(n)}(x))\,dx$.

SOLVED PROBLEMS

Problem 1: Find the first four successive approximations of the IVP $y' = 1 + xy$, $y(0) = 1$ by Picard's method.

Solution: Here, $x_0 = 0$, $y_0 = 1$ and

$$f(x, y) = 1 + xy \qquad (1)$$

The $(n + 1)$th successive approximation of y by the Picard's method is given by

$$y^{(n+1)}(x) = y_0 + \int_{x_0}^{x} f(x, y^{(n)}(x))\,dx$$

and $y^{(0)}(x) = y_0$. Therefore,

$$y^{(n+1)}(x) = 1 + \int_{x_0}^{x} \left\{ 1 + x.y^{(n)}(x) \right\} dx \qquad (2)$$

and

$$y^{(0)}(x) = 1 \;\; (y^{(0)}(x) = y_0) \qquad (3)$$

To find the first approximation of y:
Substituting $n = 0$ in (2), we get

$$y^{(1)}(x) = 1 + \int_0^x \left\{1 + xy^{(0)}(x)\right\} dx$$

and using (3) the above equation becomes

$$y^{(1)}(x) = 1 + \int_0^x \left\{1 + (x)(1)\right\} dx$$

i.e.,
$$y^{(1)}(x) = 1 + \int_0^x 1 + (x) dx = 1 + x + \frac{x^2}{2}$$

Thus,

$$y^{(1)}(x) = 1 + x + \frac{x^2}{2} \tag{4}$$

To find the second approximation of y:
Putting $n = 1$ in Eq. (2), we get

$$y^{(2)}(x) = 1 + \int_0^x \left\{1 + (x)y^{(1)}(x)\right\} dx$$

In view of (4), the above equation reduces to

$$y^{(2)}(x) = 1 + \int_0^x \left[1 + (x)\left\{1 + x + \frac{x^2}{2}\right\}\right] dx$$

$$\therefore \qquad y^{(2)}(x) = 1 + x + \frac{x^2}{2} + \frac{x^3}{3} + \frac{x^4}{8} \tag{5}$$

To find third successive approximation of y:
Putting $n = 2$ in Eq. (2) we have

$$y^{(3)}(x) = 1 + \int_0^x \left\{1 + (x)y^{(2)}(x)\right\} dx$$

In view of (5), this equation becomes

$$y^{(3)}(x) = 1 + \int_0^x \left\{1 + (x)\left(1 + x + \frac{x^2}{2} + \frac{x^3}{3} + \frac{x^4}{8}\right)\right\} dx$$

$$\therefore \qquad y^{(3)}(x) = 1 + x + \frac{x^2}{2} + \frac{x^3}{3} + \frac{x^4}{8} + \frac{x^5}{15} + \frac{x^6}{48} \qquad (6)$$

To find the fourth successive approximation of y:

Putting $n = 3$ in (2) and using (6), we get

$$y^{(4)}(x) = 1 + \int_0^x \left\{ 1 + (x)\left(1 + x + \frac{x^2}{2} + \frac{x^3}{3} + \frac{x^4}{8} + \frac{x^5}{15} + \frac{x^6}{48} \right) \right\} dx$$

$$\therefore \quad y^{(4)}(x) = 1 + x + \frac{x^2}{2} + \frac{x^3}{3} + \frac{x^4}{8} + \frac{x^5}{15} + \frac{x^6}{48} + \frac{x^7}{108} + \frac{x^8}{384} \qquad (7)$$

Thus, the first four successive approximations of y of given IVP by Picard's method are given by (4), (5), (6) and (7).

Problem 2: Find the value of y correct to 4 decimal places at $x = 0.1$ by Picard's method of Successive approximations, given that $\frac{dy}{dx} = x + y$, $y(0) = 1$.

Solution: Here, $x_0 = 0$, $y_0 = 1$ and

$$f(x, y) = x + y \qquad (1)$$

The $(n + 1)$th approximation of y by Picard's method is given by

$$y^{(n+1)}(x) = 1 + \int_0^x \left\{ x + y^{(n)}(x) \right\} dx \qquad (2)$$

and

$$y^{(0)}(x) = 1 \ (y^{(0)}(x) = y_0) \qquad (3)$$

Putting $n = 0$ in (2) and using (3), we have

$$y^{(1)}(x) = 1 + \int_0^x \left\{ x + 1 \right\} dx$$

$$\therefore \qquad y^{(1)}(x) = 1 + x + \frac{x^2}{2} \qquad (4)$$

Hence

$$y^{(1)}(0.1) = 1 + 0.1 + \frac{(0.1)^2}{2}$$

$$\therefore \qquad y^{(1)}(0.1) = 1.105.$$

Now, substituting $n = 1$ in (2) and using (4), we get

$$y^{(2)}(x) = 1 + \int_0^x \left\{ x + \left(1 + x + \frac{x^2}{2} \right) \right\} dx$$

$$\therefore \qquad y^{(2)}(x) = 1 + x + x^2 + \frac{x^3}{6} \qquad (5)$$

Hence

$$y^{(2)}(0.1) = 1 + 0.1 + (0.1)^2 + \frac{(0.1)^3}{6}$$

$$\therefore \qquad y^{(2)}(0.1) = 1.110167.$$

Again substituting $n = 2$ in Eq. (2) and using Eq. (5)

$$y^{(3)}(x) = 1 + \int_0^x \left\{ x + \left(1 + x + x^2 + \frac{x^3}{6} \right) \right\} dx$$

$$\therefore \qquad y^{(3)}(x) = 1 + x + x^2 + \frac{x^3}{3} + \frac{x^4}{24}. \qquad (6)$$

Hence, $\qquad y^{(3)}(0.1) = 1 + 0.1 + (0.1)^2 + \frac{(0.1)^3}{3} + \frac{(0.1)^4}{24}.$

$$\therefore \qquad y^{(3)}(0.1) = 1.110171167.$$

Upto four decimal places $y^{(3)}(0.1) = y^{(2)}(0.1)$.

$\therefore$ The value of y correct to four decimal places at $x = 0.1$ is 1.1101.

Problem 3: Given that $y' = x - y$ and $y(0) = 1$. Find $y(0.2)$ from successive approximation of y by Picard's method.

Solution: Here, $x_0 = 0$, $y_0 = 1$ and

$$f(x, y) = x - y \qquad (1)$$

The $(n + 1)$th successive approximation of y by Picard's method is given by

$$y^{(n+1)}(x) = 1 + \int_0^x \left\{ 1 - y^{(n)}(x) \right\} dx \qquad (2)$$

and

$$y^{(0)} = 1 \qquad (3)$$

To find the first approximation of y:
Putting $n = 0$ in (2) and using (3), we get

$$y^{(1)}(x) = 1 + \int_0^x \{x - 1\}\, dx$$

$$\therefore \qquad y^{(1)}(x) = 1 - x + \frac{x^2}{2} \qquad (4)$$

To find the second approximation of y:
Substituting $n = 1$ in (2) and using (4), we have

$$y^{(2)}(x) = 1 + \int_0^x \left\{ x - \left(1 - x + \frac{x^2}{2}\right) \right\} dx$$

$$\therefore \qquad y^{(2)}(x) = 1 + x^2 - x - \frac{x^3}{6} \qquad (5)$$

To find the third approximation of y:
Putting $n = 2$ in (2) and using (6), we obtain

$$y^{(3)}(x) = 1 + \int_0^x \left\{ x - \left(1 - x + x^2 - \frac{x^3}{6}\right) \right\} dx$$

$$\therefore \qquad y^{(3)}(x) = 1 - x + x^2 - \frac{x^3}{3} + \frac{x^4}{24} \qquad (6)$$

Hence, the value of y at $x = 0.2$ from the third successive approximation is

$$y^{(3)}(0.2) = 1 - 0.2 + (0.2)^2 - \frac{(0.2)^3}{3} + \frac{(0.2)^4}{24}$$

$$\therefore \qquad y(0.2) = 0.8374.$$

Problem 4: Given that $\dfrac{dy}{dx} = 1 + y^2$, $y(0) = 0$. Find $y(0.5)$ from third successive approximation of y applying Picard's method of successive approximation. Also, find the error in the approximation of $y(0.5)$.

Solution: Here, $x_0 = 0$, $y_0 = 0$ and

$$f(x, y) = 1 + y^2 \qquad (1)$$

The $(n + 1)$th approximation of y for the given IVP by Picard's method is

$$y^{(n+1)}(x) = \int_0^x \left[1 + \left\{y^{(n)}(x)\right\}^2\right] dx \tag{2}$$

and

$$y^{(0)}(x) = 0. \tag{3}$$

To find first approximation of y:

Substituting $n = 0$ in Eq. (2) and using (3), we get

$$y^{(1)}(x) = \int_0^x \left[1 + (0)^2\right] dx = x$$

$$\therefore \qquad y^{(1)}(x) = x. \tag{4}$$

To find second approximation of y:

Substituting $n = 1$ in the Eq. (2) and using (4), we get

$$y^{(2)}(x) = \int_0^x \left[1 + x^2\right] dx$$

$$\therefore \qquad y^{(2)}(x) = x + \frac{x^3}{3}. \tag{5}$$

To find the third approximation of y:

Substituting $n = 2$ in the Eq. (2) and using (5), we obtain

$$y^{(3)}(x) = \int_0^x \left[1 + \left\{x + \frac{x^3}{3}\right\}^2\right] dx$$

$$\therefore \qquad y^{(3)}(x) = x + \frac{x^3}{3} + 2.\frac{x^5}{15} + \frac{x^7}{63} \tag{6}$$

Hence,

$$y^{(3)}(0.5) = 0.5 + \frac{(0.5)^3}{3} + 2.\frac{(0.5)^5}{15} + \frac{(0.5)^7}{63}$$

$$\therefore \qquad y^{(3)}(0.5) = 0.545956.$$

For estimation of error in $y^{(3)}(0.5)$, we find the exact solution of given IVP. Recall the given differential equation

$$\frac{dy}{dx} = 1 + y^2$$

i.e.,
$$\frac{dy}{1+y^2} = dx.$$

Integrating, $\tan^{-1}y = x + c$. Since $y(0) = 0$,
We have
$$\tan^{-1}y = x$$

i.e.,
$$y = \tan(x).$$

$\therefore$
$$y(0.5) = 0.546302.$$

Thus, the error in $y(0.5)$ is $0.546302 - 0.545956 = 0.000346$.

Problem 5: Solve $y' = y - x^2$, $y(0) = 1$ by Picard's method and find the value of $y(0.1)$, $y(0.2)$, $y(0.3)$, $y(0.4)$ and $y(0.5)$ from fourth approximation of y.

Solution: Here, $x_0 = 0$, $y_0 = 1$ and
$$f(x, y) = y - x^2 \tag{1}$$

The $(n + 1)$th approximation of y by Picard's method for the given problem is
$$y^{(n+1)}(x) = 1 + \int_0^x \left\{ y^{(n)}(x) - x^2 \right\} dx \tag{2}$$
and
$$y^{(0)}(x) = 1. \tag{3}$$

To find the first approximation of y:
Substituting $n = 0$ in the Eq. (2) and using (3), we get
$$y^{(1)}(x) = 1 + \int_0^x \left\{ 1 - x^2 \right\} dx$$

$\therefore$
$$y^{(1)}(x) = 1 + x - \frac{x^3}{3} \tag{4}$$

To find the second approximation of y:
Substituting $n = 1$ in the Eq. (2) and using (4) we obtain

$$y^{(2)}(x) = 1 + \int_0^x \left[\left(1 + x - \frac{x^3}{3} \right) - x^2 \right] dx \quad y^{(2)}(x) = 1 + x + \frac{x^2}{2} - \frac{x^3}{3} - \frac{x^4}{12}. \tag{5}$$

To find the third approximation of y:

Substituting $n = 2$ in the Eq. (2) and using (5), we have

$$y^{(3)}(x) = 1 + \int_0^x \left[\left(1 + x + \frac{x^2}{2} - \frac{x^3}{3} - \frac{x^4}{12} \right) - x^2 \right] dx$$

$$\therefore \qquad y^{(3)}(x) = 1 + x + \frac{x^2}{2} - \frac{x^3}{6} - \frac{x^4}{12} - \frac{x^5}{60}. \qquad (6)$$

To find the fourth approximation of y:

Substituting $n = 3$ in the Eq. (2) and using (6), we get

$$y^{(4)}(x) = 1 + \int_0^x \left[\left(1 + x + \frac{x^2}{2} - \frac{x^3}{6} - \frac{x^4}{12} - \frac{x^5}{60} \right) - x^2 \right] dx$$

$$\therefore \qquad y^{(4)}(x) = 1 + x + \frac{x^2}{2} - \frac{x^3}{3} - \frac{x^4}{12} - \frac{x^5}{60} - \frac{x^6}{360}, \qquad (7)$$

which is the solution of given IVP by Picard's method upto fourth approximation.

Now, the values of $y(0.1)$, $y(0.2)$, $y(0.3)$, $y(0.4)$ and $y(0.5)$ from fourth approximation of y are computed from (7) and they are shown in the following table.

x	0.1	0.2	0.3	0.4	0.5
y	1.1048	1.2186	1.3401	1.4681	1.6010

EXERCISE 6(c)

1. Find the first four Picard's successive approximations of the IVP
   $$2y\frac{dy}{dx} - \cos x = 0, \, y(0) = 1.$$

2. Solve the IVP $\dfrac{dy}{dx} - x - y^2 = 0$, $y(0) = 1$ by Picard's method.

3. Find the value of $y(0.1)$ by Picard's method, given that $\dfrac{dy}{dx} = \dfrac{y-x}{y+x}$, $y(0) = 1$.

4. Solve $y' = x^2 + y^2$, $y(0) = 1$ using Picard's method of successive approximation.

5. Obtain the first four successive approximation of Picard's method for the IVP $\dfrac{dy}{dx} = 1 + 2xy$, $y(0) = 0$.

6. Find $y(0.1)$ and $y(0.2)$ by Picard's method given that $y' = x + y^2 + 1$, $y(0) = 0$.

ANSWERS

1. 10 993

2. $y = 1 + x + \dfrac{3}{2}x^2 + \dfrac{2}{3}x^3 + \dfrac{1}{4}x^4 + \dfrac{1}{20}x^5 + \ldots\ldots\ldots$

3. $y = 1 + x - 2\displaystyle\int_0^x \dfrac{x}{1 + 2\log(1 + x)}\,dx$

4. $y = 1 + x + x^2 + \dfrac{2}{3}x^3 + \dfrac{1}{6}x^4 + \dfrac{2}{15}x^5 + \dfrac{1}{63}x^7 + \ldots\ldots\ldots$

5. $y = x + \dfrac{2}{3}x^3 + \dfrac{1}{2}x^4 + \dfrac{2}{5}x^5 + \dfrac{1}{9}x^8 + \ldots\ldots\ldots$

6. $y(0.1) = 0.105358$, $y(0.2) = 0.223082$.

6.12   TAYLOR SERIES METHOD

In this section, y_i', y_i'', y_i''', y_i^{iv}….. etc denote the values of $\dfrac{dy}{dx}, \dfrac{d^2y}{dx^2}, \dfrac{d^3y}{dx^3}, \dfrac{d^4y}{dx^4}$….. respectively at $(x_i,\ y_i)$, i.e., $\left(\dfrac{dy}{dx}\right)_{(x_i,y_i)} = y_i'$,

$\left(\dfrac{d^2y}{dx^2}\right)_{(x_i,y_i)} = y_i''$, $\left(\dfrac{d^3y}{dx^3}\right)_{(x_i,y_i)} = y_i'''$, $\left(\dfrac{d^4y}{dx^4}\right)_{(x_i,y_i)} = y_i^{iv}$, etc.

Now, we describe the Taylor's series method to solve the initial value problem given by (6.1) and (6.2), i.e.,

$$\frac{dy}{dx} = f(x, y)$$

and

$$y(x_0) = y_0.$$

The Taylor's series expansion of $y(x)$ about $x = x_0$ is given by

$$y(x) = y(x_0) + (x - x_0)y'(x_0) + (x - x_0)^2 \frac{y''(x_0)}{\lfloor 2} + (x - x_0)^3 \frac{y'''(x_0)}{\lfloor 3} + \ldots$$

Using the notation introduced at the beginning of this section, the above equation can be written as

$$y(x) = y_0 + (x - x_0)y_0' + (x - x_0)^2 \frac{y_0''}{\lfloor 2} + (x - x_0)^3 \frac{y_0'''}{\lfloor 3} + \ldots \qquad (6.22)$$

Let h be the interval of differencing and $x_1 = x_0 + h$. Now, substituting $x_1 = x_0 + h$ for x in the Eq. (6.22), we obtain

$$y(x_1) = y_0 + hy_0' + \frac{h^2}{\lfloor 2} y_0'' + \frac{h^3}{\lfloor 3} y_0''' + \ldots \qquad (6.23)$$

The value of $y(x_1)$ will be known approximately, if we are able to compute $y_0', y_0'', y_0''', \ldots$ Note that y_0 is known from the initial condition. The values of $y_0', y_0'', y_0''' \ldots$ etc can be computed in the following manner.

Since $y' = f(x, y)$ we have

$$y_0' = f(x_0, y_0)$$

Differentiating (6.1), i.e., $\dfrac{dy}{dx} = f(x, y)$ with respect to x, we get

$$y'' = \frac{\partial f}{\partial x} + \frac{\partial f}{\partial y} y'$$

Let

$$\frac{\partial f}{\partial x} + \frac{\partial f}{\partial y} y' = g_1(x, y')$$

Then

$$y'' = g_1(x, y, y') \qquad (6.24)$$

Hence, $y_0'' = g_1(x_0, y_0, y_0')$

Again differentiating (6.24) with respect to x, we obtain

$$y''' = \frac{\partial g_1}{\partial x} + \frac{\partial g_1}{\partial y} \frac{dy}{dx} + \frac{\partial g_1}{\partial y'} \frac{\partial y'}{\partial x}$$

i.e.,
$$y''' = \frac{\partial g_1}{\partial x} + \frac{\partial g_1}{\partial y}y' + \frac{\partial g_1}{\partial y'}\cdot y''$$

Let
$$\frac{\partial g_1}{\partial x} + \frac{\partial g_1}{\partial y}y' + \frac{\partial g_1}{\partial y'}y'' = g_2(x, y, y', y'') \tag{6.25}$$

Then
$$y''' = g_2(x, y, y', y'').$$

Hence,
$$y_0''' = g_2(x_0, y_0, y_0', y_0'')$$

Similarly, we can find $y_0^{iv}, y_0^{v}, \dots$etc.

Substituting $y_0', y_0'', y_0'''\dots$ and h in the Eq. (6.23) and truncating the terms of it on and after h^k, k is a positive integer, we get the approximate value of y_1.

To compute the values of y at $x = x_0 + 2h$ approximately, we use the following Taylor series.

$$y_2 = y_1 + hy_1' + \frac{h^2}{\lfloor 2}y_1'' + \frac{h^3}{\lfloor 3}y_1''' + \dots \tag{6.26}$$

The values of $y_1', y_1'', y_1'''\dots$etc. can be computed from the following:
$$y_1' = f(x_1, y_1)$$
$$y_1'' = g_1(x_1, y_1, y_1')$$
$$y_1''' = g_2(x_1, y_1, y_1', y_1'')$$

etc., Note that $f(x, y)$ is the R.H.S. of (6.1) and $g_1, g_2\dots\dots$ are given by

$$g_1(x, y, y') = \frac{\partial f}{\partial x} + \frac{\partial f}{\partial y}y'$$

$$g_2(x, y, y', y'') = \frac{\partial g_1}{\partial x} + \frac{\partial g_1}{\partial y}y' + \frac{\partial g_1}{\partial y'}y''$$

Now, substituting $h, y_1, y_1', y_1'', y_1'''\dots$ in the Eq. (6.26), we obtain $y(x_2)$.

Continuing this process, we can find y values at $x_3, x_4, x_5\dots, x_n$. Note that

$$y_{i+1} = y_i + hy_i' + \frac{h^2}{\lfloor 2}y_i'' + \frac{h^3}{\lfloor 3}y_i''' + \frac{h^4}{\lfloor 4}y_i^{(iv)} + \dots, \tag{6.27}$$

where $i = 0, 1, 2, 3, \dots$

Note: To find the numerical solution of an IVP given by (6.1) and (6.2) by Taylor's series method, usually we compute the terms containing upto h^3 or h^4 if the number of terms to be included in the Taylor series is not specified in a problem or question.

SOLVED PROBLEMS

Problem 1: Using the Taylor's series method find $y(0.1)$ and $y(0.2)$ for the IVP $y' = x + y^2$ and $y(0) = 1$.

Solution: Let $h = 0.1$ be the interval of differencing. Since $y(0) = 1$, we have $x_0 = 0$ and $y_0 = 1$. Further,

$$y' = x + y^2 \tag{1}$$

Differentiating (1) with respect to x, we get

$$y'' = 1 + 2yy' \tag{2}$$

Again differentiating (2) with respect to x,
We obtain

$$y''' = 2yy'' + 2(y')^2 \tag{3}$$

Similarly,

$$y^{iv} = 2yy''' + 6y'y'' \tag{4}$$

Thus,

$$y'_i = x_i + y_i^2 \tag{5}$$

$$y''_i = 1 + 2y_i y'_i \tag{6}$$

$$y'''_i = 2y_i y''_i + 2(y'_i)^2 \tag{7}$$

and

$$y^{iv}_i = 2y_i y''_i + 6y'_i y''_i, \tag{8}$$

where $i = 0, 1, 2, 3,\ldots$.

To find $y\,(0.1)$ (i.e., y_1):
First, we compute y_0', y_0'', y_0''' and y_0^{iv} from the Eqs. (5) to (8):

$$y_0' = x_0 + y_0^2 = 0 + (1)^2 = 1$$

$$y_0'' = 1 + 2y_0 y_0' = 1 + 2(1)(1) = 3$$

$$y_0''' = 2y_0 y_0'' + 2(y_0')^2 = 2 \times 1 \times 3 + 2(1)^2 = 8$$

$$y_0^{iv} = 2y_0 y_0''' + 6\,y_0' y_0'' = 2 \times 1 \times 8 + 6 \times 1 \times 3 = 34$$

Now, Taylor's series expansion about x_0 is given by

$$y_1 = y_0 + \frac{h}{\lfloor 1} y_0' + \frac{h^2}{\lfloor 2} y_0'' + \frac{h^3}{\lfloor 3} y_0''' + \frac{h^4}{\lfloor 4} y_0^{(iv)} + \cdots$$

$$\therefore \qquad y_1 = 1 + \frac{0.1}{\lfloor 1}(1) + \frac{(0.1)^2}{\lfloor 2}(3) + \frac{(0.1)^3}{\lfloor 3}(8) + \frac{(0.1)^4}{\lfloor 4} \times 34 + \cdots$$

i.e., $\qquad y_1 = 1.116475.$

Thus,

$$x_1 = 0.1 \text{ and } y_1 = 1.116475.$$

To find $y\,(0.2)$ (i.e., y_2):
From (6.27), for $i = 1$, we have

$$y_2 = y_1 + \frac{h}{\lfloor 1} y_1' + \frac{h^2}{\lfloor 2} y_1'' + \frac{h^3}{\lfloor 3} y_1''' + \frac{h^4}{\lfloor 4} y_1^{(iv)} + \cdots \qquad (9)$$

Now, we compute y_1', y_1'', y_1''' and y_1^{iv} from the Eqs. (5) to (8).

$$y_1' = x_1 + y_1^2$$

$$= 0.1 + (1.116475)^2 = 1.346516$$

$$y_1'' = 1 + 2y_1 y_1'$$

$$= 1 + 2(1.116475)(1.346516) = 4.006703$$

$$y_1''' = 2y_1 y_1'' + 2(y_1')^2$$

$$= 2(1.116475)(4.006703) + 2(1.346516)^2$$

$$= 12.572978$$

$$y_1^{iv} = 2y_1 y_1''' + 6y_1' y_1''$$

$$= 2(1.116475)(12.572978) + 6(1.346516) \times (4.006703)$$

$$= 60.445369.$$

Substituting the values of h, y_1, y_1', y_1'', y_1''' and y_1^{iv} in Eq. (9), we obtain

$$y_2 = 1.116475 + 0.1(1.346516) + \frac{(0.1)^2}{\underline{|2}}(4.006703)$$

$$+ \frac{(0.1)^3}{\underline{|3}}(12.572978) + \frac{(0.1)^4}{\underline{|4}}60.445369$$

$$\therefore \qquad y_2 = 1.273507.$$

Thus, $y(0.1) = 1.116475$ and $y(0.2) = 1.273507$.

Problem 2: Using the Taylor's series method, solve the IVP $\frac{dy}{dx} = 1 + xy, y(0) = 2$ for $0.1(0.1)(0.3)$.

Solution: Here, $x_0 = 0$, $y_0 = 2$, $h = 0.1$ and $f(x, y) = 1 + xy$.
Consider

$$y' = 1 + xy \tag{1}$$

Differentiating (1) successively with respect to x two times, we get

$$y'' = xy' + y \tag{2}$$

$$y''' = xy'' + 2y' \tag{3}$$

$$\therefore \qquad y_i' = 1 + x_i y_i \tag{4}$$

$$y_i'' = x\,y_i' + y_i \tag{5}$$

$$y_i''' = x\,y_i'' + 2\,y_i' \tag{6}$$

To find $y(0.1)$ (i.e., y_1):
From (6.27) with $i = 0$, we get

$$y_1 = y_0 + \frac{h}{\underline{|1}}y_0' + \frac{h^2}{\underline{|2}}y_0'' + \frac{h^3}{\underline{|3}}y_0''' + \ldots \tag{7}$$

Now

$$y_0' = 1 + x_0 y_0 = 1 + (0)(2) = 1 \tag{8}$$

$$y_0'' = x_0 y_0' + y_0 = 0(1) + 2 = 2 \tag{9}$$

$$y_0''' = x_0 y_0'' + 2\,y_0' = 0(2) + 2 \times 1 = 2 \tag{10}$$

Substituting (8), (9), (10), h and y_0 in (7) and neglecting the terms containing h^k $(k \geq 4)$,
We obtain

$$y_1 = 2 + \frac{0.1}{\underline{|1}} \times (1) + \frac{(0.1)^2}{\underline{|2}}(2) + \frac{(0.1)^3}{\underline{|3}}(2)$$

$$= 2 + 0.1 + 0.01 + 0.000333$$

$$= 2.1103 \text{ (upto 4 decimal places).}$$

Thus,

$$x_1 = 0.1 \text{ and } y_1 = 2.1103.$$

To find y at $x = 0.2$ (i.e., y_2):
Substituting $i = 1$ in the Eq. (6.27), we get

$$y_2 = y_1 + \frac{h}{\lfloor 1} y_1' + \frac{h^2}{\lfloor 2} y_1'' + \frac{h^3}{\lfloor 3} y_1''' + \dots \tag{11}$$

From the Eqs. (4), (5) and (6) (with $i = 1$),
We have

$$y_1' = 1 + x_1 y_1 = 1 + (0.1)(2.1103) = 1.21103$$

$$y_1'' = x y_1' + y_1$$

$$= 0.1(1.21103) + 2.1103$$

$$\approx 2.2314$$

and

$$y_1''' = x y_1'' + 2 y_1'$$

$$= (0.1)(2.2314) + 2(1.21103)$$

$$\approx 2.6452$$

Substituting y_1, y_1', y_1'', y_1''' and h values in the Eq. (11),
We have,

$$y_2 = 2.1103 + \frac{0.1}{\lfloor 1} 1.21103 + \frac{(0.1)^2}{\lfloor 2}(2.2314) + \frac{(0.1)^3}{\lfloor 3}(2.6452)$$

$$\therefore \quad y_2 \approx 2.2430$$

Thus,

$$x_2 = 0.2 \text{ and } y_2 = 2.2430.$$

To find y at $x = 0.3$ (i.e., y_3):
Substituting $i = 2$ in the Eq. (6.27), we get

$$y_3 = y_2 + \frac{h}{\lfloor 1} y_2' + \frac{h^2}{\lfloor 2} y_2'' + \frac{h^3}{\lfloor 3} y_2''' + \dots \tag{12}$$

From the Eqs. (4), (5) and (6) with $i = 2$, we have

$$y_2' = 1 + x_2 y_2 = 1 + (0.2)(2.2430) \approx 1.4486$$
$$y_2'' = x y_2' + y_2 = (0.1)(1.4486) + 2.2430 \approx 2.5327$$
$$y_2''' = x_2 y_2'' + 2 y_2'$$
$$= (0.2)(2.5327) + 2 \times (1.4486)$$
$$= 0.50654 + 2.8972$$
$$\approx 3.4037.$$

Substituting y_2, y_2', y_2'', y_2''' and h values in the Eq. (12), We get,

$$y_3 = 2.2430 + \frac{0.1}{\lfloor 1} \times 1.4486 + \frac{(0.1)^2}{\lfloor 2} \times 2.5327 + \frac{(0.1)^3}{\lfloor 3} \times 3.4037$$

$$y_3 = 2.3707.$$
Thus, $x_3 = 0.3$ and $y_3 = 2.3707$.

Problem 3: Using the Taylor's series method solve the initial value problem $y' = x^2 - y$, $y(0) = 1$ for $0.1(0.1)0.4$.

Solution: Here, $h = 0.1$, $x_0 = 0$, $y_0 = 1$ and

$$y' = x^2 - y \tag{1}$$

Differentiating (1) w.r.t. x, we get
$$y'' = 2x - y' \tag{2}$$
Similarly,
$$y''' = 2 - y'' \tag{3}$$
and
$$y^{iv} = -y''' \tag{4}$$
$$\therefore \qquad y_i' = x_i^2 - y_i \tag{5}$$
$$y_i'' = 2x_i - y_i' \tag{6}$$
$$y_i''' = 2 - y_i'' \tag{7}$$
$$y_i^{iv} = -y_i''' \tag{8}$$

To find y at $x = 0.1$ (i.e., y_1):
From 6.27, with $i = 0$, we obtain

$$y_1 = y_0 + \frac{h}{2} y_0' + \frac{h^2}{\lfloor 2} y_0'' + \frac{h^3}{\lfloor 3} y_0''' + \frac{h^4}{\lfloor 4} y_0^{(iv)} + \dots \tag{9}$$

For $i = 0$, the Eqs. (5), (6), (7) and (8) become

$$y_0' = x_0^2 - y_0 = 0^2 - 1 = -1$$
$$y_0'' = 2x_0 - y_0' = 2(0) - (-1) = 1$$
$$y_0''' = 2 - y_0'' = 2 - 1 = 1$$
$$y^{iv} = -y_0''' = -1.$$

Substituting $y_0, y_0', y_0'', y_0''', y_0^{iv}$, h and neglecting the terms involving $h^k (k \geq 5)$, we get

$$y_1 = 1 + (0.1)(-1) + \frac{(0.1)^2}{\lfloor 2} (1) + \frac{(0.1)^3}{\lfloor 3} \times 1 + \frac{(0.1)^4}{\lfloor 4} (-1)$$

$$y_1 \approx 0.9052.$$

Thus,

$$x_1 = 0.1 \text{ and } y_1 = 0.9052.$$

To find y at $x = 0.2$ (i.e., y_2):
From the Eq. (6.27) with $i = 1$, we have

$$y_2 = y_1 + hy_1' + \frac{h^2}{\lfloor 2} y_1'' + \frac{h^3}{\lfloor 3} y_1''' + \frac{h^4}{\lfloor 4} y_1^{(iv)} + \dots \tag{10}$$

Now,

$$y_1' = x_1^2 - y_1 = (0.1)^2 - (0.9052) = -0.89552$$
$$y_1'' = 2x_1 - y_1' = 2(0.1) - (-0.8952) = 1.0952$$
$$y_1''' = 2 - y_1'' = 2 - 1.0952 = 0.9048$$
$$y_1^{iv} = -y_1''' = -0.9048.$$

Substituting $y_1, y_1', y_1'', y_1''', y_1^{iv}$ and h in the Eq. (10), we get

$$y_2 = 0.0952 + (0.1)(-0.8952) + \frac{(0.1)^2}{\lfloor 2} \times (1.09952)$$

$$+ \frac{(0.1)^3}{\lfloor 3} \times (0.9048) + \frac{(0.1)^4}{\lfloor 4} (-0.9048)$$

$$y_2 \approx 0.8213.$$

Thus, $\qquad x_2 = 0.2 \text{ and } y_2 = 0.8213.$

To find y at $x = 0.3$ (i.e., y_3):
From the Eq. (6.27), y_3 is given by

$$y_3 = y_2 + hy_2' + \frac{h^2}{\lfloor 2} y_2'' + \frac{h^3}{\lfloor 3} y_2''' + \frac{h^4}{\lfloor 4} y_2^{(iv)} + \dots \qquad (11)$$

Now,

$$y_2' = x_2^2 - y_2 = (0.2)^2 - 0.8213 = -0.7813$$
$$y_2'' = 2x_2 - y_2' = 2 \times 0.2 - (-0.7813) = 1.1813$$
$$y_2''' = 2 - y_2'' = 2 - 1.1813 = 0.8187$$
$$y_2^{iv} = -y_2''' = -0.8187.$$

$$\therefore \qquad y_3 = 0.8213 + 0.1(-0.7813) + \frac{(0.1)^2}{\lfloor 2} \times 1.1813$$

$$+ \frac{(0.1)^3}{\lfloor 3} \times 0.8187 + \frac{(0.1)^4}{\lfloor 4} \times (-0.8187)$$

$$y_3 \approx 0.7492.$$

Thus, $x_3 = 0.3$ and $y_3 = 0.7492$.

To find y at $x = 0.4$ (i.e., y_4):

From the Eq. (6.27), we have

$$y_4 = y_3 + hy_3' + \frac{h^2}{\lfloor 2} y_3'' + \frac{h^3}{\lfloor 3} y_3''' + \frac{h^4}{\lfloor 4} y_3^{(iv)} + \dots \qquad (12)$$

Now,

$$y_3' = x_3^2 - y_3 = (0.3)^2 - 0.7492 = -0.6592$$

$$y_3'' = 2x_3 - y_3' = 2 \times 0.3 - (-0.6592) = 1.2592$$

$$y_3''' = 2 - y_3'' = 2 - 1.2592 = 0.7408$$

$$y_3^{iv} = -y_3''' = -0.7408.$$

$$\therefore \qquad y_4 = 0.7492 + 0.1(-0.6592) + \frac{(0.1)^2}{\lfloor 2} \times 1.2592$$

$$+ \frac{(0.1)^3}{\lfloor 3} \times 0.7408 + \frac{(0.1)^4}{\lfloor 4} \times (-0.7408)$$

$$\approx 0.68969.$$

Problem 4: Using the Taylor's series method, solve the IVP $y' = x + y$, $y(1) = 0$ for $x = 1.1$ and 1.2 with $h = 0.1$.

Solution: Here, $x_0 = 1, y_0 = 0, h = 0.1$ and

$$y' = x + y \tag{1}$$

Differentiating (1) successively, we obtain

$$y'' = 1 + y'$$

$$y''' = y''$$

$$y^{iv} = y'''$$

$$\therefore \qquad y_i' = x_i + y_i \tag{2}$$

$$y_i'' = 1 + y_i' \tag{3}$$

$$y_i''' = y_i'' \tag{4}$$

$$y_i^{iv} = y_i''' \tag{5}$$

To find y at $x = 1.1$ (i.e., to find y_1):
From (6.27), we have

$$y_1 = y_0 + hy_0' + \frac{h^2}{\lfloor 2} y_0'' + \frac{h^3}{\lfloor 3} y_0''' + \frac{h^4}{\lfloor 4} y_0^{(iv)} + \ldots \tag{6}$$

Now, from the Eqs. (2), (3), (4) and (5), we have

$$\therefore \qquad y_0' = x_0 + y_0 = 1 + 0 = 1$$

$$y_0'' = 1 + y_0' = 1 + 1 = 2$$

$$y_0''' = y_0'' = 2$$

$$y_0^{iv} = y_0''' = 2$$

Now substituting the values of h, y_0, y_0', y_0'', y_0''', y_0^{iv} and neglecting the terms involving h^k $(k \geq 5)$, we have

$$y_1 = 0 + (0.1)(1) + \frac{(0.1)^2}{\lfloor 2}.2 + \frac{(0.1)^3}{\lfloor 3}.2 + \frac{(0.1)^4}{\lfloor 4}(2)$$

$$y_1 \approx 0.1103.$$

Thus, $x_1 = 0.1$ and $y_1 = 0.1103$.

To find y at $x = 1.2$ (i.e., to find $y(1.2)$):
We have y_2 from Eq. (6.27) as

$$y_2 = y_1 + hy_1' + \frac{h^2}{\underline{|2}} y_1'' + \frac{h^3}{\underline{|3}} y_1''' + \frac{h^4}{\underline{|4}} y_1^{(iv)} + \ldots \tag{7}$$

Now,

$$y_1' = x_1 + y_1 = 1.1 + 0.1103 = 1.2103$$
$$y_1'' = 1 + y_1' = 1 + 1.2103 = 2.2103$$
$$y_1''' = y_1'' = 2.2103$$
$$y_1^{iv} = y_1''' = 2.2103.$$

Substituting the above values, h and y_1 in Eq. (7)

We get

$$y_2 = 0.1103 + (0.1)(1.2103) + \frac{(0.1)^2}{\underline{|2}} \times 2.2103 + \frac{(0.1)^3}{\underline{|3}} \times 2.2103$$

$$+ \frac{(0.1)^4}{\underline{|4}} \times 2.2103$$

$$\approx 0.2427.$$

Thus, $x_2 = 1.2$ and $y_2 = 0.2427$.

Problem 5: Find $y(1.1)$ using the Taylor's series, given $y' = xy^{1/3}$, $y(1) = 1$.

Solution: Here, $x_0 = 1, y_0 = 1, h = 0.1$ and

$$y' = xy^{1/3} \tag{1}$$

Differentiating (1) w.r.t. x, we get

$$y'' = y^{1/3}(1) + x \cdot \frac{1}{3} \cdot y^{-2/3} \cdot y'$$

$$= y^{1/3} + x \cdot \frac{1}{3} \cdot y^{-2/3} \cdot x \cdot y^{1/3} \qquad (\because y' = xy^{1/3})$$

$$\therefore \qquad y'' = y^{1/3} + \frac{1}{3} x^2 \cdot y^{-1/3} \tag{2}$$

Again differentiating (2) with respect to x, we get

$$y''' = \frac{1}{3} y^{-2/3} \cdot y' + \frac{1}{3}\left[x^2\left(-\frac{1}{3}\right) y^{-4/3} y' + y^{-1/3}(2x) \right]$$

$$= \frac{1}{3}\, y^{-2/3}.\, xy^{1/3} - \frac{1}{9}.x^2 y^{-4/3}.xy^{1/3} + \frac{2x}{3} y^{-1/3}$$

$$= xy^{-1/3} - \frac{x^3}{9}\, y^{-1} \tag{3}$$

Again differentiating (3) with respect to x, we get

$$y^{iv} = x\left(-\frac{1}{3}\right).y^{-\frac{4}{3}}.y^1 + y^{-1/3} - \frac{1}{9}x^3(-1)y^{-2}y' + y^{-1}.3x^2$$

$$= -\frac{x}{3}.y^{-\frac{4}{3}}.xy^{1/3} + y^{-1/3} + \frac{1}{9}.x^3 y^{-2}(xy^{1/3}) - \frac{1}{3}x^2 y^{-1}$$

$$= -\frac{x^3}{3}y^{-1} + y^{-1/3} + \frac{1}{9}x^4.y^{-5/3} - \frac{1}{3}x^2 y^{-1}$$

$$= -\frac{2x^2}{3}y^{-1} + y^{-1/3} + \frac{1}{9}.x^4.y^{-5/3} \tag{4}$$

Now,

$$y_0' = x_0 y_0^{1/3} = 1.(1)^{1/3} = 1 \tag{5}$$

$$y_0'' = y_0^{1/3} + \frac{1}{3}\, x_0^2.\, y_0^{-1/3} = 1^{1/3} + \frac{1}{3}(1)^2(1)^{-1/3} = 4/3 \tag{6}$$

$$y_0''' = xy^{-1/3} - \frac{x^3}{9}y^{-1} = (1)(1)^{-1/3} - \frac{(1)^3}{9}(1)^{-1} = \frac{8}{9} \tag{7}$$

$$y_0^{iv} = -\frac{2x^2}{3}y^{-1} + y^{-1/3} + \frac{1}{9}x^4.y^{-5/3} = -\frac{2}{3} + 1 + \frac{1}{9} = \frac{4}{9} \tag{8}$$

From the Eq. (6.27), we have

$$y_1 = y_0 + hy_0' + \frac{h^2}{\lfloor 2}\, y_0'' + \frac{h^3}{\lfloor 3}\, y_0''' + \frac{h^4}{\lfloor 4}\, y_0^{(iv)} + ...$$

$$\therefore \qquad y_1 = 1 + (0.1)(1) + \frac{(0.1)^2}{\lfloor 2} \times \frac{4}{3} + \frac{(0.1)^3}{\lfloor 3} \times \frac{8}{9} + \frac{(0.1)^4}{\lfloor 4} \times \frac{4}{9} + ...$$

$$y_1 \approx 1.1067$$

Thus, $y(1.1) = 1.1067.$

EXERCISE 6(d)

1. Use Taylor series method to determine $y(0.2)$, given that $2y\dfrac{dy}{dx} = \cos x$, $y(0) = 1$.

2. Use Taylor series method to solve $xy' - x + y = 0$, $y(2) = 2$ for $x = 2.1(0.1)2.2$.

3. Determine $y(0.1)$ and $y(0.2)$ using the Taylor series method, given that $\dfrac{dy}{dx} = 1 - y$, $y(0) = 0$.

4. Solve $y' = x^2y - 1$, $y(0) = 1$ for $x = 0.1(0.1)0.3$ by Taylor series method.

5. Solve $y' = 2y + 3e^x$, $y(0) = 0$ Using Taylor series method for $0.1(0.1)0.4$.

6. Use Taylor series method to solve $10\ y' = x^2 + y^2$, $y(0) = 1$ for $x = 0.5(0.5)1.0$.

7. Determine y at $x = 0.1(0.1)0.5$ by Taylor series method, given that $y' + 2x + y = 0$, $y(0) = -1$.

8. Using Taylor series method, solve $y' = y\ \sin x + \cos x$, $y(0) = 0$ for $x = 0.1(0.1)0.3$.

ANSWERS

1. 1.094833
2. 2.002375, 2.009081
3. 0.095171, 0.181284
4. 0.9000, 0.801701, 0.706224
5. 0.348695, 0.811265, 1.416779, 2.201147
6. 1.056792, 1.146055
7. −0.914487, −0.856147, −0.822393, −0.810886, −0.819508
8. 0.100167, 0.201335, 0.304518.

6.13   RUNGE–KUTTA METHODS

There are various-order Runge–Kutta formulae to solve the IVP given by (6.1) and (6.2). Further, even for the same order, we have different Runge–Kutta formulas to solve the IVP. These methods were discovered by Runge, and a few years later they were extended by Kutta. The derivation of the formulas of Runge–Kutta methods are somewhat lengthy and will not be discussed in this book. Out of the different-order Runge–Kutta methods, we describe two methods of Runge–Kutta and they are Runge-Kutta second-order method and fourth-order methods to solve the IVP. $\dfrac{dy}{dx} = f(x,y)$, $y(x_0) = y_0$ in section 6.14 and 6.15. The advantage of these methods are there is no need to calculate the higher-order derivatives of $y(x)$ as in the case of obtaining the numerical solution of an IVP by Taylor series method.

6.14   RUNGE–KUTTA SECOND-ORDER METHOD

Let $\dfrac{dy}{dx} = f(x,y)$, $y(x_0) = y_0$ be an IVP.

The value of y at $x = x_0 + h$ is computed by the following formulas

$$K_1 = h f(x_0, y_0) \tag{6.28}$$
$$K_2 = hf(x_0 + h, y_0 + K_1) \tag{6.29}$$

and

$$y_1 = y_0 + \frac{K_0 + K_1}{2} \tag{6.30}$$

Thus, $x_1 = x_0 + h$ and $y_1 = y_0 + \dfrac{K_1 + K_2}{2}$,

Similarly, y value (i.e., y_2) at $x_1 = x_0 + 2h$ is computed by the following formulas

$$K_1 = hf(x_1, y_1) \tag{6.31}$$
$$K_2 = hf(x_1 + h, y_1 + K_1) \tag{6.32}$$

and

$$y_2 = y_1 + \frac{K_1 + K_2}{2} \tag{6.33}$$

Thus, $x_2 = x_0 + 2h$ and $y_2 = y_1 + \dfrac{K_1 + K_2}{2}$.

Proceeding in this manner, we can find $y_3, y_4 \ldots\ldots, y_n$ at $x = x_0 + 3h$, $x_0 + 4h, \ldots, x_0 + nh$ respectively. Note that the only change in the formulae for the calculation of y_i is the values of x and y to be substituted. Thus, to find y_n at $x = x_0 + nh$, we should have substituted x_{n-1} and y_{n-1} in the expressions of K_1 and K_2.

SOLVED PROBLEMS

Problem 1: Using the Runge–Kutta second-order method, solve the IVP $\dfrac{dy}{dx} = x + y$, $y(0) = 1$ for 0.1(0.1)(0.2).

Solution: Here, $x_0 = 0$, $y_0 = 1$, $h = 0.1$ and

$$f(x, y) = x + y \tag{1}$$

To find $y(0.1)$(i.e., y_1):

As per the Runge–Kutta second-order method,

$$K_1 = hf(x_0, y_0) = 0.1f(0.1) = 0.1\{0 + 1\} = 0.1$$
$$K_2 = hf(x_0 + h, y_0 + k)$$
$$= 0.1\, f(0 + 0.1,\ 1 + 0.1)$$
$$= 0.1\, f(0.1,\ 1.1)$$
$$= 0.1\ \{0.1 + 1.1\} = 0.1 \times 1.2 = 0.12.$$

Hence, $\qquad y_1 = y_0 + \dfrac{1}{2}\left(K_1 + K_2\right)$

$$= 1 + \dfrac{1}{2}(0.1 + 0.12)$$

$$= 1.11.$$

Thus, $x_1 = 0.1$ and $y_1 = 1.11$.

To find $y(0.2)$ (i.e., y_2):

Now $\qquad K_1 = hf(x_1, y_1)$

$$= 0.1\, f(0.1,\ 1.11)$$
$$= 0.1\ \{0.1 + 1.11\}$$
$$= 0.121.$$
$$K_2 = hf(x_1 + h, y_1 + K_1)$$
$$= 0.1\, f(0.1 + 0.1,\ 1.11 + .121)$$
$$= 0.1\, f(0.2,\ 1.231).$$
$$K_2 = 0.1\ \{0.2 + 1.231\} = 0.1431.$$

Hence $\qquad y_2 = y_1 + \dfrac{1}{2}\left(K_1 + K_2\right)$

$$= 1.11 + \dfrac{1}{2}(0.121 + 0.1431)$$

$$= 1.24205.$$

Thus, $x_2 = 0.2$ and $y_2 = 1.24205$.

Problem 2: Using Runge-Kutta second-order method, solve the IVP $\dfrac{dy}{dx} = \dfrac{y+x}{x}$, $y(2) = 2$ for $x = 2.25$ and $x = 2.5$.

Solution: Here, $h = 0.25$, $x_0 = 2$, $y_0 = 2$, and

$$f(x, y) = \frac{y+x}{x} \tag{1}$$

To find y at $x = 2.25$ (i.e., y_1):

In this case,
$$\begin{aligned}
K_1 &= hf(x_0, y_0) \\
&= 0.25\, f(2, 2) \\
&= 0.25 \left(\frac{2+2}{2} \right) = 0.5 \\
K_2 &= hf(x_1 + h, y_1 + K_1) \\
&= 0.25\, f(2 + 0.25, 2 + 0.5) \\
&= 0.25\, f(2.25, 2.5) \\
&= 0.25 \left\{ \frac{2.5 + 2.25}{2.25} \right\} \\
&= 0.527777.
\end{aligned}$$

Hence,
$$y_1 = y_0 + \frac{1}{2}(K_1 + K_2)$$

$$y_1 = 2 + \frac{1}{2}[0.5 + 0.527777]$$

$$\approx 2.513888$$

Thus, $x_1 = 2.25$ and $y_1 = 2.513888$.

To find y at $x = 2.25$ (i.e., y_2):

In this case,
$$\begin{aligned}
K_1 &= hf(x_1, y_1) \\
K_1 &= 0.25f(2.25, 2.513888) \\
&= 0.25 \left\{ \frac{2.513888 + 2.25}{2.25} \right\} \\
&= 0.529320 \\
K_2 &= hf(x_1 + h, y_1 + K_1)
\end{aligned}$$

$$= 0.25\,f\,(2.25 + 0.25,\ 2.513888 + 0.529320)$$

$$= 0.25\,f\,(2.5,\ 3.043208)$$

$$= 0.25\left\{\frac{3.043208 + 2.5}{2.5}\right\}$$

$$= 0.554320.$$

Hence
$$y_2 = y_1 + \frac{1}{2}\left(K_1 + K_2\right)$$

$$= 2.513888 + \frac{1}{2}\,[0.529320 + 0.554320]$$

$$= 3.055708.$$

Thus, $x_2 = 2.5$ and $y_2 = 3.055708$.

Problem 3: Solve the IVP $\dfrac{dy}{dx} = y - x$, $y(0) = 2$ using Runge–Kutta second-order method for 0.1(0.1)0.3.

Solution: Here, $x_0 = 0,\, y_0 = 2,\, h = 0.1$ and
$$f(x, y) = y - x \tag{1}$$

To find y at $x = 0.1$ (i.e., $y\,(0.1)$):

$$K_1 = h\,f\,(x_0, y_0) = 0.1\,f\,(0.2) = 0.1(2 - 0) = 0.2$$
$$K_2 = h\,f\,(x_0 + h,\, y_0 + K_1)$$
$$= (0.1)\,f\,(0 + 0.1, 2 + 0.2) = 0.1\,f\,(0.1,\, 2.2)$$
$$= (0.1)\{2.2 - 0.1\} = 0.1(2.1) = 0.21.$$

$$\therefore \qquad y_1 = y_0 + \frac{1}{2}\left(K_1 + K_2\right)$$

$$= 2 + \frac{1}{2}\,(0.2 + 0.21)$$

$$= 2.2050.$$

Thus, $x_1 = 0.1$ and $y_1 = 2.2050$.

To find y at $x = 0.2$ (i.e., $y(0.2)$):

$$K_1 = h\,f\,(x_1, y_1)$$
$$= 0.1f\,(0.1,\, 2.2050)$$
$$= 0.1\{2.2050 - 0.1\}$$
$$= 0.2105.$$

$$K_2 = h f(x_1 + h, y_1 + K_1)$$
$$= 0.1 f(0.1 + 0.1, 2.2050 + .2105)$$
$$= 0.1 f(0.2, 2.4155)$$
$$= 0.1\{2.4155 - 0.2\}$$
$$= 0.1(2.2155)$$
$$= .22155.$$

$$\therefore \quad y_2 = y_1 + \frac{1}{2}(K_1 + K_2)$$

$$= 2.2050 + \frac{1}{2}(0.21050 + .22155)$$

$$= 2.4210.$$

Thus,

$x_2 = 0.2$ and $y_2 = 2.4210$.

To find y at $x = 0.3$(i.e., to find $y(0.3)$):

$$K_1 = hf(x_2, y_2)$$
$$= 0.1 f(0.2, 2.4210)$$
$$= 0.1\{2.4210 - 0.2\}$$
$$= 0.22210.$$

$$K_2 = h f(x_2 + h, y_2 + K_1)$$
$$= 0.1 f(0.2 + 0.1, 2.4210 + 0.22210)$$
$$= 0.1 f(0.3, 2.6431)$$
$$= 0.1\{2.6431 - 0.3\}$$
$$= 0.23431$$

$$\therefore \quad y_3 = y_2 + \frac{1}{2}(K_1 + K_2)$$

$$= 2.4210 + \frac{1}{2}(0.22210 + 0.23431)$$

$$= 2.6492.$$

Thus, $x_3 = 0.3$ and $y_3 = 2.6492$.

Problem 4: Using Runge–Kutta second-order method, solve
$\dfrac{dy}{dx} = \dfrac{y-x}{y+x}$, $y(0) = 1$ for $x = 0.2\ (0.2)\ (0.6)$.

Solution: Here, $h = 0.2$, $x_0 = 0$, $y_0 = 1$ and

$$f(x, y) = \frac{y - x}{y + x} \tag{1}$$

To find y at $x = 0.2$:

Runge–Kutta second-order formula to compute y_1 is given by

$$y_1 = y_0 + \frac{1}{2}(K_1 + K_2) \tag{2}$$

where
$$K_1 = hf(x_0, y_0) \tag{3}$$
$$K_2 = hf(x_0 + h, y_0 + K_1) \tag{4}$$

Therfore, we compute the values of K_1 and K_2.

$$\begin{aligned}
K_1 &= hf(x_0, y_0) \\
&= 0.2 f(0,1) \\
&= 0.2 \left\{ \frac{1-0}{1+0} \right\} = 0.2
\end{aligned}$$

$$\begin{aligned}
K_2 &= h f(x_0 + h, y_0 + K_1) \\
&= 0.2 f(0 + 0.2, 1 + 0.2) \\
&= 0.2 f(0.2, 1.2) \\
&= 0.2 \left\{ \frac{1.2 - 0.2}{1.2 + 0.2} \right\} \\
&= 0.2 \times \left(\frac{1}{1.4} \right) \\
&\approx 0.1428
\end{aligned}$$

Hence,

$$y_1 = 1 + \frac{1}{2}(0.2 + 0.1482)$$

$$\approx 1.1714$$

Thus, $x_1 = 0.2$ and $y_1 = 1.1714$.

To find y at $x = 0.4$ (i.e., to find y_2):

Since $y_2 = y_1 + \frac{1}{2}(K_1 + K_2)$ and $K_1 = h f(x_1, y_1)$ and $K_2 = h f(x_1 + h, y_1 + K_1)$, first we compute K_1 and K_2 values.

$$K_1 = h f(x_1, y_1)$$

$$= 0.2\,f\,(0.2,\,1.1714)$$

$$= 0.2\left\{\frac{1.1714 - 0.2}{1.1714 + 0.2}\right\}$$

$$= 0.2\left\{\frac{0.9714}{1.3714}\right\} = 0.1416$$

$$K_2 = hf\,(x_1 + h,\, y_1 + K_1)$$

$$= 0.2\,f\,(0.2 + 0.2,\, 1.1714 + 0.1416)$$

$$= 0.2\,f\,(0.4,\, 1.313)$$

$$= 0.2\left\{\frac{1.313 - 0.4}{1.313 + 0.4}\right\}$$

$$= 0.1065.$$

$$\therefore \qquad y_2 = 1.1714 + \frac{1}{2}\,[0.1416 + 0.1065]$$

$$= 1.1714 + 0.12405$$

$$= 1.2954.$$

Thus, $x_2 = 0.4$ and $y_2 = 1.2954$.

To find y at $x = 0.6$:

In this case, $\qquad K_1 = h\,f\,(x_2,\, y_2)$

$$= 0.2\,f\,(0.4,\, 1.2954)$$

$$= 0.2\left\{\frac{1.2954 - 0.4}{1.2954 + 0.4}\right\}$$

$$= 0.1056$$

$$K_2 = hf\,(x_2 + h,\, y_2 + K_1)$$

$$= 0.2\,f\,(0.4 + 0.2,\, 1.2954 + 0.1056)$$

$$= 0.2\,f\,(0.6,\, 1.401)$$

$$= 0.2\left\{\frac{1.401 - 0.6}{1.404 + 0.6}\right\}$$

$$= 0.0800.$$

Now,

$$y_3 = y_2 + \frac{1}{2}\left(K_1 + K_2\right)$$

$$=1.2954 + \frac{1}{2}\,(0.1056 + 0.0800)$$

$$= 1.3882.$$

Thus,

$x_3 = 0.6$ and $y_3 = 1.3882$.

EXERCISE 6(e)

1. Determine $y(1.1)$ using Runge–Kutta method of second order, given that $y' = 3x + y^2$, $y(1) = 1.2$.

2. Solve $\dfrac{dy}{dx} = x - y$, $y(1) = 0.4$ for $x = 1.2$ by Runge–Kutta method of second order with $h = 0.1$.

3. Solve $\dfrac{dy}{dx} = \dfrac{y^2 - x^2}{y^2 + x^2}$, $y(0) = 1$ for $x = 0.2(0.2)0.4$ by Runge–Kutta method of second order.

4. Determine $y(0.6)$ and $y(0.8)$ using Runge–Kutta method of second order, given that $y' = \sqrt{x + y}$, $y(0.4) = 0.41$.

5. Solve $\dfrac{dy}{dx} = 2ye^x$, $y(0) = 2$ for $x = 0.1(0.1)0.3$ by Runge–Kutta method of second order.

6. Solve $\dfrac{dy}{dx} = y - x^2$, $y(0) = 1$ for $x = 0.2(0.2)0.6$ by Runge–Kutta method of second order.

ANSWERS

1. 1.722137
2. 0.52761
3. $y(0.2) = 1.194595$, $y(0.4) = 1.373717$
4. $y(0.6) = 0.609087$, $y(0.8) = 0.846678$
5. $y(0.1) = 2.465241$, $y(0.2) = 3.105352$, $y(0.3) = 4.006216$, $y(0.4) = 5.30607$
6. $y(0.2) = 1.216$, $y(0.4) = 1.46272$, $y(0.6) = 1.729318$.

6.15   RUNGE–KUTTA FOURTH ORDER METHOD

Let $\dfrac{dy}{dx} = f(x, y)$, $y(x_0) = y_0$ be an initial value problem.

In this method y value at $x = x_0 + h$ (i.e., x_1) is computed by the following formulas.

$$y_1 = y_0 + \frac{1}{6}(K_1 + 2K_2 + 2K_3 + K_4), \qquad (6.34)$$

where

$$K_1 = h f(x_0, y_0) \qquad (6.34a)$$

$$K_2 = hf\left(x_0 + \frac{h}{2}, y_0 + \frac{K_1}{2}\right) \qquad (6.34b)$$

$$K_3 = hf\left(x_0 + \frac{h}{2}, y_0 + \frac{K_2}{2}\right) \qquad (6.34c)$$

and

$$K_4 = h f(x_0 + h, y_0 + K_3) \qquad (6.34d)$$

Similarly, the y value at $x = x_0 + 2h$ (i.e., x_2) is computed by the following formulae

$$y_2 = y_1 + \frac{1}{6}(K_1 + 2K_2 + 2K_3 + K_4), \qquad (6.35)$$

where

$$K_1 = hf(x_1, y_1) \qquad (6.35a)$$

$$K_2 = hf\left(x_1 + \frac{h}{2}, y_1 + \frac{K_1}{2}\right) \qquad (6.35b)$$

$$K_3 = hf\left(x_1 + \frac{h}{2}, y_1 + \frac{K_2}{2}\right) \qquad (6.35c)$$

and $\qquad\qquad K_4 = h f(x_1 + h, y_1 + K_3) \qquad (6.35d)$

Continuing this process, we can find $y_3, y_4, \dots y_n$ at $x = x_0 + 3h$, $x_0 + 4h$, ..., $x_0 + nh$, respectively. It can be observed that the only change in the formulas for the calculations of y_i is the values of x and y to be substituted. Thus, to find y_n, we should have to substitute x_{n-1} and y_{n-1} in the expressions K_1, K_2, K_3 and K_4.

SOLVED PROBLEMS

Problem 1: Find $y(0.1)$ using Runge–Kutta fourth-order method, given that $y' = x^2 - y$ and $y(0) = 1$.

Solution: Here, $x_0 = 0$, $y_0 = 1$, $h = 0.1$ and
$$y' = x^2 - y \tag{1}$$

To find y value at $x = 0.1$:

First, we compute K_1, K_2, K_3 and K_4 values.

$$K_1 = hf(x_0, y_0) = 0.1\, f(0, 1) = 0.1\{0^2 - 1\} = -0.1$$

$$K_2 = hf\left(x_0 + \frac{h}{2}, y_0 + \frac{K_1}{2}\right)$$

$$= 0.1 f\left(0 + \frac{0.1}{2}, 1 + \left(\frac{-0.1}{2}\right)\right)$$

$$= 0.1\, f(0.05, 0.95)$$
$$= 0.1\{(0.05)^2 - 0.95\}$$
$$= -0.09475.$$

$$K_3 = hf\left(x_0 + \frac{h}{2}, y_0 + \frac{K_2}{2}\right)$$

$$= 0.1 f\left(0 + \frac{0.1}{2}, 1 + \left(\frac{-0.09475}{2}\right)\right)$$

$$= 0.1\, f(0.05, 0.0952625)$$
$$= 0.1\{(0.05)^2 - 0.0952625\}$$
$$= -0.0950125.$$

$$K_4 = hf(x_0 + h, y_0 + K_3)$$

$$= 0.1\, f(0 + 0.1, 1 - 0.0950125)$$
$$= 0.1\, f(0.1, 0.9049875)$$
$$= 0.1\{(0.1)^2 - 0.9049875\}$$
$$= 0.1\{0.01 - 0.7049875\}$$
$$= -0.08949875.$$

$$\therefore \quad y_1 = y_0 + \frac{1}{6}(K_1 + 2K_2 + 2K_3 + K_4) \quad \text{(R–K fouth order formula)}$$

$$= 1 + \frac{1}{6}\left[(-0.1) + 2(-0.09475) + 2(-0.0950125) + (-0.08949875)\right]$$

$$= 1 + \frac{1}{6}\left[-0.1 - 0.1895 - 0.190025 - 0.08949875\right]$$

$$= 0.9051627083.$$

Thus, the required value $y\,(0.1) = 0.9051627083$.

Problem 2: Solve $\dfrac{dy}{dx} = y - x$, $y\,(0) = 2$, using Runge–kutta fourth-order method for $x = 0.2$ and $x = 0.4$ by taking $h = 0.2$.

Solution: Here, $h = 0.2$, $x_0 = 0$, $y_0 = 2$ and

$$f(x, y) = y - x \tag{1}$$

To find y value at $x = 0.2$ (i.e., to find y_1):

For the computation of y_1 we require the values of K_1, K_2, K_3, and K_4, where K_1, K_2, K_3, and K_4 are given by (6.34a), (6.34b), (6.34c) and (6.34d) respectively.
Thus,

$$K_1 = h f(x_0, y_0) = 0.2 f(0, 2) = 0.2[2 - 0] = 0.4.$$

$$K_2 = h f\left(x_0 + \frac{h}{2}, y_0 + \frac{K_1}{2}\right) = 0.2 f\left(0 + \frac{0.2}{2}, 2 + \frac{0.4}{2}\right)$$

$$= 0.2 f(0.1, 2.2) = 0.2[2.2 - 0.1] = 0.42.$$

$$K_3 = h f\left(x_0 = \frac{h}{2}, y_0 + \frac{K_2}{2}\right) = 0.2 f\left(0 + \frac{0.2}{2}, 2 + \frac{0.42}{2}\right)$$

$$= 0.2 f(0.1, 2.21)$$

$$= 0.2[2.21 - 0.1] = 0.422.$$

$$K_4 = h f(x_0 + h, y_0 + K_3)$$

$$= 0.2 f(0 + 0.2, 2 + 0.422)$$

$$= 0.2 f(0.2, 2.422)$$

$$= 0.2[2.422 - 0.2]$$

$$= 0.4444.$$

Hence

$$y_1 = y_0 + \frac{1}{6}(K_1 + 2K_2 + 2K_3 + K_4)$$

$$= 2 + \frac{1}{6}[0.4+2(0.42) + 2(0.422) + 0.4444]$$

$$= 2 + \frac{1}{6}(2.5284)$$

$$= 2 + 0.4214$$

$$= 2.4214.$$

Thus, $x_1 = 0.2$ and $y_1 = 2.4214$.

To find y at $x = 0.4$:

In this case, the formula for k_1, k_2, k_3, k_4 are given by (6.35a), (6.35b), (6.35c) and (6.35d) respectively

$\therefore$
$$K_1 = hf(x_1, y_1)$$

$$= 0.2f(0.2, 2.4214)$$

$$= 0.2[2.4214 - 0.2]$$

$$= 0.44428.$$

$$K_2 = hf\left(x_1 + \frac{h}{2}, y_1 + \frac{K_1}{2}\right)$$

$$= 0.2f\left(0.2 + \frac{0.2}{2}, 2.4214 + \frac{0.44428}{2}\right)$$

$$= 0.2f(0.3, 2.64354)$$

$$= 0.2[2.64354 - 0.3]$$

$$= 0.468708.$$

$$K_3 = hf\left(x_1 + \frac{h}{2}, y_1 + \frac{K_2}{2}\right)$$

$$= 0.2f\left(0.2 + \frac{0.2}{2}, 2.4214 + \frac{0.48708}{2}\right)$$

$$= 0.2\, f\,(0.3,\, 2.655754)$$

$$= 0.2\{2.655754 - 0.3\}$$

$$= 0.47115.$$

$$K_4 = h\, f\,(x_1 + h,\, y_1 + K_3)$$

$$= 0.2 f\,(0.2 + 0.2,\, 2.4214 + 0.47115)$$

$$= 0.2\, f\,(0.4,\, 2.89255)$$

$$= 0.2[2.89255 - 0.4]$$

$$= 0.49851.$$

Hence, $\qquad\qquad y_2 = y_1 + \dfrac{1}{6}\,(K_1 + 2K_2 + 2K_3 + K_4)$

$$= 2.4214 + \frac{1}{6}\,[0.44428 + 2(0.468708) + 2(0.47115) + 0.49851]$$

$$= 2.891817667.$$

Thus, $x_2 = 0.4$ and $y_2 = 2.891817$ (up to six decimal places).

Problem 3: Solve $y' = 1 + xy$, $y\,(0) = 2$ using Runge–kutta fourth-order method for $0.1(0.1)0.3$.

Solution: Here, $h = 0.1$, $x_0 = 0$, $y_0 = 2$ and

$$f\,(x, y) = 1 + xy \qquad\qquad\qquad (1)$$

To find y at $x = 0.1$:

The formulae of K, K_2, K_3, K_4, to compute y_1 are given by 6.34a, 6.34b (6.34c) and (6.34d) respectively and

$$y_1 = y_0 + \frac{1}{6}\,(K_1 + 2K_2 + 2K_3 + K_4) \qquad\qquad (2)$$

Now, we compute K_1, K_2, K_3 and K_4.

$$K_1 = h\, f\,(x_0,\, y_0).$$

$$= 0.1\, f\,(0,\, 1) = 0.1[1 + (0)(1)] = 0.1$$

$$K_2 = h\, f\left(x_0 + \frac{h}{2},\, y_0 + \frac{K_1}{2}\right)$$

$$= 0.1 f\left(0 + \frac{0.1}{2},\, 2 + \frac{0.1}{2}\right)$$

$$= 0.1 f(0.05, 2.05)$$
$$= 0.1[1 + (0.05)\,(2.05)]$$
$$= 0.11025.$$

$$K_3 = hf\left(x_0 = \frac{h}{2},\, y_0 + \frac{K_2}{2}\right)$$

$$= 0.1 f\left(0 + \frac{0.1}{2},\, 2 + \frac{0.11025}{2}\right)$$

$$= 0.1 f(0.05, 2.055125)$$
$$= 0.1[1 + (0.05)\,(2.055125)]$$
$$= 0.110275625$$

$$K_3 \approx 0.1103.$$
$$K_4 = hf(x_0 + h,\, y_0 + K_3)$$
$$= 0.1f(0 + 0.1,\, 2 + 0.1103)$$
$$= 0.1 f(0.1, 2.1103)$$
$$= 0.1\{1 + (0.1)\,(2.1103)\}$$
$$= 0.121103.$$

From (6.34),

$$y_1 = y_0 + \frac{1}{6}(K_1 + 2K_2 + 2K_3 + K_4)$$

$$\therefore \qquad y_1 = 2 + \frac{1}{6}[0.1 + 2(.11025) + 2(.1103) + 0.121103]$$

$$= 2.110367167$$

$$\approx 2.1104.$$

Thus, $x_1 = 0.1$ and $y_1 = 2.1104$.

To find y at $x = 0.2$:

For the computation of y_2 the formulae of K_1, K_2, K_3, K_4 of Runge–kutta fourth-order method are given by (6.35a) to (6.35d)

$$\therefore \qquad K_1 = hf(x_1, y_1)$$
$$= 0.1 f(0.1, 2.1104)$$
$$= 0.1[1 + (0.1)\,(2.1104)]$$

$$= 0.1 \times [1.21104]$$

$$= 0.121104$$

$$K_1 \approx 0.1211.$$

$$K_2 = hf\left(x_1 + \frac{h}{2}, y_1 + \frac{K_1}{2}\right)$$

$$= 0.1\,f\left(0.1 + \frac{0.1}{2}, 2.1104 + \frac{0.1211}{2}\right)$$

$$= 0.1\,f(0.15,\ 2.17095)$$
$$= 0.1[1 + (0.15)\,(2.17095)]$$
$$\approx 0.1325.$$

$$K_3 = hf\left(x_1 + \frac{h}{2}, y_1 + \frac{K_2}{2}\right)$$

$$= 0.1f\left(0.1 + \frac{0.1}{2}, 2.1104 + \frac{0.1325}{2}\right)$$

$$= 0.1\,f(0.15,\ 2.17665)$$
$$= 0.1[1 + (0.15)\,(2.17665)]$$
$$\approx 0.1326.$$

$$K_4 = hf(x_1 + h, y_1 + K_3)$$
$$= 0.1f(0.1 + 0.1,\ 2.1104 + 0.1326)$$
$$= 0.1\,f(0.2,\ 2.243)$$
$$= 0.1[1 + (0.2)\,(2.243)]$$
$$= 0.1$$
$$= 0.14486.$$

$$\therefore \qquad y_2 = y_1 + \frac{1}{6}(K_1 + 2K_2 + 2K_3 + K_4)$$

i.e., $\qquad y_2 = 2.1104 + \dfrac{1}{6}[0.1211 + 2(0.1325) + 2(0.1326) + 0.14486]$

i.e., $\qquad y_2 \approx 2.2431.$

To find y at $x = 0.3$:

For the computation of y_3, the formulae of Rung-kutta fourth-order method are as follows:

$$K_1 = hf(x_2, y_2)$$

$$K_2 = hf\left(x_2 + \frac{h}{2},\, y_2 + \frac{K_1}{2}\right)$$

$$K_3 = hf\left(x_2 + \frac{h}{2},\, y_2 + \frac{K_2}{2}\right)$$

$$K_4 = hf(x_2 + h,\, y_2 + K_3)$$

and

$$y_3 = y_2 + \frac{1}{6}(K_1 + 2K_2 + 2K_3 + K_4)$$

$\therefore$

$$K_1 = hf(x_2, y_3)$$
$$= 0.1\, f(0.2, 2.2431)$$
$$= 0.1[1 + 0.2(2.2431)]$$
$$\approx 0.1448.$$

$$K_2 = hf\left(x_2 + \frac{h}{2},\, y_2 + \frac{K_1}{2}\right)$$

$$= 0.1\, f\left(0.2 + \frac{0.1}{2},\, 2.2431 + \frac{0.1448}{2}\right)$$

$$= 0.1\, f(0.25, 2.3155)$$
$$= 0.1[1 + (0.25)(2.3155)]$$
$$\approx 0.1579.$$

$$K_3 = hf\left(x_2 + \frac{h}{2},\, y_2 + \frac{K_2}{2}\right)$$

$$= 0.1f\left(0.2 + \frac{0.1}{2},\, 2.2431 + \frac{0.1579}{2}\right)$$

$$= 0.1\, f(0.25, 2.32205)$$
$$= 0.1[1 + (0.25)(2.32205)]$$
$$\approx 0.1580.$$

$$K_4 = hf(x_2 + h,\, y_2 + K_3)$$
$$= 0.1f(0.2 + 0.1,\, 2.2431 + 0.1580)$$
$$= 0.1\, f(0.3, 2.4011)$$
$$= 0.1[1 + (0.3)(2.4011)]$$
$$\approx 0.1720,$$

and

$$y_3 = y_2 + \frac{1}{6}(K_1 + 2K_2 + 2K_3 + K_4)$$

$$= 2.2431 + \frac{1}{6}\,[0.1448 + 2*(.1579) + 2(.158) + 0.1720]$$

$$= 2.2431 + 0.1581$$
$$= 2.4012.$$

Thus, $x_3 = 0.3$ and $y_3 = 2.4012$.

Problem 4: Solve $\dfrac{dy}{dx} = x\,y$, $y\,(0) = 1$ using Runge–kutta method of order four for $x = 0.1(0.1)0.2$

Solution: Here, $h = 0.1$, $x_0 = 0$, $y_0 = 1$ and
$$f(x, y) = x\,y \tag{1}$$

To find y_1 at $x_1 = x_0 + h$ i.e. $x_1 = 0.1$:
   As per the Runge–kutta method of order four,

$$K_1 = h f(x_0, y_0) = 0.1\,f(0.1)\,[(0).(1)] = 0.$$

$$K_2 = h f\left(x_0 + \frac{h}{2}, y_0 + \frac{K_1}{2}\right)$$

$$= 0.1 + (0.05,\ 1)$$

$$= 0.1[(0.05)\,(1)] = 0.005.$$

$$K_3 = h f\left(x_0 = \frac{h}{2},\ y_0 + \frac{K_2}{2}\right)$$

$$= 0.1\,f(0.05,\ 1.0025)$$

$$= 0.1[(0.05)\,(1.0025)]$$

$$= 0.005012.$$

$$K_4 = h f(x_0 + h,\ y_0 + K_3) = 0.1\,f(0.1,\ 1.005012)$$

$$= 0.1\,f[(0.1) \times (1.005012)]$$

$$\approx 0.010050.$$

Hence

$$y_1 = y_0 + \frac{1}{6}\,(K_1 + 2K_2 + 2K_3 + K_4)$$

$$= 1 + \frac{1}{6}\,[0 + 2 \times 0.005 + 2 \times 0.005012 + 0.010050]$$

$$= 1.005012$$

Thus, $x_1 = 0.1$ and $y_1 = 1.005012$.

To find y_2 at $x_2 = x_0 + 2h = 0.2$

In this case,

$$K_1 = h f(x_1, y_1) = 0.1 f(0.1, 1.005012)$$
$$= 0.1[(0.1)(1.005012)]$$
$$= 0.010050.$$

$$K_2 = h f\left(x_1 + \frac{h}{2}, y_1 + \frac{K_1}{2}\right)$$

$$= 0.1 f(0.15, 1.012587)$$
$$= 0.015150.$$

$$K_3 = h f\left(x_1 + \frac{h}{2}, y_1 + \frac{K_2}{2}\right)$$

$$= 0.1 f\left(0.1 + \frac{0.1}{2}, 1.005012 + \frac{0.015150}{2}\right)$$

$$= 0.1 f(0.15, 1.012587)$$
$$= 0.1[(0.15)(1.012587)] = 0.15188.$$

$$K_4 = h f(x_1 + h, y_1 + K_3)$$
$$= 0.1 f(0.1 + 0.1, 1.005012 + 0.015188)$$
$$= 0.1 f(0.2, 1.0202)$$
$$= 0.1[(0.2)(1.0202)]$$
$$\approx 0.0204.$$

$$\therefore \quad y_2 = y_1 + \frac{1}{6}(K_1 + 2K_2 + 2K_3 + K_4)$$

$$= 1.005012 + \frac{1}{6}[0.010050 + 2(0.01515) + 2(0.015188) + 0.0204]$$

$$\approx 1.0202.$$

Thus, $x_2 = 0.2$ and $y_2 = 1.0202$.

EXERCISE 6(f)

1. Use Runge–kutta method of fourth order to find the numerical solution at $x = 0.2$, given that $y' = x + y, y(0) = 1$.

2. Solve the IVP $y' = x^2 + y^2$, $y(0) = 0$ using Runge–Kutta method of fourth order to obtain the numerical solution at $x = 0.4$ with $h = 0.2$.

3. Determine y at $x = 0.1(0.1)0.5$ by Runge–kutta fourth-order method, given that $(x + y)\, y' = 1$, $y(0) = 2$.

4. Given that $\dfrac{dy}{dx} + \dfrac{y}{x} = \dfrac{1}{x^2}$, $y(1) = 1$, determine y value at $x = 1.1$ using Runge–kutta method of order four.

5. Determine $y(2.5)$ using Runge–kutta method of fourth order with $h = 0.5$, given that $y' = \dfrac{y + x}{x}$, $y(2) = 2$.

6. Given that $y' = y - \dfrac{2x}{y}$, $y(0) = 1$. Determine y values at $x = 0.1(0.1)0.5$ by Runge–kutta method of fourth order.

7. Solve $y' = \dfrac{1}{(x + y)}$, $y(0) = 1$ for $x = 0.5(0.5)2$ by Runge–kutta method of fourth order.

8. Determine y values at $x = 0.1$ and $x = 0.2$ using Runge–kutta method of fourth order with $h = 0.1$ given that $y' + y = 0$ and $y(0) = 1$.

ANSWERS

1. $y(0.2) = 1.2428$
2. $y(0.4) = 0.02136$
3. $y(0.1) = 2.048227$, $y(0.2) = 2.093268$, $y(0.3) = 2.135571$, $y(0.4) = 2.17549$, $y(0.5) = 2.213312$
4. $y(1.1) = 0.995737$
5. $y(2.5) = 3.057819$
6. $y(0.1) = 1.095446$, $y(0.2) = 1.183217$, $y(0.3) = 1.264912$, $y(0.4) = 1.341642$, $y(0.5) = 1.414216$
7. $y(0.5) = 1.357065$, $y(1.0) = 1.58368$, $y(1.5) = 1.755507$, $y(2.0) = 1.895639$
8. $y(0.1) = 0.904837$, $y(0.2) = 0.818731$.

6.16   MILNE'S METHOD

To solve $y' = f(x, y)$, $y(x_0) = y_0$ numerically by this method, we need y values y_1, y_2, y_3, at $x = x_0 + h$, $x_0 + 2h$ and $x_0 + 3h$ respectively. These values can be found using the methods learnt in earlier sections. For the quick calculations of these values, we can employ the Euler's method. Using these values, we find a predicted value of y_4 by the

Milne's predicted formula. The predicted value of y_4 is obtained from the Milne's predictor formula is denoted by y_4^p. The improved value of y_4^p is obtained by the Milne's corrected value of y_4 is denoted by y_4^c. The value of y_4^c is taken as y_4.

Similarly, using y_1, y_2, y_3, y_4 values we can compute y_5^p and y_5^c. Hence, $y_5 = y_5^c$ this process can be repeated to find y_i^p and y_i^c ($i = 6$, 7, 8, ... n). Note that to find y_i^p, y_i^c ($i > 4$) we need the values of y_{i-4}, y_{i-3}, y_{i-2}, y_{i-1}. Now, we proceed to derive the predictor and corrector formulas of Milne's method.

Integrating $\dfrac{dy}{dx} = f(x, y)$ from x_0 to $x_0 + 4h$, we have

$$\int_{y(x_0)}^{y(x_0 + 4h)} dy = \int_{x_0}^{x_0 + 4h} f(x, y)dx$$

i.e.,

$$y_4 = y_0 + \int_{x_0}^{x_0 + 4h} f(x, y)dx \tag{6.36}$$

By the Newton's interpolation formula, we have

$$y' = y_0' + u\Delta y_0' + \frac{u(u-1)}{\lfloor 2} \Delta^2 y_0' + \frac{u(u-1)(u-2)}{\lfloor 3} \Delta^3 y_0' + ..., \tag{6.37}$$

where

$$u = \frac{x - x_0}{u} \tag{6.38}$$

Introducing $y_0' = f_0$ and using the fact $y' = f(x, y)$, the Eq. (6.37) can be written as

$$f(x, y) = f_0 + u\Delta f_0 + \frac{u(u-1)}{\lfloor 2} \Delta^2 f_0 + \frac{u(u-1)(u-2)}{\lfloor 3} \Delta^3 f_0 + ..., \tag{6.39}$$

Since, $u = \dfrac{x - x_0}{u}$, we have $dx = hdu$ and when $x = x_0 + 4h$ the value of $u = 4$, and when $x = x_0$ the value of $u = 0$. Thus, the Eq. (6.36) becomes

$$y_4 = y_0 + \int_o^4 \left[f_0 + u\Delta f_0 + \frac{u(u-1)}{\lfloor 2} \Delta^2 f_0 + \frac{u(u-1)(u-2)}{\lfloor 3} \Delta^3 f_0 + ... \right] hdu$$

$$y_4 = y_0 + h\left[uf_0 + \frac{u^2}{\lfloor 2} \Delta f_0 + \left(\frac{u^3}{3} - \frac{u^2}{2}\right)\frac{\Delta^2 f_0}{\lfloor 2} + \left(\frac{u^4}{4} - u^3 + \frac{u^2}{2}\right)\frac{\Delta^3 y_0}{\lfloor 3} + \ldots\ldots \right]_0^4$$

i.e., $y_4 = y_0 + h[4f_0 + 8\Delta f_0 + \dfrac{20}{3}\Delta^2 f_0 + \dfrac{8}{3}\Delta^3 f_0 + \ldots .]$

Neglecting fourth- and higher-order differences of f_0, we have

$$y_4 = y_0 + \frac{4h}{3}[3f_0 + 6\Delta f_0 + 5\Delta^2 f_0 + 2\,\Delta^3 f_0]$$

i.e., $\qquad y_4 = y_0 + \dfrac{4h}{3}[3f_0 + 6(E-1)f_0 + 5(E-1)^2 f_0 + 2(E-1)^3 f_0]$

i.e., $y_4 = y_0 + \dfrac{4h}{3}[3f_0 + 6f_1 - 6f_0 + 5f_2 - 10f_1 + 5f_0 + 2f_3 - 6f_2 + 6f_1 - 2f_0]$

$$\therefore \quad y_4 = y_0 + \frac{4h}{3}[2f_1 - f_2 + 2f_3] \tag{6.40}$$

The predicted value of y_4 is computed from (6.40). Hence,

$$y_4^p = y_0 + \frac{4h}{3}[2f_1 - f_2 + 2f_3] \tag{6.41}$$

The general form of (6.41) is given by

$$y_{n+1}^p = y_{n-3} + \frac{4h}{3}[2f_{n-2} - f_{n-1} + 2f_n] \tag{6.42}$$

The formula (6.42) is known as Milne's predictor formula.

To derive the corrector formula, integrate $\dfrac{dy}{dx} = f(x, y)$ from x_0 to $x_0 + 2h$. Then

$$y_2 = y_0 + \int_{x_0}^{x_0 + 22h} f(x, y)dx \tag{6.43}$$

Now substituting (6.39) in (6.43), we get

$$y_2 = y_0 + \int_{x_0}^{x_0 + 2h}\left[f_0 + u\Delta f_0 + \frac{u(u-1)}{\lfloor 2}\Delta^2 f_0 + \frac{u(u-1)(u-2)}{\lfloor 3}\Delta^3 f_0 + \ldots \right]dx.$$

Since $u = \dfrac{x - x_0}{h}$, the above equation becomes

$$y_2 = y_0 + \int_0^2 \left[f_0 + u\Delta f_0 + \frac{u(u-1)}{\lfloor 2} \Delta^2 f_0 + \frac{u(u-1)(u-2)}{\lfloor 3} \Delta^3 f_0 + \dots \right] h.du.$$

Neglecting third-and higher-order differences of the integrand of the above equation, we have

$$y_2 = y_0 + h \int_0^2 \left[f_0 + u\Delta f_0 + \frac{u(u-1)}{\lfloor 2} \Delta^2 f_0 \right] du$$

i.e.,
$$y_2 = y_0 + h \left[u f_0 + \frac{u^2}{2} \Delta f_0 + \left(\frac{u^3}{3} - \frac{u^2}{2} \right) \frac{\Delta^2 f_0}{\lfloor 2} \right]_0^2$$

i.e.,
$$y_2 = y_0 + h \left[2 f_0 + 2\Delta y_0 + \frac{2}{3} \Delta^2 f_0 \right]$$

i.e.,
$$y_2 = y_0 + \frac{h}{3} (f_0 + 4 f_1 + f_2).$$

Operating E^2 on both sides of the above equation, we get

$$y_4 = y_2 + \frac{h}{3} (f_2 + 4 f_3 + f_4) \tag{6.44}$$

The corrected value of y_4 can be obtained from (6.44), hence

$$y_4^c = y_2 + \frac{h}{3} (f_2 + 4 f_3 + f_4^p) \tag{6.45}$$

where
$$f_4^p = f(x_4, y_4^p).$$

The general form of (6.45) is given by

$$y_{n+1}^c = y_{n-1} + \frac{h}{3} (f_{n-1} + 4 f_n + f_{n+1}^p) \tag{6.46}$$

where
$$f_{n+1}^p = f(x_{n+1}, y_{n+1}^p). \tag{6.47}$$

6.17 WORKING RULE TO IMPLEMENT THE MILNE'S METHOD TO SOLVE IVP

Step 1: Express the given differential equation in the form $\dfrac{dy}{dx} = f(x, y)$. At this step, we have the expression of $f(x, y)$.

Step 2: Note down the expression of $f(x, y)$.

Step 3: In addition to $y(x_0) = y_0$, if y values y_1, y_2, y_3 at $x = x_0 + h$, $x_0 + 2h$, $x_0 + 3h$ are given in the question then go to Step 5, otherwise go to Step 4.

Step 4: Find y values (y_1, y_2, y_3) at $x = x_0 + h$, $x_0 + 2h$, $x_0 + 3h$ using Euler's or any other method.

Step 5: Construct the following table:

x_i	y_i	$y' = f(x_i, y_i)$
x_0	y_0	$f_0 = f(x_0, y_0)$
x_1	y_1	$f_1 = f(x_1, y_1)$
x_2	y_2	$f_2 = f(x_2, y_2)$
x_3	y_3	$f_3 = f(x_3, y_3)$

Step 6: Write the predictor formula of Milne's method

$$y_4^p = y_0 + \frac{4h}{3}(2f_1 - f_2 + 2f_3). \tag{1}$$

Step 7: Substitute the values of y_0, f_1, f_2, f_3 obtained in Step 5 in (1). We get y_4^p.

Step 8: Write the corrector formula of Milne's method.

$$y_4^c = y_2 + \frac{h}{3}(f_2 + 4f_3 + f_4^p). \tag{2}$$

Step 9: Compute $f_4^p = f(x_4, y_4^p)$.

Step 10: Substitute y_2, h, f_2, f_3 obtained in Step 5 and f_4^p obtained in Step 9 in (2) to get y_4^c. At this step, we have y_4^c.

Step 11: $y_4 = y_4^c$. i.e. $y(x_4) = y_4$.

If y_5 is required then go to Step 12, otherwise terminate the process.

Step 12: Construct the following table:

x_i	y_i	$y' = f(x_i, y_i)$
x_1	y_1	$f_1 = f(x_1, y_1)$
x_2	y_2	$f_2 = f(x_2, y_2)$
x_3	y_3	$f_3 = f(x_3, y_3)$
x_4	y_4	$f_4 = f(x_4, y_4)$

Step 13: To obtain y_5^p, use the predictor formula

$$y_5^p = y_1 + \frac{4h}{3}\,(2f_2 - f_3 + 2f_4).$$

Step 14: To compute y_5^c, use the corrector formula

$$y_5^c = y_3 + \frac{h}{3}\,(y_3 + 4f_4 + f_5^p), \text{ where } f_5^p = f(x_5, y_5^n).$$

Step 15: Write $y(x_5) = y_5^c$.

Repeat Steps 12 to 17 for computing $y_6,\ y_7,\ \ldots..$ advancing the subscripts of x, y and f by one successively for each computation.

SOLVED PROBLEMS

Problem 1: Using Milne's method, find $y(0.4)$ for the differential equation $\dfrac{dy}{dx} = 1 + xy$ with $y(0) = 2$.

Solution: Let $h = 0.1$. To compute $y(0.4)$ by Milne's method, we need $y(0)$, $y(0.1)$, $y(0.2)$ and $y(0.3)$. The value $y(0)$ is known from the hypothesis. Therefore, it is required to compute $y(0.1)$, $y(0.2)$ and $y(0.3)$. We use Euler's method to find them, Note that

$$f(x, y) = 1 + xy \tag{1}$$

To find y_1 (i.e., y value at $x_1 = 0.1$):

By Euler's algorithm

$$y_1 = y_0 + h f(x_0, y_0)$$

$\therefore$
$$y_1 = 2 + 0.1\, f(0, 2) = 2.1$$

Thus,
$$x_1 = 0.1,\, y_1 = 2.1.$$

To find y_2:

Since
$$y_2 = y_1 + h f(x_1, y_1),$$
$$y_2 = 2.1 + 0.1\, f(0.1 + 2.1)$$
$$= 2.1 + 0.1\{1 + (0.1)(2.1)\}$$
$$= 2.221$$

$\therefore x_2 = 0.2$ and $y_2 = 2.221$.

To find y_3:

Since $y_3 = y_2 + h f(x_2, y_2)$ we have

$$y_3 = 2.221 + 0.1 f(0.2, 2.221)$$
$$= 2.221 + 0.1\{1 + (0.2)(2.221)\}$$
$$= 2.36542$$

Thus, $x_3 = 0.3$ and $y_3 = 2.36542$. The values of f_0, f_1, f_2 and f_3 are shown in the following table.

i	x_i	y_i	$f_i = 1 + x_i y_i$
0	0	2	$f_0 = 1$
1	0.1	2.1	$f_1 = 1.21$
2	0.2	2.221	$f_2 = 1.4442$
3	0.3	2.36542	$f_3 = 1.709626$

Milne's predictor formula for y_4 is

$$y_4^p = y_0 + \frac{4h}{3}(2f_1 - f_2 + 2f_3)$$

$\therefore$
$$y_4^p = 2 + \frac{4 \times 1}{3}[2 \times 1.21 - 1.4442 + 2 \times 1.709626]$$

$$= 2 + \frac{0.4}{3}[4.395052]$$

$$= 2.586006.$$

Now
$$f_4^p = f(x_4, y_4^p)$$

$\therefore$
$$f_4^p = f(0.4, 2.586006)$$

$$= 1 + (0.4)(2.586006)$$
$$= 2.0344024.$$

The Milne's corrector formula for y_4 is

$$y_4^c = y_2 + \frac{h}{3}(f_2 + 4f_3 + f_4^p)$$

$\therefore$
$$y_4^c = 2.221 + \frac{0.1}{3}[1.4442 + 4(1.709626) + 2.0344024]$$

$$= 2.221 + \frac{0.1}{3}[10.3171064]$$

$$= 2.5649035$$

Thus, $\qquad y_4 = 2.5649035.$

Problem 2: Given, $\dfrac{dy}{dx} = y - x^2$, $y(0) = 1$, $y(0.2) = 1.2187$, $y(0.4) = 1.4684$ and $y(0.6) = 1.7383$ find $y(0.8)$ using Milne's method.

Solution: Here, $h = 0.2$, $x_0 = 0$, $x_1 = 0.2$, $x_2 = 0.4$, $x_3 = 0.6$, $x_4 = 0.8$, $y_0 = 1$, $y_1 = 1.2187$, $y_2 = 1.4684$, $y_3 = 1.7383$ and

$$f(x, y) = y - x^2 \qquad (1)$$

We need to find y_4, i.e., $y(0.8)$. The values of f_0, f_1, f_2, f_3 are shown in the following table

i	x_i	y_i	$f_i = y_i - x_i^2$
0	0	1	$f_0 = 1$
1	0.2	1.2187	$f_1 = 1.1787$
2	1.4684	1.4684	$f_2 = 1.3084$
3	1.7383	1.7383	$f_3 = 1.3783$

The predicted value of y_4 is

$$y_4^p = y_0 + \frac{4h}{3}(2f_1 - f_2 + 2f_3)$$

$$\therefore \qquad y_4^p = 1 + \frac{4 \times 0.2}{3}[2 \times 1.1787 - 1.3084 + 2 \times 1.3783]$$

$$= 1 + \frac{0.8}{3}[3.8056]$$

$$\approx 2.5743.$$

Now

$$f_4^p = f(x_4, y_4^p)$$

$$= f(0.8, 2.5743)$$

$$= 2.5743 - (0.8)^2$$

$$= 1.9343.$$

The corrected value of y_4 is given by

$$y_4^c = y_2 + \frac{h}{3}[f_2 + 4f_3 + f_4^p]$$

$$\therefore \qquad y_4^c = 1.4684 + \frac{0.2}{3}[1.3084 + 4 \times 1.3783 + 1.9343]$$

$$= 1.4684 + \frac{0.2}{3}[8.7559]$$

$$= 2.0521.$$

Thus, $x_4 = 0.8$ and $y_4 = 2.0521$.

Problem 3: Given $10. \dfrac{dy}{dx} = x^2 + y^2$, $y(0) = 1$, $y(0.1) = 1.0101$ $y(0.2) = 1.0206$ and $y(0.3) = 1.0318$, find $y(0.4)$ using Milne's method.

Solution: Here, $h = 0.1$, $x_0 = 0$, $x_1 = 0.1$, $x_2 = 0.2$, $x_3 = 0.3$, $x_4 = 0.4$, $y_0 = 1$, $y_1 = 1.0101$, $y_2 = 1.0206$, $y_3 = 1.0318$ and

$$f(x, y) = \frac{x^2 + y^2}{10} \tag{1}$$

First, we compute the values of f_0, f_1, f_2 and f_3 using $f_i = f(x_i, y_i)$. These values are shown in the following table:

i	x_i	y_i	f_i
0	0	1	$f_0 = 0.1$
1	0.1	1.0101	$f_1 = 0.103030$
2	0.2	1.0206	$f_2 = 0.108162$
3	0.3	1.0318	$f_3 = 0.115461$

We know that

$$y_4^p = y_0 + \frac{4h}{3}[2f_1 - f_2 + 2f_3]$$

$$\therefore \qquad y_4^p = 1 + \frac{4 \times 0.1}{3}[2(0.103030) - 0.108162 + 2(0.115461)]$$

$$y_4^p \approx 1.043842.$$

Now

$$f_4^p = f(x_4, y_4^p)$$

$$\therefore \quad f_4^p = f(0.4, 1.043842)^2$$

$$= \frac{(0.4)^2 + (1.043842)^2}{10}$$

$$= 0.124960.$$

The corrected value of y_4^c is given by

$$y_4^c = y_2 + \frac{h}{3}[f_2 + 4f_3 + f_4^p]$$

$$y_4^c = 1.0206 + \frac{0.1}{3}[0.108162 + 4(0.113461) + 0.124960]$$

$$= 1.043765.$$

Hence $y_4 = 1.043765$ i.e., $y(0.4) = 1.043765$.

Problem 4: Solve $y^1 = 1 + y^2$, $y(0) = 0$ using Milne's method for 0.8(0.2)1.0.

Solution: Here, $f(x, y) = 1 + y^2$, $x_0 = 0$, $y_0 = 0$. We take $h = 0.2$. To find $y(0.8)$, we need the values of y at $x = 0, 0.2, 0.4$ and 0.6. As $y(0) = y_0 = 0$ is given in the hypothesis, it is enough to find the remaining of these values. These vales namely $y(0.2)$, $y(0.4)$ and $y(0.6)$ can be found using Euler's method. Note that we can also use the other methods such as Taylor's series method, Rurge-Kutta method, etc. to find them. First, we find the y values at $x = 0.2, 0.4$ and 0.6 by Euler's method.

To find at y_1 i.e., $y(0.2)$:

By Euler's Algorithm, we have

$$y_1 = y_0 + h f(x_0, y_0)$$

$$\therefore \quad y_1 = 0 + 0.2\,(0, 0)$$

$$= 0.2\{1 + 0^2\} = 0.2$$

Thus, $x_1 = 0.2$ and $y_1 = 0.2$.

To find at y_2, i.e., $y(0.4)$:

$$y_2 = y_1 + h f(x_1, y_1)$$

$\therefore$
$$y_2 = 0.2 + 0.2 f(0.2, 0.2)$$
$$= 0.2 + 0.2\{1 + (0.2)^2\}$$
$$= 0.408.$$

Thus, $x_2 = 0.4$ and $y_2 = 0.408$.

To find at y_3, i.e., $y(0.6)$:

$$y_3 = y_2 + h f(x_2, y_2)$$

$\therefore$
$$y_3 = 0.408 + 0.2 f(0.4, 0.408)$$
$$= 0.408 + 0.2\{1 + (.408)^2\}$$
$$\approx 0.64129.$$

Thus, $x_3 = 0.6$ and $y_3 = 0.64129$.

Now we find the values of f_0, f_1, f_2, f_3 using $f_i = 1 + y_i^2$

i	x_i	y_i	f_i
0	0	0	$f_0 = 1$
1	0.2	0.2	$f_1 = 1.04$
2	0.4	0.408	$f_2 = 1.16646$
3	0.6	0.64129	$f_3 = 1.41125$

The predicted values of y at 0.8, i.e., y_4^p is given by

$$y_4^p = y_0 + \frac{4h}{3}[2f_1 - f_2 + 2f_3]$$

i.e.,
$$y_4^p = 0 + \frac{4 \times 0.2}{3}[2 \times 1.04 - 1.16646 + 2(1.41125)]$$

$$= \frac{0.8}{3}[2.08 - 1.16646 + 2.8225]$$

$$\approx 0.99627.$$

Thus, $y_4^p \approx 0.99627$

Now
$$f_4^p = f(x_4, y_4^p)$$
$$= f(0.8, 0.99627)$$

$$= 1 + (0.99627)^2$$
$$= 1.99255.$$

The corrected value of y_4, i.e., y_4^c is given by

$$y_4^c = y_2 + \frac{h}{3}[f_2 + 4f_3 + f_4^p]$$

$$\therefore \qquad y_4^c = 0.408 + \frac{0.2}{3}[1.16646 + 4 \times 1.41125 + 1.99255]$$

$$= 0.408 + \frac{0.2}{3}[1.16646 + 5.645 + 1.99255]$$

$$= 0.408 + \frac{0.2}{3} \times 8.80401$$

$$= 0.994934.$$

Thus, $x_4 = 0.8$ and $y_4 = 0.994934$.

To find $y\,(1.0)$ i.e., y_5:

i	x_i	y_i	$f_i = 1 + y_i^2$
1	0.2	0.2	1.04
2	0.4	0.408	1.16646
3	0.6	0.64129	1.41125
4	0.8	0.994934	1.989893

Now

$$y_5^p = y_1 + \frac{4h}{3}[2f_2 - f_3 + 2f_4]$$

$$= 0.2 + \frac{4 \times 0.2}{3}[2 \times 1.6646 - 1.41125 + 2(1.989893)]$$

$$= 0.2 + \frac{0.8}{3}[4.901456]$$

$$\approx 1.507055$$

The value of f_5^p is given by

$$f_5^p = f(x_5, y_5^p)$$
$$= f(1, 1.507055) = 1 + (1.507055)^2$$
$$\approx 3.271214.$$

The corrected values of y_5^c is

$$y_5^c = 0.64129 + \frac{0.2}{3}[1.41125 + 4(1.989893) + 3.271214]$$
$$\approx 1.48409.$$

Thus, $x_5 = 1.0$ and $y_5 = 1.48409$.

EXERCISE 6(g)

1. Given that $\dfrac{dy}{dx} = \dfrac{xy}{1+x^2}$, $y(0) = 1$, $y(0.1) = 1.0049$, $y(0.2) = 1.0198$ and $y(0.3) = 1.0440$. Find y (0.4) using Milne's method, and compare the result with the value obtained for the exact solution.

2. Given that $\dfrac{dy}{dx} = \dfrac{1}{x+y}$, $y(0) = 1$, $y(0.5) = 1.3571$, $y(1) = 1.5837$, $y(1.5) = 1.7565$, Find $y(2)$ using Milne's method.

3. For the equation $y' = x^2y + x^2$, $y(1) = 1$, get the starting values by Taylor series method for $x = 1.1, 1.2$ and 1.3. Determine $y(1.4)$ by Milne's method.

4. For the equation $(1 + x^2)y' - (1 + y^2) = 0$, $y(0) = 1$, get the starting values by Euler's method for $x = 0.1(0.1)0.3$ and determine $y(0.4)$ by Milne's method.

5. Use Milne's method to find $y(0.4)$ for the IVP $\dfrac{dy}{dx} = 2 \, e^x y$. Determine the starting values by Euler's method for $x = 0.1(0.1)0.3$.

6. For the equation $y' = x^2y$, $y(0) = 1$ get starting values by Euler's method for $x = 0.2(0.2)0.6$ and obtain $y(0.8)$ by Milne's method.

ANSWERS

1. 1.077028	2. 1.895999	3. 2.575555
4. 2.203881	5. 4.987399	6. 1.16888.

6.18   ADAM–MOULTON METHOD

This method is similar to the Milne's method. It also needs y_{n-1}, y_{n-2} and y_{n-3} to find y_n of an IVP by this method. The predictor and corrector formulae of this method are derived using the Newton's backward formula of interpolation in contrast to the use of Newton's forward formula of interpolation used in the derivation of predictor and corrector formulas of Milne's method. Now, we proceed to derive the predictor and corrector formulae of Adam–Moulton method.

Integrating $y' = f(x, y)$ with respect to x from x_0 to $x_0 + h$, we get

$$\int_{y(x_0)}^{y(x_0+h)} dy = \int_{x_0}^{x_0+h} f(x, y)dx$$

i.e.,
$$y_1 = y_0 + \int_{x_0}^{x_0+h} f(x, y)dx \tag{6.48}$$

By the Newton's backward formula, we have

$$y' = y_0' + u\,\nabla\, y_0' + \frac{u(u+1)}{\lfloor 2} \nabla^2 y_0' + \dots \tag{6.49}$$

where
$$u = \frac{x - x_0}{u} \tag{6.50}$$

Introducing $y_0' = f_0$, the Eq. 6.49 becomes

$$f(x, y) = f_0 + u\nabla f_0 + \frac{u(u+1)}{\lfloor 2} \nabla^2 f_0 + \dots \tag{6.51}$$

since, $y' = f(x, y)$
Substituting (6.51) in (6.48), we get

$$y_1 = y_0 + \int_{x_0}^{x_1} \left[f_0 + u\nabla f_0 + \frac{u(u+1)}{\lfloor 2} \nabla^2 f_0 + \dots \right] dx$$

Since, $u = \dfrac{x - x_0}{h}$, we have $dx = h\,du$ and when $x = x_0$, $u = 0$ and when $x = x_0 + h$, $u = 1$.

$$\therefore \qquad y_1 = y_0 + h \int_0^1 \left[f_0 + u\nabla f_0 + \frac{u(u+1)}{\lfloor 2} \nabla^2 f_0 + \dots \right] du$$

$$= y_0 + h\left[u\nabla f_0 + \frac{\nabla^2 f_0}{2} + \left(\frac{u^3}{3} + \frac{u^2}{2}\right)\frac{\nabla^2 f_0}{\lfloor 2} + \dots \right]_0^1$$

$$= y_0 + h\left[f_0 + \frac{\nabla f_0}{2} + \frac{5}{12}\cdot\nabla^2 f_0 + \frac{3}{8}\nabla^3 f_0 + \frac{251}{750}\nabla^3 f_0 + \dots \right]$$

Neglecting fourth-and higher-order differences of ∇, we get

$$y_1 = y_0 + h\left[f_0 + \frac{\nabla f_0}{2} + \frac{5}{12}\nabla^2 f_0 + \frac{3}{8}\nabla^3 f_0 \right]$$

Simplifying the right-hand side of above equation using $\nabla = 1 - E^{-1}$, we get

$$y_1 = y_0 + \frac{h}{24}\left(55 y_0 + 59 y_{-1} + 37 y_{-2} - 9 y_{-3}\right)$$

In general,

$$y_{n+1} = y_n + \frac{h}{24}[55 f_n - 59 f_{n-1} + 37 y_{n-2} - 9 y_{n-3}] \qquad (6.52)$$

The values y_{n+1} computed from (6.52) is called predicted value of y_{n+1}, i.e., y_{n+1}^p

$$\therefore \qquad y_{n+1}^p = y_n + \frac{h}{24}[55 f_n - 59 f_{n-1} + 37 y_{n-2} - 9 y_{n-3}], \qquad (6.53)$$

which is called as Adams–Predicted formula.

A corrected formula can be derived in a similar way by using Newton's backward interpolation formula at f_1, i.e.,

$$f(x, y) = f_1 + u\nabla f_1 + \frac{u(u+1)}{\lfloor 2}\nabla^2 f_1 + \dots \qquad (6.54)$$

where

$$u = \frac{x - x_1}{h}. \qquad (6.55)$$

Substituting (6.54) in (6.48), we obtain

$$y_1 = y_0 + \int_{x_0}^{x_0+h}\left[f_1 + u\nabla f_1 + \frac{u(u+1)}{\lfloor 2}\nabla^2 f_1 + \frac{u(u+1)(u+2)}{\lfloor 3}\nabla^3 f_1 + \dots \right]dx$$

$$(6.56)$$

By (6.55), we have $u = \dfrac{x - x_1}{h}$. Therefore, $dx = hdu$. Further, when $x = x_0$ the value of $u = -1$ and when $x = x_1$ the value of $u = 0$. Now the Eq. (6.56) reduces to

$$y_1 = y_0 + h \int_{-1}^{0} \left[f_1 + u\nabla f_1 + \frac{u(u+1)}{\lfloor 2} \nabla^2 f_1 + \frac{u(u+1)(u+2)}{\lfloor 3} \nabla^3 f_1 + \ldots \right] dx$$

$$= y_0 + h \left[1 - \frac{1}{2}\nabla f_1 - \frac{1}{12}\nabla^2 f_1 - \frac{1}{24}\nabla^3 f_1 - \frac{19}{720}\nabla^4 f_1 + \ldots \right]$$

Neglecting fourth-and higher-order differences of ∇, we get

$$y_1 = y_0 + \frac{h}{24}\,(9f_1 + 19f_0 - 5f_{-1} + f_{-2})$$

Replacing f_1 by $f_1^p = f(x_1, y_1^p)$, we have

$$y_1^c = y_0 + \frac{h}{24}\,(9\,f_1^p + 19f_0 - 5f_{-1} + f_{-2}) \tag{6.57}$$

In general,

$$y_{n+1}^c = y_n + \frac{h}{24}\left[9f_{n+1}^p + 19f_n - 5f_{n-1} + f_{n-2}\right] \tag{6.58}$$

This formula is known as Adams–Moulton corrector formula.

The method described in this section to solve an IVP is known as Adams–Moulton method.

SOLVED PROBLEMS

Problem 1: Using Adam–Moulton method, find $y(1.4)$, given that $y' = x^2(1 + y)$ and $y(1) = 1$.

Solution: Let $h = 0.1$, $x_0 = 1$ and $y_0 = 1$. Then, $x_1 = 1.1$, $x_2 = 1.2$, $x_3 = 1.3$ and $x_4 = 1.4$. To find y_4, i.e., $y(1.4)$ we require the values of y_3, y_2 and y_1, i.e., $y(1.1)$, $y(1.2)$, $y(1.3)$ in order to employ the Adam–Moulton method. First, we find y_1, y_2, y_3 values using the Euler's method .

To find at y_1, i.e., $y\,(1.1)$:

By Euler's Algorithm, we have

$$y_1 = y_0 + hf(x_0, y_0)$$

$$\therefore \qquad y_1 = 1 + 0.1\, f(1, 1)$$
$$= 1 + 0.1[1^2(1+1)]$$
$$= 1.2.$$

Thus, $x_1 = 0.1$ and $y_1 = 1.2$

To find at y_2, i.e., $y(1.2)$:

$$y_2 = y_1 + h f(x_1, y_1)$$
$$\therefore \qquad y_2 = 1.2 + 0.1\, f(1.1, 1.2)$$
$$= 1.2 + 0.1[(1.1)^2\,(1 + 1.2)]$$
$$= 1.4662.$$

Thus, $x_2 = 0.2$ and $y_2 = 1.4662$

To find at y_3, i.e., $y(1.3)$:

$$y_3 = y_2 + h f(x_2, y_2)$$
$$\therefore \qquad y_3 = 1.4662 + 0.1\, f(1.2, 1.4662)$$
$$= 1.4662 + 0.1\,[(1.2)^2\{1 + 1.46623\}]$$
$$= 1.4662 + 0.1\,(3.551328)$$
$$\approx 1.8213.$$

Thus, $x_3 = 1.2$ and $y_3 = 1.8213$

We compute the values of f_i $(i = 0, 1, 2, 3)$ and they are shown in the following table:

i	x_i	y_i	$f_i = x_i^2\,(1 + y_i)$
0	1	1	$f_0 = 2$
1	1.1	1.2	$f_1 = 2.662$
2	1.2	1.4662	$f_2 \approx 3.5513$
3	1.3	1.8213	$f_3 \approx 4.7680$

The predicted value of y_4 is given by

$$y_4^p = y_3 + \frac{h}{24}[55 f_3 - 59 f_2 + 37 f_1 - 9 f_0]$$

$$\therefore y_4^p = 1.8213 + \frac{0.1}{24}[55 \times 4.7680 - 59(3.5513) + 37(2.662) - 9(2)]$$

$$\text{i.e., } y_4^p \approx 2.3763.$$

Now we compute $f_4^p = f(x_4, y_4^p)$

$$\therefore \qquad\qquad f_4^p = f(1.4, 2.3763)$$
$$= (1.4)^2\{1 + 2.3763\}$$
$$= 1.96(3.3763).$$
$$= 6.6175$$

$$\therefore \qquad\qquad f_4^p = 6.6175.$$

Adams–Moulton corrector formula is given by

$$y_4^c = y_3 + \frac{h}{24}\,[9\,f_4^p + 19\,f_3 - 5\,f_2 + f_1]$$

$$\therefore \; y_4^c = 1.8213 + \frac{0.1}{24}\,[9(6.6175) + 19(4.7680) - 5(3.5513) + 2.6662]$$

$$= 2.3840.$$

Thus, $x_4 = 1.4$ and $y_4 = 2.3840$.
Hence, the required value of $y(1.4)$ is 2.3840.

Problem 2: Given $\dfrac{dy}{dx} = x^2 - y$, $y(0.1) = 0.90516$, $y(0.2) = 0.82127$ and $y(0.3) = 0.74918$; find $y(0.4)$ using Adam–Moulton method.

Solution: Let $h = 0.1$, $x_0 = 0$, $x_1 = 0.1$, $x_2 = 0.2$, $x_3 = 0.3$, $x_4 = 0.4$, $y_0 = 1$, $y_1 = 0.90516$, $y_2 = 0.82127$ and $y_3 = 0.74918$. Here

$$f(x, y) = x^2 - y \qquad\qquad\qquad (1)$$

We have to find y_4, i.e., $y(0.4)$.
The computed values of $f_i = x_i^2 - y_i$, $i = 0, 1, 2, 3$ are shown below in the following table:

i	x_i	y_i	$f_i = x_i^2 - y_i$
0	0	1	$f_0 = 0^2 - 1 = -1$
1	0.1	0.90516	$f_1 = (0.1)^2 - 0.90516 = -0.89516$
2	0.2	0.82127	$f_2 = (0.2)^2 = -0.82127 = -0.78127$
3	0.3	0.74918	$f_3 = (0.3)^2 - 0.74918 = -0.65918$

The Adam–Moulton predictor formula is

$$y_4^p = y_3 + \frac{h}{24}[55f_3 - 59f_2 + 37f_1 - 9f_0] \tag{2}$$

Substituting the values of h, y_3, f_3, f_2, f_1 and f_0 (shown in the table) in the Eq. (2), we obtain

$$y_4^p = 0.74918 + \frac{0.1}{24}[55(-0.65918) - 59(-0.78127)$$
$$+ 37(-0.89516) - 9(-1)]$$

i.e., $\qquad y_4^p = 0.689676.$

Now $\qquad f_4^p = f(x_4, y_4^p)$

i.e. $\qquad f_4^p = f(0.4, 0.689676)$

$$= (0.4)^2 - 0.689676$$

$$= -0.529676.$$

Thus, $\qquad f_4^p = -0.529676.$

The corrector formula of Adam–Moulton method is

$$y_4^c = y_3 + \frac{h}{24}[9f_4^p + 19f_3 - 5f_2 + f_1]$$

$$\therefore \qquad y_4^c = 0.74918 + \frac{0.1}{24}[9(-0.529676) + 19(-0.65918)$$
$$-5(-0.78127) + (-0.89516)]$$

i.e., $\qquad y_4^c = 0.689679.$

Thus, $x_4 = 0.4$ and $y_4 = 0.689679.$

Problem 3: Use Adam–Moulton method to find $y(0.4)$. Given $2\dfrac{dy}{dx} = x\,y, y(0) = 1.$

Solution: Here, $x_0 = 0$ and $y_0 = 1$, Take $h = 0.1$. Then $x_1 = 0.1$, $x_2 = 0.2$, $x_3 = 0.3$ and $x_4 = 0.4$ To implement Adam–Moulton method to find $y(0.4)$, we must know the values of $y(0.3)$, $y(0.2)$, $y(0.1)$ and $y(0)$. Therefore, it is required to find $y(0.3)$, $y(0.2)$ and $y(0.1)$. We find these values by Euler's method.

Expressing the given differential equation in the form of $\dfrac{dy}{dx} = f(x, y)$, we get

$$\frac{dy}{dx} = \frac{xy}{2} \qquad (1)$$

Now, we proceed to find $y\,(0.3)$, $y\,(0.2)$ and $y(0.1)$.

To find $y\,(0.1)$, i.e., y_1:

By Euler's Algorithm, we have

$$y_1 = y_0 + h f(x_0, y_0)$$

$$\therefore \qquad y_1 = 1 + 0.1\,(0,\,1) = 1 + 0.1\left\{\frac{1}{2}(0)(1)\right\} = 1.$$

Thus, $x_1 = 0.1$ and $y_1 = 1$.

To find at $y(0.2)$, i.e., y_2:

In this case, $\qquad\qquad y_2 = y_1 + h f(x_1, y_1)$

$$\therefore \qquad\qquad y_2 = 1 + 0.1 f(0.1,\,1)$$

$$= 1 + 0.1\left\{\frac{1}{2}(0.1)(1)\right\} = 1.005.$$

Thus, $x_2 = 0.2$ and $y_2 = 1.005$.

To find $y\,(0.3)$, i.e., y_3:

As $y_3 = y_2 + h f(x_2, y_2)$ by Euler's algorithm, we have

$$y_3 = 1.005 + 0.1 f(0.2,\,1.005)$$

$$= 1.005 + 0.1\left\{\frac{1}{2}(0.2)(1.005)\right\}$$

$$= 1.01505.$$

Thus, $x_3 = 0.3$ and $y_3 = 1.01505$.

Now, we compute the values of $f_i = \dfrac{1}{2} x_i\, y_i$, $i = 0, 1, 2, 3$. These values are shown in the following table:

i	x_i	y_i	$f_i = \dfrac{x_i \, y_i}{2}$
0	0	1	$f_0 = 0$
1	0.1	1	$f_1 = 0.05$
2	0.2	1.005	$f_2 = 0.1005$
3	0.3	1.01505	$f_3 \approx 0.152257$

The predictor formula of Adam–Moulton method is given by

$$y_4^p = y_3 + \frac{h}{24}\,[55f_3 - 59f_2 + 37f_1 - 9f_0] \tag{2}$$

Substituting the values of h, y_3, f_0, f_1, f_2 and f_3 (which are shown in the table) in the Eq. (2), we obtain

$$y_4^p = 1.01505 + \frac{0.1}{24}\,[55(0.152257) - 59(0.1005) + 37(0.05) - 9(0)]$$

$$= 1.026006.$$

Thus, $y_4^p = 1.026006.$

Now $f_4^p = f(x_4, y_4^p) = f(0.4, 1.026006)$

$$= \frac{1}{2} \times 0.4 \times 1.026006 = 0.205201. \tag{3}$$

The corrector formula of Adam–Moulton method is given by

$$y_4^{(c)} = y_3 + \frac{h}{24}\,[9\,f_4^p + 19f_3 - 5f_2 + f_1]. \tag{4}$$

Substituting y_3, h, f_3, f_2, f_1 and f_4^p (given by (3)) in the Eq. (4), we have

$$y_4^c = 1.01505 + \frac{0.1}{24}\,[9(0.205201) + 19(0.152257) - 5(0.1005) + 0.05]$$

i.e., $$y_4^c = 1.032913.$$

Hence, $$y(0.4) = 1.032913.$$

EXERCISE 6(h)

1. Find $y(0.8)$, given that $y' = x + y$, $y(0) = 1$ by Adam–Moulton method. Determine the starting values by Euler's method for $x = 0.2(0.2)0.6$.

2. For the equation $(1 + x^2)\, y' - (1 + y^2) = 0$, $y(0) = 1$, get starting values by Taylor series method for $x = 0.1(0.1)0.3$ and, determine $y(0.4)$ and $y(0.5)$ by Adam–Moulton method.

3. Given that $\dfrac{dy}{dx} = \dfrac{xy}{1+x^2}$, $y(0) = 1$, $y(0.1) = 1.0049$, $y(0.2) = 1.0198$ and

   $y(0.3) = 1.0440$. Find $y(0.4)$ using Adam–Moulton method. Also, compare the result with the value obtained from the exact solution.

4. Using Adam–Moulton method find $y(0.8)$ and $y(1.0)$ given that $y' = x + y^2$, $y(0) = 0$, $y(0.2) = 0.02$, $y(0.4) = 0.0795$ and $y(0.6) = 0.1762$.

5. Given the equation $y' = y\sin(\pi x)\, y(0) = 1$, find $y(0.8)$ by Adam–Moulton method with $h = 0.2$. Obtain starting values $y(0.2)$, $y(0.4)$ and $y(0.6)$ by Runge–kutta second-order method.

6. Solve $y' = x + y + xy$, $y(0) = 1$ to find the numerical solution at $x = 0.8$ by Adam–Moulton method with $h = 0.2$ Determine starting values $y(0.2)$, $y(0.4)$ and $y(0.6)$ by Euler's method.

ANSWERS

1. 2.41925

2. 2.159969, 2.737113

3. 1.077003

4. 0.328897, 0.547071

5. 1.748623

6. 3.017565.

7

Curve Fitting

In chapter 3, we have learnt that for a given data (x_i, y_i) $(i = 0, 1, 2, \ldots n)$ how to find an interpolating polynomial $p_n(x)$ of degree less than or equal to n such that $p_n(x_i) = y_i$ $(i = 0, 1, 2, \ldots n)$. However, some scientists and engineers often wish to represent an experimental data of certain experiment in a simplest form rather than of higher-degree polynomial using some principle of mathematics. It can be done by the method of least squares that use the principle of least squares. The other methods to express the data in simplest form are group of averages and method of moments.

The objective of curve fitting is to obtain the dependent variable (y) connected by a rule that can express in x like $y = a + bx$ or $y = e^{bx}$ or $y = a + bx + cx^2$, etc; using the emperical/experimental data (x_i, y_i) $(i = 1, 2, \ldots, n)$. We can use this connected relation between x and y to predict the values of y corresponding to the values of x. The point x must either belongs to $[x_0, x_n]$ or $x < x_0 (x > x_n)$ but near to x_0 (x_n). If x is far away from x_0 or x_n, the predicted value y obtained from the relation for such value may no longer be accurate in some cases.

In this chapter, we learn how to fit a curve to the given data using the principle of least squares and method of moments.

7.1   DEFINITION

Let (x_i, y_i) $(i = 1, 2, 3, \ldots n)$ be a given set of values of x (independent variable) and y (dependent variable).

The diagram obtained by locating these points on a graph is known as scatter diagram. The scatter diagram helps to know the relationship between the variables x and y. The scatter diagram shown in Figs. 7.1 and 7.2 suggest the relation between x and y as $y = a + bx$ and $y = a + bx + cx^2$, respectively.

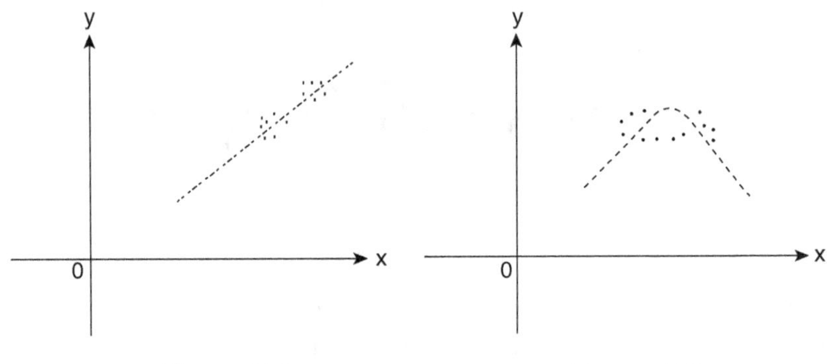

Figure 7.1	Figure 7.2

7.2 DEFINITION—RESIDUAL

Let (x_i, y_i) $(i = 1, 2, 3, ... n)$ be a given set of values x and y with $x_1 < x_2 < .... < x_n$,

Let $y = f(x)$ be the assumed relation between x and y. For example, $f(x) = a + bx$ or $f(x) = ae^{bx}$ or $f(x) = a + bx + cx^2$. The y_i given at x_i is called observed value of y and $f(x_i)$ is called expected value. Note that the parameters involved in $f(x)$ are estimated through the principle of least squares or some other methods such as methods of moments and group of averages.

The difference $y_i - f(x_i)$ is defined as residual at y_i. The residual at y_i is $P_c Q_i$, which is shown in the Fig. 7.3.

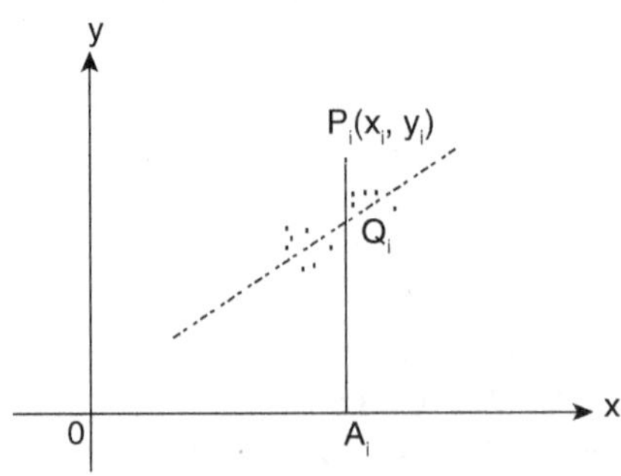

Figure 7.3

7.3 PRINCIPLE OF LEAST SQUARES

The principle of least square states that the sum of squares of residuals is minimum. The curve obtained by applying the principle of least squares is called a best fit. The method of fitting a curve using the principle of least squares is called the method of least square.

Note that fitting of a curve to the data (x_i, y_i) $(i = 1, 2, 3, \ldots n)$ means to find $y = f(x)$ such as $y = a + bx$, or $y = ab^x$ or $y = a.e^{bx}$ or $y = a + bx + cx^2$. The equation $y = a + bx$ represents a curve that is a straight line, the equation $y = a + bx + cx^2$ represents a parabola and the equation $y = a.e^{bx}$ represents an exponential curve. The parameters (or constants) a, b or a, b, c involved in $f(x)$ are estimated through certain methods. Further, note that the value estimated through $f(x)$ at x_i need not be equal to y_i, whereas if $p_x(x)$ is interpolating polynomial of $f(x)$ for this data, then $p_x(x_i) = y_i$.

7.4 FITTING OF A STRAIGHT LINE BY THE METHOD OF LEAST SQUARES

Let (x_i, y_i) $(i = 1, 2, 3, \ldots n)$ be the given data pertaining to certain entity and

$$y = a + bx \tag{7.1}$$

be the straight line to be fitted to the given data. Here, a and b are to be determined constants. The task is to estimate the parameters a, b of Equation (7.1). Since there are two parameters, we require two simultaneous equations involved with a and b to estimate them.

Suppose, E denotes the sum of squares of residuals, then

$$E = \sum_{i=1}^{n} \{y_i - f(x_i)\}^2,$$

i.e.,
$$E = \sum_{i=1}^{n} \{y_i - (a + bx_i)\}^2 \tag{7.2}$$

since,
$$f(x) = a + bx.$$

By the principle of least squares, the condition for minimum of E (from the calculus) is

$$\frac{\partial E}{\partial a} = 0 \tag{7.3}$$

and
$$\frac{\partial E}{\partial b} = 0. \tag{7.4}$$

Differentiating (7.2) partially with respect to a and using the equation (7.3), we obtain

$$\sum_{i=1}^{n} 2\{y_i - (a + bx_i)\}(-1) = 0,$$

which is equivalent to

$$\sum_{i=1}^{n} y_i = na + b \sum_{i=1}^{n} x_i. \tag{7.5}$$

Similarly, differentiating (7.2) partially swith respect to b, and using (7.4), we get

$$\sum_{i=1}^{n} x_i y_i = a \sum_{i=1}^{n} x_i + b \sum_{i=1}^{n} x_i^2. \tag{7.6}$$

Eqs. (7.5) and (7.6) are called normal equations to fit a straight line to the given data using the method of least squares.

Solving (7.5) and (7.6), we get the values of a and b. Let these values be a_f and b_f.

Now, the required fitting of straight line to the given data is
$$y = a_f + b_f x.$$

Note: We can write the normal Eqs. (7.5), (7.6) in the following way:

Let the straight line to be fitted to the given data be
$$y = a + bx; \tag{1}$$

Express it as
$$y = ax^0 + bx, \tag{2}$$

and multiplying Eq. (2) with x, we have
$$xy = ax + bx^2. \tag{3}$$

To get the normal Eqs. (7.5) and (7.6), write the suffix i for the variables x and y in (2) and (3), and taking $\sum_{i=1}^{n}$ on both sides of them, i.e.,

$$\sum_{i=1}^{n} y_i = \sum_{i=1}^{n} ax_i^0 + b \sum_{i=1}^{n} x_i$$

$$\sum_{i=1}^{n} x_i y_i = \sum_{i=1}^{n} ax_i + b \sum_{i=1}^{n} x_i^2$$

i.e.,

$$\sum_{i=1}^{n} y_i = n\ a + b\sum_{i=1}^{n} x_i \qquad \left(\because \sum_{\partial=1}^{n} x_i^{(0)} = n\right)$$

$$\sum_{i=1}^{n} x_i\, y_i = a\sum_{i=1}^{n} x_i + b\sum_{i=1}^{n} x_i^2,$$

which are normal equations to fit a straight line $y = a + bx$.

Remark: If the data values x_i or y_i of (x_i, y_i) are too big or large, we can use the transformation:

$$U = \frac{x-A}{h} \text{ and } V = \frac{y-B}{k} \text{ or } u = x - A \text{ and } V = y - B \text{ or } U = \frac{x-A}{h},$$

$V = y - b$ or $U = x - A$, $V = \dfrac{y-B}{k}$ so that U_i and V_i are small in numbers compared with x_i and y_i, respectively. In case a transformation is used, we find $V = A + BU$ to fit a straight line to the transformed data of the given data namely (U_i, V_i); substituting the transformations of U and V in $V = A + BU$, we get $y = a + bx$.

7.5 PROCEDURE TO FIT A STRAIGHT LINE TO THE GIVEN DATA (x_i, y_i) $(I = 1, 2, \ldots, n)$ BY THE METHOD OF LEAST SQUARES

Step 1: Suppose $y = a + bx$ be the required line.

Step 2: Note down the value of n, where n is number of paired observations.

Step 3: Construct the table having columns heading $x_i, y_i, x_i \cdot y_i$ and x_1^2 and having n rows below the column headings.

Step 4: Fill x_i, y_i data under column headings of x_i, y_i, respectively.

Step 5: Compute $x_i \cdot y_i, x_1^2$ and fill the columns of x_1^2 and $x_i y_i$ at corresponding places for $i = 1, 2, \ldots\ldots n$.

Step 6: Compute the sum of each columns and find

$$\sum_{i=1}^{n} x_i, \sum_{i=1}^{n} y_i, \sum_{i=1}^{n} x_i^2 \text{ and } \sum_{i=1}^{n} x_i y_i.$$

Step 7: Write down the normal equations.

$$\sum_{i=1}^{n} y_i = n\,a + b \sum_{i=i}^{n} x_i \tag{1}$$

$$\sum_{i=1}^{n} x_i\, y_i = a \sum_{i=1}^{n} x_i + b \sum_{i=1}^{n} x_i^2 \tag{2}$$

Step 8: Substitute n, $\displaystyle\sum_{i=1}^{n} x_i$, $\displaystyle\sum_{i=1}^{n} x_i^2$ and $\displaystyle\sum_{i=1}^{n} x_i\, y_i$, which are obtained in Steps 2 and 6 in (1) and (2).

Step 9: Solve the equations obtained in Step 8 for a and b; Let these values be a_f and b_f.

Step 10: Write the required straight line as $y = a_f + b_f x$.

SOLVED PROBLEMS

Problem 1: Fit a straight line for the following data choosing x as independent variable and y as dependent variable.

x	0	2	3	5
y	1	7	14	22

Solution: Let the required straight line be $y = a + bx$ $\hspace{2cm}$ (1)

Note that n (no. of paired observations) $= 4$.

Now, we construct the following table from the given data:

x_i	y_i	$x_i\, y_i$	x_i^2
0	1	0	0
2	7	14	4
3	14	42	+
5	22	110	25
$\sum x_i = 10$	$\sum y_i = 44$	$\sum x_i\, y_i = 166$	$\sum x_i^2 = 38$

The normal equations for fitting a straight line to the given data are

$$\sum_{i=1}^{4} y_i = 4a + b\sum_{i=1}^{4} x_i \tag{2}$$

$$\sum_{i=1}^{4} x_i\,y_i = a\sum_{i=1}^{4} x_i + b\sum_{i=1}^{4} x_i^2 \tag{3}$$

Now, substituting the values of $\sum x_i = 10$, $\sum x_i^2 = 38$, $\sum y_i = 44$ and $\sum x_i\,y_i = 166$ in (2) and (3) we get,

$$4a + 10b = 44. \tag{4}$$

$$10a + 38b = 166. \tag{5}$$

Solving (4) and (5), we get $a = 0.2308$ and $b = 4.3077$.
Thus, the required straight line is $y = 0.2308 + 4.3077x$.

Problem 2: Fit a straight line for the data given below:

x	0	1	2	3	4
y	1	1.9	3.4	4.6	6.2

Also find the expected values of y for $x = 4.5$ and 3.5.

Solution: Here, n (number of paired observations) $= 5$.
Let the required straight line be

$$y = a + bx. \tag{1}$$

To find $\sum x_i$, $\sum y_i$, $\sum x_i^2$, $\sum x_i \cdot y_i$, we construct the following table:

x	y	xy	x^2
0	1	0	0
1	1.9	1.9	1
2	3.4	6.8	4
3	4.6	13.8	9
4	6.2	24.8	16
$\sum x_i = 10$	$17.1 = \sum y_i$	$\sum x_i \cdot y_i = 38.6$	$\sum x_i^2 = 30$

The normal equations of (1) are

$$\sum_{i=1}^{5} y_i = 5\,a + \left(\sum_{i=1}^{5} x_i\right).b \tag{2}$$

$$\sum_{i=1}^{5} x_i\,y_i = \left(\sum_{i=1}^{5} x_i\right) a + \left(\sum_{i=1}^{5} x_i^2\right) b \qquad (3)$$

Now, substituting $\sum x_i$, $\sum x_i^2$, $\sum y_i$ and $\sum x_i \cdot y_i$ in (2) and (3), we obtain

$$5a + 10b = 17.1 \qquad (4)$$

$$0a + 30b = 38.6. \qquad (5)$$

Solving (4) and (5) for a and b, we get $a = 2.54$ and $b = 0.44$.

Thus, the required straight line is $y = 2.54 + 0.44\,x$.

The expected values of for $x = 4.35$ and $x = 3.5$ are $2.45 + 0.44(4.5) = 4.52$ and $2.54 + 0.44(3.5) = 4.08$ respectively.

Problem 3: A simply supported beam carries a concentrated load p at its mid point. Corresponding to various values of p, the maximum deflections y are measured and are as follows:

P (in lbs).	100	120	140	160	180	200
y	0.42	0.53	0.58	0.64	0.73	0.80

Fit the above data in the form of straight line $y = a + bP$.

Solution: As the squares of P are large for the given data, we use the transformation $U = \dfrac{P - 140}{20}$. Similarly, to reduce the numerical values of y, the following transformation is used:

$$V = y - 0.64.$$

The advantage of taking $V = y - 0.64$ (instead of $V = y - 0.54$) is $V = 0$ in some row other than the row in which $U = 0$ in the data table. Now, the given data of P and y in the transformed variables U and V are as follows:

U	-2	-1	0	1	2	3
V	-0.22	-0.11	-0.06	0	0.09	0.16

Let $\qquad\qquad\qquad V = a + bU \qquad\qquad\qquad (1)$

be the line to fit the transformed data; note that number of paired observations is 6.

U	V	UV	U^2
-2	-0.22	0.44	4
-1	-0.11	0.11	1
0	-0.06	0	0
1	0	0	1
2	0.09	0.18	4
3	0.16	0.48	9
$\sum U_i = 3$	$\sum V_i = -0.14$	$\sum U_i V_i = 1.21$	$\sum U_i^2 = 19$

The normal equations of (1) are

$$\sum_{i=1}^{6} V_i = 5a + \left(\sum_{i=1}^{6} U_i\right) b \tag{2}$$

$$\sum_{i=1}^{6} U_i V_i = \left(\sum_{i=1}^{6} U_i\right) a + \sum_{i=1}^{6} U_i^2 b \tag{3}$$

From table, the equation(2) and (3) reduce to

$$5a + 3b = -0.14 \tag{4}$$

$$3a + 19b = 1.21 \tag{5}$$

Solving (4) and (5) for a and b, we get $a = -0.0731$ and $b = 0.0752$.

Substituting the values of a, b, $V = y - 0.64$ and $U = \dfrac{P - 140}{20}$ in

Eq. (1), we have

$$y - 0.64 = (-0.0731) + (0.0752)\left(\frac{P - 140}{20}\right)$$

i.e., $\qquad y = 0.0349 + 0.0038\,P$

which is the required line.

Problem 4: The following table shows the amount cof weight (y in gms) of potassium bromide dissolved in 100 ml of water at temperature (x in °C); Fit a straight line $y = A + Bx$ to the given data:

x^0	10	20	30	40	50	60
y	58.9	64.8	70.4	76.2	79.6	85.2

Solution: We transform the variables x and y to the variables U and V by the following relations:

$$U = \frac{x-30}{10}, \tag{1}$$

and
$$V = y - 76.2 \tag{2}$$

x	y	U	V	UV	U^2
10	58.9	-2	-17.3	34.6	4
20	64.8	-1	-11.9	11.4	1
30	70.4	0	-5.8	0	0
40	76.2	1	0	0	1
50	79.6	2	3.4	6.8	4
60	85.2	3	9	27	9
		$\sum U_i = 3$	$\sum V_i = -22.1$	$\sum U_i V_i = 79.8$	$\sum U_i^2 = 19$

Let $V = a + bU...$(3) be the straight line to fit data in the variable U and V. Then, the normal equations of (3) are

$$\sum_{i=1}^{6} V_i = 6a + \left(\sum_{i=1}^{6} U_i \right) b$$

$$\sum_{i=1}^{6} U_i V_i = a \left(\sum_{i=1}^{6} U_i \right) + b \left(\sum_{i=1}^{6} U_i^2 \right).$$

Substituting $\sum U_i, \sum U_i^2, \sum V_i$ and $\sum U_i V_i$ in the above equations, we get

$$6a + 3b = -22.1$$

$$3a + 19b = 79.8$$

Solving the above equations for a and b, we get

$$a = -6.279 \text{ and } b = 5.191$$

The Eq. 3 becomes

$$V = -6.279 + 5.1914U$$

In view of (1) and (2), the above equation reduces to

$$y - 76.2 = -6.279 + 5.1914\left(\frac{x-30}{10} \right)$$

i.e.,
$$y = 54.3468 - 0.51914\,x,$$

which is the required straight line.

Problem 5: Below are given the production values of a fertilizer factory of certain entity for various years:

Year	2008	2009	2010	2011	2012
Production (in tonnes)	40	45	52	60	66

Find the expected production in the year 2014 after fitting the data into a straight line.

Solution: Let x be denote the year and y be the amount of production of certain entity in the year x.

Let

$$U = x - 2010 \tag{1}$$

$$V = y - 45. \tag{2}$$

Then, the given data in terms of transformed variables reduced to the values shown in column 3 and column 4 of the following table:

x	y	$U = x - 2010$	$V = y - 45$	UV	U^2
2008	40	-2	-5	10	4
2009	45	-1	0	0	1
2010	52	0	7	0	0
2011	60	1	15	15	1
2012	66	2	21	42	4
		$\sum U = 0$	$\sum V = 38$	$\sum UV = 67$	$\sum U^2 = 10$

Let the straight line to fit the data in U and V variables be

$$V = a + bU \tag{3}$$

The normal equations of (3) for the given data are

$$\sum_{i=1}^{5} V_i = 5a + \left(\sum_{i=1}^{5} U_i \right) b$$

$$\sum_{i=1}^{5} U_i V_i = \left(\sum_{i=1}^{5} U_i \right) a + \left(\sum_{i=1}^{5} U_i^2 \right) b$$

Substituting the values of $\sum_{i=1}^{5} U_i$, $\sum_{i=1}^{5} U_i^2$, $\sum_{i=1}^{5} V_i$ and $\sum_{i=1}^{5} U_i V_i$ in the above two equations, we get

$$5a + 0.b = 38$$
$$0.a + 10.b = 67$$

Thus, $a = 7.6$ and $b = 6.7$. Hence, the Eq. (3) reduces to

$$V = 7.6 + 6.7U$$

In view of (1) and (2), the above equation becomes

$$y - 45 = 7.6 + 6.7(x - 2010)$$

i.e.,
$$y = -13414 + (6.7)x,$$

which is the straight line to the given data in the variable x and y.
The expected amount of production in the year 2014 is $y = -13414 + (6.7)(2014)$

i.e.,
$$y = 79.8 \text{ tonnes.}$$

Problem 6: Given that $y = 0.7 + 1.3x$ is the straight line fitted to the data given below:

x	0	1	2	3	4
y	1	1.8	3.3	4.5	6.3

Find the sum of squares of residuals.

Solution: Let E be the sum of square of residuals. Then

$$E = \sum_{i=1}^{5} \{y_i - (0.7 + 1.3x_i)\}^2 \tag{1}$$

We construct the following table to show the values of x_i, y_i, $1.3\,x_i$
$y_i - (0.7 + 1.3\,x_i)$ and $\{y_i - (0.7 + 1.3\,x_i)\}^2$ for $i = 1, 2, 3, 4$ and 5:

x_i	x_i	$1.3\,x_i$	$0.7 + 1.3\,x_i$	$y_i - (0.7 + 1.3\,x_i)$	$\{y_i - (0.7 + 1.3\,x_i)\}^2$
0	1	0	0.7	0.3	0.09
1	1.8	1.3	2	−0.2	0.04
2	3.3	2.6	3.3	0	0
3	4.5	3.9	4.6	−0.1	0.01
4	6.3	5.2	5.9	0.4	0.16
					0.3

$$\therefore \qquad\qquad\qquad E = 0.3.$$

i.e., The sum of squares of residuals is 0.3.

7.6 FITTING OF A SECOND-DEGREE POLYNOMIAL (OR SECOND-DEGREE CURVE OR PARABOLA) TO THE GIVEN DATA BY THE METHOD OF LEAST SQUARES

Let (x_i, y_i) $(i = 1, 2, 3, \ldots\ldots n)$ be a set of observed data; suppose, we wish to fit this data to a parabola (second-degree polynomial). Let

$$y = a + bx + cx^2 \tag{7.7}$$

be the relation between x and y, where a, b and c are to be determined constants. Note that $y = ax^2 + bx + c$ represents a parabola.

Further, let E denote the sum of squares of residuals of y_i. Then

$$E = \sum_{i=1}^{n} \{y_i - (a + bx_i + c\,x_i^2)\}^2 \tag{7.8}$$

We have to choose a, b, c such that E is minimum. From the calculus, we know that the necessary conditions for E to be minimum are

$$\frac{\partial E}{\partial a} = \frac{\partial E}{\partial b} = \frac{\partial E}{\partial c} = 0.$$

Differentiating (7.8) partially with respect a, b and c respectively and using the fact $\dfrac{\partial E}{\partial a} = \dfrac{\partial E}{\partial b} = \dfrac{\partial E}{\partial c} = 0$, we have

$$\frac{\partial E}{\partial a} = \sum_{i=1}^{n} 2\{y_i - (a + bx_i + cx_i^2)\}(-1) = 0$$

$$\frac{\partial E}{\partial b} = \sum_{i=1}^{n} 2\{y_i - (a + bx_i + c\,x_i^2)\}(-x_i) = 0$$

$$\frac{\partial E}{\partial c} = \sum_{i=1}^{n} 2\{y_i - (a + bx_i + c\,x_i^2)\}(-x_i^2) = 0$$

After slight simplifications, the above equations reduce to the following:

$$\sum_{i=1}^{n} y_i = na + b\sum_{i=1}^{n} x_i + c\sum_{i=1}^{n} x_i^2 \tag{7.9}$$

$$\sum_{i=1}^{n} x_i y_i = a\sum_{i=1}^{n} x_i + b\sum_{i=1}^{n} x_i^2 + c\sum_{i=1}^{n} x_i^3 \tag{7.10}$$

$$\sum_{i=1}^{n} x_i^2 y_i = a\sum_{i=1}^{n} x_i^2 + b\sum_{i=1}^{n} x_i^3 + c\sum_{i=1}^{n} x_i^4 \tag{7.11}$$

Note that n is the number of paired observations of x_i and y_i. Equations (7.9), (7.10) and (7.11) are called normal equations of fitting a parabola to the given data. Solving these equations, we get the values of a, b and c. Let these values be a_f, b_f, c_f. Hence the required fit of parabola for the given data is

$$y = a_f + b_f x + c_f x^2.$$

Note: We can write the normal equations given by (7.9) to (7.11), following the procedure given below:

First, write the parabolic curve equation as $y = ax^0 + bx^1 + cx^2$. Multiplying it with x and x^2 successively, we get two more equations namely $yx = ax + bx^2 + cx^3$ and $x^2y = ax^2 + bx^3 + cx^3$; put the suffix i for the variables x and y in each of these three equations and take summation for $i = 1$ to n. Using the fact $\sum_{i=1}^{n} x_i^0 = n$, these equations are

$$\sum_{i=1}^{n} y_i = na + b\sum_{i=1}^{n} x_i + c\sum_{i=1}^{n} x_i^2$$

$$\sum_{i=1}^{n} x_i y_i = a\sum_{i=1}^{n} x_i + b\sum_{i=1}^{n} x_i^2 + c\sum_{i=1}^{n} x_i^3$$

$$\sum_{i=1}^{n} x_i^2 y_i = a\sum_{i=1}^{n} x_i^2 + b\sum_{i=1}^{n} x_i^3 + c\sum_{i=1}^{n} x_i^4.$$

7.7 PROCEDURE TO FIT A PARABOLA FOR THE GIVEN DATA (x_i, y_i) $(I = 1, 2, \ldots\ldots n)$

Step 1: Suppose the required fitting of parabola be $y = a + bx + cx^2$ (1)

Step 2: Note down n (number of paired observations) value.

Step 3: Construct a table having column heading x_i, y_i, $x_i y_i$, $x_i^2 y$, x_i^2, x_i^3, x_i^4 and having n rows below the column headings.

Step 4: Fill x_i, y_i, values under x_i and y_i.

Step 5: Compute $x_i y_i$, $x_i^2 y$, x_i^2, x_i^3, and x_i^4 for each i and fill the columns at corresponding places.

Step 6: Compute the sum of entries in each column and note down the values of

$$\sum x_i, \sum y_i, \sum x_i y_i, \sum x_i^2 y, \sum x_i^2, \sum x_i^3 \text{ and } \sum x_i^4$$

Step 7: Write down the normal equations of fitting a parabola namely,

$$\sum y_i = na + b\sum x_i + c\sum x_i^2 \qquad (2)$$

$$\sum x_i y_i = a\sum x_i + b\sum x_i^2 + c\sum x_i^3 \qquad (3)$$

$$\sum x_i^2 y_i = a\sum x_i^2 + b\sum x_i^3 + c\sum x_i^4 \qquad (4)$$

Step 8: Substitute $\sum x_i, \sum x_i y_i, \sum x_i^2 y_i, \sum x_i^2, \sum x_i^2, \sum x_i^4$ (obtained in Step 6) and n value (obtained in Step 2) in the Eqs. (2), (3) and (4)

Step 9: Solve the equations obtained in Step 8 for a, b and c. Let these values be a_f, b_f, and c_f.

Step 10: Required parabola is

$$y = a_f + b_f x + c_f x^2.$$

SOLVED PROBLEMS

Problem 1: Fit a parabola of second degree to the following data:

x	0	1	2	4
y	1	5	9	20

Solution: Let

$$y = a + bx + cx^2 \qquad (1)$$

be the required parabola of second degree; note that n (number of paired observations) = 4. Now, we construct a table to find the

$$\sum x_i, \sum x_i^2, \sum x_i^3, \sum x_i^4, \sum y_i, \sum x_i y_i \text{ and } \sum x_i^2 y_i : \text{values}$$

x_i	y_i	x_i^2	x_i^3	x_i^4	$x_i y_i$	$x_i^2 y_i$
0	1	0	0	0	0	0
1	5	1	1	1	5	5
2	9	4	8	16	18	36
4	20	16	64	256	80	320
$7 = \sum x_i$	$35 = \sum y_i$	$21 = \sum x_i^2$	$73 = \sum x_i^3$	$273 = \sum x_i^4$	$103 = \sum x_i y_i$	$361 = \sum x_i^2 y_i$

The normal equations of fitting a second-degree curve to the given data are

$$4a + b\sum_{i=1}^{4} x_i + c\sum_{i=1}^{4} x_i^2 = \sum_{i=1}^{4} y_i \tag{2}$$

$$a\sum_{i=1}^{4} x_i + b\sum_{i=1}^{4} x_i^2 + c\sum_{i=1}^{4} x_i^3 = \sum_{i=1}^{4} x_i y_i \tag{3}$$

$$a\sum_{i=1}^{4} x_i^2 + b\sum_{i=1}^{4} x_i^3 + c\sum_{i=1}^{4} x_i^4 = \sum_{i=1}^{4} x_i^2 y_i \tag{4}$$

Substituting the values of $\sum x_i, \sum x_i^2, \sum x_i^3, \sum x_i^4, \sum x_i y_i,$ and $\sum x_i^2 y_i$ in (2), (3) and (4), we obtain

$$4a + 7b + 21c = 35$$
$$7a + 21b + 73c = 103$$
$$21a + 73b + 273c = 360.$$

Solving the above equations for a, b and c, we get

$$a = 0.9227 \ b = 0.6886 \text{ and } c = 0.2614.$$

Hence, the required fit of parabola is

$$y = 0.9227 + (0.6886)x + (0.2614)x^2$$

Problem 2: Fit a parabola to the following data:

x	1	2	3	4	5
y	4	3	6	3	2

Solution: Let the required fit of parabola be $y = a + bx + cx^2$. $\tag{1}$

Here, $n = 5$, i.e., the number paired observations is 5.

x_i	y_i	$x_i y_i$	x_i^2	$x_i^2 y$	x_i^3	x_i^4
1	4	4	1	4	1	1
2	3	6	4	12	8	16
3	6	18	9	54	27	81
4	3	12	15	48	64	256
5	2	10	25	50	125	625
15	18	50	55	168	225	979

From the above table, we have $\sum_{i=1}^{5} x_i = 15$, $\sum_{i=1}^{5} y_i = 18$, $\sum_{i=1}^{5} x_i y_i = 50$,

$\sum_{i=1}^{5} x_i^2 = 55$, $\sum_{i=1}^{5} x_i^2 y_i = 168$, $\sum_{i=1}^{5} x_i^3 = 225$ and $\sum_{i=1}^{5} x_i^4 = 979$.

The normal equations of fitting a parabola (1) are

$$\sum_{i=1}^{5} y_i = 5a + b\sum_{i=1}^{5} x_i + c\sum_{i=1}^{5} x_i^2 \tag{2}$$

$$\sum_{i=1}^{5} x_i y_i = a\sum_{i=1}^{5} x_i + b\sum_{i=1}^{5} x_i^2 + c\sum_{i=1}^{5} x_i^3 \tag{3}$$

$$\sum_{i=1}^{5} x_i^2 y_i = a\sum_{i=1}^{5} x_i^2 + b\sum_{i=1}^{5} x_i^3 + c\sum_{i=1}^{5} x_i^4 \tag{4}$$

Substituting the values of $\sum x_i, \sum y_i, \sum x_i y_i, \sum x_i^2 y_i, \sum x_i^3, \sum x_i^4$ and n in the Eqs. (2) to (4), we get

$$5a + 15b + 55c = 18$$
$$15a + 55b + 225c = 50$$
$$55b + 225c + 979c = 168.$$

Solving the above equations for a, b and c, we get $a = 1.8$, $b = 2.1714$ and $c = -0.4286$, Thus, the required fit of parabola is

$$y = 1.8 + 2.1714x - 0.4286\,x^2.$$

Problem 3: The profit of a certain company in the xth year of its life are given below:

x	1	2	3	4	5
y (Rs in laks)	1150	1300	1650	1850	2200

Fit a parabola to the above data.

Solution: Here, $n = 5$; we use the following transformation to change the variables x, y to the variables U and V to reduce the numerical values of x and y:

$$U = x - 3 \tag{1}$$

and
$$V = \frac{y - 1300}{50}.$$
(2)

The given data in terms of U and V variables are as follows:

U	-2	-1	0	1	2
V	-3	0	7	11	18

Let the parabola to fit the data of U and V be

$$V = a + bU + cU^2$$
(3)

U	V	U^2	U^3	U^4	UV	U^2v
-2	-3	4	-8	16	6	-12
-1	0	1	-1	1	0	0
0	7	0	0	0	0	0
1	11	1	1	1	11	11
2	18	4	8	16	36	72
0	33	10	0	34	53	71

From the above table, we have

$$\sum U_i = 0, \sum V_i = 33, \sum U_i^2 = 10, \sum U_i^3 = 0,$$

$$\sum U_i^4 = 34, \sum U_i V_i = 53 \quad \text{and} \quad \sum U_i^2 V_i = 71.$$

The normal equations of parabola given by (1) are

$$\sum V_i = 5a + b\sum U_i + c\sum U_i^2$$
$$\sum U_i V_i = a\sum U_i + b\sum U_i^2 + c\sum U_i^3$$
$$\sum U_i^2 V_i = a\sum U_i^2 + b\sum U_i^3 + c\sum U_i^4$$

Substituting the values of $\sum U_i$, $\sum V_i$ ...... etc in the above equations, we get

$$5a + 0.b + 10.c = 33$$

$$0.a + 10.b + 0.c = 53$$

$$10a + 0.b + 34c = 71.$$

Solving the above equations for a, b and c, we get $a = 5.8857$ $b = 5.3000$ and $c = 0.3511$.

Hence, the equation (3) becomes

$$\frac{y-1300}{50} = 5.8857 + 5.3(x-3) + 0.3571(x-3)^2$$

$$y = 1300 + 50[5.8857 + 5.3(x-3) + 0.3571\,(x^2 - 6x + 9)]$$

$$= 1300 + 50[5.8857 + 5.3x - 5.3 \times 3 + 0.3571$$
$$x^2 - 0.3571 \times 6x + 0.3571 \times 9]$$

$$= 1300 + 50[0.3571x^2 + 3.1574x - 6.8004]$$

$$= 17.855x^2 + 157.85x + 959.98.$$

Problem 4: Fit a second-degree curve by the method of least squares for the given data:

x	20	40	60	80	100	120
y	5.5	9.1	14.9	22.8	33.3	46.0

Solution: Here, number of paired observations of x and y is 6, i.e., $n = 6$. Let

$$U = \frac{x-60}{20} \tag{1}$$

and $$V = y - 22.8 \tag{2}$$

The values of U and V for given values of x and y are shown in third and seventh column of the following table:

x	y	U	U^2	U^3	U^4	V	UV	U^2V
20	5.5	−2	4	−8	16	−17.3	+ 34.6	−69.2
40	9.1	−1	1	−1	−13.7	13.7	13.7	60
14.9	0	0	0	0	−7.9	0	0	80
22.8	1	1	4	1	0	0	0	0
100	33.3	2	4	8	16	10.5	21.0	42
120	46.0	3	9	27	81	23.2	69.6	208.8
		3	19	27	115	−5.2	138.9	167.9

Suppose the second-degree curve to be fitted to the data in U and V be

$$V = a + bU + cU^2. \tag{3}$$

The normal equations of (3) for the data in U and V are

$$\sum_{i=1}^{6} V_i = 6a + b\sum_{i=1}^{6} U_i + c\sum_{i=1}^{6} U_i^2$$

$$\sum_{i=1}^{6} U_i V_i = a\sum_{i=1}^{6} U_i + b\sum_{i=1}^{6} U_i^2 + c\sum_{i=1}^{6} U_i^2$$

$$\sum_{i=1}^{5} U_i^2 V_i = a\sum_{i=1}^{6} U_i^2 + b\sum_{i=1}^{6} U_i^3 + c\sum_{i=1}^{6} U_i^3$$

Substituting the value of $\sum U, \sum U^2 \ldots\ldots$ etc in the above equations, we get

$$6a + 3b + 19c = -5.2$$

$$3a + 19b + 27c = 138.9$$

$$19a + 27b + 115c = 16.9.$$

Solving the above equations for a, b and c, we obtain
$a = -7.9714$, $b = 6.9375$ and $c = 1.1482$.
Now, the equation (3) becomes
$V = -7.9714 + 6.9375U + 1.1482U^2$.
In view of (1) and (2), the above equation reduces to

$$y - 22.8 = -7.9714 + 6.9375\left(\frac{x-60}{20}\right) + 1.1482\left(\frac{x-60}{20}\right)^2$$

$$y = 22.8 - 7.9714 + \frac{6.9375}{20}(x-60) + \frac{1.1482}{400}(x^2 - 120x + 3600)$$

$$= 22.8 - 7.9714 + 0.3047(x-60)$$
$$+ 0.0029(x^2 - 120x + 360 + 3600)$$

$$= 22.8 - 7.914 + 0.3047x - 0.307*60 + 0.0029.x^2$$
$$- 0.0029*120x + 0.0029*3600$$

$$= 22.8 - 7.914 + 0.3047x - 18.2820 + 0.0029x^2$$
$$- 0.3480x + 10.44$$

$$= 7.040 - 0.0433x + 0.0029x^2,$$

which is the required second-degree curve to the given data in the variable x and y

7.8   FITTING THE DATA IN THE FORM $y = ae^{bx}$

Suppose the points of scatter diagram are joined by a smooth curve like an exponential curve. In such cases, one has to fit the data in the exponential form or exponential curve. Now, we learn how to fit the data in the exponential form $y = ae^{bx}$.

Suppose the exponential curve to represent the given data (x_i, y_i) $(i = 1, 2, 3, \ldots n)$ be

$$y = ae^{bx}, \tag{7.12}$$

where a and b are to be determined constants. Taking logarithms on both sides of the Eq. (7.12) with base e, we get

$$\log_e y = \log_e a + bx \tag{7.13}$$

Let $Y = \log_e y$ and $\log_e a = A$; then the Eq. (7.13) reduces to

$$Y = A + bx \tag{7.14}$$

The Eq. (7.14) is a straight line. Thus, the problem of fitting an exponential curve to the given data is reduced to fitting of straight line to the data $(x_i, \log y_i)$. Hence, the normal equations of (7.14) are

$$n A + b \sum_{i=1}^{n} x_i = \sum_{i=1}^{n} Y_i \tag{7.15}$$

and

$$A \sum_{i=1}^{n} x_i + b \sum_{i=1}^{n} x_i^2 = \sum_{i=1}^{n} x_i Y_i \tag{7.16}$$

Solving (7.15) and (7.16), we obtain A and b; suppose these values are A_f and b_f, respectively. Since $A = \log_e a$, we have the value of a say a_f. Thus, the equation of exponential curve to the given data will be

$$y = a_f e^{b_f x}.$$

7.9   FITTING OF DATA IN THE FORM $y = a.b^x$

Suppose the data (x_i, y_i) $(i = 1, 2, \ldots\ldots, n)$ to be fitted in the form

$$y = a \, b^x \tag{7.17}$$

Taking logarithms with base e on both sides of 7.17, we get

$$\log_e y = \log_e a + x \log_e b. \tag{7.18}$$

Introducing $\log_e y = Y$ and $A = \log_e a$, $B = \log_e b$.
The Eq. (7.18) can be expressed as

$$Y = A + Bx \tag{7.19}$$

Clearly, (7.19) represent a line in the variable Y and x. Therefore, we fit a straight line to the data $(x_i, \log_e y_i)(i = 1, 2, ...n)$. Thus, the normal equations of (7.19) are given by

$$n A + B \sum n_i = \sum Y_i \tag{7.20}$$

$$\left(\sum x_i\right) A + B \left(\sum x_i^2\right) = \sum x_i y_i . \tag{7.21}$$

Solving (7.20) and (7.21), we obtain the values of A and B. Let these values be A_f and B_f.
    Since $A = \log_e a$, we have $a = e^A$. The value of a (say a_f) is given by $a_f = e^{A_f}$. Similarly, $b_f = e^{B_f}$. Thus, the required curve to the given data will be

$$y = a_f e^{(b_f x)}$$

7.10   FITTING THE DATA IN THE FORM $y = a.x^b$

Suppose the data $(x_i, y_i)(i = 1, 2, ...n)$ to be fitted in the form
$$y = ax^b \tag{7.22}$$
Taking logarithm on both sides of 7.22 with base e,
we have

$$\log_e y = \log_e a + b \log_e x \tag{7.23}$$

Introducing $\log_e y = Y$, $\log_e x = X$ and $A = \log_e a$, the Eq. 7.23 reduces to

$$Y = A + bX \tag{7.24}$$

Hence, the normal equations of (7.24) are

$$nA + b \sum x_i = \sum Y_i \tag{7.25}$$

$$A\left(\sum x_i\right) + b \sum x_i^2 = \sum x_i y_i. \tag{7.26}$$

Solving (7.25) and (7.26), we get A and b substituting A value in $A = \log a$; we can find a. Let the value of a and b be a_f and b_f. Hence, the required fit will be $y = a_f\,(x)^{b_f}$.

7.11   FITTING THE DATA IN THE FORM $y = a + \dfrac{b}{x}$ OR $xy = ax + b$

Suppose the data (x_i, y_i) $(i = 1, 2, 3\ldots.. m)$ to be fitted in the form

$$y = a + \frac{b}{x} \tag{7.27}$$

where each $x_i \neq 0$.

Let $\dfrac{1}{x} = X$. Then, the Eq. 7.27reduces to $y = a + bX$. $\tag{7.28}$

Hence, the normal equations of (7.28) are

$$n\,a + b \sum X_i = \sum y_i \tag{7.29}$$

$$a \sum X_i + \sum X_i^2 = \sum X_i\, y_i \tag{7.30}$$

Solving (7.29) and (7.30), we get the values of a and b, suppose these valuse as a_f and b_f. Thus, the required fit for the data will be

$$y = a_f + b_f \left(\frac{1}{x}\right).$$

7.12   FITTING THE DATA IN THE FORM $P\,V^a = c$

Suppose the data (P_i, V_i) $(i = 1, 2, 3, \ldots\ldots.. n)$ to be fitted in the form

$$P\,V^a = c \tag{7.31}$$

where a, c are constants.

Taking logarithm with base e on both sides of (7.31), we get

$$\log_e P + a \log_e V = \log_e c. \tag{7.32}$$

Introducing $\log_e V = y$, $\log_e P = x$, $A = \dfrac{\log c}{a}$ and $B = \dfrac{-1}{a}$, the Eq. (7.32) reduces to

$$y = A + B\,x. \tag{7.33}$$

Hence, the normal equations of (7.33) are

$$\sum y_i = nA + B\sum x_i \tag{7.34}$$

$$\sum x_i y_i = A\sum x_i + 1B\sum x_1^2 \tag{7.35}$$

Solving Eqs. (7.34) and (7.35), we get the values of A and B. The values of a and c can be obtained from the relations $B = \dfrac{-1}{a}$ and $A = \dfrac{\log c}{a}$. Let the values of a and c be a_f and c_f and the required fit will be

$$P\, V^{a_f} = c_f,$$

7.13 FITTING THE DATA IN THE FORM OF $y = a + \dfrac{b}{x} + \dfrac{c}{x^2}$

Let the data (x_i, y_i) to be fitted in the form

$$y = a + \frac{b}{x} + \frac{c}{x^2} \tag{7.36}$$

where $x_i \neq 0$ $(i = 1, 2, \ldots, n)$

Introducing $\dfrac{1}{x} = X$, the Eq. (7.36) becomes

$$y = a + bX + cX^2 \tag{7.37}$$

The Eq. (7.37) represents a parabolic curve in the variable y and X. Hence, the normal equations of 7.37 are

$$\sum_{i=1}^{n} y_i = na + \left(\sum_{i=1}^{n} X_i\right) b + \left(+b\sum_{i=1}^{n} X_i^2\right) c \tag{7.38}$$

$$\sum_{i=1}^{n} X_i y_i = a\sum_{i=1}^{n} X_i + b\sum_{i=1}^{n} X_i^2 + c\sum_{i=1}^{n} X_i^3 \tag{7.39}$$

$$\sum_{i=1}^{n} X_i^2 y_i = a\sum_{i=1}^{n} X_i^2 + b\sum_{i=1}^{n} X_i^3 + c\sum_{i=1}^{n} X_i^4 \tag{7.40}$$

Solving the above equations, we get the values of a, b, and c. Let these values be a_f, b_f and c_f, respectively Thus, the required fit will be

$$y = a_f + \frac{b_f}{x} + \frac{c_f}{x^2}.$$

SOLVED PROBLEMS

Problem 1: Find the curve of the type $y = ae^{bx}$ to the following date:

x	0	1	2	3
y	1.05	2.10	3.85	8.30

Solution: The curve to be fitted to $y = a\,e^{bx}$ to the given data is equivalent to fitting the $(x_i, \log y_i)$ $(i = 1, 2, ..., n)$ in the form $Y = A + bx$.

where $A = \log_e a$ and $Y = \log_e y$.
The normal equations of $Y = A + bx$ are

$$\sum_{i=1}^{4} Y_i = 4A + b\sum_{i=1}^{4} x_i$$

$$\sum_{i=1}^{4} x_i Y_i = A\sum_{i=1}^{4} x_i + b\sum_{i=1}^{4} x_i^2,$$

where 4 is the number of paired observations. Now, we construct the following table:

x_i	y_i	$Y_i = \log y_i$	x_i^2	$x_i \cdot \log_e y_i = x_i Y_i$
0	1.05	0.0488	0	0
1	2.10	0.7420	1	0.7420
2	3.85	1.3481	4	2.6962
3	8.3	2.1163	9	6.3489
$6 = \sum x_1$	$15.3 = \sum y_1$	$4.2552 = \sum \log y_i = (Y_i)$	$\sum x_1^2 = 14$	$9.7871 = \sum x_i Y_i$

From the above table, the normal equations are

$$4A + 6\,b = 4.2552$$

$$6A + 14\,b = 9.7871$$

Solving the above equations for A and b, we get $A = 0.0425$ and $B = 0.6809$. Hence, $a = e^A (\because A = \log_e a)$ gives $a = e^{0.0425} = 1.0434$. Thus, the required fit in the form $y = a\,e^{bx}$ is $y = 1.0434\,e^{0.6809x}$.

Problem 2: Fit the following data in the form $y = ae^{bx}$

x	1	5	7	9	12
y	10	14	12	15	19

Solution: The curve to be fitted $y = ae^{bx}$ to the given date (x_i, y_i) is equivalent to fitting the data $(x_i, \log y_i)$ $(i = 1, 2, ..., n)$ in the form $y = A + bx$, where $A = \log_e a$ and $Y = \log_e y$. Now, the normal equations of $Y = A + bx$ are

$$\sum_{i=1}^{5} Y_i = 5A + b \sum_{i=1}^{5} x_i \tag{1}$$

$$\sum_{i=1}^{5} x_i Y_i = A \sum_{i=1}^{5} x_i + b \sum_{i=1}^{5} x_i^2. \tag{2}$$

We construct the following table to know the sums involved in Eqs. (1) and (2):

x	y	$Y_i = \log y_i$	x^2	$x.\log y_i = xY_i$
1	10	2.3026	1	2.3026
5	14	2.6391	25	13.1955
7	12	2.4850	49	17.3950
9	15	2.7081	81	24.3729
12	19	2.9445	144	35.3340
34	70	13.0793	300	92.6

Thus, $\sum x_i = 34$, $\sum Y_i = 13.0793$, $\sum x_i Y_i = 92.6$ and $\sum x_i^2 = 300$;

Substituting these values in (1) and (2), we have

$$5A + 34b = 13.0793$$

$$34A + 300b = 92.6.$$

Solving the above equations for A and b, we get $A = 2.2540$ and $b = 0.0532$.

Since, $\log_e a = A$, we have $\log_e a = 2.2540$,

i.e., $a = e^{2.2540} = 9.5254$. Thus, the required fit to the given data in the form $y = a\ e^{bx}$ is $y = 9.5258\ e^{(0.0532)x}$.

Problem 3: Fit a curve $y = ab^x$ to the data given below:

x	3	4	5	6	7
y	122.4	135.6	142.5	151.6	157.8

Solution: The curve to be fitted $y = ab^x$ to the given data (x_i, y_i) is equivalent to fitting the data form $Y = A + Bx$, where $A = \log_e a$, $B = \log_e b$ and $\log_e y = Y$, the normal equations of $Y = A + Bx$ are given by

$$nA + B \sum x_i = \sum Y_i \tag{1}$$

$$A \sum x_i + \sum x_i^2 = \sum x_i Y_i ; \tag{2}$$

once we find A and B from (1) and (2), we can find a and b from the relations $A = \log_e a$ and $B = \log_e b$.

x	y	$Y_i = \log_e y_i$	$x_i Y_i$	x_i^2
3	122.4	4.8073	14.4219	9
4	135.6	4.9098	19.6392	16
5	142.5	4.9594	24.797	25
6	151.6	5.0213	30.1278	36
7	157.8	5.0614	35.4298	49
25	709.9	24.7592	124.4154	135

From the above table, we have, $\sum x_i = 25$, $\sum x_i^2 = 135$, $\sum Y_i = 24.7592$, $\sum x_i Y_i = 124.4154$ and $n = 5$; substituting these values (1) and (2), we obtain

$$5A + 25B = 24.7592$$

$$25A + 135B = 124.4157.$$

Solving the above equation for A and B, we get $A = 4.6420$ and $B = 0.0620$.

Since $A = \log_e a$, we have $a = e^A = e^{4.6420} = 103.7516$; similarly, $b = e^{0.0620} = 1.064$. Thus, the required fit to the given data is $y = 103.7516(1.064)^x$.

Problem 4: Fit a curve $y = a.b^x$ to the following data:

x	2	4	6	8
y	16	23	36	45

Solution: The curve $y = a.\, b^x$ reduces to $Y = A + BX$ where $A = \log_e a$, $B = \log_e b$ and $Y = \log_e y$.

Therefore, the normal equations of $y = A + Bx$ are

$$n A + \sum x_i .B = \sum Y_i \tag{1}$$

$$A \sum x_i + B \sum x_i^2 = \sum x_i Y_i \tag{2}$$

Now, we construct the following table to know the $\sum x_i$, $\sum x_i^2$, $\sum Y_i$ and $\sum x_i Y_i$:

x_i	y_i	$Y_i = \log_e y_i$	xY_i	x_i^2
2	16	2.7726	5.5452	4
4	23	3.1355	12.5420	16
6	36	3.5836	21.5016	36
8	45	3.8067	30.4536	64
20	120	13.2984	70.0424	120

Substituting $n = 4$, $\sum x_i = 20$, $\sum x_i^2 = 120$, $\sum Y_i = 13.2984$, $\sum x_i Y_i = 70.0424$ in the normal Eqs. (1) and (2), we have

$$4A + 20B = 13.2984$$
$$20A + 120B = 70.0424$$

Solving the above two equations for A and B, we obtain $A = 2.4370$ and $B = 0.1775$

Since $A = \log_e a$, we have $a = e^{2.4370} = 11.4387$. Similarly, $b = e^{0.1775} = 1.1942$. Thus, the required curve in the form $y = a\, b^x$ is $y = 1.4387(1.1942)^x.$

Problem 5: Find the curve of fit of the type. $y = a.x^b$ to data given below:

x	1	2	3	4	5
y	4	12	27	48	100

Solution: The curve to be fitted is $y = a.x^b$, which is equavalent to $Y = A + bx$ where $Y = \log_e y$, $X = \log_e x$ and $A = \log_e a$,. The normal equations of $Y = A + bx$ are

$$n A + b \sum X_i = \sum Y_i \tag{1}$$

$$A \sum X_i + b \sum X_i^2 = \sum X_i Y_i. \tag{2}$$

x_i	y_i	$X_i = \log_e x_i$	$Y_i = \log_e y_i$	$X_i Y_i$	X_i^2
1	4	0	1.3863	0	0
2	12	0.6932	2.485	1.7227	0.4806
3	27	1.0987	3.2959	3.6213	1.2072
4	48	1.3863	3.8713	5.3668	1.9219
5	100	1.6095	4.6052	7.4121	2.5905
15	191	4.7877	15.6437	18.1229	6.2002

From the table, we have $n = 5$, $\sum X_i = 4.7877$, $\sum X_i^2 = 6.2002$, $\sum Y_i = 15.6437$ and $\sum X_i Y_i = 18.1229$. Putting these values in the normal Eqs. (1) and (2), we get

$$5A + 4.7877b = 15.643$$

$$4.7877A + 6.2002b = 18.1229.$$

Solving the above equations, we get $A = 1.2659$ and $b = 1.9455$, since $A = \log_e a$, we have $a = e^A$ i.e. $a = e^{1.2659} = 3.5463$. Thus, the required fit to the given data is $y = (3.5463)(x)^{1.9455}$.

Problem 6: Fit the following data in the form $y = a. x^b$:

x	1	3	5	8
y	3.8	35.4	98.6	254.8

Solution: The curve $y = a. x^b$ reduces to $Y = A + bX$, where $Y = \log_e y$, $X = \log_e x$ and $A = \log_e a$. The normal equations of $y = A + bx$ are

$$n A + b \sum X_i = \sum Y_i \tag{1}$$

$$A \sum X_i + b \sum X_i^2 = \sum X_i Y_i \tag{2}$$

x_i	y_i	$X_i = \log_e x_i$	$Y_i = \log_e y_i$	X_i^2	XY_i
1	3.8	0	1.3351	0	0
3	35.4	1.0987	3.5668	1.2072	3.9189
5	98.6	1.6095	4.5911	2.5905	7.3894
8	254.8	2.0795	5.5405	4.3244	11.5215
17	392.6	4.7877	15.0335	8.1221	22.8298

Now, the normal Eqs. (1) and (2) are reduced to

$$4A + 4.7877b = 15.0335$$

$$4.7877A + 8.1221b = 22.8298.$$

Solving the above equations, we get $A = 1.3382$ and $b = 2.0220$; since $A = \log_e a$ we have $a = e^A$, hence $a = e^{1.3382} = 3.8122$. Thus, the required curve is

$$y = (3.8122)\,(x)^{2.022}.$$

Problem 7: Fit the following data in the form $y = a + \dfrac{b}{x}$:

x	1	2	3	4	5	6
y	5.63	6.42	8.41	10.23	12.61	14.38

Solution: As stated in (7.11), assuming $X = \dfrac{1}{x}$, the equation of $y = a + \dfrac{b}{x}$ reduces to $y = a + bX$ and its normal equations are

$$\sum_{i=1}^{n} y_i = na + b\sum_{i=1}^{n} X_i \tag{1}$$

$$\sum_{i=1}^{n} X_i y_i = a\sum_{i=1}^{n} X_i + b\sum_{i=1}^{n} X_i^2 \tag{2}$$

Note the number of paired observation of the given data is 6, i.e., $n = 6$.

x_i	y_i	$X_i = \dfrac{1}{x_i}$	X_i^2	$X_i Y_i$
1	563	1	1	5.63
2	6.42	0.5	0.25	3.2100
3	8.41	0.3334	0.1112	2.8039

4	10.23	0.25	0.0625	2.5575
5	12.61	0.2	0.04	2.5220
6	14.38	0.1667	0.0278	2.3972
21	57.68	2.4501	1.4915	19.1206

Substituting n, $\sum X_i$, $\sum X_i^2$, $\sum y_i$ and $\sum X_i y_i$ in the normal Eqs. (1) and (2), we get

$$6a + 2.4501b = 57.68$$

$$2.4501a + 1.4915b = 19.1206.$$

Solving the above two equations for a and b, we obtain $a = 13.3001$ and $b = -9.0285$, Hence the required fit in given by

$$y = (13.3001) - (9.0285)\left(\frac{1}{x}\right).$$

Problem 8: The pressures (P) of the gas corresponding to various volumes (V) are given in the following table:

V (cm³)	40	50	60	70	80
P (kg/cm³)	67.3	63.4	50.2	40.6	30.4

If V and P are related by $PV^a = c$. Find a and c using the method of least squares.

Solution: The relation $PV^a = c$ reduces to the following form by taking logarithms on both sides of it with base e

$$y = A + Bx \tag{1},$$

where

$$y = \log_e V,\ x = \log_e P,\ A = \frac{\log_e}{a}\ \text{and}\ B = \frac{-1}{a}. \tag{2}$$

Further, the normal equations of (1) are

$$\sum y_i = nA + \left(\sum x_i\right)B \tag{3}$$

$$\sum x_i y_i = \left(\sum x_i\right)A + \left(\sum x_i^2\right)B \tag{4}$$

Here, $n = 5$, i.e., the number of paired observations is 5.

V	P	$x = \log_e P$	$y = \log V$	x^2	xy
40	67.3	3.6889	4.2092	13.608	15.5274
50	63.4	3.9121	4.1495	15.3046	16.2333
60	50.2	4.0944	3.9161	16.7642	16.0341
70	40.6	4.2485	3.7038	18.0498	15.7356
80	30.4	4.3821	3.4145	19.2029	14.9627
		20.326	19.3931	82.9295	78.4931

Thus the normal equations are

$$5A + 20.326B = 19.3931$$

$$20.326A + 82.9295B = 78.4931$$

Solving the above equations, we have

$$A = 8.5326 \text{ and } B = -1.1448$$

Since $a = \dfrac{-1}{B}$, we have $B = 0.8735$ and $A = \dfrac{\log_e c}{a}$ implies $\log_e c = aA =$

$(0.8735)(8.5326) = 7.4532.$

Hence, $c = e^{7.4532} = 17 = 5.4$. Thus , the required relation is

$$P\,V^{\,0.8735} = 1725.4,$$

and the values of a and c are 0.8735 and 1725.4, respectively.

Problem 9: The time taken by a runner to cover same fixed distance for four consecutive days are given below:

day(x)	1	2	3	4
Time(y)	15.2	14.8	14.2	14

Fit the above data in the form $y = a + \dfrac{b}{x} + \dfrac{c}{x^2}$. Also find the expected time to be taken by the runner on fifth day to run the same distance.

Solution: Here, $n = 4$. Introducing $X = \dfrac{1}{x}$, the relation $y = a + \dfrac{b}{x} + \dfrac{c}{x^2}$ takes the form $y = a + bX + cX^2$ and its normal equations are the normal equations of second-degree curve, namely

$$na + b\sum X_i + c\sum X_i^2 = \sum y_i \qquad (1)$$

$$a\sum X_i + b\sum X_i^2 + c\sum X_i^3 = \sum X_i y_i \qquad (2)$$

$$a\sum X_i^2 + b\sum X_i^3 + c\sum X_i^4 = \sum X_i^2 y_i \qquad (3)$$

x_i	y_i	$X_i = \dfrac{1}{x_i}$	X_i^2	X_i^3	X_i^4	$X_i Y_i$	$X_i^2 Y_i$
1	15.2	1	1	1	1	15.2	15.2
2	14.8	0.5	0.25	0.125	0.0625	7.4	3.7
3	14.2	0.3334	0.1112	0.0371	0.0124	4.7343	1.5791
4	14	0.25	0.0625	0.0157	0.0004	3.5	0.875
10	58.2	2.0834	1.4237	1.1778	1.0789	30.8343	21.3541

Thus, the normal equations become

$$4a + 2.0834b + 1.4237c = 58.2$$

$$2.0834a + 1.4237b + 1.1778c = 30.8343$$

$$1.4237a + 1.1778b + 1.0789c = 21.351$$

Solving the above equations, we get $a = 12.6466$, $b = 6.0435$ and $c = -3.4962$.

Thus, the required curve to the given data is $y = 12.6466 + 6.0435 \left(\dfrac{1}{x}\right) - 3.4962 \left(\dfrac{1}{x}\right)^2$.

The expected time taken by the runner on the fifth day is 13.7155.

EXERCISE 7(a)

1. Fit a straight line to the following data in the form $y = a + bx$

x	0	1	2	3	4
y	1	1.6	3.1	4.3	6.1

2. Given that

x	1	2	3	4	5
y	14	27	38	54	64

Fit above data in the form $y = a + bx$ by the method of least squares.

3. Fit a straight line of the form $y = a + bx$ to the below given data using the method of least squares

x	0	1	2	3
y	2	5	8	12

4. Find the values of a and b to that $y = a + bx$ is the best fit to the following data

x	5	10	15	20	25	30
y	14	19	25	29	34	38

5. Fit a second-degree curve to the following data

x	1	2	4	6	8	9
y	2	6	8	11	10	9

6. Fit the following data in the form of $y = a + bx + cx^2$

x	2	4	6	8	10
y	3.07	12.82	31.45	57.36	90.24

7. Fit a curve of the form $y = a\, b^x$ to the following data

x	1	2	3	4	5	6
y	151	100	61	50	20	8

8. Fit a curve of the type $y = a\, e^{bx}$ to the data given below

x	1	4	6	8	10
y	10	14	13	15	20

9. Fit the following data in the form $y = a.x^b$

x	2	2.5	3	3.5	4
y	0.83	0.94	1.02	1.23	1.31

10. Fit a curve to the data given below in the form $y = a + \dfrac{b}{x}$

x	1	4	6	8	9	10
y	42	34	25	18	15	12

11. Find the least-squares straight line for the data points (−2, 10) (0, 9), (1, 11), (3, 4), (4, 3) and (5, 1) where first co-ordinate represents x and second co-ordinate represents y.

12. Fit a curve $y = a\, e^{bx}$ to the data given below

x	0	1	2	3	4	5	6
y	20	30	52	77	135	211	626

13. Let P represents the pull required to life a load W. Fit the following data in the form $P = a + bW$

W (kg)	50	70	100	110
P	10	13	20	23

Find the expected amount of pull required to lift the load of 150 kg.

14. The weight of a calf taken at bi-weekly intervals are given below. Fit a straight line using the method of least squares.

Age in weeks x	2	4	6	8	10
weight	54.6	68.4	79.4	92.6	104.8

15. Fit an exponential curve of the form $y = a\, b^x$ to the following data

x	1	2	3	4	5	6	7	8
y	1.0	1.3	1.9	2.7	3.6	4.8	6.9	9.5

16. Fit a second-degree curve to the following data (x, y) by the method of least squares, where x represents the year and y represents the population of a town in lakhs.

year (x)	1950	1960	1970	1980	1990	2000
population: (y)	25.6	29.8	34.2	38.4	41.3	46.4

17. Fit the following data in the form (i) $y = a + bx$ (ii) $y = a + bx + cx^2$:

x	1	2	3	4	5
y	2	5	10	6	3

18. The profits of a certain company in x^{th} year of its life are as follows:

x	1	2	3	4	5	6
y (Rs. in lakhs)	2000	2500	2700	3200	3400	3900

Fit the above data in the form of $y = a + bx + cx^2$.

19. The pressruns of the gas corresponding to various volumes V are given below

V (cm^3)	20	40	60	80	100
P (kg/cm^3)	69.3	62.7	51.3	22.4	16.2

Fit the data in the form $PV^a = c$.

20. Fit the following data in the form $y = a + \dfrac{b}{x} + \dfrac{c}{x^2}$

x	1	2	3	4	5
y	14.3	14.1	14	13.5	13.2

21. The production of urea by a fertilizer company for various years are given below:

year:	2006	2008	2010	2012
Production (in tonnes)	40	45	52	61

Fit the above data in the form $y = a + bx + cx^2$. Also find the expected production for the year 2014.

22. The sales of c company in lakhs of rupees for the years 2008 to 2012 are given below:

year (x)	2008	2009	2010	2011	2012
Sales (y)	40	45	52	60	66

Fit the above data in the form $y = a + bx$. Also find the expected sales in the year 2014.

ANSWERS 7 (a)

1. $y = 0.64 + 1.29\,x$ 2. $y = 1.3 + 12.7\,x$ 3. $y = 1.1 + 3.6x$

4. $y = 9.6 + 0.965714\,x$ 5. $y = -0.7929 + 3.4749\,x - 0.2648\,x^2$

6. $y = 0.94 - 0.528857\,x + 0.956071\,x^2$ 7. $y = 316.7816 \times (0.5693)^x$

8. $y = 9.5429 \times e^{0.0666x}$ 9. $y = 0.5084 \times (x)^{0.679}$

10. $y = 16.2107 + \dfrac{27.8047}{x}$ 11. $y = 11.4393 - 2.785x$

12. $y = 17.4087 \times e^{0.5424x}$ 13. $P = -1.4505 + 0.2176\,w$

14. $w = 42.5 + 6.23\,x$ 15. $y = 1.0908 \times (1.3169)^x$

16. $y = -0.00026786\,x^2 + 1.4657\,x - 1815$

17. (i) $y = 4.3 + 0.3\,x$ (ii) $y = -6.2 + 9.3\,x - 1.5\,x^2$

18. $y = 1600 + 457.142857\,x - 14.285714\,x^2$

19. $PV^{1.1797} = 4045.3$

20. $y = 12.310564 + \dfrac{5.675633}{x} - \dfrac{3.697828}{x^2}$

21. $y = 52 + 8\left(\dfrac{x - 2010}{2}\right) + \left(\dfrac{x - 2010}{2}\right)^2,\ 72$

22. $y = 52.171429 + 6.7\,(x - 2010) + 0.21429\,(x - 2010)^2,\ 82.4.$

7.14 METHODS OF MOMENTS

Let $y_1, y_2, \ldots\ldots y_n$ be a set of observations of a function y at equidistant points $x_1, x_2, x_3, \ldots\ldots x_n$, respectively with the interval of differencing h and $x_1 < x_2 < x_3 < \ldots\ldots < x_n$.

In this method, the moments m_k of the observed values are defined as follows:

$$m_k = h \sum_{i=1}^{n} x_i^{k-1} \cdot y_i \ (k = 1, 2, \ldots, p), \tag{7.41}$$

where p is a positive integer.

We call m_k as kth moment. The moments m_1, m_2, ...m_n are called the moments of the observed values.

Let the curve to be fitted for the given data be $y = f(x)$. Note that $f(x)$ involves some constants that are to be determined constants. For example, if $f(x)$ form is $a + bx$, then a, and b are to be determined. If $f(x)$ form is assumed as $a + bx + cx^2$, then a, b and c are to be determined constant. Now, the moments M_k of the expected values of y are defined as follows:

$$\mu_k = \int_{l}^{m} x^{k-1} \cdot y \, dx \ (k = 1, 2, \ldots\ldots, p), \tag{7.42}$$

where

$$l = x_1 - \frac{h}{2} \text{ and } m = x_n + \frac{h}{2} \tag{7.43}$$

The method of moments for curve fitting is based on the principle that the moments of observed values are respectively equal to the moments of expected values of y. i.e..,

$$m_k = \mu_k \ (k = 1, 2 \, .. \, p). \tag{7.44}$$

These equations are called observation equations. We can find the constants involved in the expression of $f(x)$ from the Eq. (7.44). If the expression of $f(x)$ involves two constants, we consider $m_1 = M_1$ and $m_2 = M_2$ and in case of $f(x)$ has 3 constants we consider $m_1 = M_1$, $m_2 = M_2$, and $m_3 = M_3$ to find them.

7.15   FITTING OF A STRAIGHT LINE $y = a + bx$ BY THE METHOD OF MOMENTS

Let (x_i, y_i) $(i = 1, 2, \ldots, n)$ be the set of n observations and x_i s are equidistants with h as the interval of differencing; we wish to fit the data in the form $y = f(x)$, where $f(x) = a + bx$ where a and b are constants.

In this case, the observation equations are

$$m_1 = \mu_1 \tag{7.45}$$

and

$$m_2 = \mu_2 \tag{7.46}$$

In view of (7.41) and (7.42), the Eqs. (7.45) and (7.46) reduce to

$$h \sum_{i=1}^{n} y_i = \int_{l}^{m} (a + bx)\,dx \tag{7.47}$$

and

$$h \sum_{i=1}^{n} x_i y_i = \int_{l}^{m} x(a + bx)\,dx. \tag{7.48}$$

where $l = x - h/2$ and $m = x_n + h/2$ the Eqs. (7.47) and (7.48) are given by

$$a\,(m - l) + \frac{b}{2}\,(m^2 - l^2) = h \sum_{i=1}^{n} y_i \tag{7.49}$$

and

$$\frac{a}{2}\,(m^2 - l^2) + \frac{b}{3}\,(m^3 - l^3) = h \sum_{i=1}^{n} x_i y_i. \tag{7.50}$$

Substituting the values of l, m, h, $\displaystyle\sum_{i=1}^{n} y_i$, and $\displaystyle\sum_{i=1}^{n} x_i y_i$ in (7.49) and (7.50) and solving them, we get the values of a and b. Let these values be a_f and b_f. The required fit to the data by the method of moments in $y = a_f + b_f x$.

SOLVED PROBLEMS

Problem 1: Fit a straight line to the following data by the method of moments:

x	2	3	4	5
y	5	7	9	11

Solution: Here, $x_1 = 2$, $x_2 = 3$, $x_3 = 4$, $x_4 = 5$, $y_1 = 5$, $y_2 = 7$, $y_3 = 9$, $y_4 = 11$, $n = 1$ and $n = 4$. Further $l = x_1 - \dfrac{h}{2} = 2 - \dfrac{1}{2} = 1.5$ and $m = x_n + \dfrac{h}{2} = 5 + \dfrac{1}{2} = 5.5$.

Let the line of fit to the given data by the method of moments be

$$y = a + bx \tag{1}$$

The observation equations are

$$a(m - l) + \frac{b}{2}(m^2 - l^2) = h\sum_{i=1}^{n} y_i \tag{2}$$

$$\frac{a}{2}(m^2 - l^2) + \frac{b}{3}(m^3 - l^3) = h\sum_{i=1}^{n} x_i y_i \tag{3}$$

x_i	y_i	$x_i y_i$
2	5	10
3	7	21
4	9	36
5	11	55
14	32	122

From the above table, we have $\sum x_i = 14$, $\sum y_i = 32$ and $\sum x_i y_i = 122$.

Now, Substituting the values of m, l , h, $\sum y_i$ and $\sum x_i y_i$ in (2) and (3), we have

$$a\{5.5 - 1.5\} + \frac{b}{2}\{(5.5)^2 - (1.5)^2\} = 1 \times 32$$

$$\frac{a}{2}\{(5.5)^2 - (1.5)^2\} + \frac{b}{3}\{(5.5)^3 - (1.5)^3\} = 1 \times 122,$$

i.e.,

$$2a + 7b = 16 \tag{4}$$

$$42a + 163b = 366 \tag{5}$$

Solving the above equations, we get $a = 1.4375$ and $b = 1.8750$.

Thus, the required straight line of fit by the method of least squares is

$$y = (1.4375) + (1.8750)x.$$

Problem 2: Fit a straight line to the data given below using the method of moments.

x	1	2	3	4	5
y	4	8	10	12	16

Solution: Here, $x_1 = 1$, $x_2 = 2$, $x_3 = 3$, $x_4 = 4$, $x_5 = 5$, $y_1 = 4$, $y_2 = 8$, $y_3 = 10$, $y_4 = 12$, $y_5 = 16$, $h = 1$ and n (number of data points) $= 5$. The values of l and m are 0.5 and 5.5 respectively. Let

$$y = a + bx \qquad (1)$$

be the required fit to the given data by method of moments. The observation equations for fitting a line by the method of moments are

$$a(m - l) + \frac{b}{2}(m^2 - l^2) = h\sum_{i=1}^{n} y_i \qquad (2)$$

$$\frac{a}{2}(m^2 - l^2) + \frac{b}{3}(m^3 - l^3) = h\sum_{i=1}^{n} x_i y_i \qquad (3)$$

x_i	y_i	$x_i y_i$
1	4	4
2	8	16
3	10	30
4	12	48
5	16	80
15	50	178

From the above table, $\sum y_i = 50$ and $\sum x_i y_i = 178$ and $\sum x_i = 15$.

Now, substituting m, l, m, $\sum y_i$ and $\sum x_i y_i$ in the Eqs. (2) and (3), we have

$$a\{5.5 - 0.5\} + \frac{b}{2}\{(5.5)^2 - (0.5)^2\} = 50, \qquad (4)$$

and $\qquad \dfrac{a}{2}\{(5.5)^2 - (0.5)^2\} + \dfrac{b}{3}\{(5.5)^3 - (0.5)^3\} = 178, \qquad (5)$

i.e., $\qquad\qquad\qquad\qquad\qquad\qquad a + 3b = 10 \qquad (6)$

$$45a + 163.25 = 53. \qquad (7)$$

Solving (6) and (7), we have $a = 1.0796$ and $b = 1.0796 + 2.9735x$.

Problem 3: Fit a straight line to the given data by (i) the method of moments and (ii) the methods of least sequences.

x	1	3	5	7
y	2.5	2.9	3.6	4.8

Also, find y value at $x = 4$ by each of these methods.

Solution: The following table is constructed to know the $\sum x_i, \sum y_i,$ $\sum x_i y_i$ and $\sum x_i^2$.

x_i	y_i	x_i^2	$x_i y_i$
1	2.5	1	2.5
3	2.9	9	8.7
5	3.6	25	18.0
7	4.8	49	33.6
16	13.8	24	62.8

Thus, $\sum x_i = 16, \sum y_i = 13.8, \sum x_i^2 = 84$ and $\sum x_i y_i = 62.8$.

(i) **To find $y = a + bx$ by the method of moments:**

For the given problem $h = 2$, $l = x_i - \dfrac{h}{2} = 1 - \dfrac{2}{2} = 0$

and $m = x_n + \dfrac{h}{2} = 7 + \dfrac{2}{2} = 8.$

The observation equations for fitting of a straight line $y = a + bx$ by the method of moments are

$$a(m - l) + \frac{b}{2}(m^2 - l^2) = h \sum y_i$$

$$\frac{a}{2}(m^2 - l^2) + \frac{b}{3}(m^3 - l^3) = h \sum x_i y_i.$$

Sustituting the relevant values in the above equations, we get

$$8a + 32b = 27.6$$

$$96a + 512b = 376.8$$

Solving the above equations, we get $a = 7.9125$ and $b = -1.1156$

$$\therefore \qquad y = 7.9125 - 1.1156x$$

is the required straight line by the method of y movements. The value of y at $x = 4$ by this method is $7.9125 - 1.1156 \times 4$ $= 3.45$.

(ii) **To find $y = a + bx$ by the method of least squares:**
Suppose $y = a + bx$ be the straight line to be fitted to the given data by the method of least squares. Then, its normal equations are

$$4a + 16b = 13.8$$

and $$16a + 84b = 62.8$$

Solving the above equations, we get $a = 1.93$ and $b = 0.38$. The required straight line by the method of least squares is $y = 1.93 + 0.38x$. Thus, the value of y at $n = 4$ by the method of least square estimate is 3.45.

7.16 FITTING OF CURVE OF THE FORM $y = a + bx + cx^2$ TO THE GIVEN DATA BY THE METHOD OF MOMENTS

Suppose $x_1, x_2, \ldots\ldots\ldots x_n$ $(x_1 < x_2 < \ldots\ldots < x_n)$ be n equispaced points with the interval of differencing h, and $y_1, y_2, \ldots\ldots\ldots y_n$ be the values of a function at these points respectively. We wish to fit the data in the form $y = a + bx + cx^2$, where a, b and c are to be determined constants. Since it involves three constants, the observation equations are

$$m_1 = \mu_1, \tag{7.51}$$
$$m_2 = \mu_2, \tag{7.52}$$
and $$m_3 = \mu_3, \tag{7.53}$$

In view of (7.41) and (7.42), the Eqs. (7.51), (7.52) and (7.53) are reduced to the following equations:

$$a(m - l) + \frac{b}{2}(m^2 - l^2) + \frac{c}{3}(m^3 - l^3) = h \sum y_i \tag{7.54}$$

$$\frac{a}{2}(m^2 - l^2) + \frac{b}{3}(m^3 - l^3) + \frac{c}{4}(m^4 - l^4) = h \sum x_i y_i \tag{7.55}$$

$$\frac{a}{3}(m^3 - l^3) + \frac{b}{4}(m^4 - l^4) + \frac{c}{5}(m^5 - l^5) = h \sum x_i^2 y_i, \tag{7.56}$$

where $l = x_1 - \dfrac{h}{2}$ and $m = x_n + \dfrac{h}{2}$.

Solving the above equations, we get the values of a, b and c Let these values be a_f, b_f and c_f.

Thus, the required fit of parabola ($y = a + bx + cx^2$) to the given data by the method of moments is $y = a_f + b_f\, x + c_f x^2$.

SOLVED PROBLEMS

Problem 1: Fit a parabola (or second-degree curve) to the data given below using the method of moments:

x	2	4	6	8
$f(x)$	2	3	5	8

Solution: Here, n (the number of data points) $= 4$, $h = 2$, $x_1 = 2$, $x_2 = 4$, $x_3 = 6$, $x_4 = 8$, $y_1 = 2$, $y_2 = 3$, $y_3 = 5$, $y_4 = 8$, $l = x_1 - \dfrac{h}{2} = 2 - \dfrac{2}{2} = 1$ and $m = x_n + \dfrac{h}{2} = 8 + \dfrac{2}{2} = 9$.

The observation equations for fitting a parabola $y = a + bx + cx^2$ to the given data are given by (from (7.54) to (7.56))

$$a(m - l) + \frac{b}{2}(m^2 - l^2) + \frac{c}{3}(m^3 - l^3) = h \sum_{i=1}^{n} y_i \tag{1}$$

$$\frac{a}{2}(m^2 - l^2) + \frac{b}{3}(m^3 - l^3) + \frac{c}{4}(m^4 - l^4) = h \sum_{i=1}^{n} x_i\, y_i \tag{2}$$

$$\frac{a}{3}(m^3 - l^3) + \frac{b}{4}(m^4 - l^4) + \frac{c}{5}(m^5 - l^5) = h \sum_{i=1}^{n} x_i^2\, y_i \tag{3}$$

x_i	y_i	x_i^2	$x_i y_i$	$x_i^2 y_i$
2	2	4	4	8
4	3	16	12	48
6	5	36	30	180
8	8	64	64	512
20	18	120	110	748

From the above table, we have $\sum y_i = 18$, $\sum x_i y_i = 110$ and $\sum x_i^2 y_i = 748$.

Substituting l, m, n, $\sum y_i$, $\sum x_i y_i$ and $\sum x_i^2 y_i$ in (1), (2) and (3), we get

$$8a + \frac{b}{2}\,80 + \frac{c}{3}\,(728) = 2 \times (18)$$

$$\frac{a}{2}\,(80) + \frac{b}{3}\,(728) + \frac{c}{4}\,(6560) = 2 \times (110)$$

$$\frac{a}{3}\,(728) + \frac{b}{4}\,(6560) + \frac{c}{5}\,(59048) = 2 \times (748),$$

i.e.,

$$8a + 40b + 242.66667\, c = 36 \tag{4}$$

$$40a + 242.6667b + 1640\, c = 220 \tag{5}$$

$$242.6667\, a + 1640\, b + 11809.6\, c = 1496. \tag{6}$$

Solving the above equations for a, b and c, we get

$$a = 0.2447,\ b = 0.7148,\ \text{and}\ c = 0.0220$$

Thus required fit of parabola by the method of moments is

$$y = (0.2447) + (0.7148)x + (0.0220)x^2$$

Problem 2: Fit a parabola of the form $y = a + bx + cx^2$ to the data given below:

x	1	2	3	4
y	0.30	0.64	1.32	5.3

Solution: Here, $n = 4$, $x_1 = 1$, $x_2 = 2$, $x_3 = 3$, $x_4 = 4$, $y_1 = 0.30$, $y_2 = 0.64$, $y_3 = 1.32$, $y_4 = 5.3$, $h = 2$, $l = x_1 - \dfrac{h}{2} = 1 - \dfrac{1}{2} = 0.5$ and $m = x_n + \dfrac{h}{2} = 4 + \dfrac{1}{2} = 4.5$.

The observation equations for fitting a parabola

$$y = a + bx + cx^2 \tag{1}$$

to the given data are given by

$$a(m - l) + \frac{b}{2}(m^2 - l^2) + \frac{c}{3}(m^3 - l^3) = h\sum_{i=1}^{n} y_i \qquad (2)$$

$$\frac{a}{2}(m^2 - l^2) + \frac{b}{3}(m^3 - l^3) + \frac{c}{4}(m^4 - l^4) = h\sum_{i=1}^{n} x_i\, y_i \qquad (3)$$

$$\frac{a}{3}(m^3 - l^3) + \frac{b}{4}(m^4 - l^4) + \frac{c}{5}(m^5 - l^5) = h\sum_{i=1}^{n} x_i^2\, y_i \qquad (4)$$

x_i	y_i	x_i^2	$x_i\, y_i$	$x_i^2\, y_i$
1	0.30	1	0.30	0.30
2	0.64	4	1.28	2.56
3	1.32	9	3.96	11.88
4	5.30	16	21.2	84.80
10	7.56	30	26.74	99.54

From the above table, $\sum_{i=1}^{4} y_i = 7.26$, $\sum_{i=1}^{4} x_i y_i = 26.74$ and $\sum_{i=1}^{4} x_i^2 y_i = 99.54$; substituting the values of $n, h, l, m, \sum_{i=1}^{4} y_i, \sum_{i=1}^{4} x_i y_i, \sum_{i=1}^{4} x_i^2 y_i$ in (2), (3), and (4), we get

$$a\{4.5 - 0.5\} + \frac{b}{2}\{(4.5)^2 - (0.5)^2\} + \frac{c}{3}\{(4.5)^3 - (0.5)^3\} = 1 \times (7.56)$$

$$\frac{a}{2}\{(4.5)^2 - (0.5)^2\} + \frac{b}{3}\{(4.5)^3 - (0.5)^3\} + \frac{c}{4}\{(4.5)^4 - (0.5)^4\} = 1 \times (26.74)$$

$$\frac{a}{3}\{(4.5)^3 - (0.5)^3\} + \frac{b}{4}\{(4.5)^4 - (0.5)^4\} + \frac{c}{5}\{(4.5)^5 - (0.5)^5\} = 1 \times (99.54),$$

i.e.,
$$4a + 10b + 30.3333c = 7.56$$
$$10a + 30.3333b + 102.5c = 26.74$$
$$30.3333a + 102.5b + 369.05c = 99.54.$$

Solving the above equations for a, b and c, we get a

$$a = 0.8167, b = -1.1758 \text{ and } c = 0.5292$$

Thus, the required fit of parabola by the method of moments is

$$y = 0.8167 - 1.1758x + 0.5292\ x_i^2.$$

EXERCISE 7(b)

1. The amounts of export in lakhs made by an industry for various years are given below:

Year:	2008	2009	2010	2011
Amount of Export (in Rs. lakhs)	25.4	28.2	29.5	31.6

   Fit the above data in the form $y = a + bx$ by the method of moments and by the method of least squares.

2. Fit the following data to a straight line by the method of moments:

x	2	4	6	8	10
y	7	9	13	14	21

3. Fit the data given below in the form $y = a + bx$.

x	1.2	1.4	1.6	1.8	2.0
y	2	4	5	7	8

4. Fit the data given below in the form $y = a + bx + cx^2$ by the method of moments:

x	0	1	2	3	4
y	1	1.8	1.3	2.5	2.3

5. Fit the data given below in the form $y = a + bx + cx^2$ by the method of moments and also by the method of least squares.

x	1	2	3	4	5
y	2	4	5	8	6

ANSWERS 7(b)

1. (i) $y = -3720.3 + 1.8656\,x$ (ii) $y = -3970.23 + 1.99\,x$

2. $y = 3.0320 + 1.7280\,x$

3. $y = -6.32 + 9.2\,x$

4. $y = 1.0314 + 0.5568\,x - 0.06\,x^2$

5. (i) $y = -1.6764 + 3.9456\,x - 0.04656\,x^2$

   (ii) $y = -1.6 + 3.771429\,x - 0.42857\,x^2.$

8

Approximations of Functions

There are two situations of approximations of functions. One among them is approximating a function $y = f(x)$ *by a* polynomial or some other forms like $y = ab^x$ or $y = a + \dfrac{b}{x}$ for a given empirical data (x_i, y_i) ($i = 1, 2, 3, .., n$) of $y = f(x)$. In chapter 4, we have learnt how a given empirical data $y = f(x)$ is approximated with a polynomial using Newton's forward difference interpolation formula, Newton's divided difference formula of interpolation and spline functions. Further, in Chapter 7, we have learnt how a given data (x_i, y_i) ($i = 1, 2, 3, .., n$) of $y = f(x)$ is approximated with a polynomial form $y = a + bx$, $y = a + bx + cx^2$ or a function of the form $y = ab^x$ or $y = a + \dfrac{b}{x}$ using the method of least squares. The other situation is approximating a continuous function on $[a, b]$ that matches some other types of functions with a minimum error. For example, an exponential function $y = e^x$ is approximated with a function of the form $y = c_0 + c_1 x$ with a minimum error on an interval $[0, 2]$.

In this chapter, we discuss the approximation of a continuous function of $y = f(x)$ on $[a, b]$ with an approximating function $\phi(x)$ on $[a, b]$ with a minimum error. Usually, the approximating functions $\phi(x)$ are polynomial functions or trigonometric functions or rational functions. However, from the application point of view, the polynomial functions are mostly used. The existence of a polynomial function $P_n(x)$ that approximates any continuous functions $f(x)$ on $[a, b]$ is guaranteed by the Weierstrass approximation theorem (Section 0.22).

8.1   DEFINITIONS

Let $f(x)$ be a continuous function on $[a, b]$ and $\int_a^b |f(x)|^p dx$ $(p \geq 1)$ exists. Then

(i) $\|f\| = \left[\int_a^b w(x)|f(x)|^p dx \right]^{1/p}$ (where $w(x)$ is the weight function)

is called L_p-norm.

(ii) $\|f\| = \left[\int_a^b w(x) f^2(x) dx \right]^{1/2}$ is called L_2-norm or Euclidean norm.

(iii) $\|f\| = \max_{a \leq x \leq b} |f(x)|$ is called L_∞-norm.

An approximation in which the L_2-norm is minimised is called least square approximation.

8.2   CO-ORDINATE FUNCTIONS

Suppose, we wish to approximate a continuous function $y = f(x)$ on $[a, b]$ in the form

$$f(x) \approx c_0 \phi_0(x) + c_1 \phi_1(x) + \ldots + c_n \phi_n(x), \tag{8.1}$$

where $\phi_i(x)$ $(i = 0, 1, 2, .., n)$ are linearly independent functions and c_i $(i = 0, 1, 2, .., n)$ are to be determined constants/parameters. The functions $\phi_0(x)$, $\phi_1(x)$, $\ldots$ $\phi_x(x)$ are called coordinate functions.

The error of the approximation is defined as

$$E(f) = \left\| f(x) - \sum_{i=0}^{n} c_i \phi_i(x) \right\|, \tag{8.2}$$

where $\|\cdot\|$ is a well-defined norm.

8.3   NORMAL EQUATIONS

For the least square approximations, the norm in 8.2 is L_2-norm or Euclidean norm. Therefore, the problem of approximation is to determine the parameters $c_0, c_1, c_2, \ldots c_n$ such that

$$I(c_0, c_1, .., c_n) = \int_a^b w(x) \left\{ f(x) - \sum_{i=0}^{n} c_i \phi_i(x) \right\}^2 dx \tag{8.3}$$

is minimum, where $w(x) > 0$ is the weight function.

The necessary conditions for $I(c_0, c_1, c_2, \ldots c_n)$ (given by 8.3) to have a minimum value are

$$\frac{\partial I}{\partial c_i} = 0 \ (i = 0, 1, 2, .., n) \tag{8.4}$$

i.e.,
$$\frac{\partial I}{\partial c_i} = \int_a^b w(x)\left\{ f(x) - \sum_{i=0}^{n} c_i \phi_i(x) \right\} \phi_j(x) = 0, \tag{8.5}$$

where $j = 0, 1, 2, .., n$. Eq. (8.5) gives a system of $(n + 1)$ linear equations in $(n + 1)$ unknown parameters c_0, c_1, .. and c_n. These equations are called normal equations. If $w(x) = 1$, then Eq. (8.5) reduces to

$$\frac{\partial I}{\partial c_i}(c_0, c_1, .., c_n) = \int_a^b \left\{ f(x) - \sum_{i=0}^{n} c_i \phi_i(x) \right\} \phi_j(x)dx = 0 \tag{8.6}$$

8.4 LEAST SQUARE APPROXIMATION OF A FUNCTION WITH A POLYNOMIAL FUNCTION

In case of approximation of function $y = f(x)$ with a polynomial function $c_0\phi_0(x) + c_1\phi_1(x) + \ldots + c_n\phi_n(x)$, usually the coordinate functions $\phi_i(x)$ are taken as $\phi_i(x) = x^i$ $(i = 0, 1, 2, .., n)$ and weight function $w(x) = 1$. Hence, Eq. (8.3) reduces to

$$I(c_0, c_1, .., c_n) = \int_a^b \left[f(x) - \sum_{i=0}^{n} c_i x^i \right]^2 dx \tag{8.7}$$

and the normal Eq. (8.6) is reduced to

$$\frac{\partial I}{\partial c_i} = \int_a^b \left\{ f(x) - \sum_{i=0}^{n} c_i x_i \right\} x^j = 0, \tag{8.8}$$

where $j = 0, 1, 2, .., n$.

8.5 REMARK

When $w(x) \neq 1$, $f(x) \approx c_0\phi_0(x) + c_1\phi_1(x) + \ldots + c_n\phi_n(x)$ and $\phi_i(x) = x^i$ $(i = 0, 1, 2, .., n)$, the normal equations are

$$\int_a^b w(x)\left\{ f(x) - \sum_{i=0}^{n} c_i x^i \right\} x^j = 0 \tag{8.9}$$

Note:

(i) When a continuous function $f(x)$ on $[a, b]$ is to be approximated by a linear polynomial function $c_0 + c_1 x$ (i.e., $f(x) \approx c_0 + c_1 x$ on $[a, b]$) by the least square approximation, the normal equations are

$$\int_a^b \left[f(x) - (c_0 + c_1 x) \right](1)dx = 0 \tag{8.10a}$$

and

$$\int_a^b \left[f(x) - (c_0 + c_1 x) \right](x)dx = 0 \tag{8.10b}$$

(ii) When a continuous function $f(x)$ to be approximated by a second degree polynomial function $c_0 + c_1 x + c_2 x^2$ by the least square approximation, the normal equations are

$$\int_a^b \left[f(x) - (c_0 + c_1 x + c_2 x^2) \right](1)dx = 0 \tag{8.11a}$$

$$\int_a^b \left[f(x) - (c_0 + c_1 x + c_2 x^2) \right](x)dx = 0 \tag{8.11b}$$

and

$$\int_a^b \left[f(x) - (c_0 + c_1 x + c_2 x^2) \right](x^2)dx = 0 \tag{8.11c}$$

8.6   PROCEDURE TO FIND THE APPROXIMATE POLYNOMIAL FUNCTION $y = f(x)$ ON $[a, b]$ BY THE METHOD OF LEAST SQUARE APPROXIMATION WITH $w(x) = 1$

Step 1: Note down the function $f(x)$ which is to be approximated with a polynomial function and the interval $[a, b]$.

Step 2: To approximate $f(x)$ with k^{th} degree polynomial function, suppose as

$$P_k(x) = c_0 + c_1 x + \ldots + c_k x^k$$

i.e.,
$$f(x) \approx c_0 + c_1 x + \ldots + c_k x^k$$

Step 3: Form the equation

$$I(c_0, c_1, .., c_n) = \int_a^b \left[f(x) - \left\{ c_0 + c_1 x + ... + c_k x^k \right\} \right]^2 dx \qquad (1)$$

Where $f(x)$, a, b are noted in Step 1.

Step 4: Write the normal equations

$$\frac{\partial I}{\partial c_0} = \int_a^b \left[f(x) - \left(c_0 + c_1 x + ... + c_k x^k \right) \right](1) dx = 0 \qquad (A_0)$$

$$\frac{\partial I}{\partial c_1} = \int_a^b \left[f(x) - \left(c_0 + c_1 x + ... + c_k x^k \right) \right](x) dx = 0 \qquad (A_1)$$

$$\frac{\partial I}{\partial c_2} = \int_a^b \left[f(x) - \left(c_0 + c_1 x + ... + c_k x^k \right) \right](x^2) dx = 0 \qquad (A_2)$$

$$..........................$$

$$\frac{\partial I}{\partial c_k} = \int_a^b \left[f(x) - \left(c_0 + c_1 x + ... + c_k x^k \right) \right](x^k) dx = 0 \qquad (A_k)$$

The equations (A_0), (A_1), .., (A_k) give a system of linear equations in c_0, c_1, ... c_k. Note that $f(x)$, a, b are known as function and real numbers respectively.

Step 5: Solve the equations obtained in Step 4 for c_0, c_1, .., c_k, let these values be $c_0{}^f$, $c_1{}^f$,..., $c_k{}^f$.

Step 6: The required approximation of $f(x)$ is given by
$$f(x) \approx c_0{}^f + c_1{}^f x + c_2{}^f x^2 ..... + c_k{}^f x^k.$$

SOLVED PROBLEMS

Problem 1: Obtain a linear polynomial function that is an approximation of the function $f(x) = \sqrt{x}$ on $[0, 1]$ using the least square approximation.

Solution: Here $f(x) = \sqrt{x}$ and the interval is $[0, 1]$. Therefore, $a = 0$ and $b = 1$. Let the required linear polynomial function on $[0, 1]$ be

$$P_1(x) = c_0 + c_1 x \qquad (1)$$

The normal equations $\dfrac{\partial I}{\partial c_0} = 0$ and $\dfrac{\partial I}{\partial c_1} = 0$ are

$$\int_0^1 \left\{ \sqrt{x} - (c_0 + c_1 x) \right\} (1)\, dx = 0 \tag{2}$$

and

$$\int \left\{ \sqrt{x} - (c_0 + c_1 x) \right\} (x)\, dx = 0 \tag{3}$$

where

$$I(c_0, c_1) = \int_0^1 \left\{ x^{\frac{1}{2}} - (c_0 + c_1 x) \right\}^2 dx \tag{4}$$

Evaluating the integral of Eq. (2), we have

$$\left[\frac{2}{3} x^{\frac{3}{2}} - \left(c_0 x + c_1 \frac{x^2}{2} \right) \right]_0^1 = 0$$

i.e.,
$$c_0 + \frac{c_1}{2} = \frac{2}{3} \tag{5}$$

Similarly, from Eq. (3), we have

$$\frac{c_0}{2} + \frac{c_1}{3} = \frac{2}{5} \tag{6}$$

Solving (5) and (6) for c_0 and c_1, we obtain $c_0 = \dfrac{4}{15}$ and $c_1 = \dfrac{4}{5}$. Thus, the required linear polynomial which is an approximation of $f(x) = \sqrt{x}$ on [0, 1] is $\dfrac{4}{15}(1 + 3x)$.

i.e.,
$$f(x) = \sqrt{x} \approx \frac{4}{15}(1 + 3x)$$

Problem 2: Obtain a linear polynomial function so that it is an approximation of the function $f(x) = 20x^3$ on [0, 1] using the least square approximation.

Solution: Here, $f(x) = 20x^3$ and the interval is [0, 1]. Therefore, $a = 0$ and $b = 1$. Let the required linear polynomial function be

$$P_1(x) = c_0 + c_1 x \tag{1}$$

Then,

$$I(c_0, c_1) = \int_0^1 \left\{ 20x^3 - (c_0 + c_1 x) \right\}^2 dx \tag{2}$$

The normal equations $\dfrac{\partial I}{\partial c_0} = 0$ and $\dfrac{\partial I}{\partial c_1} = 0$ are

$$\int_0^1 \left\{ 20x^3 - (c_0 + c_1 x) \right\}(1)dx = 0 \tag{3}$$

$$\int_0^1 \left\{ 20x^3 - (c_0 + c_1 x) \right\}(x)dx = 0 \tag{4}$$

From Eq. (3), we have

$$\left[20\frac{x^4}{4} - c_0 x - c_1 \frac{x^2}{2} \right]_0^1 = 0$$

i.e.,
$$5 - c_0 - \frac{c_1}{2} = 0$$

i.e.,
$$c_0 + \frac{c_1}{2} = 5 \tag{5}$$

Similarly, from Eq. (4), we obtain

$$\frac{c_0}{2} + \frac{c_1}{3} = 4 \tag{6}$$

Solving (5) and (6), we get $c_0 = -4$ and $c_0 = 18$. Thus, the required linear polynomial function is $-4 + 18x$. i.e.,

$$f(x) = 20x^3 \approx -4 + 18x \text{ on } [0, 1].$$

Problem 3: Find the least square line for $f(x) = x^3$ on the interval $[-1, 1]$.

Solution: Here $f(x) = x^3$ and the interval is $[-1, 1]$. Therefore, $a = -1$ and $b = 1$. Let the required least square line for $f(x) = x^3$ on $[-1, 1]$ be

$$P_1(x) = c_0 + c_1 x \tag{1}$$

From (8.10a) and (8.10b), the normal equations are

$$\int_{-1}^1 \left\{ x^3 - (c_0 + c_1 x) \right\}(1)dx = 0 \tag{2}$$

and

$$\int_{-1}^{1}\{x^3 - (c_0 + c_1 x)\}(x)dx = 0 \tag{3}$$

Of course, these normal equations can be obtained from $\dfrac{\partial I}{\partial c_0} = 0$ and $\dfrac{\partial I}{\partial c_1} = 0$, where

$$I(c_0, c_1) = \int_{-1}^{1}\left[x^3 - (c_0 + c_1 x)\right]^2 dx.$$

From Eq. (2), we have

$$\left[\frac{x^4}{4} - c_0 x - c_1 \frac{x^2}{2}\right]_{1}^{-1} = 0$$

$$-2c_0 = 0$$

i.e., $$c_0 = 0$$

Similarly, from Eq. (3), we have $c_1 = \dfrac{3}{5}$. Thus, the required least square

line is $\dfrac{3}{5}x$. i.e., $x^3 \approx \dfrac{3}{5}x$ on $[-1, 1]$.

Problem 4: Obtain the linear polynomial function for $f(x) = 2x^{1/2}$ on $[0, 2]$ so that it is an approximation function of $f(x)$ on $[0, 2]$.

Solution: Here, $f(x) = 2x^{1/2}$, a = 0 and b = 2. Let the required polynomial function be

$$P_1(x) = c_0 + c_1 x \tag{1}$$

From (8.10a) and (8.10b), the normal equations are

$$\int_{0}^{2}\left[2\sqrt{x} - (c_0 + c_1 x)\right](1)dx = 0 \tag{2}$$

and

$$\int_{0}^{2}\left[2\sqrt{x} - (c_0 + c_1 x)\right](x)dx = 0 \tag{3}$$

From Eq. (2), we have

$$\left[2 \cdot \frac{2}{3} \cdot x^{3/2} - c_0 x - c_1 \frac{x^2}{2}\right]_{0}^{2} = 0$$

i.e.,
$$c_0 + c_1 = \frac{4\sqrt{2}}{3} \tag{4}$$

Similarly, from Eq. (3), we have

$$\left[\frac{4}{5}\cdot x^{5/2} - \frac{c_0 x^2}{2} - \frac{c_1 x^3}{3} \right]_0^2 = 0$$

i.e.,
$$2c_0 + \frac{8}{3}c_1 = \frac{16\sqrt{2}}{5} \tag{5}$$

Solving Eqs. (4) and (5) for c_0 and c_1, we get $c_0 = \dfrac{8\sqrt{2}}{15}$ and $c_1 = \dfrac{4\sqrt{2}}{5}$.

Thus, the required polynomial function is $\dfrac{8\sqrt{2}}{15} + \dfrac{4\sqrt{2}}{5}x = \dfrac{4\sqrt{2}}{15}(2+3x)$,

i.e., $f(x) \approx \dfrac{4\sqrt{2}}{15}(2+3x)$.

Problem 5: Obtain the least square polynomial function of second degree for $f(x) = \sqrt{x}$ on $[0, 1]$ using the method of least square approximation.

Solution: Here, $f(x) = \sqrt{x}$, $a = 0$ and $b = 1$. Let the required second-degree polynomial function be

$$P_2(x) = c_0 + c_1 x + c_2 x^2 \tag{1}$$

By (8.11a), (8.11b) and (8.11c), the normal equations of least square approximation are

$$\int_0^1 \left[\sqrt{x} - (c_0 + c_1 x + c_2 x^2) \right](1)dx = 0 \tag{2}$$

$$\int_0^1 \left[\sqrt{x} - (c_0 + c_1 x + c_2 x^2) \right](x)dx = 0 \tag{3}$$

and

$$\int_0^1 \left\{ \sqrt{x} - (c_0 + c_1 x + c_2 x^2) \right\}(x^2)dx = 0 \tag{4}$$

Evaluating the integrals of Eqs. (2), (3) and (4), we obtain

$$c_0 + \frac{c_1}{2} + \frac{c_2}{3} = \frac{2}{3} \tag{5}$$

$$\frac{c_0}{2} + \frac{c_1}{3} + \frac{c_2}{4} = \frac{2}{5} \qquad (6)$$

$$\frac{c_0}{3} + \frac{c_1}{4} + \frac{c_2}{5} = \frac{2}{7} \qquad (7)$$

Solving Eqs. (5), (6) and (7) for c_0, c_1 and c_2, we get

$$c_0 = \frac{6}{35}, c_1 = \frac{48}{35} \text{ and } c_2 = \frac{-20}{35}.$$

Thus, the required second degree polynomial function is

$\frac{6}{35} + \frac{48}{35}x - \frac{20}{35}x^2$. i.e.,

$$f(x) \approx \frac{2}{35}(3 + 24x - 10x^2).$$

To facilitate further discussion on approximation of functions with Chebyshev polynomials by the least square approximation, we give the definitions of the orthogonal functions, Chebyseve polynomials and their properties in Sections (8.7 and 8.8) The discussion on approximations of functions restored from the section (8.9) onwards.

8.7   DEFINITIONS

Two functions $\phi_1(x)$ and $\phi_2(x)$ are said to be orthogonal on an interval $[a, b]$ with respect to weight function $w(x) > 0$ if

$$\int_a^b w(x)\phi_1(x)\phi_2(x) = 0 \qquad (8.12)$$

Further, a set of functions $\{\phi_0(x), \phi_1(x), \phi_2(x), \ldots\}$ is said to be orthogonal on $[a, b]$ with respect to weight function $w(x) > 0$ if

$$\int_a^b w(x)\phi_i(x)\phi_j(x)dx = 0 \text{ if } i \neq j \qquad (8.13)$$

For example, $\phi_1(x) = x$ and $\phi_2(x) = \dfrac{3x^2 - 1}{2}$ are orthogonal functions on $[-1, 1]$ with respect to weight function $w(x) = 1$ since $\displaystyle\int_{-1}^{1} x\left(\frac{3x^2 - 1}{2}\right)dx = 0$ as the integrand is an odd function and the lower and upper limits of definite integral are equal in magnitude but opposite in signs. Another example is the set $\{\sin x, \sin 2x, \sin 3x, \ldots\}$ is orthogonal on $[0, 2\pi]$ since

$$\int_{0}^{2\pi} \sin mx \sin nx = 0 \qquad \text{for } m \neq n.$$

i.e., any two distinct function of the set $\{\sin x, \sin 2x, \sin 3x, \ldots\}$ are orthogonal on $[0, 2\pi]$.

8.8 CHEBYSHEV POLYNOMIALS AND THEIR PROPERTIES

The Chebyshev polynomials are denoted by $T_n(x)$ and they are defined as

$$T_n(x) = \cos\,(n\cos^{-1} x), \tag{8.14}$$

where n is an integer and $-1 \leq x \leq 1$. Clearly, $T_n(x) = T_{-n}(x)$, $T_0(x) = 1$ and $T_1(x) = x$.

Suppose $x = \cos \theta$ then $T_n(x) = \cos n\theta$, where $\theta = \cos^{-1} x$. Now consider the trigonometric identity

$$\cos\,(n+1)\,\theta + \cos\,(n-1)\,\theta = 2\cos\theta \,.\cos n\theta$$

Hence, $\qquad\qquad\qquad T_{n+1}(x) + T_{n-1}(x) = 2xT_n(x)$

It can be used to find $T_{n+1}(x)$ using known $T_n(x)$ and $T_{n-1}(x)$ functions. Thus, we have the following recurrence relation:

$$T_{n+1}(x) = 2xT_n(x) - T_{n-1}(x), \tag{8.15}$$

with $T_0(x) = 1$ and $T_1(x) = x$.

From the recurrence relation (8.15), the Chebyshev polynomials $T_2(x), T_3(x), T_4(x), \ldots$ etc are as follows:

$$T_2(x) = 2x^2 - 1; \ T_3(x) = 4x^3 - 3x; \ T_4(x) = 8x^4 - 8x^2 + 1;$$

$$T_5(x) = 16x^5 - 20x^3 + 5x; \ T_6(x) = 32x^6 - 48x^4 + 18\,x^2 - 1 \tag{8.16}$$

$$T_7(x) = 64x^7 - 112x^5 + 26\,x^3 - 7x.$$

The graphs of $T_0(x), T_1(x), T_2(x)$ and $T_3(x)$ are shown in Fig. 8.1.

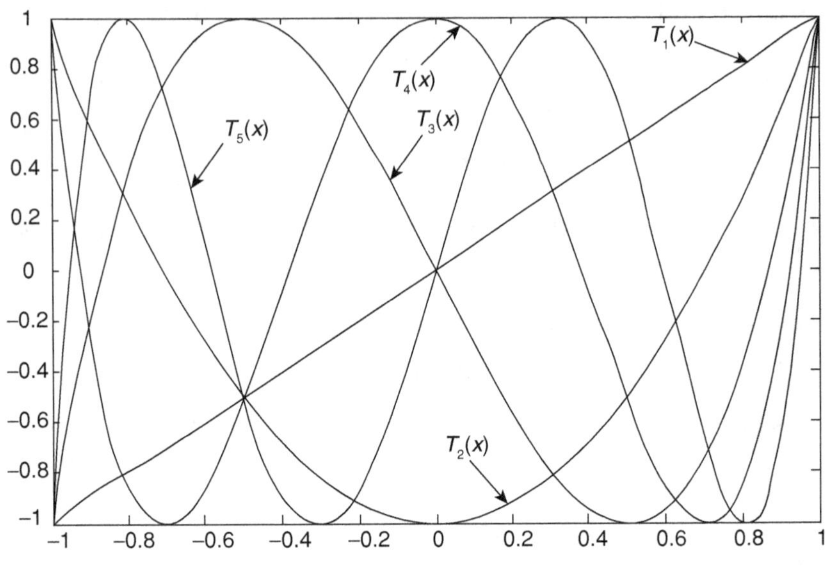

Figure 8.1

The various powers of x can be expressed as a linear combination of Chebyshev polynomials using (8.16), $T_0(x) = 1$ and $T_1(x) = x$. Some powers of x in terms of Chebyshev polynomials are given below:

$$x^0 = 1 = T_0(x)$$

$$x = T_1(x)$$

$$x^2 = \frac{1}{2}(T_0 + T_2)$$

$$x^3 = \frac{1}{4}(3T_1 + T_3)$$

$$x^4 = \frac{1}{8}(3T_0 + 4T_2 + T_4) \qquad (8.17)$$

$$x^5 = \frac{1}{16}(10T_1 + 5T_3 + T_5)$$

$$x^6 = \frac{1}{32}(10T_0 + 15T_2 + 6T_4 + T_6)$$

$$x^7 = \frac{1}{64}(35T_1 + 21T_3 + 7T_5 + T_7)$$

The following are some of the properties of Chebyshev polynomials:

   (i) $T_n(x)$ is a polynomial of degree n.

  (ii) If n is even, then $T_n(x)$ is a polynomial in x of even degree and it contains only even powers of x.

 (iii) If n is odd, then $T_n(x)$ is a polynomial function in x of odd degree and it contains only odd powers of x.

 (iv) $T_n(x)$ has n simple zeros and they are given by

$$x_k = \cos\left\{\frac{(2k-1)\pi}{2n}\right\}, \quad k = 1, 2, 3, \ldots, n \text{ on the domain of}$$

Chebyshev polynomials.

  (v) $\left|T_n(x)\right| \le 1 \quad \forall\ x \in [-1, 1]$ and for every n.

 (vi) $T_n(x)$ attains extreme values at $(n + 1)$ points and they are given by

$$x_k = \cos\left(\frac{k\pi}{n}\right), \quad k = 0, 1, 2, \ldots, n.$$

and the value of extreme at x_k is $(-1)^k$.

 (vii) The set of Chebyshev polynomials $\{T_0(x),\ T_1(x),\ T_2(x),\ \ldots\}$ are orthogonal on $[-1,\ 1]$ with respect to weight function

$$w(x) = \frac{1}{\sqrt{1-x^2}}, \text{ i.e.,}$$

$$\int_{-1}^{1} \frac{1}{\sqrt{1-x^2}} T_m(x) T_n(x)\, dx = 0 \qquad \text{if } m \neq n \qquad (8.18)$$

(viii) $\displaystyle\int_{-1}^{1} \frac{T_i(x) . T_j(x)}{\sqrt{1-x^2}} dx = \begin{cases} 0 & \text{if } i \neq j \\ \pi & \text{if } i = j = 0 \\ \dfrac{\pi}{2} & \text{if } i = j \neq 0 \end{cases} \qquad (8.19)$

8.9 APPROXIMATION OF FUNCTIONS WITH CHEBYSHEV POLYNOMIALS BY THE LEAST SQUARES APPROXIMATION

Let $f(x)$ be a continuous function defined on $[-1, 1]$.

Suppose, we wish to approximate the function $f(x)$ with Chebyshev polynomials of the following form

$$f(x) \approx c_0 T_0(x) + c_1 T_1(x) + \ldots + c_n T_n(x) \tag{8.20}$$

where $c_0, c_1, \ldots, c_n$ are to be determined parameters. From Eq. (8.3), we have

$$I(c_0, c_1, \ldots, c_n) = \int_{-1}^{1} \frac{1}{\sqrt{1-x^2}} \left[f(x) - \sum_{i=0}^{n} c_i T_i(x) \right]^2 dx \tag{8.21}$$

Note that the weight function $w(x) = \dfrac{1}{\sqrt{1-x^2}}$. Now, we have to choose $c_0, c_1, \ldots, c_n$ such that the error function with L_2-norm is minimum. The necessary conditions for $I(c_0, c_1, \ldots, c_n)$ is minimum are $\dfrac{\partial I}{\partial c_0} = \dfrac{\partial I}{\partial c_1} = \ldots = \dfrac{\partial I}{\partial c_n} = 0$, which are the normal equations. Using (8.21), we have the following normal equations:

$$\int_{-1}^{1} \frac{1}{\sqrt{1-x^2}} \left[f(x) - \sum_{i=0}^{n} c_i T_i(x) \right] T_j(x) dx = 0 \tag{8.22}$$

$$(j = 0, 1, 2, \ldots, n)$$

It can be shown that

$$c_0 = \frac{1}{\pi} \int_{-1}^{1} \frac{f(x) T_0(x)}{\sqrt{1-x^2}} dx \tag{8.23}$$

and

$$c_k = \frac{2}{\pi} \int_{-1}^{1} \frac{f(x) T_k(x)}{\sqrt{1-x^2}} dx \qquad \text{for } k \neq 0 \tag{8.24}$$

Thus $f(x) \approx c_0 T_0(x) + c_1 T_1(x) + \ldots + c_n T_n(x)$, where $c_0, c_1, \ldots c_n$ are given by (8.23) and (8.24).

Note:

(i) To have an approximation of a continuous function $f(x)$ on $[a, b]$, the function $f(x)$ has to be transformed to a function $F(t)$ so that its domain is $[-1, 1]$. The reason for this is as c_0 and c_k given by (8.23) and (8.24) contains the lower limit and upper limit of the definite integral are -1 and 1 respectively.

(ii) A function $f(x)$ defined on $[a, b]$ can be converted in to a function with domain $[-1, 1]$ by the transformation

$$t = \frac{2}{b-a} x - \left(\frac{b+a}{b-a} \right), \quad (b \neq a).$$

8.10 SOME USEFUL RESULTS

(i) $\displaystyle\int_{-1}^{1}\frac{x^{k}}{\sqrt{1-x^{2}}}dx=\int_{0}^{\pi}\cos^{k}\theta\,d\theta$

(ii) $\displaystyle\int_{-1}^{1}\frac{x^{k}}{\sqrt{1-x^{2}}}dx=\begin{cases}\pi & \text{if } k=0\\ 0 & \text{if } k \text{ is odd}\\ \dfrac{\pi}{2^{k}}\,{}^{k}C_{\frac{k}{2}} & \text{if } k \text{ is even}\end{cases}$ (8.25)

This formula is useful in the approximation of functions with Chebyshev polynomials when a polynomial function is to be approximated by a function involving Chebyshev polynomials.

SOLVED PROBLEMS

Problem 1: A function $f(x)=x^{2}$ is defined on [0, 1]. Transform the function $f(x)$ into a function $g(\cdot)$ so that the domain of $g(\cdot)$ is [−1, 1].

Solution: By note 8.9 (ii), the required transformation is $t = 2x - 1$ or $x = \dfrac{(t+1)}{2}$. Clearly, when $x = 0$, $t = -1$ and when $x = 1$, $t = 1$. The transformed function is $g(t)=\left(\dfrac{t+1}{2}\right)^{2}=\dfrac{t^{2}+2t+1}{4}$ and its domain is [−1, 1].

Problem 2: Express $P_{3}(x) = x^{3} - 4x^{2} + 3$ into a function involving Chebyshev polynomials.

Solution: Since $1 = T_{0}(x)$, $x^{2}=\dfrac{1}{2}(T_{0}+T_{2})$ and $x^{3}=\dfrac{1}{4}(3T_{1}+T_{3})$, we have

$$P_{3}(x)=\frac{1}{4}(3T_{1}+T_{3})-4\cdot\frac{1}{2}(T_{0}+T_{2})+3T_{0}(x)$$

i.e.,

$$P_{3}(x)=\frac{1}{4}\left[(3T_{1}+T_{3})-8(T_{0}+T_{2})+12T_{0}\right]$$

i.e.,
$$P_3(x) = \frac{1}{4}\left[T_3 - 8T_2 + 3T_1 + 4T_0\right]$$

Problem 3: Using the Chebyshev polynomials obtain the least square approximation of second degree for $f(x) = x^4$ on $[-1, 1]$.

Solution: Here $f(x) = x^4$ and the interval is $[-1, 1]$. Suppose

$$f(x) \approx c_0 T_0(x) + c_1 T_1(x) + c_2 T_2(x)$$

Then, by (8.23) and (8.24), the values of c_0, c_1 and c_2 can be computed. The values of c_0, c_1 and c_2 are given by

$$c_0 = \frac{1}{\pi}\int_{-1}^{1} \frac{x^4 T_0(x)}{\sqrt{1-x^2}}\, dx$$

$$c_1 = \frac{2}{\pi}\int_{-1}^{1} \frac{x^4 T_1(x)}{\sqrt{1-x^2}}\, dx$$

and
$$c_2 = \frac{2}{\pi}\int_{-1}^{1} \frac{x^4 T_2(x)}{\sqrt{1-x^2}}\, dx$$

Since $T_0(x) = 1$, $T_1(x) = x$, and $T_2(x) = 2x^2 - 1$, we have

$$c_0 = \frac{1}{\pi}\int_{-1}^{1} \frac{x^4}{\sqrt{1-x^2}}\, dx$$

$$c_1 = \frac{2}{\pi}\int_{-1}^{1} \frac{x^4 (x)}{\sqrt{1-x^2}}\, dx$$

and
$$c_2 = \frac{2}{\pi}\int_{-1}^{1} \frac{x^4 \{2x^2 - 1\}}{\sqrt{1-x^2}}\, dx$$

Using Eq. (8.28), we have

$$c_0 = \frac{1}{\pi}\cdot\frac{\pi}{2^4}\,^4C_2 = \frac{6}{16} = \tfrac{3}{8}$$

$$c_1 = 0 \qquad\qquad\qquad (\because \text{integral is odd})$$

and

$$c_2 = \frac{2}{\pi} \int_{-1}^{1} \frac{2x^6 - x^4}{\sqrt{1-x^2}} dx$$

$$= \frac{2}{\pi} \left[2 \cdot \int_{-1}^{1} \frac{x^6}{\sqrt{1-x^2}} dx - \int_{-1}^{1} \frac{x^4}{\sqrt{1-x^2}} dx \right]$$

$$= \frac{2}{\pi} \left[2 \cdot \frac{\pi}{2^6} \cdot 6C_3 - \frac{\pi}{2^4} \cdot 4C_2 \right]$$

$$= \frac{2}{\pi} \left[\frac{2}{64} \cdot \frac{6 \cdot 5 \cdot 4}{1 \cdot 2 \cdot 3} - \frac{6}{16} \right] \pi$$

$$= 2 \left[\frac{5}{8} - \frac{3}{8} \right] = \frac{1}{2}$$

Thus, $c_0 = \dfrac{3}{8}$, $c_1 = 0$ and $c_2 = \dfrac{1}{2}$. Hence, the required approximation of

$f(x)$ is $\dfrac{3}{8} T_0 + \dfrac{1}{2} T_2$. i.e.,

$$f(x) \approx \frac{3}{8} T_0(x) + \frac{1}{2} T_2(x)$$

Alternate solution:

Suppose $f(x) \approx c_0 T_0 + c_1 T_1 + c_2 T_2$. Then, the error function with L_2-norm is

$$I(c_0, c_1, c_2) = \int_{-1}^{1} \frac{1}{\sqrt{1-x^2}} \left\{ x^4 - (c_0 T_0 + c_1 T_1 + c_2 T_2) \right\}^2 dx,$$

where $w(x) = \dfrac{1}{\sqrt{1-x^2}}$ is the weight function. The necessary condi-

tions for $I(c_0, c_1, c_2)$ to be minimum are $\dfrac{\partial I}{\partial c_0} = \dfrac{\partial I}{\partial c_1} = \dfrac{\partial I}{\partial c_2} = 0$. Therefore,

$$\frac{\partial I}{\partial c_0} = -2 \int_{-1}^{1} \frac{\left\{ x^4 - (c_0 T_0 + c_1 T_1 + c_2 T_2) \right\}}{\sqrt{1-x^2}} T_0(x) dx = 0$$

$$\frac{\partial I}{\partial c_1} = -2 \int_{-1}^{1} \frac{\left\{ x^4 - (c_0 T_0 + c_1 T_1 + c_2 T_2) \right\}}{\sqrt{1-x^2}} T_1(x) dx = 0$$

$$\frac{\partial I}{\partial c_2} = -2\int_{-1}^{1} \frac{\left\{x^4 - \left(c_0 T_0 + c_1 T_1 + c_2 T_2\right)\right\}}{\sqrt{1-x^2}} T_2(x)\,dx = 0.$$

The above three equations are reduced to the following:

$$c_0 \int_{-1}^{1} \frac{T_0^2\,dx}{\sqrt{1-x^2}} + c_1 \int_{-1}^{1} \frac{T_0 T_1}{\sqrt{1-x^2}}\,dx + c_2 \int_{-1}^{1} \frac{T_0 T_2}{\sqrt{1-x^2}}\,dx = \int_{-1}^{1} \frac{x^4 T_0}{\sqrt{1-x^2}}\,dx \qquad (1)$$

$$c_0 \int_{-1}^{1} \frac{T_0 T_1}{\sqrt{1-x^2}}\,dx + c_1 \int_{-1}^{1} \frac{T_1^2\,dx}{\sqrt{1-x^2}} + c_2 \int_{-1}^{1} \frac{T_0 T_3}{\sqrt{1-x^2}}\,dx = \int_{-1}^{1} \frac{x^4 T_1}{\sqrt{1-x^2}}\,dx \qquad (2)$$

$$c_0 \int_{-1}^{1} \frac{T_0 T_2}{\sqrt{1-x^2}}\,dx + c_1 \int_{-1}^{1} \frac{T_1 T_2}{\sqrt{1-x^2}}\,dx + c_2 \int_{-1}^{1} \frac{T_2^2\,dx}{\sqrt{1-x^2}} = \int_{-1}^{1} \frac{x^4 T_2}{\sqrt{1-x^2}}\,dx \qquad (3)$$

We know that

$$\int_{-1}^{1} \frac{T_m T_n}{\sqrt{1-x^2}}\,dx = \begin{cases} 0 & \text{if } m \neq n \\[2mm] \dfrac{\pi}{2} & \text{if } m = n \neq 0 \\[2mm] \pi & \text{if } m = n = 0 \end{cases} \qquad (4)$$

In view of (4), Eqs. (1), (2) and (3) reduces to

$$c_0 \pi + 0 + 0 = \int_{-1}^{1} \frac{x^4 T_0}{\sqrt{1-x^2}}\,dx$$

$$0 + c_1 \frac{\pi}{2} + 0 = \int_{-1}^{1} \frac{x^4 T_1}{\sqrt{1-x^2}}\,dx$$

$$0 + 0 + c_2 \frac{\pi}{2} = \int_{-1}^{1} \frac{x^4 T_2}{\sqrt{1-x^2}}\,dx$$

Thus,

$$c_0 = \frac{1}{\pi} \int_{-1}^{1} \frac{x^4 T_0}{\sqrt{1-x^2}}\,dx \qquad (5)$$

$$c_1 = \frac{2}{\pi} \int_{-1}^{1} \frac{x^4 T_1}{\sqrt{1-x^2}}\,dx \qquad (6)$$

$$c_2 = \frac{2}{\pi} \int_{-1}^{1} \frac{x^4 T_2}{\sqrt{1-x^2}}\,dx. \qquad (7)$$

The function x^4 in terms of Chebyshev polynomial is

$$x^4 = \frac{1}{8}\{T_4 + 4T_2 + 3T_0\}.$$

Now, we evaluate the definite integrals of Eqs. (5), (6) and (7). Consider

$$c_0 = \frac{1}{\pi}\int_{-1}^{1}\frac{x^4 T_0(x)}{\sqrt{1-x^2}}\,dx$$

$$= \frac{1}{\pi}\int_{-1}^{1}\frac{\frac{1}{8}\{T_4 + 4T_2 + 3T_0\}T_0}{\sqrt{1-x^2}}\,dx$$

$$= \frac{1}{8\pi}\int_{-1}^{1}\frac{3\cdot T_0^2}{\sqrt{1-x^2}}\,dx$$

$$= \frac{3}{8\pi}\times\pi = \frac{3}{8}$$

Similarly

$$c_1 = \frac{2}{\pi}\int_{-1}^{1}\frac{\frac{1}{8}\{T_4 + 4T_2 + 3T_0\}T_1}{\sqrt{1-x^2}}\,dx = 0$$

and

$$c_2 = \frac{2}{\pi}\int_{-1}^{1}\frac{\frac{1}{8}\{T_4 + 4T_2 + 3T_0\}T_2}{\sqrt{1-x^2}}\,dx$$

$$= \frac{2}{8\pi}\int_{-1}^{1}\frac{4\cdot T_2^2}{\sqrt{1-x^2}}\,dx$$

$$= \frac{1}{\pi}\int_{-1}^{1}\frac{T_2^2}{\sqrt{1-x^2}}\,dx$$

$$= \frac{1}{\pi}\times\frac{\pi}{2} = \frac{1}{2}$$

Hence, $f(x) \approx \frac{3}{8}T_0 + \frac{1}{2}T_2(x)$.

Problem 4: Approximate the function $f(x) = 3x^2 - 2x - 5$ on $[-1, 1]$ in terms of Chebyshev polynomials.

Solution: Since $f(x)$ is a polynomial function of second degree, we suppose

$$f(x) \approx c_0 T_0(x) + c_1 T_1(x) + c_2 T_2(x) \tag{1}$$

Then, by the least square approximation, we have

$$c_0 = \frac{1}{\pi} \int_{-1}^{1} \frac{3x^2 - 2x - 5}{\sqrt{1-x^2}} T_0 \, dx \tag{2}$$

$$c_1 = \frac{2}{\pi} \int_{-1}^{1} \frac{(3x^2 - 2x - 5) T_1}{\sqrt{1-x^2}} dx \tag{3}$$

and

$$c_2 = \frac{2}{\pi} \int_{-1}^{1} \frac{(3x^2 - 2x - 5) T_2}{\sqrt{1-x^2}} dx . \tag{4}$$

Now we find c_0, c_1 and c_2 values. Consider

$$c_0 = \frac{1}{\pi} \int_{-1}^{1} \frac{3x^2 - 2x - 5}{\sqrt{1-x^2}} \, dx \qquad\qquad (\because T_0 = 1)$$

$$= \frac{1}{\pi} \left[3 \int_{-1}^{1} \frac{x^2}{\sqrt{1-x^2}} dx - 2 \int_{-1}^{1} \frac{x}{\sqrt{1-x^2}} dx - 5 \int_{-1}^{1} \frac{dx}{\sqrt{1-x^2}} \right]$$

$$= \frac{1}{\pi} \left[3 \cdot \frac{\pi}{3^2} \cdot {}^2C_1 - 0 - 5\pi \right]$$

$$= \frac{-7}{2} . \qquad\qquad (\text{by } (8.25))$$

Similarly,

$$c_1 = \frac{2}{\pi} \int_{-1}^{1} \frac{(3x^2 - 2x - 5) x}{\sqrt{1-x^2}} dx$$

$$= \frac{2}{\pi} \left[3 \int_{-1}^{1} \frac{x^3}{\sqrt{1-x^2}} dx - 2 \int_{-1}^{1} \frac{x^2}{\sqrt{1-x^2}} dx - 5 \int_{-1}^{1} \frac{x}{\sqrt{1-x^2}} dx \right]$$

$$= \frac{2}{\pi} \left[0 - 2 \cdot \frac{\pi}{2^2} \cdot {}^2C_1 - 0 \right] = -2$$

and
$$c_2 = \frac{2}{\pi}\int_{-1}^{1}\frac{(3x^2-2x-5)(2x^2-1)}{\sqrt{1-x^2}}dx$$

$$= \frac{2}{\pi}\int_{-1}^{1}\frac{(6x^4-4x^3-13x^2+2x+5)}{\sqrt{1-x^2}}dx$$

$$= \frac{2}{\pi}\left[6\int_{-1}^{1}\frac{x^4}{\sqrt{1-x^2}}dx-0-13\int_{-1}^{1}\frac{x^2}{\sqrt{1-x^2}}dx+0+5\int_{-1}^{1}\frac{dx}{\sqrt{1-x^2}}\right]$$

$$= \frac{2}{\pi}\left[6\cdot\frac{\pi}{2^4}\cdot{}^4C_2-0-13\cdot\frac{\pi}{2^2}\cdot{}^2C_1+0+5\pi\right]=\frac{3}{2}$$

Thus, the required approximation $f(x) = 3x^2 - 2x - 5$ on $[-1, 1]$ is $\frac{-7}{2}T_0 - 2T_1 + \frac{3}{2}T_2$, i.e.,

$$f(x) \approx \frac{-7}{2}T_0 - 2T_1 + \frac{3}{2}T_2.$$

Problem 5: If $f(x) = \cos^{-1} x$, $-1 \le x \le 1$, then show that

$$f(x) \approx \frac{\pi}{2} - \frac{4}{\pi}T_1(x) - \frac{4}{9\pi}T_3(x) - \frac{4}{25\pi}T_5(x) - \frac{4}{49\pi}T_7(x).$$

Solution: Let $f(x) \approx c_0 T_0 + c_1 T_1 + c_2 T_2 + \ldots + c_7 T_7$. Then, the coefficients of $c_0, c_1, c_2, \ldots, c_7$ are given by

$$c_0 = \frac{1}{\pi}\int_{-1}^{1}\frac{\cos^{-1}x\cdot T_0}{\sqrt{1-x^2}}dx$$

$$c_k = \frac{2}{\pi}\int_{-1}^{1}\frac{\cos^{-1}x\cdot T_k}{\sqrt{1-x^2}}dx \qquad (k = 1, 2, 3, .., 7)$$

Consider

$$c_0 = \frac{1}{\pi}\int_{-1}^{1}\frac{\cos^{-1}x\cdot T_0(x)}{\sqrt{1-x^2}}dx$$

$$= \frac{-1}{\pi}\left[\frac{(\cos^{-1}x)^2}{2}\right]_{-1}^{1}$$

$$= \frac{-1}{\pi}\left[\frac{-\pi^2}{2}\right]=\frac{\pi}{2}.$$

Now, we compute c_k values for $k = 1, 2, 3, .., 7$.

$$c_k = \frac{2}{\pi} \int_{-1}^{1} \frac{\cos^{-1} x \cdot T_k}{\sqrt{1-x^2}} dx .$$

Put $x = \cos \theta$; then $dx = -\sin\theta \, d\theta$ and the limits of θ are $-\pi$ to 0.

$$\therefore \qquad c_k = \frac{2}{\pi} \int_{-\pi}^{0} \frac{\theta \cdot T_k (\cos \theta)}{\sqrt{1-\cos^2 \theta}}(-\sin \theta)d\theta$$

i.e., $$c_k = \frac{2}{\pi} \int_{0}^{\pi} \theta \cdot \cos k\theta \, d\theta$$

$$= \frac{2}{\pi k^2}\left\{(-1)^k -1\right\} \quad k = 1, 2, 3, \ldots, 7$$

Thus, $c_1 = \dfrac{-4}{\pi}$, $c_2 = 0$, $c_3 = \dfrac{-4}{9\pi}$, $c_4 = 0$, $c_5 = \dfrac{-4}{25\pi}$, $c_6 = 0$, $c_7 = \dfrac{-4}{49\pi}$.

Hence, the required approximation of $f(x)$ is

$$f(x) \approx \frac{\pi}{2} - \frac{4}{\pi}T_1(x) - \frac{4}{9\pi}T_3(x) - \frac{4}{25\pi}T_5 - \frac{4}{49\pi}T_7 .$$

Problem 6: Using the Chebyshev polynomials, find the least square approximation of a polynomial of degree one for $f(x) = x^2$ over the interval $[0, 1]$ using the weighted function $w(x) = \dfrac{1}{\sqrt{1-x^2}}$ on $[-1, 1]$.

Solution: Given function $f(x) = x^2$ is defined on $[0, 1]$; hence it is required to transform a function whose domain is $[-1, 1]$. Suppose $t = 2x - 1 \left(x = \dfrac{(t+1)}{2} \right)$. Then, when $x = 0$, $t = -1$ and when $x = 1$, $t = 1$. The transformed function is $g(t) = \left(\dfrac{t+1}{2} \right)^2 = \dfrac{1}{4}(t^2 + 2t + 1)$ and $-1 \le t \le 1$.

Now, we find the least square approximation of $g(t)$ using the Chebyshev polynomial with weighted function $w(t) = \dfrac{1}{\sqrt{1-t^2}}, -1 \le t \le 1$.

Let $g(t) \approx c_0 T_0(t) + c_1 T_1(t)$. Then

$$c_0 = \frac{1}{\pi} \int_{-1}^{1} \frac{\frac{1}{4}(t^2 + 2t + 1)T_0}{\sqrt{1-t^2}} dt$$

and

$$c_1 = \frac{1}{\pi}\int_{-1}^{1}\frac{\frac{1}{4}(t^2 + 2t + 1)T_1}{\sqrt{1 - t^2}}\,dt$$

It can be shown that $c_0 = \dfrac{3}{8}$ and $c_1 = \dfrac{1}{2}$. Thus,

$$g(t) = \frac{3}{8}T_0(t) + \frac{1}{2}T_1(t) = \frac{3}{8} + \frac{1}{2}t.$$

The approximation function in terms of the variable x is

$$f(x) \approx \frac{3}{8} + \frac{1}{2}(2x - 1) = \frac{1}{8}(8x - 1).$$

8.11   LANCZOS ECONOMIZATION

Consider the power series expansion of $f(x)$ in the form

$$f(x) = A_0 + A_1 x + A_2 x^2 + \ldots + A_n x^n \text{ where } -1 \le x \le 1. \tag{8.26}$$

The polynomials $1, x, .., x^n$ can be expressed in chebyshev polynomials $T_0, T_1, T_2, \ldots$ and T_n. Substituting $1, x, .., x^n$ in terms of $T_0, T_1, T_2, \ldots, T_n$ in (8.26), we get

$$f(x) = B_0 T_0(x) + B_1 T_1(x) + \ldots + B_n T_n(x) \tag{8.27}$$

For large values of n, the expansion (8.27) converges more rapidly than the power series given by (8.26). For a given tolerance of error $\in$, the number of terms of (8.27) may be less than the number of terms of (8.26). Neglecting the terms of (8.27) whose absolute value is less than tolerance of error, the resultant series is known as Lanczos economization of power series.

SOLVED PROBLEMS

Problem 1: Find a polynomial approximation of degree 4 or less to e^x on $[-1, 1]$ using Lanczos economization with a tolerance of error $\in = 0.02$.

Solution: The power series expansion of e^x with the given error of tolerance is given

$$f(x) = e^x \simeq 1 + x + \frac{x^2}{2} + \frac{x^3}{6} + \frac{x^4}{24} \tag{1}$$

Since $\left|\dfrac{x^k}{\lfloor k}\right| \le 0.00833 \,\forall k \ge 5$

Expressing 1, x, x^2, x^3 and x^4 in terms of Chebyshev polynomials, Eq. (1) becomes

$$f(x) \approx T_0 + T_1 + \frac{1}{2}\left(\frac{T_0 + T_2}{2}\right) + \frac{1}{6}\left(\frac{3T_1 + T_3}{4}\right) + \frac{1}{24}\left(\frac{3T_0 + 4T_2 + T_4}{8}\right)$$

i.e., $\qquad f(x) \approx \dfrac{81}{64}T_0 + \dfrac{9}{8}T_1 + \dfrac{13}{48}T_2 + \dfrac{1}{24}T_3 + \dfrac{1}{192}T_4$ $\qquad\qquad$ (2)

Now, $\qquad \left|\dfrac{1}{192}T_4\right| \le \dfrac{1}{192}$ $\qquad\qquad\qquad (\because |T_n| \le 1)$

$$= 0.00515.$$

Therefore, ignoring the term involving T_4 $\left(\because \left|\dfrac{1}{192}T_4\right| < 0.02\right)$, we have the Lanczos economization power series

$$f(x) \approx \frac{81}{64}T_0 + \frac{9}{8}T_1 + \frac{13}{48}T_2 + \frac{1}{24}T_3$$ $\qquad\qquad$ (3)

Note that the number of terms in (3) is four, i.e., one term less than number of terms of (1). Now,

$$f(x) = \frac{81}{64}(1) + \frac{9}{8}x + \frac{13}{48}(2x^2 - 1) + \frac{1}{24}(4x^3 - 3x)$$

i.e., $\qquad f(x) = \dfrac{1}{6}x^3 + \dfrac{13}{24}x^2 + x + \dfrac{191}{192}.$

Problem 2: Find a polynomial approximation of degree 2 to $5x^3 + 3x^2 - 1$ on $[-1, 1]$ using Lanczos economization of power series.

Solution: Let $f(x) = 5x^3 + 3x^2 - 1$. Expressing it in Chebyshev polynomials, we get

$$f(x) = 5 \cdot \frac{1}{4}(T_3 + 3T_1) + 3 \cdot \frac{1}{2}(T_0 + T_2) - T_0$$

i.e.,
$$f(x) = \frac{T_0}{2} + \frac{15}{4}T_1 + \frac{3}{2}T_2 + \frac{5}{4}T_3$$

Lanczos economization of power series of second degree will be
$$f(x) \approx \frac{T_0}{2} + \frac{15}{4}T_1 + \frac{3}{2}T_2$$

i.e.,
$$f(x) \approx \frac{1}{2} + \frac{15}{4}x + \frac{3}{2}(2x^2 - 1)$$

i.e.,
$$f(x) = 3x^2 + \frac{15}{4}x - 1.$$

Problem 3: Express $\sin x = x - \dfrac{x^3}{6} + \dfrac{x^5}{120}$ in terms of Chebyshev series. Economize the result to give a third-degree polynomial using Lanczos economization of power series.

Solution: Let $f(x) = \sin x = x - \dfrac{x^3}{6} + \dfrac{x^5}{120}$. Now, expressing x, x^3 and x^5 in terms of Chebyshev polynomials, we obtain

$$f(x) = \sin x = T_1(x) - \frac{1}{24}[3T_1(x) + T_3(x)] + \frac{1}{120} \cdot$$
$$\frac{1}{16}[10T_1(x) + 5T_3(x) + T_5(x)]$$
$$= \frac{169}{192}T_1(x) - \frac{5}{128}T_3(x) + \frac{1}{1920}T_5(x) \tag{1}$$

Economized series of (1) as a third-degree polynomial by Lanczos economization, we have

$$f(x) \approx \frac{169}{192}T_1(x) - \frac{5}{128}T_3(x)$$
$$= \frac{169}{192}x - \frac{5}{128}(4x^3 - 3x)$$
$$= \frac{383}{384}x - \frac{5}{32}x^3.$$

8.12   APPROXIMATIONS WITH TRIGONOMETRIC FUNCTION

In the study of periodic motion in acoustics, optics, electrodynamics, the theory of heat and some engineering problems it may require to express a function in a series of sines and cosines. The features of these functions are:

    (i) Periodic functions;
    (ii) Can be easily computed;
    (iii) They have orthogonal properties;
    (iv) Sum of series in sines and cosines converges rapidly;
    (v) Their successive derivatives are again sines or cosines;
    (vi) The repeated integrals are again sines or cosines.

Thus, when a function is periodic, it is desirable to express it in sines and cosines. In this section, we study the approximation of functions by sine and cosine functions.

If a function $f(x)$ is periodic with a period $2l$, then the function $f(x)$ in $(c, c + 2l)$ is approximated by

$$f(x) = \frac{a_0}{2} + \sum_{n=1}^{\infty} a_n \cos \frac{n\pi x}{l} + \sum_{n=1}^{\infty} b_n \sin \frac{n\pi x}{l} \qquad (8.28)$$

The sum on the right–hand side of (8.28) is known as Fourier series of $f(x)$ and, a_n and b_n are called Fourier coefficients, they are given by

$$a_0 = \frac{1}{l} \int_{c}^{c+2l} f(x)dx \qquad (8.29a)$$

$$a_n = \frac{1}{l} \int_{c}^{c+2l} f(x) \cos \frac{2\pi x}{l} dx \qquad (8.29b)$$

and

$$b_n = \frac{1}{l} \int_{c}^{c+2l} f(x) \sin \frac{n\pi x}{l} dx, \qquad (8.29c)$$

where c is any point belongs to the domain of the function $f(x)$.

The Fourier series of $f(x)$ with period $2l$ in the interval $(0, 2l)$ is given by

$$f(x) = \frac{a_0}{2} + \sum_{n=1}^{\infty} a_n \cos \left(\frac{n\pi x}{l} \right) + \sum_{n=1}^{\infty} b_n \sin \left(\frac{n\pi x}{l} \right),$$

where the Fourier coefficients are

$$a_0 = \frac{1}{l} \int_0^{2l} f(x)dx \tag{8.30a}$$

$$a_n = \frac{1}{l} \int_0^{2l} f(x)\cos\frac{n\pi x}{l}\,dx \tag{8.30b}$$

$$b_n = \frac{1}{l} \int_0^{2l} f(x)\sin\frac{n\pi x}{l}\,dx \tag{8.30c}$$

Further, if $f(x)$ is periodic with period $2l$, then the Fourier series of $f(x)$ in the interval $(-l, l)$ is

$$f(x) = \frac{a_0}{2} + \sum_{n=1}^{\infty} a_n \cos\frac{n\pi x}{l} + \sum_{n=1}^{\infty} b_n \sin\frac{n\pi x}{l},$$

where

$$a_0 = \frac{1}{l} \int_{-l}^{l} f(x)dx \tag{8.31a}$$

$$a_n = \frac{1}{l} \int_{-l}^{l} f(x)\cos\frac{n\pi x}{l}\,dx \tag{8.31b}$$

$$b_n = \frac{1}{l} \int_{-l}^{l} f(x)\sin\frac{n\pi x}{l}\,dx \tag{8.31c}$$

Note that, the Fourier series is the approximation of function $f(x)$ in trigonometric functions namely, sines and cosines.

SOLVED PROBLEMS

Problem 1: If $f(x) = x^2$ is periodic function with period 2, then find the Fourier series of $f(x)$ on the interval $(0, 2)$.

Solution: Here, the period of $f(x) = x^2$ is given as 2. Therefore, $2l = 2$, hence $l = 1$. Now the Fourier series of $f(x)$ is given by (from Eq. 8.28)

$$f(x) = \frac{a_0}{2} + \sum_{n=1}^{\infty} a_n \cos n\pi x + \sum_{n=1}^{\infty} b_n \sin n\pi x \tag{1}$$

where (from Eqs. (8.30a) to (8.30c))

$$a_0 = \int_0^2 x^2\, dx$$

$$a_n = \int_0^2 x^2 \cos n\pi x\, dx$$

and

$$b_n = \int_0^2 x^2 \sin n\pi x\, dx$$

Now, we proceed to find the Fourier coefficients a_0, a_n and b_n.

$$a_0 = \int_0^2 x^2\, dx = \left[\frac{x^3}{3}\right]_0^2 = \frac{8}{3}$$

$$a_n = \int_0^2 x^2 \cos n\pi x\, dx$$

$$= \left[\frac{x^2 \sin n\pi x}{n\pi} + \frac{2x \cos n\pi x}{n^2 \pi^2} + \frac{2}{n^3 \pi^3} \sin n\pi x\right]_0^2$$

$$= \frac{4}{n^2 \pi^2}$$

$$b_n = \int_0^2 x^2 \sin n\pi x\, dx$$

$$= \left[\frac{-x^2 \cos n\pi x}{n\pi} + \frac{2x}{n^2 \pi^2} \sin n\pi x + \frac{2}{n^3 \pi^3} \cos n\pi x\right]_0^2$$

$$= -\frac{4}{n\pi}$$

Substituting the values of a_0, a_n and b_n in (1), we get

$$f(x) = \frac{4}{3} + \frac{4}{\pi^2} \sum_{n=1}^{\infty} \frac{\cos(n\pi x)}{n^2} - \frac{4}{11} \sum_{n=1}^{\infty} \frac{\sin(n\pi x)}{n}$$

i.e.,

$$f(x) = \frac{4}{3} + \frac{4}{\pi^2}\left[\frac{\cos \pi x}{1^2} + \frac{\cos 2\pi x}{2^2} + \frac{\cos 3\pi x}{3^2} + \ldots\right]$$

$$- \frac{4}{\pi}\left[\sin \pi x + \frac{1}{2}\sin(2\pi x) + \frac{1}{3}\sin(3\pi x) + \ldots\right]$$

Problem 2: Find the Fourier series of the periodic function $f(x)$ with period 2π, where $f(x) = \begin{cases} -1 \text{ for } -\pi \le x \le 0 \\ 1 \text{ for } 0 \le x \le \pi. \end{cases}$

Solution: Here, period $2l = 2\pi$, and therefore $l = \pi$. From (8.28), the Fourier series of $f(x)$ is given by

$$f(x) = \frac{a_0}{2} + \sum_1^\infty a_n \cos nx + \sum_1^\infty b_n \sin nx, \qquad (1)$$

where the Fourier coefficients are (from Eqs. 8.31a, 8.31b and 8.31c)

$$a_0 = \frac{1}{\pi} \int_{-\pi}^{\pi} f(x) dx \qquad (2)$$

$$a_n = \frac{1}{\pi} \int_{-\pi}^{\pi} f(x) \cos nx dx \qquad (3)$$

and

$$b_n = \frac{1}{\pi} \int_{-\pi}^{\pi} f(x) \sin nx dx \qquad (4)$$

Now, we determine the Fourier coefficients a_0, a_n and b_n. Recalling Eq. (2)

$$a_0 = \frac{1}{\pi} \int_{-\pi}^{\pi} f(x) dx$$

$$= \frac{1}{\pi} \left[\int_{-\pi}^{0} (-1) dx + \int_{0}^{\pi} (1) dx \right]$$

$$= \frac{1}{\pi} \left[\{-x\}_{-\pi}^{0} + \{x\}_{0}^{\pi} \right]$$

$$= 0.$$

Recalling Eq. (3)

$$a_n = \frac{1}{\pi} \int_{-\pi}^{\pi} f(x) \cos nx dx$$

$$= \frac{1}{\pi} \left[\int_{-\pi}^{0} (-1) \cos nx dx + \int_{0}^{\pi} (1) \cos nx dx \right]$$

$$= \frac{1}{\pi}\left[\left\{\frac{-\sin nx}{n}\right\}_{-\pi}^{0} + \left\{\frac{\sin nx}{n}\right\}_{0}^{\pi}\right]$$

$$= 0.$$

Finally,

$$b_n = \frac{1}{\pi}\int_{-\pi}^{\pi} f(x)\sin nx\,dx$$

$$= \frac{1}{\pi}\left[\int_{-\pi}^{0}(-1)\sin nx\,dx + \int_{0}^{\pi}(1)\sin nx\,dx\right]$$

$$= \frac{1}{\pi}\left[\left\{\frac{\cos nx}{n}\right\}_{-\pi}^{0} + \left\{\frac{-\cos nx}{n}\right\}_{0}^{\pi}\right]$$

$$= \frac{2}{n\pi}(1 - \cos n\pi)$$

$$= \begin{cases} 0 & \text{if } n \text{ is even} \\ \dfrac{4}{n\pi} & \text{if } n \text{ is odd} \end{cases}$$

Thus, the Fourier series of $f(x)$ is

$$f(x) = \frac{4}{\pi}\left[\sin x + \frac{\sin 3x}{3} + \frac{\sin 5x}{5} + \ldots\right].$$

Problem 3: Find the Fourier series of $f(x) = \begin{cases} 1 \text{ if } 0 \le x \le \pi \\ -1 \text{ if } \pi \le x \le 2\pi \end{cases}$ over the interval $(0, 2\pi)$ if it is periodic with period 2π.

Solution: Here, the period of given function is 2π. Therefore, $l = \pi$. From (8.28), and the Fourier series of $f(x)$ is given by

$$f(x) = \frac{a_0}{2} + \sum_{n=1}^{\infty} a_n \cos nx + \sum_{n=1}^{\infty} b_n \sin nx$$

where

$$a_0 = \frac{1}{\pi}\int_{0}^{2\pi} f(x)\,dx$$

$$a_n = \frac{1}{\pi} \int_0^{2\pi} f(x)\cos nx\, dx$$

$$b_n = \frac{1}{\pi} \int_0^{2\pi} f(x)\sin nx\, dx$$

(from Eqs. 8.30a, 8.30b and 8.30c). Consider

$$a_0 = \frac{1}{\pi} \int_0^{2\pi} f(x)\, dx$$

$$= \frac{1}{\pi}\left[\int_0^{\pi}(1)\, dx + \int_{\pi}^{2\pi}(-1)\, dx \right]$$

$$= \frac{1}{\pi}[\pi - \pi] = 0.$$

$$a_n = \frac{1}{\pi}\left[\int_0^{\pi} 1.\cos nx\, dx + \int_{\pi}^{2\pi} \cos nx\, dx \right]$$

$$= \frac{1}{\pi}\left[\left\{ \frac{\sin nx}{n} \right\}_0^{\pi} + \left\{ \frac{\sin nx}{n} \right\}_0^{\pi} \right]$$

$$= 0.$$

Similarly,

$$b_n = \frac{1}{\pi}\left[\int_0^{\pi} \sin nx\, dx + \int_{\pi}^{2\pi} -\sin nx\, dx \right]$$

$$= \frac{1}{\pi}\left[\left\{ \frac{-\cos nx}{n} \right\}_0^{\pi} + \left\{ \frac{\cos nx}{n} \right\}_{\pi}^{2\pi} \right]$$

$$= \frac{1}{\pi}\left[\left\{ \left(\frac{-\cos n\pi}{n} \right) - \left(\frac{-1}{n} \right) \right\} + \left\{ \frac{\cos 2n\pi}{n} - \frac{\cos n\pi}{n} \right\} \right]$$

$$= \frac{1}{\pi}\left[\frac{2}{n} - \frac{2\cos n\pi}{n} \right]$$

$$= \frac{2}{n\pi}[1 - \cos n\pi]$$

$$= \begin{cases} 0 & \text{if } n \text{ is even} \\ \dfrac{4}{n\pi} & \text{if } n \text{ is odd} \end{cases}$$

Thus,

$$f(x) = \frac{4}{\pi}\left[\sin x + \frac{\sin 3x}{3} + \frac{\sin 5x}{5} + \dots \right].$$

Problem 4: Find the Fourier series expansion of $f(x) = x^2 - 2$ in the interval $(-2, 2)$ if it is periodic function with period 4.

Solution: Here $2l = 4$, hence $l = 2$. The Fourier series of $f(x)$ by (8.28) is given by

$$f(x) = \frac{a_0}{2} + \sum_{n=1}^{\infty} a_n \cos \frac{n\pi x}{l} + \sum_{n=1}^{\infty} b_n \sin \frac{n\pi x}{l}$$

where

$$a_0 = \frac{1}{2} \int_{-2}^{2} (x^2 - 2)\,dx$$

$$a_n = \frac{1}{2} \int_{-2}^{2} (x^2 - 2)\cos \frac{n\pi x}{2}\,dx$$

$$b_n = \frac{1}{2} \int_{-2}^{2} (x^2 - 2)\sin \frac{n\pi x}{2}\,dx$$

It can be shown that $a_0 = \dfrac{-4}{3}$, $a_n = (-1)^n \dfrac{16}{n^2 \pi^2}$ and $b_n = 0$. Thus, required Fourier series of $f(x)$ is

$$f(x) = \frac{-2}{3} + \frac{16}{\pi^2} \sum_{n=1}^{\infty} \frac{(-1)^n}{n^2} \cos \frac{n\pi x}{2}.$$

Problem 5: Find the Fourier series expansion of $f(x) = \dfrac{x}{4}$ in the interval $-\pi < x < \pi$, if the period of $f(x)$ is 2π.

Solution: Here, $2l = 2\pi$, hence $l = \pi$. The Fourier series of $f(x)$ is given by

$$f(x) = \frac{a_0}{2} + \sum_1^\infty a_n \cos nx + \sum_1^\infty b_n \sin nx \tag{1}$$

where

$$a_0 = \frac{1}{\pi} \int_{-\pi}^{\pi} \frac{x}{4} \, dx$$

$$a_n = \frac{1}{\pi} \int_{-\pi}^{\pi} \frac{x}{4} \cos nx \, dx$$

and

$$b_n = \frac{1}{\pi} \int_{-\pi}^{\pi} \frac{x}{4} \sin nx \, dx$$

Clearly, $a_0 = 0$, $a_n = 0$ as $\dfrac{x}{4}$ and $\dfrac{x}{4}\cos nx$ are odd functions. Now,

$$b_n = \frac{1}{4\pi} \left[2 \times \int_0^{\pi} x \sin nx \, dx \right]$$

$$= \frac{1}{2\pi} \left[\left\{ \frac{-x \cos nx}{n} \right\}_0^{\pi} + \left\{ \frac{\sin nx}{n^2} \right\}_0^{\pi} \right]$$

$$= \frac{1}{2\pi} \times \left\{ \frac{-\pi \cos n\pi}{n} \right\} = \frac{-\cos n\pi}{2n}$$

Therefore, the required Fourier series of $f(x)$ is

$$f(x) = \sum_{n=1}^\infty \frac{-\cos n\pi}{2n} \sin nx = \frac{1}{2}\left[\frac{1}{1}\sin x - \frac{1}{2}\sin 2x + \frac{1}{3}\sin 3x... \right].$$

Problem 6: If the period of $f(x) = e^{-x}$ is 2π, then find its Fourier series in $(0, 2\pi)$.

Solution: Since $2l = 2\pi$, and we have $l = \pi$. The Fourier series of $f(x)$ is

$$f(x) = \frac{a_0}{2} + \sum_1^\infty a_n \cos nx + \sum_1^\infty b_n \sin nx \tag{1}$$

where

$$a_0 = \frac{1}{\pi} \int_0^{2\pi} e^{-x} dx = \frac{1 - e^{-2\pi}}{\pi}$$

$$a_n = \frac{1}{\pi} \int_0^{2\pi} e^{-x} \cos nx\, dx$$

$$= \frac{1}{\pi} \cdot \frac{1}{(n^2 + 1)} \left[e^{-x}(-\cos nx + n\sin nx) \right]_0^{2\pi}$$

$$= \frac{1}{(n^2 + 1)\pi}(1 - e^{-2\pi})$$

$$b_n = \frac{1}{\pi} \int_0^{2\pi} e^{-x} \sin nx\, dx$$

$$= \frac{1}{\pi} \cdot \frac{1}{(n^2 + 1)} \left[e^{-x}(-\sin nx - n\cos nx) \right]_0^{2\pi}$$

$$= \frac{n}{(n^2 + 1)\pi}(1 - e^{-2\pi})$$

Thus, required Fourier series of $f(x)$ is

$$f(x) = \frac{1}{2\pi}(1 - e^{-2\pi}) + \frac{(1 - e^{-2\pi})}{\pi} \sum_{n=1}^{\infty} \frac{1}{(n^2 + 1)} \cos nx$$

$$+ \frac{(1 - e^{-2\pi})}{\pi} \sum_{n=1}^{\infty} \frac{n}{(n^2 + 1)} \sin nx.$$

Problem 7: Find a Fourier series of $f(x)$ in the interval $-\pi < x < \pi$

when $f(x) = \begin{cases} 0 & \text{if } -\pi < x \le 0 \\ \dfrac{\pi x}{4} & \text{if } 0 < x < \pi. \end{cases}$

Solution: Here, the period of $f(x)$ $2l = 2\pi$. Therefore, $l = \pi$. By (8.28), the Fourier series of $f(x)$ is given by

$$f(x) = \frac{a_0}{2} + \sum_{1}^{\infty} a_n \cos n\pi x + \sum_{n=1}^{\infty} b_n \sin n\pi x \qquad (1)$$

where

$$a_0 = \frac{1}{\pi}\int_{-\pi}^{\pi} f(x)dx = \frac{1}{\pi}\left[\int_{-\pi}^{0} 0\cdot dx + \int_{0}^{\pi}\frac{\pi x}{4}dx\right] = \frac{\pi^2}{8}$$

$$a_n = \frac{1}{\pi}\int_{-\pi}^{\pi} f(x)\cos nx\,dx$$

$$= \frac{1}{\pi}\left[\int_{-\pi}^{0} 0\cdot\cos nx\,dx + \int_{0}^{\pi}\frac{\pi x}{4}\cos nx\,dx\right]$$

$$= \frac{1}{4n^2}(-1+\cos n\pi)$$

$$b_n = \frac{1}{\pi}\int_{-\pi}^{\pi} f(x)\sin nx\,dx$$

$$= \frac{1}{\pi}\left[\int_{-\pi}^{0} 0\cdot\sin nx\,dx + \int_{0}^{\pi}\frac{\pi x}{4}\sin nx\,dx\right]$$

$$= \frac{1}{\pi}\times\frac{\pi}{4}\int_{0}^{\pi} x\sin nx\,dx$$

$$= \frac{-\pi}{4n}\cos n\pi$$

Thus,

$$f(x) = \frac{\pi}{16} + \frac{1}{4}\sum_{n=1}^{\infty}\frac{(\cos n\pi - 1)}{n^2}\cos nx - \frac{\pi}{4}\sum_{n=1}^{\infty}\frac{\cos n\pi}{n}\sin nx$$

i.e., $$f(x) = \frac{\pi^2}{16} - \frac{1}{2}\left[\cos x + \frac{\cos 3x}{3^2} + \frac{\cos 3x}{5^2} + \ldots\right]$$

$$-\frac{\pi}{4}\left[-\sin x + \frac{1}{2}\sin 2x - \frac{\sin 3x}{3} + \frac{\sin 4x}{4} + \ldots\right].$$

8.13   APPROXIMATION OF A FUNCTION WITH RATIONAL FUNCTION

We call the approximation of a function with rational function as rational approximation of a function for convenience. A rational approximation of $f(x)$ is supposed in the form

$$f(x) \approx R_N = \frac{a_0 + a_1 x + a_2 x^2 + \dots + a_n x^n}{1 + b_1 x + b_2 x^2 + \dots + b_m x^m} \tag{8.32}$$

where $N = m + n$; the approximation of a function of type (8.32) is also called Pade approximation. Let the Maclaurin series of $f(x)$ be

$$f(x) = \sum_{j=0}^{\infty} c_j x^j \tag{8.33}$$

where c_j are known constants. Now,

$$f(x) - R_N = \sum_{j=0}^{\infty} c_j x^j - \frac{\displaystyle\sum_{i=0}^{n} a_i x^i}{\displaystyle\sum_{k=0}^{m} b_k x^k} \qquad \text{(with } b_0 = 1\text{)}$$

i.e., $\quad f(x) - R_N(x) = \dfrac{\left(\displaystyle\sum_{j=0}^{\infty} c_j x^j\right)\left(\displaystyle\sum_{k=0}^{m} b_k x^k\right) - \displaystyle\sum_{i=0}^{n} a_i x^i}{\displaystyle\sum_{k=0}^{m} b_k x^k}. \tag{8.34}$

The $(n + m + 1)$ constants namely a_i $(i = 0, 1, 2,\dots, n)$ and b_k $(k = 1, 2, 3,\dots, m)$ are determined equating the coefficients of x^l $(l = 0, 1, 2, \dots, (m + n))$ in the numerator of the right–hand sides of Eq. (8.34) to zero. Now, we illustrate the rational approximation of a function.

SOLVED PROBLEMS

Problem 1: Obtain the rational approximation of $f(x) = e^x$ in the form $\dfrac{a_0 + a_1 x}{1 + b_1 x}$.

Solution: Here, $n = 1$ and $m = 1$, hence $N = n + m = 1 + 1 = 2$.

The Macluarin's expansion of $f(x) = e^x$ is

$$e^x = 1 + x + \frac{x^2}{\lfloor 2} + \frac{x^3}{\lfloor 3} + \ldots = \sum_{i=0}^{\infty} \frac{x^i}{\lfloor i}$$

Let
$$R_N = \frac{a_0 + a_1 x}{1 + b_1 x}$$

Therefore,
$$e^x - R_N = e^x - \frac{a_0 + a_1 x}{1 + b_1 x}$$

$$= \left(\sum_{i=0}^{\infty} \frac{x^i}{\lfloor i} \right) - \frac{a_0 + a_1 x}{1 + b_1 x}$$

$$= \frac{\left(\sum_{i=0}^{\infty} \frac{x^i}{\lfloor i} \right)(1 + b_1 x) - (a_0 + a_1 x)}{(1 + b_1 x)}$$

$$= \frac{\left(1 + x + \frac{x^2}{\lfloor 2} + \frac{x^3}{\lfloor 3} + \ldots \right)(1 + b_1 x) - (a_0 + a_1 x)}{(1 + b_1 x)}$$

$$= \frac{(1 + b_1 x + x + b_1 x^2 + \frac{x^2}{2} + \frac{b_1 x^3}{2} + \ldots - a_0 - a_1 x)}{(1 + b_1 x)}$$

$$= \frac{(1 - a_0) + x(b_1 + 1 - a_1) + x^2(b_1 + \frac{1}{2}) + \ldots}{(1 + b_1 x)} \tag{1}$$

Equating the coefficients of x^0, x^1 and x^2 in the numerator of right hand side of Eq. (1) to zero, we get

$$1 - a_0 = 0 \Rightarrow a_0 = 1$$

$$b_1 + 1 - a_1 = 0 \tag{2}$$

$$b_1 + \tfrac{1}{2} = 0 \tag{3}$$

From (3) $b_1 = -\tfrac{1}{2}$ and from (2) and (3), we have $a_1 = \tfrac{1}{2}$. Thus, $a_0 = 1$, $a_1 = \tfrac{1}{2}$, $b_1 = -\tfrac{1}{2}$. Hence, the required rational approximation of $f(x) = e^x$ is $\dfrac{1 + x/2}{1 - x/2}$.

Problem 2: Obtain the rational approximation of $f(x) = e^x$ in the form $\dfrac{a_0 + a_1 x + a_2 x^2}{1 + b_1 x + b_2 x^2}$.

Solution: Here $n = 2$, $m = 2$ and hence $N = n + m = 2 + 2 = 4$. Let $R_N = \dfrac{a_0 + a_1 x + a_2 x^2}{1 + b_1 x + b_2 x^2}$. The Maclaurin's expansion of $f(x) = e^x$ is

$$e^x = 1 + x + \frac{x^2}{2} + \frac{x^3}{6} + \frac{x^4}{24} + \dots.$$

Now, consider $f(x) - R_N(x)$, i.e.,

$$f(x) - R_N(x) = \left(1 + x + \frac{x^2}{2} + \frac{x^3}{6} + \frac{x^4}{24} + \dots.\right) - \frac{a_0 + a_1 x + a_2 x^2}{1 + b_1 x + b_2 x^2}$$

$$= \frac{\left(1 + x + \dfrac{x^2}{2} + \dfrac{x^3}{6} + \dfrac{x^4}{24} + \dots.\right)\left(1 + b_1 x + b_2 x^2\right) - \left(a_0 + a_1 x + a_2 x^2\right)}{1 + b_1 x + b_2 x^2}$$

$$= \frac{\left[(1 - a_0) + (b_1 + 1 - a_1)x + (b_1 + b_2 + \dfrac{1}{2} - a_2)x^2 + (b_2 + \dfrac{b_1}{2} + \dfrac{1}{6})x^3 + (\dfrac{1}{24} + \dfrac{b_1}{6} + \dfrac{b_2}{2})x^4 + \dots.\right]}{(1 + b_1 x + b_2 x^2)} \tag{1}$$

Equating the coefficients of x^0, x^1, x^2, x^3 and x^4 in the numerator of right-hand side of Eq. (1) to zero, we get

$$1 - a_0 = 0 \tag{2}$$

$$b_1 + 1 - a_1 = 0 \tag{3}$$

$$b_1 + b_2 + \tfrac{1}{2} - a_2 = 0 \tag{4}$$

$$b_2 + \frac{b_1}{2} + \frac{1}{6} = 0 \tag{5}$$

$$\frac{1}{24} + \frac{b_1}{6} + \frac{b_2}{2} = 0 \tag{6}$$

From (5) and (6), we get $b_1 = -\frac{1}{2}$ and $b_2 = \dfrac{1}{12}$. In the view of $b_2 = \dfrac{1}{12}$

and $b_1 = -\frac{1}{2}$ from Eq. (4), we have $a_2 = \dfrac{1}{12}$. From (3), $a_1 = 1 + b_1$ and

hence $a_1 = 1 - \frac{1}{2} = \frac{1}{2}$. From Eq. (2), we have $a_0 = 1$. Thus, the required

rational approximation of $f(x) = e^x$ is $\dfrac{1 + \dfrac{x}{2} + \dfrac{1}{12}x^2}{1 - \dfrac{1}{2}x + \dfrac{1}{12}x^2}$.

Problem 3: Obtain the rational approximation of $f(x) = \log(1+x)$ in the form $\dfrac{a_0 + a_1 x + a_2 x^2 + a_3 x^3}{1 + b_1 x + b_2 x^2 + b_3 x^3}$.

Solution: Here $n = 3$, $m = 3$ and hence $N = 3 + 3 = 6$.

Let $R_N(x) = \dfrac{a_0 + a_1 x + a_2 x^2 + a_3 x^3}{1 + b_1 x + b_2 x^2 + b_3 x^3}$. We have to find $(N+1) = (6+1) = 7$

coefficients namely $a_0, a_1, a_2, a_3, b_1, b_2$ and b_3. Now, consider

$$f(x) - R_N(x) = \log(1+x) - \frac{a_0 + a_1 x + a_2 x^2 + a_3 x^3}{1 + b_1 x + b_2 x^2 + b_3 x^3}.$$

Using the Maclaurin's expansion of $\log(1+x)$,

i.e.,
$$\log(1+x) = x - \frac{x^2}{2} + \frac{x^3}{3} - \frac{x^4}{4} + \frac{x^5}{5} - \frac{x^6}{6} + \frac{x^7}{7} - \cdots,$$

we have

$$f(x) - R_N(x) = \left(x - \frac{x^2}{2} + \frac{x^3}{3} - \frac{x^4}{4} + \frac{x^5}{5} - \frac{x^6}{6} + \frac{x^7}{7} - \cdots \right)$$
$$- \frac{a_0 + a_1 x + a_2 x^2 + a_3 x^3}{1 + b_1 x + b_2 x^2 + b_3 x^3}$$

$$= \frac{\left[\left\{ x - \frac{x^2}{2} + \frac{x^3}{3} - \frac{x^4}{4} + \frac{x^5}{5} - \frac{x^6}{6} + \frac{x^7}{7} - \cdots \right\} \left\{ 1 + b_1 x + b_2 x^2 + b_3 x^3 \right\} \right.}{1 + b_1 x + b_2 x^2 + b_3 x^3}$$
$$\left. - \left(a_0 + a_1 x + a_2 x^2 + a_3 x^3 \right) \right]$$

i.e.,

$$f(x) - R_N(x) = \frac{\begin{bmatrix} -a_0 + (1-a_1)x + (b_1 - \dfrac{1}{2} - a_2)x^2 + (b_2 - \dfrac{b_1}{2} + \dfrac{1}{3} - a_3)x^3 \\[2mm] + (b_3 - \dfrac{b_2}{2} + \dfrac{b_1}{3} - \dfrac{1}{4})x^4 + (-\dfrac{b_3}{2} + \dfrac{b_2}{3} - \dfrac{b_1}{4} + \dfrac{1}{5})x^5 \\[2mm] + (\dfrac{b_3}{3} - \dfrac{b_2}{4} + \dfrac{b_1}{5} - \dfrac{1}{6})x^6 + \ldots \end{bmatrix}}{(1 + b_1 x + b_2 x^2 + b_3 x^3)} \tag{1}$$

Equating the coefficients of x^0, x^1 x^2, x^3, x^4, x^5 and x^6 in the numerator of right-hand side of Eq. (1) to zero, we obtain the following linear equations in the unknowns a_0, a_1, a_2, a_3, b_1, b_2 and b_3.

$$-a_0 = 0 \tag{2}$$

$$1 - a_1 = 0 \tag{3}$$

$$b_1 - \tfrac{1}{2} - a_2 = 0 \tag{4}$$

$$b_2 - \frac{b_1}{2} + \frac{1}{3} - a_3 = 0 \tag{5}$$

$$b_3 - \frac{b_2}{2} + \frac{b_1}{3} - \frac{1}{4} = 0 \tag{6}$$

$$-\frac{b_3}{2} + \frac{b_2}{3} - \frac{b_1}{4} + \frac{1}{5} = 0 \tag{7}$$

$$\frac{b_3}{3} - \frac{b_2}{4} + \frac{b_1}{5} - \frac{1}{6} = 0 \tag{8}$$

Note that Eqs. (6), (7) and (8) are linear equations in b_1, b_2 and b_3 only. Solving them for b_1, b_2 and b_3, we obtain $b_1 = \dfrac{3}{2}$, $b_2 = \dfrac{3}{5}$ and $b_3 = \dfrac{1}{20}$. From (2) and (3), we have $a_0 = 0$ and $a_1 = 1$. Since $b_1 = \dfrac{3}{2}$, from (4), we have $a_2 = 1$. Further substituting $b_1 = \dfrac{3}{2}$ and $b_2 = \dfrac{3}{5}$ in (5), we get $a_3 = \dfrac{11}{60}$. Thus, the required rational approximation of

$$\log(1+x) \text{ is } \dfrac{x + x^2 + \dfrac{11}{60}x^3}{1 + \dfrac{3}{2}x + \dfrac{3}{5}x^2 + \dfrac{1}{20}x^3}.$$

i.e.,
$$f(x) = \log(1+x) \approx \dfrac{x + x^2 + \dfrac{11}{60}x^3}{1 + \dfrac{3}{2}x + \dfrac{3}{5}x^2 + \dfrac{1}{20}x^3}.$$

EXERCISE 8

1. Obtain a linear polynomial approximation of the function $f(x) = x^2 - x$ on the interval $[0, 1]$ using the least square approximation with weight function $w(x) = 1$.

2. Obtain a least square polynomial approximation of degree two for $f(x) = e^{-x}$ on $[0, 1]$ with weight function $w(x) = 1$.

3. Obtain a linear polynomial approximation for $f(x) = \sin x$ on the interval $[0, 1]$ using the least square approximation with weight function $w(x) = 1$.

4. Obtain a least square polynomial approximation of degree 3 for $f(x) = e^x$ on $[0, 1]$ with weight function $w(x) = 1$.

5. Obtain a least square polynomial approximation of degree two for $f(x) = 2x^3$ on the interval $[-1, 1]$ with weight function $w(x) = 1$.

6. Find the least square line for $f(x) = 3x^2 - 2x + 5$ on $[-1, 1]$ with weight function $w(x) = \dfrac{1}{\sqrt{1-x^2}}$.

7. Obtain the least square polynomial of second degree for $f(x) = x^3$ on $[-1, 1]$ with weight function $w(x) = \dfrac{1}{\sqrt{1-x^2}}$.

8. Show that $\sin 3x$ and $\sin 4x$ are orthogonal functions on $[0, 2\pi]$ with weight function $w(x) = 1$.

9. Show that x^2 and $(2x^3 - 7x)$ are orthogonal functions on $[-1, 1]$ with weight function $w(x) = 1$.

10. Obtain the transform function $g(\cdot)$ of $f(x) = x^3$ defined on $[2, 7]$ so that the domain of $g(\cdot)$ is $[-1, 1]$.

11. Let $f(x) = x + e^x$ is defined on $[-2, 2]$. Obtain the transform function $g(\cdot)$ of $f(x)$ so that the domain of $g(\cdot)$ is $[-1, 1]$.

12. Express $f(x) = x^3$ in Chebyshev polynomials.

13. Express $f(x) = 4x^3 - 3x + 3$ in Chebyshev polynomials.

14. Express $e^x \approx 1 + x + \dfrac{x^2}{2} + \dfrac{x^3}{6}$ in terms of Chebyshev polynomials.

15. Express $\cos x \approx 1 - \dfrac{x^2}{2} + \dfrac{x^4}{24} - \dfrac{x^6}{720}$ in terms of Chebyshev polynomials.

16. Using the Chebyshev polynomials obtain the least square approximation of second degree for $f(x) = x^4 + 2$ on $[-1, 1]$.

17. Using the Chebyshev polynomials, obtain the least square approximation of second degree for $f(x) = e^x$ on $[-1, 1]$.

18. Obtain Chebyshev series of $\sin^{-1} x$.

19. If $\log(1 + x) \approx x - \dfrac{x^2}{2} + \dfrac{x^3}{6}$, then find the Chebyshev series for $\log(1 + x)$ on $[-1, 1]$.

20. If $e^{-x} \approx 1 - x + \dfrac{x^2}{2} - \dfrac{x^3}{6} + \dfrac{x^4}{24}$ then find the Chebyshev series of e^{-x} on $[-1, 1]$.

21. Show that the coefficient of x^n in $T_n(x)$ is 2^{n-1}.

22. Express $T_8(x)$ as a polynomial of degree 8.

23. Economize $\tan^{-1} x = x - \dfrac{x^3}{3} + \dfrac{x^5}{5}$ to give a third-degree polynomial.

24. Economize $e^{-x} = 1 - x + \dfrac{x^2}{2} - \dfrac{x^3}{6} + \dfrac{x^4}{24}$ to give a third-degree polynomial.

25. Find the Fourier series for $f(x) = |x|$, $-\pi \le x \le \pi$ if the period of $f(x)$ is 2π.

26. Find the Fourier series for $f(x) = \pi x$ in $0 \le x \le 2$ if the period of $f(x)$ is 2.

27. Obtain the Fourier series of $f(x) = 2x - x^2$ in $(0, 3)$ if its period is 3.

28. Find the Fourier series for $f(x) = x$ in $-\pi \le x \le \pi$ if the period of $f(x)$ is 2π.

29. Obtain the Fourier series for $f(x) = x^2 - 2$ in the interval $(-2, 2)$ if its period is 4.

30. Given that the period of $f(x)$ is 8, where $f(x) = \begin{cases} 0 & \text{if } -4 \le x \le 0 \\ 1 & \text{if } \ \ 0 \le x \le 4 \end{cases}$, Obtain the Fourier series of $f(x)$ in the interval $(-4, 4)$.

31. Find the Fourier series for $f(x) = x + x^2$ in the interval $-\pi \le x \le \pi$ if its period is 2π.

32. Find Pade approximation of $f(x) = e^{-x}$ with numerator and denominator each of degree 2.

33. Obtain the rational approximation of $\sin x$ in the form $\dfrac{a_0 + a_1 x}{1 + b_1 x}$.

34. Obtain the rational approximation of $\cos x$ in the form $\dfrac{a_0 + a_1 x}{1 + b_1 x + b_2 x^2}$.

35. Obtain the rational approximation of $\cos x + x^2$ in the form $\dfrac{\left(a_0 + a_1 x + a_2 x^2\right)}{\left(1 + b_1 x + b_2 x^2\right)}$.

ANSWERS

1. $-0.8333 - 0.125x$

2. $0.9945 - 0.9305 + 0.3087x^2$

3. $-3.6458 + 6.3722x$

4. $16.9992 - 118.9828x + 240.4238x^2 - 139.7231x^3$

5. $f(x) \approx 1.2x$

6. $f(x) \approx \dfrac{13}{2} T_0 - T_1 + \dfrac{3}{2} T_0$

7. $f(x) \approx \dfrac{3}{4} T_1$

10. $g(t) = \dfrac{1}{8}\left[125t^3 + 675t^2 + 1215t + 729\right]$, $t\varepsilon[-1, 1]$

11. $g(t) = 2t + e^{2t}$, $t\varepsilon[-1, 1]$

12. $\dfrac{1}{4}\left[T_2(x) + 3T_1(x)\right]$

13. $T_3(x) + 3T_0(x)$

14. $\dfrac{1}{24} T_3 + \dfrac{1}{2} T_2(x) + \dfrac{9}{8} T_1(x) + \dfrac{3}{2} T_0(x)$

15. $\cos x \approx 0.7652\, T_0(x) - 0.2298\, T_2(x) + 0.0049\, T_4(x) - 0.000043403\, T_6(x)$

16. $x^4 + 2 \approx \dfrac{19}{8} T_0(x) + \dfrac{1}{2} T_2(x)$

17. $e^x \approx \dfrac{5}{4} T_0(x) + T_1(x) + \dfrac{1}{4} T_2(x)$

18. $\sin^{-1}(x) \approx \dfrac{4}{\pi}\left[T_1(x) - \dfrac{1}{9} T_3(x) + \dfrac{1}{25} T_5(x) + \dfrac{1}{49} T_7(x) + \ldots\ldots\right]$

19. $\log(1+x) \approx -\dfrac{1}{4}T_0(x) + \dfrac{5}{4}T_1(x) - \dfrac{1}{4}T_2(x) + \dfrac{1}{12}T_3(x)$

20. $e^x \approx 1.2656\,T_0(x) - 1.125\,T_1(x) + 0.2708T_2(x) - 0.0417\,T_3(x) + 0.0052T_4(x)$

22. $T_8(x) = 128\,x^8 - 256\,x^6 + 160\,x^4 - 32\,x^2 + 1$

23. $\mathrm{Tan}^{-1}(x) \approx -0.0208\,T_3(x) + 0.875\,T_1(x)$

24. $1.375\,T_0(x) - 1.125\,T_1(x) + 0.2708T_2(x) - 0.0417\,T_3(x)$

25. $\dfrac{\pi}{2} - \dfrac{4}{\pi}\left(\cos(x) + \dfrac{\cos(3x)}{3^2} + \dfrac{\cos(5x)}{5^2} + \dfrac{\cos(7x)}{7^2} + \ldots\ldots\ldots \right)$

26. $-\dfrac{2}{\pi}\left[\dfrac{\sin \pi x}{1} + \dfrac{\sin 2\pi x}{2} + \dfrac{\sin \pi x}{3} + \ldots\ldots\ldots \right]$

27. $-\dfrac{9}{\pi^2}\sum_{n=1}^{\infty}\dfrac{1}{n^2}Cos\left(\dfrac{2n\pi x}{3} \right) + \dfrac{3}{\pi}\sum_{n=1}^{\infty}\dfrac{1}{n}Sin\left(\dfrac{2n\pi x}{3} \right)$

28. $2\left[\dfrac{\sin x}{1} - \dfrac{\sin 2x}{2} + \dfrac{\sin 3x}{3} - \dfrac{\sin 4x}{4} + \ldots\ldots\ldots \right]$

29. $-\dfrac{2}{3} - \dfrac{16}{\pi^2}\left[\cos\left(\dfrac{\pi x}{2} \right) - \dfrac{1}{2^2}\cos(\pi x) + \dfrac{1}{3^2}\cos\left(\dfrac{3\pi x}{2} \right) - \ldots\ldots\ldots\ldots \right]$

30. $\dfrac{1}{2} + \dfrac{2}{\pi}\left[\sin\left(\dfrac{\pi x}{4} \right) + \sin\left(\dfrac{3\pi x}{4} \right) + \sin\left(\dfrac{5\pi x}{4} \right) + \ldots\ldots\ldots\ldots \right]$

31. $\dfrac{\pi^2}{3} - 4\left[\dfrac{\cos x}{1^2} - \dfrac{\cos 2x}{2^2} + \dfrac{\cos 3x}{3^2} - \ldots\ldots \right]$

$\quad + 2\left[\dfrac{\sin x}{1} - \dfrac{\sin 2x}{2} + \dfrac{\sin 3x}{3} - \dfrac{\sin 4x}{4} + \ldots\ldots\ldots \right]$

32. $\dfrac{1 - 0.5x + 0.08332x^2}{1 + 0.5x + 0.0833x^2}$

33. $\dfrac{x}{1}$ 34. $\dfrac{2}{2+x}$ 35. $\dfrac{12 + 5x^2}{12 - x^2}$.

9

Solutions of Algebraic and Transcendental Equations

9.1 NUMERICAL METHODS OF SOLVING ALGEBRAIC AND TRANSCENDENTAL EQUATIONS

We are often encounter problems of determining the roots of an equation of the form $f(x) = 0$ in applied mathematics, physics and engineering. For example, the polynomial equation $x^6 - 27x^5 + 105x^4 - 140x^3 + 81x^2 - 21x + 2 = 0$ was originated while studying a problem "Dynamic forces in aircraft engines" by *E.C.* Carter. To analyze the problem of this type, we need to know the roots of algebraic equation.

It is known in the algebra that $x = \dfrac{-b}{a}$ is the solution of $ax + b = 0$ $(a \neq 0)$ and $\dfrac{-b \pm \sqrt{b^2 - 4ac}}{2a}$ are the roots of $ax^2 + bx + c = 0$ $(a \neq 0)$. A cubic equation $x^3 + ax^2 + bx + c = 0$ is solved by Cardon's method and using transformation of an equation. A formula for the roots of biquadratic equation is also available. However, there is no standard formula to find the roots of $P_n(x) = 0$ when $P_n(x)$ is nth-degree polynomial and $n \geq 5$. In such cases, numerical methods can be employed to find the approximate roots of $P_n(x) = 0$ up to a desired degree of accuracy. Of course, these methods can also be used to find the roots of cubic and biquadratic equations. Moreover, these methods can also be used to find an approximate root of non-algebraic equations like $x + \tan x = 0$, $x \log_e x - 3 = 0$ and $\sinh x - x = 0$. In this chapter, we learn some numerical methods for finding the roots of an equation $f(x) = 0$.

Also, we discuss the convergence and order of convergence of some special methods.

9.2 DEFINITION: ALGEBRAIC EQUATION

If $f(x)$ is an algebraic function, then $f(x) = 0$ is called an algebraic equation.

The polynomial $P_n(x) = a_0x^n + a_1x^{n-1} + \dots + a_n$ are the members of the set of all algebraic functions. Therefore, $x^3 - 5x^2 + 7x - 8 = 0$ is an example of an algebraic equation.

A function which is not algebraic is called a transcendental function or non-algebraic function. For example $e^{-x} + \sin x - 3$ is a transcendental function.

9.3 DEFINITION: TRANSCENDENTAL EQUATION

If $f(x)$ is a transcendental or non-algebraic function, then $f(x) = 0$ is called a transcendental or non-algebraic equation.

$2\log_{10}x - 1.2 = 0$, $e^x + \tan x - x^2 = 0$ and $x\sin x + 2 = 0$ are examples of transcendental equations. Usually transcendental equations involve one or more elementary functions like e^x, $\sin x$, $\cos x$, $\log x$ etc.

9.4 DEFINITION: SOLUTION OR ROOT OF AN EQUATION

A number a is said to be a solution of an equation $f(x) = 0$ if $f(a) = 0$. Such a solution is also called a root or zero of $f(x) = 0$. For example, $x = 1$ is a root of $x^3 + x - 2 = 0$.

Note: Geometrically a root of $f(x) = 0$ is the abscissa (x-coordinate) of a point of intersection of the graph $y = f(x)$ and x-axis. For example, $y = x^2 - 5x + 6$ and x-axis intersect at the points (2, 0) and (3, 0). The numbers 2 and 3, which are abscissa of points of intersection, are the roots of $x^2 - 5x + 6 = 0$.

9.5 DEFINITIONS: DIRECT AND ITERATIVE METHODS

If a solution of a problem is obtained by a method in a finite number of steps and it is exact, then the method is called a direct method. For example, obtaining the roots of $ax^2 + bx + c = 0\,(a \neq 0)$ by $\dfrac{-b \pm \sqrt{b^2 - 4ac}}{2a}$ is a direct method. On the other hand, a method is said to be an iterative method if the solution of a problem is obtained by the method is limiting value of successive approximations. For example, suppose a is a root of an equation $f(x) = 0$. By assuming one or more approximations of the root a, we obtain a sequence of approximations $\{x_k\}$ by a method. If the sequence $\{x_k\}$ converges to a, then the method through which the sequence of approximations is obtained is called an iterative method. The successive approximations obtained by the method are called its iterates.

9.6 DEFINITIONS: BRACKETING AND OPEN METHODS

A method of finding an approximation of a root of $f(x) = 0$ is said to be a bracketing method if it needs two initial guess values a_0 and b_0 such that a root a of $f(x) = 0$ lies between a_0 and b_0. The successive approximations c_k of the root a are obtained such that $a \leq a_n < c_k < b_n \leq b$, where a_n and b_n are suggested by bracketing method.

A method of finding an approximation of a root a of $f(x) = 0$ is said to be an open method if it needs a single guess x_0 to get successive approximations $\{c_k\}$ of the root by it.

9.7 DEFINITION: CONVERGENCE OF A METHOD

Let $\{x_k\}_{k=1}^{\infty}$ be a sequence of approximations of a root a of an equation $f(x) = 0$ are obtained by a method. The method is said to be convergent if

$$\lim_{n \to \infty} |x_n - a| = 0 \tag{9.1}$$

9.8 DEFINITION: ORDER OF CONVERGENCE OF AN ITERATION METHOD

Suppose a is a root of $f(x) = 0$ and $\{x_k\}_{k=1}^{\infty}$ be successive approximations of a root a obtained by an iterative method. Let $x_n = a + \varepsilon_n$ and $x_{n+1} = a + \varepsilon_{n+1}$, where ε_n, ε_{n+1} are errors in nth, $(n + 1)$th approximations of the root a respectively. The iterative method is said to has the rate of convergence p if p is the largest positive real number such that

$$\left| \varepsilon_{n+1} \right| \leq |c| \left| \varepsilon_n \right|^p \tag{9.2}$$

where $c \neq 0$ is a constant.

Note:

1. If $p = 1$, the convergence of the iterative method is said to be linear.
2. If $p = 2$, the convergence of the iterative method is said to be quadratic.
3. The order of convergence of an iterative method need not be an integer.

Now, we proceed to learn some iterative methods to find approximations of a root of an equation $f(x) = 0$ where it is an algebraic or non-algebraic equation.

9.9   BISECTION METHOD

Suppose $f(x)$ is continuous function on an interval $[a, b]$ and two real numbers a_0 and b_0 are such that $a \leq a_0 < b_0 \leq b$, and $f(a_0) f(b_0) < 0$. Then, the equation $f(x) = 0$ has at least one root in the interval (a_0, b_0).

Let $c_1 = \dfrac{a_0 + b_0}{2}$. Then c_1 is the midpoint of the interval (a_0, b_0). If $f(c_1) = 0$, then c_1 is the root $f(x) = 0$; otherwise the root of $f(x) = 0$ lies in the interval (a_0, c_1) or (c_1, b_0). Let $I = (a_1, b_1)$ where $a_1 = a_0$, $b_1 = c_1$ if $f(a_0) f(c_1) < 0$ and $a_1 = c_1$, $b_1 = b_0$ if $f(c_1) f(b) < 0$.

Again compute the midpoint of (a_1, b_1). Suppose $c_2 = \dfrac{a_1 + b_1}{2}$. If $f(c_2) = 0$, then c_2 is a root, otherwise the root lies in the interval (a_1, c_2) or (c_2, b_1). Suppose $I_2 = (a_2, b_2)$ where $a_2 = a_1$, $b_2 = c_2$ if $f(a_1) f(c_2) < 0$ and $a_2 = c_2$, $b_2 = b_1$ if $f(c_2) f(b_1) < 0$.

Continuing this process, we have successive approximations of the root of $f(x) = 0$. The approximations $c_1, c_2, c_3, \ldots$ are called iterative values or iterates of the root. The iterative formula of bisection method is

$$c_{k+1} = \frac{1}{2}(a_k + b_k) \tag{9.3}$$

where

$$\left(a_{k+1}, b_{k+1}\right) = \begin{cases} \left(a_k, c_{k+1}\right) & \text{if } f\left(a_k\right) f\left(c_{k+1}\right) < 0 \\ \left(c_{k+1}, b_k\right) & \text{if } f\left(b_k\right) f\left(c_{k+1}\right) < 0. \end{cases}$$

Among the computed c_1, c_2, ..., c_l, (l is a positive integer), we take c_l as an approximation to the root of $f(x) = 0$. The number l depends on the desired degree of accuracy. The process of finding an approximation of a root of $f(x) = 0$ by bisection iterative formula is called a bisection method.

Note:

1. The error in the computation of real root of $f(x) = 0$ after n iterations of bisection method is less than $\dfrac{\left(b_0 - a_0\right)}{2^n}$, where $f(a_0) f(b_0) < 0$ and a_0, b_0 are the initial iterates for computing successive approximations by bisection method.

2. Bisection method is a bracketing method as it needs two initial guesses of approximations of a root to find successive approximations by it.

9.10 PROCEDURE TO FIND ITERATES OF A REAL ROOT OF $f(x) = 0$ BY BISECTION METHOD

Step 1: Find two values a_0 and b_0 such that $f(a_0) f(b_0) < 0$.

Step 2:

   (i) The first iterative approximation of the root c_1 is obtained by

$$c_1 = \frac{1}{2}(a_0 + b_0)$$

   (ii) Find $f(c_1)$.

   (iii) If $f(a_0) f(c_1) < 0$, then set $a_1 = a_0$ and $b_1 = c_1$, else set $a_1 = c_1$ and $b_1 = b_0$.

   (iv) The root lies between a_1 and b_1.

Step 3:

   (i) The second iterative approximation of the root c_2 is given by

$$c_2 = \frac{1}{2}(a_1 + b_1).$$

(ii) Find $f(c_2)$.

(iii) If $f(a_1)\,(c_2) < 0$, then set $a_2 = a_1$ and $b_2 = c_2$, or else set $c_2 = c_2$ and $b_2 = b_1$.

(iv) Now, the root lies between a_2 and b_2.
Continue this process till we get the desired accuracy of the root.

SOLVED PROBLEMS

Problem 1: Find a real root of the equation $x^3 - 3x + 1 = 0$ lying between 1 and 2 correct to two places of decimal by bisection method.

Solution: Given that $a_0 = 1$ and $b_0 = 2$.

Let $f(x) = x^3 - 3x + 1 = 0$. Then $f(a_0) = f(1) = -1$ and $f(b_0) = f(2) = 3$. Clearly, $f(1)\,f(2) < 0$.

Iteration 1: By Eq. (9.3), with $k = 0$, we have

$$c_1 = \frac{a_0 + b_0}{2} = \frac{1+2}{2} = 1.5$$

$$f(1.5) = (1.5)^3 - 3(1.5) + 1 = -0.125$$

Thus $f(2)\,f(1.5) < 0$. Therefore, the required root lies between 1.5 and 2.

Iteration 2: Let $a_1 = 1.5$ and $b_1 = 2$. Again by Eq. (9.3), we have

$$c_2 = \frac{1.5 + 2}{2} = 1.75$$

$$f(1.75) = (1.75)^3 - 3(1.75) - 1 = 1.1095.$$

Thus, $f(1.5)\,f(1.75) < 0$. Therefore, the root lies between 1.5 and 1.75.

Iteration 3: Let $a_2 = 1.5$ and $b_2 = 1.75$. Then

$$c_3 = \frac{1.5 + 1.75}{2} = 1.625$$

$$f(1.625) = (1.625)^3 - 3(1.625) + 1 = 0.416015624$$

Now, the root lies between 1.5 and 1.625, since $f(1.5)\,f(1.625) < 0$.

Iteration 4: Let $a_3 = 1.5$ and $b_3 = 1.625$. Then,

$$c_4 = \frac{1.5 + 1.625}{2} = 1.5625,$$

and

$$f(1.5625) = (1.5625)^3 - 3(1.5625) + 1 = 0.127197265.$$

As $f(1.5) f(1.5625) < 0$, the root lies between 1.5 and 1.5625.

Iteration 5: Let $a_4 = 1.5$ and $b_4 = 1.5625$. Then,

$$c_5 = \frac{1.5 + 1.5625}{2} = 1.53125$$

and

$$f(1.53125) = -0.00387452.$$

Since $f(1.5625) f(1.53125) < 0$, the root lies between 1.53125 and 1.5625.

Iteration 6: Let $a_5 = 1.53125$ and $b_5 = 1.5625$. Then

$$c_6 = \frac{1.53125 + 1.5625}{2} = 1.546875$$

and

$$f(1.546875) = 0.060771941.$$

The required root lies between 1.53125 and 1.546875, since $f(1.53125) f(1.546875) < 0$.

Iteration 7: Let $a_6 = 1.53125$ and $b_6 = 1.546875$. Then, bisection formula (9.3) is

$$c_7 = \frac{1.53125 + 1.546875}{2} = 1.5390625,$$

and

$$f(1.5390625) = 0.028410434.$$

Iteration 8: Let $a_7 = 1.53125$ and $b_7 = 1.5390625$. Then,

$$c_8 = \frac{1.53125 + 1.5390625}{2} = 1.53515625.$$

Up to two decimal places c_7 and c_8 are equal. Hence, 1.53 is the approximate root of $x^3 - 3x + 1 = 0$ correct to two decimal places.

Problem 2: Find a positive real root of $x \log_{10} x = 1.2$ using bisection method in four iterations.

Solution: Let $f(x) = x \log_{10} x - 1.2$. Then, we have to find a positive root of $f(x) = 0$. Therefore, we have to search two positive numbers a_0 and b_0 (these are the two initial guesses) such that $f(a_0) f(b_0) < 0$. These two guesses can be found by trail and error method. For $x = 1, 2, 3$ the values of $f(x)$ are

$$f(1) = 1\log_{10} 1 - 1.2 = -1.2$$
$$f(2) = 2\log_{10} 2 - 1.2 = -0.5979$$
$$f(3) = 3\log_{10} 3 - 1.2 = 0.2313 \quad .$$

Since $f(2) f(3) < 0$, a positive root of $f(x) = 0$ lies between 2 and 3.

Iteration 1: Let $a_0 = 2$ and $b_0 = 3$. Then, by Eq. (9.3), we have

$$c_1 = \frac{a_0 + b_0}{2} = \frac{2 + 3}{2} = 2.5.$$

Now, $f(2.5) = 2.5 \log_{10} 2.5 - 1.2 = -0.2051$. As $f(2.5) f(3) < 0$, the required root lies between 2.5 and 3. Note that the length of $(2.5, 3)$ is half the length of the interval $(2, 3)$.

Iteration 2: Let $a_1 = 2.5$ and $b_1 = 3$. Then,

$$c_2 = \frac{2.5 + 3}{2} = 2.75$$

and $f(2.75) = 2.75\log_{10} 2.75 - 1.2 = 0.0081$.

Since, $f(2.5) f(2.75) < 0$, the required root lies between 2.5 and 2.75.

Iteration 3: Let $a_2 = 2.5$ and $b_2 = 2.75$. Then,

$$c_3 = \frac{2.5 + 2.75}{2} = 2.625$$

and $f(2.625) = -0.0997$.

As $f(2.625) f(2.75) < 0$, the root lies between 2.625 and 2.75.

Iteration 4: Let $a_3 = 2.625$ and $b_3 = 2.75$. Then,

$$c_4 = \frac{2.625 + 2.75}{2} = 2.6875$$

After four iterations, the approximate positive root of $x\log x = 1.2$ is 2.6875.

Problem 3: Perform three iterations of bisection method to find a root of $3\sin x + 4x - 5 = 0$.

Solution: Here $f(x) = 3\sin x + 4x - 5$. For trial values of $x = 0$ and 1 the values of $f(x)$ are

$$f(0) = -5$$
$$f(1) = 1.5244.$$

Therefore, a root of $f(x) = 0$ lies between 0 and 1. Note that, the units of these numbers are radians i.e., the values of x should be in radians.

Iteration 1: Let $a_0 = 0$ and $b_0 = 1$. Then,

$$c_1 = \frac{a_0 + b_0}{2} = \frac{0+1}{2} = 0.5$$

and

$$f(0.5) = -1.5617.$$

Since, $f(0.5) f(1) < 0$, the root lies between 0.5 and 1.

Iteration 2: Let $a_1 = 0.5$ and $b_1 = 1$. Then,

$$c_2 = \frac{0.5+1}{2} = 0.75$$

and $f(0.75) = 0.0449$.
Since, $f(0.5) f(0.75) < 0$, the root of $f(x) = 0$ lies between 0.5 and 0.75.

Iteration 3: Let $a_2 = 0.5$ and $b_2 = 0.75$. Then,

$$c_2 = \frac{0.5+0.75}{2} = 0.625$$

Thus, the first three iterates of a root of $3\sin x + 4\cos n - 5 = 0$ are 0.5, 0.75 and 0.625.

Problem 4: Find a negative root of $x^3 - 4x + 8 = 0$ correct to 2 decimal places using bisection method.

Solution: Let $f(x) = x^3 - 4x + 8$. Now,

$$f(-1) = -1 + 4 + 8 = 11 > 0$$
$$f(-2) = -8 + 8 + 8 = 8 > 0$$
$$f(-3) = -27 + 12 + 8 = -7 > 0.$$

Therefore, a negative root lies between -3 and -2.

Iteration 1: For $a_0 = -3$ and $b_0 = 2$ we have $c_1 = -2.5$ and $f(-2.5) = (-2.5)^3 - 4(-2.5) + 8 = 2.375 > 0$. Hence, the root lies between -3 and -2.5.

Iteration 2: For $a_1 = -3$ and $b_1 = -2.5$ we have

$$c_2 = \frac{-3 - 2.5}{2} = -2.75$$

and $f(-2.75) = -1.7968$. Since $f(-2.75)\,f(-2.5) < 0$, the required root lies between -2.75 and -2.5.

Iteration 3: For $a_2 = -2.75$ and $b_2 = -2.5$, we have

$$c_3 = \frac{-2.75 - 2.5}{2} = -2.625$$

and $f(-2.625) = 0.4142$. Since $f(-2.75)\,f(-2.625) < 0$, the required root lies between -2.75 and -2.625.

Iteration 4: For $a_3 = -2.75$ and $b_3 = -2.625$, we have

$$c_4 = \frac{-2.75 - 2.625}{2} = -2.6875$$

and $f(-2.6875) = -0.6608$. Now, the root lies between -2.625 and -2.6875.

Iteration 5: For $a_4 = -2.6875$ and $b_4 = -2.625$, we get

$$c_5 = \frac{-2.6875 - 2.625}{2} = -2.65625,$$

$f(-2.65625) = 0.1166$ and the required root lies between -2.625 and -2.65625, since $f(-2.625)\,f(-2.65625) < 0$.

Iteration 6: For $a_5 = -2.65625$ and $b_5 = -2.625$

$$\text{The value of } c_6 = \frac{-2.65625 - 2.625}{2} = -2.640625$$

and $f(-2.640625) = 0.1496849$.

Iteration 7: For $a_6 = -2.65625$ and $b_6 = -2.640625$,

$$\text{The value of } c_7 = \frac{-2.65625 - 2.640625}{2} = -2.6484375$$

and $f(-2.6484375) = 0.0170$.

Iteration 8: For $a_7 = -2.65625$ and $b_7 = -2.6484375$,

$$\text{The value of } c_8 = \frac{-2.65625 - 2.6484375}{2} = -2.65234375$$

and $f(c_8) = 0.0496$.

Since, $f(-2.6484375) \, f(-2.65234375) < 0$, the root lies between -2.6484375 and -2.65234375.

Iteration 9: For $a_8 = -2.65234375$ and $b_8 = -2.6484375$,
We have

$$c_9 = \frac{(-2.65234375 - 2.6484375)}{2} = -2.65039062$$

and $f(c_9) = -0.0162932$. Therefore, the root lies between -2.65039625 and -2.6484375.

Iteration 10: For $a_9 = -2.65039625$ and $b_9 = -2.6484375$, the value of

$$c_{10} = \frac{(a_9 + b_9)}{2} = -2.64941406$$

Similarly, we can obtain $c_{11} = -2.64990234$, $c_{12} = -2.64965820$. Hence, the negative root of given equation correct to 2 decimal places is -2.64.

Notation:

We denote the length of an interval I by $|I|$.

Theorem:

The successive approximations of a root of $f(x) = 0$ obtained by bisection method converges.

Proof: Suppose a root α of $f(x) = 0$ lies between a_0 and b_0 with $a_0 < b_0$.

Let $I_0 = (a_0, b_0)$. Then $|I_0| = b_0 - a_0$. If $c_1 = \dfrac{a_0 + b_0}{2}$ is the first iterative value of α, then

$$|c_1 - \alpha| < \frac{(b_0 - a_0)}{2} \tag{9.4}$$

Suppose $f(c_1) \neq 0$. If $f(a_0) \, f(c_1) < 0$, then set $a_1 = a_0$, $b_1 = c_1$ else set $a_1 = c_1$ and $b_1 = b_0$. Let $I_1 = (a_1, b_1)$. The next approximation or iterative value of the root a by bisection method is $c_2 = \dfrac{(a_1 + b_1)}{2}$. Now, $|c_2 - a| < \dfrac{(b_1 - a)}{2}$. Note that $|I_1| = \dfrac{1}{2}I_0$. Hence,

$$|c_2 - a| < \frac{(b_0 - a_0)}{2^2} \tag{9.5}$$

Again suppose $f(c_2) \neq 0$. Then we have

$$|c_3 - a| < \frac{(b_0 - a_0)}{2^3}$$

where c_3 is the 3rd approximation of the root a. In general if c_n denotes nth iterate of the root a obtained by bisection method, we have

$$|c_n - a| < \frac{(b_0 - a_0)}{2^n} \tag{9.6}$$

Clearly, as $n \to$, $|c_n - a| \to 0$ i.e.,

$$\underset{n \to \infty}{\text{Lt}}\ c_n = a$$

Hence, the successive approximations (or iterates) of a root of $f(x) = 0$ obtained by bisection method converges.

Note that if $f(c_m) = 0$ for some positive integer m, then $c_m = a$. In such case, we set $a_m = c_m$ and $b_m = c_m$, and we have successive approximations.

$$c_{m+1} = c_{m+2} = c_{m+3} = \ldots\ldots\ldots\ldots = c_m(a)$$

Clearly, the sequence $\{c_1, c_2, c_{m-1}, a, a, a, \ldots\}$ converges to a.

Note: The convergence of bisection method is very slow. To get a root upto the desired degree of accuracy in a lesser number of iterations, choose a_0 and b_0 such that they are very close and $f(a_0)f(b_0) < 0$. For example, to find a root of $f(x) = x^2 - 17 = 0$. Choose the initial guesses as $a_0 = 4.1$ and $b_0 = 4.2$ instead of $a_0 = 4$ and $b_0 = 5$. Selection of two such closed values are not necessary in case the problem to solved on a computer through a program written for the implementation of bisection method.

9.11 SECANT METHOD

Secant method is also a bracketing method as it needs two initial guesses of a root to find the iterates of it. Suppose x_0 and x_1 are two distinct approximations of a root a of $f(x) = 0$.

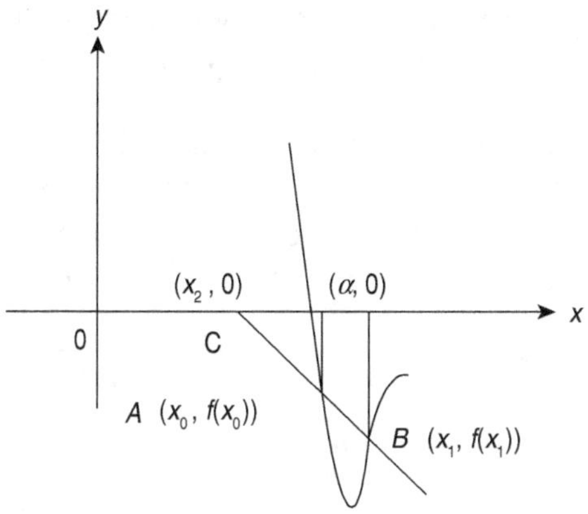

Figure 9.1

The equation of chord joining the points $A(x_0, f\,x_0)$ and $B(x_1, f(x_1))$ (Fig. 9.1) is

$$y - f(x_0) = \frac{f(x_1) - f(x_0)}{x_1 - x_0}(x - x_0) \tag{9.7}$$

Note that the variable x and y involved in (9.7) are the coordinates of any point on the line containing the chord AB.

Let the chord AB meets the x-axis at $C(x_2, 0)$.

Since $(x_2, 0)$ is a point of the chord AB, we have

$$0 - f(x_0) = \frac{f(x_1) - f(x_0)}{x_1 - x_0}(x_2 - x_0)$$

$$\text{i.e., } x_2 = \frac{x_0 f(x_1) - x_1 f(x_0)}{f(x_1) - f(x_0)} \tag{9.8}$$

Similarly, proceeding with $(x_1, f(x_1))$ and $(x_2, f(x_2))$, we have next approximation x_3 of the root α as

$$x_3 = \frac{x_1 f(x_2) - x_2 f(x_1)}{f(x_2) - f(x_1)}$$

In general, the sequence of approximations $\{x_n\}_{n=2}^{\infty}$ are computed by the following iteration formula:

$$x_{n+1} = \frac{x_{n-1} f(x_n) - x_n f(x_{n-1})}{f(x_n) - f(x_{n-1})} \ (n = 1, 2, ...) \tag{9.9}$$

The Eq. (9.9) is known as secant iterative formula and it is used to find iterates of a root of $f(x) = 0$.

9.12 PROCEDURE TO FIND AN APPROXIMATION TO A ROOT OF $f(x) = 0$ BY SECANT METHOD

Step 1: First choose two approximations x_0 and x_1 to a root a of $f(x) = 0$ such that x_0 and x_1 are close to a. For example, if there is a root of $f(x) = 0$ between 2 and 3, then we can take initial approximation to a as $x_0 = 2.5$ and $x_1 = 2.6$.

$$\text{Compute } x_2 = \frac{x_0 f(x_1) - x_1 f(x_0)}{f(x_1) - f(x_0)}$$

Step 2: Compute x_3 by substituting $n = 2$ in secant iterative formula. i.e.,

$$x_3 = \frac{x_1 f(x_2) - x_2 f(x_1)}{f(x_2) - f(x_1)}$$

Step 3: Repeat the Step 2 to compute $x_4, x_5, x_6, \ldots$ using secant iterative formula by taking $n = 3, 4, 5,\ldots$ till we get the desired degree of accuracy of the root or tolerance of the error.

SOLVED PROBLEMS

Problem 1: Given that a real root of the equation $f(x) = x^3 - 6x + 2 = 0$ lies in the interval $(0, 1)$. Perform four iterations of secant method to find the approximation of the root of $f(x) = 0$.

Solution: Choose $x_0 = 0$ and $x_1 = 1$. Then $f(0) = 2$ (positive) and $f(1) = 1 - 6 + 2 = -3$ (negative). By secant formula,

$$x_2 = \frac{x_0 f(x_1) - x_1 f(x_0)}{f(x_1) - f(x_0)}$$

$$x_2 = \frac{0 \cdot f(1) - 1 f(0)}{f(1) - f(0)}$$

$$= \frac{(0)(-3) - 1(2)}{-3 - 2} = \frac{-2}{-5} = 0.4$$

and $f(x_2) = f(0.4) = -0.33600003$. By secant formula $(n = 2)$,

$$x_3 = \frac{x_1 f(x_2) - x_2 f(x_1)}{f(x_2) - f(x_1)}$$

$$\therefore \quad x_3 = \frac{(1) f(0.4) - (0.4) f(1)}{f(0.4) - f(1)}$$

$$= \frac{(1)(-.33600003) - 0.4(-3)}{-.33600003 - (-3)}$$

$$= 0.34247$$

and $f(x_3) = -0.01413$. Similarly,

$$x_4 = \frac{x_2 f(x_3) - x_3 f(x_2)}{f(x_3) - f(x_2)}$$

$$\therefore x_4 = \frac{0.4 f(0.34247) - 0.34247\ f(0.4)}{f(0.34247) - f(0.4)}$$

$$= 0.33998$$

$$f(x_4) = -0.00058.$$

Again by secant formula with $n = 4$,

$$x_5 = \frac{x_3 f(x_4) - x_4 f(x_3)}{f(x_4) - f(x_3)}$$

$$= \frac{(0.34247)(-0.000058) - (0.33998)(-0.01413)}{(-0.000058) - (-0.01463)} = 0.33988.$$

After performing 4 iterations of the secant method, the approximation to a root of given equation is 0.33988.

Problem 2: Compute 8 iterations of Secant method to find an approximation to a root of $f(x) = e^x - 3\sin x - 2 = 0$.

Solution: Since $f(1) = -1.80613$ and $f(2) = 2.66116$, we choose $x_0 = 1$ and $x_1 = 2$. Now, by Secant formula (9.9) with $n = 1$,

$$x_2 = \frac{x_0 f(x_1) - x_1 f(x_0)}{f(x_1) - f(x_0)}$$

i.e., $x_2 = \dfrac{(1) f(2) - (2) f(1)}{f(2) - f(1)} = \dfrac{(1)(2.66116) - 2(-1.80613)}{(2.66116) - (-1.80613)} = 1.40430$

and $f(x_2) = f(1.40430) = -0.88584$.

Similarly, with $n = 2$, the secant formula is given by

$$x_3 = \frac{x_1 f(x_2) - x_2 f(x_1)}{f(x_2) - f(x_1)}$$

$$= \frac{2 f(1.40430) - 1.40430 f(2)}{f(1.40430) - f(2)}$$

$$= \frac{2(-0.88584) - 1.40430(2.66116)}{-.88584 - 2.66116} = 1.55307$$

and $f(x_3) = f(1.55307) = -0.27356$.
Now,

$$x_4 = \frac{x_2 f(x_3) - x_3 f(x_2)}{f(x_3) - f(x_2)}$$

$$= \frac{(1.40430) f(1.55307) - 1.55307 f(1.40430)}{f(1.55307) - f(1.40430)}$$

$$= \frac{(1.40430)(-0.27356) - 1.55307(-0.88584)}{(-0.27536) - (-0.88584)} = 1.59473$$

and $f(1.59473) = -0.07213$.
Similarly, $x_5 = 1.60543$, $x_6 = 1.60811$, $x_7 = 1.60878$, and $x_8 = 1.60894$.

Problem 3: Find a real root of $x - e^{-x} = 0$ correct 3 decimal places.

Solution: Let $f(x) = x - e^{-x} = 0$. Then, $f(0) = -1$ and $f(1) = 0.63212$.
Let $x_0 = 0$ and $x_1 = 1$. With $n = 1$, the Secant iterative formula becomes

$$x_2 = \frac{x_0 f(x_1) - x_1 f(x_0)}{f(x_1) - f(x_0)}$$

$$\therefore x_2 = \frac{0 f(1) - 1 f(0)}{f(1) - f(0)} = \frac{0(0.63212) - 1(-1)}{(63212) - (-1)} = .61270$$

Substituting $n = 2$ in Secant iterative formula, we have

$$x_3 = \frac{x_1 f(x_2) - x_2 f(x_1)}{f(x_2) - f(x_1)}$$

$$= \frac{1 f(\cdot 61270) - (\cdot 61270) f(1)}{f(.61270) - f(1)}$$

$$= \frac{(1) \cdot (0.07081) - (.61270)(.63212)}{0.07081 - 0.63212}$$

$$= \frac{-0.31648}{-0.56131} = 0.56384$$

Similarly, we have $x_4 = .56717$, $x_5 = .56714$.
The desired root correct to three decimal places is 0.567.

9.13   REGULA FALSI METHOD

The Secant method which we learnt in Section 9.12 is used to find the iterates of a root of $f(x) = 0$ by Eq. 9.9, and for each computation of x_{n+1}, we did not have any condition like $f(x_{n-1}) \, f(x_n) < 0$. The secant method is improved by imposing a logical condition for each computation of successive approximations. The improved method is known as Regula Falsi method or method of false position. Now, we describe the Regula Falsi method to determine a root of $f(x) = 0$.

We suppose $f(x)$ is a continuous function. Let a_0 and b_0 are two real numbers such that $f(a_0) f(b_0) < 0$. Then a root of $f(x) = 0$ lies between a_0 and b_0. (See Fig. 9.2).

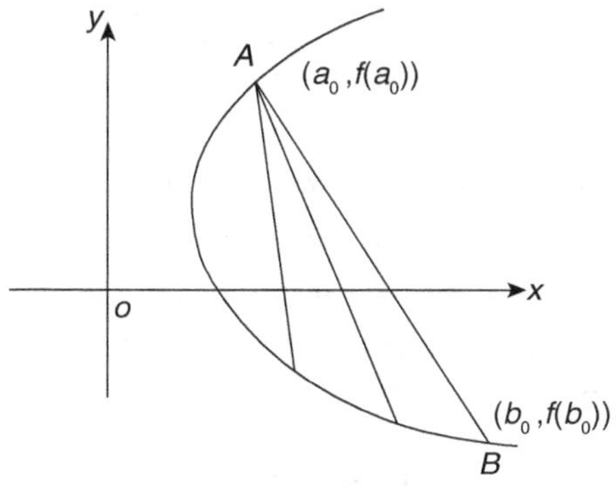

Figure 9.2

Let $A(a_0, f(a_0))$ and $B(b_0, f(b_0))$ be two points lying on the curve $y = f(x)$. Then the equation of secant containing AB is

$$y - f(a_0) = \frac{f(b_0) - f(a_0)}{b_0 - a_0}(x - a_0) \qquad (9.10)$$

Let $(c_1, 0)$ be the point of intersection of the secant line with x-axis. As the point $(c_1, 0)$ lies on the secant, we have

$$0 - f(b_0) = \frac{f(b_0) - f(a_0)}{b_0 - a_0}(c_1 - a_0)$$

i.e.,
$$c_1 = \frac{a_0 f(b_0) - b_0 f(a_0)}{f(b_0) - f(a_1)}$$

If $f(a_0) f(c_1) < 0$, then set $a_1 = a_0$ and $b_1 = c_1$ else set $a_1 = c_1$ and $b_1 = b_0$. Next approximation c_2 to the root of $f(x) = 0$ lies between a_1 and b_1, and it is given by

$$c_2 = \frac{a_1 f(b_1) - b_1 f(a_1)}{f(b_1) - f(a_1)}.$$

If $f(a_1) f(c_2) < 0$, then set $a_2 = a_1$ and $b_2 = c_2$ else set $a_2 = c_2$ and $b_2 = b_1$. The third approximation c_3 is given by

$$c_3 = \frac{a_2 \, f(b_2) - b_2 f(a_2)}{f(b_2) - f(a_2)}.$$

Continuing this process, we have successive approximations $c_4, c_5, c_6, \ldots$ to the root of $f(x) = 0$. The iterative formula of Regula–Falsi method is

$$c_{k+1} = \frac{a_k f(b_k) - b_k f(a_k)}{f(b_k) - f(a_k)}, \qquad (9.11)$$

where

(i) $a_{k+1} = a_k$ and $b_{k+1} = c_{k+1}$ if $f(a_k) f(c_{k+1}) < 0$.

(ii) $a_{k+1} = c_{k+1}$ and $b_{k+1} = b_k$ if $f(a_k) f(c_{k+1}) > 0$.

The method is shown graphically in Fig (9.2). The successive approximations $c_1, c_2, c_3, \ldots$ are approaching the exact solution of $f(x) = 0$. Note that Regula Falsi method is an open method.

9.14   PROCEDURE TO FIND THE APPROXIMATIONS TO A ROOT OF F (X) = 0 BY REGULA FALSI METHOD

Step 1: Find two values a_0 and b_0 such that $f(a_0) f(b_0) < 0$.

Step 2: Using Regula Falsi formula (with $n = 0$) compute c_1 where

$$c_1 = \frac{a_0 f(b_0) - b_0 f(a_0)}{f(b_0) - f(a_0)}.$$

Step 3: If $f(a_0) f(c_1) < 0$, then set $a_1 = a_0$, $b_1 = c_1$ else set $a_1 = c_1$ and $b_1 = b_0$.

Step 4: Repeat step 2 and step 3 with appropriate suffixes till we get the desired accuracy or error tolerance of the root of $f(x) = 0$.

SOLVED PROBLEMS

Problem 1: Perform six iterations of Regula Falsi method to find an approximate root of $x^3 - x - 11 = 0$.

Solution: Let $f(x) = x^3 - x - 11$. Then, $f(2) = -5$ and $f(3) = 13$. Hence, a root of $f(x) = 0$ lies between 2 and 3. Let $a_0 = 2$ and $b_0 = 3$.

With $k = 0$, the Regula Falsi formula becomes

$$c_1 = \frac{a_0 f(b_0) - b_0 f(a_0)}{f(b_0) - f(a_0)}$$

$$\therefore c_1 = \frac{2 f(3) - 3 f(2)}{f(3) - f(2)}$$

$$= \frac{2(13) - 3 f(-5)}{(13) - (-5)} = 2.27778$$

and $f(c_1) = f(2.27778) = -1.46005$.

Now, the root lies between 2.27778 and 3. Let $a_1 = 2.27778$ and $b_1 = 3$.

$$\therefore c_2 = \frac{2.27778 f(3) - 3 f(2.27778)}{f(3) - f(2.27778)}$$

$$= 2.35070$$

and $f(c_2) = f(2.35070) = -0.36120$.

As the root lies between 2.35070 and 3, suppose $a_2 = 2.35070$ and $b_2 = 3$.

$$c_3 = \frac{2.35070 f(3) - 3 f(2.35070)}{f(3) - f(2.35070)}$$

$$= 2.36825$$

and $f(c_3) = f(2.36825) = -0.08559$.

Now, the root lies between 2.36825 and 3. Let $a_3 = 2.36825$ and $b_3 = 3$. Then,

$$c_4 = \frac{2.36825\,f(3) - 3f(2.36825)}{f(3) - f(2.36825)}$$

$$= 2.37239$$

and $f(c_4) = f(2.3729) = -0.02008$.

Continuing this process, we get $c_5 = 2.7335$, $c_6 = 2.37358$, $c_7 = 2.37363$ and $c_8 = 2.37365$.

Problem 2: Find a real root correct to three decimal places of $f(x) = xe^x - 3 = 0$ by Regula Falsi method.

Solution: Since $f(x) = xe^x - 3$, we have $f(0) = -3$, $f(1) = e - 3 = -0.281718$ and $f(2) = 2e^2 - 3 = 11.778112$. Therefore, a root of given equation lies between 1 and 2. Let $a_0 = 1$, $b_0 = 2$. By Regula Falsi iterative formula for $k = 0$ is

$$c_1 = \frac{a_0 f(b_0) - b_0 f(a_0)}{f(b_0) - f(a_0)}$$

i.e.,
$$c_1 = \frac{(1)\, f(2) - 2f(1)}{f(2) - f(1)}$$

$$= \frac{1(11.778112) - 2(-0.281718)}{(11.778112) - (-0.281718)} = 1.023360$$

and $f(c_1) = f(1.02336) = -0.152472$. Root of given equation lies between 1.02336 and 2. Let $a_1 = 1.02336$ and $b_1 = 2$. Thus,

$$c_2 = \frac{a_1 f(b_1) - b_1 f(a_1)}{f(b_1) - f(a_1)}$$

$$= \frac{1.02336\,f(2) - 2f(1.02336)}{f(2) - f(1.02336)} = 1.035841$$

and $f(c_2) = f(1.035841) = -0.081542$.

Repeating the same process, we get $c_3 = 1.04247$, $c_4 = 1.045980$, $c_5 = 1.047835$, $c_6 = 1.048815$, $c_7 = 1.049332$ and $c_8 = 1.049604$. We can conclude that a root of given equation correct to three decimal places is 1.049.

Problem 3: Perform three iterations of Regula Falsi method to find an approximate root of $e^x \sin x - 1 = 0$ lying between 0.4 and 0.6.

Solution: Let $f(x) = e^x \sin x - 1$, $a_0 = 0.4$ and $b_0 = 0.6$. Then, $f(a_0) = -0.41906$ and $f(b_0) = 0.02885$.

Iteration 1: With $k = 1$, Regula Falsi formula becomes

$$c_1 = \frac{a_0 f(b_0) - b_0 f(a_0)}{f(b_0) - f(a_0)}$$

$$\therefore c_1 = \frac{0.4 f(0.6) - 0.6 f(0.4)}{f(0.6) - f(0.4)}$$

$$= \frac{0.4(0.02885) - 0.6(-0.41906)}{(0.02885) - (-0.41906)} = 0.58712$$

$\therefore f(c_1) = f(0.58712) = -0.00353$.

Now, the root lies between 0.58712 and 0.6.

Let $a_1 = 0.58712$ and $b_1 = 0.6$.

Iteration 2: With $k = 1$, Regula Falsi formula gives

$$c_2 = \frac{a_1 f(b_1) - b_1 f(a_1)}{f(b_1) - f(a_1)}$$

$$c_2 = \frac{0.58712 f(0.6) - 0.6 f(0.58712)}{f(0.6) - f(0.58712)} = 0.58852$$

The value of $f(x)$ at $x = c_2$ is -0.00002, i.e., $f(0.58852) = 0.00002$.

Now, the root lies between 0.58852 and 0.6. Let $a_2 = 0.58852$ and $b_2 = 0.6$.

Iteration 3: With $k = 2$, Regula Falsi formula reduces to

$$c_3 = \frac{a_2 f(b_2) - b_2 f(a_2)}{f(b_2) - f(a_2)}$$

$$= \frac{0.58852 f(0.6) - 0.6 f(0.58852)}{f(0.6) - f(0.58852)} = .58853$$

The approximate root of $e^x \sin x - 1 = 0$ is 0.58853.

Problem 4: Perform five iterations of Regula Falsi method to find an approximate root of $f(x) = x - 2\sin x = 0$ lying between 1.5 and 2.

Solution: Let $a_0 = 1.5$ and $b_0 = 2$. Then $f(a_0) = f(1.5) = -0.49499$ and $f(b_0) = f(2) = 0.18141$. Therefore, root of $f(x) = 0$ lies between 1.5 and 2.

Iteration 1: $c_1 = \dfrac{a_0 f(b_0) - b_0 f(a_0)}{f(b_0) - f(a_0)}$

$$c_1 = \frac{1.5 f(2) - 2 f(1.5)}{f(2) - f(1.5)} = 1.86590$$

and $f(c_1) = -0.04764$.

Iteration 2: Here $a_1 = 1.86590$ and $b_1 = 2.0$.

$$\therefore c_2 = \frac{1.86590\, f(2) - 2 f(1.86590)}{f(2) - f(1.86590)} = 1.89379$$

and $f(c_2) = -0.00278$.

Iteration 3: Here $a_2 = 1.89379$ and $b_2 = 2$ since $f(a_2) f(b_2) < 0$.

$$\therefore c_3 = \frac{1.89379\, f(2) - 2 f(1.89379)}{f(2) - f(1.89379)} = 1.89540$$

and $f(c_3) = f(1.89540) = -0.00016$.

Iteration 4: Here $a_3 = 1.89540$ and $b_3 = 2$.

$$\therefore c_4 = \frac{1.89540\, f(2) - 2. f(1.89540)}{f(2) - f(1.89540)} = 1.89549$$

and $f(c_4) = f(1.89549) = -0.00001$.

Iteration 5: Here $a_4 = 1.89549$ and $b_4 = 2$.

$$\therefore c_5 = \frac{1.89549\, f(2) - 2 f(1.89549)}{f(2) - f(1.89549)} = 1.89549$$

9.15   ORDER OF CONVERGENCE OF SECANT AND REGULA FALSI METHOD

Either in the secant method or in Regula Falsi method, the formula used in each method for computations of successive approximations of a root of $f(x) = 0$ is of the form

$$x_{n+1} = \frac{x_{n-1} f(x_n) - x_n f(x_{n-1})}{f(x_n) f(x_{n-1})} \tag{9.12}$$

This formula is applicable for Regula Falsi method if $f(x_{n-1})f(x_n) < 0$

Suppose the sequence of approximations $\{x_n\}$ of a root a of $f(x) = 0$ are obtained from either of these methods. Let $x_{n-1} = a + \varepsilon_{n-1}$, $x_n = a + \varepsilon_n$ and $x_{n+1} = a + \varepsilon_{n+1}$ where ε_{n-1}, ε_n, ε_{n+1} are errors in $(n-1)$th, nth and $(n+1)$th approximations of the root of $f(x) = 0$. Now, substituting $x_{n-1} = a + \varepsilon_{n-1}$, $x_n = a + \varepsilon_n$ and $x_{n+1} = a + \varepsilon_{n+1}$ in (9.12), we get

$$a + \varepsilon_{n+1} = \frac{(a + \varepsilon_{n-1})f(a + \varepsilon_n) - (a + \varepsilon_n)f(a + \varepsilon_{n-1})}{f(a + \varepsilon_n) - f(a + \varepsilon_{n-1})}. \tag{9.13}$$

Applying Taylor's series for $f(a + \varepsilon_{n-1})$ and $f(a + \varepsilon_n)$, Eq. (9.13) reduces to

$$\varepsilon_{n+1} = \frac{\begin{bmatrix} \varepsilon_{n-1}\left\{ f(a) + \varepsilon_n f'(a) + \dfrac{\varepsilon_n^2}{\lfloor 2} f''(a) + \dfrac{\varepsilon_n^3}{\lfloor 3} f'''(a) + \dots \right\} \\ -\varepsilon_n \left\{ f(a) + \varepsilon_{n-1} f'(a) + \dfrac{\varepsilon_{n-1}^2}{\lfloor 2} f''(a) + \dfrac{\varepsilon_{n-1}^3}{\lfloor 3} f'''(a) + \dots \right\} \end{bmatrix}}{\begin{bmatrix} \left\{ f(a) + \varepsilon_n f'(a) + \dfrac{\varepsilon_n^2}{\lfloor 2} f''(a) + \dfrac{\varepsilon_n^3}{\lfloor 3} f'''(a) + \dots \right\} \\ -\left\{ f(a) + \varepsilon_{n-1} f'(a) + \dfrac{\varepsilon_{n-1}^2}{\lfloor 2} f''(a) + \dfrac{\varepsilon_{n-1}^3}{\lfloor 3} f'''(a) + \dots \right\} \end{bmatrix}}$$

As a being the root of $f(x) = 0$, we have $f(a) = 0$. Therefore,

$$\varepsilon_{n+1} = \frac{\left[\left(\varepsilon_{n-1}\varepsilon_n^2 - \varepsilon_n\varepsilon_{n-1}^2\right) \dfrac{f''(a)}{\lfloor 2} + \left(\varepsilon_{n-1}\varepsilon_n^3 - \varepsilon_n\varepsilon_{n-1}^3\right) \cdot \dfrac{f'''(a)}{\lfloor 3} + \dots \right]}{\left[\left(\varepsilon_n - \varepsilon_{n-1}\right)f'(a) + \left(\varepsilon_n^2 - \varepsilon_{n-1}^2\right)\dfrac{f''(a)}{\lfloor 2} + \left(\varepsilon_n^3 - \varepsilon_{n-1}^3\right)\dfrac{f'''(a)}{\lfloor 3} + \dots \right]}$$

$$= \frac{\left[\varepsilon_n\varepsilon_{n-1}\left(\varepsilon_n - \varepsilon_{n-1}\right)\dfrac{f''a}{\lfloor 2} + \varepsilon_n\varepsilon_{n-1}\left(\varepsilon_n^2 - \varepsilon_{n-1}^2\right)\dfrac{f'''a}{\lfloor 3} + \dots \right]}{\left[\left(\varepsilon_n - \varepsilon_{n-1}\right)f'(a) \times \right.}$$

$$\left. \left\{ 1 + \left(\varepsilon_n + \varepsilon_{n-1}\right)\dfrac{f''(a)}{2!\,f'(a)} + \left(\varepsilon_n^2 + \varepsilon_n\varepsilon_{n-1} + \varepsilon_{n-1}^2\right)\dfrac{f'''(a)}{3!\,f'(a)} + \dots \right\} \right]}$$

$$\left[\left\{\varepsilon_n\varepsilon_{n-1}\frac{f''(a)}{\lfloor 2}+\varepsilon_n\varepsilon_{n-1}\cdot\left(\varepsilon_n+\varepsilon_{n-1}\right)\frac{f'''(a)}{\lfloor 3}+\ldots\right\}\times\right.$$

$$=\frac{\left.\left\{1+\left(\varepsilon_n+\varepsilon_{n-1}\right)\frac{f''(a)}{\lfloor 2f'(a)}+\frac{\varepsilon_n^2+\varepsilon_n\varepsilon_{n-1}+\varepsilon_{n-1}^2}{\lfloor 3f'(a)}\frac{f'''(a)}{}+\ldots\right\}^{-1}\right]}{f'(a)}$$

$$=\frac{1}{f'(a)}\left[\varepsilon_n\varepsilon_{n-1}\frac{f''(a)}{\lfloor 2}+\varepsilon_n\varepsilon_{n-1}\left(\varepsilon_n+\varepsilon_{n-1}\right)\frac{f'''(a)}{\lfloor 3}+\ldots\right].$$

$$\left[1-\left\{\left(\varepsilon_n+\varepsilon_{n-1}\right)\frac{f''(a)}{\lfloor 2f'(a)}+\left(\varepsilon_n^2+\varepsilon_n\varepsilon_{n-1}+\varepsilon_{n-1}^2\right)\frac{f'''(a)}{\lfloor 3f'(a)}+\ldots\right\}+\right.$$

$$\left.\left\{\left(\varepsilon_n+\varepsilon_{n-1}\right)\frac{f''(a)}{\lfloor 2f'(a)}+\left(\varepsilon_n^2+\varepsilon_n\varepsilon_{n-1}+\varepsilon_{n-1}^2\right)\frac{f'''(a)}{\lfloor 3f'(a)}+\ldots\right\}^2+\ldots\right]$$

$$=\frac{\varepsilon_n\varepsilon_{n-1}}{2f'(a)}f''(a)+O_3\left(\varepsilon_n,\varepsilon_{n-1}\right)\tag{9.14}$$

where $O_3\left(\varepsilon_n,\varepsilon_{n-1}\right)$ is the sum of terms involving $\varepsilon_n^r\varepsilon_{n-1}^s$ such that $r+s\geq 3$. Neglecting $O_3\left(\varepsilon_n,\varepsilon_{n-1}\right)$ Eq. (9.14) reduces to

$$\varepsilon_{n+1}=A\cdot\varepsilon_n\varepsilon_{n-1}$$

$$\text{where } A=\frac{f''(a)}{2f'(a)}\tag{9.15}$$

Let m be the order of convergence of the Regula–Falsi/Secant method. Then

$$\varepsilon_{i+1}=C\varepsilon_i^m\tag{9.16}$$

where C is a constant.

For $i=n$ and $(n-1)$, Eq. (9.16) gives

$$\varepsilon_{n+1}=C\varepsilon_n^m\tag{9.17}$$

$$\varepsilon_n=C\varepsilon_{n-1}^m\tag{9.18}$$

From Eq. (9.18), we get

$$\varepsilon_{n-1} = \frac{1}{C^{\frac{1}{m}}} \cdot \varepsilon_n^{\frac{1}{m}} \qquad\qquad (9.19)$$

Substituting (9.17) and (9.19) in (9.15), we get

$$C\varepsilon_n^m = A\varepsilon_n \cdot \frac{1}{C^{\frac{1}{m}}} \cdot \frac{1}{\varepsilon_n^m}$$

$$\varepsilon_n^m = \frac{A}{C^{1+\frac{1}{m}}} \cdot \varepsilon_n^{1+\frac{1}{m}}$$

Comparing the powers of ε_m, on both sides of above equation we have

$$m = 1 + \frac{1}{m}$$

i.e., $\qquad\qquad m^2 - m - 1 = 0$

$$\therefore m = \frac{1}{2}\left(1 + \sqrt{5}\right)$$

Thus, the rate of convergence of the Secant/Regula Falsi method is $\dfrac{1+\sqrt{5}}{2}$ or 1.618.

9.16   NEWTON–RAPHSON METHOD

Newton–Raphson method is an open method, which uses a single guess to obtain the iterates of a root of an equation $f(x) = 0$. Now, we proceed to describe the Newton–Raphson method.

Let x_0 be an approximation of a non-repeated root a of the equation $f(x) = 0$. Suppose that the exact solution is $x_1 = x_0 + h$ where h is very small. Then,

$$f(x_0 + h) = 0$$

Expanding $f(x_0 + h)$ by Taylor series about x_0, we get

$$f\left(x_0\right) + hf'\left(x_0\right) + \frac{h^2}{\lfloor 2} f''\left(x_0\right) + \ldots = 0$$

If all the terms of second-and higher-order powers of h are neglected, we have

$$f(x_0) + hf'(x_0) = 0$$

$$\therefore h = -\frac{f(x_0)}{f'(x_0)}$$

The improved value or next approximation x_1 of the initial approximation x_0 of the root is given by

$$x_1 = x_0 - \frac{f(x_0)}{f'(x_0)}$$

Repeating the above process with x_1 instead of x_0, we get next approximation of the root of $f(x) = 0$ and it is given by

$$x_2 = x_1 - \frac{f(x_1)}{f'(x_1)}$$

Continuing this process, we have a general formula

$$x_{n+1} = x_n - \frac{f(x_n)}{f'(x_n)} \tag{9.20}$$

and it is known as Newton–Raphson formula. The process of finding a root by (9.20) is called a Newton–Raphson method.

Note:

1. The Newton–Raphson method may fail to find an approximate root of $f(x) = 0$ upto desired accuracy when $\dfrac{f(x)}{f'(x)}$ is very large at the points lying in the neighbourhood of the root.

2. The Newton–Raphson method is not applicable to find a root a of $f(x) = 0$ when a is a multiple root or $f(x) = 0$ has another root near to (close to) a. However, the Newton–Raphson method is applicable to find the other real roots of it ($f(x) = 0$) if they are not multiple roots and no two roots are close to one another.

9.17 GEOMETRICAL INTERPRETATION OF NEWTON–RAPHSON METHOD

Let $y = f(x)$ and x_0 be an initial approximation to a root of $f(x) = 0$. We know that $f'(x_0)$ represents the slope of the tangent drawn to the curve $y = f(x)$ at $(x_0, f(x_0))$. Consider a point $A(x_0, f(x_0)$ that lies on the curve $y = f(x)$. The equation of the tangent to the curve $y = f(x)$ at A is given by

$$y - f(x_0) = f'(x_0)(x - x_0) \tag{9.21}$$

Suppose $B_1(x_1, 0)$ be the point of intersection of the tangent with x-axis. Then,

$$0 - f(x_0) = f'(x_0)(x_1 - x_0)$$

i.e.,
$$x_1 = x_0 - \frac{f(x_0)}{f'(x_0)}$$

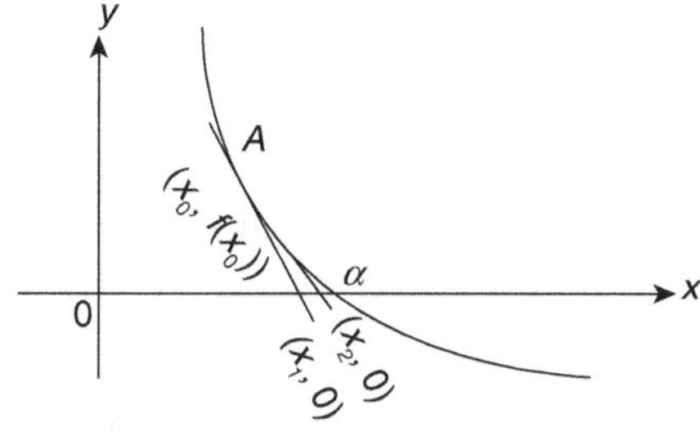

Figure 9.3

Let $B_2 = (x_1, f(x_1))$ and the tangent drawn to the curve at B_2 meet the x-axis at $(x_2, 0)$. It can be shown that

$$x_2 = x_1 - \frac{f(x_1)}{f'(x_1)}.$$

Continuing this process, we have the sequence of approximations $\{x_n\}$, where $x_{n+1} = x_n - \frac{f(x_n)}{f'(x_n)}$. The points $(x_1, 0)$, $(x_2, 0)$, $(x_3, 0)$, ... approach to $(a, 0)$, where a is the exact root of $f(x) = 0$.

9.18 PROCEDURE TO FIND AN APPROXIMATE ROOT OF $f(x) = 0$ BY NEWTON–RAPHSON METHOD

Step 1: First trace two values a and b such that $f(a) f(b) < 0$.
Step 2: Assume $x_0 \in (a, b)$ as an initial approximation.
Step 3: Find the derivative of $f(x)$.

Step 4: Write the Newton–Raphson formula

$$x_{n+1} = x_n - \frac{f(x_n)}{f'(x_n)} \quad (n = 0, 1, 2, 3,...),$$

where $f(x)$ is given function and $f'(x)$ obtained in step 3.

Step 5: For $n = 0$, the formula becomes

$$x_1 = x_0 - \frac{f(x_0)}{f'(x_0)}.$$

Substitute x_0 in the R.H.S. of above equation to obtain x_1.

Step 6: Repeat the Step 5 with appropriate changes in suffixes to get the root upto a desired degree of accuracy.

SOLVED PROBLEMS

Problem 1: Find a real root of the equation $x^3 - x - 2 = 0$ using Newton–Raphson method, correct to four decimal places.

Solution: Let $f(x) = x^3 - x - 2$. Then $f(0) = -2$ (negative), $f(1) = -2$ (negative) and $f(2) = 4$ (positive). Since $f(1)\,f(2) < 0$, a root of $f(x) = 0$ lies between 1 and 2. Choose $x_0 = 1$ as the initial approximation to the root of $f(x) = 0$. The derivative of $f(x)$ is $f'(x) = 3x^2 - 1$. Therefore, the Newton–Raphson iterative formula for the equation $f(x) = x^3 - x - 2 = 0$ becomes

$$x_{n+1} = x_n - \frac{x_n^3 - x_n - 2}{3x_n^2 - 1} \quad (n = 0, 1, 2, 3...) \tag{1}$$

Substituting $n = 0$ in (1), we have

$$x_1 = x_0 - \frac{x_0^3 - x_0 - 2}{3x_0^2 - 1}$$

$$\therefore x_1 = 1 - \frac{1^3 - 1 - 2}{3(1)^2 - 1} = 2$$

Substituting $n = 1$ in (1), we have

$$x_2 = x_1 - \frac{x_1^3 - x_1 - 2}{3x_1^2 - 1} = 2 - \frac{(2)^3 - 2 - 2}{3(2)^2 - 1}$$

$$= 2 - \frac{4}{11} = 1.63636$$

Similarly, the values of $x_{n+1} = x_n - \dfrac{x_n^3 - x_n - 2}{3x_n^2 - 1}$ for $n = 2, 3, 4, 5$ are

$$x_3 = 1.63636 - \frac{(0.74530)}{7.03306} = 1.53039$$

$$x_4 = 1.53039 - \frac{0.05394}{6.02630} = 1.52144$$

$$x_5 = 1.52144 - \frac{0.00037}{5.94435} = 1.52138$$

$$\text{and } x_6 = 1.52138 - \frac{0}{5.94379} = 1.52138$$

Therefore, the root of given equation correct to four decimal places is 1.5213.

Problem 2: Perform five iterations of Newton–Raphson method to find an approximate root of $f(x) = x - e^{-x} = 0$.

Solution: For the given function $f(0) = -1$(negative) and $f(1) = 1 - e^{-1} = 0.63212$. Therefore, a root of $f(x) = 0$ lies between 0 and 1. Choose $x_0 = 0$ as the initial approximation to root of $f(x) = 0$. The derivative of $f(x)$ is $f'(x) = 1 + e^{-x}$. Thus,

$$x_{n+1} = x_n - \frac{x_n - e^{-x_n}}{1 + e^{-x_n}}$$

is the Newton–Raphson iterative formula for the given function $f(x) = x - e^{-x}$. The successive iterative of a root lying between 0 and 1 are as follows.

$$x_1 = x_0 - \frac{x_0 - e^{-x_0}}{1 + e^{-x_0}}$$

$$= 0 - \frac{(-1)}{2} = 0.5$$

$$x_2 = 0.5 - \frac{(-0.10653)}{1.60653} = 0.56631,$$

$$x_3 = 0.56631 - \frac{(-0.00130)}{1.56762} = 0.56714$$

$$x_4 = 0.56714 - \frac{(-0.0000)}{1.56714} = 0.56714$$

$$x_5 = x_6 = \ldots = 0.56714$$

Problem 3: Find a real root of $f(x) = xe^x - 2 = 0$ by Newton–Raphson method.

Solution: Here $f(0) = -2$ (negative) and $f(1) = 1 - 2 = 0.71828$. Further $f'(x) = (x + 1)e^x$

Hence,

$$x_{n+1} = x_n - \frac{x_n e^{x_n} - 2}{(x_n + 1)e^{x_n}}$$

We choose $x_0 = 1$ as the initial approximation of a root $f(x) = 0$, then x_1, x_2, x_3, x_4 and x_5 are given by

$$x_1 = x_0 - \frac{x_0 e^{x_0} - 2}{(x_0 + 1)e^{x_0}} = 1 - \frac{0.71828}{5.43656} = 0.86788$$

$$x_2 = x_1 - \frac{x_1 e^{x_1} - 2}{(x_1 + 1)e^{x_1}} = 0.86788 - \frac{0.06716}{4.44902} = 0.85278$$

$$x_3 = x_2 - \frac{x_2 e^{x_2} - 2}{(x_2 + 1)e^{x_2}} = 0.85278 - \frac{(0.00077)}{4.34694} = 0.85261$$

$$x_4 = x_3 - \frac{x_3 e^{x_3} - 2}{(x_3 + 1)e^{x_3}} = 0.85261 - \frac{0}{4.34575} = 0.85261$$

and $\quad x_5 = x_4 - \dfrac{x_4 e^{x_4} - 2}{(x_4 + 1)e^{x_4}} = 85261 - \dfrac{0}{4.34575} = 0.85261$

Problem 4: Find an iterative formula to find the square root of a number N.

Solution: Let $x = \sqrt{N}$. Then, the square root of N is the solution or root of $x^2 - N = 0$.

Suppose $f(x) = x^2 - N$. Then, $f'(x) = 2x$. Therefore, the Newton–Raphson iterative formula for the function $f(x) = x^2 - N$ becomes

$$x_{n+1} = x_n - \frac{x_n^2 - N}{2x_n}$$

$$\text{i.e., } x_{n+1} = \frac{1}{2}\left[x_n + \frac{N}{x_n}\right]$$

Problem 5: Find $\sqrt{12}$ by an iterative formula.

Solution: In Problem 4, we have shown that the iterative formula for $\sqrt{N}$ is

$$x_{n+1} = \frac{1}{2}\left[x_n + \frac{N}{x_n}\right]$$

Therefore, the iterative formula for $\sqrt{12}$ is

$$x_{n+1} = \frac{1}{2}\left[x_n + \frac{12}{x_n}\right] \tag{1}$$

Since $9 < 12 < 16$, we have $3 < \sqrt{12} < 4$. It shows that $\sqrt{12}$ lies between 3 and 4. Choose an initial approximation of $\sqrt{12}$ as $x_0 = 3.5$. Now,

$$x_1 = \frac{1}{2}\left[3.5 + \frac{12}{3.5}\right] = 3.46429$$

$$x_2 = \frac{1}{2}\left[x_1 + \frac{12}{x_1}\right] = \frac{1}{2}\left[3.46429 + \frac{12}{3.46429}\right] = 3.46410$$

$$x_3 = \frac{1}{2}\left[x_2 + \frac{12}{x_2}\right] = \frac{1}{2}\left[3.46410 + \frac{12}{3.46410}\right] = 3.46410$$

Since $x_2 = x_3$, we can say that the value of $\sqrt{12}$ correct to five iterative places is 3.46410.

Problem 6: Find an iterative formula for pth root of N.

Solution: Let $x = N^{1/p}$. Then $N^{1/p}$ is a root of the equation $x^P - N = 0$. Suppose $f(x) = x^P - N$. Then $f'(x) = p.x^{P-1}$. By Newton–Raphson formula, the iterative formula of $N^{1/p}$

$$x_{n+1} = x_n + \frac{x_n^p - N}{p \cdot x_n^{p-1}}$$

$$x_{n+1} = \frac{1}{p}\left[(p-1)x_n + \frac{N}{x_n^{p-1}}\right]$$

Problem 7: Find an iterative formula to compute $\dfrac{1}{N}$.

Solution: Let $x = \dfrac{1}{N}$. Then, $N = \dfrac{1}{x}$ or $N - \dfrac{1}{x} = 0$.

We suppose $f(x) = N - \dfrac{1}{x}$. Then, $f'(x) = \dfrac{1}{x^2}$ and by Newton–Raphson formula to compute $\dfrac{1}{N}$ is given by

$$x_{n+1} = x_n - \frac{N - \dfrac{1}{x_n}}{\dfrac{1}{x_n^2}}$$

or

$$x_{n+1} = x_n\left(2 - Nx_n\right)$$

Problem 8: Find $\dfrac{1}{32}$ by an iterative formula.

Solution: The iterative formula for $\dfrac{1}{N}$ is

$$x_{n+1} = x_n\left(2 - Nx_n\right).$$

Since $10 < 32 < 100$, we have $0.01 < \dfrac{1}{32} < 0.1$, and choose $x_0 = 0.001$.

The successive approximations of $\dfrac{1}{32}$ are

$$x_1 = x_0(2 - 32x_0) = 0.00197,$$
$$x_2 = x_1(2 - 32x_1) = 0.00381,$$

$$x_3 = x_2(2 - 32x_2) = 0.00716$$
$$x_4 = -x_3(2 - 32x_3) = 0.001268$$
$$x_5 = x_4(2 - 32x_4) = 0.02021$$
$$x_6 = x_5(2 - 32x_5) = 0.02735$$
$$x_7 = x_6(2 - 32x_6) = 0.03076$$
$$x_8 = x_7(2 - 32x_7) = 0.03124$$
$$x_9 = x_8(2 - 32x_8) = 0.03125$$
$$x_{10} = x_9(2 - 32x_9) = 0.03125$$

$\therefore$ The value of $\dfrac{1}{32}$ is 0.03125.

Problem 9: Find an iterative formula to determine $\left(\dfrac{1}{N}\right)^{\frac{1}{P}}$.

Solution: Let $x = \left(\dfrac{1}{N}\right)^{\frac{1}{P}}$. Taking p power on both sides, we have $x^P = \dfrac{1}{N}$. It can be expressed as $N = \dfrac{1}{x^P}$.

Let $f(x) = N - \dfrac{1}{x^P}$. Then $f'(x) = +\dfrac{p}{x^{P-1}}$. Thus the iterative formula for $\left(\dfrac{1}{N}\right)^{\frac{1}{P}}$ is

$$x_{n+1} = x_n - \frac{N - x_n^{-P}}{p \cdot x_n^{-P-1}}$$

or

$$x_{n+1} = \frac{x_n}{p}\left[(p+1) - Nx_n^P\right]$$

$$n = 0, 1, 2, 3, \ldots\ldots$$

Problem 10: Find the value of $\left(\dfrac{1}{12}\right)^{\frac{1}{3}}$.

Solution: The iterative formula for $\left(\dfrac{1}{12}\right)^{\frac{1}{3}}$ can be obtained from Problem 9 with $p = 3$ and $N = 12$ and it is given by

$$x_{n+1} = \frac{x_n}{3}\left[4 - 12x_n^3\right]$$

Choose $x_0 = 0.2$. Now,

$$x_1 = \frac{x_0}{3}\left[4 - 12x_0^3\right] = \frac{0.2}{3}\left[4 - (0.2)^3\right] = 0.26027$$

Similarly, $x_2 = 0.32867$, $x_3 = 0.39155$, $x_4 = 0.42805$, $x_5 = 0.43644$, $x_6 = .43679$, $x_7 = 0.43679$. Thus, the approximate value of $\left(\dfrac{1}{12}\right)^{\frac{1}{3}}$ is 0.43679 correct to five decimal places.

9.19   ORDER OF CONVERGENCE OF NEWTON–RAPHSON METHOD

Let a be a root of $f(x) = 0$ and, x_n and x_{n+1} be two successive approximations of a root a obtained from Newton–Raphson method. Further let ϵ_i be the error in x_i. Then $x_n = a + \varepsilon_n$, and $x_{n+1} = a + \varepsilon_{n+1}$. Substituting these in Newton–Raphson formula (9.20), we get

$$a + \varepsilon_{n+1} = a + \varepsilon_n - \frac{f\left(a + \varepsilon_n\right)}{f'\left(a + \varepsilon_n\right)}$$

$$\varepsilon_{n+1} = \varepsilon_n - \frac{f\left(a + \varepsilon_n\right)}{f'\left(a + \varepsilon_n\right)}$$

or
$$\varepsilon_{n+1} = \frac{\varepsilon_n f'\left(a + \varepsilon_n\right) - f\left(a + \varepsilon_n\right)}{f'\left(a + \varepsilon_n\right)} \tag{9.22}$$

Expanding $f(a + \varepsilon_n)$ and $f'(a + \varepsilon_n)$ in Taylor series about a, we have

$$\varepsilon_{n+1} = \frac{\left[\varepsilon_n\left\{f'(a) + \varepsilon_n f''(a) + \varepsilon_n^2 \dfrac{f'''(a)}{\lfloor 2} + \ldots\right\} - \left\{f(a) + \varepsilon_n f'(a) + \dfrac{\varepsilon_n^2}{\lfloor 2} f''(a) + \ldots\right\}\right]}{\left[f'(a) + \varepsilon_n f''(a) + \dfrac{\varepsilon_n^2}{\lfloor 2} f'''(a) + \ldots\right]}$$

As a being a root of $f(x) = 0$, we have $f(a) = 0$. Using this fact, the above equation can be written as

$$\varepsilon_{n+1} = \frac{\left[\left\{\varepsilon_n^2 f''(a) + \dfrac{\varepsilon_n^3}{\lfloor 2} f'''(a) + \ldots\right\} - \left\{\dfrac{\varepsilon_n^2}{\lfloor 2} f''(a) + + \dfrac{\varepsilon_n^3}{\lfloor 3} f'''(a) + \ldots\right\}\right]}{f'(a)\left[1 + \varepsilon_n \dfrac{f''(a)}{f'(a)} + \varepsilon_n^2 \dfrac{f'''(a)}{f'(a)} + \ldots\right]}$$

$$= \frac{1}{f'(a)}\left[\frac{1}{2}\varepsilon_n^2 f''(a) + \frac{1}{3}\varepsilon_n^3 f'''(a) + \ldots\right]$$

$$\left[1 + \left\{\varepsilon_n \frac{f''(a)}{f'(a)} + \frac{\varepsilon_n^2}{\lfloor 2}\frac{f'''(a)}{f'(a)} + \ldots\right\}\right]^{-1}$$

$$= \frac{1}{f'(a)}\left[\frac{1}{2}\varepsilon_n^2 f''(a) + \frac{\varepsilon_n^3}{3} f'''(a) + \ldots\right]$$

$$\times\left[1 - \left\{\varepsilon_n \frac{f''(a)}{f'(a)} + \frac{\varepsilon_n^2}{\lfloor 2}\frac{f'''(a)}{f'(a)} + \ldots\right\} + \left\{\varepsilon_n \frac{f''(a)}{f'(a)} + \frac{\varepsilon_n^2}{\lfloor 2}\frac{f'''(a)}{f'(a)} + \ldots\right\}^2 - \ldots\right]$$

$$= \frac{1}{f'(a)}\left[\frac{1}{2}\varepsilon_n^2 f''(a) + 0_3\left(\varepsilon_n\right)\right]$$

where $O_3(\varepsilon_n)$ is the sum of terms involving ε_n^k, $k \geq 3$. Neglecting $O_3(\varepsilon_n)$, we obtain

$$\varepsilon_{n+1} \approx \frac{f''(a)}{2f'(a)}\varepsilon_n^2 \tag{9.23}$$

Hence, the order of convergence of Newton–Raphson method is 2.

9.20   NEWTON–RAPHSON METHOD FOR REPEATED ROOTS

In this section, we discuss the Newton–Raphson method to find an approximate root a of $f(x) = 0$ where a is a repeated root of $f(x) = 0$. Suppose a is a root of $f(x) = 0$ with multiplicity p. The generalized Newton–Raphson iterative formula to determine an approximation of the root a is given by

$$x_{n+1} = x_n - p\frac{f(x_n)}{f'(x_n)} \tag{9.24}$$

If the multiplicity of a root of $f(x) = 0$ is not known in advance, then we have to proceed as follows to know its multiplicity. For some positive integer m, compute $x_0 - m\dfrac{f(x_0)}{f'(x_0)}$, $x_0 - \dfrac{(m-1)f'(x_0)}{f''(x_0)}$, ..., $x_0 - \dfrac{f^{(n-1)}_{(x_0)}}{f^{(n)}_{(x_0)}}$, and if they are equal or nearly equal, then $f(x) = 0$ has a root of multiplicity m. For example, the roots of the equation $f(x) = x^3 - 4x^2 + 5x - 2 = 0$ are 1, 1 and 2. Therefore, 1 is a double root of it. For $x_0 = 0.9$, the values of

$$x_0 - 2\frac{f(x_0)}{f'(x_0)} \text{ and } x_0 - \frac{f'(x_0)}{f''(x_0)} \text{ are}$$

$$0.9 - 2\frac{f(0.9)}{f'(0.9)} = 0.99565$$

and

$$0.9 - \frac{f'(0.9)}{f''(0.9)} = 0.98846,$$

respectively. As the values 0.99565 and 0.98846 are close, we can say that $f(x) = 0$ has a double root near unity.

SOLVED PROBLEMS

Problem 1: Given that there is a double root of $f(x) = x^3 - 5x^2 + 8x - 4 = 0$ near 1.9. Find the double root of $f(x) = 0$.

Solution: Choose $x_0 = 1.9$ as the initial approximation of the double root of $f(x) = 0$. The Newton's generalized formula for a double root becomes

$$x_{n+1} = x_n - 2\frac{f(x_n)}{f'(x_n)},$$

i.e.,

$$x_{n+1} = x_n - 2 \cdot \frac{x_n^3 - 5x_n^2 + 8x_n - 4}{3x_n^2 + 10x_n + 8}$$

Now,

$$x_1 = 1.9 - 2\frac{\left[(1.9)^3 - 5(1.9)^2 + 8(1.9) - 4\right]}{\left[3(1.9)^2 - 10(1.9) + 8\right]} = 2.0058$$

Similarly, the values of x_2, x_3, x_4 are $x_2 = 2.00002$, $x_3 = 2.0000$ and $x_4 = 2.0000$. Thus the required double root of given equation is 2.

Problem 2: The equation $f(x) = x^3 - 7x^2 + 16x - 12 = 0$ has a double root near $x = 1.9$. Find the double root correct to four decimal places.

Solution: Choose $x_0 = 1.5$ as an initial approximation of the double root of $f(x) = 0$. The generalized Newton–Raphson formula for $p = 2$ becomes

$$x_{n+1} = x_n - 2\frac{f(x_n)}{f'(x_n)}$$

i.e.,
$$x_{n+1} = x_n - 2\frac{\left\{x_n^3 - 7x_n^2 + 16x_n - 12\right\}}{\left\{3x_n^2 - 14x_n + 16\right\}} \tag{1}$$

For $n = 0$, Eq. (1) reduces to

$$x_1 = x_0 - 2\frac{\left\{x_0^3 - 7x_0^2 + 16x_0 - 12\right\}}{\left\{3x_0^2 - 14x_0 + 16\right\}}$$

Substituting $x_0 = 1.5$, we get

$$x_1 = 1.92857$$

and similarly the values of x_2, x_3, x_4 are $x_2 = 1.92857$, $x_3 = 1.99770$, $x_4 = 2$ and $x_5 = 2$. Thus, the double root of given equation is 2.

Problem 3: Show that the order of convergence of the method

$$x_{n+1} = x_n - 2\frac{f(x_n)}{f'(x_n)} \text{ is two.}$$

Solution: Note that given method is nothing but a Newton–Raphson generalized method for $p = 2$. Let a be a double root of $f(x) = 0$ such that nth approximation of the root x_n is near to a.

Suppose $x_n = a + \varepsilon_n$ and $x_{n+1} = a + \varepsilon_{n+1}$ where ε_n and ε_{n+1} are the errors in the approximations x_n and x_{n+1} respectively. Now, substituting these values in the given method, we get

$$a + \varepsilon_{n+1} = a + \varepsilon_n - 2 \cdot \frac{f(a + \varepsilon_n)}{f'(a + \varepsilon_n)} \tag{1}$$

Since a is a double root we have $f(a) = 0$ and $f'(a) = 0$. In view of these, the Taylors expansions of $f(a + \varepsilon_n)$ and $f'(a + \varepsilon_n)$ become

$$f(a + \varepsilon_n) = \frac{\varepsilon_n^2}{\lfloor 2} f''(a) + \frac{\varepsilon_n^3}{\lfloor 3} f'''(a) + \dots.$$

and

$$f'(a + \varepsilon_n) = \varepsilon_n f''(a) + \frac{\varepsilon_n^2}{\lfloor 2} f'''(a) + \dots.$$

Further

$$\frac{f(a + \varepsilon_n)}{f'(a + \varepsilon_n)} = \frac{\varepsilon_n^2 \left\{ \dfrac{f''(a)}{\lfloor 2} + \dfrac{\varepsilon_n}{\lfloor 3} f'''(a) + \dots \right\}}{\left\{ \varepsilon_n f''(a) + \dfrac{\varepsilon_n}{\lfloor 2} f'''(a) + \dots \right\}}$$

i.e.,

$$\frac{f(a + \varepsilon_n)}{f'(a + \varepsilon_n)} = \frac{\varepsilon_n \left\{ \dfrac{f''(a)}{\lfloor 2} + \dfrac{\varepsilon_n f'''(a)}{\lfloor 3} + \dots \right\}}{\left\{ f''(a) + \dfrac{\varepsilon_n}{\lfloor 2} f'''(a) + \dots \right\}} \tag{2}$$

From (1) and (2),

$$\varepsilon_{n+1} = \varepsilon_n - 2\frac{\varepsilon_n \left\{ \dfrac{f''(a)}{\lfloor 2} + \dfrac{\varepsilon_n}{\lfloor 3} f'''(a) + \dots. \right\}}{\left\{ f''(a) + \dfrac{\varepsilon_n}{\lfloor 2} f'''(a) + \dots. \right\}}$$

$$= \frac{\varepsilon_n \left\{ f''(a) + \dfrac{\varepsilon_n}{\lfloor 2} f'''(a) + \ldots \right\} - 2\varepsilon_n \left\{ \dfrac{f''(a)}{\lfloor 2} + \dfrac{\varepsilon_n}{\lfloor 3} f'''(a) + \ldots \right\}}{f''(a) \left\{ 1 + \dfrac{\varepsilon_n}{\lfloor 2} \dfrac{f'''(a)}{f''(a)} + \dfrac{\varepsilon_n^2}{\lfloor 3} \dfrac{f'^v(a)}{f''(a)} + \ldots \right\}}$$

$$= \frac{\varepsilon_n^2 \left(\dfrac{1}{2} - \dfrac{1}{3} \right) f'''(a) + O_3(\varepsilon_n)}{f''(a) \left\{ 1 + \dfrac{\varepsilon_n}{\lfloor 2} \dfrac{f'''(a)}{f''(a)} + \dfrac{\varepsilon_n^2}{\lfloor 3} \dfrac{f'^v(a)}{f''(a)} + \ldots \right\}}$$

$$= \frac{\left[\varepsilon_n^2 \times \dfrac{1}{\lfloor 3} \cdot f'''(a) + O_3(\varepsilon_n) \right] \left[1 + \left\{ \dfrac{\varepsilon_n}{\lfloor 2} \dfrac{f'''(a)}{f''(a)} + \dfrac{\varepsilon_n^2}{\lfloor 3} \dfrac{f'^v(a)}{f''(a)} \ldots \right\} \right]^{-1}}{f''(a)}$$

$$= \frac{1}{f''(a)} \left[\varepsilon_n^2 \frac{f'''(a)}{\lfloor 3} + O_3(\varepsilon_n) \right] \left[1 - \left\{ \frac{\varepsilon_n}{\lfloor 2} \frac{f'''(a)}{f''(a)} + \frac{\varepsilon_n^2}{\lfloor 3} \frac{f'^v(a)}{f''(a)} \ldots \right\} \right.$$

$$\left. + \left\{ \frac{\varepsilon_n}{\lfloor 2} \frac{f'''(a)}{f''(a)} + \frac{\varepsilon_n^2 f'^v(a)}{f''(a)} + \cdots \right\}^2 + \ldots\ldots \right]$$

$$= \frac{1}{f''(a)} \left[\varepsilon_n^2 \frac{f'''(a)}{\lfloor 3} + O_3(\varepsilon_n) \right]$$

where $O_3(\varepsilon_n)$ is the sum of infinite terms involving ε_n^k, $k \geq 3$. Neglecting $O_3(\varepsilon_n)$, we have

$$\varepsilon_{n+1} \approx \frac{\varepsilon_n^2 f'''(a)}{\lfloor 3\ f''(a)}.$$

The order of convergence of given method is 2.

9.21 METHOD OF ITERATION OR FIXED-POINT ITERATION METHOD

Let $f(x) = 0$ be an equation. We rearrange $f(x) = 0$ in the form $x = g(x)$. If $a = g(a)$, then a is a root of $f(x) = 0$. Let x_0 be an approximation to a root of $f(x) = 0$. Then, it is also an approximate root of $x - g(x) = 0$. We find successive approximations $x_1, x_2, x_3, \ldots\ldots$ by

$$x_{n+1} = g(x_n), \qquad (9.25)$$

which is known as iterative formula of method of iteration or fixed-point iteration method. A number a is said to be a fixed point of the function $g(x)$ if $g(a) = a$. The process of finding the iterates of a root of $f(x) = 0$ using (9.25) is called method of iteration.

Note:

1. An equation $f(x) = 0$ can be written in the form $x = g(x)$ in several ways. For example, $f(x) = x^2 - 3x + 4 = 0$ can be expressed as $x = \dfrac{1}{3}\left(x^2 + 4\right)$ or $x = 3 - \dfrac{4}{x}$ or $x = \sqrt{3x - 4}$.

2. The approximations of a root of $f(x) = 0$ found by the method of iteration need not converge. For example, if we consider $x_0 = 1$ as initial approximation of a root of $x = 6 - x^2$, then the successive approximations of the root are 5, -19, -355, -126019 ...... and the sequence of these approximations does not converge.

We establish the sufficient condition for the convergence of sequence of successive approximations of a root of $x = g(x)$ in the next section.

9.22   SUFFICIENT CONDITION FOR CONVERGENCE OF THE METHOD OF ITERATION

Let $a \in [a, b]$ be a root of $f(x) = 0$ where the interval $[a, b]$ is the domain of the function. Let $f(x) = 0$ can be thrown in the form $x = g(x)$. Then, a satisfies $x = g(x)$. Therefore,

$$a = g(a) \qquad (9.26)$$

Suppose x_0 be an initial approximation to the root a. Then,

$$x_1 = g(x_0) \qquad (9.27)$$

Subtracting 9.27 from (9.26), we get

$$a - x_1 = g(a) - g(x_0). \qquad (9.28)$$

By Lagrange's mean value theorem, we have

$$\frac{g(a) - g(x_0)}{a - x_0} = g'(a_0) \quad \text{where } x_0 < a_0 < a$$

i.e.,

$$g(a) - g(x_0) = (a - x_0)\, g'(a_0). \qquad (9.29)$$

Substituting (9.29) in (9.28), we obtain

$$a - x_1 = (a - x_0)\, g'(a_0). \qquad (9.30\,a_1)$$

Similarly, it can be shown that

$$(a - x_2) = (a - x_1)\, g'\,(a_1) \tag{9.30 a_2}$$
$$(a - x_3) = (a - x_2)\, g'\,(a_2) \tag{9.30 a_3}$$
$$(a - x_{n+1}) = (a - x_n)\, g'\,(a_n) \tag{9.30 a_n}$$

Multiplying together 9.30 a_1 to 9.30 a_n, we get

$$(a - x_1)\,(a - x_2)\,(a - x_3)\,\ldots(a - x_{n+1})$$
$$= (a - x_0)\,(a - x_1)\,\ldots(a - x_n)\, g'\,(a_0)\, g'\,(a_1)\,\ldots\, g'\,(a_n)$$

i.e.,
$$(a - x_{n+1}) = (x - x_0)\, g'\,(a_0)\, g'\,(a_1)\,\ldots\, g'\,(a_n). \tag{9.31}$$

Suppose $g'|(a_i)| \le k < 1$ for $i = 1, 2, \ldots, n$; Then,

$$\left|(a - x_{n+1})\right| \le \left|(a - x_0)\right| k^n$$

Since $k < 1$, we have $\underset{n\to\infty}{\mathrm{Lt}}\ k^n = 0$. Hence, $\underset{n\to\infty}{\mathrm{Lt}}\ x_{n+1} = a$. Thus, if $|g'(x)| < 1$ for all x in the neighborhood of a, then the successive approximations of the root a of $x = g(x)$ converges.

9.23 ORDER OF CONVERGENCE OF METHOD OF ITERATION

Suppose $x = g(x)$ is a rearrangement of equation $f(x) = 0$ such that $|g'(x)| < 1$ for all x in the vicinity of a root a of $f(x) = 0$ or $x = g(x)$. Since a is a root of $x = g(x)$, we have

$$a = g(a) \tag{9.32}$$

Let $x_{n+1} = a + \varepsilon_{n+1}$ and $x_n = a + \varepsilon_n$ where ε_{n+1} and ε_n are errors in nth and $(n + 1)$th approximations x_n and x_{n+1} of the root a respectively. Then,

$$x_{n+1} = g(x_n)$$

becomes

$$a + \varepsilon_{n+1} = g(a + \varepsilon_n)$$

Using Taylor series, it reduces

$$a + \varepsilon_{n+1} = g(a) + \varepsilon_n g'(a) + \frac{\varepsilon_n^2}{\lfloor 2} g''(a) + \cdots$$

i.e.,
$$\varepsilon_{n+1} = \varepsilon_n g'(a) + \frac{\varepsilon_n^2}{\lfloor 2} g''(a) + \cdots$$

i.e.,
$$\varepsilon_{n+1} = \varepsilon_n g'(a) + O_2(\varepsilon_n), \tag{9.33}$$

where $O_2(\varepsilon_n)$ is sum of infinite terms involving ε_n^k, $k \geq 2$. Neglecting $O_2(\varepsilon_n)$ in (9.33), we get

$$\varepsilon_{n+1} \approx \varepsilon_n g'(a).$$

Thus, the order of convergence of the method of iteration is linear.

SOLVED PROBLEMS

Problem 1: Find a real root correct to four decimal places of $x^2 - 2x - 4 = 0$.

Solution: Let $f(x) = x^2 - 2x - 4$. Then, $f(0) = -4$ (negative), $f(1) = -5$ (negative), $f(2) = -4$ (negative), $f(3)) = -1$ (negative) and $f(4) = 4$ (positive). Thus, $f(x)$ has a root between 3 and 4. The equation $x^2 - 2x - 4 = 0$ can be rearranged as

$$x = \sqrt{2x + 4} \tag{1}$$

so that $g'(x) = \dfrac{1}{\sqrt{2x+4}}$, where $g(x) = \sqrt{2x+4}$. Choose the initial approximation $x_0 = 3$. Before the computations of successive approximations of the root, we verify the condition $|g'(x)| < 1$. For

$$|g'(x)| = \left| \frac{1}{\sqrt{2x+4}} \right|$$

Clearly, $|g'(x)| < 1$ for all $x \in (3, 4)$. Hence, the method iteration $x_{n+1} = g(x_n)$ convergence for the function $g(x) = \sqrt{2x+4}$. Now,

$$x_1 = g(x_0) = g(3) = \sqrt{2 \times 3 + 4} = 3.162278$$

$$x_2 = g(x_1) = \sqrt{2x_1 + 4} = 3.213185$$

$$x_3 = g(x_2) = \sqrt{2x_2 + 4} = 3.228989$$

$$x_4 = g(x_3) = \sqrt{2x_3 + 4} = 3.233880$$

$$x_5 = g(x_4) = \sqrt{2x_4 + 4} = 3.235392$$

$$x_6 = g(x_5) = \sqrt{2x_5 + 4} = 3.235859$$

$$x_7 = g(x_6) = \sqrt{2x_6 + 4} = 3.236003$$

$$x_8 = g(x_7) = \sqrt{2x_7 + 4} = 3.236048$$

Thus, a real root correct to 4 decimal places of $x^2 - 2x - 4 = 0$ is 3.2360.

Problem 2: Find a real root of $x^3 + x^2 - 1 = 0$ by the method of iteration correct to three decimal places.

Solution: Let $f(x) = x^3 + x^2 - 1$. Then, $f(0) = -1$ (negative) and $f(1) = 1$ (positive). Therefore, a real root lies in the interval (0, 1). Now, given $f(x) = 0$ can be rearranged in the form

$$x = \frac{1}{\sqrt{x+1}}. \tag{1}$$

Let $g(x) = \dfrac{1}{\sqrt{x+1}}$. Then $g'(x) = \dfrac{-1}{2(x+1)^{3/2}}$.

$$\left| g'(x) \right| = \left| \frac{-1}{2(x+1)^{3/2}} \right| < 1 \ \ \forall x \in (0, 1).$$

Choose the initial approximation of the root lying between 0 and 1 as $x_0 = 0.5$. Then,

$$x_1 = g(x_0) = g(0.5) = \frac{1}{\sqrt{0.5 + 1}} = 0.816497$$

$$x_2 = g(x_1) = \frac{1}{\sqrt{x_1 + 1}} = 0.741964$$

$$x_3 = g(x_2) = \frac{1}{\sqrt{x_2 + 1}} = 0.757671$$

$$x_4 = g(x_3) = \frac{1}{\sqrt{x_3 + 1}} = 0.754278$$

$$x_5 = g(x_4) = \frac{1}{\sqrt{x_4 + 1}} = 0.755007$$

$$x_6 = g(x_5) = \frac{1}{\sqrt{x_5 + 1}} = 0.754850$$

$$x_7 = g(x_6) = \frac{1}{\sqrt{x_6 + 1}} = 0.754884$$

Thus, 0.754 is a root of given equation correct to three decimal places.

Problem 3: Find a real root of $3x = \cos x + 1$ by the method of iteration, correct to three decimal places.

Solution: Let $f(x) = 3x - \cos x - 1$. Then, $f(0) = -2$ (negative) and $f(1) = 1.459697692$ (positive). Therefore, a root of given equation lies between 0 and 1. The given equation can be rearranged in the form of $x = g(x)$ as follows:

$$x = \frac{1}{3}(\cos x + 1) \tag{1}$$

Hence, $g(x) = \frac{1}{3}(\cos x + 1)$. Now,

$$\left|g'(x)\right| = \frac{|\sin x|}{3} < 1 \; \forall \; x \in (0, 1]$$

Therefore, the method of iteration is applicable for $x = g(x)$ to find its real root. Choose $x_0 = 0.5$. Then,

$$x_1 = g(x_0) = g(0.5) = \frac{1}{3}(1 + \cos(0.5)) = 0.625861$$

$$x_2 = g(x_1) = \frac{1}{3}(1 + \cos x_1) = 0.603486$$

$$x_3 = g(x_2) = \frac{1}{3}(1 + \cos x_2) = 0.607787$$

$$x_4 = g(x_3) = \frac{1}{3}(1 + \cos x_3) = 0.606971$$

$$x_5 = g(x_4) = \frac{1}{3}(1 + \cos x_4) = 0.607126$$

$$x_6 = g(x_5) = \frac{1}{3}(1 + \cos x_5) = 0.607097$$

Thus, 0.607 is a real root of $3x - \cos x - 1 = 0$ correct to 3 decimal places.

Problem 4: Find a real root of $e^x - 3x^2 = 0$ by the method of iteration. Perform ten iterations.

Solution: Let $f(x) = e^x - 3x^2$. Then $f(0) = 1$ (positive) and $f(1) = -0.281718171$ (negative) therefore, a root of $f(x) = 0$ lies between 0 and 1. Choose the initial approximation of a root of $e^x - 3x^2 = 0$ as $x_0 = 0$. The given equation can be expressed as $x = \sqrt{\dfrac{e^x}{3}}$ so that

$g(x) = \dfrac{e^{\frac{x}{2}}}{\sqrt{3}}$, and now $g'(x) = \dfrac{1}{2\sqrt{3}} e^{\frac{x}{2}}$. It can be shown that

$$|g'(x)| = \left| \frac{e^{\frac{x}{2}}}{2\sqrt{3}} \right| < 1 \,\forall\, x \in (0,1)$$

Therefore, the method of iteration is applicable for the problem at our disposal. As $x_0 = 0$ is chosen,

$$x_1 = g(x_0) = g(0) = \sqrt{\frac{e^0}{3}} = 0.577350$$

$$x_2 = g(x_1) = \sqrt{\frac{e^{x_1}}{3}} = 0.770565$$

$$x_3 = g(x_2) = \sqrt{\frac{e^{x_2}}{3}} = 0.848722$$

$$x_4 = g(x_3) = \sqrt{\frac{e^{x_3}}{3}} = 0.882545$$

$$x_5 = g(x_4) = \sqrt{\frac{e^{x_4}}{3}} = 0.897598$$

$$x_6 = g(x_5) = \sqrt{\frac{e^{x_5}}{3}} = 0.904378$$

$$x_7 = g(x_6) = \sqrt{\frac{e^{x_6}}{3}} = 0.907450$$

$$x_8 = g(x_7) = \sqrt{\frac{e^{x_7}}{3}} = 0.908845$$

$$x_9 = g(x_8) = \sqrt{\frac{e^{x_8}}{3}} = 0.909479$$

$$x_{10} = g(x_9) = \sqrt{\frac{e^{x_9}}{3}} = 0.909767$$

After ten iterations, the approximate root of given equation is 0.909767.

9.24 AITKEN'S Δ^2-METHOD

Suppose x_{n-1}, x_n and x_{n+1} be three successive approximations of a desired root a of an equation $f(x) = 0$. Let ε_{n-1}, ε_n, ε_{n+1} be errors in the approximations x_{n-1}, x_n and x_{n+1} respectively. Assume that each error in an approximation of the root is proportional to its error in its immediate preceding approximation. Then $\varepsilon_n = k\,\varepsilon_{n-1}$ are $\varepsilon_{n+1} = k\,\varepsilon_n$. Therefore,

$$\frac{\varepsilon_{n+1}}{\varepsilon_n} = \frac{k.\varepsilon_n}{k.\varepsilon_{n-1}}$$

$$\frac{\varepsilon_{n+1}}{\varepsilon_n} = \frac{\varepsilon_n}{\varepsilon_{n-1}}$$

i.e.,

$$\frac{a - x_{n+1}}{a - x_n} = \frac{a - x_n}{a - x_{n-1}}.$$

It gives

$$x_{n+1}\, x_{n-1} - x_n^2 = (x_{n+1} - 2x_n + x_{n-1})a$$

or

$$a = \frac{x_{n+1}\, x_{n-1} - x_n^2}{(x_{n+1} - 2x_n + x_{n-1})}$$

$$= \frac{x_{n+1}(x_{n+1} - 2x_n + x_{n-1}) - x_{n+1}^2 + 2x_n x_{n+1} - x_n^2}{(x_{n+1} - 2x_n + x_{n-1})}$$

$$= x_{n+1} - \frac{(x_{n+1}^2 - 2x_n x_{n+1} - x_n^2)}{x_{n+1} - 2x_n + x_{n-1}}$$

$$\therefore \qquad a = x_{n+1} - \frac{(\Delta x_n)^2}{\Delta^2 x_n} \qquad\qquad (9.34)$$

The convergence of approximations accelerate by computing the approximations by Eq. (9.34). This acceleration is known as Aitken's acceleration. Note that it can be applied to any iterative method. It is also to be noted that the values x_{n-1}, x_n and x_{n+1} need not be in equidistant.

SOLVED PROBLEMS

Problem 1: Perform three iterations of the method of iteration followed by one iteration of Aitken's Δ^2-method to find a root of $x^3 + x^2 - 1 = 0$. Choose $x_0 = 0.5$ as initial approximation.

Solution: The equation $x^3 + x^2 - 1 = 0$ can be rearranged as $x = \sqrt{\dfrac{1}{(x+1)}}$ so that $g(x) = \dfrac{1}{\sqrt{x+1}}$. Now,

$$x_1 = g(x_0) = g(0.5) = 0.816497$$

$$x_2 = g(x_1) = g(0.816497) = 0.741964$$

$$x_3 = g(x_2) = g(0.741964) = 0.757671$$

Let x_3^A be the value to be determined by Aitken's method using the approximations x_1, x_2 and x_3. By Aitken's Δ^2-method,

$$x_3^A = x_1 - \frac{(x_2 - x_1)^2}{(x_3 - 2x_2 + x_1)}$$

$$\therefore\ x_3^A = 0.816497 - \frac{(0.741964 - 0.816497)}{(0.757671 - 2(0.741964) + 0.816497)}$$

$$= 0.7549$$

Problem 2: Perform three iterations of a bisection method starting with two initial guesses 0 and 1 to determine the approximate root of $f(x) = x^3 + x - 1 = 0$. Using these values, compute the approximate root of $f(x) = 0$ by Aitken's Δ^2-method.

Solution: Since $x_0 = 0$ and $b_0 = 1$.

The first approximation $c_1 = \dfrac{0+1}{2} = 0.5$

$$f(0.5) = (0.5)^3 + 0.5 - 1$$
$$= 0.125 + 0.5 - 1$$
$$= -0.375$$

The root lies between 0.5 and 1

$$c_2 = \frac{0.5 + 1}{2} = 0.75$$

$f(0.75) = (0.75)^3 + 0.75 - 1 = 0.171875$

The root lies between 0.5 and 0.75;

$$\therefore \qquad c_3 = \frac{0.75 + 0.5}{2} = 0.625$$

The next approximation c_4^A by Aitken's Δ^2-method is

$$c_4^A = c_3 - \frac{(c_2 - c_1)^2}{c_3^2 - 2c_2 + c_1}$$

$$= 0.625 - \frac{(0.75 - 0.5)^2}{(0.625)^2 - 2 \times 0.75 + 0.5}$$

$$= 0.625 - \frac{0.0625}{0.390625 - 1.5 + 0.5} = 0.727564$$

9.25 MULLER'S METHOD

The Muller's method of solving an equation $f(x) = 0$ is based on approximating $f(x)$ by a second-degree curve such that the abscissas of the points of the intersection of second-degree curve and the curve $y = f(x)$ are close to a root of $f(x) = 0$. To approximate a function by a second-degree curve (a curve of second-degree polynomial), we need three points. Suppose that $A(x_{n-2}, f(x_{n-2}))$, $B(x_{n-2}, f(x_{n-1}))$ and $C(x_n, f(x_n))$ are three points on the curve $y = f(x)$ such that x_{n-2}, x_{n-1} and x_n are close to a root a of $f(x) = 0$. Let $p_2(x)$ be a second define curve which is an approximation of $y = f(x)$ and passing through the points A, B and C. Then

$$p_2(x_{n-2}) = f(x_{n-2})$$
$$p_2(x_{n-1}) = f(x_{n-1}) \tag{9.35}$$
$$p_2(x_n) = f(x_n)$$

Now suppose

$$p_2(x) = a_0(x - x_{n-1})(x - x_n) + a_1(x - x_{n-2})(x - x_n)$$
$$+ a_2(x - x_{n-2})(x - x_{n-1}) \tag{9.36}$$

Where a_0, a_1 and a_2 are to be determined constants and it satisfies the equations given by (9.35). Substituting $x = x_{n-2}$ in (9.36) we get

$$p_2(x_{n-2}) = a_0(x_{n-2} - x_{n-1})(x_{n-2} - x_n)$$

In view of 1st equation of (9.35), it reduces to

$$f(x_{n-2}) = a_0(x_{n-2} - x_{n-1})(x_{n-2} - x_n)$$

$$\therefore \qquad a_0 = \frac{f(x_{n-2})}{(x_{n-2} - x_{n-1})(x_{n-2} - x_n)} \tag{9.37}$$

Similarly by substituting $x = x_{n-1}$ and x_n in (9.36) and using the second and third equations of (9.35) we have

$$a_1 = \frac{f(x_{n-2})}{(x_{n-1} - x_{n-2})(x_{n-1} - x_n)} \qquad (9.38)$$

$$a_2 = \frac{f(x_n)}{(x_n - x_{n-2})(x_n - x_{n-1})} \qquad (9.39)$$

Thus,

$$p_2(x) = \frac{(x - x_{n-1})(x - x_n)}{(x_{n-2} - x_{n-1})(x_{n-2} - x_n)} f_{n-2} +$$

$$\frac{(x - x_n)(x - x_{n-2})}{(x_{n-1} - x_n)(x_{n-1} - x_{n-2})} f_{n-1} + \frac{(x - x_{n-2})(x - x_{n-1})}{(x_n - x_{n-2})(x_n - x_{n-1})} f_n \qquad (9.40)$$

where $f(x_k) = f_k$. The absicissas of the point of intersection of $y = p_2(x)$ with x-axis is an approximate root of $f(x) = 0$ and it is one of the root of $p_2(x) = 0$.

$$\frac{(x - x_{n-1})(x - x_n)}{(x_{n-2} - x_{n-1})(x_{n-2} - x_n)} f_{n-2} + \frac{(x - x_n)(x - x_{n-2})}{(x_{n-1} - x_n)(x_{n-1} - x_{n-2})} f_{n-1}$$

$$+ \frac{(x - x_{n-2})(x - x_{n-1})}{(x_n - x_{n-2})(x_n - x_{n-1})} f_n = 0 \qquad (9.41)$$

We express (9.40) in a suitable form to obtain a root of equation (9.41) by introducing

$$h = x - x_n$$

$$h_n = x_n - x_{n-1}$$

$$\text{and } h_{n-1} = x_{n-1} - x_{n-2} \qquad (9.42)$$

In view of (9.42) the equation given by (9.41) becomes

$$\frac{h(h + h_n)}{h_{n-1}(h_{n-1} + h_n)} f_{n-2} - \frac{h(h + h_n + h_{n-1})}{h_{n-1} h_n} f_{n-1}$$

$$+ \frac{(h + h_n + h_{n-1})(h + h_n)}{(h_n + h_{n-1}) h_n} f_n = 0 \qquad (9.43)$$

Further introducing

$$\lambda = \frac{h}{h_n},\; \lambda_n = \frac{h_n}{h_{n-1}}\; \text{and}\; \delta_n = 1 + \lambda_n\; \text{ the equation (9.43) reduces to}$$

$$\lambda(\lambda+1)\lambda_n^2 \delta_n^{-1} f_{n-2} - \lambda[(1+\lambda)\lambda_n + 1]f_{n-1}$$

$$+ (\lambda+1)\delta_n^{-1}[1 + \lambda_n(\lambda+1)]f_n = 0 \tag{9.44}$$

Since

$$\frac{h(h+h_n)}{h_{n-1}(h_{n-1}+h_n)} = \frac{\dfrac{h}{h_n}\left(\dfrac{h}{h_n}+1\right)}{\dfrac{h_{n-1}}{h_n}\left(\dfrac{h_{n-1}}{h_n}+1\right)} = \frac{\lambda(\lambda+1)}{\dfrac{1}{\lambda_n}\left(\dfrac{1}{\lambda_n}+1\right)}$$

$$= \frac{\lambda_n^2(\lambda)(\lambda+1)}{(1+\lambda_n)} = \lambda(\lambda+1)\lambda_n^2\delta_n^{-1}$$

$$\frac{h(h+h_n+h_{n-1})}{h_{n-1}h_n} = \frac{\dfrac{h}{h_n}\left(\dfrac{h}{h_n}+1+\dfrac{h_{n-1}}{h_n}\right)}{\dfrac{h_{n-1}}{h_n}} \cdot (1)$$

$$= \frac{\lambda(\lambda+1+\lambda_n^{-1})}{\dfrac{1}{\lambda_n}} = \lambda[1 + \lambda_n(\lambda+1)]$$

and

$$\frac{(h+h_n+h_{n-1})(h+h_n)}{(h_n+h_{n-1})h_n} = \frac{\left(\dfrac{h}{h_n}+1+\dfrac{h_{n-1}}{h_n}\right)\left(\dfrac{h}{h_n}+1\right)}{1+\dfrac{h_{n-1}}{h_n}} \cdot (1)$$

$$\frac{\left(\lambda+1+\dfrac{1}{\lambda n}\right)(\lambda+1)}{\left(1+\dfrac{1}{\lambda n}\right)} = \frac{[\lambda_n+(\lambda+1)+1][\lambda+1]}{(1+\lambda_n)}$$

$$= (\lambda+1)[\lambda_n(\lambda+1)+1]\delta_n^{-1}.$$

Now the equation (9.44) gives

$$\lambda_n[\lambda_n f_{n-2} - \delta_n f_{n-1} + f_n]\lambda^2$$
$$+\left[\lambda_n^2 f_{n-2} - \delta_n^2 f_{n-1} + f_n(\lambda_n + \delta_n)\right]\lambda + f_n\delta_n = 0 \tag{9.45}$$

Let

$$c_n = \lambda_n(f_{n-2}\lambda_n - f_{n-1}\delta_n + f_n) \tag{9.46}$$

and

$$g_n = f_{n-2}\lambda_n^2 - \delta_n^2 f_{n-1} + f_n(\lambda_n + \delta_n) \tag{9.47}$$

In view of (9.46) and (9.47) the equation (9.45) reduces to

$$c_n\lambda^2 + g_n\lambda + \delta_n f_n = 0 \tag{9.48}$$

or

$$\delta_n f_n \cdot \frac{1}{\lambda^2} + g_n \frac{1}{\lambda} + c_n = 0$$

Hence

$$\frac{1}{\lambda} = \frac{-g_n \pm \sqrt{g_n^2 - 4\delta_n c_n f_n}}{2\delta_n f_n}$$

$$\therefore \qquad \lambda = \frac{-2\delta_n f_n}{g_n \mp \sqrt{g_n^2 - 4\delta_n c_n f_n}} \tag{9.49}$$

The sign in the denominator is chosen so that λ has smallest absolute value, we have to take negative sign if g_n is negative otherwise we take positive sign in the equation (9.49). Let such value of λ be λ_{n+1}. Then

$$\lambda_{n+1} = \frac{h_{n+1}}{h_n} = \frac{x_{n+1} - x_n}{x_n - x_{n-1}}$$

Hence,

$$x_{n+1} = x_n + (x_n - x_{n-1})\lambda_{n+1}$$

i.e.,
$$x_{n+1} = x_n + (x_n - x_{n-1}).\left\{\frac{-2\delta_n f_n}{g_n \mp \sqrt{g_n^2 - 4\delta_n c_n f_n}}\right\} \tag{9.50}$$

The said procedure can be repeated with $(x_{n-1}, f(x_{n-1}))$, $(x_n, f(x_n))$ and $(x_{n+1}, f(x_{n+1}))$ to have a better approximation x_{n+2} of the root of $f(x) = 0$. This process can be repeated until the desired degree of accuracy of the root is obtained. The process of finding a root by (9.50) is known as Muller's method.

9.26   GEOMETRICAL INTERPRETATION OF MULLER'S METHOD

Let $f(x) = 0$ be an equation and the curve $y = f(x)$ crosses the x–axis at $A(a, 0)$.

Suppose $B(x_{n-2}, f(x_{n-2}))$, $C(x_{n-1}, f(x_{n-1}))$ and $D(x_n, f(x_n))$ be any three points on the curve so that x_{n-2}, x_{n-1} and x_n are close to a. Let $p_2(x)$ be the second-degree curve passing through B, C and D (see Fig. (9.4)). Now, x_{n+1} of the Muller's method is the abscissa of the point of intersection of the curve $y = p_2(x)$ and x-axis.

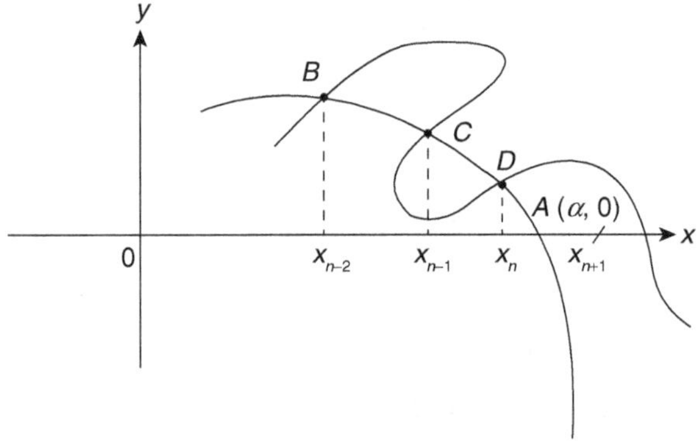

Figure 9.4

9.27   CONVERGENCE OF MULLER'S METHOD

The approximate root of $f(x) = 0$ is obtained by a second-degree curve

$$p_2(x) \approx f(x) = \frac{(x - x_{n-1})(x - x_n)}{(x_{n-2} - x_{n-1})(x_{n-2} - x_n)} f_{n-2}$$

$$+ \frac{(x - x_{n-2})(x - x_n)}{(x_{n-1} - x_{n-2})(x_{n-1} - x_n)} f_{n-1} + \frac{(x - x_{n-2})(x - x_{n-1})}{(x_n - x_{n-1})(x_n - x_{n-2})} f_n \;, \quad (9.51)$$

which is an approximation of $y = f(x)$ and passing through the points $(x_{n-2}, f(x_{n-2}))$, $(x_{n-1}, f(x_{n-1}))$ and $(x_n, f(x_n))$. We compute the rate of convergence of Muller's methods replacing $f(x)$ in Newton–Raphson formula by R.H.S. of Eq. (9.51). Differentiating (9.51) with respect to x, and substituting $x = x_n$, we get

$$f'(x_n) = \frac{1}{(x_n - x_{n-2})}\left[f_n + \frac{(x_n - x_{n-1})}{(x_{n-1} - x_{n-2})} f_{n-2} \right]$$
$$+ \frac{1}{(x_n - x_{n-1})}\left[f_n + \frac{(x_n - x_{n-2})}{(x_{n-2} - x_{n-1})} f_{n-1} \right]$$

or

$$f'(x_n) = \frac{1}{(x_n - x_{n-2})}\left[f(x_n) + \frac{(x_n - x_{n-1})}{(x_{n-1} - x_{n-2})} f(x_{n-2}) \right]$$
$$+ \frac{1}{(x_n - x_{n-1})}\left[f(x_n) + \frac{(x_n - x_{n-2})}{(x_{n-2} - x_{n-1})} f(x_{n-1}) \right]$$

since $f_k = f(x_k)$.

Let $x_{n-2} = a + \varepsilon_{n-2}$, $x_{n-1} = a + \varepsilon_{n-1}$ and $x_n = a + \varepsilon_n$ where ε_{n-2}, ε_{n-1} and ε_n are the errors in the approximations x_{n-2}, x_{n-1} and x_n, respectively.

Now, $\quad f'(a + \varepsilon_n) = \dfrac{1}{(\varepsilon_n - \varepsilon_{n-2})}\left[f(a + \varepsilon_n) + \dfrac{\varepsilon_n - \varepsilon_{n-1}}{\varepsilon_{n-1} - \varepsilon_{n-2}} f(a + \varepsilon_{n-2}) \right]$

$$+ \frac{1}{(\varepsilon_n - \varepsilon_{n-1})}\left[f(a + \varepsilon_n) + \frac{\varepsilon_n - \varepsilon_{n-2}}{\varepsilon_{n-2} - \varepsilon_{n-1}} f(a + \varepsilon_{n-1}) \right]$$

$$= \frac{1}{(\varepsilon_n - \varepsilon_{n-2})}\left[\left\{ \varepsilon_n f'(a) + \frac{\varepsilon_n^2}{\underline{2}} f''(a) + \frac{\varepsilon_n^3}{\underline{3}} f'''(a) + \cdots \right\} \right.$$
$$\left. + \frac{\varepsilon_n - \varepsilon_{n-1}}{\varepsilon_{n-1} - \varepsilon_{n-2}} \left\{ \varepsilon_{n-2} f'(a) + \frac{\varepsilon_{n-2}^2}{\underline{2}} f''(a) + \frac{\varepsilon_{n-1}^3}{\underline{3}} f'''(a) \cdots \right\} \right]$$

$$+ \frac{1}{(\varepsilon_n - \varepsilon_{n-1})}\left[\left\{ \varepsilon_n f'(a) + \frac{\varepsilon_n^2}{\underline{2}} f''(a) + \frac{\varepsilon_n^3}{\underline{3}} f'''(a) + \cdots \right\} \right.$$
$$\left. + \frac{\varepsilon_n - \varepsilon_{n-2}}{\varepsilon_{n-2} - \varepsilon_{n-1}} \left\{ \varepsilon_{n-1} f'(a) + \frac{\varepsilon_{n-1}^2}{\underline{2}} f''(a) + \frac{\varepsilon_{n-1}^3}{\underline{3}} f'''(a) + \cdots \right\} \right]$$

since $f(a) = 0$. Now, we wish to collect to the coefficients of $f'(a)$, $f''(a)$ etc. in $f'(a + \varepsilon_n)$. Now,

$$f'(a + \varepsilon_n) = \frac{1}{(\varepsilon_n - \varepsilon_{n-2})}\left[\varepsilon_n f'(a) + \frac{\varepsilon_n^2}{\lfloor 2} f''(a) + \frac{\varepsilon_n^3}{\lfloor 3} f'''(a) \cdots \right]$$

$$+ \frac{\varepsilon_n - \varepsilon_{n-1}}{(\varepsilon_n - \varepsilon_{n-2})(\varepsilon_{n-1} - \varepsilon_{n-2})}\left[\varepsilon_{n-2} f'(a) + \frac{\varepsilon_{n-2}^2}{\lfloor 2} f''(a) + \frac{\varepsilon_{n-2}^3}{\lfloor 3} f'''(a) \cdots \right]$$

$$+ \frac{1}{(\varepsilon_n - \varepsilon_{n-1})}\left\{ \varepsilon_n f'(a) + \frac{\varepsilon_n^2}{\lfloor 2} f''(a) + \frac{\varepsilon_n^3}{\lfloor 3} f'''(a) \cdots \right\}$$

$$+ \frac{(\varepsilon_n - \varepsilon_{n-2})}{(\varepsilon_n - \varepsilon_{n-1})(\varepsilon_{n-2} - \varepsilon_{n-1})}\left[\varepsilon_{n-1} f'(a) + \frac{\varepsilon_{n-2}^2}{\lfloor 2} f''(a) + \frac{\varepsilon_{n-1}^3}{\lfloor 3} f'''(a) \cdots \right]$$

$$= f'(a)\left[\frac{\varepsilon_n}{(\varepsilon_n - \varepsilon_{n-2})} + \frac{(\varepsilon_n - \varepsilon_{n-1})\varepsilon_{n-2}}{(\varepsilon_n - \varepsilon_{n-2})(\varepsilon_{n-1} - \varepsilon_{n-2})} + \frac{\varepsilon_n}{(\varepsilon_n - \varepsilon_{n-1})} + \frac{\varepsilon_{n-1}(\varepsilon_n - \varepsilon_{n-1})}{(\varepsilon_n - \varepsilon_{n-1})(\varepsilon_{n-2} - \varepsilon_{n-1})} \right]$$

$$+ \frac{f''(a)}{\lfloor 2}\left[\frac{\varepsilon_n^2}{(\varepsilon_n - \varepsilon_{n-2})} + \frac{\varepsilon_{n-2}^2(\varepsilon_n - \varepsilon_{n-1})}{(\varepsilon_n - \varepsilon_{n-2})(\varepsilon_{n-1} - \varepsilon_{n-2})} + \frac{\varepsilon_n^2}{(\varepsilon_n - \varepsilon_{n-1})} + \frac{\varepsilon_{n-1}^2(\varepsilon_n - \varepsilon_{n-1})}{(\varepsilon_n - \varepsilon_{n-1})(\varepsilon_{n-2} - \varepsilon_{n-1)})} \right]$$

$$+ \frac{f'''(a)}{\lfloor 3}\left[\frac{\varepsilon_n^3}{(\varepsilon_n - \varepsilon_{n-2})} + \frac{\varepsilon_{n-2}^3(\varepsilon_n - \varepsilon_{n-1})}{(\varepsilon_n - \varepsilon_{n-2})(\varepsilon_{n-1} - \varepsilon_{n-2})} + \frac{\varepsilon_n^3}{(\varepsilon_n - \varepsilon_{n-1})} + \frac{\varepsilon_{n-1}^3(\varepsilon_n - \varepsilon_{n-1})}{(\varepsilon_n - \varepsilon_{n-1})(\varepsilon_{n-2} - \varepsilon_{n-1})} \right] + \cdots$$

$$= f'(a) \left[\frac{1}{(\varepsilon_n - \varepsilon_{n-2})} \left\{ \varepsilon_n + \frac{\varepsilon_{n-2}(\varepsilon_n - \varepsilon_{n-1})}{\varepsilon_{n-1} - \varepsilon_{n-2}} \right\} + \frac{1}{(\varepsilon_n - \varepsilon_{n-1})} \left\{ \varepsilon_n + \frac{\varepsilon_{n-1}(\varepsilon_n - \varepsilon_{n-2})}{(\varepsilon_{n-2} - \varepsilon_{n-1})} \right\} \right]$$

$$+ \frac{f''(a)}{\lfloor 2} \left[\frac{1}{(\varepsilon_n - \varepsilon_{n-2})} \left\{ \varepsilon_n^2 + \frac{\varepsilon_{n-2}^2(\varepsilon_n - \varepsilon_{n-1})}{(\varepsilon_{n-1} - \varepsilon_{n-2})} \right\} + \frac{1}{\varepsilon_n - \varepsilon_{n-1}} \left\{ \varepsilon_n^2 + \frac{\varepsilon_{n-1}^2(\varepsilon_n - \varepsilon_{n-2})}{(\varepsilon_{n-2} - \varepsilon_{n-1})} \right\} \right]$$

$$+ \frac{f'''(a)}{\lfloor 3} \left[\frac{1}{(\varepsilon_n - \varepsilon_{n-1})} \left\{ \varepsilon_n^3 + \frac{\varepsilon_{n-2}^3(\varepsilon_n - \varepsilon_{n-1})}{(\varepsilon_{n-1} - \varepsilon_{n-2})} \right\} + \frac{1}{(\varepsilon_n - \varepsilon_{n-1})} \left\{ \varepsilon_n^3 + \frac{\varepsilon_{n-1}^3(\varepsilon_n - \varepsilon_{n-2})}{(\varepsilon_{n-2} - \varepsilon_{n-1})} \right\} \right] + \dots$$

$$= f'(a) \left[\frac{\varepsilon_{n-1}}{\varepsilon_{n-1} - \varepsilon_{n-2}} - \frac{\varepsilon_{n-2}}{(\varepsilon_{n-1} - \varepsilon_{n-2})} \right]$$

$$+ \frac{f''(a)}{\lfloor 2} \left[\frac{\varepsilon_{n-1}(\varepsilon_n^2 - \varepsilon_{n-2}^2) - \varepsilon_n \varepsilon_{n-2}(\varepsilon_n - \varepsilon_{n-2})}{(\varepsilon_n - \varepsilon_{n-2})(\varepsilon_{n-1} - \varepsilon_{n-2})} + \frac{\varepsilon_{n-2}(\varepsilon_n^2 - \varepsilon_{n-1}^2) - \varepsilon_n \varepsilon_{n-1}(\varepsilon_n - \varepsilon_{n-1})}{(\varepsilon_n - \varepsilon_{n-1})(\varepsilon_{n-2} - \varepsilon_{n-1})} \right]$$

$$+ \frac{f'''(a)}{\lfloor 3} \left[\frac{\varepsilon_{n-1}(\varepsilon_n^3 - \varepsilon_{n-2}^3) - \varepsilon_n \varepsilon_{n-2}(\varepsilon_n^2 - \varepsilon_{n-2}^2)}{(\varepsilon_n - \varepsilon_{n-2})(\varepsilon_{n-1} - \varepsilon_{n-2})} + \frac{\varepsilon_{n-2}(\varepsilon_n^3 - \varepsilon_{n-1}^3) - \varepsilon_n \varepsilon_{n-1}(\varepsilon_n^2 - \varepsilon_{n-1}^2)}{(\varepsilon_n - \varepsilon_{n-1})(\varepsilon_{n-2} - \varepsilon_{n-1})} \right] + \dots$$

$$= f'(a)(1) + \frac{f''(a)}{\lfloor 2} \left[\left\{ \frac{\varepsilon_{n-1}(\varepsilon_n + \varepsilon_{n-2}) - \varepsilon_n \varepsilon_{n-2}}{(\varepsilon_{n-1} - \varepsilon_{n-2})} \right\} + \left\{ \frac{\varepsilon_{n-2}(\varepsilon_n + \varepsilon_{n-1}) - \varepsilon_n \varepsilon_{n-1}}{(\varepsilon_{n-2} - \varepsilon_{n-1})} \right\} \right]$$

$$+ \frac{f'''(a)}{\lfloor 3} \left[\left\{ \frac{\varepsilon_{n-1}(\varepsilon_n^2 + \varepsilon_n \varepsilon_{n-2} + \varepsilon_{n-2}^2) - \varepsilon_n \varepsilon_{n-2}(\varepsilon_n + \varepsilon_{n-2})}{(\varepsilon_{n-1} - \varepsilon_{n-2})} \right\} + \left\{ \frac{\varepsilon_{n-2}(\varepsilon_n^2 + \varepsilon_n \varepsilon_{n-1} + \varepsilon_{n-1}^2) - \varepsilon_n \varepsilon_{n-1}(\varepsilon_n + \varepsilon_{n-1})}{(\varepsilon_{n-2} - \varepsilon_{n-1})} \right\} \right]$$

Thus

$$f'(a + \varepsilon_n) = f'(a) + \varepsilon_n f''(a) + \frac{f'''(a)}{\lfloor 3} \times$$

$$\left\{ 2\varepsilon_n^2 - \varepsilon_{n-1}\varepsilon_{n-2} + \varepsilon_n \varepsilon_{n-1} + \varepsilon_n \varepsilon_{n-2} \right\} + \cdots \tag{9.52}$$

Substituting $f(x_n) = f(a + \varepsilon_n)$, $f'(x_n) = f'(a + \varepsilon_n)$ and $x_{n+1} = a + \varepsilon_{n+1}$ in Newton–Raphson formula, we get

$$a + \varepsilon_{n+1} = a + \varepsilon_n - \frac{f(a + \varepsilon_n)}{f'(a + \varepsilon_n)} \,.$$

Using Taylor series of $f(a + \varepsilon_n)$ and (9.52), the above equation reduces to

$$\varepsilon_{n+1} = \varepsilon_n - \frac{\left\{ f'(a)\varepsilon_n + \dfrac{\varepsilon_n^2 f''(a)}{\lfloor 2} + \dfrac{\varepsilon_n^3 f'''(a)}{\lfloor 3} + \cdots \right\}}{\left[f'(a) + \varepsilon_n f''(a) + (2\varepsilon_n^2 - \varepsilon_{n-1}\varepsilon_{n-2} + \varepsilon_n \varepsilon_{n-1} + \varepsilon_n \varepsilon_{n-2}) \times \dfrac{f'''(a)}{\lfloor 3} + \cdots \right]}$$

$$= \frac{\left[\varepsilon_n \left\{ f'(a) + \varepsilon_n f''(a) + (2\varepsilon_n^2 - \varepsilon_{n-1}\varepsilon_{n-2} + \varepsilon_n \varepsilon_{n-1} + \varepsilon_n \varepsilon_{n-2}) \dfrac{f'''(a)}{\lfloor 3} + \cdots \right\} - \left\{ f'(a)\varepsilon_n + \dfrac{\varepsilon_n^2}{\lfloor 2} f''(a) + \dfrac{\varepsilon_n^3}{\lfloor 3} f'''(a) + \cdots \cdots \right\} \right]}{\left[f'(a) + \varepsilon_n f''(a) + (2\varepsilon_n^2 - \varepsilon_{n-1}\varepsilon_{n-2} + \varepsilon_n \varepsilon_{n-1} + \varepsilon_n \varepsilon_{n-2}) \dfrac{f'''(a)}{\lfloor 3} + \cdots \right]}$$

$$= \frac{\dfrac{\varepsilon_n^2}{\lfloor 2} f''(a) + (\varepsilon_n^3 - \varepsilon_n \varepsilon_{n-1} \varepsilon_{n-2} + \varepsilon_n^2 \varepsilon_{n-1} + \varepsilon_n^2 \varepsilon_{n-2}) \dfrac{f'''(a)}{\lfloor 3} + \cdots}{f'(a) \left[1 + \dfrac{f''(a)}{f'(a)} \varepsilon_n + \dfrac{f'''(a)}{\lfloor 3\, f'(a)} \times (2\varepsilon_n^2 - \varepsilon_{n-1}\varepsilon_{n-2} + \varepsilon_n\varepsilon_{n-1} + \varepsilon_n\varepsilon_{n-2}) + \cdots \right]}$$

$$= \frac{1}{f'(a)} \left\{ \frac{\varepsilon_n^2}{\lfloor 2} f''(a) + (\varepsilon_n^3 - \varepsilon_n\varepsilon_{n-1}\varepsilon_{n-2} + \varepsilon_n^2\varepsilon_{n-1} + \varepsilon_n^2\varepsilon_{n-2}) \frac{f'''(a)}{\lfloor 3} + \cdots \right\} \times$$

$$\left[1 - \left\{ \varepsilon_n \frac{f''(a)}{f'(a)} + \frac{f'''(a)}{\lfloor 3\, f'(a)} \times (2\varepsilon_n^2 - \varepsilon_{n-1}\varepsilon_{n-2} + \varepsilon_n\varepsilon_{n-1} + \varepsilon_n\varepsilon_{n-2}) + \cdots \right\} \right.$$

$$\left. + \left\{ \varepsilon_n \frac{f''(a)}{f'(a)} + \frac{f'''(a)}{\lfloor 3\, f'(a)} \times (2\varepsilon_n^2 - \varepsilon_{n-1}\varepsilon_{n-2} + \varepsilon_n\varepsilon_{n-2} + \varepsilon_n\varepsilon_{n-1} + \varepsilon_n\varepsilon_{n-2}) + \cdots \right\}^2 \cdots\cdots \right]$$

Neglecting all the terms except the term involving $\varepsilon_n \varepsilon_{n-1} \varepsilon_{n-2}$, We have

$$\varepsilon_{n+1} = -\frac{f'''(a)}{\lfloor 3\, f'(a)} \varepsilon_n \varepsilon_{n-1} \varepsilon_{n-2} \tag{9.53}$$

Let $\varepsilon_k = c\, \varepsilon_{k-1}^p$, where c and p are constants.

Then,

$$\varepsilon_{k-1} = \varepsilon_k^{1/p} \cdot c^{-1/p} \tag{9.54}$$

Substituting $k = n,\ n-1$ in (9.54), we get

$$\therefore\ \varepsilon_{n-1} = \varepsilon_n^{1/p} \cdot c^{-1/p} \tag{9.55}$$

and

$$\varepsilon_{n-2} = \varepsilon_{n-1}^{1/p} c^{-1/p}. \tag{9.56}$$

Further,

$$\varepsilon_{n-2} = \left\{ \varepsilon_n^{+1/p} \cdot c^{-1/p} \right\}^{+1/p} \cdot c^{-1/p}$$

$$= \varepsilon_n^{1/p^2} \cdot c^{-1/p^2} \cdot c^{-1/p}$$

i.e.,
$$\varepsilon_{n-2} = \varepsilon_n^{\frac{1}{p^2}} \cdot c^{-\frac{1}{p^2}-\frac{1}{p}} \tag{9.57}$$

and

$$\varepsilon_{n+1} = c\,\varepsilon_n^p. \tag{9.58}$$

Substituting ε_{n-2}, ε_{n-1}, ε_{n+1} in, (9.53)
we get

$$c \cdot \varepsilon_n^p = -\frac{f'''(a)}{\lfloor 3\, f'(a)} \cdot \varepsilon_n \cdot \varepsilon_n^{\frac{1}{p}} \cdot c^{-\frac{1}{p}} \cdot \varepsilon_n^{\frac{1}{p^2}} \cdot c^{-\frac{1}{p}-\frac{1}{p^2}}$$

$$\varepsilon_n^p = A c^{-\left(1+\frac{2}{p}+\frac{1}{p^2}\right)} \cdot \varepsilon_n^{1+\frac{1}{p}+\frac{1}{p^2}},$$

where

$$A = -\frac{f'''(a)}{\lfloor 3\, f'(a)}. \tag{9.59}$$

Equating the powers of ε_n on both sides, we obtain

$$p = 1 + \frac{1}{p} + \frac{1}{p^2}$$

i.e.,
$$p^3 - p^2 - p - 1 = 0 \tag{9.60}$$

Let
$$g(p) = p^3 - p^2 - p - 1 = 0. \text{ Then,}$$

$$g(1) = -2 \text{ (negative) and } g(2) = 1 \text{ (positive).}$$

Therefore, a root of (9.60) lies between 1 and 2. The iterative formula
for (9.60) is

$$p_{k+1} = p_k - \frac{p_k^3 - p_k^2 - p_k - 1}{3p_k^2 - 2p_k - 1}$$

Choose $p_0 = 1.6$. Then,

$$p_1 = 1.6 - \frac{(1.6)^3 - (1.6)^2 - (1.6) - 1}{3(1.6)^2 - 2(1.6) - 1}$$

$$= 1.90575$$

Similarly, we can find successive approximations and their values
are as shown below,

$$p_2 = 1.84266$$
$$p_3 = 1.83930$$
$$p_4 = 1.83929$$
$$p_5 = 1.83929$$

Hence, the order of convergence of Muller's method is 1.84.

SOLVED PROBLEMS

Problem 1: Perform three iterations of the Muller's method to find a real root of $x^3 - 3x - 5 = 0$. Use the initial approximations 2, 2.5 and 3.

Solution: Let $x_0 = 2$, $x_1 = 2.5$ and $x_2 = 3$.

Iteration 1: $h_2 = x_2 - x_1 = 0.5$, $h_1 = x_1 - x_0 = 0.5$, $\lambda_2 = \dfrac{h_2}{h_1} = 1$,

$$\delta_2 = 2, \lambda_2 + \delta_2 = 3,$$

$$g_2 = \lambda_2^2 f_0 - \delta_2^2 f_1 + (\lambda_2 + \delta_2)f_2 = 23.5,$$

and

$$c_2 = \lambda_2(\lambda_2 f_0 - \delta_2 f_1 + f_2) = 3.75$$

By Muller's method

$$x_3 = x_2 + (x_2 - x_1).\left\{ \frac{-2\delta_2 f_2}{g_2 + \sqrt{g_2^2 - 4\delta_2 f_2 c_2}} \right\} \tag{1}$$

Note that as $g_2 > 0$, we have taken positive sign.
   Substituting x_1, x_2, δ_2, f_2, g_2 and c_2 value in (1) we get

$$x_3 = 2.282516.$$

Iteration 2: Now, we take $x_1 = 2.5$, $x_2 = 3$ and $x_3 = 2.282516$ as the initial approximations of the root.

$$h_3 = x_3 - x_2 = -0.717484, h_2 = x_2 - x_1 = 0.5$$

$$\lambda_3 = \frac{h_3}{h_2} = -1.434968, \delta_3 = 1 + \lambda_3 = -0.434968$$

$$\lambda_3 + \delta_3 = -1.869936, g = 3.892793$$

and $c_3 = -1.742617$.

By Muller's method,

$$x_4 = x_3 + (x_3 - x_2) \cdot \frac{-2\delta_3 f_3}{g_3 + \sqrt{g_3^2 - 4\delta_3 f_3 c_3}}$$

$\therefore x_4 = 2.278974$.

Iteration 3: In this iteration we take $x_2 = 3$ $x_3 = 2.282516$ and $x_4 = 2.27894$ as the initial approximation to the root of given equation

$$h_4 = x_4 - x_3 = -0.003542, h_3 = x_3 - x_2 = -0.717484$$

$$\lambda_4 = \frac{h_4}{h_3} = 0.004937, \delta_4 = 1 + \lambda_4 = 1.004937$$

$$\lambda_4 + \delta_4 = 1.009874, g_4 = -0.044774$$

$c_4 = 0.000095$. Since $g_4 < 0$, by the Muller's method

$$x_5 = x_4 + (x_4 - x_3) \cdot \left\{ \frac{-2\delta_4 f_4}{g_4 - \sqrt{g_4^2 - 4g_4 f_4 c_4}} \right\}$$

Substituting $x_3, x_4, \delta_4, f_4, g_4$ and c_4 values in the above equation. We get
$$x_5 = 2.279019.$$

The approximate root of given equation is 2.279019.

Problem 2: Perform three iterations of the Muller's method to find a real root of $2x^3 - 1 = 0$. Use the initial approximations 0, 1 and 0.5.

Solution: Let $x_0 = 0$, $x_1 = 1$ and $x_2 = 0.5$.

Iteration 1: $h_2 = x_2 - x_1 = -0.5, h_1 = x_1 - x_0 = 1, \lambda_2 = \frac{h_2}{h_1} = -0.5$

$$\delta_2 = (1 + \lambda_2) = 0.5, \lambda_2 + \delta_2 = 0, g_2 = -0.5 \text{ and } c_2 = 0.375$$

since $g_2 < 0$,

$$x_3 = x_2 + (x_2 - x_1) \left\{ \frac{-2\delta_2 f_2}{g_2 - \sqrt{g_2^2 - 4\delta_2 f_2 c_2}} \right\}$$

Substituting the relevant values, we get

$$x_3 = 0.767592.$$

Iteration 2: Here, we take $x_1 = 1$, $x_2 = 0.5$ and $x_3 = 0.767592$ as initial approximations of the root

$$h_3 = x_3 - x_2 = 0.267592,\ h_2 = x_2 - x_1 = -0.5,$$

$$\lambda_3 = \frac{h_3}{h_2} = -0.535184,\ \delta_3 = 1 + \lambda_3 = 0.464816$$

$$\lambda_3 + \delta_3 = -0.070368,\ g_3 = 0.455181 \text{ and } c_3 = 0.150946.$$

$$x_4 = x_3 + (x_3 - x_2) \cdot \left\{ \frac{-2\delta_3 f_3}{g_3 + \sqrt{g_3^2 - 4\delta_3 f_3 c_3}} \right\}$$

Substituting the relevant values,

$$x_4 = 0.792888.$$

Iteration 3: In this iteration, $x_2 = 0.5$, $x_3 = 0.767592$ and $x_4 = 0.79288$ are as three initial approximations of the root. Then, $h_4 = x_4 - x_3 = 0.025296$, $h_3 = x_3 - x_2 = 0.267592$

$$\lambda_4 = \frac{h_4}{h_3} = 0.094532,\ \delta_4 = 1 + \lambda_4 = 1.094532$$

$$\lambda_4 + \delta_4 = 1.189064,\ g_4 = 0.104026 \text{ and } c_4 = 0.002886$$

By the Muller's method,

$$x_5 = x_4 + (x_4 - x_3) \cdot \left\{ \frac{-2\delta_4 f_4}{g_4 + \sqrt{g_4^2 - 4\delta_4 f_4 c_4}} \right\}$$

Substituting the relevant values, we get

$$x_5 = 0.793704.$$

Problem 3: Perform three iterations of the Muller's method to find a real root of $\sin x - 0.5 = 0$ use the initial approximations 1.5, 2 and 2.5

Solution: Let $x_0 = 1.5$, $x_1 = 2$ and $x_2 = 2.5$.

Iteration 1: $h_2 = x_2 - x_1 = 0.5$, $h_1 = x_1 - x_0 = 0.5$

$$\lambda_2 = \frac{h_2}{h_1} = 1,\ \delta_2 = 1 + \lambda_2 = 2,\ \delta_2 + \lambda_2 = 3$$

$$g_2 = -0.844278 \text{ and } c_2 = -0.222628.$$

By Muller's method,

$$x_3 = x_2 + (x_2 - x_1)\left\{ \frac{-2\delta_2 f_2}{g_2 - \sqrt{g_2^2 - 4\delta_2 f_2 c_2}} \right\}$$

(since g is negative)

Substituting the relevant values, we have

$$x_3 = 2.610227.$$

Iteration 2: Here, we take $x_1 = 2$, $x_2 = 2.5$ and $x_3 = 2.610227$ as initial approximations of the root of given equation. Then,

$$h_3 = x_3 - x_2 = 0.110227$$

$$h_2 = x_2 - x_1 = 0.5$$

$$\lambda_3 = 0.220454, \delta_4 = (1 + \lambda_4) = 1.220454$$

$$\lambda_3 + \delta_4 = 1.440908, g_4 = -0.117113$$

and

$$c_4 = -0.005123$$

$$\therefore x_4 = x_3 + (x_2 - x_1)\left\{ \frac{-2\delta_3 f_3}{g_3 - \sqrt{g_3^2 - 4\delta_3 f_3 c_3}} \right\}$$

$$x_4 = 2.617913.$$

Iteration 3: In this iteration, we take $x_2 = 2.5$, $x_3 = 2.610227$ and $x_4 = 2.617913$ as the initial approximations of the root of the given equation. Then,

$$h_4 = x_4 - x_3 = 0.007686$$

$$h_3 = x_3 - x_2 = .110227$$

$$\lambda_4 = \frac{h_4}{h_3} = 0.069725$$

$$\delta_4 = (1 + \lambda_4) = 1.069725$$

$$\lambda_4 + \delta_4 = 1.139450$$

$$g_4 = -0.007121$$

and

$$c_4 = -0.000017.$$

By Muller's method,

$$\therefore x_5 = x_4 + (x_4 - x_3)\left\{\frac{-2\delta_4 f_4}{g_4 - \sqrt{g_4^2 - 4g_4 f_4 c_4}}\right\}$$

Substituting the relevant values, we get

$$x_5 = 2.617994.$$

Problem 4: Perform two iterations of Muller's method to find a root of $x^3 - x - 1 = 0$ use the initial approximations -0.5, 1 and 1.5.

Solution: Let $x_0 = 0.5$, $x_1 = 1$ and $x_2 = 1.5$.

Iteration 1: $h_2 = x_2 - x_1 = 0.5$

$$h_1 = x_1 - x_0 = 0.5$$

$$\lambda_2 = \frac{h_2}{h_1} = 1$$

$$\delta_2 = 1 + \lambda_2 = 2$$

$$\lambda_2 + \delta_2 = 3$$

$$g_2 = 5.25$$

$$c_2 = 1.5$$

By Muller's method

$$x_3 = x_2 + (x_2 - x_1) \cdot \left\{\frac{-2\delta_2 f_2}{g_2 - \sqrt{g_2^2 - 4\delta_2 f_2 c_2}}\right\}$$

Substituting the values of x_1, x_2, δ_2, f_2, g_2 and c_2 we get

$$\therefore x_3 = 1.313446.$$

Iteration 2: In this iteration $x_1 = 1$, $x_2 = 1.5$ and $x_3 = 1.313446$. Hence,

$$h_3 = x_3 - x_2 = -0.186554$$

$$h_2 = x_2 - x_1 = 0.5$$

$$\lambda_3 = \frac{h_3}{h_2} = -0.373107$$

$$\delta_3 = 1 + \lambda_3 = 0.626893$$

$$\delta_3 + \lambda_3 = 0.253786$$

$$g_3 = -0.495151$$

and $$c_3 = 0.083199$$

By Muller's method

$$x_4 = x_3 + (x_3 - x_2)\left\{\frac{-2\delta_3 f_3}{g_3 - \sqrt{g_3^2 - 4\delta_3 f_3 c_3}}\right\}$$

Substituting the relevant values, we have

$$x_4 = 1.324569.$$

9.28   BIRGE–VIETA METHOD

This method is used to find a real number p such that $(x - p)$ will be a factor of

$$P_n(x) = x^n + a_1 x^{n-1} + \ldots + a_{n-1} x + a_n = 0, \qquad (9.61)$$

where a_i $(i = 1, 2, \ldots, n)$ are real numbers. If P_n (a) P_n $(b) < 0$ for some real numbers a and b, then p is a point in the interval (a, b). We start with an approximate value p_0 of p such that $a < p_0 < b$ and its values are improved by repeated application of this method so that $(x - p_k)$ is a exact factor or an approximate linear factor of $P_n(x)$. Now, we proceed to describe Birge–Vieta method.

Suppose we divide the polynomial $P_n(x)$ given by (9.61) by $(x - p)$; then we have quotient

$$Q_{n-1}(x) = x^{n-1} + b_1 x^{n-2} + b_2 x^{n-3} + \ldots \ldots + b_{n-1} \qquad (9.62)$$

of degree $(n - 1)$ and a remainder R, which is a constant. Now, $P_n(x)$ can be expressed as

$$P_n(x) = (x - p) Q_{n-1}(x) + R. \qquad (9.63)$$

Note that R and the coefficients of various powers of x in $Q_{n-1}(x)$ depends on the value p, i.e., for different values of p, we have different values of R and the coefficients of various powers of x namely b_1, b_2, ., b_{n-1}. If $R = 0$ is zero, then $(x - p)$ is a factor and p is a root of (9.61). In general, with initial value p_0 of p, the remainder R may not be zero. The problem is to improve the value of p_0 so that R is zero or close to zero for improved value of p_0.

Let p_0 be the initial value of p and $p_0 + \Delta p$ be the exact root of (9.61). Then,

$$P_n (p_0 + \Delta p) = 0$$

Expanding in Taylor series, we obtain

$$P_n(p_0) + \Delta p\ P_n'(p_0) + \frac{(\Delta p)^2}{\lfloor 2} P_n''(p_0) + \cdots = 0$$

Neglecting second- and higher-order powers of Δp, we get

$$P_n(p_0) + \Delta p\ P_n'(p_0) = 0$$

$$\therefore\ \Delta p = \frac{-P_n(p_0)}{P_n'(p_0)}$$

Let $p_1 = p_0 + \Delta p$. Then,

$$p_1 = p_0 - \frac{P_n(p_0)}{P_n'(p_0)} \tag{9.64}$$

Evaluating $P_n(p_0)$ and $P_n'(p_0)$, we can find p_1 value. But this method gives a different approach to have $P_n(p_0)$ and $P_n'(p_0)$.

These two can be obtained using the synthetic division.

Now, equating the coefficients of various powers of x in (9.63), we get

$$a_1 = b_1 - p$$
$$a_2 = b_2 - p\ b_1$$
$$a_3 = b_3 - p\ b_2$$
$$- - - - - -$$
$$a_k = b_k - p\ b_{k-1}$$
$$- - - - - -$$
$$a_{n-1} = b_{n-1} - p\ b_{n-2} \tag{9.65}$$
$$a_n = R - p\ b_{n-1}$$

From equations of (9.65), we have

$$b_1 = a_1 + p$$
$$b_2 = a_2 + p\ b_1$$
$$b_3 = a_2 + p\ b_2$$
$$- - - - - -$$
$$b_k = a_k + p\ b_{k-1}$$
$$- - - - - -$$
$$b_{n-1} = a_{n-1} + p\ b_{n-1}$$
$$R = a_n + p\ b_{n-1} \tag{9.66}$$

Note that b_k values are determined using the values of a_k. Let us define the following recurrence relation by introducing b_n

$$b_k = a_k + p\, b_{k-1} \quad (k = 1, 2, 3, \ldots, n) \tag{9.67}$$

with $b_0 = 1$. Note that

$$b_n = R \tag{9.68}$$

Since $R = a_n + p\, b_{n-1}$
From (9.63), we have

$$P_n(p) = R \tag{9.69}$$

Differentiating (9.67) w. r. t. p, we get

$$\frac{db_k}{dp} = b_{k-1} + p \cdot \frac{db_{k-1}}{dp} \tag{9.70}$$

Let us introduce

$$\frac{db_k}{dp} = c_{k-1} \quad (k = 1, 2, n-1) \tag{9.71}$$

with $c_0 = 1$. Then, Eq. (9.70) becomes

$$c_{k-1} = b_{k-1} + p\, c_{k-2}$$

or

$$c_k = b_k + p\, c_{k-1} \quad (k = 1, 2, \ldots, n-1)$$

Now, differentiating (9.69) w. r. t. p, we get

$$P_n'(p) = \frac{dR}{dp}$$

Since $R = b_n$, we have $\dfrac{dR}{dp} = \dfrac{db_n}{dp}$. using (9.68),

$$\frac{dR}{dp} = \frac{db_n}{dp} = c_{n-1}. \text{ Thus,}$$

$$P_n'(p) = \frac{dR}{dp} = c_{n-1} \tag{9.72}$$

Thus, c_k are determined from b_k. Had we started with p_0 instead of p, we would have

$$P_n(p_0) = b_n \tag{9.73}$$

and

$$P_n'(p_0) = c_{n-1} \tag{9.74}$$

In view of (9.73) and (9.74), Eq. 9.64 becomes

$$p_1 = p_0 - \frac{b_n}{c_{n-1}} \tag{9.75}$$

Since $P_n(x)$ is a polynomial, the values of b_k's and c_k's can be obtained from the following synthetic division.

p_0	1	a_1	a_2	......	a_{n-2}	a_{n-1}	a_n
		p_0	$p_0 b_1$		$p_0 b_{n-3}$	$p_0 b_{n-2}$	$p_0 b_{n-1}$
p_0	1	b_1	b_2		b_{n-2}	b_{n-1}	$b_n = R$
		p	pc_1		$p c_{n-3}$	$p c_{n-2}$	
	1	c_1	c_2		c_{n-2}	$c_{n-1} = \dfrac{dR}{dp}$	

The general iteration formula for Birge–Vieta method is given by

$$p_{k+1} = p_k - \frac{b_n}{c_{n-1}} \quad (k = 0, 1, 2, \cdots),$$

where b_n and c_{n-1} are last entries of 3rd and 5th row of synthetic division table. Note that n is the degree of polynomial.

SOLVED PROBLEMS

Problem 1: Perform four iterations of the Birge–Vieta method to find a root of $x^3 - 11x^2 + 32x - 22 = 0$.

Solution: The Birge–Vieta method is given by

$$p_{k+1} = p_k - \frac{b_n}{c_{n-1}} \quad (k = 0, 1, 2, 3, \cdots). \; ,$$

where n is the degree of the polynomial. The given polynomial is of third degree, and therefore,

$$p_{k+1} = p_k - \frac{b_3}{c_2} \quad (k = 0, 1, 2, 3, \cdots). \tag{1}$$

Let $p_0 = 0.5$ be an initial approximation.

Iteration 1:

0.5	1.00000	−11.00000	32.00000	−22.00000
		0.50000	−5.25000	13.37500
0.5	1.00000	−10.50000	26.75000	$-8.62500 = b_3$
		0.50000	−5.00000	
	1.00000	−10.00000	$21.75000 = c_2$	

$$p_1 = p_0 - b_3/c_2$$
$$p_1 = 0.50000 - (-8.62500)/(21.75000) = 0.89655$$

Iteration 2:

0.89655	1.00000	−11.00000	32.00000	−22.00000
		0.89655	−9.05826	20.56845
0.89655	1.00000	−10.10345	22.94174	$-1.43155 = b_3$
	−	0.89655	−8.25446	
	1.00000	−9.20690	$14.68728 = c_2$	

$$p_2 = p_1 - b_3/c_2$$
$$p_2 = 0.89655 - (-1.43155)/(14.68728) = 0.99402$$

Iteration 3:

0.99402	1.00000	−11.00000	32.00000	−22.00000
		0.99402	−9.94615	21.92198
0.99402	1.00000	−10.00595	22.05385	$-0.07802 = b_3$
		0.99402	−8.95807	
	1.00000	−9.01196	$13.09578 = c_3$	

$$p_3 = 0.99402 - (-0.07802)/(13.09578) = 0.99998.$$

Iteration 4:

0.99998	1.00000	−11.00000	32.00000	−22.00000
		0.99998	−9.99980	21.99972

0.99998	1.00000	−10.00002	22.00020	−0.00028 = b_3
		0.99998	−8.99985	
	1.00000	−9.00004	13.00035 = c_2	

$$P_4 = P_3 - b_3/c_2$$

$$P_4 = 0.99998 - (-0.00028)/(13.00035) = 1.00000$$

Thus, 1 is a root of given equation.

Problem 2: Perform four iteration of the Birge–Vieta method to find a root of $x^4 + 7x^3 + 24x^2 - 15 = 0$.

Solution: Let $P_4(x) = x^4 + 7x^3 + 24x^2 - 15$. Then, $P_4(0) = -15$ and $P_4(1) = 17$. Therefore, a root of $P_4(x) = 0$ lies between an interval (0, 1) to choose $p_0 = 0.4$. The Birge–Vieta method is given by

$$p_{k+1} = p_k - \frac{b_n}{c_{n-1}} \quad (k = 0, 1, 2, \cdots)$$

and n is the degree of given polynomial

$$\therefore \ p_{k+1} = p_k - \frac{b_4}{c_3}.$$

Since the degree of given polynomial is 4,
Iteration 1: $p_0 = 0.4$

0.4	1.00000	7.00000	24.00000	0.00000	−15.00000
		0.40000	2.96000	10.78400	4.31360
0.4	1.00000	7.40000	26.96000	10.78400	−10.68640 = b_4
		0.40000	3.12000	12.03200	
	1.00000	7.80000	30.08000	22.81600 = c_3	

$$P_1 = P_0 - b_4/c_3$$

$$P_1 = 0.40000 - (-10.68640)/(22.81600) = 0.86837.$$

Iteration 2:

0.86837	1.00000	7.00000	24.00000	0.00000	−15.00000
		0.86837	6.83268	26.77427	23.25006

0.86837	1.00000	7.86837	30.83268	26.77427	$8.25006 = b_4$
		0.86837	7.58675	33.36240	
	1.00000	8.73675	38.41944	$60.13667 = c_3$	

$$p_2 = p_1 - \frac{b_4}{c_3}$$

$$p_2 = 0.86837 - (8.25006)/(60.13667) = 0.73118.$$

Iteration 3:

0.73118	1.00000	7.00000	24.00000	0.00000	−15.00000
		0.73118	5.65292	21.68176	15.85337
0.73118	1.00000	7.73118	29.65292	21.68176	$0.85337 = b_4$
		0.73118	6.18755	26.20600	
	1.00000	8.46237	35.84048	$47.88776 = c_3$	

$$p_3 = p_2 - \frac{b_4}{c_3}$$

$$p_3 = 0.73118 - (0.85337)/(47.88776) = 0.71336.$$

Iteration 4:

0.71336	1.00000	7.00000	24.00000	0.00000	−15.00000
		0.71336	5.50244	21.04599	15.01346
0.71336	1.00000	7.71336	29.50244	21.04599	$0.01346 = b_4$
		0.71336	6.01133	25.33426	
	1.00000	8.42673	35.51377	$46.38025 = c_3$	

$$p_4 = p_3 - \frac{b_4}{c_3}$$

$$p_4 = 0.71336 - (0.01346)/(46.38025) = 0.71307.$$

Thus, one root of given equation is 0.71307.

9.29 BAIRSTOW'S METHOD

A polynomial is said to be a real polynomial equation if the coefficients of various powers of x in it are real numbers. For example, $x^4 - \sqrt{3}x^3 + x^2 + 6x - 4 = 0$ is real polynomial equation. With regard to computational point of view, finding a complex root of such equations by the methods that we learnt in earlier sections may be tedious in some situations. The quadratic factor corresponding to two real roots or conjugate complex roots of real polynomial equations can be in the form of $x^2 + p_n + q$. Note that the complex roots of real polynomial equations occur in pairs. If a quadratic factor is found for real polynomial equations, then its roots can be found by using $\dfrac{-b \pm \sqrt{b^2 - 4ac}}{2a}$ or factorizing the quadratic factor. Bairstow's method is a method to extract a quadratic factor of real polynomial equation. We now proceed to describe Bairstow's method.

Let $P_n(x) = x^n + a_1 x^{n-1} + a_2 x^{n-2} + \ldots\ldots + a_{n-2}x^2 + a_{n-1}x + a_n = 0$ (9.76)

be a real polynomial equation of degree n and $x^2 + ux + v$ be a quadratic factor of $P_n(x)$.

Dividing $P_n(x)$ by $x^2 + ux + v$ we obtain quotient $Q_{n-2}(x)$ a polynomial of degree n–2 and a remainder that is linear in x.Let

$$Q_{n-2}(x) = x^{n-2} + b_1 x^{n-3} + b_2 x^{n-4} + \ldots\ldots\ldots + b_{n-3}x + b_{n-2} \qquad (9.77)$$

and remainder $Rx + S$.Then

$$P_n(x) = (x^2 + ux + v)Q_{n-2}(x) + (Rx + S) \qquad (9.78)$$

Comparing the coefficients of like powers of x in Eq.(9.78),we have

$$a_1 = b_1 + u$$
$$a_2 = b_2 + ub_1 + v \qquad (9.79)$$
$$a_3 = b_3 + ub_2 + vb_1$$

$$\cdots\cdots\cdots\cdots\cdots\cdots\cdots\cdots$$

$$a_k = b_k + ub_{k-1} + vb_{k-2}$$

$$\cdots\cdots\cdots\cdots\cdots\cdots\cdots\cdots$$

$$a_{n-2} = b_{n-2} + ub_{n-3} + vb_{n-4}$$
$$a_{n-1} = R + ub_{n-2} + vb_{n-3}$$
$$a_n = S + vb_{n-2}$$

From the above equations, we have

$$b_1 = a_1 - u$$
$$b_2 = a_2 - ub_1 - v$$
$$b_3 = a_3 - ub_2 - vb_1 \qquad (9.80)$$

$$\cdots\cdots\cdots\cdots\cdots\cdots\cdots$$

$$b_k = a_k - ub_{k-1} - vb_{k-2}$$

$$\cdots\cdots\cdots\cdots\cdots\cdots\cdots$$

$$b_{n-2} = a_{n-2} - ub_{n-3} - vb_{n-4}$$
$$R = a_{n-1} - ub_{n-2} - vb_{n-3}$$
$$S = a_n - vb_{n-2}$$

Now, we introduce a recursive formula

$$b_k = a_k - ub_{k-1} - vb_{k-2} \quad (k = 1,2,3,\ldots,n) \qquad (9.81)$$

where $b_0 = 1$ *and* $b_1 = 0$

In view of Eq. (9.81), the last two equations of Eq. (9.80) become

$$R = b_{n-1} \qquad (9.82)$$

and

$$S = b_n + ub_{n-1} \qquad (9.83)$$

Note that R and S are functions of u and v. From Eq. (9.78), it is evident that $x^2 + ux + + v$ is a quadratic factor of $P_n(x)$ provided R(u,v) = 0 and S(u,v) = 0. Thus we need to find u, v such that R(u,v) = s(u,v) = 0. Suppose u_0 and v_0 are trial solutions of u and v *and* $u_0 + \Delta u$, $v_0 + \Delta v$ are exact solutions.

Then,

$$R\,(u_0 + \Delta u,\, v_0 + \Delta v) = 0$$
$$S\,(u_0 + \Delta u,\, v_0 + \Delta v) = 0.$$

Now, expanding $R\,(u_0 + \Delta u,\, v_0 + \Delta v)$ and $S\,(u_0 + \Delta u,\, v_0 + \Delta v)$ in Taylor series about the point (u_0, v_0), we have

$$R(u_0,v_0) + \left[\Delta u\, R_u(u_0,v_0) + \Delta v R_v(u_0,v_0)\right]$$
$$+ \frac{1}{\lfloor 2}\left[(\Delta u)^2 R_{uu}(u_0,v_0) + 2\Delta u \Delta v R_{uv}(u_0,v_0) + (\Delta v)^2 R_{vv}(u_0,v_0)\right] + \cdots = 0$$

and

$$S(u_0, v_0) + \left[\Delta u\, S_u(u_0, v_0) + \Delta v\, S_v(u_0, v_0) \right]$$

$$+ \frac{1}{\lfloor 2} \left[(\Delta u)^2 S_{uu}(u_0, v_0) + 2\Delta u\, \Delta v S_{uv}(u_0, v_0) + (\Delta v)^2 S_{vv}(u_0, v_0) \right] + \cdots\cdots = 0,$$

where $R_u = \dfrac{\partial R}{\partial u}, R_u = \dfrac{\partial R}{\partial u}, R_{uv} = \dfrac{\partial^2 R}{\partial u\, \partial v}, R_{uu} = \dfrac{\partial R^2}{\partial u^2}$ and $R_{vv} = \dfrac{\partial^2 R}{\partial v^2}$ etc.

Neglecting second- and higher-order terms of Δu and Δv, we obtain

$$R(u_0, v_0) + \Delta u\, R_u(u_0, v_0) + \Delta v\, R_v(u_0, v_0) = 0$$
$$S(u_0, v_0) + \Delta u\, S_u(u_0, v_0) + \Delta v\, S_v(u_0, v_0) = 0$$

Solving the above two system of linear equations, we get

$$\Delta u = \frac{S\, R_u - R\, S_v}{R_u\, S_v - S_u\, R_v} \tag{9.84}$$

and

$$\Delta v = \frac{R\, S_u - S\, R_u}{R_u S_v - S_u R_v} \tag{9.85}$$

Bairstow's method gives a mechanism to get R.H.S of (9.84) and (9.85) in terms of quantities, which can be had from the synthetic division in a systematic manner. Differentiating (9.81) partially with respect to u and v, we get

$$\frac{\partial b_k}{\partial u} = -b_{k-1} - u\frac{\partial b_{k-1}}{\partial u} - v\frac{\partial b_{k-2}}{\partial u}, \frac{\partial b_0}{\partial u} = \frac{\partial b_{-1}}{\partial u} = 0 \tag{9.86}$$

$$\frac{\partial b_k}{\partial v} = -b_{k-1} - u\frac{\partial b_{k-1}}{\partial v} - v\frac{\partial b_{k-2}}{\partial v}, \frac{\partial b_0}{\partial v} = \frac{\partial b_{-1}}{\partial v} = 0$$

Introducing

$$\frac{\partial b_k}{\partial u} = -c_{k-1}, \; k = 1, 2, \cdots, n$$

and

$$\frac{\partial b_k}{\partial v} = -c_{k-2} \left(\text{i.e.} \frac{\partial b_k}{\partial v} = \frac{\partial b_{k-1}}{\partial u} = -c_{k-2} \right)$$

the two equations of (9.86) become

$$c_{k-1} = b_{k-1} - u\, c_{k-2} - v\, c_{k-3} \tag{9.87}$$

and

$$c_{k-2} = b_{k-2} - u\, c_{k-3} - v\, c_{k-4}.$$

Thus, we obtain a recurrence relation for c_k from Eq. (9.87) and it is given by

$$c_k = b_k - u\, c_{k-1} - v\, c_{k-2} \quad (k=1, 2,\cdots, n-1) \tag{9.88}$$

with $c_0 = 0,\ c_{-1} = 0$.
Recall

$$\frac{\partial b_k}{\partial v} = \frac{\partial b_{k-1}}{\partial u} = -c_{k-2} \tag{9.89}$$

Put $k = n$ in (9.89)

$$-c_{n-2} = \frac{\partial b_n}{\partial v} = \frac{\partial b_{n-1}}{\partial u}$$

But $b_{n-1} = R$ (Eq. 9.82), therefore,

$$R_u = -c_{n-2} \tag{9.90}$$

Similarly $R_v = -c_{n-3}$,

$$S_u = (b_{n-1} - c_{n-1}) - u\, c_{n-2}$$
$$S_v = - (c_{n-2} + u\, c_{n-2})$$

Substituting the values $R = b_{n-1}$ (Eq. 9.82) $S = b_n + u\, b_{n-1}$ (Eq. 9.83) and the values R_u, R_v, S_u and S_v in (9.84) and (9.85), we have

$$\Delta u = -\frac{b_n c_{n-3} - b_{n-1}c_{n-2}}{c_{n-2}^2 - c_{n-3}(c_{n-1} - b_{n-1})} \tag{9.91}$$

and

$$\Delta v = -\frac{b_{n-1}(c_{n-1} - b_{n-1}) - b_n c_{n-2}}{c_{n-2}^2 - c_{n-3}(c_{n-1} - b_{n-1})} \tag{9.92}$$

Hence, the improved values of u and v are given by

$$u_1 = u_0 + \Delta u$$
$$v_1 = v_0 + \Delta v.$$

Now, taking u_1, v_1 as the approximations to u and v, and repeating the above process, we can obtain further better approximations to u and v.

The values of b_k's and c_k's are determined with the help of synthetic division in a systematic manner, which is as follows:

	1	a_1	a_2	a_3	$\ldots$	a_{n-2}	a_{n-1}	a_n
$-u$	$-$	$+(-u)1$	$+(-u)b_1$	$+(-u)b_2$		$+(-u)\,b_{n-3}$	$+(-u)\,b_{n-2}$	$+(-u)\,b_{n-1}$
$-v$	$-$	$-$	$+(-v)(1)$	$+(-u)b_1$		$+(-u)\,b_{n-4}$	$+(-v)\,b_{n-3}$	$+(-v)\,b_{n-2}$
	1	b_1	b_2	b_3	$\ldots$	b_{n-2}	$b_{n-1}=R$	b_n
$-u$	$-$	$+(-u)1$	$+(-u)c_1$	$+(-u)c_2$	$\ldots$	$+(-u)\,c_{n-3}$	$+(-u)\,c_{n-2}$	
$-v$	$-$	$-$	$+(-v)(1)$	$+(-v)c_1$	$\ldots$	$+(-v)\,c_{n-4}$	$+(-v)\,c_{n-3}$	
	1	c_1	c_2	c_3		c_{n-2}	c_{n-1}	

The values b_{n-1}, b_{n-2}, c_{n-3}, c_{n-2} and c_{n-1} can be obtained from the above table.

SOLVED PROBLEMS

Problem 1: Perform five iteration of the Bairstow's method to extract approximate quadratic factor of $P_3(x) = x^3 - 8x^2 + 17x - 10 = 0$.

Solution: Choose the initial approximate of u and v as $u_0 = 0.8$ and $v_0 = .8$. The successive iterations u and v are computed as follows.

Iteration 1:

-0.8	1.00000	-8.00000	17.00000	-10.00000
-0.8		-0.80000	7.04000	-18.59200
			-0.80000	7.04000
-0.8	1.00000	-8.80000	23.24000	-21.55200
-0.8		-0.80000	7.68000	
			-0.80000	
	1.00000	-9.60000	30.12000	

$$\Delta u = -\frac{b_3 c_0 - b_2 c_1}{c_1^2 - c_0(c_2 - b_2)} = -(201.55200)/(85.28001) = -2.36341$$

$$\Delta v = -\frac{b_2(c_2 - b_2) - b_3 c_1}{c_1^2 - c_0(c_2 - b_2)} = -(-47.00798)/(85.28001) = 0.55122$$

$$u_1 = u_0 + \Delta u = 0.80000 + (-2.36341) = -1.56341$$

$$v_1 = v_0 + \Delta v = 0.80000 + (0.55122) = 1.35122.$$

Iteration 2:

1.56341	1.00000	−8.00000	17.00000	−10.00000
−1.35122		1.56341	−10.06305	8.73281
			−1.35122	8.69724
1.56341	1.00000	−6.43659	5.58573	7.43005
−1.35122		1.56341	−7.61879	
			−1.35122	
	1.00000	−4.87317	−3.38428	

$$\Delta u = -\frac{b_3 c_0 - b_2 c_1}{c_1^2 - c_0(c_2 - b_2)} = -(34.65026)/(32.71780) = -1.05906$$

$$\Delta v = -\frac{b_2(c_2 - b_2) - b_3 c_1}{c_1^2 - c_0(c_2 - b_2)} = -(-13.89612)/(32.71780) = 0.42473$$

$$u_2 = u_1 + \Delta u = -1.56341 + (-1.05906) = -2.62248$$

$$v_2 = v_1 + \Delta v = 1.35122 + (0.42473) = 1.77595.$$

Iteration 3:

2.62248	1.00000	−8.00000	17.00000	−10.00000
−1.77595		2.62248	−14.10244	2.94142
			−1.77595	9.55019
2.62248	1.00000	−5.37752	1.12162	2.49161
−1.77595		2.62248	−7.22504	
			−1.77595	
	1.00000	−2.75504	−7.87937	

$$\Delta u = -\frac{b_3 c_0 - b_2 c_1}{c_1^2 - c_0(c_2 - b_2)} = -(5.58171)/(16.59124) = -0.33643$$

$$\Delta v = -\frac{b_2(c_2 - b_2) - b_3 c_1}{c_1^2 - c_0(c_2 - b_2)} = -(-3.23119)/(16.59124) = 0.19475$$

$$u_3 = u_2 + \Delta u = -2.62248 + (-0.33643) = -2.95890$$

$$v_3 = v_2 + \Delta v = 1.77595 + (0.19475) = 1.97070.$$

Iteration 4:

2.95890	1.00000	−8.00000	17.00000	−10.00000
−1.9070		2.95890	−14.91612	0.33489
			−1.97070	9.93448
2.95890	1.00000	−5.04110	0.11318	0.26937
−1.9070		2.95890	−6.16101	
			−1.97070	
	1.00000	−2.08219	−8.01852	

$$\Delta u = -\frac{b_3 c_0 - b_2 c_1}{c_1^2 - c_0(c_2 - b_2)} = -(0.50504)/(12.46723) = -0.04051$$

$$\Delta v = -\frac{b_2(c_2 - b_2) - b_3 c_1}{c_1^2 - c_0(c_2 - b_2)} = -(-0.35947)/(12.46723) = 0.02883$$

$$u_4 = u_3 + \Delta u = -2.95890 + (-0.04051) = -2.99941$$

$$v_4 = v_3 + \Delta v = 1.97070 + (0.02883) = 1.99953.$$

Iteration 5:

2.99941	1.00000	−8.00000	17.00000	−10.00000
−1.99953		2.99941	−14.99883	0.00492
			−1.99953	9.99883
2.99941	1.00000	−5.00059	0.00164	0.00375
−1.99953		2.99941	−6.00235	
			−1.99953	
	1.00000	−2.00117	−8.00024	

$$\Delta u = -\frac{b_3 c_0 - b_2 c_1}{c_1^2 - c_0(c_2 - b_2)} = -(0.00703)/(12.00657) = -0.00059$$

$$\Delta v = -\frac{b_2(c_2 - b_2) - b_3 c_1}{c_1^2 - c_0(c_2 - b_2)} = -(-0.00562)/(12.00657) = 0.00047$$

$$u_5 = u_4 + \Delta u = -2.99941 + (-0.00059) = -3.00000$$

$$v_5 = v_4 + \Delta v = 1.99953 + (0.00047) = 2.00000.$$

Hence, the required quadratic factor is $x^2 - 3x + 2$.

Problem 2: Extract the quadratic factor of the polynomial $p_3(x) = x^3 - 2x^2 + x - 2$. Perform two iterations.

Solution: Choose $u_0 = -0.5$ and $v_0 = 1$.

Iteration 1:

0.5	1.0	−2.0	1.0	−2.0
−1	−	0.5	−0.750	−0.37500
	−	−	−1.0	1.5
0.5	1.0	−1.5	$-0.75 = b_2$	$-0.875 = b_3$
−1	−	0.5	−0.5	
	−		−1.0	
	$1 = c_0$	$-1.00 = c_1$	$-2.25 = c_2$	

$$\Delta u = -\frac{b_3 c_0 - b_2 c_1}{c_1^2 - c_0(c_2 - b_2)} = 0.65$$

$$\Delta v = -\frac{b_2(c_2 - b_2) - b_3 c_1}{c_1^2 - c_0(c_2 - b_2)} = -0.1$$

$$\therefore u_1 = -0.5 + 0.65 = .15$$

$$u_2 = 1 - 0.1 = 0.9.$$

Iteration 2:

−0.15	1	−2.0	1.0	−2.0
−0.9	−	−0.15	0.3225	−0.06337
	−	−	−0.9	1.935
−0.15	1.0	−2.5	$0.4225 = b_2$	$-0.12837 = b_3$
−0.9	−	−0.15	0.34500	
	−	−	−0.9	
	$1.0 = c_0$	$-2.3 = c_1$	$-0.13250 = c_2$	

$$\Delta u = -0.14429$$

$$\Delta v = 0.09063$$

$$\therefore \quad u_2 = u_1 + \Delta u = 0.15 - 0.14429$$

$$= 0.00571$$

$$v_2 = v_1 + \Delta v = 0.9 + 0.09063$$
$$= 0.99063$$

Hence, the required quadratic factor is $x^2 + 0.0057x + 0.99063$.

Problem 3: Find a quadratic facto of $x^3 - 2x - 5 = 0$ by Bairstow's method. Perform 3 iterations.

Solution: Choose $u_0 = 2$ and $v_0 = 2.5$.

Iteration 1:

-2	1	.0	-2	-5
-2.5	$-$	-2	$+4$	1
	$-$	$-$	-2.5	5
-2	1	-2	$-0.5 = b_2$	$1 = b_3$
-2.5	$-$	-2	8	
	$-$	$-$	-2.5	
	$1 = c_0$	$-4 = c_1$	$5 = c_2$	

$$\Delta u = \; = 0.09524$$
$$\Delta v = \; = -0.11905$$
$$\therefore \quad u_1 = u_0 + \Delta u = 2.09524$$
$$v_1 = v_0 + \Delta v = 2.38095.$$

Iteration 2:

-2.09524	1.0	0.0	-2.0	-5.0
-2.38095	$-$	-2.09524	4.39002	-0.01901
	$-$	$-$	-2.38095	4.98866
-2.09524	1.0	-2.09524	0.00907	-0.03034
-2.38095	$-$	-2.09524	8.78005	
	$-$	$-$	-2.38095	
	1.0	-4.19048	6.40816	

$$\Delta u = \; = -0.00069$$
$$\Delta v = \; = 0.00069$$
$$\therefore \quad u_2 = u_1 + \Delta u = 2.09524 - 0.00069$$

$$= 2.09455$$
$$v_2 = v_1 + \Delta v = 2.38095 + 0.00619$$
$$= 2.38714.$$

Iteration 3:

-2.09455	1	0	-2	-5
-2.38714	$-$	-2.09455	4.38714	0
	$-$	$-$	-2.38714	5.0
-2.09455	1	-2.09455	$0 = b_2$	$-0 = b_3$
-2.38714	$-$	-2.09455	8.7729	
	$-$	$-$	-2.38714	
	1	$-4.18910 = c_1$	$6.38714 = c_2$	

$$\Delta u = \ = 0$$
$$\Delta v = \ = 0$$
$$\therefore \qquad u_3 = u_2 = 2.09455$$
$$v_3 = v_2 = 2.38714.$$

The required quadratic factor is $x^2 + 2.09455x + 2.38714$.

Problem 4: Perform three iteration of Bairstow's method to find a quadratic factor of $x^4 + x^3 + 2x^2 + x + 1$.

Solution: Choose $u_0 = 0.5$ and $v_0 = 0.5$.

Iteration 1:

-0.5	1.0	1.0	2.0	1.0	1.0
-0.5	$-$	-0.5	-0.25	-0.625	-0.06250
	$-$		-0.5	-0.25	-0.625
-0.5	1	0.5	1.25	$0.125 = b_3$	$0.3125 = b_4$
-0.5	$-$	-0.5	-0	-0.375	
	$-$		-0.5	-0.0	
	$1 = c_0$	$0 = c_1$	$0.75 = c_2$	$-2.25 = c_3$	

$$\Delta u = \ = 0.16667$$
$$\Delta v = \ = 0.50000$$
$$\therefore \qquad u_1 = u_0 + \Delta u = 0.066667$$
$$v_1 = v_0 + \Delta v \ = 1.0.$$

Iteration 2:

-0.66667	1.0	1.0	2.0	1.0	1.0
-1	$-$	-0.66667	-0.22222	-0.51852	-0.09877
	$-$	$-$	-1	-0.33333	-0.77778
-0.66667	1	-0.33333	0.77778	0.14815	0.12346
-1	$-$	-0.66667	0.22222	-0	
	$-$	$-$	-1	0.33333	
	1	-0.33333	0	0.48148	

$$\Delta u = \; = 0.37037$$
$$\Delta v = \; = -0.4444$$
$$u_2 = 1.03704$$
$$v_2 = 0.55556.$$

∴

Iteration 3:

-1.03704	1.0	1.0	2.0	1.0	1.0
-0.55556	$-$	-1.03704	0.03841	-1.53777	0.53635
	$-$	$-$	-0.5556	0.02058	-0.82381
-1.03704	1	-0.03704	1.48285	-0.51720	0.71255
-0.55556	$-$	-1.03704	1.11385	-2.11675	
		$-$	-0.5556	0.59671	
	1	-1.07407	2.04115	-2.03724	

$$\Delta u = \; = -0.11460$$
$$\Delta v = \; = 0.26375$$
$$u_3 = u_2 + \Delta u = 0.92244$$
$$v_3 = v_2 + \Delta v = 0.81930$$

∴

The required quadratic function is $x^2 + 0.92244x + 0.81930$.

9.30   GRAFFE'S ROOT-SQUARING METHOD

This method is used to find all the roots of a polynomial equation if all roots of it are real and distinct.

Let a_1, a_2,..... a_n be n real and distinct roots of the polynomial equation of degree n.

$$Pn(x) = a_0 x^n + a_1 x^{n-1} + \dots + a_n = 0 \qquad (9.93)$$

where $a_0, a_1, \ldots a_n$ are real numbers and $a_0 \neq 0$, separating the even and odd power of x in (9.81) and squaring, we obtain

$$a_0^2 x^{2n} - (a_1^2 - 2a_0 a_2) x^{2n-2} + (a_2^2 - 2a_1 a_3 + 2a_0 a_4) x^{2n-4} + \ldots + a_n^2 = 0$$

Suppose $-x^2 = z$, then the above $(2n)^{\text{th}}$-degree polynomial becomes

$$a_0^2 z^n + (a_1^2 - 2a_0 a_2) z^{n-1} + (a_2^2 - 2a_1 a_3 + 2a_0 a_4) z^{n-2} + \ldots + a_n^2 = 0$$

This equation can be expressed as

$$b_0 z^n + b_1 z^{n-1} + b_2 z^{n-2} + \ldots + b_n = 0, \tag{9.94}$$

where

$$b_0 = a_0^2, \, b_1 = a_1^2 - 2a_0 a_2, \, b_2 = a_2^2 - 2a_1 a_3 + 2a_0 a_4 \ldots, \, b_n = a_n^2 \tag{9.95}$$

Note that the roots of (9.94) are $-a_1^2, -a_2^2, \ldots, -a_n^2$.

This procedure can be repeated m times and we obtain an equation of the form

$$L_0 u^n + L_1 u^{n-1} + L_2 u^{n-2} + \ldots + L_{n-1} u + L_n = 0, \tag{9.96}$$

whose rots $R_1, R_2, R_3, \ldots, R_n$ where

$$R_k = -\alpha_c^{2^m}.$$

If we assume $|a_1| > |a_2| > \ldots > |a_n|$, then

$$|R_1| \gg |R_2| \gg \ldots \gg |R_n|.$$

Since the roots of (9.93) are distinct, the (2^m)th power roots are widely separated for large values m where m is a positive integer. For example, if 1, 2, 3, are roots of $P_3(x) = 0$, then (2^3)th powers of 1, 2, 3 are 1,256,6561. These values are separated much more than 1, 2, 3. Since $R_1, R_2, \ldots, R_n$ are roots of (9.96), we have

$$\sum R_i = \frac{-L_1}{L_0}$$

$$\sum R_i R_j = \frac{L_2}{L_0}$$

$$\sum R_i R_j R_k = \frac{-L_3}{L_0}$$

$$\ldots \ldots \ldots \ldots \ldots \ldots$$

$$R_1 R_2 \ldots R_n = (-1)^n L_n$$

Consider

$$\sum R_i = \frac{-L_1}{L_0}$$

It can be written as

$$R_1\left(1 + \frac{R_2}{R_1} + \frac{R_3}{R_1} + \dots + \frac{R_n}{R_1}\right) = \frac{-L_1}{L_0}$$

when the roots are widely separated $\dfrac{R_2}{R_1}, \dfrac{R_3}{R_1}, \dots, \dfrac{R_n}{R_1}$ and are very very small or close to zero,

$$\therefore \quad R_1 = \frac{-L_1}{L_0}$$

Similarly, $R_1 R_2 = \dfrac{L_2}{L_0}$, $R_1 R_2 R_3 = \dfrac{-L_3}{L_0}, \dots, R_1 R_2 \dots R_n = (-1)^n \dfrac{L_n}{L_0}$.

From these relations, we can find R_1, $R_2,\dots,R_n$, and using $\left|R_i\right| = \left|-a_i^{2^m}\right|$, we can have $\left|a_1\right|,\left|a_2\right|,\dots,\left|a_n\right|$. Hence, the roots of Eq. 9.93 can be determined by substituting $\left|a_i\right|$ or $-\left|a_i\right|$ in Eq. 9.93 and verifying $P_n\left(\left|a_i\right|\right) = 0$ or $P_n\left(-\left|a_i\right|\right) = 0$

Note: The value of L_0, $L_1,\dots,L_n$ can be obtained from the synthetic division in a systematic manner, which is as shown below:

a_0	a_1	a_2	a_3	$\dots$	a_{n-2}	a_{n-1}	a_n
a_0^2	a_1^2 $-2a_0a_2$	a_2^2 $-2a_1a_3$ $+2a_0a_4$	a_3^2 $-2a_2a_4$ $+2a_1a_5$		a_{n-2}^2 $-$ $-$	a_{n-1}^2 $-$ $-$	a_n^2 $-$ $-$
			$\cdot$ $\cdot$ $\cdot$ $\cdot$				$\cdot$ $\cdot$ $\cdot$
b_0	b_1	b_2	b_3		b_{n-2}	b_{n-1}	b_n

Note that is the polynomial is of nth degree; then, there will $(n-1)$ rows between the two horizontal lines. Note that there will no entry at first and last of 2nd rows between the horizontal lines. There will be no first two entries at the beginning and last for third row between horizontal lines. For third-degree polynomial, there will be only two rows between horizontal lines for each squaring operation in the table of synthetic division.

SOLVED PROBLEMS

Problem 1: Using Graffe's root-squaring method find all the roots of the equation $x^3 - 8x^2 + 17x - 10 = 0$.

Solution: Here, $a_0 = 1$, $a_1 = -8$, $a_2 = 17$ and $a_3 = -10$. Let a_1, a_2 and a_3 be roots of given polynomial equation. These roots can be obtained from the table of synthetic division, which is short-cut method to find them by Graeffe's root-squaring method.

	1	−8	17	−10
	1	64	289	100
	−	+(−2)(1)(17)	+(−2)(−8)(−10)	−
1st square	1	30	129	100
	1	900	16641	104
	−	+(−2)(1)(129)	+(−2)(30)(100)	−
2nd square	1	642	10641	104
	1	412164	113230881	108
	1	+(−2)(1)(10641)	+(−2)(642)	−
3rd square	1	390882	100390881	108

Now $\quad |a_1| = \sqrt[8]{390882} = 5.00041 \approx 5$

$$|a_2| = \sqrt[8]{\frac{100390881}{390882}} = 2.00081 \approx 2$$

$$|a_3| = \sqrt[8]{\frac{10^8}{100390881}} = 0.99951 \approx 1$$

We can verify that 5, 2 and 1 are roots of given polynomial equation.

Problem 2: Find all the roots of $x^3 - 3x^2 + 2x = 0$ by Graeffe's root-squaring method.

Solution: Here, $a_0 = 1$, $a_1 = -3$, $a_2 = 2$ and $a_3 = 0$. Let a_1, a_2 and a_3 be roots of given polynomial function. The coefficients of each polynomial obtained from each squaring process are shown in the following table:

	1	−3	2	0
	1	9	4	0
		−4	0	
First square	1	5	4	0
	1	25	16	0
		−8	0	
Second square	1	17	16	0
	1	289	256	0
		−32	0	
Third square	1	257	256	0

$$\text{Now} \quad |a_1| = \sqrt[8]{\frac{257}{1}}\, 2.00097 \approx 2$$

$$|a_2| = \sqrt[8]{\frac{256}{257}}\, 0.99951 \approx 1$$

$$|a_3| = \sqrt[8]{\frac{0}{256}} = 0$$

Hence, the roots are 0, 1 and 2.

Problem 3: Find all the roots of $x^3 - 4x^2 + 3x + 1 = 0$ using Graeffe's root-squaring method.

Solution: Here $a_0 = 1$, $a_1 = -4$, $a_2 = 3$ and $a_3 = 1$. Let a_1, a_2 and a_3 be roots of the given polynomial equation. Similar to earlier problems, we construct synthetic division table, which show the coefficients of polynomial on each squaring process:

	1	−4	3	1
	1	16	9	1
		−6	8	

First square	1	10	17	1
	1	100	289	1
		−34	−20	
Second square	1	66	269	1
	1	4356	72361	1
		−538	−132	
Third square	1	3818	72229	1

$$|a_1| = \sqrt[8]{\frac{3818}{1}} = 2.803687$$

$$|a_2| = \sqrt[8]{\frac{72229}{3819}} = 1.444141$$

$$|a_3| = \sqrt[8]{\frac{1}{72229}} = 0.24698$$

It can be easily verified that 0.24698, 1.444141 and 2.803687 are roots of the given polynomial equation.

9.31   SYSTEM OF NON-LINEAR EQUATIONS

In this section, we shall learn to find the solution of a system of two non-linear equations in two variables by Newton–Raphson method. Let

$$f(x, y) = 0$$

and

$$g(x, y) = 0 \qquad (9.97)$$

be a system of two non-linear equations in x and y. Suppose $x = \alpha$ and $y = \beta$ be an exact solution of (9.97). Let x_0 and y_0 be a suitable approximations to α and β, respectively. Let Δx be an increment in x_0 and Δy be an increment in y_0 such that $(x_0 + \Delta x, y_0 + \Delta y)$ is an exact solution of (9.97). Then,

$$f(x_0 + \Delta x, y_0 + \Delta y) = 0 \qquad (9.98)$$

and

$$g(x_0 + \Delta x, y_0 + \Delta y) = 0 \qquad (9.99)$$

Expanding $f(x_0 + \Delta x, y_0 + \Delta y)$ and $g(x_0 + \Delta x, y_0 + \Delta y)$ in Taylor's series about the point (x_0, y_0), we have

$$f(x_0,y_0)+\left[\Delta x\, f_x(x_0,y_0)+\Delta y\, f_y(x_0,y_0)\right]$$

$$+\frac{1}{\lfloor 2}\left[(\Delta x)^2 f_{xx}(x_0,y_0)+2\Delta x\Delta y\, f_{xy}(x_0,y_0)+(\Delta y)^2 f_{yy}(x_0,y_0)\right]+\cdots=0$$

and

$$g(x_0,y_0)+\left[\Delta x\, g_x(x_0,y_0)+\Delta y\, g_y(x_0,y_0)\right]$$

$$+\frac{1}{\lfloor 2}\left[(\Delta x)^2 g_{xx}(x_0,y_0)+2\Delta x\Delta y\, g_{xy}(x_0,y_0)+(\Delta y)^2 g_{yy}(x_0,y_0)\right]$$

$$+\cdots=0,$$

where, $f_x=\dfrac{\partial f}{\partial x}, f_{xy}=\dfrac{\partial^2 f}{\partial x\partial y}$ etc. Neglecting the terms involving the second- and higher-order powers of Δx, Δy and, the product Δx, and Δy, we have

$$f(x_0,y_0)+\Delta x\, f_x(x_0,y_0)+\Delta y\, f_y(x_0,y_0)=0, \tag{9.100}$$

$$g(x_0,y_0)+\Delta x\, g_x(x_0,y_0)+\Delta y\, g_y(x_0,y_0)=0. \tag{9.101}$$

If $J=f_x(x_0,y_0)\,g_y(x_0,y_0)-f_y(x_0,y_0)\,g_x(x_0,y_0)\neq 0$, then Eqs. (9.100) and (9.101) have unique solution. Assuming $J\neq 0$, we have

$$\Delta x=-\frac{f(x_0,y_0)g_y(x_0,y_0)-g(x_0,y_0)f_y(x_0,y_0)}{J}$$

and

$$\Delta y=-\frac{g(x_0,y_0)f_x(x_0,y_0)-f(x_0,y_0)g_x(x_0,y_0)}{J}$$

here $J=\begin{vmatrix}\dfrac{\partial f}{\partial x} & \dfrac{\partial f}{\partial y}\\[2ex] \dfrac{\partial g}{\partial x} & \dfrac{\partial g}{\partial y}\end{vmatrix}_{(x_0,y_0)}$ is the value of Jacobian of functions f and g at (x_0,y_0). The new approximation (x_1,y_1) to the solution of (9.97) is given by

$$x_1=x_0+\Delta x=x_0-\frac{f(x_0,y_0)g_y(x_0,y_0)-g(x_0,y_0)f_y(x_0,y_0)}{J}$$

and

$$y_1 = y_0 + \Delta y = y_0 - \frac{g(x_0,y_0)f_x(x_0,y_0) - f(x_0,y_0)g_x(x_0,y_0)}{J}$$

Continuing the above process with (x_1, y_1), we have another approximation (x_2, y_2) and so on. Thus, the Newton's iterative formulae to find the solutions of (9.97) are

$$x_{n+1} = x_n - \frac{f(x_n,y_n)g_y(x_n,y_n) - g(x_n,y_n)f_y(x_0,y_0)}{J} \qquad (9.102)$$

$$y_{n+1} = y_n - \frac{g(x_0,y_0)f_x(x_0,y_0) - f(x_0,y_0)g_x(x_0,y_0)}{J} \qquad (9.103)$$

where,

$$J = f_x(x_n, y_n)\,g_y(x_n, y_n) - f_y(x_n, y_n)\,g_x(x_n, y_n). \qquad (9.104)$$

SOLVED PROBLEMS

Problem 1: Find a solution of the following system of equations $x^2 - y^2 = 5$ and $x^2 + y^2 = 13$. Choose the initial approximations of x and y as 2.5 and 1.5.

Solution: Let $f(x, y) = x^2 - y^2 - 5$, $g(x, y) = x^2 + y^2 - 13$, $x_0 = 2.5$ and $y_0 = 1.5$.

By the Newton–Raphson formula for the system of two equations in two variables x and y are given by

$$x_{n+1} = x_n - \frac{f(x_n,y_n)g_y(x_n,y_n) - g(x_n,y_n)f_y(x_n,y_n)}{J} \qquad (1)$$

$$y_{n+1} = y_n - \frac{g(x_n,y_n)f_x(x_n,y_n) - f(x_n,y_n)g_x(x_n,y_n)}{J}, \qquad (2)$$

where,

$$J = f_x(x_n, y_n)\,g_y(x_n, y_n) - f_y(x_n, y_n)\,g_x(x_n, y_n). \qquad (3)$$

In the problem at our disposal, $f_x = 2x$, $f_y = -2y$, $g_x = 2x$ and $g_y = 2y$ substituting $n = 0$ in (1), (2) and (3), we have

$$x_1 = x_0 - \frac{f(x_0,y_0)g_y(x_0,y_0) - g(x_0,y_0)f_y(x_0,y_0)}{J} \qquad (4)$$

$$y_1 = y_0 - \frac{g(x_0,y_0)f_x(x_0,y_0) - f(x_0,y_0)g_x(x_0,y_0)}{J} \qquad (5)$$

and

Iteration 1: $\qquad J = f_x(x_0, y_0)\,g_y(x_0, y_0) - g_x(x_0, y_0)\,f_y(x_0, y_0) \qquad (6)$

Now, we find the quantities appeared in (4), (5) and (6) for $x_0 = 2.5$ and $y_0 = 1.5$.

$$f(x_0, y_0) = f(2.5, 1.5) = (2.5)^2 - (1.5)^2 - 5 = -1$$
$$g(x_0, y_0) = g(2.5, 1.5) = (2.5)^2 + (1.5)^2 - 13 = -4.5$$
$$f_x(x_0, y_0) = f_x(2.5, 1.5) = 2 \times 2.5 = 5$$
$$f_y(x_0, y_0) = f_y(2.5, 1.5) = -2 \times 1.5 = -3$$
$$g_x(x_0, y_0) = g_x(2.5, 1.5) = 2 \times 2.5 = 5$$
$$g_y(x_0, y_0) = g_y(2.5, 1.5) = 2 \times 1.5 = 3$$
$$J = f(2.5, 1.5)\,g_y(2.5, 1.5) - g_x(2.5, 1.5)\,f_y(2.5, 1.5) = 30$$

The first approximations to x and y are

$$x_1 = 2.5 - \frac{(-16.5)}{30} = 3.05$$
$$y_1 = 1.5 - \frac{(-17.5)}{30} = 2.083333$$

Iteration 2: The values of f, g, f_x etc. at x_1 and y_1 are

$$f(x_1, y_1) = -0.03778,\ g(x_1, y_1) = 0.64278$$
$$f_x(x_1, y_1) = 6.1,\ f_y(x_1, y_1) = -4.16667$$
$$g_x(x_1, y_1) = 6.1,\ g_y(x_1, y_1) = 4.16667 \text{ and } J = 5083333.$$

Hence,

$$x_2 = 3.05 - \left(\frac{2.52083}{50.83333}\right) = 3.000410$$

$$y_2 = 2.083333 - \frac{(4.15138)}{50.83333} = 2.001667.$$

Iteration 3: The value of $f, g, f_x, f_y \ldots$ etc. at $x_2 = 3.000410$ and $y_2 = 2.001667$ are as follows:

$$f(x_2, y_2) = -0.00421,\ g(x_2, y_2) = 0.00913$$
$$f_x(x_2, y_2) = 6.00082,\ f_y(x_2, y_2) = -4.00333$$
$$g_x(x_2, y_2) = 6.00082,\ g_y(x_2, y_2) = 4.00333$$
$$J = 48.04656.$$

Thus

$$x_3 = 3.000410 - \frac{(+0.01969)}{48.04656} = 3.00000$$

and

$$y_3 = 2.001667 - \left(\frac{+0.08004}{48.04656}\right) = 2.00001$$

The solution of given system of equations is $x = 3$ and $y = 2$.

Problem 2: Use Newton–Raphson method to find a solution of the following system of equations.

$$f(x, y) = x^3 - 2y^2 - 1 = 0$$

and
$$g\,(x, y) = 5y^3 + x^2 - 2xy - 4 = 0.$$

Choose $x_0 = -0.5$ and $y_0 = 0.8$ as initial approximations of x and y. Perform four iterations.

Solution: Differentiating f and g partially w.r.t. x and y, we have

$$f_x = 3x^2, f_y = -4y$$

$$g_x = 2x - 2y,\ g_y = 15y^2 - 2x.$$

The iteration of Newton–Raphson method for system of two equations in two variables x and y are given by

$$x_{n+1} = x_n - \frac{f(x_n, y_n)g_y(x_n, y_n) - g(x_n, y_n)f_y(x_n, y_n)}{J}$$

$$y_{n+1} = y_n - \frac{f(x_n, y_n)g(x_n, y_n) - f(x_n, y_n)g_x(x_n, y_n)}{J}$$

where

$$J = f_x\,(x_n, y_n)\,g_y\,(x_n, y_n) - f_y\,(x_n, y_n)\,g_x\,(x_n, y_n)$$

For $n = 0$, we have

$$x_1 = x_0 - \frac{f(x_0, y_0)g_y(x_0, y_0) - g(x_0, y_0)f_y(x_0, y_0)}{J}$$

$$y_1 = y_0 - \frac{g(x_0, y_0)f_x(x_0, y_0) - f(x_0, y_0)g_x(x_0, y_0)}{J}$$

and

$$J = f_x(x_0, y_0)g_y(x_0, y_0) - g_x(x_0, y_0)g_y(x_0, y_0)$$

Since $x_0 = -0.5$, $y_0 = 0.8$

$$f(x_0, y_0) = f(-0.6, 0.8) = 0.064$$

$$g(x_0, y_0) = f(-0.6, 0.8) = -0.1200$$

$$f_x(x_0, y_0) = f_x(-0.6, 0.8) = 1.08$$

$$f_y(x_0, y_0) = f_y(-0.6, 0.8) = 3.2$$

$$g_x(x_0, y_0) = g_x(-0.6, 0.8) = -2.8$$

$$g_y(x_0, y_0) = g_y(-0.6, 0.8) = 10.8$$

$$J = 20.62400$$

$$\therefore x_1 = -0.6 - \frac{1.07520}{20.62400} = -0.652133$$

$$y_1 = 0.8 - \frac{(0.04960)}{2062400} = 0.797595.$$

Iteration 2 : The values of f, g, f_x, f_y, g_x and g_y at (x_1, y_1) are

$$f(x_1, y_1) = -0.00502, \quad g(x_1, y_1) = 0.00254$$

$$f_x(x_1, y_1) = 1.27583, \quad g_x(x_1, y_1) = -2.89946$$

$$f_y(x_1, y_1) = 3.19038, \quad g_y(x_1, y_1) = 10.84663$$

$$J = 23.08887$$

$$\therefore \quad x_2 = -0.652133 - \frac{(-0.06257)}{23.08887} = -0.649424$$

$$y_2 = 0.797595 - \frac{(-0.0113)}{23.08887} = 0.798086$$

Similarly, it can be shown that

$$x_3 = -0.649416$$

$$y_3 = 0.798087.$$

EXERCISE 9

1. Find a real root of $x^3 - x - 1 = 0$ by bisection method.

2. Find a real root of $x\log_{10} x - 1.2 = 0$ by bisection method.

3. Find a root of $2x - 3\sin x - 5 = 0$ by bisection method.

4. Find a root correct the three decimal places of $x^3 - x = 40$ by (i) Secant method and (ii) Regula Falsi method.

5. Find a root of the equation correct the three decimal places of $x^3 - 3x - 5 = 0$ by Regula Falsi method.

6. Find a root of $x\,e^x - \cos x = 0$ by Newton–Raphson method.

7. Find a root of $x - e^{-x} = 0$ by Newton–Raphson method correct to three decimal places.

8. Obtain $\sqrt{17}$ using Newton–Raphson method

9. Obtain $\sqrt[3]{12}$ using Newton–Raphson method

10. Find a root of $\sin x - \dfrac{x}{2} = 0$ by Muller's method.

11. Find the iteration formula based on Newton–Raphson method for finding $N^{1/4}$.

12. Use Muller's method to find a root of $x^3 - x^2 - x + 1 = 0$.

13. Find a quadratic factor of $x^4 - 3x^3 + 20x^2 + 44x + 54 = 0$ by Bairstow's method.

14. Find a quadratic factor of $2x^4 + 7x^3 - 4x^2 + 29x + 14 = 0$.

15. Find all the roots of $x^3 - 4x^2 + 3x + 1 = 0$ by Graeffes's root squaring method.

16. Find all the root of $x^3 + 3x^2 - 4 = 0$ by Graeffe's root-squaring method.

17. Find a solution of $x^2 + y^2 = 1.12$ and $xy = 0.23$ by Newton–Raphson method.

18. Find a solution of $x^3 - 3xy^2 + 1 = 0$ and $3x^2 - y^2 = 0$.

19. Find a root of $x^3 - x^2 - 2x + 1 = 0$ by Birga–Vieta method.

20. Find a root of $x^3 + 1.2x^2 - 4x - 4.8 = 0$ by Birga–Vieta method.

ANSWERS

1. 1.3247

2. 2.74064

3. 2.88323

4. 3.5174, 3.5174

5. 2.2790

6. 0.51775736

7. 0.56714329

8. 4.12310563

9. 2.28942849

10. 0.52359878

11. $x_{n+1} = \dfrac{3x_n^4 + N}{4x_n^3}$

12. 1.83928676

13. $x^2 + 1.9412x + 1.95377$

14. $x^2 + 5x + 2$, $2x^2 - 3x + 7$

15. 2.8019, 1.4450, −0.2470

16. −2 −2 ,1

17. (1.035, 0.222), (−1.035, −0.222)

18. $x = 0.5086, y = 0.8411$

19. 1.8019 or −1.2470 or 0.445

20. −1.2 or −2 or 2.

10

Solution of System of Linear Equations

Some problems in the fields of engineering, elasticity and statistics yield a system of linear equations. For example, in the study of mechanical vibrations of n degrees of freedom, we get a system of linear equations. To know the natural frequencies, we need the solution of the system of linear equations. Representing the system of linear equation in matrices simplified the task of finding the solution of system of linear equations. In this chapter, we discuss some of the numerical methods to solve the system of linear equations.

10.1 DEFINITION: SYSTEM OF LINEAR EQUATIONS

The following equations are defined as a system of m linear equations in n unknowns (m, n are positive integers):

$$a_{11} x_1 + a_{12} x_2 + \ldots\ldots + a_{1n} x_n = b_1$$

$$a_{21} x_1 + a_{22} x_2 + \ldots\ldots + a_{2n} x_n = b_2 \qquad (10.1)$$

$$\ldots\ldots\ldots\ldots\ldots\ldots\ldots\ldots\ldots$$

$$a_{m1} x_1 + a_{m2} x_2 + \ldots\ldots + a_{mn} x_n = b_n,$$

where a_{ij} ($i = 1, 2\ldots, m$; $j = 1, 2\ldots, n$) are known constants, b_i ($i = 1, 2, \ldots, m$) are also known constants and x_i ($i = 1, 2, \ldots, n$) are called unknowns.

For example, $2x + 3y + 4z = 6$, $x - y + 3z = 4$ is a system of two linear equations in three unknowns.

Note: In this chapter, we restrict our study to the system of n linear equations in n unknown only. Further, we also assume that such system have a unique solution.

10.2   DEFINITION: SOLUTION OF SYSTEM OF LINEAR EQUATIONS

We say that a_1, a_2, ..., a_n is a solution of the system of n linear equations in n unknowns:

$$a_{11} x_1 + a_{12} x_2 + ... + a_{1n} x_n = b_1$$
$$a_{21} x_1 + a_{22} x_2 + ... + a_{2n} x_n = b_2 \qquad (10.2)$$
$$\cdots\cdots\cdots\cdots\cdots\cdots\cdots\cdots$$
$$a_{n1} x_1 + a_{n2} x_2 + ... + a_{nn} x_n = b_n$$

if

$$a_{11} a_1 + a_{12} a_2 + ... + a_{1n} a_n = b_1$$
$$a_{21} a_1 + a_{22} a_2 + ... + a_{2n} a_n = b_2$$
$$\cdots\cdots\cdots\cdots\cdots\cdots\cdots\cdots$$
$$a_{n1} a_1 + a_{n2} a_2 + ... + a_{nn} a_n = b_n$$

For example, $x_1 = 2$ and $x_2 = 1$ is a solution of system of equations $x_1 + 2x_2 = 4$ and $4x_1 - x_2 = 7$.

10.3   REPRESENTATION OF SYSTEM OF LINEAR EQUATIONS IN A MATRIX FORM

The system of linear equations given by (10.2) can be written as

$$AX = B, \qquad (10.3)$$

where

$$A = \begin{bmatrix} a_{11} & a_{12}.... & a_{1n} \\ a_{21} & a_{22}.... & a_{2n} \\ ...... & .... & ..... \\ a_{n1} & a_{n2}.... & a_{nn} \end{bmatrix}, \ X = \begin{bmatrix} x_1 \\ x_2 \\ ..... \\ x_n \end{bmatrix} \text{and } B = \begin{bmatrix} b_1 \\ b_2 \\ .... \\ b_n \end{bmatrix}$$

The matrix A is defined as coefficient matrix of the system of Eq. (10.2) and matrix X is defined as unknown matrix of it. Further the matrix B is defined as known constant matrix of the system of Eq. (10.2). The matrix

$$[A:B] = \begin{bmatrix} a_{11} & a_{12}.... & a_{1n} & b_1 \\ a_{21} & a_{22}.... & a_{2n} & b_2 \\ ...... & .... & .... & .... \\ a_{n1} & a_{n2}..... & a_{nn} & b_n \end{bmatrix} \qquad (10.4)$$

is called augmented matrix of the system of equations given by (10.2).

Note that the system of n linear equations in n unknowns given by (10.2) is equivalent to a single matrix Eq. (10.3). Throughout this chapter, we assume that determinant of the coefficient matrix different from zero to guarantee the unique solution of the system of linear equations.

10.4   UPPER AND LOWER TRIANGULAR SYSTEM OF LINEAR EQUATIONS

The system of equations of the following type is called the upper triangular system of linear equations:

$$a_{11}x_1 + a_{12}x_2 + a_{13}x_3 + ... + a_{1n}x_n = b_1$$

$$a_{22}x_2 + a_{23}x_3 + ... + a_{2n}x_n = b_2$$

$$a_{33}x_3 + ... + a_{3n}x_n = b_3 \qquad (10.5)$$

$$..$$

$$a_{nn}x_n = b_n$$

The Eq. (10.5) can be expressed as

$$UX = B,$$

where

$$U = \begin{bmatrix} a_{11} & a_{12}.... & a_{1n} \\ 0 & a_{22}.... & a_{2n} \\ 0 & 0 & a_{nn} \end{bmatrix}, \quad X = \begin{bmatrix} x_1 \\ x_2 \\ x_n \end{bmatrix} \text{and } B = \begin{bmatrix} b_1 \\ b_2 \\ b_n \end{bmatrix}$$

Note that U is an upper triangular matrix.

Similarly, the system of linear equations of the following form is called the lower triangular system of linear equations:

$$a_{11}x_1 = b_1$$

$$a_{21}x_1 + a_{22}x_2 = b_2 \qquad (10.6)$$

$$a_{31}x_1 + a_{32}x_2 + a_{33}x_3 = b_3$$

$$\cdots\cdots\cdots\cdots\cdots\cdots\cdots\cdots\cdots\cdots$$

$$a_{n1}x_1 + a_{n2}n_2 + a_{n3}n_3 + \ldots + a_{nn}x_n = b_n$$

and Eq. (10.6) can be expressed as

$$LX = B,$$

where

$$L = \begin{bmatrix} a_{11} & 0 & 0\ldots\ldots & 0 & 0 \\ a_{21} & a_{22} & 0\ldots\ldots & 0 & 0 \\ \multicolumn{5}{c}{\cdots\cdots\cdots\cdots\cdots\cdots\cdots\cdots\cdots\cdots} \\ a_{n1} & a_{n2} & a_{n3}\ldots & a_{n(n-1)} & a_{nn} \end{bmatrix}, \ X = \begin{bmatrix} x_1 \\ x_2 \\ \cdots \\ x_n \end{bmatrix} \text{and } B = \begin{bmatrix} b_1 \\ b_2 \\ \cdots \\ b_n \end{bmatrix}$$

Clearly, L is a lower triangular matrix.

10.5   DEFINITIONS

(i) *Direct methods*

A method to obtain the solution of (10.2) is said to be a direct method if its solution is obtained in a finite number of steps and it is an exact solution.

If a number is rounded in the intermediate steps of a direct method, the solution may not be exact by the direct method. Still, the method is regarded as a direct method.

(ii) *Iteration method*

A method to obtain the solution of (10.2) is said to be an iterative method if it give a sequence of approximate solutions which converges to exact solution when the number of steps tends to very very large or infinite. The successive approximations are called iterates.

10.6   DIRECT METHODS

In this section, we discuss the direct methods of solving the system of linear Eq. (10.2), and in Section (10.7), we introduce the iteration methods of solving Eq. (10.2).

10.6.1 Method of solving upper triangular system of linear equations

The upper triangular system of equations are given by (10.5) .i.e.,

$$a_{11}x_1 + a_{12}x_2 + \ldots + a_{1(n-1)}x_{n-1} + a_{1n}x_n = b_1 \qquad (A_1)$$

$$a_{22}x_2 + \ldots + a_{2(n-1)}x_{n-1} + a_{2n}x_n = b_2 \qquad (A_2)$$

$$a_{33}x_3 + \ldots + a_{3(n-1)}x_{n-1} + a_{3n}x_n = b_3 \qquad (A_3)$$

$$\ldots\ldots\ldots\ldots\ldots\ldots\ldots\ldots\ldots\ldots\ldots\ldots$$

$$a_{nn}x_n = b_n \qquad (A_n)$$

From the equation (A_n), we can find x_n. Substituting x_n $\left(\text{i.e. } x_n = \dfrac{b_n}{a_{nn}}\right)$ in the equation (A_{n-1}), we get x_{n-1}. Similarly, we can find x_{n-2}, x_{n-3}, ..., x_3, x_2, and x_1. The unknowns are solved by back substitution, and hence this method is called back-substitution method

10.6.2 Method of solving lower triangular linear system of equations

The lower triangular system of linear equation is given by

$$a_{11}x_1 = b_1 \qquad (B_1)$$

$$a_{21}x_1 + a_{22}x_2 = b_2 \qquad (B_2)$$

$$\ldots\ldots\ldots\ldots\ldots\ldots\ldots\ldots\ldots\ldots\ldots$$

$$a_{n1}x_1 + a_{n2}x_2 \ldots\ldots + a_{nn} = b_n \qquad (B_n)$$

Solving the equation (B_1) for x_1, we obtain $x_1 = \dfrac{b_1}{a_{11}}$; substituting the known value x_1 in the equation (B_2), we get x_2; similarly, we can find x_3, x_4, ... and x_n. Here, the unknown are solved by forward substitution, and hence this method is called forward-substitution method.

SOLVED PROBLEMS

Problem 1: Solve $3x_1 + 4x_2 - 2x_3 = 9$, $3x_2 + 2x_3 = 4$ and $2x_3 = 1$.

Solution: The given system of linear equations is

$$3x_1 + 4x_2 - 2x_3 = 9 \qquad (1)$$

$$3x_2 + 2x_3 = 4 \tag{2}$$

$$2x_3 = 1 \tag{3}$$

From Eq. (3), we have $x_3 = \dfrac{1}{2}$ and substituting it in (2), we get

$$3x_2 + 2\left(\dfrac{1}{2}\right) = 4, \text{ i.e., } x_2 = 1.$$

Now, substituting $x_2 = 1$ and $x_3 = \dfrac{1}{2}$ in (1), we obtain

$$3x_1 + 4(1) - 2\left(\dfrac{1}{2}\right) = 9 \text{ i.e., } x_1 = 2. \text{ Thus, } x_1 = 2, x_2 = 1 \text{ and } x_3 = \dfrac{1}{2} \text{ is}$$

the solution of the given system of linear equations.

Problem 2: Solve $4x_1 = 1, 3x_1 + 5x_2 = \dfrac{13}{4}$ and $x_1 + x_2 + x_3 = \dfrac{3}{2}$.

Solution: The given system of linear equations is

$$4x_1 = 1 \tag{1}$$

$$3x_1 + 5x_2 = \dfrac{13}{4} \tag{2}$$

$$x_1 + x_2 + x_3 = \dfrac{3}{2} \tag{3}.$$

From Eq. (1), we have $x_1 = \dfrac{1}{4}$; substituting $x_1 = \dfrac{1}{4}$ in (2), we obtain

$$3\left(\dfrac{1}{4}\right) + 5x_2 = \dfrac{13}{4} \text{ i.e., } x_2 = \dfrac{1}{2}. \text{ Finally, substituting } x_1 = \dfrac{1}{4} \text{ and } x_2 = \dfrac{1}{2}$$

in (3), we get $x_3 = \dfrac{3}{4}$. Thus, $x_1 = \dfrac{1}{4}, x_2 = \dfrac{1}{2}$ and $x_3 = \dfrac{3}{4}$ is the solution of given system of linear equations.

Problem 3: Solve $x_1 - 0.7x_2 + 0.3x_3 + 0.5x_4 = 0.6$, $x_2 + 0.63636$ $x_3 - 0.27273x_4 = -0.72727$, $x_3 - 0.02247x_4 = -7.02247$ and $10.37079\, x_4 = 10.37079$.

Solution: The given system of equations is

$$x_1 - 0.7\,x_2 + 0.3x_3 + 0.5\,x_4 = 0.6 \tag{1}$$

$$x_2 + 0.63636\, x_3 - 0.27273\, x_4 = -0.72727 \tag{2}$$

$$x_3 - 0.02247\, x_4 = -7.02247 \tag{3}$$

$$10.37079\, x_4 = 10.37079. \tag{4}$$

Clearly, $x_4 = 1$, substituting $x_4 = 1$ in (3), we get $x_3 - (0.02247)$ (1) $= -7.02247$, i.e., $x_3 = -7$. Now, substituting $x_4 = 1$ and $x_3 = -7$ (2), we obtain $x_2 + (0.63636)$ $(-7) - 0.27273$ (1) $= -0.72727$, i.e., $x_2 = -0.72727 + (0.63636)$ $(7) + 0.27273$ i.e., $x_2 = 3.99998$ and finally, substituting $x_2 = 3.99998$, $x_3 = -7$ and x_4 in Eq. (1), we obtain

$x_1 - 0.7$ (3.99998) $+ 0.3$ $(-7) + 0.5$ (1) $= 0.6$, i.e., $x_1 = 0.7$ (3.99998) $+ 0.3$ (7) $- 0.5 + 0.6 = 4.999986$. Thus, the solution of the given system of equations is $x_1 = 4.999986$, $x_2 = 3.99998$, $x_3 = -7$ and $x_4 = 1$.

10.6.3 Cramer's rule

We present the proof of it for a system of three linear equations in three unknowns. It can be extended to n linear equations in n unknowns. Let

$$a_1 x_1 + b_1 x_2 + c_1 x_3 = d_1$$

$$a_2 x_1 + b_2 x_2 + c_2 x_3 = d_2 \tag{10.7}$$

$$a_3 x_1 + b_3 x_3 + c_3 x_3 = d_3$$

be a system of linear equations in three unknowns

$$\text{Let } \Delta = \begin{vmatrix} a_1 & b_1 & c_1 \\ a_2 & b_2 & c_2 \\ a_3 & b_3 & c_3 \end{vmatrix}, \quad \Delta_1 = \begin{vmatrix} d_1 & b_1 & c_1 \\ d_2 & b_2 & c_2 \\ d_3 & b_3 & c_3 \end{vmatrix},$$

$$\Delta_2 = \begin{vmatrix} a_1 & d_1 & c_1 \\ a_2 & d_2 & c_2 \\ a_3 & d_3 & c_3 \end{vmatrix}, \quad \Delta_4 = \begin{vmatrix} a_1 & b_1 & d_1 \\ a_2 & b_2 & d_2 \\ a_3 & b_3 & d_3 \end{vmatrix} \text{ and } \Delta \neq 0$$

Now, consider

$$\Delta_1 = \begin{vmatrix} d_1 & b_1 & c_1 \\ d_2 & b_2 & c_2 \\ d_3 & a_3 & a_3 \end{vmatrix}$$

i.e.,
$$\Delta_1 = \begin{vmatrix} a_1x_1 + b_1x_2 + c_1x_3 & b_1 & c_1 \\ a_2x_1 + b_2x_2 + c_2x_3 & b_2 & c_2 \\ a_3x_1 + b_3x_2 + c_3x_3 & b_3 & c_3 \end{vmatrix} \qquad \text{(by 10.7)}$$

Since first column elements of Δ_1 are sum of three elements, by the properties of the determinants

$$\Delta_1 = \begin{vmatrix} a_1x_1 & b_1 & c_1 \\ a_2x_1 & b_2 & c_2 \\ a_3x_1 & b_3 & a_3 \end{vmatrix} + \begin{vmatrix} b_1x_2 & b_1 & c_1 \\ b_2x_2 & b_2 & c_2 \\ b_3x_2 & b_3 & c_3 \end{vmatrix} + \begin{vmatrix} c_1x_3 & b_1 & c_1 \\ c_2x_3 & b_2 & c_2 \\ c_3x_3 & b_3 & c_3 \end{vmatrix}$$

$$\Delta_1 = x_1\,\Delta + 0 + 0$$

$$x_1 = \frac{\Delta_1}{\Delta}.$$

Similarly, it can be shown that $x_2 = \dfrac{\Delta_2}{\Delta}$ and $= x_3 = \dfrac{\Delta_3}{\Delta}$. Thus, the Cramer's rule is given by

$$\frac{x_1}{\begin{vmatrix} d_1 & b_1 & c_1 \\ d_2 & b_2 & c_2 \\ d_3 & b_3 & c_3 \end{vmatrix}} = \frac{x_2}{\begin{vmatrix} a_1 & d_1 & c_1 \\ a_2 & d_2 & c_2 \\ a_3 & d_3 & c_3 \end{vmatrix}} = \frac{x_3}{\begin{vmatrix} a_1 & b_1 & d_1 \\ a_2 & b_2 & d_2 \\ a_3 & b_3 & d_3 \end{vmatrix}} = \frac{1}{\begin{vmatrix} a_1 & a_2 & a_3 \\ b_1 & b_2 & b_3 \\ c_1 & c_2 & c_3 \end{vmatrix}}$$

Note: The Cramer's rule for a system of four linear equations in four unknowns:

$$a_1 x_1 + b_1 x_2 + c_1 x_3 + d_1 x_4 = e_1$$

$$a_2 x_1 + b_2 x_2 + c_2 x_3 + d_2 x_4 = e_2$$

$$a_3 x_1 + b_3 x_2 + c_3 x_3 + d_3 x_3 = e_3$$

$$a_4 x_1 + b_4 x_2 + c_4 x_3 + d_4 x_4 = e_4$$

is given by

$$x_1 = \frac{\Delta_1}{\Delta},\ x_2 = \frac{\Delta_2}{\Delta},\ x_3 = \frac{\Delta_3}{\Delta} \text{ and } x_4 = \frac{\Delta_4}{\Delta},$$

where

$$\Delta_1 = \begin{vmatrix} e_1 & b_1 & c_1 & d_1 \\ e_2 & b_2 & c_2 & d_2 \\ e_3 & b_3 & c_3 & d_3 \\ e_4 & b_4 & c_4 & e_4 \end{vmatrix}, \quad \Delta_2 = \begin{vmatrix} a_1 & e_1 & c_1 & d_1 \\ a_2 & e_2 & c_2 & d_2 \\ a_3 & e_3 & c_3 & d_3 \\ a_4 & e_4 & c_4 & d_4 \end{vmatrix}$$

$$\Delta_3 = \begin{vmatrix} a_1 & b_1 & e_1 & d_1 \\ a_2 & b_2 & e_2 & d_2 \\ a_3 & b_3 & e_3 & d_3 \\ a_4 & b_4 & e_4 & d_4 \end{vmatrix}, \quad \Delta_4 = \begin{vmatrix} a_1 & b_1 & c_1 & e_1 \\ a_2 & b_2 & c_2 & e_2 \\ a_3 & b_3 & c_3 & e_3 \\ a_4 & b_4 & c_4 & e_4 \end{vmatrix}$$

and

$$\Delta = \begin{vmatrix} a_1 & b_1 & c_1 & d_1 \\ a_2 & b_2 & c_2 & d_2 \\ a_3 & b_3 & c_3 & d_3 \\ a_4 & b_4 & c_4 & e_4 \end{vmatrix} \neq 0$$

SOLVED PROBLEMS

Problem 1: Solve $2x_1 + 3x_2 = 5$ and $3x_1 - x_2 = 2$ by Cramer's rule.

Solution: The given system of linear equation is in two unknowns and two equations. Therefore, by the Cramer's rule

$$\frac{x_1}{\begin{vmatrix} 5 & 3 \\ 2 & -1 \end{vmatrix}} = \frac{x_2}{\begin{vmatrix} 2 & 5 \\ 3 & 2 \end{vmatrix}} = \frac{1}{\begin{vmatrix} 2 & 3 \\ 3 & -1 \end{vmatrix}}$$

i.e.,
$$\frac{x_1}{-11} = \frac{x_2}{-11} = \frac{1}{-11}$$

$$\therefore \qquad x_1 = \frac{-11}{-11} = 1 \text{ and } x_2 = \frac{-11}{-11} = 1.$$

Thus, $x_1 = 1$ and $x_2 = 1$ is the solution of given system of linear equations.

Problem 2: Solve $x + y + z = 4$, $2x - y + z = 4$ and $x + 2y + 3z = 7$ using Cramer's rule.

Solution: Here,

$$\Delta = \begin{vmatrix} 1 & 1 & 1 \\ 2 & -1 & 1 \\ 1 & 2 & 3 \end{vmatrix} = -5,\ \Delta_1 = \begin{vmatrix} 4 & 1 & 1 \\ 4 & -1 & 1 \\ 7 & 2 & 3 \end{vmatrix} = -10,$$

$$\Delta_2 = \begin{vmatrix} 1 & 4 & 1 \\ 2 & 4 & 1 \\ 1 & 7 & 3 \end{vmatrix} = -5 \text{ and } \Delta_3 = \begin{vmatrix} 1 & 1 & 4 \\ 2 & -1 & 4 \\ 1 & 2 & 7 \end{vmatrix} = -5$$

Hence,

$$x = \frac{\Delta_1}{\Delta} = \frac{-10}{-5} = 2,$$

and

$$y = \frac{\Delta_2}{\Delta} = \frac{-5}{-5} = 1,$$

$$z = \frac{\Delta_3}{\Delta} = \frac{-5}{-5} = 1,$$

Thus, $x = 2$, $y = 1$ and $z = 1$ is the solution of given system of linear equations.

Problem 3: Solve the following equations by the Cramer's rule.

$$10x_1 - 7x_2 + 3x_3 + 5x_4 = 6$$
$$-6x_1 + 8x_2 - x_3 - 4x_4 = 5$$
$$3x_1 + x_2 + 4x_3 + 11x_4 = 2$$
$$5x_1 - 9x_2 - 2x_3 + 4x_4 = 7$$

Solution: Here, $\Delta = \begin{vmatrix} 10 & -7 & 3 & 5 \\ -6 & 8 & -1 & -4 \\ 3 & 1 & 4 & 11 \\ 5 & -9 & -2 & 4 \end{vmatrix}$

$$\Delta_1 = \begin{vmatrix} 6 & -7 & 3 & 5 \\ 5 & 8 & -1 & -4 \\ 2 & 1 & 4 & 11 \\ 7 & -9 & -2 & 4 \end{vmatrix},\ \Delta_2 = \begin{vmatrix} 10 & 6 & 3 & 5 \\ -6 & 5 & -1 & -4 \\ 3 & 2 & 4 & 11 \\ 5 & 7 & -2 & 4 \end{vmatrix}$$

$$\Delta_3 = \begin{vmatrix} 10 & -7 & 6 & 5 \\ -6 & 8 & 5 & -4 \\ 3 & 1 & 2 & 11 \\ 5 & -9 & 7 & 4 \end{vmatrix} \text{ and } \Delta_4 = \begin{vmatrix} 10 & -7 & 3 & 6 \\ -6 & 8 & -1 & 5 \\ 3 & 1 & 4 & 2 \\ 5 & -9 & -2 & 7 \end{vmatrix}$$

By evaluating the determinants, we have $\Delta = 923$, $\Delta_1 = 4615$, $\Delta_2 = 3692$, $\Delta_3 = 6461$ and $\Delta_4 = 923$.

Hence,

$$x_1 = \frac{\Delta_1}{\Delta} = \frac{4615}{923} = 5$$

$$x_2 = \frac{\Delta_2}{\Delta} = \frac{3692}{923} = 4$$

$$x_3 = \frac{\Delta_3}{\Delta} = \frac{-6461}{923} = -7$$

and
$$x_4 = \frac{\Delta_4}{\Delta} = \frac{923}{923} = 1.$$

Thus, $x_1 = -5$, $x_2 = 4$, $x_3 = -7$ and $x_4 = 1$ is the solution of given system of equations.

10.6.4 Matrix inversion method

Let $AX = B$ be the matrix equation representing the system of linear equations given by (10.2) and det $(A) \neq 0$. Then

$$A^{-1}(AX) = A^{-1}B$$

i.e.,
$$\left(A^{-1}A\right)X = A^{-1}B$$

i.e.,
$$IX = A^{-1}B$$

i.e.,
$$X = A^{-1}B$$

Thus, the unknown matrix of $AX = B$ is the product A^{-1} and B. The column elements of $A^{-1}B$ will be the solution of the system of Eq. (10.2). This method is known as matrix inversion method.

SOLVED PROBLEMS

Problem 1: Solve $3x + 2y = 12$ and $2x - y = 1$ by matrix inversion method.

Solution:

Here
$$A = \begin{bmatrix} 3 & 2 \\ 2 & -1 \end{bmatrix}, X = \begin{bmatrix} x \\ y \end{bmatrix} \text{ and } B = \begin{bmatrix} 12 \\ 1 \end{bmatrix}$$

Now, we find A^{-1}, we know that $A^{-1} = \dfrac{\text{adj } A}{\det A}$

$$\det A = \begin{vmatrix} 3 & 2 \\ 2 & -1 \end{vmatrix} = -7 \text{ and adj } (A) = \begin{bmatrix} -1 & -2 \\ -2 & 3 \end{bmatrix}$$

$$\therefore \qquad A^{-1} = \frac{1}{-7}\begin{bmatrix} -1 & -2 \\ -2 & 3 \end{bmatrix}.$$

Now, $X = A^{-1} B$ reduces to

$$X = \frac{1}{(-7)} \cdot \begin{bmatrix} -1 & -2 \\ -2 & 3 \end{bmatrix}\begin{bmatrix} 12 \\ 1 \end{bmatrix}$$

i.e.,
$$X = \frac{1}{(-7)}\begin{bmatrix} -14 \\ -21 \end{bmatrix}$$

i.e.,
$$X = \begin{bmatrix} 2 \\ 3 \end{bmatrix}.$$

Thus, the solution of given system of linear equations is $x = 2$ and $y = 3$.

Problem 2: Solve $x_1 + x_2 + x_3 = 4$, $2x_1 - x_2 + x_3 = 4$ and $x_1 + 2x_2 + 3x_3 = 7$ by matrix inversion method.

Solution: Here,
$$A = \begin{bmatrix} 1 & 1 & 1 \\ 2 & -1 & 1 \\ 1 & 2 & 3 \end{bmatrix}, X = \begin{bmatrix} x_1 \\ x_2 \\ x_3 \end{bmatrix} B = \begin{bmatrix} 4 \\ 4 \\ 7 \end{bmatrix} \text{ Now,}$$

$$\det A = \begin{vmatrix} 1 & 1 & 1 \\ 2 & -1 & 1 \\ 1 & 2 & 3 \end{vmatrix} = -5 \text{ and}$$

$$\operatorname{adj}\, A = \begin{bmatrix} -5 & -1 & 2 \\ -5 & 2 & 1 \\ 5 & -1 & -3 \end{bmatrix}$$

Hence,

$$A^{-1} = \frac{1}{(-5)} \begin{bmatrix} -5 & -1 & 2 \\ -5 & 2 & 1 \\ 5 & -1 & -3 \end{bmatrix}$$

Thus, $\qquad X = A^{-1}B$

$$= \frac{1}{(-5)} \begin{bmatrix} -5 & -1 & 2 \\ -5 & 2 & 1 \\ 5 & -1 & -3 \end{bmatrix} \begin{bmatrix} 4 \\ 4 \\ 7 \end{bmatrix}$$

$$= \frac{1}{(-5)} \begin{bmatrix} -10 \\ -5 \\ -5 \end{bmatrix} = \begin{bmatrix} 2 \\ 1 \\ 1 \end{bmatrix}$$

$\therefore x_1 = 2,\ x_2 = 1$ and $x_3 = 1$ which is the solution of given system of linear equations.

10.6.5 Gauss elimination method

In this method, following a certain process (which is going to be described shortly), the given system of n linear equations in n unknowns is reduced to a upper triangular system of equations that can be solved by back-substitution method. Now, we proceed to describe the process of reducing the given system of linear equations to a upper triangular system of equations. Let the system of n linear equations in n unknowns is given by (10.2). i.e.,

$$a_{11}x_1 + a_{12}x_2 + \ldots\ldots + a_{1n}x_n = b_1 \qquad \left(A_1^{(1)}\right)$$

$$a_{21}x_1 + a_{22}x_2 + \ldots\ldots + a_{2n}x_n = b_2 \qquad \left(A_2^{(1)}\right)$$

$$\ldots\ldots\ldots\ldots\ldots\ldots\ldots\ldots\ldots\ldots\ldots\ldots\ldots\ldots\ldots$$

$$a_{n1}x_1 + a_{n2}x_2 + \ldots\ldots + a_{nn}x_n = b_n \qquad \left(A_n^{(1)}\right)$$

Let $a_{11} \neq 0$. Multiplying the equation $\left(A_1^{(1)}\right)$ by $\dfrac{a_{21}}{a_{11}}$ and subtracting from $\left(A_2^{(1)}\right)$, multiplying the equation $A_1^{(1)}$ by $\dfrac{a_{31}}{a_{11}}$ and subtracting from $\left(A_3^{(1)}\right)$ .... multiplying $\left(A_1^{(1)}\right)$ by $\dfrac{a_{n1}}{a_{11}}$ and subtracting from $\left(A_n^{(1)}\right)$, we obtain

$$a_{22}^{(1)}x_2 + a_{23}^{(1)}x_3 + \ldots\ldots + a_{2n}^{(1)}x_n = b_2^{(1)} \qquad \left(A_2^{(2)}\right)$$

$$a_{32}^{(1)}x_2 + a_{33}^{(1)}x_3 + \ldots\ldots + a_{3n}^{(1)}x_n = b_3^{(1)} \qquad \left(A_3^{(2)}\right)$$

$$a_{n2}^{(1)}x_2 + a_{n3}^{(1)}x_3 + \ldots\ldots + a_{nn}^{(1)}x_n = b_n^{(1)} \qquad \left(A_2^{(2)}\right)$$

Suppose ${a^{(1)}}_{22} \neq 0$. Now, multiplying the equation $\left(A_2^{(2)}\right)$ by $\dfrac{a_{32}^{(1)}}{a_{22}^{(1)}}$ and subtracting from the equation $\left(A_3^{(2)}\right)$, multiplying $(A_2^{(2)})$ by $\dfrac{a_{42}^{(1)}}{a_{22}}$ and subtracting from $A_3^{(2)}$, ..., multiplying the equation $A_2^{(2)}$ by $\dfrac{a_{n2}^{(1)}}{a_{22}^{(1)}}$ and subtracting from $A_3^{(2)}$, we get

$$a_{33}^{(2)}x_3 + a_{34}^{(2)}x_4 + \ldots\ldots + a_{3n}^{(2)}x_n = b_3^{(2)} \qquad \left(A_3^{(3)}\right)$$

$$a_{43}^{(2)}x_3 + a_{44}^{(2)}x_4 + \ldots\ldots + a_{4n}^{(2)}x_n = b_4^{(2)} \qquad \left(A_4^{(3)}\right)$$

$$a_{n3}^{(2)}x_3 + a_{n4}^{(2)}x_4 + \ldots\ldots + a_{nn}^{(2)}x_n = b_n^{(2)} \qquad \left(A_n^{(3)}\right)$$

Assuming $a_{33}^{(2)} \neq 0$, $a_{44}^{(2)} \neq 0$,..., $a_{nn}^{(n-1)} \neq 0$, and repeating the above process, the given system of Eq. (10.2) reduces to the following:

$$a_{11}x_1 + a_{12}x_2 + \ldots\ldots + a_{1n}x_n = b_1$$

$$a_{22}^{(1)}x_2 + \ldots\ldots + a_{2n}^{(1)}x_n = b_2^{(1)}$$

$$a_{33}^{(2)}x_3 + \ldots\ldots + a_{3n}^{(2)}x_n = b_3^{(2)}$$

$$\ldots\ldots\ldots\ldots\ldots\ldots\ldots\ldots\ldots\ldots\ldots$$

$$a_{nn}^{(n-1)}x_n = b_n^{(n-1)}$$

Clearly, the above equations are upper triangular system of n equations in n unknowns, which can be solved by back-substitution method. The method of obtaining the solution of (10.2), by the process described here is known as Gauss elimination method. The elements a_{11}, $a_{22}^{(1)}$, $a_{33}^{(2)}$, ... and $a_{nn}^{(n-1)}$ are called pivot elements.

SOLVED PROBLEMS

Problem 1: Using the Gauss elimination method, solve the following equations:

$$5x_1 + 2x_2 = 19 \tag{1}$$
$$3x_1 + 4x_2 = 17 \tag{2}$$

Solution: Multiplying (1) by $\dfrac{3}{5}$ and subtracting from (2), we have

$\left(4 - \dfrac{6}{5}\right) x_2 = \left(17 - \dfrac{57}{5}\right)$, i.e., $\dfrac{14}{5} x_2 = \dfrac{28}{5}$. Thus, the given Eqs. (1) and (2) are reduced to

$$5x_1 + 2x_2 = 19$$
$$\frac{14}{5} x_2 = \frac{28}{5} \tag{3}$$

From (3), we get $x_2 = 2$ and substituting it in (1), we have $x_1 = 3$. Thus, $x_1 = 3$ and $x_2 = 2$ is the solution of given system of equations.

Problem 2: Using the Gauss elimination method, solve the following system of equations:

$$10x_1 - x_2 + 2x_3 = 4 \tag{1}$$
$$x_1 + 10x_2 - x_3 = 3 \tag{2}$$
$$2x_1 + 3x_2 + 20x_3 = 7 \tag{3}$$

Solution: Multiplying (1) by $\dfrac{1}{10}$, subtracting from (2), and multiplying (1) by $\dfrac{2}{10}$ and subtracting from (3), we get

$$\left(10+\frac{1}{10}\right)x_2 - \left(1+\frac{2}{10}\right)x_3 = \left(3-\frac{4}{10}\right)$$

and

$$\left(3+\frac{2}{10}\right)x_2 + \left(20-\frac{4}{10}\right)x_3 = \left(7-\frac{8}{10}\right)$$

i.e.,
$$10.1x_2 - 1.2x_3 = 2.6 \tag{4}$$

$$3.2x_2 + 19.6x_3 = 6.2. \tag{5}$$

Now, we eliminate x_2 from Eqs. (4) and (5). Multiplying Eq. (4) by $\dfrac{3.2}{10.1}$ and subtracting from Eq. (5), we obtain

$$\left(19.6 + \frac{3.2 \times 1.2}{10.1}\right)x_3 = 6.2 - \left(\frac{(2.6)(3.2)}{10.1}\right)$$

$$(19.6 + 0.3802)x_3 = 6.2 - 0.8238$$

$$(19.9802)x_3 = 5.3762 \tag{6}$$

The given system of equations are reduced to

$$10x_1 - x_2 + 2x_3 = 4 \tag{1}$$
$$3.2x_2 + 19.6x_3 = 6.2 \tag{5}$$

$$(19.9802)x_3 = 5.3762 \tag{6}$$

From (6), we get $x_3 = 0.26907$. Substituting $x_3 = 0.26907$ in (5), we get $x_2 = 0.28944$. Further, substituting $x_3 = 0.26907$ and $x_2 = 0.28944$ in (1), we get $x_1 = 0.37513$.

Problem 3: Using Gauss elimination method, solve the following equations:

$$2x_1 + x_2 + x_3 = 10 \tag{1}$$
$$3x_1 + 2x_2 + 3x_3 = 18 \tag{2}$$
$$x_1 + 4x_2 + 9x_3 = 16 \tag{3}$$

Solution: Multiplying Eq. (1) by $\dfrac{3}{2}$, subtracting from (2) and multiplying the Eq. (1) by $\dfrac{1}{2}$ and subtracting from (3), we get

and
$$\frac{1}{2}x_2 + \frac{3}{2}x_3 = 3 \,,$$

$$\frac{7}{2}x_2 + \frac{17}{2}x_3 = 11$$

i.e.,
$$x_2 + 3x_3 = 6 \tag{4}$$

$$7x_2 + 17x_3 = 22 \,. \tag{5}$$

Now, multiplying Eq. (4) by 7 and subtracting from (5), we obtain
$$-4x_3 = -20 \tag{6}$$

The given system of equation is reduced to
$$2x_1 + x_2 + x_3 = 0 \tag{1}$$
$$x_2 + 3x_3 = 6 \tag{4}$$
$$-4x_3 = -20 \tag{6}$$

By back-substitution method, we get $x_1 = 7$, $x_2 = -9$ and $x_3 = 5$.

Problem 4: Using Gauss elimination method, solve

$$10x_1 + 7x_2 + 8x_3 + 7x_4 = 32 \tag{1}$$

$$7x_1 + 5x_2 + 6x_3 + 5x_4 = 23 \tag{2}$$

$$8x_1 + 6x_2 + 10x_3 + 9x_4 = 33 \tag{3}$$

$$7x_1 + 5x_2 + 9x_3 + 10x_4 = 31 \tag{4}$$

Solution: The first task of Gauss elimination method is to eliminate x_1 from Eqs. (2), (3) and (4). In contrast to the procedure followed in the solution of earlier problems, it can also be done by making the coefficient of x_1 is 1 in Eq. (1) by dividing it by 10 and multiplying it with appropriate numbers and subtracting from the remaining equations. The same may be followed in the elimination of other unknowns. We adopt this technique for quick and easier calculations.

Dividing the Eq. (1) by 10, we have

$$x_1 + 0.7x_2 + 0.8x_3 + 0.7x_3 = 3.2 \tag{5}$$

Now, multiplying Eq. (5) by 7, subtracting from(2), multiplying Eq. (5) by 8, subtracting from (3) and multiplying Eq. (5) by 7, and subtracting from (4), we obtain

$$0.1x_2 + 0.4x_3 + 0.1x_4 = 0.6 \tag{6}$$

$$0.4x_2 + 3.6x_3 + 3.4x_4 = 7.4 \tag{7}$$

$$0.1x_2 + 3.4x_3 + 5.1x_4 = 8.6 \tag{8}$$

Dividing the Eq. (6) by 0.1, we have

$$x_2 + 4x_3 + x_4 = 6 \tag{9}$$

Now, we eliminate x_2 from Eq. (7) and (8) using the Eq. (9). To do so, multiply Eq. (9) by 0.4; subtract it from Eq. (7); and multiply Eq. (9) by 0.1; and subtract from Eq. (8). Then, we have

$$2x_3 + 3x_4 = 5 \tag{10}$$

$$3x_3 + 5x_4 = 8 \tag{11}$$

Dividing the Eq. (10) by 2, we obtain

$$x_3 + 1.5x_4 = 2.5 \tag{12}$$

To eliminate x_3 from (11), multiplying (12) by 3 and subtracting from (11), we get

$$0.5x_4 = 0.5 \tag{3}$$

Thus, the given system of linear equations is reduced to the following upper tri-angular system of equations.

$$x_1 + 0.7x_2 + 0.8x_3 + 0.7x_3 = 3.2 \tag{5}$$

$$x_2 + 4x_3 + x_4 = 6 \tag{9}$$

$$x_3 + 1.5x_4 = 2.5 \tag{12}$$

$$0.5x_4 = 0.5 \tag{13}$$

Solving these equations by back-substitution method, we have $x_4 = 1, x_3 = 1, x_2 = 1$ and $x_1 = 1$. Hence, the solution of given system of linear equations is $x_1 = 1, x_2 = 1, x_3 = 1$ and $x_4 = 1$.

Remark: If the pivot element is zero or nearly zero (i.e., a very very small number), then Gauss elimination method may give no solution or provide an incorrect solution due to round off the solution of an unknown at the intermediate steps of Gauss elimination method.

10.6.6 An example

Consider the following system of two linear equations in two unknowns.

$$10^{-6} x_1 + x_2 = 1 \tag{1}$$

$$x_1 + x_2 = 2 \tag{2}$$

Let the solution of this system is required upto five decimal places by Gauss elimination method. Eliminating x_1 from (2) following the Gauss elimination method, we have

$$\left(1 - \frac{1}{10^{-6}}\right) x_2 = \left(2 - \frac{1}{10^{-6}}\right)$$

$$999999 \, x_2 = 0.999998$$

$$x_2 = 0.999999.$$

If we make $x_2 = 1$, we obtain $x_1 = 0$ from the Eq. (1), which is incorrect solution of given system of equations.

10.6.7 Partial pivoting and complete pivoting

A slight modification of the Gauss elimination method is the partial pivoting.

In the partial pivoting, the first colum elements $(a_{11}, a_{21}, a_{31}, ..., a_{n})$ are searched for the largest element is magnitude and brought as first pivot (a_{11}) by interchanging the first equation with the equation having the largest-element in magnitude. Changing the order of equations is known as pivoting. After this interchange of equations, x_1, is eliminated from the remaining $(n-1)$ equations following the procedure of Gauss elimination method. The same procedure is continued for the elimination of x_2 from the equations involving $x_2, x_3, ..., x_n$ only (i.e., equations obtained by eliminating x_1) and x_3 from the equations involving $x_3, x_4, ..., x_n$ only and so on until the given system of equation reduced the upper-triangular system of equations in $x_1, x_2, ..., x_n$. This process is known as partial pivoting.

In the case of complete pivoting, the elements of the coefficient matrix A are searched for the largest element in magnitude and brought to it as first pivot. To do so, it requires not only an interchange of equations but also an an interchange of unknowns. In this process, the positions of unknowns also get changed. The partial pivoting is preferred rather than complete pivoting as the later is complicated in view of computations.

Example

We illustrate the partial pivoting to solve the following system of equations:

$$2x_1 - 3x_2 + x_3 = -1 \tag{1}$$

$$x_1 + 4x_2 + 5x_3 = 25 \tag{2}$$

$$3x_1 - 4x_2 + 5x_3 = 1 \tag{3}$$

The largest element in the first column (i.e., a_{11}, a_{21}, a_{31}) of the coefficient matrix is $a_{31} = 3$. Therefore, we interchange Eqs. (1) and (3)

$$3x_1 - 4x_2 + 5x_3 = 2 \tag{1a}$$

$$x_1 + 4x_2 + 5x_3 = 25 \tag{2}$$

$$2x_1 - 3x_2 + x_3 = -1 \tag{3a}$$

Using Eq. (1a) and eliminating x_1 from Eqs. (2) and 3 (a), we obtain

$$\frac{16}{3}x_2 + \frac{10}{3}x_3 = \frac{73}{3} \tag{4}$$

$$-\frac{1}{3}x_2 - \frac{7}{3}x_3 = \frac{-7}{3} \tag{5}$$

For Eqs. (4) and (5), the largest element of the coefficient of x_2 is $\frac{16}{3}$; therefore, there is no need of interchange of the equations. Eliminating x_2 from (5) using (4), we get

$$\frac{-17}{8}x_3 = \frac{-13}{16}. \tag{6}$$

Thus, the given equations are reduced to the following upper triangular system of equations.

$$3x_1 - 4x_2 + 5x_3 = 2$$

$$\frac{16}{3}x_2 + \frac{10}{3}x_3 = \frac{73}{3}$$

$$\frac{-17}{8}x_3 = \frac{-13}{6}$$

Hence, solving these equations by back-substitution method, we obtain $x_3 = 0.3824$, $x_2 = 4.3235$ and $x_1 = 5.794$.

10.6.8 Ill-Conditioned equations

Very small changes in the elements of a coefficient matrix, we may arrive at a solution which is far different from the actual solution. Such system of equations are called ill-conditioned ones. For example, the equations $2x_1 + x_2 = 2$ and $2x_1 + 1.01x_2 = 2.01$ are ill-conditioned. The solutions of these equations are $x_1 = 1/2$ and $x_2 = 1$. With a slight changes in the elements of second equation, we consider $2x_1 + x_2 = 2$ and $2.01x_1 + x_2 = 2.05$. The solution, of this system is $x_1 = 5$ and $x_2 = -8$. Thus, the very small change in the coefficient of unknown in the equation $2x_1 + 1.01x_2 = 2.01$ has given a solution with a great change in their solutions.

10.6.9 Notations

We introduce the following notations for the elementary row transformations:

(i) $R_i \leftarrow kR_i$ denotes the elements of i th row are replaced by k times the corresponding elements of it, where k is a non-zero real number.

(ii) R_{ij} denotes the interchange of ith and j^{th} row elements.

(iii) $R_i \leftarrow R_i - kR_j$ denotes the replacements of elements of ith row by corresponding (i^{th} row element $- k$ times j^{th} row element). i.e., j^{th} row elements are multiplied by k and subtracted from the elements of ith row. Similarly $R_i \leftarrow R_i + kR_j$ denotes replacements of elements of i th row by corresponding (i^{th} row element $+ k$ times j^{th} row element).

10.6.10 Gauss–Jordan method

Consider the system of n linear equations in n unknowns given (10.2), i.e.,

$$a_{11} x_1 + a_{12} x_2 + \ldots + a_{1n} x_n = b_1$$

$$a_{21} x_1 + a_{22} x_2 + \ldots + a_{2n} x_n = b_2$$

$$\ldots\ldots\ldots\ldots\ldots\ldots\ldots\ldots\ldots\ldots\ldots\ldots$$

$$a_{n1} x_1 + a_{n2} x_2 + \ldots + a_{nn} x_n = b_n$$

In this method, first we form an augmented matrix of (10.2), i.e.,

$$[A:B] = \begin{bmatrix} a_{11} & a_{12}.... & a_{1n} & b_1 \\ a_{21} & a_{22}.... & a_{2n} & b_2 \\ .... & .... & .... & .... \\ a_{n1} & a_{n2}.... & a_{nn} & b_n \end{bmatrix}$$

Next, using the elementary row transfrmations, the augmented matrix will be reduced to the following form:

$$\begin{bmatrix} 1 & 0 & 0.... & 0 & 0 & a_1 \\ 0 & 1 & 0.... & 0 & 0 & a_2 \\ 0 & 0 & 1.... & 0 & 0 & a_3 \\ .... & .... & .... & .... & .... & .... \\ 0 & 0 & 0.... & 0 & 1 & a_n \end{bmatrix}$$

The solution of the system of Eqs. (10.2) is $x_1 = a_1, x_2 = a_2, \ldots,$ and $x_n = a_n$. This process of obtaining the solution of (10.2) is known as Gauss–Jordan method.

Note: The elementary row transformations on the augmented matrix do not effect the solution of a system of linear equations.

10.6.11 Procedure to find the solution of a system of 3 linear equations in 3 unknowns by Gauss–Jordan method

Suppose the system of 3 linear equations in 3 unknown be

$$a_{11} x_1 + a_{12} x_2 + a_{13} x_3 = b_1 \tag{1}$$

$$a_{21} x_1 + a_{22} x_2 + a_{23} x_3 = b_2 \tag{2}$$

$$a_{31} x_1 + a_{32} x_2 + a_{33} x_3 = b_3 \tag{3}$$

and it has a unique solution, i.e., det $A \neq 0$.

Step 1: Express the given Eqs. (1) to (3) in the form $AX = B$ where

$$A = \begin{bmatrix} a_{11} & a_{12} & a_{13} \\ a_{21} & a_{22} & a_{23} \\ a_{31} & a_{32} & a_{33} \end{bmatrix} X = \begin{bmatrix} x_1 \\ x_2 \\ x_3 \end{bmatrix} \text{and } B = \begin{bmatrix} b_1 \\ b_2 \\ b_3 \end{bmatrix}$$

Step 2: Form an augmented matrix

$$[A:B] = \begin{bmatrix} a_{11} & a_{12} & a_{13} & b_1 \\ a_{21} & a_{22} & a_{23} & b_2 \\ a_{31} & a_{32} & a_{33} & b_3 \end{bmatrix}$$

Suppose $a_{11} \neq 0$. If $a_{11} = 0$, then using the elementary transformation R_{ij}, it is possible to make $a_{11} \neq 0$. For example, if

$$[A:B] = \begin{bmatrix} 0 & 3 & 3 & 5 \\ 6 & 2 & 1 & 11 \\ 2 & 4 & 6 & 14 \end{bmatrix}$$

then applying R_{12}, we have

$$[A:B] \sim \begin{bmatrix} 6 & 2 & 1 & 11 \\ 0 & 3 & 3 & 5 \\ 2 & 4 & 6 & 14 \end{bmatrix}$$

so that $a_{11} \neq 0$. The symbol $\sim$ denotes equivalent.

Step 3: Applying the elementary transformation, $\dfrac{1}{a_{11}} R_1$ we have

$$\begin{bmatrix} 1 & a_{12}^{\,1} & a_{13}^{\,1} & b_1^{\,1} \\ a_{21} & a_{22} & a_{23} & b_2 \\ a_{31} & a_{32} & a_{33} & b_3 \end{bmatrix}$$

where $a_{12}^{\,1} = \dfrac{a_{12}}{a_{11}}$, $a_{13}^{\,1} = \dfrac{a_{13}}{a_{11}}$, and $b_1^{\,1} = \dfrac{b_1}{a_{11}}$.

Step 4: Applying $R_2 \leftarrow R_2 - a_{21} R_1$, $R_3 \leftarrow R_3 - a_{31} R_1$, the augmented matrix reduces to

$$[A:B] \sim \begin{bmatrix} 1 & a_{12}^{\,1} & a_{13}^{\,1} & b_1^{\,1} \\ 0 & a_{22}^{\,1} & a_{23}^{\,1} & b_2^{\,1} \\ 0 & a_{32}^{\,1} & a_{33}^{\,1} & b_3^{\,1} \end{bmatrix} \tag{4}$$

Step 5: If $a_{22}^{\,1} = 0$, then apply R_{23} so that 2nd row and 2nd column element of augmented matrix is non-zero.

If $a_{22}^{\,1} \neq 0$, then augmented matrix is given by (4) only.

Step 6: We suppose the matrix obtained in Step 5 be

$$[A:B] \sim \begin{bmatrix} 1 & c_{12} & c_{13} & l_1 \\ 0 & c_{22} & c_{23} & l_2 \\ 0 & c_{32} & c_{33} & l_3 \end{bmatrix} \qquad (5)$$

Where $c_{22} \neq 0$

Step 7: Applying $R_2 \leftarrow \dfrac{R_2}{c_{22}}$ on (5), we get

$$[A:B] \sim \begin{bmatrix} 1 & c_{12} & c_{13} & l_1 \\ 0 & 1 & c_{23}^{\;1} & l_2^{\;1} \\ 0 & c_{32} & c_{33} & l_3 \end{bmatrix} \qquad (6)$$

Step 8: Applying $R_1 \leftarrow R_1 - c_{12} R_2$, $R_3 \leftarrow R_3 - c_{32} R_2$ on (6), we obtain

$$[A:B] \sim \begin{bmatrix} 1 & 0 & c_{13}^{\;1} & l_1^{\;1} \\ 0 & 1 & c_{23}^{\;1} & l_2^{\;1} \\ 0 & 0 & c_{33}^{\;1} & l_3^{\;1} \end{bmatrix} \qquad (7)$$

Note that $c_{33}^{\;1} \neq 0$, since det $(A) \neq 0$

Step 9: Applying $\dfrac{1}{c_{33}^{\;1}} R_3$ on (7), we have

$$[A:B] \sim \begin{bmatrix} 1 & 0 & c_{13}^{\;1} & l_1^{\;1} \\ 0 & 1 & c_{23}^{\;1} & l_2^{\;1} \\ 0 & 0 & 1 & l_3^{\;11} \end{bmatrix} \qquad (8)$$

Step 10: Applying $R_1 \leftarrow R_1 - c_{13}^{\;1} R_3$ and $R_2 \leftarrow R_2 - c_{23}^{\;1} R_3$ on (8), we get

$$[A:B] \sim \begin{bmatrix} 1 & 0 & 0 & a_1 \\ 0 & 1 & 0 & a_2 \\ 0 & 0 & 1 & a_3 \end{bmatrix}$$

Step 11: The solution of (1), (2) and (3) is given by $x_1 = a_1$, $x_2 = a_2$ and $x_3 = a_3$.

SOLVED PROBLEMS

Problem 1: Solve the following system of equations by Gauss–Jordan method.

$$2x_1 + x_2 + x_3 = 10 \tag{1}$$
$$3x_1 + 2x_2 + 3x_3 = 18 \tag{2}$$
$$x_1 + 4x_2 + 9x_3 = 16 \tag{3}$$

Solution: The given system of Eqs. (1) to (3) can be expressed as

$$\begin{bmatrix} 2 & 1 & 1 \\ 3 & 2 & 3 \\ 1 & 4 & 9 \end{bmatrix} \begin{bmatrix} x_1 \\ x_2 \\ x_3 \end{bmatrix} = \begin{bmatrix} 10 \\ 18 \\ 16 \end{bmatrix}$$

Hence, the augmented matrix

$$[A:B] = \begin{bmatrix} 2 & 1 & 1 & 10 \\ 3 & 2 & 3 & 18 \\ 1 & 4 & 9 & 16 \end{bmatrix} \tag{4}$$

Applying $R_1 \leftarrow \dfrac{1}{2} R_1$ on (4), we obtain

$$[A:B] \sim \begin{bmatrix} 1 & \dfrac{1}{2} & \dfrac{1}{2} & 5 \\ 3 & 2 & 3 & 18 \\ 1 & 4 & 9 & 16 \end{bmatrix} \tag{5}$$

Now, applying $R_2 \leftarrow R_2 - 3R_1,\ R_3 \leftarrow R_3 - R_1$ on (5), we have

$$[A:B] \sim \begin{bmatrix} 1 & \dfrac{1}{2} & \dfrac{1}{2} & 5 \\ 0 & \dfrac{1}{2} & \dfrac{3}{2} & 3 \\ 0 & \dfrac{7}{2} & \dfrac{17}{2} & 11 \end{bmatrix} \tag{6}$$

Applying $R_2 \leftarrow 2R_2$ on (6), we obtain

$$[A:B] \sim \begin{bmatrix} 1 & \dfrac{1}{2} & \dfrac{1}{2} & 5 \\ 0 & 1 & 3 & 6 \\ 0 & \dfrac{7}{2} & \dfrac{17}{2} & 11 \end{bmatrix} \qquad (7)$$

Now applying $R_1 \leftarrow R_1 - \dfrac{1}{2} R_2$ and $R_3 \leftarrow R_3 - \dfrac{7}{2} R_2$ on (7), we get

$$[A:B] \sim \begin{bmatrix} 1 & 0 & -1 & 2 \\ 0 & 1 & 3 & 6 \\ 0 & 0 & -2 & -10 \end{bmatrix} \qquad (8)$$

Applying on $R_3 \leftarrow \left(-\dfrac{1}{2} \right) R_3$ on (8), we have

$$[A:B] \sim \begin{bmatrix} 1 & 0 & -1 & 2 \\ 0 & 1 & 3 & 6 \\ 0 & 0 & 1 & 5 \end{bmatrix} \qquad (9)$$

Finally, applying $R_1 \leftarrow R_1 + R_3$, $R_2 \leftarrow R_2 - 3R_3$ on (9), we get

$$[A:B] \sim \begin{bmatrix} 1 & 0 & 0 & 7 \\ 0 & 1 & 0 & -9 \\ 0 & 0 & 1 & 5 \end{bmatrix}$$

Thus, $x_1 = 7$, $x_2 = -9$ and $x_3 = 5$ is the solution of given system of linear equations.

Problem 2: Solve

$$x_1 + x_2 + x_3 = 9 \qquad (1)$$
$$2x_1 - 3x_2 + 4x_3 = 13 \qquad (2)$$
$$3x_1 + 4x_2 + 5x_3 = 40 \qquad (3)$$

by Gauss–Jordan method.

Solution: For the given system of equations, the augmented matrix will be

$$[A:B] = \begin{bmatrix} 1 & 1 & 1 & 9 \\ 2 & -3 & 4 & 13 \\ 3 & 4 & 5 & 40 \end{bmatrix} \qquad (4)$$

Applying $R_2 \leftarrow R_2 - 2R_1$, $R_3 \leftarrow R_3 - 3R_1$ on (4), we get

$$[A:B] \sim \begin{bmatrix} 1 & 1 & 1 & 9 \\ 0 & -5 & 2 & -5 \\ 0 & 1 & 2 & 13 \end{bmatrix} \qquad (5)$$

Now, applying $R_2 \leftarrow \left(\dfrac{-1}{5} \right) R_2$, on (5), we obtain

$$[A:B] \sim \begin{bmatrix} 1 & 1 & 1 & 9 \\ 0 & 1 & \dfrac{-2}{5} & 1 \\ 0 & 1 & 2 & 13 \end{bmatrix} \qquad (6)$$

On (6), performing $R_1 \leftarrow R_1 - R_2$, $R_3 \leftarrow R_3 - R_2$, we have

$$[A:B] \sim \begin{bmatrix} 1 & 0 & \dfrac{7}{5} & 8 \\ 0 & 1 & \dfrac{-2}{5} & 1 \\ 0 & 0 & \dfrac{12}{5} & 12 \end{bmatrix} \qquad (7)$$

Applying $R_3 \leftarrow \dfrac{5}{12} R_3$ on (7), we get

$$[A:B] \sim \begin{bmatrix} 1 & 0 & \dfrac{7}{5} & 8 \\ 0 & 1 & \dfrac{-2}{5} & 1 \\ 0 & 0 & 1 & 5 \end{bmatrix} \qquad (8)$$

Finally, performing $R_1 \leftarrow R_1 - \dfrac{7}{5} R_3$ and $R_2 \leftarrow R_2 + \dfrac{2}{5} R_3$ on (8), we get

$$[A:B] \sim \begin{bmatrix} 1 & 0 & 0 & 1 \\ 0 & 1 & 0 & 3 \\ 0 & 0 & 1 & 5 \end{bmatrix}$$

Hence, $x_1 = 1$, $x_2 = 3$ and $x_3 = 5$ is the solution of (1), (2) and (3).

Problem 3: Solve $2x_1 + x_2 = 3$ and $3x_1 + 2x_2 = 5$ by Gauss–Jordan method.

Solution: For the given system of equations, the augmented matrix is

$$[A:B] = \begin{bmatrix} 2 & 1 & 3 \\ 3 & 2 & 5 \end{bmatrix} \tag{1}$$

Applying $R_1 \leftarrow \dfrac{1}{2} R_1$ on (1), we get

$$[A:B] \sim \begin{bmatrix} 1 & \dfrac{1}{2} & \dfrac{3}{2} \\ 3 & 2 & 5 \end{bmatrix} \tag{2}$$

Now, performing $R_2 \leftarrow R_2 - 3R_1$ on (2), we get

$$[A:B] \sim \begin{bmatrix} 1 & \dfrac{1}{2} & \dfrac{3}{2} \\ 0 & \dfrac{1}{2} & \dfrac{1}{2} \end{bmatrix} \tag{3}$$

Applying $R_2 \leftarrow 2R_2$ on (3) we obtain

$$[A:B] \sim \begin{bmatrix} 1 & \dfrac{1}{2} & \dfrac{3}{2} \\ 0 & 1 & 1 \end{bmatrix} \tag{4}$$

Finally, applying $R_1 \leftarrow R_1 - \dfrac{1}{2} R_2$ on (4), we have

$$[A:B] \sim \begin{bmatrix} 1 & 0 & 1 \\ 0 & 1 & 1 \end{bmatrix}$$

Hence, $x_1 = 1$ and $x_2 = 1$ is the solution of given system of equations.

10.6.12 Escalator method

This method is based on the partition of matrices. The escalator method is used to find the inverse of a matrix B of type $(n + 1) \times (n + 1)$

when the inverse of a A of type $n \times n$ is known where the matrix A of type $n \times n$ is obtained by the deleting bottom row elements and extreme right column elements of B of type $(n + 1) \times (n + 1)$.

For example, if we know the inverse of $\begin{bmatrix} a_{11} & a_{12} \\ a_{21} & a_{22} \end{bmatrix}$ then using this

method, we can find the inverse of

$$\begin{bmatrix} a_{11} & a_{12} & a_{13} \\ a_{21} & a_{22} & a_{23} \\ a_{31} & a_{32} & a_{33} \end{bmatrix}$$

Now we describe the escalator method to find the inverse of a matrix. Let

$$A = \begin{bmatrix} a_{11} & a_{12}\dots & a_{1n} & a_{1(n+1)} \\ a_{21} & a_{22}\dots & a_{2n} & a_{2(n+1)} \\ a_{n1} & a_{n2}\dots & a_{nn} & a_{n(n+1)} \\ a_{(n+1)1} & a_{(n+1)2}\dots & a_{(n+1)n} & a_{(n+1)(n+1)} \end{bmatrix} \qquad (10.8)$$

Suppose

$$A_1 = \begin{bmatrix} a_{11} & \dots & a_{1n} \\ \dots & \dots & \dots \\ a_{n1} & \dots & a_{nn} \end{bmatrix}, \ A_2 = \begin{bmatrix} a_{1(n+1)} \\ a_{2(n+1)} \\ \vdots \\ a_{n(n+1)} \end{bmatrix}$$

$$A_3^T = \begin{bmatrix} a_{(n+1)1} & a_{(n+1)2}\dots & a_{(n+1)n} \end{bmatrix} \text{ and } a = a_{(n+1)(n+1)}$$

We express the matrix A as partition of matrices

$$A = \begin{bmatrix} A_1 & \vdots & A_2 \\ \dots & \dots & \dots \\ A^T_{\ 3} & \vdots & a \end{bmatrix} \qquad (10.9)$$

Let the inverse of A be

$$A^{-1} = \begin{bmatrix} X_1 & \vdots & X_2 \\ .... & .... & .... \\ X^T_3 & \vdots & x \end{bmatrix} \qquad (10.10)$$

and suppose A_1^{-1} is known. The matrix X_2 is a column vector and x is a real number. Now, $AA^{-1} = I$ gives

$$\begin{bmatrix} A_1 X_1 + A_2 X_3^T & \vdots & A_1 X_2 + A_2 x \\ .... & .... & .... \\ A_3^T x_1 + a X^T_3 & \vdots & A_3^T X_2 + ax \end{bmatrix} = I \quad (10.11)$$

Therefore, we have

$$A_1 X_1 + A_2 X_3^T = I \qquad (10.12)$$
$$A_1 X_2 + A_2 x = 0 \qquad (10.13)$$
$$A_3^T X_1 + ax_3^T = 0 \qquad (10.14)$$
$$A_3^T X_2 + ax = 1 \qquad (10.15)$$

From the Eq. (10.13), we have

$$A_1 X_2 = - A_2 x$$
$$\therefore \qquad X_2 = - A_1^{-1} A_2 x \qquad (10.16)$$

Substituting X_2 in (10.15), we get

$$A^T_3 \{ -A_1^{-1} A_2 x \} + ax = 1$$
$$\therefore \qquad x(a - A^T_3 A_1^{-1} A_2) = 1 \qquad (10.17)$$

This equation gives the value of x as a, A^T_3, A_1^{-1} and A_2 are known. Substituting x in (10.16), we obtain X_2. Now, from Eq. (10.12), we have

$$X_1 = A_1^{-1}(I - A_2 X_3^T),$$
$$\text{i.e., } X_1 = A_1^{-1} - A_1^{-1} A_2 X_3^T \qquad (10.18)$$

Substituting X_1 of (10.18) in (10.14), we get

$$A_3^T \{ A_1^{-1} - A_1^{-1} A_2 X_3^T \} + a X_3^T = 0$$
$$\text{i.e., } \qquad A^T_3 A_1^{-1} - A^T_3 A_1^{-1} A_2 X^T_3 + a X^T_3 = 0$$

i.e., $\qquad aX^T_3 - A^T_3 A_1^{-1} A_2 X^T_3 = - A^T_3 A_1^{-1}$

i.e., $\qquad (a - A^T_3 A_1^{-1} A_2) X^T_3 = - A^T_3 A_1^{-1}$

i.e., $\qquad X^T_3 = - x A^T_3 A_1^{-1} \qquad$ by (10.17) (10.19)

Thus, the unknown matrices X_1, X_2, X_3^T and the element x can be computed from the following relations:

$$(a - A^T_3 A_1^{-1} A_2) x = 1$$

$$X_2 = - x A_1^{-1} A_2$$

$$X_3^T = - x A^T_3 A_1^{-1}$$

$$X_1 = A_1^{-1} (I - A_2 X^T_3)$$

Note: To find $X = A^{-1}B$ of a system of equations $AX = B$, we find the A^{-1} by escalator method and using $X = A^{-1}B$, we can find the solution of $AX = B$.

SOLVED PROBLEMS

Problem 1: Find the inverse of

$$A = \begin{bmatrix} 2 & 1 & 2 \\ 1 & 1 & 2 \\ 1 & 4 & 3 \end{bmatrix}$$

by escalator method and solve $2x_1 + x_2 + 2x_3 = 7$, $x_1 + x_2 + 2x_3 = 6$ and $x_1 + 4x_2 + 2x_3 = 11$.

Solution: Let $A = \begin{bmatrix} A_1 & \vdots & A_2 \\ \cdots & \cdots & \cdots \\ A^T_3 & \vdots & a \end{bmatrix}$ where

$$A_1 = \begin{bmatrix} 2 & 1 \\ 1 & 1 \end{bmatrix}, A_2 = \begin{bmatrix} 2 \\ 2 \end{bmatrix}, A_3^T = \begin{bmatrix} 1 & 4 \end{bmatrix} \text{ and } a = 3.$$

Suppose
$$A^{-1} = \begin{bmatrix} X_1 & \vdots & X_2 \\ \cdots & \vdots & \cdots \\ X^T{}_3 & \vdots & x \end{bmatrix}$$

Clearly, the inverse of A_1 is $\begin{bmatrix} 1 & -1 \\ -1 & 2 \end{bmatrix}$ i.e.,

$$A_1^{-1} = \begin{bmatrix} 1 & -1 \\ -1 & 2 \end{bmatrix}$$

To find x:

The value of x can be found from the relation (by 10.17)

$$(a - A^T{}_3\, A_1^{-1} A_2)\, x = 1$$

$$\therefore \left\{ 3 - \begin{bmatrix} 1 & 4 \end{bmatrix} \begin{bmatrix} 1 & -1 \\ -1 & 2 \end{bmatrix} \begin{bmatrix} 2 \\ 2 \end{bmatrix} \right\} x = 1$$

It gives $x = \dfrac{-1}{5}$

To find X_2:

Since $X_2 = -x\, A_1^{-1} A_2$ (by 10.16), we have

$$X_2 = -\left(\frac{-1}{5}\right) \begin{bmatrix} 1 & -1 \\ -1 & 2 \end{bmatrix} \begin{bmatrix} 2 \\ 2 \end{bmatrix}$$

i.e.,
$$X_2 = \begin{bmatrix} 0 \\ \dfrac{2}{5} \end{bmatrix}$$

To find $X_3{}^T$:

By Eq. (10.19), $X^T{}_3 = -x A^T{}_3\, A_1^{-1}$

$$\therefore \qquad X^T{}_3 = -\left(\frac{-1}{5}\right) \begin{bmatrix} 1 & 4 \end{bmatrix} \begin{bmatrix} 1 & -1 \\ -1 & 2 \end{bmatrix}$$

i.e.,
$$X_3^T = \begin{bmatrix} \dfrac{-3}{5} & \dfrac{7}{5} \end{bmatrix}$$

To find X_1:

Again by (10.18), we have $X_1 = A^{-1}_{\ 1}(I - A_2 X_3^T)$.

Therefore,

$$X_1 = \begin{bmatrix} 1 & -1 \\ -1 & 2 \end{bmatrix} \left\{ \begin{bmatrix} 1 & 0 \\ 0 & 1 \end{bmatrix} - \begin{bmatrix} 2 \\ 2 \end{bmatrix} \begin{bmatrix} \dfrac{-3}{5} & \dfrac{7}{5} \end{bmatrix} \right\}$$

$$X_1 = \begin{bmatrix} 1 & -1 \\ \dfrac{1}{5} & \dfrac{-4}{5} \end{bmatrix}$$

i.e.,

Thus, the inverse of the matrix A is

$$A^{-1} = \begin{bmatrix} 1 & -1 & 0 \\ \dfrac{1}{5} & \dfrac{-4}{5} & \dfrac{2}{5} \\ \dfrac{-3}{5} & \dfrac{7}{5} & \dfrac{-1}{5} \end{bmatrix}$$

Therefore, the solution of given system of equations is

$$\begin{bmatrix} x_1 \\ x_2 \\ x_3 \end{bmatrix} = \begin{bmatrix} 1 & -1 & 0 \\ \dfrac{1}{5} & \dfrac{-4}{5} & \dfrac{2}{5} \\ \dfrac{-3}{5} & \dfrac{7}{5} & \dfrac{-1}{5} \end{bmatrix} \begin{bmatrix} 7 \\ 6 \\ 11 \end{bmatrix}$$

$$= \begin{bmatrix} 7 - 6 + 0 \\ \dfrac{7}{5} - \dfrac{24}{5} + \dfrac{22}{5} \\ \dfrac{-21}{5} + \dfrac{42}{5} - \dfrac{11}{5} \end{bmatrix} = \begin{bmatrix} 1 \\ 1 \\ 2 \end{bmatrix}$$

Thus, the solution of given system of equations is $x_1 = 1$, $x_2 = 1$ and $x_3 = 2$.

Problem 2: Find the inverse of the following matrix by escalator method:

$$A = \begin{bmatrix} 3 & -1 & 10 & 2 \\ 5 & 1 & 20 & 3 \\ 9 & 7 & 39 & 4 \\ 1 & -2 & 2 & 1 \end{bmatrix}$$

Solution: We find the inverse of given matrix by escalator method and first we apply it to 3×3 and later to 4×4 matrix (i.e., A). Suppose

$$B = \begin{bmatrix} 3 & -1 & 10 \\ 5 & 1 & 20 \\ 9 & 7 & 39 \end{bmatrix}$$

and

$$B = \begin{bmatrix} B_1 & \vdots & B_2 \\ \cdots & \cdots & \cdots \\ B^T_{\ 3} & \vdots & b \end{bmatrix}$$

where

$$B_1 = \begin{bmatrix} 3 & -1 \\ 5 & 1 \end{bmatrix}, B_2 = \begin{bmatrix} 10 \\ 20 \end{bmatrix}, B_3^T = \begin{bmatrix} 9 & 7 \end{bmatrix} \text{ and } b = 39. \text{ Clearly,}$$

$$B_1^{-1} = \begin{bmatrix} 0.1250 & 0.1250 \\ -0.6250 & 0.3750 \end{bmatrix}.$$

Suppose

$$B^{-1} = \begin{bmatrix} Y_1 & \vdots & Y_2 \\ \cdots & \cdots & \cdots \\ Y_3^{\ T} & \vdots & y \end{bmatrix}$$

The matrices Y_1, Y_2, Y_3^T and the element y can be found from the following relations:

$$(b - B_3^T B_1^{-1} B_2) y = 1$$
$$Y_2 = -y B_1^{-1} B_2$$
$$Y_3^T = -y B_3^T B_1^{-1}$$
$$Y_1 = B_1^{-1} (I - B_2 Y_3^T),$$

which are the relations from (10.16) to (10.19) in place of Y_i is X_i, a is b, A_i is B_i and x is y. Therefore,

$$Y_2 = \begin{bmatrix} 1.0714 \\ 0.3571 \end{bmatrix}, Y_3^T = \begin{bmatrix} -0.9286 & 1.0714 \end{bmatrix}, Y_1 = \begin{bmatrix} 3.6071 & -3.8929 \\ 0.5357 & -0.9643 \end{bmatrix}$$

and
$$y = -0.2857.$$

Thus,

$$B^{-1} = \begin{bmatrix} 3.6071 & -3.8929 & 1.0714 \\ 0.5357 & -0.9643 & 0.3571 \\ -0.9286 & 1.0714 & -0.2857 \end{bmatrix} \qquad (1)$$

Now, suppose the matrix A as

$$A = \begin{bmatrix} A_1 & \vdots & A_2 \\ \dots & \dots & \dots \\ A_3^T & \vdots & a \end{bmatrix}$$

where
$$A_1 = B, A_2 = \begin{bmatrix} 2 \\ 3 \\ 4 \end{bmatrix}, A^T_3 = \begin{bmatrix} 1 & -2 & 2 \end{bmatrix}$$

and $a = 1$ and the inverse of A as

$$A^{-1} = \begin{bmatrix} X_1 & \vdots & X_2 \\ \dots & \dots & \dots \\ X_3^T & \vdots & x \end{bmatrix}$$

The inverse A_1 is (by (1))

$$A_1^{-1} = \begin{bmatrix} 3.6071 & -3.8929 & 1.0714 \\ 0.5357 & -0.9643 & 0.3571 \\ -0.9286 & 1.0714 & -0.2857 \end{bmatrix}$$

The matrices X_1, X_2, X_3^T and the number x can be found from the relations (10.16) to (10.19). These matrices and the number x are as follows:

$$X_1 = \begin{bmatrix} 7 & -3 & 0 \\ 8 & 1 & -2 \\ -5 & 0 & 1 \end{bmatrix}, X_2 = \begin{bmatrix} -5 \\ -11 \\ 6 \end{bmatrix}, X_3^T = \begin{bmatrix} 19 & 5 & 6 \end{bmatrix} \text{ and } x = -28$$

Therefore ,the inverse of the given matrix is

$$A^{-1} = \begin{bmatrix} 7 & -3 & 0 & -5 \\ 8 & 1 & -2 & -11 \\ -5 & 0 & 1 & 6 \\ 19 & 5 & 6 & -28 \end{bmatrix}$$

10.6.13 Cholesky method

This method is used to solve $AX = B$ when the coefficient matrix A is symmetric and positive definite. In the Cholesky method, the coefficient matrix A is decomposed as

$$A = LL^T \tag{10.20}$$

where L is lower triangular matrix, $L = [l_{ij}]$ where $l_{ij} = 0 \; for \; i < j$. Now,

$$A^{-1} = (LL^T)^{-1}$$

$$= (L^T)^{-1}.L^{-1}$$

$$= (L^{-1})^T.L^{-1}.$$

Thus, we can determine A^{-1} by computing the inverse of L. The solution of $AX = B$ can be found *by* $X = A^{-1}B$. This method of finding X of $AX = B$ is known as Cholesky method. This method is also known as the Square–root method.

10.6.14   The relations between the elements of *A* and *L* when *A* = *LL*ᵀ

We present these relations for 3×3 coefficient matrix. This procedure can be extended to higher order matrices.

$$A = \begin{bmatrix} a_{11} & a_{12} & a_{13} \\ a_{21} & a_{22} & a_{23} \\ a_{31} & a_{32} & a_{33} \end{bmatrix} \text{ and } L = \begin{bmatrix} l_{11} & 0 & 0 \\ l_{21} & l_{22} & 0 \\ l_{31} & l_{32} & l_{33} \end{bmatrix}$$

Then, $A = L\,L^T$ becomes

$$\begin{bmatrix} a_{11} & a_{12} & a_{13} \\ a_{21} & a_{22} & a_{23} \\ a_{31} & a_{32} & a_{33} \end{bmatrix} = \begin{bmatrix} l_{11} & 0 & 0 \\ l_{21} & l_{22} & 0 \\ l_{31} & l_{32} & l_{33} \end{bmatrix}\begin{bmatrix} l_{11} & l_{21} & l_{31} \\ 0 & l_{22} & l_{32} \\ 0 & 0 & l_{33} \end{bmatrix}$$

$$= \begin{bmatrix} l^2_{11} & l_{11}l_{21} & l_{11}l_{31} \\ l_{21}l_{11} & l^2_{21} + l^2_{22} & l_{21}l_{31} + l_{22}l_{32} \\ l_{31}l_{11} & l_{31}l_{21} + l_{32}l_{22} & l^2_{31} + l^2_{32} + l^2_{33} \end{bmatrix}$$

Comparing the corresponding elements of A and $L.L^T$, we obtain

$$\begin{aligned} l^2_{11} &= a_{11} & l_{11}l_{21} &= a_{12} & l_{11}l_{31} &= a_{13} \\ l_{21}l_{11} &= a_{21} & l^2_{21} + l^2_{22} &= a_{22} & l_{21}l_{31} + l_{22}l_{32} &= a_{23} \\ l_{31}l_{11} &= a_{31} & l_{31}l_{21} + l_{32}l_{22} &= a_{32} & l^2_{31} + l^2_{32} + l^2_{33} &= a_{33} \end{aligned} \qquad (10.21)$$

Solving the above equations for $l_{11}, l_{21}, l_{22}, l_{31}, l_{32},$ and l_{33} we obtain

$$l_{11} = \sqrt{a_{11}}\,,\ l_{21} = \frac{a_{21}}{\sqrt{a_{11}}}\,, l_{31} = \frac{a_{31}}{\sqrt{a_{11}}}\left(\frac{a_{13}}{\sqrt{a_{11}}} \quad \because a_{13} = a_{31} \right)$$

$$l_{22} = \sqrt{a_{22} - \frac{a_{21}^2}{a_{11}}}\,, l_{32} = \frac{\left[a_{23} - \dfrac{a_{21}.a_{31}}{a_{11}} \right]}{\sqrt{a_{22} - \dfrac{a_{21}^{\,2}}{a_{11}}}} \qquad (10.22)$$

$$\text{and } l_{33} = \sqrt{a_{33} - \frac{a_{31}^2}{a_{11}} - \frac{\left[a_{23} - \dfrac{a_{21}.a_{13}}{a_{11}} \right]^2}{\left(a_{22} - \dfrac{a_{21}^2}{a_{11}} \right)}}$$

SOLVED PROBLEMS

Problem 1: Using Cholesky method, solve the following system of equations:

$$x_1 + 2x_2 + 3x_3 = 5$$
$$2x_1 + 8x_2 + 22x_3 = 6$$
$$3x_1 + 22x_2 + 82x_3 = -10$$

Solution: Here the coefficient matrix A is

$$A = \begin{bmatrix} 1 & 2 & 3 \\ 2 & 8 & 22 \\ 3 & 22 & 82 \end{bmatrix}, X = \begin{bmatrix} x_1 \\ x_2 \\ x_3 \end{bmatrix} \text{ and } B = \begin{bmatrix} 5 \\ 6 \\ -10 \end{bmatrix}$$

Therefore, we suppose

$$\begin{bmatrix} 1 & 2 & 3 \\ 2 & 8 & 22 \\ 3 & 22 & 82 \end{bmatrix} = \begin{bmatrix} l_{11} & 0 & 0 \\ l_{21} & l_{22} & 0 \\ l_{31} & l_{32} & l_{33} \end{bmatrix} \begin{bmatrix} l_{11} & l_{21} & l_{31} \\ 0 & l_{22} & l_{32} \\ 0 & 0 & l_{33} \end{bmatrix} \qquad (1)$$

Using (10.22), we have

$$l_{11} = \sqrt{a_{11}} = \sqrt{1} = 1$$

$$l_{21} = \frac{a_{21}}{\sqrt{a_{11}}} = \frac{2}{1} = 2$$

$$l_{31} = \frac{a_{31}}{\sqrt{a_{11}}} = \frac{3}{1} = 3$$

$$l_{22} = \sqrt{a_{22} - \frac{a_{21}^{2}}{a_{11}}} = \sqrt{8 - \frac{4}{1}} = 2$$

$$l_{32} = \frac{a_{23} - \dfrac{a_{21} a_{31}}{a_{11}}}{\sqrt{a_{22} - \dfrac{a_{21}^{2}}{a_{11}}}} = \frac{22 - \dfrac{(2)(3)}{(1)}}{\sqrt{8 - \dfrac{4}{1}}} = 8$$

and

$$l_{33} = \sqrt{a_{33} - \frac{a_{31}^2}{a_{11}} - \frac{\left[a_{23} - \dfrac{a_{21}a_{13}}{a_{11}}\right]^2}{\left(a_{22} - \dfrac{a_{21}^2}{a_{11}}\right)}}$$

$$= \sqrt{82 - \frac{9}{1} - \frac{[22-(2)(3)]^2}{(8-4)}} = \sqrt{82-9-64} = \sqrt{9} = 3$$

Thus,

$$L = \begin{bmatrix} 1 & 0 & 0 \\ 2 & 2 & 0 \\ 3 & 8 & 3 \end{bmatrix}$$

Computing the inverse of L, we obtain

$$L^{-1} = \begin{bmatrix} 1 & 0 & 0 \\ -1 & \dfrac{1}{2} & 0 \\ \dfrac{5}{3} & \dfrac{-4}{3} & \dfrac{1}{3} \end{bmatrix}$$

Thus,

$$A^{-1} = \left(L^{-1}\right)^T L^{-1} = \begin{bmatrix} 1 & -1 & \dfrac{5}{3} \\ 0 & \dfrac{1}{2} & \dfrac{-4}{3} \\ 0 & 0 & \dfrac{1}{3} \end{bmatrix} \begin{bmatrix} 1 & 0 & 0 \\ -1 & \dfrac{1}{2} & 0 \\ \dfrac{5}{3} & \dfrac{-4}{3} & \dfrac{1}{3} \end{bmatrix}$$

i.e.,

$$A^{-1} = \begin{bmatrix} \dfrac{43}{9} & \dfrac{-49}{18} & \dfrac{5}{9} \\ \dfrac{-49}{18} & \dfrac{73}{36} & \dfrac{-4}{9} \\ \dfrac{5}{9} & \dfrac{-4}{9} & \dfrac{1}{9} \end{bmatrix}$$

Now, $X = A^{-1}B$ becomes

$$X = \begin{bmatrix} \dfrac{43}{9} & \dfrac{-49}{18} & \dfrac{5}{9} \\ \dfrac{-49}{18} & \dfrac{73}{36} & \dfrac{-4}{9} \\ \dfrac{5}{9} & \dfrac{-4}{9} & \dfrac{1}{9} \end{bmatrix} \begin{bmatrix} 5 \\ 6 \\ -10 \end{bmatrix} = \begin{bmatrix} 1 \\ 3 \\ -1 \end{bmatrix}$$

Hence $x_1 = 1, x_2 = 3, x_3 = -1$ is the solution of given system of equations

Alternate method:

Since $A = LL^T$, the matrix equation $AX = B$ becomes

$$LL^T X = B \tag{2}$$

Let
$$L^T X = Y, \tag{3}$$

where
$$Y^T = \begin{bmatrix} y_1 & y_2 & y_3 \end{bmatrix}$$

Then, Eq. (2) becomes

$$LY = B \tag{4}$$

$$\text{i.e.;} \quad \begin{bmatrix} 1 & 0 & 0 \\ 2 & 2 & 0 \\ 3 & 8 & 3 \end{bmatrix} \begin{bmatrix} y_1 \\ y_2 \\ y_3 \end{bmatrix} = \begin{bmatrix} 5 \\ 6 \\ -10 \end{bmatrix}$$

It gives the following equations:

$$y_1 = 5 \tag{5}$$
$$2y_1 + 2y_2 = 6 \tag{6}$$
$$3y_1 + 8y_2 + 3y_3 = -10 \tag{7}$$

Solving Eqs. (5), (6) and (7) for y_1, y_2 and y_3, we obtain $y_1 = 5, y_2 = -2$ and $y_3 = -3$

Thus, $Y^T = \begin{bmatrix} 5 & -2 & -3 \end{bmatrix}$; Now, Eq. (3) reduces to

$$\begin{bmatrix} 1 & 2 & 3 \\ 0 & 2 & 8 \\ 0 & 0 & 3 \end{bmatrix} \begin{bmatrix} x_1 \\ x_2 \\ x_3 \end{bmatrix} = \begin{bmatrix} 5 \\ -2 \\ -3 \end{bmatrix}. \text{ Hence,}$$

$$x_1 + 2x_2 + 3x_3 = 5 \tag{8}$$
$$2x + 8x_3 = -2 \tag{9}$$
$$3x_3 = -3 \tag{10}$$

Solving Eqs. (8), (9) and (10) by method of back substitution, we obtain $x_1 = 2, x_2 = 3$ and $x_3 = -1$.

Problem 2: Solve the following system of equations by Cholesky's method.

$$x_1 + 2x_2 + 6x_3 = 4 \tag{1}$$
$$2x_1 + 5x_2 + 15x_3 = 12 \tag{2}$$
$$6x_1 + 15x_2 + 46x_3 = 36 \tag{3}$$

Solution: Here $a_{11} = 1, a_{12} = 2, a_{13} = 6, a_{21} = 2, a_{22} = 5, a_{23} = 15, a_{31} = 6, a_{32} = 15$ and $a_{33} = 46$. The Eq. (1) to (3) can be expressed as $AX = B$, where

A is the coefficient matrix, $X = \begin{bmatrix} x_1 \\ x_2 \\ x_3 \end{bmatrix}$ and $B = \begin{bmatrix} 4 \\ 12 \\ 36 \end{bmatrix}$

We suppose $A = LL^T$ where $L = \begin{bmatrix} l_{11} & 0 & 0 \\ l_{21} & l_{22} & 0 \\ l_{31} & l_{32} & l_{33} \end{bmatrix}$

Then

$$l_{11} = \sqrt{a_{11}} = \sqrt{1} = 1, \quad l_{21} = \frac{a_{21}}{\sqrt{a_{11}}} = \frac{2}{1} = 2$$

$$l_{31} = \frac{a_{31}}{\sqrt{a_{11}}} = \frac{6}{1} = 6, \quad l_{22} = \sqrt{a_{22} - \frac{a_{21}^2}{a_{11}}} = \sqrt{5 - \frac{2^2}{1}} = 1$$

$$l_{32} = \frac{a_{23} - \dfrac{(a_{12})(a_{13})}{a_{11}}}{\sqrt{a_{22} - \dfrac{a_{21}^2}{a_{11}}}} = \frac{15 - (2)(6)}{\sqrt{5 - \dfrac{2^2}{1}}} = 3$$

$$l_{33} = \sqrt{a_{33} - \frac{a_{31}^2}{a_{11}} - \frac{\left\{a_{23} - (a_{21})(a_3)\right\}}{\left(a_{22} - \dfrac{a_{21}^2}{a_{11}}\right)}}$$

$$= \sqrt{46 - 36 - \frac{\left[\dfrac{15-(2)(6)}{1}\right]}{\left(5-\dfrac{4}{1}\right)}} = 1$$

Thus, $L = \begin{bmatrix} 1 & 0 & 0 \\ 2 & 1 & 0 \\ 6 & 3 & 1 \end{bmatrix}$, hence $L^T = \begin{bmatrix} 1 & 2 & 6 \\ 0 & 1 & 3 \\ 0 & 0 & 1 \end{bmatrix}$

Now, $AX = B$ reduces to

$$LL^T X = B \tag{4}$$

Let $L^T X = Y$

where
$$Y = \begin{bmatrix} y_1 \\ y_2 \\ y_3 \end{bmatrix} \tag{5}$$

Then, (4) becomes

$$LY = B$$

i.e.,
$$\begin{bmatrix} 1 & 0 & 0 \\ 2 & 1 & 0 \\ 6 & 3 & 1 \end{bmatrix} \begin{bmatrix} y_1 \\ y_2 \\ y_3 \end{bmatrix} = \begin{bmatrix} 4 \\ 12 \\ 36 \end{bmatrix}$$

It gives

$$y_1 = 4 \tag{6}$$

$$2y_1 + y_2 = 12 \tag{7}$$

$$6y_1 + 3y_2 + y_3 = 36 \tag{8}$$

Solving Eqs. (6), (7) and (8) by forward substitution, we have $y_1 = 4$, $y_2 = 4$ and $y_3 = 0$. Hence,

$$Y = \begin{bmatrix} 4 \\ 4 \\ 0 \end{bmatrix} \tag{9}$$

Now, Eq. (5) becomes

$$\begin{bmatrix} 1 & 2 & 6 \\ 0 & 1 & 3 \\ 0 & 0 & 1 \end{bmatrix} \begin{bmatrix} x_1 \\ x_2 \\ x_3 \end{bmatrix} = \begin{bmatrix} 4 \\ 4 \\ 0 \end{bmatrix}$$

Hence,

$$x_1 + 2x_2 + 6x_3 = 4 \tag{10}$$

$$x_2 + 3x_3 = 4 \tag{11}$$

$$x_3 = 0 \tag{12}$$

Solving Eqs. (10), (11) and (12) by back substitution, we obtain

$$x_1 = -4, x_2 = 4 \text{ and } x_3 = 0$$

10.6.15 Crout's method

Let the system of n linear equations in n unknowns be given by (10.2), i.e.,

$$a_{11}x_1 + \ldots\ldots + a_{1n}x_n = b_1$$

$$a_{21}x_1 + \ldots\ldots + a_{2n}x_n = b_2$$

$$a_{n1}x_1 + \ldots\ldots + a_{nn}x_n = b_n$$

It is equivalent to $AX = B$, where A is the coefficient matrix of (10.2) $X^T = \begin{bmatrix} x_1, & x_2 \ldots x_n \end{bmatrix}$ and $B^T = \begin{bmatrix} b_1, b_2 \ldots b_n \end{bmatrix}$. In this method, first we find lower triangular L and upper triangular matrices U such that

$$A = LU, \tag{10.23}$$

where

$$L = \begin{bmatrix} l_{11} & 0 & 0 & 0 & 0 \\ l_{21} & l_{22} & 0 & 0 & 0 \\ \ldots & \ldots & \ldots & \ldots & \ldots \\ l_{n1} & l_{n2} & l_{n3} & l_{n(n-1)} & l_{nn} \end{bmatrix} \tag{10.24}$$

and

$$U = \begin{bmatrix} 1 & u_{12} & u_{13} & \ldots & u_{1(n-1)} & u_{1n} \\ 0 & 1 & u_{23} & \ldots & u_{2(n-1)} & u_{2n} \\ \ldots & \ldots & \ldots & \ldots & \ldots \\ 0 & 0 & 0 & \ldots & u_{n(n-1)} & u_{nn} \end{bmatrix} \tag{10.25}$$

Now the matrix equation $AX = B$ reduces to

$$LUX = B \tag{10.26}$$

Next, we suppose

$$UX = Y, \tag{10.27}$$

where

$$Y = \begin{bmatrix} y_1 \\ y_2 \\ y_n \end{bmatrix} \tag{10.28}$$

In view of (10.27), Eq. (10.26) reduces to

$$LY = B \tag{10.29}$$

The equations of (10.29) are solved for $y_1, y_2, \ldots y_n$ using forward substitution. These solutions give the matrix $Y = \begin{bmatrix} y_1 \\ y_2 \\ y_n \end{bmatrix}$. Now,

Eq. (10.27), i.e., $UX = Y$ can be solved for X by backward substitution. The method of obtaining the solution of $AX = B$ by the procedure described here is called Crout's method. It is one of the triangularisation method or decomposition method. Further, the coefficient matrix can also be expressed as $A = LU$, where

$$L = \begin{bmatrix} 1 & 0 & 0 & \ldots & 0 & 0 \\ l_{21} & 1 & 0 & \ldots & 0 & 0 \\ l_{31} & l_{32} & 1 & \ldots & & \\ l_{n1} & l_{n2} & l_{n3} & l_{n(n-1)} & 1_{nn} \end{bmatrix} \text{ and } U = \begin{bmatrix} u_{11} & u_{12} & u_{13} & \ldots u_{1n} \\ 0 & u_{22} & u_{23} & \ldots u_{2n} \\ \ldots & \ldots & \ldots & \ldots \\ 0 & 0 & 0 & \ldots u_{nn} \end{bmatrix}$$
$$\tag{10.30}$$

The matrix equation $AX = B$ can also be solved by expressing $AX = B$ as $LUX = B$, (where L and U are given by (10.30)) following the procedure described above. This method is known as Doolittle's method.

10.6.16 Formulae to find the elements of lower and upper triangular matrices *L* and *U* when *A* = *LU*

We present the formulae to compute the elements of lower triangular matrix L and upper triangular matrix U when $A = LU$, where

$$A = \begin{bmatrix} a_{11} & a_{12} & a_{13} \\ a_{21} & a_{22} & a_{23} \\ a_{31} & a_{32} & a_{33} \end{bmatrix}, L = \begin{bmatrix} l_{11} & 0 & 0 \\ l_{21} & l_{22} & 0 \\ l_{31} & l_{32} & l_{33} \end{bmatrix} \text{ and }$$

$$U = \begin{bmatrix} 1 & u_{12} & u_{13} \\ 0 & 1 & u_{23} \\ 0 & 0 & 1 \end{bmatrix}$$

First column elements of L:

$$l_{11} = a_{11} \tag{10.31}$$

$$l_{21} = a_{21} \tag{10.32}$$

$$l_{31} = a_{31} \tag{10.33}$$

First row elements of U:

$$u_{12} = \frac{a_{12}}{a_{11}} \tag{10.34}$$

$$u_{13} = \frac{a_{13}}{a_{11}} \tag{10.35}$$

Second column elements of L:

$$l_{22} = a_{22} - l_{21}u_{12}, \tag{10.36}$$

where l_{21} and u_{12} are given by (10.32) and (10.34), respectively.

$$l_{32} = a_{32} - l_{31}u_{12}, \tag{10.37}$$

where l_{31} and u_{12} are given by (10.33) and (10.34), respectively.

Second row elements of U:

$$u_{23} = \frac{a_{23} - l_{21}u_{13}}{l_{22}}, \tag{10.38}$$

where l_{21}, u_{13} and l_{22} are given by (10.32), (10.35) and (10.36), respectively.

Third column element of L:

$$l_{33} = a_{33} - l_{31}u_{13} - l_{32}u_{23}, \tag{10.39}$$

where l_{31}, u_{13}, l_{32} and u_{23} are given by (10.33), (10.35), (10.37) and (10.38), respectively.

Similar formulae can be derived when the coefficient matrix A is of order 4.

SOLVED PROBLEMS

Problem 1: Using Crout's method, solve the following system of linear equations:

$$x_1 + 2x_2 + 3x_3 = 14 \tag{1}$$
$$2x_1 + 5x_2 + 2x_3 = 18 \tag{2}$$
$$3x_1 + x_2 + 5x_3 = 20 \tag{3}$$

Solution: Here, $A = \begin{bmatrix} 1 & 2 & 3 \\ 2 & 5 & 2 \\ 3 & 1 & 5 \end{bmatrix}$, $X = \begin{bmatrix} x_1 \\ x_2 \\ x_3 \end{bmatrix}$ and $B = \begin{bmatrix} 14 \\ 18 \\ 20 \end{bmatrix}$

Let
$$A = \begin{bmatrix} l_{11} & 0 & 0 \\ l_{21} & l_{22} & 0 \\ l_{31} & l_{32} & l_{33} \end{bmatrix} \begin{bmatrix} 1 & u_{12} & u_{13} \\ 0 & 1 & u_{23} \\ 0 & 0 & 1 \end{bmatrix} \tag{4}$$

From Eqs. (10.31) to (10.39), we have

$$l_{11} = a_{11} = 1 \tag{5}$$

$$l_{21} = a_{21} = 2 \tag{6}$$

$$l_{31} = a_{31} = 3 \tag{7}$$

$$u_{12} = \frac{a_{12}}{a_{11}} = \frac{2}{1} = 2 \tag{8}$$

$$u_{13} = \frac{a_{13}}{a_{11}} = \frac{3}{1} = 3 \tag{9}$$

$$l_{22} = a_{22} - l_{21}u_{12} = 5 - (2)(2) = 1 \tag{10}$$

$$l_{32} = a_{32} - l_{31}u_{12} = 1 - (3)(2) = -5 \tag{11}$$

$$u_{23} = \frac{a_{23} - l_{21}u_{13}}{l_{22}} = \frac{2 - (2)(3)}{(1)} = -4 \tag{12}$$

$$u_{33} = a_{33} - l_{31}u_{13} - l_{32}u_{23} = 5 - (3)(3) - (-5)(-4) = -24$$

Thus,
$$A = \begin{bmatrix} 1 & 0 & 0 \\ 2 & 1 & 0 \\ 3 & -5 & -24 \end{bmatrix} \begin{bmatrix} 1 & 2 & 3 \\ 0 & 1 & -4 \\ 0 & 0 & 1 \end{bmatrix}$$

Therefore, the matrix equation $AX = B$ reduces to

$$\begin{bmatrix} 1 & 0 & 0 \\ 2 & 1 & 0 \\ 3 & -5 & -24 \end{bmatrix} \begin{bmatrix} 1 & 2 & 3 \\ 0 & 1 & -4 \\ 0 & 0 & 1 \end{bmatrix} X = \begin{bmatrix} 14 \\ 18 \\ 20 \end{bmatrix} \tag{13}$$

Let
$$UX = Y \tag{14}$$

where
$$Y = \begin{bmatrix} y_1 \\ y_2 \\ y_3 \end{bmatrix} \tag{15}$$

Then, (13) reduces to

$$\begin{bmatrix} 1 & 0 & 0 \\ 2 & 1 & 0 \\ 3 & -5 & -24 \end{bmatrix} \begin{bmatrix} y_1 \\ y_2 \\ y_3 \end{bmatrix} = \begin{bmatrix} 14 \\ 18 \\ 20 \end{bmatrix}$$

Hence,

$$y_1 = 14 \tag{16}$$
$$2y_1 + y_2 = 18 \tag{17}$$
$$3y_1 - 5y_2 - 24y_3 = 20 \tag{18}$$

Solving Eqs. (16) to (18) for y_1, y_2 and y_3 by forward substitution, we get $y_1 = 14$, $y_2 = -10$ and $y_3 = 3$

i.e.,
$$Y = \begin{bmatrix} 14 \\ -10 \\ 3 \end{bmatrix}$$

Now, from Eq. (14), we have

$$\begin{bmatrix} 1 & 2 & 3 \\ 0 & 1 & -4 \\ 0 & 0 & 1 \end{bmatrix} \begin{bmatrix} x_1 \\ x_2 \\ x_3 \end{bmatrix} = \begin{bmatrix} 14 \\ -10 \\ 3 \end{bmatrix}$$

It gives the following system of equations:

$$x_1 + 2x_2 + 3x_3 = 14 \tag{19}$$

$$x_2 - 4x_3 = -10 \tag{20}$$

$$x_3 = 3 \tag{21}$$

Solving (19) to (21) for x_1, x_2 and x_3 by back substitution, we get $x_1 = 1$, $x_2 = 2$ and $x_3 = 3$, which is the solution of given system of equations.

Problem 2: Solve the following equations by Crout's method.

$$2x_1 - 6x_2 + 8x_3 = 24 \tag{1}$$

$$5x_1 + 4x_2 - 3x_3 = 2 \tag{2}$$

$$3x_1 + x_2 + 2x_3 = 6 \tag{3}$$

Solution: The given system of Eqs. (1) to (3) can be expressed as

$$AX = B, \tag{4}$$

where

$$A = \begin{bmatrix} 2 & -6 & 8 \\ 5 & 4 & -3 \\ 3 & 1 & 2 \end{bmatrix}, \ X = \begin{bmatrix} x_1 \\ x_2 \\ x_3 \end{bmatrix} \text{ and } B = \begin{bmatrix} 24 \\ 2 \\ 6 \end{bmatrix}$$

Let $A = LU$, where

$$A = \begin{bmatrix} l_{11} & 0 & 0 \\ l_{21} & l_{22} & 0 \\ l_{31} & l_{32} & l_{33} \end{bmatrix} \text{ and } U = \begin{bmatrix} 1 & u_{12} & u_{13} \\ 0 & 1 & u_{23} \\ 0 & 0 & 1 \end{bmatrix}$$

Now l_{11}, l_{21}, l_{22}, l_{31}, l_{32}, l_{33}, u_{12}, u_{13} and u_{23} are found from the relations (10.31) to (10.39). The values of these elements are $l_{11} = 2, l_{21} = 5$,

$l_{31} = 3, u_{12} = -3, l_{22} = 19, l_{32} = 10, u_{13} = 4, u_{23} = \dfrac{-23}{19}$ and $l_{33} = \dfrac{40}{19}$

Thus,

$$L = \begin{bmatrix} 2 & 0 & 0 \\ 3 & 19 & 0 \\ 3 & 10 & \dfrac{40}{19} \end{bmatrix} \text{ and } U = \begin{bmatrix} 1 & -3 & 4 \\ 0 & 1 & \dfrac{-23}{19} \\ 0 & 0 & 1 \end{bmatrix}$$

Now, Eq. (4) becomes

$$\begin{bmatrix} 2 & 0 & 0 \\ 3 & 19 & 0 \\ 3 & 10 & \dfrac{40}{19} \end{bmatrix}\begin{bmatrix} 1 & -3 & 4 \\ 0 & 1 & \dfrac{-23}{19} \\ 0 & 0 & 1 \end{bmatrix}\begin{bmatrix} x_1 \\ x_2 \\ x_3 \end{bmatrix} = \begin{bmatrix} 24 \\ 2 \\ 16 \end{bmatrix} \tag{5}$$

Let

$$\begin{bmatrix} 1 & -3 & 4 \\ 0 & 1 & \dfrac{-23}{19} \\ 0 & 0 & 1 \end{bmatrix}\begin{bmatrix} x_1 \\ x_2 \\ x_3 \end{bmatrix} = \begin{bmatrix} y_1 \\ y_2 \\ y_3 \end{bmatrix} \tag{6}$$

Then, Eq. (5) reduces to

$$\begin{bmatrix} 2 & 0 & 0 \\ 3 & 19 & 0 \\ 3 & 10 & \dfrac{40}{19} \end{bmatrix}\begin{bmatrix} y_1 \\ y_2 \\ y_3 \end{bmatrix} = \begin{bmatrix} 24 \\ 2 \\ 16 \end{bmatrix}$$

Therefore,

$$2y_1 = 24 \tag{7}$$

$$3y_1 + 19y_2 = 2 \tag{8}$$

$$3y_1 + 10y_2 + \frac{40}{19}y_3 = 16 \tag{9}$$

Solving (7), (8) and (9) for y_1, y_2 and y_3, we obtain

$$y_1 = 12, \; y_2 = \frac{-58}{19} \text{ and } y_3 = 5$$

Hence,

$$\begin{bmatrix} y_1 \\ y_2 \\ y_3 \end{bmatrix} = \begin{bmatrix} 12 \\ \dfrac{-58}{19} \\ 5 \end{bmatrix} \tag{10}$$

In view of (10), Eq. (6) reduces to

$$\begin{bmatrix} 1 & -3 & 4 \\ 0 & 1 & \dfrac{-23}{19} \\ 0 & 0 & 1 \end{bmatrix} \begin{bmatrix} x_1 \\ x_2 \\ x_3 \end{bmatrix} = \begin{bmatrix} 12 \\ \dfrac{-58}{19} \\ 5 \end{bmatrix}$$

$$x_1 - 3x_2 + 4x_3 = 12 \tag{11}$$

$$x_2 - \frac{23}{19}x_3 = \frac{-58}{19} \tag{12}$$

$$x_3 = 5. \tag{13}$$

Solving (11) to (13) by back-substitution method, we get $x_3 = 5$, $x_2 = 3$ and $x_1 = 1$, which is the solution of Eqs. (1), (2) and (3).

Problem 3: Find the inverse of $A = \begin{bmatrix} 1 & 1 & 1 \\ 4 & 3 & -1 \\ 3 & 5 & 3 \end{bmatrix}$ using LU decomposition method.

Solution: Suppose

$$A = LU,$$

where

$$L = \begin{bmatrix} l_{11} & 0 & 0 \\ l_{21} & l_{22} & 0 \\ l_{31} & l_{32} & l_{33} \end{bmatrix} \text{ and } U = \begin{bmatrix} 1 & u_{12} & u_{13} \\ 0 & 1 & u_{23} \\ 0 & 0 & 1 \end{bmatrix}$$

Using the relations (10.31) to (10.39), we can find the elements of lower triangular matrix L and upper triangular matrix U. The matrices L and U are

$$L = \begin{bmatrix} 1 & 0 & 0 \\ 4 & -1 & 0 \\ 3 & 2 & -10 \end{bmatrix} \text{ and } U = \begin{bmatrix} 1 & 1 & 1 \\ 0 & 1 & 5 \\ 0 & 0 & 1 \end{bmatrix}$$

It can be shown that

$$L^{-1} = \begin{bmatrix} 1 & 0 & 0 \\ 4 & -1 & 0 \\ 1 & -0.2 & -0.1 \end{bmatrix} \text{ and } U^{-1} = \begin{bmatrix} 1 & -1 & 4 \\ 0 & 1 & -5 \\ 0 & 0 & 1 \end{bmatrix}$$

Now, since $A = LU$, we have $A^{-1} = U^{-1}L^{-1}$ i.e.,

$$A^{-1} = \begin{bmatrix} 1 & -1 & 4 \\ 0 & 1 & -5 \\ 0 & 0 & 1 \end{bmatrix} \begin{bmatrix} 1 & 0 & 0 \\ 4 & -1 & 0 \\ 1 & -0.2 & -0.1 \end{bmatrix}$$

$$= \begin{bmatrix} 1.0 & 0.2 & -0.4 \\ -1.5 & 0 & 0.5 \\ 1.1 & -0.2 & -0.1 \end{bmatrix}$$

Problem 4: Decompose the matrix $A = \begin{bmatrix} 5 & -2 & 1 \\ 7 & 1 & -5 \\ 3 & 7 & 4 \end{bmatrix}$ into $A = LU$

and find A^{-1} Hence solve

$$5x_1 - 2x_2 + x_3 = 4 \tag{1}$$

$$7x_1 + x_2 - 5x_3 = 8 \tag{2}$$

$$3x_1 + 7x_2 + 4x_3 = 10 \tag{3}$$

Solution: Suppose $A = LU$, where

$$L = \begin{bmatrix} l_{11} & 0 & 0 \\ l_{21} & l_{22} & 0 \\ l_{31} & l_{32} & l_{33} \end{bmatrix} \text{ and } U = \begin{bmatrix} 1 & u_{12} & u_{13} \\ 0 & 1 & u_{23} \\ 0 & 0 & 1 \end{bmatrix}$$

Using the relations (10.31) to (10.39), we can find the elements of L and U. The matrices L and U are given by

$$L = \begin{bmatrix} 5 & 0 & 0 \\ 7 & 3.8 & 0 \\ 3 & 8.2 & 17.2105 \end{bmatrix} \text{ and } U = \begin{bmatrix} 1 & -0.4 & 0.2 \\ 0 & 1 & -1.6842 \\ 0 & 0 & 1 \end{bmatrix}$$

The inverses of L and U are given by

$$L^{-1} = \begin{bmatrix} 0.2 & 0 & 0 \\ -0.3684 & 0.2632 & 0 \\ 0.1407 & -0.1254 & 0.058 \end{bmatrix} \text{ and } U^{-1} = \begin{bmatrix} 1.0 & 0.4 & 0.4737 \\ 0 & 1 & 1.6842 \\ 0 & 0 & 1 \end{bmatrix}$$

Thus, $A^{-1} = U^{-1}L^{-1} = \begin{bmatrix} 0.1193 & 0.0459 & 0.0275 \\ -0.1315 & 0.0520 & 0.0979 \\ 0.1407 & -0.1254 & 0.0581 \end{bmatrix}$

Now $X = A^{-1}B$ becomes

$$\begin{bmatrix} x_1 \\ x_2 \\ x_3 \end{bmatrix} = \begin{bmatrix} 0.1193 & 0.0459 & 0.0275 \\ -0.1315 & 0.0520 & 0.0979 \\ 0.1407 & -0.1254 & 0.0581 \end{bmatrix} \begin{bmatrix} 4 \\ 8 \\ 10 \end{bmatrix}$$

$$= \begin{bmatrix} 1.1193 \\ 0.8685 \\ 0.1407 \end{bmatrix}$$

Hence, $x_1 = 1.1193$, $x_2 = 0.8685$ and $x_3 = 0.1407$ is the solution of Eqs. (1) to (3).

10.7 ITERATIVE METHODS

So far, the methods considered in section 10.6 are direct methods to solve a system of n linear equations in n unknowns. In this section, we introduce the iterative methods namely Jacobi's method and Gauss–Seidel method to solve such equations.

10.7.1 Definitions

An $n \times n$ matrix A is said to be diagonally dominant if the absolute value of each diagonal elements is greater than or equal to the sum of absolute values of the remaining elements of the row in which the diagonal element occurs. For example, $A = \begin{bmatrix} 10 & -5 & -2 \\ 5 & 12 & 4 \\ 1 & 6 & 14 \end{bmatrix}$ is a diagonally dominant matrix, since

$$|a_{11}| = 10 \geq |a_{12}| + |a_{13}| = |-5| + |-2| = 7$$

$$|a_{22}| = 12 \geq |a_{21}| + |a_{23}| = |5| + |4| = 9$$

$$|a_{33}| = 14 \geq |a_{31}| + |a_{32}| = 1 + 6 = 7$$

whereas $B = \begin{bmatrix} 2 & 3 & -1 \\ 5 & 8 & -4 \\ 1 & 1 & 1 \end{bmatrix}$ is not diagonally dominant matrix, since

$$|b_{22}| = 8 \not\geq |a_{21}| + |a_{23}| = |5| + |-4| = 5 + 4 = 9$$

An $n \times n$ matrix is said to be strictly diagonally dominant if the absolute value of each diagonal element is greater than the sum of absolute value of remaining elements of the row in which diogonal element occurs.

The system of equations given by $AX = B$ is said to be diagonally dominant if A is diagonally dominant and it is said to be strictly diagonally dominant if A is strictly diagonal dominant. For example, $10x_1 + x_2 + x_3 = 12$, $x_1 + 10x_2 + x_3 = 12$ and $x_1 + x_2 + 10x_3 = 12$ is a diagonal dominant.

Note that given equations may be reduced to a diagonal system by changing the order of equations if the given equations are not in a diagonal system.

It can also be done by interchanging the columns of coefficient matrix and making corresponding changes in the unknowns in X.

SOLVED PROBLEMS

Problem 1: Is the following system of equations diagonal dominant?

$$2x_1 + 3x_2 - x_3 = 6$$
$$4x_1 + 8x_2 + 13x_3 = 8$$
$$10x_1 + x_2 + x_3 = 6$$

Solution: For the given system of equations the coefficient matrix is

$$A = \begin{bmatrix} 2 & 3 & -1 \\ 4 & 8 & 13 \\ 10 & 1 & 1 \end{bmatrix}$$

Therefore $|a_{11}| = 2$ Now

$$|a_{11}| = 2 \not\geq |a_{12}| + |a_{13}| = |3| + |-1| = 4$$

Similarly, $\qquad |a_{22}| = 8 \not\ge |a_{21}| + |a_{23}| = 17$ and

$$|a_{33}| = 1 \not\ge |a_{31}| + |a_{33}| = |10| + |1| = 11$$

Hence the given system is not a diagonal dominant.

Problem 2: Check whether the system of equations $x_1 + 6x_2 - 2x_3 = 5,$ $4x_1 + x_2 + x_3 = 6$ and $-3x_1 + x_2 + 7x_3 = 5$ is a diagonal dominant? If not reduces it to a diagonal dominant.

Solution: Here the coefficient matrix

$$A = \begin{bmatrix} 1 & 6 & -2 \\ 4 & 1 & 1 \\ -3 & 1 & 7 \end{bmatrix}, \text{ Since } |a_{11}| = 1 \not\ge |a_{12}| + |a_{13}| = 6 + 2 = 8. \text{ The given}$$

system is not a diagonal system. If we rearrange the given equations in the following form:

$$6x_2 + x_1 - 2x_3 = 5$$
$$x_2 + 4x_1 + x_3 = 6$$
$$x_2 - 3x_1 + 7x_3 = 5$$

then the transformed system is a diagonal dominant whose matrix equation is $CY = D$

$$\text{where } C = \begin{bmatrix} 6 & 1 & -2 \\ 1 & 4 & 1 \\ 1 & -3 & 7 \end{bmatrix} Y = \begin{bmatrix} x_2 \\ x_1 \\ x_3 \end{bmatrix} \text{ and } D = \begin{bmatrix} 5 \\ 6 \\ 5 \end{bmatrix}$$

Clearly, C is a diagonally dominant matrix. Hence $CY = D$ is a diagonal dominant.

Note: We introduce the notation that $x_i^{(j)}$ where $i = 1, 2, 3,..., n$ and $j = 1, 2, 3, ...$, to denote j^{th} approximation of unknown x_i and $x_i^{(0)}$ $(i = 1, 2, n)$ to denote the initial approximation of x_i. For example, $x_1^{(1)}$ $x_2^{(1)}$, $x_3^{(1)},...x_n^{(1)}$, are the first approximations of $x_1, x_2, x_3...$ and x_n respectively.

10.7.2 Jacobi's method

In this method, an initial approximate solution to a given system of equations is assumed. Following certain process (which is going to describe shortly), the approximate solution of the given system of equations is improved successively so that the approximate solutions approach the exact solution of the given system of equations. Now we describe the Jacobi's method of solving a system of n linear equations in n knowns.

Suppose the system of n linear equations in n unknown be

$$a_{11}x_1 + a_{12}x_2 + \ldots\ldots + a_{1n}x_n = b_1$$
$$a_{21}x_1 + a_{22}x_2 + \ldots\ldots + a_{2n}x_n = b_2 \tag{10.40}$$
$$\ldots\ldots\ldots\ldots\ldots\ldots\ldots\ldots\ldots\ldots\ldots\ldots\ldots$$
$$a_{x1}x_1 + a_{x2}x_2 + \ldots\ldots + a_{xn}x_n = b_n$$

where the coefficient matrix of above system of equations is diagonally dominant. This system of equations can be expressed as

$$x_1 = \frac{1}{a_{11}}\left[b_1 - a_{12}x_2 - a_{13}x_3 - \ldots\ldots - a_{1n}x_n\right]$$

$$x_2 = \frac{1}{a_{22}}\left[b_2 - a_{21}x_1 - a_{23}x_3 - \ldots\ldots - a_{2n}x_n\right] \tag{10.41}$$

$$\ldots\ldots\ldots\ldots\ldots\ldots\ldots\ldots\ldots\ldots\ldots\ldots$$

$$x_n = \frac{1}{a_{nn}}\left[b_n - a_{n1}x_1 - a_{n2}x_2 - \ldots\ldots - a_{n(n-1)}x_{n-1}\right]$$

First we suppose $x_1^{(0)} = 0$, $x_2^{(2)} = 0$, $\ldots x_n^{(0)} = 0$ as initial approximate solution of (10.41). Next, we substitute these values for x_1, x_2, $\ldots x_n$ in the RHS of (10.41). The values obtained after this substitution are known as first approximate solution of the system (10.41). We denote them by $x_i^{(1)}$ $(i = 1, 2, \ldots n)$. Now substitute $x_1^{(1)}$, $x_2^{(1)}$, $\ldots, x_n^{(1)}$ for x_1, x_2, $\ldots x_n$ in the RHS of (10.41). The values obtained are known as 2nd approximate solution of (10.41) We denote them by $x_i^{(2)}$ $(i = 1, 2, \ldots n)$. This process is continued to get $x_i^{(3)}$, $x_i^{(4)}$, $x_i^{(5)}$, $\ldots$ $(i = 1, 2, \ldots n)$ and it is terminated when the desired degree of accuracy of the solution is obtained.

The general formula to find $(i + 1)$ th approximation of x_1, x_2, $\ldots, x_n$ are

$$x_1^{(i+1)} = \frac{1}{a_{11}}\left(b_1 - a_{12}x_2^{(i)} - \ldots\ldots - a_{in}x_n^{(i)}\right)$$

$$x_2^{(i+1)} = \frac{1}{a_{22}}\left(b_2 - a_{21}x_1^{(i)} - a_{23}x_3^{(i)} \ldots\ldots - a_{2n}x_n^{(i)}\right) \tag{10.42}$$

$$\ldots\ldots\ldots\ldots\ldots\ldots\ldots\ldots\ldots\ldots\ldots\ldots$$

$$x_n^{(i+1)} = \frac{1}{a_{nn}}\left(b_n - a_{n1}x_1^{(i)} - a_{n2}x_2^{(i)} - \ldots\ldots - a_{n(n-1)}x_{n-1}^{(i)}\right)$$

The process of obtaining the solution of (10.41) by the method describe above is called Jacobi's method.

Note: In the Jacobi's method of solving a system of linear equations, usually the initial approximations $x_i^{(0)}$ ($i = 1, 2, \ldots n$) are assumed to zero. It is also permissible to assume the initial approximation x_i other than zero. For example, we can assume

$$x_1^{(0)} = 1, x_2^{(0)} = 1, x_3^{(0)} = 1, \ldots, x_x^{(0)} = 1$$

SOLVED PROBLEMS

Problem 1: Using Jacobi's method, find the first eight successive approximations of x_1, x_2 and x_3 for the following equations, assuming $x_1^{(0)} = x_2^{(0)} = x_3^{(0)} = 0$.

$$27x_1 + 6x_2 - x_3 = 85 \tag{1}$$
$$6x_1 + 15x_2 + 2x_3 = 72 \tag{2}$$
$$x_1 + x_2 + 54x_3 = 110 \tag{3}$$

Solution: The given system of equations can be expressed as

$$x_1 = \frac{1}{27}\left(85 - 6x_2 + x_3\right) \tag{4}$$

$$x_2 = \frac{1}{15}\left(72 - 6x_1 - 2x_3\right) \tag{5}$$

$$x_3 = \frac{1}{54}\left(110 - x_2 - x_3\right) \tag{6}$$

Given that $x_1^{(0)} = x_2^{(0)} = x_3^{(0)} = 0$. The $(i+1)$ approximations of x_1, x_2 and x_3 are (from Eq. 10.42)

$$x_1^{(i+1)} = \frac{1}{27}\left(85 - 6x_2^{(i)} + x_3^{(i)}\right) \tag{7}$$

$$x_2^{(i+1)} = \frac{1}{15}\left(72 - 6x_1^{(i)} - 2x_3^{(i)}\right) \tag{8}$$

$$x_3^{(i+1)} = \frac{1}{54}\left(110 - x_2^{(i)} - x_3^{(i)}\right) \tag{9}$$

To find $x_1^{(1)}$, $x_2^{(1)}$ and $x_3^{(1)}$:

Putting $i = 0$ in (7), (8) and (9), we get

$$x_1^{(1)} = \frac{1}{27}\left(85 - 6x_2^{(0)} + x_3^{(0)}\right)$$

$$x_2^{(1)} = \frac{1}{15}\left(72 - 6x_1^{(0)} - 2x_3^{(0)}\right)$$

$$x_3^{(1)} = \frac{1}{54}\left(110 - x_2^{(0)} - x_3^{(0)}\right)$$

Substituting $x_1^{(0)} = 0$, $x_2^{(0)} = 0$ and $x_3^{(0)} = 0$ in the above equations, we get

$$x_1^{(1)} = \frac{85}{27} = 3.148148$$

$$x_2^{(1)} = \frac{72}{15} = 4.8 \tag{10}$$

$$x_3^{(1)} = \frac{110}{54} = 2.037037$$

To find $x_1^{(2)}$, $x_2^{(2)}$ and $x_3^{(2)}$:

Substituting $i = 1$ in (7), (8) and (9), we get

$$x_1^{(2)} = \frac{1}{27}\left(87 - 6x_2^{(1)} + x_3^{(1)}\right)$$

$$x_2^{(2)} = \frac{1}{15}\left(72 - 6x_1^{(1)} - 2x_3^{(1)}\right) \tag{11}$$

$$x_3^{(2)} = \frac{1}{54}\left(110 - x_2^{(1)} - x_3^{(1)}\right)$$

Putting the values of $x_1^{(1)}, x_2^{(1)}, x_3^{(1)}$ given in Eq. (10) in equations (11), we get

$$x_1^{(2)} = \frac{1}{27}\left[87 - 6(4.8) + 2.037037\right]$$

$$x_2^{(2)} = \frac{1}{15}\left[72 - 6(3.148148) - 2(2.037037)\right]$$

$$x_3^{(2)} = \frac{1}{54}\left[110 - (4.8) - 2.037037\right]$$

i.e.,

$$x_1^{(2)} = 3.148148 - 1.066667 + 0.075446 = 2.156927$$

$$x_2^{(2)} = 4.8 - 1.259259 - 0.271605 = 3.269136 \qquad (12)$$

$$x_3^{(2)} = 2.037037 - 0.058299 - 0.088889 = 1.889849$$

To find $x_1^{(3)}$, $x_2^{(3)}$, $x_3^{(3)}$:

Substituting $i = 2$ in Eqs. (7), (8) and (9), we have

$$x_1^{(3)} = \frac{1}{27}\left(87 - 6x_2^{(2)} + x_3^{(2)}\right)$$

$$x_2^{(3)} = \frac{1}{15}\left(72 - 6x_1^{(2)} - 2x_3^{(2)}\right) \qquad (13)$$

$$x_3^{(3)} = \frac{1}{54}\left(110 - x_2^{(2)} - x_3^{(2)}\right)$$

Substituting the values of $x_1^{(2)}, x_2^{(2)}$ and $x_3^{(2)}$ in Eqn. (13), we get

$$x_1^{(3)} = \frac{1}{27}\left[85 - 6(2.156927) + 1.889849\right]$$

$$x_2^{(3)} = \frac{1}{15}\left[72 - 6(3.269136) - 2(1.889849)\right]$$

$$x_3^{(3)} = \frac{1}{54}\left[110 - (3.269136) - 1.889849\right]$$

i.e., $$x_1^{(3)} = 3.148148 - 0.726475 + 0.069994 = 2.491668$$

$$x_2^{(3)} = 4.8 - 0.862771 - 0.25198 = 3.685249$$

$$x_3^{(3)} = 2.037037 - 0.039943 - 0.060540 = 1.936554$$

Continuing this process, the values of $x_1^{(4)}, x_2^{(4)}, x_3^{(4)}, x_1^{(5)}, x_2^{(5)}, x_3^{(5)},$ ....$x_1^{(8)}, x_2^{(8)}, x_3^{(8)}$ are as follows:

$$x_1^{(4)} = 3.148148 - 0.818944 + 0.071724 = 2.400928$$

$$x_2^{(4)} = 4.8 - 0.996667 - 0.258207 = 3.545126$$

$$x_3^{(4)} = 2.037037 - 0.046142 - 0.068245 = 1.922650,$$

$$x_1^{(5)} = 3.148148 - 0.787806 + 0.071209 = 2.431552$$

$$x_2^{(5)} = 4.800000 - 0.960371 - 0.256353 = 3.583275$$

$$x_3^{(5)} = 2.037037 - 0.044462 - 0.065650 = 1.926925$$

$$x_1^{(6)} = 3.148148 - 0.796283 + 0.071368 = 2.423232$$

$$x_2^{(6)} = 4.800000 - 0.972621 - 0.256923 = 3.570456$$

$$x_3^{(6)} = 2.037037 - 0.045029 - 0.066357 = 1.925651$$

$$x_1^{(7)} = 3.148148 - 0.793435 + 0.071320 = 2.426034$$

$$x_2^{(7)} = 4.800000 - 0.969293 - 0.256754 = 3.573954$$

$$x_3^{(7)} = 2.037037 - 0.044875 - 0.066120 = 1.926043$$

$$x_1^{(8)} = 3.148148 - 0.794212 + 0.071335 = 2.425271$$

$$x_2^{(8)} = 4.800000 - 0.970414 - 0.256806 = 3.572781$$

$$x_3^{(8)} = 2.037037 - 0.044927 - 0.066184 = 1.925926$$

Problem 2: Solve the following linear equations by Jacobi's method performing five iterations

$$10x_1 + x_2 + x_3 = 12 \tag{1}$$

$$x_1 + 10x_2 + x_3 = 12 \tag{2}$$

$$x_1 + x_2 + 10x_3 = 12 \tag{3}$$

Solution: Given equations can be expressed as

$$x_1 = \frac{1}{10}\left[12 - x_2 - x_3\right]$$

$$x_2 = \frac{1}{10}\left[12 - x_1 - x_3\right]$$

and

$$x_3 = \frac{1}{10}\left[12 - x_1 - x_2\right]$$

Hence, by Jacobi's method, we have

$$x_1^{(i+1)} = \frac{1}{10}\left[12 - x_2^{(i)} - x_3^{(i)}\right]$$

$$x_2^{(i+1)} = \frac{1}{10}\left[12 - x_1^{(i)} - x_3^{(i)}\right] \qquad (4)$$

$$x_3^{(i+1)} = \frac{1}{10}\left[12 - x_1^{(i)} - x_2^{(i)}\right],$$

where $i = 0, 1, 2, 3,\ldots$ Suppose $x_1^{(0)} = x_2^{(0)} = x_3^{(0)} = 0$

To find $x_1^{(1)}, x_2^{(1)}, x_3^{(1)}$:

For $i = 0$, Eq. (4) reduces to

$$x_1^{(1)} = \frac{1}{10}\left[12 - x_2^{(0)} - x_3^{(0)}\right]$$

$$x_2^{(1)} = \frac{1}{10}\left[12 - x_1^{(0)} - x_3^{(0)}\right] \qquad (5)$$

$$x_3^{(1)} = \frac{1}{10}\left[12 - x_1^{(0)} - x_2^{(0)}\right]$$

Substituting the values of $x_1^{(0)} = 0, x_2^{(0)} = 0, x_3^{(0)} = 0$ in the Eq. (5), we get

$$x_1^{(1)} = 1.2, \ x_2^{(1)} = 1.2, \text{ and } x_3^{(1)} = 1.2$$

To find $x_1^{(2)}, x_2^{(2)}$ and $x_3^{(2)}$:

Substituting $i = 1$ in the equations given by (4), we have:

$$x_1^{(2)} = \frac{1}{10}\ [12 - x_2^{(1)} - x_3^{(1)}]$$

$$x_2^{(2)} = \frac{1}{10}\ [12 - x_1^{(1)} - x_3^{(1)}] \qquad (6)$$

$$x_3^{(2)} = \frac{1}{10}\ [2 - x_1^{(1)} - x_2^{(1)}]$$

Now substituting $x_1^{(1)}, x_2^{(1)}$ and $x_3^{(1)}$ values in (6), we get

$$x_1^{(2)} = 1.2 - 0.12 - 0.12 = 0.96$$

$$x_2^{(2)} = 1.2 - 0.12 - 0 - 12 = 0.96$$

$$x_3^{(2)} = 1.2 - 0.12 - 0.12 = 0.96$$

To find $x_1^{(3)}$, $x_2^{(3)}$, and $x_3^{(3)}$:

Substituting $i = 2$ in the equations given by (4), we obtain

$$x_1^{(3)} = \frac{1}{10}\,[12 - x_2^{(2)} - x_3^{(2)}]$$

$$x_2^{(3)} = \frac{1}{10}\,[12 - x_1^{(2)} - x_3^{(2)}] \tag{7}$$

$$x_3^{(3)} = \frac{1}{10}\,[12 - x_1^{(2)} - x_2^{(2)}]$$

Substituting the values of $x_1^{(2)} = x_2^{(2)} = x_3^{(2)} = 0.\,96$ in (7), we get

$$x_1^{(3)} = 1.2 - 0.096 - 0.096 = 1.0\overline{0}8$$

$$x_2^{(3)} = 1.2 - 0.096 - 0.096 = 1.008$$

$$x_3^{(3)} = 1.2 - 0.096 - 0.096 = 1.008$$

Containing this process, we get

$$x_1^{(4)} = 1.2 - 0.1008 - 0.1008 = 0.9984$$

$$x_2^{(4)} = 1.2 - 0.1008 - 0.1008 = 0.9984$$

$$x_3^{(4)} = 1.2 - 0.008 - 0.1008 = 0.9984$$

and

$$x_1^{(5)} = 1.2 - 0.09984 - 0.09984 = 1.000320$$

$$x_2^{(5)} = 1.2 - 0.09984 - 0.09984 = 1.000320$$

$$x_3^{(5)} = 1.2 - 0.09984 - 0.09984 = 1.000320$$

Hence, the approximate solutions of the given equations are $x_1 = 1.000320$, $x_2 = 1.000320$ and $x_3 = 1.000230$.

Problem 3: Solve the following equations by Jacobi's method. Assume the initial approximations of x_1, x_2 and x_3 as $x_1^{(0)} = x_2^{(0)} = x_3^{(0)} = 0$.

$$10\,x_1 + 3x_2 + x_3 = 14 \tag{1}$$

$$2\,x_1 - 10\,x_2 + 3\,x_3 = -5 \tag{2}$$

$$x_1 + 3\,x_2 + 10\,x_3 = 14 \tag{3}$$

Solution: As per the Jacobi's method, Eqs. (1) to (3) are expressed in the following:

$$x_1 = \frac{1}{10}\,[14 - 3\,x_2 - x_3] \tag{4}$$

$$x_2 = \frac{-1}{10}\,[-5 - 2\,x_1 - 3\,x_3]$$

$$x_3 = \frac{1}{10}\,[14 - x_1 - 3\,x_2]$$

and the $(i + 1)$ the approximations of x_1, x_2 and x_3 in terms of their i^{th} approximations are given by

$$x_1^{(i+1)} = \frac{1}{10}\,[14 - 3\,x_2^{(i)} - x_3^{(i)}]$$

$$x_2^{(i+1)} = \frac{-1}{10}\,[-5 - 2x_1^{(i)} - 3x_3^{(i)}] \tag{5}$$

$$x_3^{(i+1)} = \frac{1}{10}\,[14 - x_1^{(i)} - 3x_2^{(i)}]$$

where $i = 0, 1, 2$ .... Note that $x_1^{(0)} = x_2^{(0)} = x_3^{(0)} = 0$ is the hypothesis. Now, we proceed to find the successive approximations of x_1, x_2, and x_3.

To find $x_1^{(1)}$, $x_2^{(1)}$ and $x_3^{(1)}$:

Substituting $i = 0$ in Eqs. of (5) we obtain:

$$x_1^{(1)} = \frac{1}{10}\,[14 - 3x_2^{(0)} - x_3^{(0)}]$$

$$x_2^{(1)} = \frac{-1}{10}\,[-5 - 2x_1^{(0)} - x_3^{(0)}] \tag{6}$$

$$x_3^{(1)} = \frac{1}{10}\,[14 - x_1^{(0)} - x_2^{(0)}]$$

Now, substituting the values of $x_1^{(0)} = x_2^{(0)} = x_3^{(0)} = 0$ in the equations of (6), we get $x_1^{(1)} = 1.4$, $x_2^{(1)} = 0.5$ and $x_3^{(1)} = 1.4$

To find $x_1^{(2)}$, $x_2^{(2)}$ and $x_3^{(2)}$

Substituting $i = 1$ in the equations of (5) we get

$$x_1^{(2)} = \frac{1}{10}\,[14 - 3\,x_2^{(1)} - x_3^{(1)}]$$

$$x_2^{(2)} = \frac{-1}{10}\,[-5 - 2x_1^{(1)} - 3x_3^{(1)}] \tag{7}$$

$$x_3^{(2)} = \frac{1}{10}\left[14 - x_2^{(1)} - x_3^{(1)}\right]$$

Now, substituting $x_1^{(1)} = 1.4$, $x_2^{(1)} = 0.5$, $x_3^{(1)} = 1.4$ in the equations of (7), we have

$$x_1^{(2)} = 1.4 - 0.15\ 0.14 = 1.11$$

$$x_2^{(2)} = 0.5 + 0.28 + 0.42 = 1.2$$

$$x_3^{(2)} = 1.4 - 0.14 - 0.15 = 1.11$$

To find $x_1^{(3)}$, $x_2^{(3)}$ and $x_3^{(3)}$:

Substituting $i = 2$ in the equations of (5), we get:

$$x_1^{(3)} = \frac{1}{10}\left[14 - 3x_2^{(2)} - x_3^{(2)}\right]$$

$$x_2^{(3)} = \frac{-1}{10}\left[-5 - 2x_1^{(2)} - 3x_3^{(2)}\right] \tag{8}$$

$$x_3^{(3)} = \frac{1}{10}\left[14 - x_2^{(2)} - x_3^{(2)}\right]$$

Putting $x_1^{(2)}$, $x_2^{(2)}$ and $x_3^{(2)}$ values in (8), we get

$$x_1^{(3)} = 1.4 - 0.36 - 0.111 = 0.929$$

$$x_2^{(3)} = 0.5 = 0.222 + 0.333 = 1.055$$

$$x_3^{(3)} = 1.4 - 0.111 - 0.36 = 0.929$$

Containing this process, we get

$$x_1^{(4)} = 1.400000 - 0.316500 - 0.092900 = 0.990600$$

$$x_2^{(4)} = 0.500000 + 0.185800 + 0.278700 = 0.964500$$

$$x_3^{(4)} = 1.400000 - 0.092900 - 0.316500 = 0.990600$$

$$x_1^{(5)} = 1.400000 - 0.289350 - 0.099060 = 1.011590$$

$$x_2^{(5)} = 0.500000 + 0.198120 + 0.297180 = 0.995300$$

$$x_3^{(5)} = 1.400000 - 0.099060 - 0.289350 = 1.011590$$

$$x_1^{(6)} = 1.400000 - 0.298590 - 0.101159 = 1.000251$$

$$x_2^{(6)} = 0.500000 + 0.202318 + 0.303477 = 1.0005795$$

$$x_3^{(6)} = 1.400000 - 0.101159 - 0.298590 = 1.000251$$

$$x^{(7)} = 1.400000 - 0.301739 - 0.100025 = 0.998236$$

$$x_2^{(7)} = 0.500000 + 0.200050 + 0.300075 = 1.000126$$

$$x_3^{(7)} = 1.400000 - 0.100025 - 0.301739 = 0.998236$$

and

$$x_1^{(8)} = 1.400000 - 0.300038 - 0.099824 = 1.000139$$

$$x_2^{(8)} = 0.500000 - 0.199647 + 0.299471 = 0.999118$$

$$x_3^{(8)} = 1.400000 - 0.099824 - 0.300038 = 1.000139$$

Hence, the solution of the given linear equations is $x_1 = 1$, $x_2 = 1$, $x_3 = 1$ (rounding the $x_1^{(8)}$, $x_2^{(8)}$, $x_3^{(8)}$ values).

Problem 4: Perform six iterations of Jacobi's method to solve the following linear equations.

$$10x_1 + x_2 - x_3 = 8 \tag{1}$$

$$x_1 + 12x_2 + x_3 = 39.66 \tag{2}$$

$$3x_1 + 4x_2 + 15x_3 = 55 \tag{3}$$

Solution: First, we express Eqs. (1) to (3) in the following form for the implementation of the Jacobi's method.

$$x_1 = \frac{1}{10}[8 - x_2 + x_3]$$

$$x_2 = \frac{1}{12}[39.66 - x_1 - x_3] \tag{4}$$

$$x_3 = \frac{1}{15}[55 - 3x_1 - 4x_2]$$

Now, $(i + 1)^{\text{th}}$ the approximations of unknowns x_1, x_2, and x_3 are (as per Jacobi's method)

$$x_1^{(i+1)} = \frac{1}{10}[8 - x_1^{(i)} + x_3^{(i)}]$$

$$x_2^{(i+1)} = \frac{1}{12}[39.66 - 3x_1^{(i)} - 4x_2^{(i)}] \tag{5}$$

$$x_3^{(i+1)} = \frac{1}{15}[55 - 3x_1^{(i)} - 4x_2^{(i)}],$$

where $i = 0, 1, 2, 3, \ldots\ldots$ Suppose the initial approximations $x_1^{(0)} = 0$, $x_2^{(0)} = 0$ and $x_3^{(0)} = 0$.

To find $x_1^{(1)}$, $x_2^{(1)}$ and $x_3^{(1)}$:

Substituting $i = 0$ in (5), we obtain

$$x_1^{(1)} = \frac{1}{10}[8 - x_1^{(0)} + x_3^{(0)}]$$

$$x_2^{(1)} = \frac{1}{12}[39.66 - 3x_1^{(0)} - 4x_2^{(0)}]$$

$$x_3^{(1)} = \frac{1}{15}[55 - 3x_1^{(0)} - 4x_2^{(0)}]$$

Putting the initial approximations $x_1^{(0)}$, $x_2^{(0)}$ and $x_3^{(0)}$ in the above equations, we get

$$x_1^{(1)} = 0.8, \ x_2^{(1)} = 3.305 \text{ and } x_3^{(1)} = 3.6666667.$$

To find $x_1^{(2)}$, $x_2^{(2)}$ and $x_3^{(2)}$:

Substituting $i = 1$ is in (5), we obtain

$$x_1^{(2)} = \frac{1}{10}[8 - x_2^{(1)} + x_1^{(3)}]$$

$$x_2^{(2)} = \frac{1}{12}[39.66 - 3x_1^{(1)} - x_2^{(1)}]$$

$$x_3^{(2)} = \frac{1}{15}[55 - 3x_1^{(1)} - 4x_2^{(1)}]$$

Putting the values of $x_1^{(1)}, x_2^{(1)}$, and $x_3^{(1)}$ in the above equations, we obtain

$$x_1^{(2)} = 0.800000 - 0.330500 + 0.366667 = 0.836167$$

$$x_2^{(2)} = 3.305000 - 0.066667 - 0.305556 = 2.932778$$

$$x_3^{(2)} = 3.666667 - 0.160000 - 0.881333 = 2.625333$$

To find $x_1^{(3)}$, $x_2^{(3)}$ and $x_3^{(3)}$:

Substituting $i = 2$ in (5), we get

$$x_1^{(3)} = \frac{1}{10}[8 - x_2^{(2)} + x_3^{(2)}]$$

$$x_2^{(3)} = \frac{1}{12}[39.66 - 3x_1^{(2)} - x_2^{(2)}]$$

$$x_3^{(3)} = \frac{1}{15}[\, 55 - 3\,x_1^{(2)} - 4\,x_2^{(2)}\,]$$

Now, putting the values of $x_1^{(2)}$, $x_2^{(2)}$ and $x_3^{(2)}$ in the above equations, we get

$$x_1^{(3)} = 0.800000 - 0.293278 + 0.262533 = 0.769256$$

$$x_2^{(3)} = 3.305000 - 0.069681 - 0.218778 = 3.016542$$

$$x_3^{(3)} = 3.666667 - 0.167233 - 0.782074 = 2.717359$$

To find $x_1^{(14)}, x_2^{(14)}, x_3^{(3)}$:

Substituting $i = 3$, in (5) and substituting the values of $x_1^{(3)}$, $x_2^{(3)}$ and $x_3^{(3)}$ we obtain

$$x_1^{(14)} = 0.800000 - 0.301654 - 0.271736 = 0.77082$$

$$x_2^{(14)} = 3.305000 - 0.064105 - 0.226447 = 3.014449$$

$$x_3^{(14)} = 3.666667 - 0.153851 - 0.804411 = 2.708404$$

To find $x_1^{(15)}, x_2^{(15)}$ and $x_3^{(15)}$:

Substituting $i = 4$ in (5) and putting the values of $x_1^{(4)}$, $x_2^{(4)}$ and $x_3^{(4)}$, we get

$$x_1^{(5)} = 0.800000 - 0.301445 + 0.270840 = 0.769396$$

$$x_2^{(5)} = 3.305000 - 0.064173 - 0.225700 = 3.015126$$

$$x_3^{(5)} = 3.666667 - 0.154016 - 0.803853 = 2.708797$$

To find $x_1^{(6)}, x_2^{(6)}$ and $x_3^{(6)}$:

Again substituting $i = 5$, in (5) and putting the values of $x_1^{(15)}$, $x_2^{(15)}$, $x_3^{(15)}$) in the resulting equations, we have

$$x_1^{(6)} = 0.800000 - 0.301513 + 0.270880 = 0.769367$$

$$x_2^{(6)} = 3.305000 - 0.064116 - 0.225733 = 3.015151$$

$$x_3^{(6)} = 3.666667 - 0.153879 - 0.804034 = 2.708754$$

Thus, sixth iterates are given by $x_1^{(6)} = 0.769367$, $x_2^{(6)} = 3.015151$ and $x_3^{(6)} = 2.708754$.

Problem 5: Solve the following equations by Jacobi's method to perform five iterations assuming the initial approximations of each unknown as zero.

$$20\,x_1 + 2x_2 - x_3 - x_4 = 19 \tag{1}$$

$$3\,x_1 + 15x_2 - x_3 + 2x_4 = 21 \tag{2}$$

$$x_1 + 3x_2 - 20x_3 + x_4 = -14 \tag{3}$$

$$x_1 + x_2 + x_3 + 6x_4 = 15 \tag{4}$$

Solution: The $(i+1)^{\text{th}}$ approximations of unknowns x_1, x_2, x_3, and x_4 are given by

$$x_1^{(i+1)} = \frac{1}{20}\,[19 - 2\,x_2^{(i)} + x_3^{(i)} + x_4^{(i)}]$$

$$x_2^{(i+1)} = \frac{1}{15}[21 - 3\,x_1^{(i)} + x_3^{(i)} - 2\,x_4^{(i)}] \tag{5}$$

$$x_3^{(i+1)} = \frac{1}{(-20)}[-14 - x_1^{(i)} - 3\,x_2^{(i)} - x_4^{(i)}]$$

$$x_4^{(i+1)} = \frac{1}{6}[15 - x_1^{(i)} - x_2^{(i)} - x_3^{(i)}],$$

where $i = 0, 1, 2, \ldots$ Suppose $x_1^{(0)} = x_2^{(0)} = x_3^{(0)} = x_4^{(0)} = 0$; substituting $i = 0$ in (5), we get

$$x_1^{(1)} = \frac{1}{20}[19 - x_2^{(0)} + x_3^{(0)} + x_4^{(0)}]$$

$$x_2^{(1)} = \frac{1}{15}[21 - 3\,x_1^{(0)} + x_3^{(0)} - 2\,x_4^{(0)}] \tag{6}$$

$$x_3^{(1)} = \frac{1}{(-20)}[-14 - x_1^{(0)} - 3\,x_2^{(0)} - x_4^{(0)}]$$

$$x_4^{(1)} = \frac{1}{6}[15 - x_1^{(0)} - x_2^{(0)} - x_3^{(0)}]$$

Substituting $x_1^{(0)} = x_2^{(0)} = x_3^{(0)} = x_4^{(0)} = 0$ in Eqs. (6), we get

$x_1^{(1)} = 0.95$, $x_2^{(1)} = 1.4$, $x_3^{(1)} = 0.7$ and $x_4^{(1)} = 2.5$. Similarly, substituting $i = 1, 2, 3, 4$ and 5 in (5) and putting the values of $x_1^{(1)}$, $x_2^{(1)}$, $x_3^{(1)}$, $x_4^{(1)}$ 1, $x_1^{(2)}$, $x_2^{(2)}$, $x_3^{(2)}$, $x_4^{(2)}$: $x_1^{(3)}$, $x_2^{(3)}$, $x_3^{(3)}$, $x_4^{(3)}$ and $x_1^{(4)}$, $x_2^{(4)}$, $x_3^{(4)}$, and $x_4^{(4)}$ in the corresponding equations, we obtain

$$x_1^{(2)} = 0.950000 + - 0.000000 + 0.000000 + 0.000000 = 0.950000$$

$$x_2^{(2)} = 1.400000 + -0.000000 + 0.000000 + 0.000000 = 1.400000$$

$$x_3^{(2)} = 0.700000 + -0.000000 + 0.000000 + 0.000000 = 0.700000$$

$$x_4^{(2)} = 2.500000 + -0.000000 + 0.000000 + 0.000000 = 2.500000$$

$$x_1^{(3)} = 0.950000 + 0.140000 + 0.035000 + 0.125000 = 0.970000$$

$$x_2^{(3)} = 1.400000 + 0.190000 + 0.046667 + 0.333333 = 0.923333$$

$$x_3^{(3)} = 0.700000 + 0.047500 + 0.210000 + 0.125000 = 1.082500$$

$$x_4^{(3)} = 2.500000 + 0.158333 + 0.233333 + 0.116667 = 1.991667$$

$$x_1^{(4)} = 0.950000 + 0.092333 + 0.054125 + 0.099583 = 1.011375$$

$$x_2^{(4)} = 1.400000 + 0.194000 + 0.072167 + 0.265556 = 1.012611$$

$$x_3^{(4)} = 0.700000 + 0.048500 + 0.138500 + 0.099583 = 0.986583$$

$$x_4^{(4)} = 2.500000 + 0.161667 + 0.153889 + 0.180417 = 2.004028$$

and

$$x_1^{(5)} = 0.950000 + 0.101261 + 0.049329 + 0.100201 = 0.998269$$

$$x_2^{(5)} = 1.400000 + 0.202275 + 0.065772 + 0.267204 = 0.996294$$

$$x_3^{(5)} = 2.500000 + 0.168562 + 0.168769 + 0.164431 = 1.998238$$

Thus, the solution of the given equations is $x_1 = 1$, $x_2 = 1, x_3 = 1$ and $x_4 = 2$.

Problem 6: Solve the following equations by Jacobi's method; Assume the initial approximations of each unknown as zero.

$$4x_1 + x_3 + x_4 = 10 \qquad (1)$$

$$4x_2 + x_4 = 20 \qquad (2)$$

$$x_1 + 4x_3 = 30 \qquad (3)$$

$$x_1 + x_2 + 4x_4 = 40 \qquad (4)$$

Solution: The $(i+1)$ th approximations of the unknowns x_1, x_2, x_3 and x_4 are given by (as per Jacobi's method)

$$x_1^{(i+1)} = \frac{1}{4}[10 - x_3^{(i)} - x_4^{(i)}]$$

$$x_2^{(i+1)} = \frac{1}{4}[20 - x_4^{(i)}]$$

$$x_3^{(i+1)} = \frac{1}{4}[30 - x_4^{(i)}]$$

$$x_4^{(i+1)} = \frac{1}{4}[40 - x_1^{(i)} - n_2^{(i)}] \qquad (5)$$

Given that $x_1^{(0)} = x_2^{(0)} = x_3^{(0)} = x_4^{(0)} = 0$. Therefore, substituting $i = 0$ in (5) and putting $x_1^{(0)} = x_2^{(0)} = x_3^{(0)} = x_4^{(0)} = 0$, we obtain

$x_1^{(1)} = 2.5, x_2^{(1)} = 5, x_3^{(1)} = 7.5$, and $x_4^{(1)} = 10$. Next substituting $i = 1$ in (5), we obtain

$$x_1^{(2)} = \frac{1}{4}[10 - x_3^{(1)} - x_4^{(1)}]$$

$$x_2^{(2)} = \frac{1}{4}[20 - x_4^{(i)}]$$

$$x_3^{(2)} = \frac{1}{4}[30 - x_1^{(i)}] \qquad (6)$$

$$x_4^{(2)} = \frac{1}{4}[40 - x_1^{(i)} - x_2^{(i)}]$$

Now, substituting the values of $x_1^{(1)}, x_2^{(i)}, x_3^{(i)}$ and $x_4^{(i)}$ in (6), we obtain

$$x_1^{(2)} = 2.500000 + 0.000000 + 1.875000 + 2.500000 = -1.875000$$

$$x_2^{(2)} = 5.000000 + 0.000000 + 0.000000 + 2.500000 = 2.500000$$

$$x_3^{(2)} = 7.500000 + 0.625000 + 0.000000 + 0.000000 = 6.875000$$

$$x_4^{(2)} = 10.000000 + 0.625000 + 1.250000 + 0.000000 = 8.125000$$

Containing this process, we obtain

i	$x_1^{(i)}$	$x_2^{(i)}$	$x_3^{(i)}$	$x_4^{(i)}$
3	−1.25	2.96875	7.968750	9.84375
4	−1.953125	2.539063	7.8125	9.570313
5	−1.845703	2.607422	7.988281	9.853516
6	−1.960449	2.536621	7.961426	9.80957
7	−1.942749	2.547607	7.9990112	9.855957
8	−1.961517	2.536011	7.985687	9.848785

Thus, the approximate solution of given linear equations is (upto 8 iterations)

$$x_1 = -1.961517, \; x_2 = 2.536011, \; x_3 = 7.985687 \text{ and } x_4 = 9.848785.$$

10.7.3 Gauss–Seidel method

Gauss–Seidel method is a slight modification of Jacobi's method. In the Jacobi's method, the $(i + 1)$th approximations of unknowns are computed using their i^{th} approximations only, whereas in Gauss–Seidel method, to compute $(i + 1)$th approximations of unknown x_k i.e. $x_k^{(i+1)}$ we use the approximations $x_1^{(k+1)}$, $x_2^{(k+1)}$, ... $x_{(k-1)}^{(k+1)}$ (as these are obtained before the computation of $x_k^{(i+1)}$) and $x_{k+1}^{(i)}$ .... $x_n^{(i)}$. Now, we describe this method of solving the linear equations in n unknowns given by (10.2),

We suppose that the coefficient matrix of (10.2) is diagonally dominant and $a_{ii} \neq 0$ $(i = 1, 2, ..., n)$. This system of equations can be expressed as

$$x_1 = \frac{1}{a_{11}}[b_1 - a_{12}x_2 - a_{13}x_3 - \ldots\ldots\ldots - a_{1n}x_n]$$

$$x_2 = \frac{1}{a_{22}}[b_2 - a_{21}x_1 - a_{23}x_3 - \ldots\ldots\ldots - a_{2n}x_n] \qquad (10.43)$$

$$x_3 = \frac{1}{a_{33}}[b_3 - a_{31}x_1 - a_{32}x_2 - a_{34}x_4 - \ldots\ldots\ldots - a_{3n}x_n]$$

$$\ldots\ldots\ldots\ldots\ldots\ldots\ldots\ldots\ldots\ldots\ldots\ldots\ldots\ldots$$

$$x_n = \frac{1}{a_{nn}}[b_n - a_{n1}x_1 - a_{n2}x_2 - \ldots\ldots\ldots\ldots - a_{n(n-1)}x_{n-1}]$$

The $(i + 1)$th approximations are computed from the following equations:

$$x_1^{(i+1)} = \frac{1}{a_{11}}[b_1 - a_{12}x_2^{(i)} - a_{13}x_3^{(i)} - \ldots\ldots\ldots\ldots - a_{1n}x_n^{(i)}]$$

$$x_2^{(i+1)} = \frac{1}{a_{22}}[b_2 - a_{21}x_1^{(i+1)} - a_{13}x_3^{(i)} - \ldots\ldots\ldots\ldots - a_{1n}x_n^{(i)}]$$

$$x_3^{(i+1)} = \frac{1}{a_{33}}[b_3 - a_{31}x_1^{(i+1)} - a_{32}x_2^{(i+1)} - a_{34}x_4^{(i)} - \ldots\ldots\ldots - a_{3n}a_n^{(i)}]$$

$$\ldots\ldots\ldots\ldots\ldots\ldots\ldots\ldots\ldots\ldots\ldots\ldots\ldots\ldots\ldots\ldots\ldots\ldots \quad (10.44)$$

$$x_n^{(i+1)} = \frac{1}{a_{nn}}\left[b_n - a_{n1}x_1^{(i+1)} - a_{n2}x_2^{(i+1)}\ldots\ldots - a_{n(n-1)}x_{n-1}^{(i+1)}\right]$$

where $i = 0, 1, 2, \ldots\ldots\ldots$ Usually the initial approximations $x_k^{(0)} = 0$ $(k = 2, 3, \ldots.. n)$ using Eq. (10.44), we find the successive approximations of x_1, x_2, ..., and x_n. As an illustration we consider a system of linear 3 equations in 3 unknowns. Suppose,

$$a_{11}x_1 + a_{12}x_2 + a_{13}x_3 = b_1$$

$$a_{21}x_2 + a_{22}x_2 + a_{23}x_3 = b_2 \quad (10.45)$$

$$a_{31}x_1 + a_{32}x_2 + a_{33}x_3 = b_3$$

be a system of equations in three linear equations in three unknowns, where the coefficient matrix of (10.45) is diagonally dominant and $a_{ii} \neq 0 (i = 1, 2, 3)$, Eq. (10.45) can be expressed as

$$x_1 = \frac{1}{a_{11}}[b_1 - a_{12}x_2 - a_{13}x_3]$$

$$x_2 = \frac{1}{a_{22}}[b_2 - a_{21}x_1 - a_{23}x_3] \quad (10.46)$$

$$x_3 = \frac{1}{a_{33}}[b_3 - a_{31}x_1 - a_{32}x_2]$$

The $(i + 1)$ approximations of x_1, x_2 and x_3 are computed by the following relations:

$$x_1^{(i+1)} = \frac{1}{a_{11}}[b_1 - a_{12}x_2^{(i)} - a_{13}x_3^{(i)}]$$

$$x_2^{(i+1)} = \frac{1}{a_{22}}[b_2 - a_{21}x_1^{(i+1)} - a_{23}x_3^{(i)}] \qquad (10.47)$$

$$x_3^{(i+1)} = \frac{1}{a_{33}}[b_3 - a_{31}x_1^{(i+1)} - a_{32}x_2^{(i+1)}]$$

Note that, as soon as the improved approximations of unknown are available, they are used in the subsequent calculations of the approximations of unknowns.

Suppose $x_2^{(0)}$, $x_3^{(0)}$ be the initial approximations of x_2 and x_3, respectively, then,

$$x_1^{(1)} = \frac{1}{a_{11}}[b_{11} - a_{12}x_2^{(0)} - a_{13}x_3^{(0)}]$$

$$x_2^{(1)} = \frac{1}{a_{22}}[b_2 - a_{21}x_1^{(1)} - a_{23}x_3^{(0)}] \qquad (10.48)$$

$$x_3^{(1)} = \frac{1}{a_{33}}[b_3 - a_{31}x_1^{(1)} - a_{32}x_2^{(1)}]$$

Note that Eqs. (10.48) are obtained from (10.47) with $i = 0$. The second approximations $x_1^{(2)}, x_2^{(2)}, x_3^{(2)}$, are obtained from the following equations:

$$x_1^{(2)} = \frac{1}{a_{11}}[b^1 - a_{12}x_2^{(1)} - a_{13}x_3^{(1)}]$$

$$x_2^{(2)} = \frac{1}{a_{22}}[b_2 - a_{21}x^{(2)} - a_{31}x_3^{(1)}] \qquad (10.49)$$

$$x_3^{(2)} = \frac{1}{a_{33}}[b_3 - a_{31}x_1^{(2)} - a_{32}x_3^{(2)}]$$

Continuing their process, we can obtain the successive approximations of the solutions of unknowns x_1, x_2 and x_3. This process is continued till the solutions of (10.45) are obtained upto desired degree of accuracy.

SOLVED PROBLEMS

Problem 1: Perform five iterations of Gauss–Seidel method for the following equations to obtain the approximate solutions.

$$10x_1 + x_2 + x_3 = 12 \tag{1}$$

$$2x_1 + 10x_2 + x_3 = 13 \tag{2}$$

$$2x_1 + 2x_2 + 10x_3 = 14 \tag{3}$$

Solution: The given equations can be expressed as

$$x_1 = \frac{1}{6}[12 - x_2 - x_3] \tag{4}$$

$$x_2 = \frac{1}{10}[13 - 2x_1 - x_3] \tag{5}$$

$$x_3 = \frac{1}{10}[14 - 2x_1 - 2x_2] \tag{6}$$

Hence, $(i+1)$ th approximations of x_1, x_2, and x_3 are given by (as per Gauss–Seidel method)

$$x_1^{(i+1)} = \frac{1}{12}[12 - x_2^{(i)} - x_3^{(i)}] \tag{7}$$

$$x_2^{(i+1)} = \frac{1}{10}[13 - 2x_1^{(i+1)} - x_3^{(i)}] \tag{8}$$

$$x_3^{(i+1)} = \frac{1}{10}[14 - 2x_1^{(i+1)} - 2x_2^{(i+1)}] \tag{9}$$

Now, we take the initial approximations of x_2 and x_3 as zero, i.e., $x_2^{(0)} = x_3^{(0)} = 0$. For $i = 0$ in (7), (8), (9) we have

$$x_1^{(1)} = \frac{1}{10}[12 - x_2^{(0)} - x_3^{(0)}] \tag{10}$$

$$x_2^{(1)} = \frac{1}{10}[13 - 2x_1^{(1)} - x_3^{(0)}] \tag{11}$$

$$x_3^{(1)} = \frac{1}{10}[14 - 2x_1^{(1)} - 2x_2^{(1)}] \tag{12}$$

Substituting $x_2^{(0)} = x_3^{(0)} = 0$ in (10), we get

$$x_1^{(1)} = 1.2$$

Now, substituting $x_1^{(1)} = 1.2$, and, $x_3^{(0)} = 0$ in Eq. (11), we get

$$x_2^{(1)} = 1.3 - 0.24 - 0 = 1.06$$

Finally, substituting $x_1^{(1)} = 1.2$ and $x_2^{(1)} = 1.06$ in Eq. (12), we get

$$x_3^{(1)} = \frac{1}{10}[14 - 2(1-2) - 2(1.06)] \text{ i.e., } x_3^{(1)} = 0.948.$$

The values of $x_1^{(2)}, x_2^{(2)}$ and $x_3^{(2)}$ are computed as follows:

Substituting $i = 1$ in Eqs. (7), (8) and (9), we have

$$x_1^{(2)} = \frac{1}{10}[12 - x_2^{(1)} - x_3^{(1)}] \tag{11}$$

$$x_2^{(2)} = \frac{1}{10}[13 - 2x_1^{(2)} - x_3^{(1)}] \tag{12}$$

$$x_3^{(2)} = \frac{1}{10}[14 - 2x_1^{(2)} - 2x_{2}^{(2)}] \tag{13}$$

Putting $x_2^{(1)} = 1.06$ and $x_3^{(1)} = 0.948$ in Eq. (11), we get

$$x_1^{(2)} = 1.2 - 0.106 - 0.0948 = 0.9992. \text{ Next, substituting}$$

$x_1^{(2)} = 0.9992$ and $x_3^{(1)} = 0.948$ in (12) we get

$$x_2^{(2)} = 1.3 - 0.19984 - 0.0948 = 1.000536$$

Finally, substituting $x_1^{(2)} = 0.9992$ and $x_2^{(2)} = 1.000536$ in Eq. (13), we obtain.

$$x_3^{(2)} = 1.4 - 0.19984 - 2.01072 = 0.999088$$

Thus, $x_1^{(2)} = 0.99992$, $x_2^{(2)} = 1.000536$ and $x_3^{(2)} = 0.999088$.

Continuing this process, we get

$$x_1^{(3)} = 0.999555, x_2^{(3)} = 1.00018, x_3^{(3)} = 1.000053,$$

$$x_1^{(4)} = 0.999977, x_2^{(4)} = 0.999999, x_3 = 1.00005, \text{ and}$$

$$x_1^{(5)} = 1.0, x_2^{(5)} = 1.0, \text{ and } x_3 = 1.0$$

Thus, the solution of given equations is $x_1 = 1.0, x_2 = 1.0$ and $x_3 = 1.0$.

Problem 2: Solve the following equations by Gauss–Seidel method upto three-decimal accuracy of the unknowns.

$$27x_1 + 6x_2 - x_3 = 85 \tag{1}$$

$$6x_1 + 15x_2 + 2x_3 = 72 \tag{2}$$

$$x_1 + x_2 + 54x_3 = 110 \tag{3}$$

Solution: To implement Gauss–Seidel method, first we express the given Eqs. (1), (2) and (3) as

$$x_1 = \frac{1}{27}[85 - 6x_2 + x_3]$$

$$x_2 = \frac{1}{15}[72 - 6x_1 - 2x_3]$$

$$x_3 = \frac{1}{54}\left[110 - x_1 - x_2\right]$$

Hence, $(i + 1)$ the approximations of x_1, x_2 and x_3 are given by,

$$x_1^{(i+1)} = \frac{1}{27}[85 - 6x_2^{(i)} + x_3^{(i)}] \tag{4}$$

$$x_2^{(i+1)} = \frac{1}{15}[72 - 6x_1^{(i+1)} - 2x_3^{(i)}] \tag{5}$$

$$x_3^{(i+1)} = \frac{1}{54}[110 - x_1^{(i+1)} - x_2^{(i+1)}] \tag{6}$$

where $i = 0, 1, 2, 3\ldots\ldots$

Starting with the initial approximations of x_2 and x_3 as zero, i.e., $x_2^{(0)} = x^{(3)} = 0$, we get $x_1^{(1)}, x_2^{(1)}$ and $x_3^{(i)}$ as follows:

Substituting $i = 0$ in (4), we obtain

$$x_1^{(1)} = \frac{1}{27}[85 - 6x_2^{(0)} + x_3^{(1)}]$$

$$x_1^{(1)} = \frac{1}{27}[85 - 6(0) + 0] = 3.148148$$

Next, substituting $i = 0$ in (5), we obtain

$$x_2^{(1)} = \frac{1}{15}[72 - 6x_1^{(1)} - 2x_3^{(0)}]$$

$$x_2^{(1)} = \frac{1}{15}[72 - 6(3.148148) - 2(0)] = 3.540741$$

Finally, substituting $i = 0$ in (6), we get

$$x_3^{(1)} = \frac{1}{54}[110 - x_1^{(1)} - x_2^{(1)}]$$

$$x_3^{(1)} = \frac{1}{54}[110 - 3.148148 - 3.540741] = 1.913369$$

Thus, $x_1^{(1)} = 3.148148, x_2^{(1)} = 3.540741$ and $x_3^{(1)} = 3.540741$.

Now, substituting $i = 1$ in Eqs. (4) to (6), we have

$$x_1^{(2)} = \frac{1}{27}[85 - 6x_2^{(1)} + x_3^{(1)}] \tag{7}$$

$$x_2^{(2)} = \frac{1}{15}[72 - 6x_1^{(2)} + 2x_3^{(1)}] \tag{8}$$

$$x_3^{(2)} = \frac{1}{54}[110 - x_1^{(2)} + x_2^{(2)}] \tag{9}$$

Substituting $x_2^{(1)}$ and $x_3^{(1)}$ values in Eq. (7), we get

$$x_1^{(2)} = 3.148148 - 0.786831 + 0.070858 = 2.432175$$

Next, substituting $x_1^{(2)}$ and $x_3^{(1)}$ values in Eq. (8), we obtain

$$x_2^{(2)} = 48 - 0.9722870 - 0.255089 = 3.572041$$

Finally, substitutions $x_1^{(2)}$ and $x_2^{(2)}$ values in Eq. (9), we obtain.

$$x_3^{(2)} = 2.037037 - 0.045040 - 0.066149 = 1.925848$$

Thus, $x_1^{(2)} = 2.432175, x_2^{(2)} = 3.572041$ and $x_3^{(2)} = 1.925848$.

To find $x_1^{(3)}, x_2^{(3)}$ and $x_3^{(3)}$ substitute $i = 2$ in Eqs. (4), (5) and (6), we have

$$x_1^{(3)} = \frac{1}{27}[85 - 6x_2^{(2)} + x_3^{(2)}] \tag{10}$$

$$x_2^{(3)} = \frac{1}{15}[72 - 6x_1^{(3)} - 2x_3^{(2)}] \tag{11}$$

$$x_3^{(3)} = \frac{1}{54}[110 - x_1^{(3)} - x_2^{(3)}] \tag{12}$$

Substituting $x_2^{(2)}$ and $x_3^{(2)}$ values in Eq. (10), we get

$$x_1^{(3)} = \frac{1}{27}[85 - 6(3.572041) + 1.925848]$$

i.e., $x_1^{(3)} = 2.425689.$

Next, substituting $x_1^{(3)}$ and $x_3^{(2)}$ values in Eq. (11), we get

$$x_2^{(3)} = \frac{1}{15}[72 - 6(2.425689 - 2(1.925848)]$$

i.e., $x_2^{(3)} = 3.572945$

Finally, substituting $x_1^{(3)}$ and $x_2^{(3)}$ values in Eq. (12), we get

$$x_3^{(3)} = \frac{1}{54}[110 - 2.425689 - 3.572945]$$

i.e., $x_3^{(3)} = 1.925951.$

Thus,

$$x_1^{(3)} = 2.425689, \; x_2^{(3)} = 3.572945 \text{ and } x_3^{(3)} = 1.925951$$

Now, substituting $i = 3$ in Eqs. (4), (5) and (6), we get

$$x_1^{(4)} = \frac{1}{54}[85 - 6x_2^{(3)} + x_3^{(3)}] \tag{13}$$

$$x_2^{(4)} = \frac{1}{15}[72 - 6x_1^{(4)} - 2x_3^{(3)}] \tag{14}$$

$$x_3^{(4)} = \frac{1}{54}[110 - x_1^{(4)} - x_2^{(4)}] \tag{15}$$

Substituting $x_2^{(3)}$ and $x_3^{(3)}$ values in Eqs. (13), we get

$$x_1^{(4)} = \frac{1}{27}[85 - 6(0.425689) + 1.925951]$$

$$= 3.148148 - 0.793988 + 0.071332$$

i.e., $x_1^{(4)} = 2.425492.$

Next, substituting $x_1^{(4)}, x_3^{(3)}$ values in Eq. (14), we have

$$x_2^{(4)} = \frac{1}{15}[72 - 6(2.425492) + 1.925951] = 4.8 - 0.970197 - 0.256793$$

i.e., $$x_2^{(4)} = 3.573010.$$

Finally, substituting $x_1^{(4)}$ and $x_2^{(4)}$ values in Eqs. (15), we get

$$x_3^{(4)} = \frac{1}{54}[110 - 2.425492 - 3.573010]$$

$$= 2.037037 - 0.044917 - 0.066167 = 1.925954$$

Thus, $x_1^{(4)} = 2.4255492$, $x_2^{(4)} = 3.573010$ and $x_3^{(4)} = 1.925954$.

Now, substituting $i = 4$ in Eqs. (4). (5) and (6), we obtain

$$x_1^{(5)} = \frac{1}{27}[85 - 6x_2^{(4)} + x_3^{(4)}] \tag{16}$$

$$x_2^{(5)} = \frac{1}{15}[72 - 6x_1^{(5)} + 2x_3^{(4)}] \tag{17}$$

$$x_3^{(5)} = \frac{1}{54}[110 - x_1^{(5)} - x_2^{(5)}] \tag{18}$$

Substituting $x_2^{(4)}$ and $x_3^{(4)}$ values in Eq. (6), we get

$$x_1^{(5)} = \frac{1}{27}[85 - 6(3.573010) + 1.925954)$$

$$= 3.148148 - 0.794002 + 0.071332$$

i.e., $$x_1^{(5)} = 2.425478$$

Next, substituting $x_1^{(5)}$ and $x_3^{(14)}$ values in (17), we have

$$x_2^{(5)} = \frac{1}{15}[72 - 6(2.425478) - 2(1.925954)]$$

$$= 4.8 - 0.970191 - 0.256794$$

i.e., $$x_2^{(5)} = 3.573015.$$

Finally, substituting $x_1^{(5)}$ and $x_2^{(5)}$ values in Eq. (18), we get

$$x_3^{(5)} = \frac{1}{54}[110 - 2.425478 - 3.573015]$$

$$= 2.037037 - 0.044916 - 0.066167$$

i.e., $$x_3^{(5)} = 1.925954.$$

Thus, upto three decimal places the solution of given system of equation is

$$x_1 = 2.425,\ x_2 = 3.573 \text{ and } x_3 = 1.925.$$

Problem 3: Perform five iterations of Gauss–Seidel method to obtain the approximate solution of the following equations:

$$20x_1 + x_2 - 2x_3 = 17 \tag{1}$$
$$3x_1 + 20x_2 - x_3 = -18 \tag{2}$$
$$2x_1 - 3x_2 + 20x_3 = 25 \tag{3}$$

Solution: The $(i+1)$th approximations of x_1, x_2 and x_3 are given by (as per Gauss–Seidel method)

$$x_1^{(i+1)} = \frac{1}{20}[17 - x_2^{(i)} + 2x_3^{(i)}] \tag{4}$$

$$x_2^{(i+1)} = \frac{1}{20}\left[-18 - 3x_1^{(i+1)} + x_3^{(i)}\right] \tag{5}$$

$$x_3^{(i+1)} = \frac{1}{20}[25 - 2x_1^{(i+1)} + 3x_2^{(i+1)}] \tag{6}$$

where $i = 0, 1, 2,\ldots\ldots$

We take the initial approximations $x_2^{(0)}$, $x_3^{(0)}$ as zero, i.e., $x_2^{(0)} = 0$, $x_3^{(0)} = 0$

To find $x_1^{(1)}$, $x_2^{(1)}$, and $x_3^{(1)}$:

Substituting $i = 0$ in (4), (5) and (6), we get

$$x_1^{(1)} = \frac{1}{20}[17 - x_2^{(0)} + 2x_3^{(0)}] \tag{7}$$

$$x_2^{(1)} = \frac{1}{20}[-18 - 3x_1^{(1)} + x_3^{(0)}] \tag{8}$$

$$x_3^{(1)} = \frac{1}{20}[25 - 2x_1^{(1)} - 3x_2^{(1)}] \tag{9}$$

Substituting $x_2^{(0)} = 0$ and $x_3^{(2)} = 0$ in (7), we get $x_1^{(1)} = 0.85$

Next substituting $x_1^{(1)} = 0.85$ and $x_3^{(0)} = 0$ in (8), we get

$$x_2^{(2)} = -0.9 - 0.1275 + 0 = -1.0275$$

Finally, substituting $x_1^{(1)} = 0.85$ and $x_2^{(1)} = -1.0275$ in Eq. (9), we obtain

$$x_3^{(1)} = 1.25 - 0.085 - 0.154125 = 1.010875.$$

To find $x_1^{(2)}, x_2^{(2)},$ and $x_3^{(2)}$:

Substituting $i = 1$ in (4) (5) and (6), we have

$$x_1^{(2)} = \frac{1}{20}[17 - x_2^{(1)} + 2x_3^{(1)}] \tag{10}$$

$$x_2^{(2)} = \frac{1}{20}[-18 - 3x_1^{(2)} + x_3^{(1)}] \tag{11}$$

$$x_3^{(2)} = \frac{1}{20}[25 - 2x_1^{(2)} + 3x_2^{(2)}] \tag{12}$$

Substituting $x_2^{(1)}$ and $x_3^{(1)}$ values in (10), we have

$$x_1^{(2)} = 0.85 + 0.051375 + 0.101088 = 1.002463$$

Next, substituting $x_1^{(2)}$ and $x_3^{(1)}$ values in (11), we get

$$x_2^{(2)} = -0.9 - 0.150369 + 0.050544 = -0.999826$$

Finally, substituting $x_1^{(2)}$ and $x_2^{(2)}$ values in (12), we have

$$x_3^{(2)} = 1.25 - 0.100246 - 0.149974 = 0.999780$$

Continuing this process, we obtain

$$x_1^{(3)} = 0.999969, x_2^{(3)} = -1.000006, x_3^{(3)} = 1.000002:$$

$$x_1^{(4)} = 1.000001, x_2^{(4)} = -1.000000, x_3^{(4)} = 1.000000;$$

and

$$x_1^{(5)} = 1.0, x_2^{(5)} = -1.0 \text{ and } x_3^{(5)} = 1.$$

Problem 4: Solve the following equations by Gauss–Seidel method.

$$10x_1 + x_2 + x_3 + x_4 = 13 \tag{1}$$

$$x_1 + 10x_2 + x_3 + x_3 = 13 \tag{2}$$

$$x_1 + x_2 + 10x_3 + x_4 = 13 \tag{3}$$

$$x_1 + x_2 + x_3 + 10x_4 = 13 \tag{4}$$

Solution: The $(i+1)^{th}$ approximations of x_i ($i = 1, 2, 3, 4$) are to be computed from the following equation (as per Gauss–Seidel method):

$$x_1^{(i+1)} = \frac{1}{10}[13 - x_2^{(i)} - x_3^{(i)} - x_4^{(i)}] \tag{5}$$

$$x_2^{(i+1)} = \frac{1}{10}[13 - x_1^{(i+1)} - x_3^{(i)} - x_4^{(i)}] \tag{6}$$

$$x_3^{(i+1)} = \frac{1}{10}[13 - x_1^{(i+1)} - x_2^{(i+1)} - x_4^{(i)}] \tag{7}$$

$$x_4^{(i+1)} = \frac{1}{10}[13 - x_1^{(i)} - x_2^{(i)} - x_3^{(i)}] \tag{8}$$

We take the initial approximations $x_2^{(0)} = x_3^{(0)} = x_4^{(0)} = 0$

To find first approximations of $x_i\,(i = 1, 2, 3, 4)$:

Putting $i = 0$ in (5), we have

$$x_1^{(1)} = \frac{1}{10}[13 - x_2^{(0)} - x_3^{(0)} - x_4^{(0)}]$$

Now, substituting $x_2^{(0)} = 0, x_3^{(0)} = 0$ and $x_4^{(0)} = 0$ in the above equation, we obtain

$$x_1^{(1)} = 1.3$$

Next, substitute $i = 0$ in Eq. (6), we have

$$x_2^{(1)} = \frac{1}{10}[13 - x_1^{(1)} - x_3^{(0)} - x_4^{(0)}]$$

$$x_2^{(1)} = \frac{1}{10}[13 - 1.3 - 0 - 0] = 1.17$$

Further, for $i = 0$, the Eq. (7) reduces to

$$x_3^{(1)} = \frac{1}{10}[13 - x_1^{(1)} - x_2^{(1)} - x_4^{(0)}]$$

i.e.,
$$x_3^{(1)} = \frac{1}{10}[13 - 1.3 - 1.17 - 0] = 1.053000.$$

Finally, putting $i = 0$ in (8), we have

$$x_4^{(1)} = \frac{1}{10}[10 - x_1^{(1)} - x_2^{(1)} - x_3^{(1)}]$$

$$= \frac{1}{10}[10 - 1.3 - 1.17 - 1.053] = 0.9477$$

To find second approximations of $x_i (i = 1, 2, 3, 4)$:

for $i = 1$, Eqs. (5) to (8) are reduce to

$$x_1^{(2)} = \frac{1}{10}[13 - x_2^{(1)} - x_3^{(1)} - x_4^{(1)}] \tag{9}$$

$$x_2^{(2)} = \frac{1}{10}[13 - x_1^{(2)} - x_3^{(1)} - x_4^{(1)}] \tag{10}$$

$$x_3^{(2)} = \frac{1}{10}[13 - x_1^{(2)} - x_2^{(2)} - x_4^{(1)}] \tag{11}$$

$$x_4^{(2)} = \frac{1}{10}[13 - x_1^{(2)} - x_2^{(2)} - x_3^{(2)}] \tag{12}$$

Substituting $x_2^{(1)}, x_3^{(1)}$ and $x_4^{(1)}$ values (9) we get

$$x_1^{(2)} = 1.3 - 0.117 - 0.1053 - 0.09477 = 0.982930$$

Next, substituting $x_1^{(2)}$, $x_3^{(1)}$ and $x_4^{(1)}$ values in Eq. (10) we get

$$x_2^{(2)} = 1.3 - 0.098293 - 0.105300 - 0.1053 = 1.001637$$

Putting the values of $x_1^{(2)}$, $x_2^{(2)}$ and $x_4^{(1)}$ in Eq. (11), we obtain

$$x_3^{(2)} = 1.3 - 0.098293 - 0.100164 - 0.100677 = 1.006773$$

Finally, substituting $x_1^{(1)}, x_2^{(1)}$ and $x_3^{(1)}$ values in Eq. (12), we obtain

$$x_4^{(2)} = 1.3 - 0.098293 - 0.100164 - 0.100677 = 1.000866$$

Continuing this process, we can computed the successive approximations of ($k = 1, 2, 3, 4$). These values are shown in the following table:

i	$x_1^{(i)}$	$x_2^{(i)}$	$x_3^{(i)}$	$x_4^{(i)}$
3	0.999072	0.999329	1.000073	1.000153
4	1.000045	0.999973	0.999983	1.0
5	1.000004	1.000001	0.999999	0.999999
6	1.0	1.0	1.0	1.0
7	1.0	1.0	1.0	1.0

Therefore, the solution of given system of equations is $x_1 = x_2 = x_3 = x_4 = 1$.

Problem 5: Solve the following equation by Gauss–Seidel method. (Perform 8 iterations):

$$x_1 - \frac{1}{4}x_2 - \frac{1}{4}x_3 = \frac{1}{2} \tag{1}$$

$$-\frac{1}{4}x_1 + x_2 - \frac{1}{4}x_4 = \frac{1}{2} \tag{2}$$

$$-\frac{1}{4}x_1 + x_3 - \frac{1}{4}x_4 = \frac{1}{4} \tag{3}$$

$$-\frac{1}{4}x_2 - \frac{1}{4}x_4 + x_4 = \frac{1}{4} \tag{4}$$

Solution: The $(i+1)^{\text{th}}$ approximations of $x_k\,(k=1,\,2,\,3,\,4)$ have to computed from the following:

$$x_1^{(i+1)} = 0.5 + 0.25x_2^{(i)} + 0.25x_3^{(i)} \tag{5}$$

$$x_2^{(i+1)} = 0.5 + 0.25x_1^{(i+1)} + 0.25x_4^{(i)} \tag{6}$$

$$x_3^{(i+1)} = 0.25 + 0.25x_1^{(i+1)} + 0.25x_4^{(i)} \tag{7}$$

$$x_4^{(i+1)} = 0.25 + 0.25x_2^{(i+1)} + 0.25x_3^{(i+1)} \tag{8}$$

we take the initial approximations $x_2^{(0)} = 0, x_3^{(0)} = 0$ and $x_4^{(0)} = 0$

To find $x_k^{(1)}\,(k=1,2,3,4)$:

For $i = 0$, Eqs. (5) to (8) reduce to

$$x_1^{(1)} = 0.5 + 0.25x_2^{(0)} + 0.25x_3^{(0)} \tag{9}$$

$$x_2^{(1)} = 0.5 + 0.25x_1^{(1)} + 0.25x_4^{(0)} \tag{10}$$

$$x_3^{(1)} = 0.25 + 0.25x_1^{(1)} + 0.25x_4^{(0)} \tag{11}$$

$$x_4^{(1)} = 0.25 + 0.25x_2^{(1)} + 0.25x_3^{(1)} \tag{12}$$

Now, substituting $x_2^{(0)} = 0$, $x_3^{(0)} = 0$ in Eq. (9) we get $x_1^{(1)} = 0.5$. Next, substituting $x_1^{(1)}$ and $x_4^{(0)}$ values in Eq. (10), we have $x_2^{(1)} = 0.625$. Further substituting $x_1^{(1)}$ and $x_4^{(0)}$ values in (11), we have $x_3^{(1)} = 0.375$. Finally, putting $x_2^{(1)}$ and $x_3^{(1)}$ values in (12), we obtain $x_4^{(1)} = 0.5$.

Repeating this process, we obtain $x_1^{(i)}$, $x_2^{(i)}$, $x_3^{(i)}$ and $x_4^{(i)}$, where $i = 2$, 3, 4, 5, 6, 7 and 8. These values are shown in the following table:

i	$x_1^{(i)}$	$x_2^{(i)}$	$x_3^{(i)}$	$x_4^{(i)}$
2	0.75	0.8125	0.5625	0.59375
3	0.843750	0.859375	0.609375	0.617188
4	0.867188	0.871094	0.621094	0.623047
5	0.873047	0.874023	0.624023	0.624512
6	0.874512	0.874756	0.624756	0.624878
7	0.874878	0.874939	0.624939	0.624969
8	0.874969	0.874985	0.624985	0.624992

Upto three decimal places, the solution of given system of equation is $x_1 = 0.874$, $x_2 = 0.874$, $x_3 = 0.624$ and $x_4 = 0.624$.

Problem 6: Solve the following equations by performing eight iterations of Gauss–Seidel method.

$$10x_1 - 2x_2 - x_3 - x_4 = 3 \tag{1}$$
$$-2x_1 + 10x_2 - x_3 - x_4 = 15 \tag{2}$$
$$-x_1 - x_2 - 10x_3 - 2x_4 = 27 \tag{3}$$
$$-x_1 - x_2 - 2x_3 + 10x_4 = -9 \tag{4}$$

Solution: As per Gauss–Seidel method, the $(i+1)^{\text{th}}$ approximations of x_1, x_2, x_3 and x_4 are as follows:

$$x_1^{(i+1)} = \frac{1}{10}[3 + 2x_2^{(i)} + x_3^{(i)} + x_4^{(i)}] \tag{5}$$

$$x_2^{(i+1)} = \frac{1}{10}[15 + 2x_1^{(i+1)} + x_3^{(i)} + x_4^{(i)}] \tag{6}$$

$$x_3^{(i+1)} = \frac{1}{10}[27 + x_1^{(i+1)} + x_2^{(i+1)} + 2x_4^{(i)}] \tag{7}$$

$$x_4^{(i+1)} = \frac{1}{10}[-9 + x_1^{(i+1)} + x_2^{(i+1)} + 2x_3^{(i+1)}] \tag{8}$$

We take initial approximations $x_2^{(0)} = x_3^{(0)} = x_4^{(0)} = 0$

To find $x_k^{(1)}(k = 1, 2, 3, 4)$:

For $i = 0$, Eqs. (5) to (8) reduce to

$$x_1^{(1)} = \frac{1}{10}[3 + 2x_2^{(0)} + x_3^{(0)} + x_4^{(0)}] \qquad (9)$$

$$x_2^{(1)} = \frac{1}{10}[15 + 2x_1^{(1)} + x_3^{(0)} + x_4^{(0)}] \qquad (10)$$

$$x_3^{(1)} = \frac{1}{10}[27 + x_1^{(1)} + x_2^{(1)} + 2x_4^{(0)}] \qquad (11)$$

$$x_4^{(1)} = \frac{1}{10}[-9 + x_1^{(1)} + x_2^{(1)} + 2x_3^{(1)}] \qquad (12)$$

Now substituting $x_2^{(0)} = x_3^{(0)} = x_4^{(0)} = 0$ in Eq. (9), we get $x_1^{(1)} = 0.3$. Next, substituting $x_1^{(1)}$, $x_3^{(0)}$ and $x_4^{(0)}$ values in Eq. (10), we get $x_2^{(1)} = 1.56$. Further, substituting $x_1^{(1)}$, $x_2^{(1)}$ and $x_4^{(0)}$ values in Eq. (11), we get $x_3^{(1)} = 2.886$. Finally, substituting $x_1^{(1)}, x_2^{(1)}$ and $x_3^{(1)}$ values in Eq. (12), we get

$$x_4^{(1)} = -0.1368.$$

Continuing this process, we obtain $x_1^{(i)}$, $x_2^{(i)}$, $x_3^{(i)}$ and $x_4^{(i)}$ for $i = 2, 3, 4, 5, 6, 7$ and 8. These values are shown in the following table:

i	$x_1^{(i)}$	$x_2^{(i)}$	$x_3^{(i)}$	$x_4^{(i)}$
2	0.886920	1.952304	2.956562	−0.024765
3	0.983641	1.989908	2.992402	−0.004165
4	0.990805	1.998185	2.998666	−0.000768
5	0.999427	1.999675	2.999757	−0.000138
6	0.999897	1.999941	2.999956	−0.000025
7	0.999981	1.999989	2.999992	−0.000005
8	0.999997	1.999998	2.999999	−0.000001

Thus, the solution of given system of equation is $x_1 = 1$, $x_2 = 2$, $x_3 = 3$ and $x_4 = 0$.

10.7.4 Relaxation method

We describe this method for three linear equations in three unknowns. Let

$$a_{11}x_1 + a_{12}x_2 + a_{13}x_3 = b_1 \tag{10.50}$$

$$a_{21}x_1 + a_{23}x_2 + a_{23}x_3 = b_2 \tag{10.51}$$

$$a_{31}x_1 + a_{33}x_3 + a_{33}x_3 = b_3 \tag{10.52}$$

be a system of three linear equations in three unknowns, where the coefficient matrix of this is diagonally dominant. The residuals R_1, R_2 and R_3 are defined by the following relations:

$$R_1 = b_1 - a_{11}x_1 - a_{12}x_1 - a_{13}x_3 \tag{10.53}$$

$$R_2 = b_2 - a_{21}x_1 - a_{22}x_2 - a_{23}x_3 \tag{10.54}$$

$$R_3 = b_3 - a_{31}x_1 - a_{33}x_3 - a_{33}x_3 \tag{10.55}$$

Note that for the values of x_1, x_2 and x_3, which are the solutions of (10.50) to (10.52), the values of R_1, R_2 and R_3 are zero. The basic idea of this method is to improve the approximations of x_1, x_2 and x_3 by making R_1, R_2 and R_3 to zero or near to zero through certain process. Now, we proceed to describe the process of obtaining the solution of the system of equations (10.50 to 10.52).

We introduce the notation $R_k^{(i)}$ denote the residuals of R_k at i^{th} iteration (k = 1, 2, 3) where i is a non-negative integer. Of course $R_k^{(0)}$ ($k = 1,2,3$) denote the residuals for the initial approximations of $x_k^{(0)}$ ($k = 1,2,3$). The ith approximations of R_1, R_2 and R_3 are given by

$$R_1^{(i)} = b_1 - a_{11}x_1^{(i)} - a_{12}x_1^{(i)} - a_{13}x_3^{(i)} \tag{10.56}$$

$$R_2^{(i)} = b_2 - a_{21}x_1^{(i)} - a_{22}x_2^{(i)} - a_{23}x_3^{(i)} \tag{10.57}$$

$$R_3^{(i)} = b_3 - a_{31}x_1^{(i)} - a_{33}x_2^{(i)} - a_{33}x_3^{(i)} \tag{10.58}$$

We take the initial approximations $x_1^{(0)} = x_2^{(0)} = x_3^{(0)} = 0$. Substituting these values in (10.56) to (10.58) with $i = 0$, we get $R_1^{(0)}$, $R_2^{(0)}$ and $R_3^{(0)}$. Next, we identify the numerically largest residual among $R_1^{(0)}$, $R_2^{(0)}$ and $R_3^{(0)}$. If

(i) $R_1^{(0)}$ is the numerically largest then x_1 is increased by $\dfrac{R_1^{(0)}}{a_{11}}$.

The increment in x_1 is denoted by $\delta x_1 = \dfrac{R_1^{(0)}}{a_{11}}$.

(ii) $R_2^{(0)}$ is numerically largest in magnitude; then x_2 is increased

by $\dfrac{R_2^{(0)}}{a_{22}}$ i.e., $\delta x_2 = \dfrac{R_2^{(0)}}{a_{22}}$.

(iii) $R_3^{(0)}$ is numerically largest in magnitude; then x_3 is increased

by $\dfrac{R_3^{(0)}}{a_{33}}$ i.e., $\delta x_3 = \dfrac{R_3^{(0)}}{a_{33}}$.

The first approximations of x_1, x_2 and x_3 are

(a) $x_1^{(1)} = x_1^{(0)} + \delta x_1$, $x_2^{(1)} = x_2^{(0)}$, $x_3^{(1)} = x_3^{(0)}$ if $R_1^{(0)}$ is the numerically largest in magnitude among $R_1^{(0)}$, $R_2^{(0)}$ and $R_3^{(0)}$.

(b) $x_1^{(1)} = x_1^{(0)}$, $x_2^{(1)} = x_2^{(0)} + \delta x_2$, $x_3^{(1)} = x_3^{(0)}$ if $R_2^{(0)}$ is the numerically largest in magnitude among $R_1^{(0)}$, $R_2^{(0)}$ and $R_3^{(0)}$.

(c) $x_1^{(1)} = x_1^{(0)}$, $x_2^{(1)} = x_2^{(0)}$, $x_3^{(1)} = x_3^{(0)} + \delta x_3$ if $R_3^{(0)}$ is the numerically largest in magnitude among $R_1^{(0)}$, $R_2^{(0)}$ and $R_3^{(0)}$.

With the values of $x_1^{(1)}$, $x_2^{(1)}$ and $x_3^{(1)}$, again we find $R_1^{(1)}$, $R_2^{(1)}$ and $R_3^{(1)}$ from (10.56) to (10.58) with $i = 1$ and repeat the above process until all the residuals are zero or very very small. The solution of the system of Eqs. (10.50) to (10.52) is $x_1 = $ Sum of the increments of x_1 of all iterations performed, $x_2 = $ sum of increments of x_2 of all the iterations performed and $x_3 = $ Sum of all the increments of x_3 of all the iterations performed. This process of finding the solutions of linear equations is called relaxation methods and it is due to R.V. Southwell.

SOLVED PROBLEMS

Problem 1: Solve the following system of equations by the relaxation method.

$$6x_1 - 3x_2 + x_3 = 11 \tag{1}$$
$$x_1 - 7x_2 + x_3 = 10 \tag{2}$$
$$2x_1 + x_2 - 8x_3 = -15 \tag{3}$$

Solution: The coefficient matrix of Eqs. (1) to (3) is diagonally dominant. Here, $a_{11} = 6$, $a_{22} = -7$ and $a_{33} = -8$. The residuals $R_1^{(i)}$, $R_2^{(i)}$ and $R_3^{(i)}$ are

$$R_1^{(i)} = 11 - 6x_1^{(i)} + 3x_2^{(i)} - x_3^{(i)} \tag{4}$$
$$R_2^{(i)} = 10 - x_1^{(i)} + 7x_2^{(i)} - x_3^{(i)} \tag{5}$$
$$R_3^{(i)} = -15 - 2x_1^{(i)} - x_2^{(i)} + 8x_3^{(i)} \tag{6}$$

where $i = 0, 1, 2, 3 \ldots\ldots$
Initial approximations of x_1, x_2, x_3:
We suppose the initial approximations of x_1, x_2 and x_3 as $x_1^{(0)} = 0$, $x_2^{(0)} = 0$ and $x_3^{(0)} = 0$

To find increment in unknown $x_1^{(1)}$, $x_2^{(1)}$ and $x_3^{(1)}$:
Putting $i = 0$ in (4) to (6), we have

$$R_1^{(0)} = 11 - 6x_1^{(0)} + 3x_2^{(0)} - x_3^{(0)}$$
$$R_2^{(0)} = 10 - x_1^{(0)} + 7x_2^{(0)} - x_3^{(0)}$$
$$R_3^{(0)} = -15 - 2x_1^{(0)} - x_2^{(0)} + 8x_3^{(0)}$$

Substituting $x_1^{(0)} = x_2^{(0)} = x_3^{(0)} = 0$ in the above equations, we get

$$R_1^{(0)} = 11; \ R_2^{(0)} = 10; \ R_3^{(0)} = -15$$

Among the $R_1^{(0)}, R_2^{(0)}$ and $R_3^{(0)}$ the numerically largest magnitude is $R_3^{(0)} = -15$. Therefore, the increment in x_3 is $\delta x_3 = \dfrac{-15}{-8} = 1.875$. Hence

$$x_1^{(1)} = x_1^{(0)}, x_2^{(1)} = x_2^{(0)}, x_3^{(1)} = x_3^{(0)} + \delta x_3$$

$$x_1^{(1)} = 0, x_2^{(1)} = 0, x_3^{(1)} = 1.875$$

To find increment in unknown $x_1^{(2)}, x_2^{(2)}$ and $x_3^{(2)}$:

for $i = 1$, Eqs. (4), (5) and (6) reduce to

$$R_1^{(1)} = 11 - 6x_1^{(1)} + 3x_2^{(1)} - x_3^{(1)}$$

$$R_2^{(1)} = 10 - x_1^{(1)} + 7x_2^{(1)} - x_3^{(1)}$$

$$R_3^{(1)} = -15 - 2x_1^{(1)} - x_2^{(1)} + 8x_3^{(1)}$$

Substituting $x_1^{(1)}, x_2^{(1)}, x_3^{(1)}$ values in the above equations, we get

$$R_1^{(1)} = 9.125; \ R_2^{(1)} = 8.125; \ R_3^{(1)} = 0$$

Clearly, $R_1^{(1)}$ is the numerically greatest one among $R_1^{(1)}, R_2^{(1)}$ and $R_3^{(1)}$.

Hence, the increment in x_1 is $\delta x_1 = \dfrac{9.125}{6} = 1.520833$. Thus,

$$x_1^{(2)} = x_1^{(1)} + \delta x_1 = 1.520833$$

$$x_2^{(2)} = x_2^{(1)} = 0$$

$$x_3^{(2)} = x_3^{(1)} = 1.875$$

To find increment in unknown $x_1^{(3)}, x_2^{(3)}$ and $x_3^{(3)}$:

for $i = 2$, Eqs. (4), (5) and (6) reduce to

$$R_1^{(2)} = 11 - 6x_1^{(2)} + 3x_2^{(2)} - x_3^{(2)}$$

$$R_2^{(2)} = 10 - x_1^{(2)} + 7x_2^{(2)} - x_3^{(2)}$$

$$R_3^{(2)} = -15 - 2x_1^{(2)} - x_2^{(2)} + 8x_3^{(2)}$$

Substituting $x_1^{(1)}, x_2^{(1)}, x_3^{(1)}$ values in the above equations, we get

$$R_1^{(2)} = 0; \ R_2^{(2)} = 6.604167; \ R_3^{(2)} = -3.041667.$$

Now $R_2^{(2)}$ is the numerically largest magnitude among $R_1^{(2)}, R_2^{(2)}$ and $R_3^{(2)}$.

$$\delta x_2 = \frac{6.604167}{-7} = -0.943452.$$

Thus,

$$x_1^{(3)} = 1.520833$$

$$x_2^{(3)} = -0.943452$$

$$x_3^{(3)} = 1.875$$

To find increment in unknown $x_1^{(4)}$, $x_2^{(4)}$ and $x_3^{(4)}$:

Since,

$$R_1^{(3)} = 11 - 6x_1^{(3)} + 3x_2^{(3)} - x_3^{(3)}$$

$$R_2^{(3)} = 10 - x_1^{(3)} + 7x_2^{(3)} - x_3^{(3)}$$

$$R_3^{(3)} = -15 - 2x_1^{(3)} - x_2^{(3)} + 8x_3^{(3)}$$

Substituting $x_1^{(3)}$, $x_2^{(3)}$ and $x_3^{(3)}$ values in the above equations, we get

$$R_1^{(3)} = -2.8303857; \; R_2^{(3)} = 0; \; R_3^{(2)} = -2.098214.$$

Numerically, largest magnitude among $R_1^{(3)}$, $R_2^{(3)}$ and $R_3^{(3)}$ is $R_1^{(3)}$. Hence

$$\delta x_1 = \frac{-2.830357}{6} = -0.471726.$$

Thus,

$$x_1^{(4)} = x_1^{(3)} + \delta x_1 = 1.049107$$

$$x_2^{(4)} = x_2^{(3)} = -0.943452$$

$$x_3^{(4)} = x_3^{(3)} = 1.875.$$

To find increment in unknown $x_1^{(5)}$, $x_2^{(5)}$ and $x_3^{(5)}$:

Substituting $i = 4$ in Eqs. (4) to (6), we get

$$R_1^{(4)} = 11 - 6x_1^{(4)} + 3x_2^{(4)} - x_3^{(4)}$$

$$R_2^{(4)} = 10 - x_1^{(4)} + 7x_2^{(4)} - x_3^{(4)}$$

$$R_3^{(4)} = -15 - 2x_1^{(4)} - x_2^{(4)} + 8x_3^{(4)}$$

Substituting $x_1^{(4)}$, $x_2^{(4)}$ and $x_3^{(4)}$ values in the above equations, we have

$$R_1^{(4)} = 0; \ R_2^{(4)} = 0.471726; \ R_3^{(4)} = -1.154762$$

Numerically, largest magnitude is $R_3^{(4)}$ among $R_1^{(4)}$, $R_2^{(4)}$ and $R_3^{(4)}$. Hence, the increment in x_3 is

$$\delta x_3 = \frac{-1.154762}{(-8)} = 0.144345.$$

Thus,

$$x_1^{(5)} = 1.049107, \ x_2^{(5)} = -0.943452, \ x_3^{(5)} = x_3^{(4)} + \delta x_3 = 2.019345.$$

To find increment in unknown $x_1^{(6)}, x_2^{(6)}$ and $x_3^{(6)}$:

Substituting $i = 5$ in Eqs. (4) to (6), we have

$$R_1^{(5)} = 11 - 6x_1^{(5)} + 3x_2^{(5)} - x_3^{(5)}$$

$$R_2^{(5)} = 10 - x_1^{(5)} + 7x_2^{(5)} - x_3^{(5)}$$

$$R_3^{(5)} = -15 - 2x_1^{(5)} - x_2^{(5)} + 8x_3^{(5)}$$

Now, Substituting the values $x_1^{(5)}$, $x_2^{(5)}$ and $x_3^{(5)}$ in the above equations, we get

$$R_1^{(5)} = -0.014345; \ R_2^{(5)} = 0.327381; \ R_3^{(5)} = 0$$

Numerically largest magnitude is among $R_1^{(5)}$, $R_2^{(5)}$ and $R_3^{(5)}$. Therefore, the increment to be made in the unknown in x_2 is

$$\delta x_2 = \frac{0.327381}{(-7)} = -0.046769.$$

Therefore,

$$x_1^{(6)} = x_1^{(5)} = 1.049107$$

$$x_2^{(6)} = x_2^{(5)} + \delta x_2 = -0.990221$$

$$x_3^{(6)} = x_3^{(5)} = 2.019345$$

To find increment in unknown $x_1^{(7)}, x_2^{(7)}$ and $x_3^{(7)}$:

Substituting $i = 6$ in Eqs. (4) to (6), we get

$$R_1^{(6)} = 11 - 6x_1^{(6)} + 3x_2^{(6)} - x_3^{(6)}$$
$$R_2^{(6)} = 10 - x_1^{(6)} + 7x_2^{(6)} - x_3^{(6)}$$
$$R_3^{(6)} = -15 - 2x_1^{(6)} - x_2^{(6)} + 8x_3^{(6)}.$$

Now, Substituting the values $x_1^{(6)}$, $x_2^{(6)}$ and $x_3^{(6)}$ in the above equations, we get

$$R_1^{(6)} = -0.284651; \ R_2^{(6)} = 0; \ R_3^{(5)} = 0.046769.$$

The Numerically largest magnitude is among $R_1^{(6)}$, $R_2^{(6)}$ and $R_3^{(6)}$ is $R_1^{(6)}$. Hence, the increment to be made in the unknown in x_1 is

$$x_1 = \frac{-0.0284651}{6} = -0.047442.$$

Therefore,

$$x_1^{(7)} = 1.001665, \ x_2^{(7)} = -0.990221, \ x_3^{(7)} = 2.019345$$

Continuing this process, it can be shown that

$$x_1^{(8)} = 1.001665, \ x_2^{(8)} = -0.990221, \ x_3^{(8)} = 2.001639$$

$$x_1^{(9)} = 1.001665, \ x_2^{(9)} = -0.999528 \text{ and } x_3^{(9)} = 2.001639. \text{ Note that}$$
$$\Sigma \, \delta x_1 = x_1^{(9)}, \ \Sigma \, \delta x_2 = x_2^{(9)} \text{ and } \Sigma \, \delta x_3 = x_3^{(9)}.$$

Thus, the approximate solution of given system of equation is $x_1 = 1.001665$, $x_2 = -0.999528$ and $x_3 = 2.001639$. The exact solution of the given system of equations is $x_1 = 1$, $x_2 = -1$ and $x_3 = 2$. Of course the approximations can be improved by further computations of iterations.

Problem 2: Solve $9x_1 - x_2 + 2x_3 = 9$, $x_1 + 10x_2 - 2x_3 = 15$ and

$2x_1 - 2x_2 - 13x_3 = 17$ by relaxation method.

Solution: The coefficient matrix of given system of equations is diagonally dominant.

Here $a_{11} = 9$, $a_{22} = 10$ and $a_{33} = -13$. Now, the residuals $R_1^{(i)}$, $R_2^{(i)}$ and $R_3^{(i)}$ are

$$R_1^{(i)} = 9 - 9x_1^{(i)} + x_2^{(i)} - 2x_3^{(i)} \tag{1}$$

$$R_2^{(i)} = 15 - x_1^{(i)} - 10x_2^{(i)} + 2x_3^{(i)} \tag{2}$$

$$R_3^{(i)} = 17 - 2x_1^{(i)} - 2x_2^{(i)} + 13x_3^{(i)} \tag{3}$$

where $i = 0, 1, 2, 3 \ldots\ldots$

Suppose $x_1^{(0)} = 0$, $x_2^{(0)} = 0$ and $x_3^{(0)} = 0$. Following the procedure described in the earlier problem, the iterates of the unknown x_1, x_2 and x_3 can be found. They are shown in the following table given overleaf.

The approximate solution of given system of equation is $x_1 = 0.917$, $x_2 = 1.647$ and $x_3 = 1.195$.

Iteration number (i)	$x_1^{(i)}$	$x_2^{(i)}$	$x_3^{(i)}$	$R_1^{(i)}$	$R_2^{(i)}$	$R_3^{(i)}$	Numericals largest $R_1^{(i)}$/$R_2^{(i)}$/$R_3^{(i)}$ and increment to be made in the unknown	Amount of increment δx_1/δx_2/δx_3
0	0	0	0	9	15	−17	$R_3^{(0)}, x_3$	$\delta x_3 = 1.307692$
1	0	0	1.307692	6.384615	17.615385	0	$R_2^{(1)}, x_2$	$\delta x_2 = 1.761538$
2	0	1.761538	1.37692	8.146154	0	3.523077	$R_1^{(2)}, x_1$	$\delta x_1 = 0.905128$
3	0.905128	1.761538	1.37692	0	−0.905128	1.172821	$R_3^{(3)}, x_3$	$\delta x_3 = -0.131755$
4	0.905128	1.761538	1.175937	0.263511	−1.168639	0	$R_2^{(4)}, x_2$	$\delta x_2 = -0.116864$
5	0.905128	1.644675	1.175937	0.146647	0	−0.233728	$R_3^{(5)}, x_3$	$\delta x_3 = 0.017979$
6	0.905128	1.644675	1.193916	0.110689	0.035958	0	$R_1^{(6)}, x_1$	$\delta x_1 = 0.012299$
7	0.917427	1.644675	1.193916	0	0.023659	−0.024598	$R_3^{(7)}, x_3$	$\delta x_3 = 0.001892$
8	0.917427	1.644675	1.193916	−0.003784	0.027444	0	$R_2^{(8)}, x_2$	$\delta x_2 = 0.002744$
9	0.917427	1.647419	1.195805	−0.001040	0	0.05489	$R_3^{(9)}, x_3$	$\delta x_3 = -0.000422$

EXERCISE 10

1. Solve the following upper triangular system of equations:
$$2x_1 + x_2 + 3x_3 = 9$$
$$3x_2 + 5x_3 = 13$$
$$6x_3 = 12$$

2. Solve the following upper triangular system of equations:
$$4x_1 + 2x_2 + x_3 + x_4 = 16$$
$$5x_2 + 3x_3 + 4x_4 = 18$$
$$4x_3 + x_4 = 16$$
$$2x_4 = 3$$

3. Solve the following lower triangular system of equations:
$$2x_3 = 7$$
$$5x_2 + 4x_3 = 10$$
$$4x_1 + 7x_2 + x_3 = 15$$

4. Solve the following lower triangular system of equations:
$$8x_4 = 12$$
$$4x_3 - x_4 = -14$$
$$2x_2 + x_3 + x_4 = 18$$
$$2x_1 + 4x_2 + 3x_3 + 5x_4 = 14$$

5. Solve $2x_1 + 5x_2 = 12$ and $-4x_1 + 7x_2 = 10$ by Cramer's rule.

6. Solve $3x_1 + x_2 + 2x_3 = 3$, $2x_1 - 3x_2 - x_3 = -3$ and $x_1 + 2x_2 + x_3 = 4$ by Cramer's rule.

7. Solve $3x_1 + 2x_2 + 4x_3 = 7$, $2x_1 + x_2 + x_3 = 7$ and $x_1 + 3x_2 + 5x_3 = 2$ by Cramer's rule.

8. Solve the following system of equations by matrix inversion method:

   (i) $3x_1 + 4x_2 + 5x_3 = 18$
   $$2x_1 - x_2 + 8x_3 = 13$$
   $$5x_1 - 2x_2 + 7x_3 = 20.$$

   (ii) $x_1 + x_2 + x_3 = 1$
   $$2x_1 + 2x_2 + 3x_3 = 6$$
   $$x_1 + 4x_2 + 9x_3 = 3.$$

(iii) $2x_1 + x_2 + x_3 = 10$

$3x_1 + 2x_2 + 3x_3 = 18$

$x_1 + 4x_2 + 9x_3 = 16.$

9. Solve the following system of equations by Gauss elimination method:

(i) $20x_1 + x_2 - 2x_3 = 17,\ \ 3x_1 + 20x_2 - x_3 = -18,$

$2x_1 - 3x_2 + 20x_3 = 25$

(ii) $x_1 + x_2 + 2x_3 = 9$, $x_1 + 3x_2 + 2x_3 = 13$, $3x_1 + x_2 + 3x_3 = 14$

(iii) $10x_1 - 2x_2 + 3x_3 = 205,\ -2x_1 + 10x_2 - 2x_3 = 154,$

$-2x_1 - x_2 + 10x_3 = 120$

(iv) $5x_1 + 2x_2 + x_3 = 12$, $x_1 + 4x_2 + 2x_3 = 15$, $x_1 + 2x_2 + 5x_3 = 20.$

10. Implementing the partial pivoting solve the following equations:

(i) $x_1 + 8x_2 - x_3 = 3$

$2x_1 + 3x_2 + 5x_3 = -1$

$13x_1 + 2x_2 - 3x_3 = 29.$

(ii) $12x_1 - 3x_2 + x_3 = -2$

$x_1 + 4x_2 + 5x_3 = 26$

$3x_1 - 14x_2 + 5x_3 = 5.$

11. Solve the following equations by Gauss–Jordan method:

(i) $2x_1 - x_2 + 3x_3 = 9$, $x_1 + x_2 + x_3 = 6$, $x_1 - x_2 + x_3 = 2.$

(ii) $3x_1 + 2x_2 + 6x_3 = 4$, $2x_1 - 4x_2 + 2x_3 = -1$, $7x_1 + 6x_2 + 2x_3 = 6.$

(iii) $x_1 - x_2 + x_3 + x_4 = 3$

$x_1 + x_2 - x_3 + x_4 = -3$

$x_1 + x_2 + x_3 - x_4 = 5$

$x_1 + x_2 + x_3 + x_4 = 1.$

12. Find the inverse of $A = \begin{bmatrix} 3 & 2 & 4 \\ 2 & 1 & 1 \\ 1 & 3 & 5 \end{bmatrix}$ by escalator method and solve

$3x_1 + 2x_2 + 4x_3 = 152$

$2x_1 + x_2 + x_3 = 56$

$x_1 + 3x_2 + 5x_3 = 176.$

13. Find the inverse of $A = \begin{bmatrix} 1 & 1 & -1 \\ 4 & 3 & -1 \\ 3 & 5 & 3 \end{bmatrix}$ by escalator method and solve

$$x_1 + x_2 + x_3 = 4$$
$$4x_1 + 3x_2 - x_3 = 4.5$$
$$3x_1 + 5x_2 + 3x_3 = 15.$$

14. Solve the following equations by Cholesky method:

  (i) $x_1 + 2x_2 + 6x_3 = 8$

   $$2x_1 + 5x_2 + 15x_3 = 19$$

   $$6x_1 + 15x_2 + 46x_3 = 58.$$

  (ii) $x_1 + 2x_2 + 3x_3 = 2$

   $$2x_1 + 8x_2 + 22x_3 = -16$$

   $$3x_1 + 22x_2 + 82x_3 = -92.$$

15. Solve the following equations by Crout's method:

  (i) $2x_1 + 4x_2 + x_3 = 5$

   $$4x_1 + 4x_2 + 3x_3 = 6$$
   $$4x_1 + 8x_2 + x_3 = 9.$$

  (ii) $x_1 + x_2 - 2x_3 = 3$

   $$4x_1 - 2x_2 + x_3 = 5$$

   $$3x_1 - x_2 + 3x_3 = 8.$$

16. Solve the following equations by Jacobi's method:

  $9x_1 + 4x_2 + x_3 = -17.$
  $x_1 + 6x_2 = 4$
  $x_1 - 2x_2 - 6x_3 = 14.$

17. Perform three iterations of Jacobi's method for the following equations:

  $6x_1 - 2x_2 + x_3 = 11$

  $-2x_1 + 7x_2 + 2x_3 = 5$

  $x_1 + 2x_2 - 5x_3 = -1.$

18. Solve the following equations by Jacobi's method:

  $12x_1 + x_2 + 2x_3 + x_4 = 26$
  $x_1 - 10x_2 + x_3 + 2x_4 = -4$

$$4x_1 + 3x_2 + 13x_3 - 3x_4 = -11$$

$$2x_1 + 2x_2 - x_3 + 11x_4 = 40.$$

19. Perform five iterations of Gauss-Seidel method to find the approximate solution of the following equations:

$$10x_1 + 2x_2 + x_3 = 9$$

$$2x_1 + 20x_2 - 2x_3 = -44$$

$$-2x_1 + 3x_2 + 10x_3 = 22.$$

20. Solve the following equations by Gauss–Seidel method:

(i) $8x_1 - x_2 + x_3 = 18$, $2x_1 + 5x_2 - 2x_3 = 3$, $x_1 + x_2 - 3x_3 = -6$.

(ii) $10x_1 - 2x_2 - x_3 - x_4 = 3$, $-2x_1 + 10x_2 - x_3 - x_4 = 15$,

$$-x_1 - x_2 + 10x_3 - 2x_4 = 27, \ -x_1 - x_2 - 2x_3 + 10x_4 = -9.$$

(iii) $2x_1 + 2x_2 + x_3 + 2x_4 = 7$, $-x_1 + 2x_2 + x_4 = -2$,

$$-3x_1 + x_2 + 2x_3 + x_4 = -3, \ -x_1 + 2x_4 = 0.$$

21. Solve the following equations by the relaxation method:

(i) $10x_1 + 2x_2 + x_3 = 9$

$$x_1 + 10x_2 - x_3 = 22$$

$$-2x_1 + 3x_2 + 10x_3 = 22.$$

(ii) $10x_1 - 2x_2 + x_3 = 14$

$$x_1 + 9x_2 - x_3 = 14$$

$$2x_1 - x_2 + 11x_3 = 16.$$

ANSWERS

1. $x_1 = 1, x_2 = 1, x_3 = 2$

2. $x_1 = 2.6063, x_2 = 0.2250, x_3 = 3.6250, x_4 = 1.5$

3. $x_1 = 4.275, x_2 = -0.8, x_3 = 3.5$

4. $-11.6875, 9.8125, -3.125, 1.5$

5. $x_1 = 1, x_2 = 2$ 6. $x_1 = 1, x_2 = 2, x_3 = -1$

7. $\dfrac{9}{4}, \dfrac{9}{8}, \dfrac{5}{8}$ 8. (i) $3, 1, 1$ (ii) $7, -10, 4$ (iii) $7, -9, 5$

9. (i) $1, -1, 1$ (ii) $1, 2, 3$ (iii) $32, 26, 21$ (iv) $1, 2, 3$

10. (i) $2, 0, 3$ (ii) $0.2558, 2.6889, 2.9977$

11. (i) $1, 2, 3$ (ii) $0.28, 0.56, 0.34$ (iii) $2, -1, 2, -2$

12. $\begin{pmatrix} 0.25 & 0.25 & -0.25 \\ -1.125 & 1.375 & 0.625 \\ 0.625 & -0.875 & -0.125 \end{pmatrix}$

13. $\begin{pmatrix} 1.4 & 0.2 & -0.4 \\ -1.5 & 0 & 0.5 \\ 1.1 & -0.2 & -0.1 \end{pmatrix}$

14. (i) $2, 0, 1$ (ii) $2, 3, 3$ 15.   (i) $\frac{1}{2}, \frac{3}{4}, 1$ (ii) $\dfrac{22}{19}, 3, \dfrac{11}{9}$

16. $-2, 1, 3$ 17. $2, 1, 1$

18. $2, 1, -1, 3$ 19. $2, 1, 3$

20. (i) $2, 1, 3$ (ii) $1, 2, 3, 0$ (iii) $2.1622, -0.4595, 1.4324, 1.0811$

21. (i) $0.2787, 2.3279, 1.5574$ (ii) $1.5745, 1.5258, 1.3070$.

11

Eigen Values and Eigen Vectors

Computation of eigen values and eigen vectors of a matrix is necessary in some engineering problems and problems in applied mathematics. For example, in solid mechanics computation of principal strains and principal stresses, there is a need to find the eigen values of a matrix formed with corresponding problem. In the study of mechanical vibrations also, they are necessary. In this chapter, we discuss some important methods of finding the eigen values and eigen vectors of a matrix.

11.1 DEFINITIONS

Let $A = [a_{ij}]$ be a square matrix of order n. Suppose there exists a scalar λ and a non-zero vector $x = \begin{bmatrix} x_1 \\ x_2 \\ \vdots \\ x_n \end{bmatrix}$ such that

$$AX = \lambda X \tag{11.1}$$

Then, λ is called an eigen value and X is called an eigen vector corresponding to the eigen value λ.

For example, if $A = \begin{bmatrix} 5 & 4 \\ 1 & 2 \end{bmatrix}$, we have $A\begin{bmatrix} 4 \\ 1 \end{bmatrix} = 6\begin{bmatrix} 4 \\ 1 \end{bmatrix}$. Therefore, 6 is an eigen value and $\begin{bmatrix} 4 \\ 1 \end{bmatrix}$ is the eigen vector corresponding to the eigen value $\lambda = 6$.

11.2   NOTATION

Here after, we denote the column matrix (vector) $X = \begin{bmatrix} x_1 \\ x_2 \\ \vdots \\ x_n \end{bmatrix}$ as

$X = \begin{bmatrix} x_1 & x_2 & \text{........} & x_n \end{bmatrix}^T$ for convenience.

11.3   CHARACTERISTIC EQUATION OF A SQUARE MATRIX *A*

The matrix Eq. (11.1) represents the following system of equations:

$$(a_{11} - \lambda)x_1 + a_{12}\, x_2 + a_{13}\, x_3 + \text{........} + a_{1n}x_n = 0$$

$$a_{21}x_1 + (a_{22} - \lambda)x_2 + a_{23}x_3 + \text{..........} + a_{2n}x_n = 0 \qquad (11.2)$$

$$\text{...}$$

$$a_{n1}x_1 + a_{n2}x_2 + a_{n3}x_3 + \text{...............} + (a_{nn} - \lambda)x_n = 0$$

A non-trivial solution of (11.2) exists if and only if

$$\begin{vmatrix} a_{11} - \lambda & a_{12} & a_{13} & \text{.........} & a_{1x} \\ a_{21} & a_{22} - \lambda & a_{23} & \text{.........} & a_{2x} \\ \text{.........} & \text{.........} & \text{.........} & \text{.........} & \text{.........} \\ a_{n1} & a_{n2} & a_{n3} & \text{.........} & a_{nn} - \lambda \end{vmatrix} = 0 \qquad (11.3)$$

Eq. (11.3) is called characteristic equation of the matrix A and it is a polynomial of degree n in λ. Hence, it has n roots.

If λ_k is a root of characteristic equation of the matrix A, then it is an eigen value of the matrix A. A vector X is said to be eigen vector corresponding to the eigen value λ_k if there exists a non-zero vector X such that $AX = \lambda_k X$. Note that the eigen values of a matrix may be distinct or some of them may be equal. All the eigen values of a matrix A are distinct if the roots of its characteristic equation are distinct and some of the eigen values of A are equal if its characteristic equation has repeated roots. To each eigen value, we have an eigen vector corresponding to it. The computation of eigen vector corresponding to repeated roots are not discussed in this chapter.

11.4 NORMALIZED VECTOR

A vector $X = \begin{bmatrix} x_1 & x_2 & x_3 & \ldots\ldots & x_n \end{bmatrix}^T$ is said to be normalized if $x_1^2 + x_2^2 + \ldots\ldots + x_n^2 = 1$.

For example, the vector $x = \begin{bmatrix} \dfrac{1}{\sqrt{6}} & \dfrac{2}{\sqrt{6}} & \dfrac{1}{\sqrt{6}} \end{bmatrix}^T$ is a normalized vector of three dimensions.

11.5 SOME RESULTS PERTAINING TO EIGEN VALUES

(i) The square matrix A and A^T have same eigen values.

(ii) If $\lambda \neq 0$ is an eigen value of a non-singular matrix A, then $\dfrac{1}{\lambda}$ is an eigen value of A^{-1}.

(iii) The sum of the eigen values of a square matrix A is equal to its trace.

(iv) The eigen values of a tri-diagonal square matrix A are the diagonal element of the matrix A.

A square matrix is said to be tri-diagonal matrix or triple-diagonal matrix if non-zero elements appear only on the principal diagonal and two of its adjacent diagonals. For example, $\begin{bmatrix} 1 & 2 & 0 \\ 2 & 1 & 10 \\ 0 & 6 & 3 \end{bmatrix}$ is a tri-diagonal matrix of 3×3.

(v) If $\lambda_1, \lambda_2, \ldots\ldots, \lambda_n$ are eigen values of a square matrix A, then $\lambda_1^k, \lambda_2^k, \ldots\ldots, \lambda_n^k$ are the eigen values of A^k.

(vi) The product of eigen values of a square matrix is equal to the determinant of the matrix A.

(vii) If λ is an eigen value of an orthogonal matrix A, then $\dfrac{1}{\lambda}$ is also an eigen value of A provided $\lambda \neq 0$.

(viii) Eigen vectors corresponding to the distinct eigen values of a square matrix are linearly independent vectors.

(ix) If a matrix B is such that $B = S^{-1} A S$ where S is a non-singular matrix, then the eigen values of A and B are same, i.e., Any similar transformation applied on a square matrix A leaves its eigen values unchanged.

(x) If a square matrix A of order n has n distinct eigen values, then there exists a similar transformation $P^{-1} A P = D$, where D is a diagonal matrix whose diagonal elements are the eigen values of the matrix A, of course; here P is a non-singular matrix.

(xi) If X is an eigen vector corresponding to eigen value λ of a square matrix A, then $k X\,(k \neq 0)$ is also an eigen vector corresponding to the eigen value λ of A.

SOLVED PROBLEMS

Problem 1: Find the eigen values of $A = \begin{bmatrix} 1 & 3 \\ 2 & 6 \end{bmatrix}$.

Solution: The characteristic equation of the given matrix A is

$$\begin{vmatrix} 1-\lambda & 3 \\ 2 & 6-\lambda \end{vmatrix} = 0$$

i.e., $\qquad\qquad (1 - \lambda)(6 - \lambda) - 6 = 0$

i.e., $\qquad\qquad 6 - 7\lambda + \lambda^2 - 6 = 0$

i.e., $\qquad\qquad \lambda^2 - 7\lambda = 0$

i.e., $\qquad\qquad \lambda(\lambda - 7) = 0$

$\therefore$ The eigen values of given matrix are 0 and 7.

Problem 2: Find the eigen values and eigen vectors of $\begin{bmatrix} 5 & 4 \\ 1 & 2 \end{bmatrix}$.

Solution: Let $A = \begin{bmatrix} 5 & 4 \\ 1 & 2 \end{bmatrix}$ Then its characteristic equation is

$$\begin{vmatrix} 5-\lambda & 4 \\ 1 & 2-\lambda \end{vmatrix} = 0$$

i.e., $$(\lambda - 1)(\lambda - 6) = 0$$

Hence, the eigen values of given matrix are 1 and 6.

To find the eigen vector corresponding to the eigen value 1:

Let $X = \begin{bmatrix} x \\ y \end{bmatrix}$ be the eigen vector corresponding to the eigen value $\lambda = 1$. Then

$$AX = 1.X$$

i.e.,
$$\begin{bmatrix} 5 & 4 \\ 1 & 2 \end{bmatrix}\begin{bmatrix} x \\ y \end{bmatrix} = 1.\begin{bmatrix} x \\ y \end{bmatrix}$$

$$\begin{bmatrix} 5x & + & 4y \\ x & + & 2y \end{bmatrix} = \begin{bmatrix} x \\ y \end{bmatrix}$$

$$\begin{bmatrix} 5x & + & 4y \\ x & + & 2y \end{bmatrix} - \begin{bmatrix} x \\ y \end{bmatrix} = 0$$

$$\begin{bmatrix} 4x & + & 4y \\ x & + & y \end{bmatrix} = 0$$

i.e., $$4x + 4y = 0 \tag{1}$$

and $$x + y = 0 \tag{2}$$

Clearly, the solution of (1) and (2) is $x = k$ and $y = -k$. Hence, the eigen vector corresponding to the eigen value is $[k, -k]^T$ or $[1, -1]^T$ (by 11.5, *xi*).

To find the eigen vector corresponding to the eigen value $\lambda = 6$:

Again suppose $X = [x\ y]^T$ be an eigen value corresponding to the eigen value $\lambda = 6$. Then

$$AX = 6X$$

i.e.,
$$\begin{bmatrix} 5 & 4 \\ 1 & 2 \end{bmatrix}\begin{bmatrix} x \\ y \end{bmatrix} = 6\begin{bmatrix} x \\ y \end{bmatrix}$$

i.e.,
$$\begin{bmatrix} 5x & + & 4y \\ x & + & 2y \end{bmatrix} - \begin{bmatrix} 6x \\ 6y \end{bmatrix} = 0$$

i.e.,
$$\begin{bmatrix} -x & + & 4y \\ x & - & 4y \end{bmatrix} = \begin{bmatrix} 0 \\ 0 \end{bmatrix}$$

Hence, we have

$$-x + 4y = 0 \tag{3}$$

$$x - 4y = 0 \tag{4}$$

Clearly, the solution of (3) and (4) is $x = 4k$ and $y = k$, Hence, the eigen vectors corresponding to $\lambda = 6$ are

$$X = \begin{bmatrix} 4k & k \end{bmatrix}^T \text{ or } X = \begin{bmatrix} 4 & 1 \end{bmatrix}^T.$$

Problem 3: Find the eigen values and eigen vector of the matrix
$$A = \begin{bmatrix} 1 & 1 & 3 \\ 1 & 5 & 1 \\ 3 & 1 & 1 \end{bmatrix}.$$

Solution: The characteristic equation of A is

$$\begin{vmatrix} 1-\lambda & 1 & 3 \\ 1 & 5-\lambda & 1 \\ 3 & 1 & 1-\lambda \end{vmatrix} = 0,$$

i.e.,
$$\lambda^3 - 7\lambda^2 + 36 = 0 \tag{1}$$

It is a cubic equation. Hence, it will have one real root. We can use trick-and-trial method to find this root. Of course, when this method fails, one can use Bisection method or Newton–Raphson method to find an approximate root or Cordon's method to find the exact root. Let

$$f(\lambda) = \lambda^3 - 7\lambda^2 + 36 = 0 \tag{2}$$

Then,
$$f(1) = 30 \neq 0, \ f(-1) = 28 \neq 0, \ f(-2) = 0.$$

Therefore, $\lambda = -2$ is a root of (1). Now, Eq. (1) can be expressed as

$$(\lambda + 2)(\lambda^2 - 9\lambda + 18) = 0,$$

i.e.,
$$(\lambda + 2)(\lambda - 3)(\lambda - 6) = 0 \tag{3}$$

Therefore, the roots of Eq. (1) are -2, 3 and 6. Hence, the eigen values of A are -2, 3 and 6.

To find the eigen vector corresponding to $\lambda = -2$:

Let $X = \begin{bmatrix} x & y & z \end{bmatrix}^T$ be the eigen vector corresponding to $\lambda = -2$. Then

$$AX = -2X$$

i.e.,
$$\begin{bmatrix} 1 & 1 & 3 \\ 1 & 5 & 1 \\ 3 & 1 & 1 \end{bmatrix}\begin{bmatrix} x \\ y \\ z \end{bmatrix} = -2\begin{bmatrix} x \\ y \\ z \end{bmatrix}$$

This matrix equation gives the following system of equations:

$$3x + y + 3z = 0 \tag{4}$$
$$x + 7y + z = 0 \tag{5}$$
$$3x + y + 3z = 0 \tag{6}$$

Applying the cross-multiplication rule for the Eqs. (4) and (5), we have

$$\frac{x}{1-21} = \frac{y}{3-3} = \frac{z}{21-1} \quad \left(\frac{1}{7} \diagdown \frac{3}{1} \diagdown \frac{3}{1} \diagdown \frac{1}{7} \right),$$

i.e.,
$$\frac{x}{-20} = \frac{y}{0} = \frac{z}{20},$$

i.e.,
$$\frac{x}{-1} = \frac{y}{0} = \frac{z}{1}$$

Let $\dfrac{x}{-1} = \dfrac{y}{0} = \dfrac{z}{1} = k$. Then $x = -k$, $y = 0$ and $z = k$. These values satisfy the Eq. (6). Thus, the non-trivial solution of system of Eqs. (4), (5) and (6) is $x = -k$, $y = 0$ and $z = k$; hence, the eigen vector corresponding to $\lambda = -2$ is $[-k \;\; 0 \;\; k]^T$ or $[-1 \;\; 0 \;\; 1]^T$ and its normalized vector is $\begin{bmatrix} \dfrac{-1}{\sqrt{2}} & 0 & \dfrac{1}{\sqrt{2}} \end{bmatrix}^T$.

The eigen vector corresponding to the eigen value $\lambda = 3$.

In this case, we have $AX = 3X$

i.e.,
$$\begin{bmatrix} 1 & 1 & 3 \\ 1 & 5 & 1 \\ 3 & 1 & 1 \end{bmatrix}\begin{bmatrix} x \\ y \\ z \end{bmatrix} = 3\begin{bmatrix} x \\ y \\ z \end{bmatrix}$$

The above matrix equation given the following system of equations

$$-2x + y + 3z = 0 \tag{7}$$
$$x + 2y + z = 0 \tag{8}$$
$$3x + y - 2z = 0 \tag{9}$$

From Eqs. (7) and (8), using the cross-multiplication rule, we have

$$\frac{x}{1-6} = \frac{y}{3+2} = \frac{z}{-4-1},$$

i.e.,

$$\frac{x}{-5} = \frac{y}{5} = \frac{z}{-5},$$

i.e.,

$$\frac{x}{1} = \frac{y}{-1} = \frac{z}{1}$$

Let $\dfrac{x}{1} = \dfrac{y}{-1} = \dfrac{z}{1} = s$. Then $x = s$, $y = -s$ and $z = s$. These values satisfy Eq. (9). The non-trivial solution of (7), (8) and (9) is $x = s$, $y = -s$ and $z = s$ $(s \neq 0)$. Thus, the eigen vector corresponding to the eigen value $\lambda = 3$ is $[s \ -s \ s]^T$ or $[1 \ -1 \ 1]^T$. The normalized eigen vector of $[1 \ -1 \ 1]^T$ is $\left[\dfrac{1}{\sqrt{3}} \ \dfrac{-1}{\sqrt{3}} \ \dfrac{1}{\sqrt{3}}\right]^T$.

Eigen vector corresponding to the eigen value $\lambda = 6$.

Since $\lambda = 6$ is an eigen value of A, we have $AX = 6X$ where $X = [x \ y \ z]^T$.

Therefore, $AX = 6X$ gives the following system of equations:

$$-5x + y + 3z = 0 \tag{10}$$
$$x - y + z = 0 \tag{11}$$
$$3x + y - 5z = 0 \tag{12}$$

From Eqs. (10) and (11), using the cross-multiplication rule

$$\frac{x}{4} = \frac{y}{8} = \frac{z}{4}, \quad \text{i.e.,} \quad \frac{x}{1} = \frac{y}{2} = \frac{z}{1}$$

Let each ratio of $\dfrac{x}{1}, \dfrac{y}{2}$ and $\dfrac{z}{1}$ be t.

Then $x = t$, $y = 2t$ and $z = t$. These values namely $x = t$, $y = 2t$ and $z = t$. satisfies Eq. (12). Thus, the non-trivial solution of system of Eqs. (10), (11) and (12) is $x = t$, $y = 2t$ and $z = t(t \neq 0)$. Therefore, the eigen vector corresponding to the eigen value $\lambda = 6$ of the matrix A is $[t\ 2t\ t]^T$ or $[1\ 2\ 1]^T$. This vector in normalized form is $\left[\dfrac{1}{\sqrt{6}}\ \dfrac{2}{\sqrt{6}}\ \dfrac{1}{\sqrt{6}}\right]^T$.

Problem 4: Find the eigen values and eigen vector of the matrix

$$A = \begin{bmatrix} 1 & 2 & 3 \\ 0 & -4 & 2 \\ 0 & 0 & 7 \end{bmatrix}.$$

Solution: The characteristic equation of the matrix A is

$$|A - \lambda I| = 0 \tag{1}$$

i.e.,

$$\begin{vmatrix} 1-\lambda & 2 & 3 \\ 0 & -4-\lambda & 2 \\ 0 & 0 & 7-\lambda \end{vmatrix} = 0$$

i.e.,

$$(7-\lambda)\begin{vmatrix} 1-\lambda & 2 \\ 0 & -4-\lambda \end{vmatrix} = 0$$

i.e.,

$$(7-\lambda)[(1-\lambda)(-4-\lambda) - 0] = 0$$

Hence, the roots of characteristic equation are 1, −4 and 7. Thus, the eigen values of given matrix A are 1, −4 and 7.

To find the eigen vector corresponding to the eigen value $\lambda = 1$.

Let $X = \begin{bmatrix} x \\ y \\ z \end{bmatrix}$ be the eigen vector corresponding to $\lambda = 1$. Then

$$AX = 1.\,X$$

i.e.,

$$\begin{bmatrix} 1 & 2 & 3 \\ 0 & -4 & 2 \\ 0 & 0 & 7 \end{bmatrix}\begin{bmatrix} x \\ y \\ z \end{bmatrix} = \begin{bmatrix} x \\ y \\ z \end{bmatrix} \tag{1}$$

i.e.,

$$\begin{bmatrix} x & + & 2y & + & 3z \\ & - & 4y & + & 2z \\ & & & & 7z \end{bmatrix} = \begin{bmatrix} x \\ y \\ z \end{bmatrix}$$

i.e.,
$$\begin{bmatrix} 2y & + & 3z \\ - & 5y & + & 2z \\ & & 6z \end{bmatrix} = \begin{bmatrix} 0 \\ 0 \\ 0 \end{bmatrix}$$

Hence, $2y + 3z = 0$, $-5y + 2z = 0$ and $6z = 0$. Clearly, the solution of these equation is $x = k$, $y = 0$ and $z = 0$. Hence, the eigen vector corresponding to the eigen value $\lambda = 1$ is

$$\begin{bmatrix} k \\ 0 \\ 0 \end{bmatrix} \text{ or } \begin{bmatrix} 1 \\ 0 \\ 0 \end{bmatrix}.$$

To find the eigen vector corresponding to the eigen value $\lambda = -4$:

Let $X = \begin{bmatrix} x \\ y \\ z \end{bmatrix}$ be the eigen vector corresponding to the eigen value

$\lambda = -4$. Then

$$AX = -4\,X$$

i.e.,
$$\begin{bmatrix} 1 & 2 & 3 \\ 0 & -4 & 2 \\ 0 & 0 & 7 \end{bmatrix}\begin{bmatrix} x \\ y \\ z \end{bmatrix} = \begin{bmatrix} -4x \\ -4y \\ -4z \end{bmatrix}$$

i.e.,
$$\begin{bmatrix} 5x & + & 2y & + & 3z \\ & & & & 2z \\ & & & & 11z \end{bmatrix} = \begin{bmatrix} 0 \\ 0 \\ 0 \end{bmatrix}$$

Hence, we have

$$5x + 2y + 3z = 0$$
$$2z = 0$$
$$11z = 0$$

Clearly, $z = 0$. The equation $5x + 2y + 3z = 0$ reduces to $5x + 2y = 0$. Suppose $y = k$, then $x = \dfrac{-2k}{5}$.

Thus, the eigen vector corresponding to $\lambda = -4$ is $\left[\dfrac{-2k}{5}, 0, 0\right]^T$ or $[-2, 0, 0]^T$.

To find the eigen vector corresponding to the eigen value $\lambda = 7$:

In this case, we have $AX = 7X$ i.e.,

$$\begin{bmatrix} -6x & + & 2y & + & 3z \\ & & - & 11y & + & 2z \\ & & & & 0 \end{bmatrix} = \begin{bmatrix} 0 \\ 0 \\ 0 \end{bmatrix}.$$

Thus, $-6x + 2y + 3z = 0$, $-11y + 2z = 0$. Let $z = k$ (k is arbitrary). Then $y = \dfrac{2k}{11}$ and $x = \dfrac{37k}{66}$, Hence, the eigen vector corresponding to $\lambda = 7$ is

$$\begin{bmatrix} \dfrac{37k}{66} & \dfrac{2k}{11} & 0 \end{bmatrix}^T \text{ or } \begin{bmatrix} 37 & 12 & 0 \end{bmatrix}^T.$$

11.6 POWER METHOD

This method is used to determine the numerically approximate largest eigen value and hence, the corresponding eigen vector.

Let A be a square matrix of order n and $\lambda_1, \lambda_2, \ldots\ldots\ldots, \lambda_n$ be its eigen values such that $|\lambda_1| > |\lambda_2| \geq |\lambda_3| \geq \ldots\ldots\ldots \geq |\lambda_n|$. Suppose $X_1, X_2, \ldots\ldots\ldots, X_n$ be n independent eigen vectors corresponding to the eigen values of the matrix A. Then

$$A X_i = \lambda_i X_i \ (i = 1, 2, \ldots\ldots\ldots, n) \tag{11.4}$$

Let X be an arbitrary vector. Then, X can be expressed as linear combination of n independent vectors $X_1, X_2, \ldots\ldots\ldots, X_n$. Hence, there exist $c_1, c_2, \ldots\ldots\ldots, c_n$ constants not all zero such that

$$X = c_1 X_1 + c_2 X_2 + \ldots\ldots\ldots + c_n X_n.$$

Now, operating A on both sides of above equation, we have

$$AX = A\{c_1 X_1 + c_2 X_2 + \ldots\ldots\ldots\ldots + c_n X_n\}$$

i.e., $$AX = c_1 AX_1 + c_2 AX_2 + \ldots\ldots\ldots\ldots + c_n AX_n \quad \text{(by 11.4)}$$

i.e., $$AX = c_1 \lambda_1 X_1 + c_2 \lambda_2 X_2 + \ldots\ldots\ldots\ldots + c_n \lambda_n X_n \quad (11.5)$$

Again operating A on both sides of (11.5), we obtain

$$A^2 X = c_1 \lambda_1^2 X_1 + c_2 \lambda_2^2 X_2 + \ldots\ldots\ldots\ldots + c_n \lambda_n^2 X_n \quad (11.6)$$

Repeating this process, we get,

$$A^r X = c_1 \lambda_1^r X_1 + c_2 \lambda_2^r X_2 + \ldots\ldots\ldots\ldots + c_n \lambda_n^r X_n$$

$$A^r X = \lambda_1^r \left\{ c_1 X_1 + c_2 \left(\frac{\lambda_r}{\lambda_1}\right)^r X_2 + \ldots\ldots\ldots\ldots + c_n \left(\frac{\lambda_n}{\lambda_n}\right)^r X_n \right\} \quad (11.7)$$

and $$A^{r+1} X = \lambda_1^{r+1} \left\{ c_1 X_1 + \left(\frac{\lambda_2}{\lambda_1}\right)^{r+1} X_2 + \ldots\ldots\ldots + c_n \left(\frac{\lambda_n}{\lambda_1}\right)^{r+1} X_n \right\} \quad (11.8)$$

As $r \to \infty$, the right-hand side of (11.7) and (11.8) tend to $\lambda_1^r c_1 X_1$ and $\lambda_1^{r+1} c_1 X_1$ since $\left|\frac{\lambda_2}{\lambda_1}\right| < 1, \left|\frac{\lambda_3}{\lambda_1}\right| < 1, \ldots\ldots\ldots, \left|\frac{\lambda_n}{\lambda_1}\right| < 1.$ The vector $c_1 X_1$ is the eigen vector corresponding to λ_1. Further, the eigen value λ_1 is obtained as the ratio of corresponding component of $A^{r+1}X$ and $A^r X$ as $r \to \infty$

i.e., $$\lambda_1 = \lim_{r \to \infty} \frac{\left(A^{r+1} X\right)r}{\left(A^r X\right)r} \qquad r = (1, 2, 3, \ldots\ldots\ldots, n),$$

where the suffix r denotes the rth component of the vector. The process of finding the largest eigen value by the method described above is known as power method.

Remark: If A is non-singular matrix, then the smallest eigen value of A is the largest eigen value of A^{-1}.

11.7   NOTATION

We denote $X^{(0)}$ to represent the initial guess eigen vector and $X^{(k)}$ denote the kth approximation of eigen vector corresponding to k^{th} approximation of eigen value $\lambda^{(k)}$.

11.8 PROCEDURE TO FIND THE LARGEST EIGEN VALUE OF A GIVEN MATRIX BY POWER METHOD

Step 1: Choose the initial (guess) eigen vector $X^{(0)}$. For example, we can choose

$$X^{(0)} = \begin{bmatrix} 1 \\ 0 \\ 0 \end{bmatrix} \text{ or } X^{(0)} = \begin{bmatrix} 1 \\ 1 \\ 1 \end{bmatrix} \text{ etc. In case given matrix is of order 3.}$$

Step 2: compute $AX^{(0)}$ and express it as

$$AX^{(0)} = \lambda^{(1)} X^{(1)},$$

where the elements of $X^{(1)}$ are obtained by dividing the element of $X^{(0)}$ by its largest element.

Step 3: Next compute $AX^{(1)}$. Again express it as $AX^{(1)} = \lambda^{(2)}X^{(2)}$, where the elements of $X^{(2)}$ are obtained by dividing the element of $X^{(1)}$ by its largest value.

Step 4: Repeat the Step 3 to get $AX^{(3)} = \lambda^{(4)} X^{(4)}$, $AX^{(4)} = \lambda^{(5)} X^{(5)}$ etc.

Step 5: Step 4 is terminated when the elements of the $X^{(r+1)} - X^{(r)}$ are either of the following:

(i) Zero (ii) near to zero (iii) upto desired degree of accuracy.

Step 6: After the termination of Step 4, we have

$$AX^{(r)} = \lambda^{(r+1)} X^{(r+1)}$$

and the largest eigen value of given matrix is $\lambda^{(n+1)}$.

SOLVED PROBLEMS

Problem 1: Find the largest eigen value and its eigen vector of $\begin{bmatrix} 1 & 2 \\ 3 & 4 \end{bmatrix}$ using the power method.

Solution: Let $A = \begin{bmatrix} 1 & 2 \\ 3 & 4 \end{bmatrix}$ and $X^{(0)} = \begin{bmatrix} 1 \\ 1 \end{bmatrix}$ be the initial eigen vector. Then,

$$AX^{(0)} = \begin{bmatrix} 1 & 2 \\ 3 & 4 \end{bmatrix}\begin{bmatrix} 1 \\ 1 \end{bmatrix} = \begin{bmatrix} 3 \\ 7 \end{bmatrix}$$

$$AX^{(0)} = 7\begin{bmatrix} 0.428571 \\ 1 \end{bmatrix}$$

Comparing $AX^{(0)}$ with $\lambda^{(1)}X^{(1)}$, we have

$\lambda^{(1)} = 7$ and
$$X^{(1)} = \begin{bmatrix} 0.428571 \\ 1 \end{bmatrix}$$

Now we compute $AX^{(1)}$.

$$AX^{(1)} = \begin{bmatrix} 1 & 2 \\ 3 & 4 \end{bmatrix}\begin{bmatrix} 0.428571 \\ 1 \end{bmatrix}$$

i.e.,
$$AX^{(1)} = \begin{bmatrix} 2.428571 \\ 5.285713 \end{bmatrix}$$

i.e.,
$$AX^{(1)} = 5.285713\begin{bmatrix} 0.459459 \\ 1 \end{bmatrix}$$

Thus, $\lambda^{(2)} = 5.285713$ and $X^{(2)} = \begin{bmatrix} 0.459459 \\ 1 \end{bmatrix}$

Again compute $AX^{(2)}$.

$$AX^{(2)} = \begin{bmatrix} 1 & 2 \\ 3 & 4 \end{bmatrix}\begin{bmatrix} 0.459459 \\ 1 \end{bmatrix} = \begin{bmatrix} 2.459459 \\ 5.378377 \end{bmatrix}$$

i.e.,
$$AX^{(2)} = 5.378377\begin{bmatrix} 0.457286 \\ 1 \end{bmatrix}$$

Thus, $\lambda^{(3)} = 5.378377$ and $X^{(3)} = \begin{bmatrix} 0.457286 \\ 1 \end{bmatrix}$

Similarly, we compute $A X^{(3)}$, $A X^{(4)}$ and $A X^{(5)}$ to get $\lambda^{(4)}$, $\lambda^{(5)}$, $\lambda^{(6)}$. The computations of these are shown below.

$$AX^{(3)} = \begin{bmatrix} 1 & 2 \\ 3 & 4 \end{bmatrix}\begin{bmatrix} 0.457286 \\ 1 \end{bmatrix} = \begin{bmatrix} 2.457286 \\ 5.371856 \end{bmatrix}$$

i.e.,
$$AX^{(3)} = 5.371856\begin{bmatrix} 0.457437 \\ 1 \end{bmatrix}$$

Thus, $\lambda^{(4)} = 5.371856$ and $X^{(4)} = \begin{bmatrix} 0.457437 \\ 1 \end{bmatrix}$

Now,
$$AX^4 = \begin{bmatrix} 1 & 2 \\ 3 & 4 \end{bmatrix}\begin{bmatrix} 0.457437 \\ 1 \end{bmatrix} = \begin{bmatrix} 2.457437 \\ 5.372311 \end{bmatrix}$$

i.e.,
$$AX^{(4)} = 5.372311\begin{bmatrix} 0.457426 \\ 1 \end{bmatrix}$$

Thus, $\lambda^{(5)} = 5.372311$ and $X^{(5)} = \begin{bmatrix} 0.457426 \\ 1 \end{bmatrix}$

Now,
$$AX^{(5)} = \begin{bmatrix} 1 & 2 \\ 3 & 4 \end{bmatrix}\begin{bmatrix} 0.457426 \\ 1 \end{bmatrix}$$

$$= \begin{bmatrix} 2.457426 \\ 5.372278 \end{bmatrix}$$

$$= 5.372278\begin{bmatrix} 0.457427 \\ 1 \end{bmatrix}$$

Thus, $\lambda^{(6)} = 5.372278$ and $X^{(6)} = \begin{bmatrix} 0.457427 \\ 1 \end{bmatrix}$.

Clearly,
$$X^{(6)} - X^{(5)} = \begin{bmatrix} 0.000001 \\ 1 \end{bmatrix} \approx \begin{bmatrix} 0 \\ 0 \end{bmatrix}$$

Therefore, the largest eigen value is 5.372278 and its eigen vector corresponding to it is $\begin{bmatrix} 0.457427 \\ 1 \end{bmatrix}$.

Problem 2: Using the power method find the largest eigen value of the matrix $A = \begin{bmatrix} 2 & -1 & 0 \\ -1 & 2 & -1 \\ 0 & -1 & 2 \end{bmatrix}$.

Solution: Suppose the initial vector $X^{(0)} = [1\ 0\ 0]^T$.

Now we compute $AX^{(0)}$, i.e.,

$$AX^{(0)} = \begin{bmatrix} 2 & -1 & 0 \\ -1 & 2 & -1 \\ 0 & -1 & 2 \end{bmatrix} \begin{bmatrix} 1 \\ 0 \\ 0 \end{bmatrix} = \begin{bmatrix} 2 \\ -1 \\ 0 \end{bmatrix}$$

This matrix $AX^{(0)}$ can be expressed as

$$AX^{(0)} = 2 \begin{bmatrix} 1 \\ -0.5 \\ 0 \end{bmatrix}$$

Hence,
$$\lambda^{(1)} = 2 \text{ and } X^{(1)} = \begin{bmatrix} 1 \\ -0.5 \\ 0 \end{bmatrix}$$

Again, we compute $AX^{(1)}$

$$AX^{(1)} = \begin{bmatrix} 2 & -1 & 0 \\ -1 & 2 & -1 \\ 0 & -1 & 2 \end{bmatrix} \begin{bmatrix} 1 \\ -0.5 \\ 0 \end{bmatrix}$$

$$= \begin{bmatrix} 2.5 \\ -2 \\ 0.5 \end{bmatrix}$$

i.e.,
$$AX^{(1)} = 2.5 \begin{bmatrix} 1 \\ -0.8 \\ 0.2 \end{bmatrix}$$

hence,
$$\lambda^{(2)} = 2 \text{ and } X^{(2)} = \begin{bmatrix} 1 \\ -0.8 \\ 0.2 \end{bmatrix}.$$

Repeating this process, we have

$$AX^{(2)} = 2.8 \begin{bmatrix} 1 \\ -1 \\ 0.43 \end{bmatrix}$$

$$\lambda^{(3)} = 2.8 \ \text{ and } \ X^{(3)} = \begin{bmatrix} 1 \\ -1 \\ 0.43 \end{bmatrix},$$

$$AX^{(3)} = 3.43 \begin{bmatrix} 0.87 \\ -1 \\ 0.54 \end{bmatrix},$$

$$\lambda^{(4)} = 3.43 \ \text{ and } \ X^{(4)} = \begin{bmatrix} 0.87 \\ -1 \\ 0.54 \end{bmatrix}$$

$$AX^{(4)} = 3.41 \begin{bmatrix} 0.80 \\ -1 \\ 0.61 \end{bmatrix}$$

$$\lambda^{(5)} = 3.41 \ \text{ and } \ X^{(5)} = \begin{bmatrix} 0.80 \\ -1 \\ 0.61 \end{bmatrix}$$

$$AX^{(5)} = 3.41 \begin{bmatrix} 0.76 \\ -1 \\ 0.65 \end{bmatrix}$$

So that
$$\lambda^{(6)} = 3.41 \ \text{ and } \ X^{(6)} = \begin{bmatrix} 0.76 \\ -1 \\ 0.65 \end{bmatrix}$$

$$AX^{(6)} = 3.41 \begin{bmatrix} 0.74 \\ -1 \\ 0.67 \end{bmatrix}$$

So that
$$\lambda^{(7)} = 3.41 \ \text{ and } \ X^{(7)} = \begin{bmatrix} 0.74 \\ -1 \\ 0.67 \end{bmatrix}$$

As $X^{(7)} - X^{(6)} \approx \begin{bmatrix} 0 \\ 0 \\ 0 \end{bmatrix}$, the largest eigen value is 3.41 and eigen vector

is $\begin{bmatrix} 0.74 \\ -1 \\ 0.67 \end{bmatrix}$.

Problem 3: Using the power matrix, find the largest eigen value and

corresponding eigen vector of $\begin{bmatrix} 1 & 6 & 1 \\ 1 & 2 & 0 \\ 0 & 0 & 3 \end{bmatrix}$.

Solution: Let $A = \begin{bmatrix} 1 & 6 & 1 \\ 1 & 2 & 0 \\ 0 & 0 & 3 \end{bmatrix}$ Suppose the initial eigen vector

$X^{(0)} = \begin{bmatrix} 1 \\ 0 \\ 0 \end{bmatrix}$.

Now, we compute

$$AX^{(0)} = \begin{bmatrix} 1 & 6 & 1 \\ 1 & 2 & 0 \\ 0 & 0 & 3 \end{bmatrix} \begin{bmatrix} 1 \\ 0 \\ 0 \end{bmatrix}$$

i.e., $$AX^{(0)} = 1.\begin{bmatrix} 1 \\ 1 \\ 0 \end{bmatrix}$$

Hence, $$\lambda^{(1)} = 1 \text{ and } X^{(1)} = \begin{bmatrix} 1 \\ 1 \\ 0 \end{bmatrix}$$

Now $$AX^{(1)} = \begin{bmatrix} 1 & 6 & 1 \\ 1 & 2 & 0 \\ 0 & 0 & 3 \end{bmatrix} \begin{bmatrix} 1 \\ 1 \\ 0 \end{bmatrix} = \begin{bmatrix} 7 \\ 3 \\ 0 \end{bmatrix}$$

i.e.,
$$AX^{(1)} = 7\begin{bmatrix} 1 \\ 0.4286 \\ 0 \end{bmatrix}$$

Hence,
$$\lambda^{(2)} = 7 \text{ and } X^{(2)} = \begin{bmatrix} 1 \\ 0.4286 \\ 0 \end{bmatrix}$$

Again
$$AX^{(2)} = \begin{bmatrix} 1 & 6 & 1 \\ 1 & 2 & 0 \\ 0 & 0 & 3 \end{bmatrix}\begin{bmatrix} 1 \\ 0.4286 \\ 0 \end{bmatrix} = \begin{bmatrix} 3.5714 \\ 1.8572 \\ 0 \end{bmatrix}$$

i.e.,
$$AX^{(2)} = 3.5714\begin{bmatrix} 1 \\ 0.52 \\ 0 \end{bmatrix}$$

Thus,
$$\lambda^{(3)} = 3.5714 \text{ and } X^{(3)} = \begin{bmatrix} 1 \\ 0.52 \\ 0 \end{bmatrix}$$

Repeating the above process, we have

$$AX^{(3)} = 4.12\begin{bmatrix} 1 \\ 0.4951 \\ 0 \end{bmatrix}$$

So that
$$\lambda^{(4)} = 4.12 \text{ and } X^{(4)} = \begin{bmatrix} 1 \\ 0.4951 \\ 0 \end{bmatrix},$$

$$AX^{(4)} = 3.9706\begin{bmatrix} 1 \\ 0.5012 \\ 0 \end{bmatrix}$$

Hence, $\qquad \lambda^{(5)} = 3.9706$ and $X^{(5)} = \begin{bmatrix} 1 \\ 0.5012 \\ 0 \end{bmatrix}$,

$$AX^{(5)} = 4.0072 \begin{bmatrix} 1 \\ 0.4997 \\ 0 \end{bmatrix}$$

Hence, $\qquad \lambda^{(6)} = 4.0072$ and $X^{(6)} = \begin{bmatrix} 1 \\ 0.4997 \\ 0 \end{bmatrix}$

$$AX^{(6)} = 3.9982 \begin{bmatrix} 1 \\ 0.5 \\ 0 \end{bmatrix}$$

So that $\qquad \lambda^{(7)} = 3.9982$ and $X^{(7)} = \begin{bmatrix} 1 \\ 0.5 \\ 0 \end{bmatrix}$

and

$$AX^{(7)} = 4 \begin{bmatrix} 1 \\ 0.5 \\ 0 \end{bmatrix}$$

So that $\qquad \lambda^{(8)} = 4$ and $X^{(8)} = \begin{bmatrix} 1 \\ 0.5 \\ 0 \end{bmatrix}$

Since $X^{(8)} - X^{(7)} = \begin{bmatrix} 0 \\ 0 \\ 0 \end{bmatrix}$, the largest eigen value of the given matrix is

4 and the eigen vector corresponding to it is $\begin{bmatrix} 1 \\ 0.5 \\ 0 \end{bmatrix}$.

Problem 4: Find the longest eigen value and the corresponding eigen vector of the matrix $A = \begin{bmatrix} 2 & 3 & 2 \\ 4 & 3 & 5 \\ 3 & 2 & 9 \end{bmatrix}$ by power method, starting with $X^{(0)} = \begin{bmatrix} 1 \\ 1 \\ 1 \end{bmatrix}$.

Solution: First we compute $AX^{(0)}$ to obtain $\lambda^{(1)}$ and $X^{(1)}$. Now,

$$AX^{(0)} = \begin{bmatrix} 2 & 3 & 2 \\ 4 & 3 & 5 \\ 3 & 2 & 9 \end{bmatrix} \begin{bmatrix} 1 \\ 1 \\ 1 \end{bmatrix} = \begin{bmatrix} 7 \\ 12 \\ 14 \end{bmatrix}$$

We express it as $A X^{(0)} = \lambda^{(1)} X^{(1)}$ where $\lambda^{(1)}$ numerically largest component of the vector $\begin{bmatrix} 7 \\ 12 \\ 14 \end{bmatrix}$ and the components of $X^{(1)}$ are obtained by dividing the components of $\begin{bmatrix} 7 \\ 12 \\ 14 \end{bmatrix}$ by 14.

Thus, $$AX^{(0)} = 14 \begin{bmatrix} 0.5 \\ 0.857142 \\ 1 \end{bmatrix}$$

Thus, $\lambda^{(1)} = 14$ and $X^{(1)} = \begin{bmatrix} 0.5 \\ 0.857142 \\ 1 \end{bmatrix}$

Next we compute $AX^{(1)}$, i.e.,

$$AX^{(1)} = \begin{bmatrix} 2 & 3 & 2 \\ 4 & 3 & 5 \\ 3 & 2 & 9 \end{bmatrix} \begin{bmatrix} 0.5 \\ 0.857142 \\ 1 \end{bmatrix}$$

$$= \begin{bmatrix} 5.571426 \\ 9.571426 \\ 12.214284 \end{bmatrix}$$

$$= 12.214284 \begin{bmatrix} 0.456140 \\ 0.783625 \\ 1 \end{bmatrix}$$

Hence, $\lambda^{(2)} = 12.214284$ and $X^{(2)} = \begin{bmatrix} 0.456140 \\ 0.783625 \\ 1 \end{bmatrix}$. Continuing the

above process, we have

$$AX^{(2)} = \begin{bmatrix} 2 & 3 & 2 \\ 4 & 3 & 5 \\ 3 & 2 & 9 \end{bmatrix} \begin{bmatrix} 0.456140 \\ 0.783625 \\ 1 \end{bmatrix}$$

$$= \begin{bmatrix} 5.263155 \\ 9.175435 \\ 11.935670 \end{bmatrix}$$

$$= 11.935670 \begin{bmatrix} 0.440960 \\ 0.768740 \\ 1.0 \end{bmatrix}$$

Hence, $\lambda^{(3)} = 11.935670$ and $X^{(3)} = \begin{bmatrix} 0.440960 \\ 0.768740 \\ 1.0 \end{bmatrix}$,

$$AX^{(3)} = \begin{bmatrix} 2 & 3 & 2 \\ 4 & 3 & 5 \\ 3 & 2 & 9 \end{bmatrix} \begin{bmatrix} 0.440960 \\ 0.768740 \\ 1.0 \end{bmatrix}$$

$$= \begin{bmatrix} 5.18814 \\ 9.07006 \\ 11.86036 \end{bmatrix}$$

$$= 11.86036 \begin{bmatrix} 0.437435 \\ 0.764737 \\ 1 \end{bmatrix}$$

Hence, $\lambda^{(4)} = 11.86036$ and $X^{(4)} = \begin{bmatrix} 0.437435 \\ 0.764737 \\ 1 \end{bmatrix}$,

$$AX^{(4)} = \begin{bmatrix} 2 & 3 & 2 \\ 4 & 3 & 5 \\ 3 & 2 & 9 \end{bmatrix} \begin{bmatrix} 0.437435 \\ 0.764737 \\ 1 \end{bmatrix}$$

$$= \begin{bmatrix} 5.169081 \\ 9.043951 \\ 11.841779 \end{bmatrix} = 11.841779 \begin{bmatrix} 0.436512 \\ 0.763732 \\ 1 \end{bmatrix}$$

Hence, $\lambda^{(5)} = 11.841779$ and $X^{(5)} = \begin{bmatrix} 0.436512 \\ 0.763732 \\ 1 \end{bmatrix}$,

$$\text{and } AX^{(5)} = \begin{bmatrix} 2 & 3 & 2 \\ 4 & 3 & 5 \\ 3 & 2 & 9 \end{bmatrix} \begin{bmatrix} 0.436512 \\ 0.763732 \\ 1 \end{bmatrix}$$

$$= \begin{bmatrix} 5.164220 \\ 9.037244 \\ 11.837 \end{bmatrix}$$

$$= 11.837 \begin{bmatrix} 0.436277 \\ 0.763474 \\ 1 \end{bmatrix}$$

Hence, $\lambda^{(6)} = 11.837$ and $X^{(6)} = \begin{bmatrix} 0.436277 \\ 0.763474 \\ 1 \end{bmatrix}$

Upto 3 decimal places $X^{(5)} - X^{(6)} \approx \begin{bmatrix} 0 \\ 0 \\ 0 \end{bmatrix}$ and therefore, the larg-

est eigen value is 11.837 and the eigen vector corresponding to it is

$\begin{bmatrix} 0.436 \\ 0.763 \\ 1 \end{bmatrix}$.

Problem 5: Find the smallest eigen value of the matrix $A = \begin{bmatrix} 1 & 6 & 1 \\ 1 & 2 & 0 \\ 0 & 0 & 3 \end{bmatrix}$.

Solution: We know that if λ is largest eigen value of A^{-1}, then $\dfrac{1}{\lambda}$ will

be the smallest eigen value of λ ($\lambda \neq 0$). Therefore, first we find the inverse of the matrix A and later its (A^{-1}) largest eigen value.

$$\det A = \begin{vmatrix} 1 & 6 & 1 \\ 1 & 2 & 0 \\ 0 & 0 & 3 \end{vmatrix} = -12 \text{ and}$$

$$\text{Adjoint of } A = \begin{bmatrix} +\begin{vmatrix} 2 & 0 \\ 0 & 3 \end{vmatrix} & -\begin{vmatrix} 1 & 0 \\ 0 & 3 \end{vmatrix} & +\begin{vmatrix} 1 & 2 \\ 0 & 0 \end{vmatrix} \\ -\begin{vmatrix} 6 & 1 \\ 0 & 3 \end{vmatrix} & +\begin{vmatrix} 1 & 1 \\ 0 & 3 \end{vmatrix} & -\begin{vmatrix} 1 & 6 \\ 0 & 0 \end{vmatrix} \\ +\begin{vmatrix} 6 & 1 \\ 2 & 0 \end{vmatrix} & -\begin{vmatrix} 1 & 1 \\ 1 & 0 \end{vmatrix} & +\begin{vmatrix} 1 & 6 \\ 1 & 2 \end{vmatrix} \end{bmatrix}^T$$

$$= \begin{bmatrix} 6 & -3 & 0 \\ -18 & 3 & 0 \\ -2 & 1 & -4 \end{bmatrix}^T = \begin{bmatrix} 6 & -18 & -2 \\ -3 & 3 & 1 \\ 0 & 0 & -4 \end{bmatrix}$$

$$\therefore \qquad A^{-1} = \frac{1}{-12} \begin{bmatrix} 6 & -18 & -2 \\ -3 & 3 & 1 \\ 0 & 0 & -4 \end{bmatrix}$$

i.e.,
$$A^{-1} = \begin{bmatrix} -0.5 & 1.5 & 1.6667 \\ 0.25 & -0.25 & 0.0833 \\ 0 & 0 & 0.3333 \end{bmatrix}$$

starting with initial vector $X^{(0)} = \begin{bmatrix} 1 \\ 0 \\ 0 \end{bmatrix}$

we have

$$A^{-1} X^{(0)} = \begin{bmatrix} -0.5 & 1.5 & 1.6667 \\ 0.25 & -0.25 & 0.0833 \\ 0 & 0 & 0.3333 \end{bmatrix} \begin{bmatrix} 1 \\ 0 \\ 0 \end{bmatrix}$$

$$= \begin{bmatrix} -0.5 \\ 0.25 \\ 0 \end{bmatrix} = -0.5 \begin{bmatrix} 1 \\ -0.5 \\ 0 \end{bmatrix}$$

Hence, $\qquad \lambda^{(1)} = -0.5$ and $X^{(1)} = \begin{bmatrix} 1 \\ -0.5 \\ 0 \end{bmatrix}$

Now we compute $AX^{(1)}$ i.e.,

$$AX^{(1)} = \begin{bmatrix} -0.5 & 1.5 & 1.6667 \\ 0.25 & -0.25 & 0.0833 \\ 0 & 0 & 0.3333 \end{bmatrix} \begin{bmatrix} 1 \\ -0.5 \\ 0 \end{bmatrix}$$

$$= \begin{bmatrix} -1.25 \\ 0.375 \\ 0 \end{bmatrix} = -1.25 \begin{bmatrix} 1 \\ -0.3 \\ 0 \end{bmatrix}$$

Hence, $$\lambda^{(2)} = -1.25 \text{ and } X^{(2)} = \begin{bmatrix} 1 \\ -0.3 \\ 0 \end{bmatrix}$$

Continuing the above process, we have

$$AX^{(2)} = \begin{bmatrix} -0.5 & 1.5 & 1.6667 \\ 0.25 & -0.25 & 0.0833 \\ 0 & 0 & 0.3333 \end{bmatrix} \begin{bmatrix} 1 \\ -0.3 \\ 0 \end{bmatrix}$$

$$= \begin{bmatrix} -0.95 \\ .325 \\ 0 \end{bmatrix} = -0.95 \begin{bmatrix} 1 \\ -0.3421 \\ 0 \end{bmatrix},$$

Hence, $$\lambda^{(3)} = -0.95 \text{ and } X^{(3)} = \begin{bmatrix} 1 \\ -0.3421 \\ 0 \end{bmatrix},$$

$$AX^{(3)} = \begin{bmatrix} -0.5 & 1.5 & 1.6667 \\ 0.25 & -0.25 & 0.8333 \\ 0 & 0 & 0.3333 \end{bmatrix} \begin{bmatrix} 1 \\ -0.3421 \\ 0 \end{bmatrix}$$

$$= \begin{bmatrix} -1.01315 \\ 0.3355 \\ 0 \end{bmatrix} = -1.01315 \begin{bmatrix} 1 \\ -0.3311 \\ 0 \end{bmatrix}$$

Hence, $$\lambda^{(4)} = -1.01315 \text{ and } X^{(4)} = \begin{bmatrix} 1 \\ -0.3311 \\ 0 \end{bmatrix}$$

$$AX^{(4)} = \begin{bmatrix} -0.5 & 1.5 & 1.6667 \\ 0.25 & -0.25 & 0.0833 \\ 0 & 0 & 0.3333 \end{bmatrix} \begin{bmatrix} 1 \\ -0.3311 \\ 0 \end{bmatrix}$$

$$= \begin{bmatrix} -0.99665 \\ 0.3327 \\ 0 \end{bmatrix}$$

$$= -0.99665 \begin{bmatrix} 1 \\ -0.3338 \\ 0 \end{bmatrix}$$

Hence, $\qquad \lambda^{(5)} = -0.99665$ and $X^{(5)} = \begin{bmatrix} 1 \\ -0.3338 \\ 0 \end{bmatrix}$ and

$$AX^{(5)} = \begin{bmatrix} -0.5 & 1.5 & 1.6667 \\ 0.25 & -0.25 & 0.0833 \\ 0 & 0 & 0.3333 \end{bmatrix} \begin{bmatrix} 1 \\ -0.3338 \\ 0 \end{bmatrix}$$

$$= \begin{bmatrix} -1.00007 \\ 0.3333 \\ 0 \end{bmatrix} = -1.00007 \begin{bmatrix} 1 \\ 0.33327 \\ 0 \end{bmatrix}$$

Hence, $\qquad \lambda^{(6)} = -1.0007 \approx -1$ and $X^{(6)} = \begin{bmatrix} 1 \\ 0.33327 \\ 0 \end{bmatrix}$

Upto 2 decimal places $X^{(5)} = X^{(6)}$. Therefore, the larger eigen value

of A^{-1} is -1; hence, the smallest eigen value of given matrix is $\dfrac{1}{-1} = -1$.

Note: The following results and definitions are useful in further discussion of eigen values of a squre matrix A.

(i) A real square matrix U is said to be unitary if $U^T U = U U^T = I$.

(ii) A real unitary matrix is called an orthogonal matrix.

(iii) A transformation $U A$ is known as orthogonal transformation if U is an orthogonal. If $B = UA$, then we say that the matrix B is obtained by orthogonal transformation on A.

(iv) For any non-singular matrix U, the formation of the matrix $U^{-1}AU$ is called similar transformation. A result from linear algebra states that the eigen value of A and $B = U^{-1}AU$ are same.

11.9   JACOBI'S METHOD TO FIND EIGEN VALUES OF A SYMMETRIC MATRIX

Let A be a real symmetric matrix. Then, there exist an orthogonal matrix U such that $U^{-1}AU$ is a diagonal matrix D whose diagonal elements are the characteristic roots of the matrix A. Further, the eigen values are real. Because of these facts, in Jacobi's method, the matrix A is diagonalized by applying a series of orthogonal transformation U_1, U_2, ... and U_n.

Now, we describe the method of finding the orthogonal matrix U by a series of orthogonal transformations.

Among the off-diagonal elements, select numerically the largest element and suppose that $|a_{ik}|$ be the numerical of largest element ($i \neq k$). From a 2×2 matrix A_1 with element a_{ii}, a_{ik}, a_{ki} and a_{kk}, i.e.,

$$A_1 = \begin{bmatrix} a_{ii} & a_{ik} \\ a_{ki} & a_{kk} \end{bmatrix}$$

Let $B = \begin{bmatrix} \cos\theta & -\sin\theta \\ \sin\theta & \cos\theta \end{bmatrix}$ be an orthogonal matrix such that $B^{-1} A_1 B$ is a diagonal matrix. It is possible to find θ such that $B^{-1} A_1 B$ is a diagonal matrix. The procedure to find θ is as follows:

Clearly, $$B^{-1} = \begin{bmatrix} \cos\theta & \sin\theta \\ -\sin\theta & \cos\theta \end{bmatrix}$$

Therefore, $$B^{-1} A_1 B = \begin{bmatrix} \cos\theta & \sin\theta \\ -\sin\theta & \cos\theta \end{bmatrix} \begin{bmatrix} a_{ii} & a_{ik} \\ a_{ki} & a_{kk} \end{bmatrix} \begin{bmatrix} \cos\theta & -\sin\theta \\ \sin\theta & \cos\theta \end{bmatrix}$$

$$= \begin{bmatrix} a_{ii}\cos^2\theta + a_{ik}\sin 2\theta + a_{kk}\sin^2\theta & (a_{kk} - a_{ii})\sin\theta\cos\theta + a_{ik}\cos 2\theta \\ (a_{kk} - a_{ii})\sin\theta\cos\theta + a_{ik}\cos 2\theta & a_{ii}\sin^2\theta - a_{ik}\sin 2\theta + a_{kk}\cos 2\theta \end{bmatrix}$$

It reduces to a diagonal form if we choose a such that

$$(a_{kk} - a_{ii})\sin\theta\cos\theta + a_{ik}\cos 2\theta = 0$$

i.e., $$\tan 2\theta = \frac{2a_{ik}}{a_{ii} - a_{kk}}$$

i.e., $$\theta = \frac{1}{2}\tan^{-1}\frac{2a_{ik}}{a_{ii} - a_{kk}} \qquad (11.9)$$

To ensure smallest rotation, we chose θ such that $\dfrac{-\pi}{4} \leq \theta \leq \dfrac{\pi}{4}$.

For the largest off-diagonal element $|a_{ik}|$, the orthogonal matrix U_1 is such that

(i) The ith row, i^{th} column element of U_1 is $\cos\theta$.

(ii) The ith row, k^{th} column element of U_1 is $-\sin\theta$.

(iii) k^{th} row, i^{th} column element of U_1 is $\sin\theta$.

(iv) k^{th} row, k^{th} column element of U_1 is $\cos\theta$.

(v) Each of the remaining principal diagonal element of U_1 is one.

(vi) All other element of U_1 are zero. The matrix U_1 will be the following form:

$$U_1 = \begin{bmatrix} 1 & & & & & & & \\ & 1 & & & & & & \\ & & \ddots & & & & & \\ & & & \cos\theta & \cdots & \sin\theta & & \\ & & & \cdots & & \cdots & & \\ & & & \sin\theta & & \cos\theta & & \\ & & & & \ddots & & & \\ & & & & & 1 & & \\ & & & & & & 1 & \\ & & & & & & & 1 \end{bmatrix}$$

Suppose $T_1 = U_1^{-1} AU$; then T_1 is the matrix obtained through the orthogonal transformation. Note that the elements of off-diagonal elements a_{ik} and a_{ki} of T_1 are zero.

Next, we find the numerically off-diagonal element in rotated (or transformed) matrix T_1 and repeat the procedure described above. Let T_2 be the matrix obtained through the orthogonal transformation, Thus, $T_2 = U_2^{-1} T_1 U_2$.

After performing r such transformation, we get

$$T_r = U_r^{-1} U_{r-1}^{-1} \ldots\ldots U_1^{-1}, A \, U_1 \, U_2 \, U_3 \ldots\ldots U_{r-1} U_r \tag{11.10}$$

Let $$U = U_1 \, U_2 \, \text{---} U_r, \tag{11.11}$$

Then $$T_r = U^{-1} A \, U. \tag{11.12}$$

For large number r, the transformed matrix T_r approached a diagonal matrix whose elements are eigen values of the given matrix.

Note:

(i) To obtain the approximate eigen values of a given matrix A by Jacobi's method, we perform a finite number of transformations described in (11.11) so that off-diagonal elements are zero or nearly zero.

(ii) We denote a diagonal matrix as D $(a_1, a_2, \ldots\ldots a_n)$ where $a_1, a_2, \ldots\ldots a_n$ are real numbers.

(iii) If the final transformed matrix is $T_k = U^{-1}A\, U = D\,(\lambda_1, \lambda_2, \ldots\ldots \lambda_n)$, then $\lambda_1, \lambda_2, \ldots\ldots \lambda_n$ are eigen values of matrix A and the column elements of U are corresponding eigen vectors. Note that
$$U = U_1\, U_2\, U_3 \ldots\ldots U_k.$$

11.10 PROCEDURE TO FIND EIGEN VALUES OF A SYMMETRIC MATRIX BY JACOBI'S METHOD

Step 1: Let A be the given symmetric matrix of order n.

Step 2: First find the numerically greatest element among the off-diagonal element of A.

Suppose a_{23} is the numerically largest off-diagonal element.

Step 3: Note down the values of $a_{22}, a_{23}, a_{32}, a_{33}$.

Step 4: Find $\theta = \dfrac{1}{2}\tan^{-1}\dfrac{2a_{23}}{a_{22} - a_{33}}$ and let this angle be θ_0. Of course

θ_0 should be chosen such that $\dfrac{-\pi}{4} \le \theta_0 \le \dfrac{\pi}{4}$. The general

formula to find θ is $\theta = \dfrac{1}{2}\tan^{-1}\left\{\dfrac{2a_{ik}}{(a_{ii} - a_{kk})}\right\}$, where a_{ik} is the

numerically greatest off-diagonal element of A.

Step 5: Write the matrix U_1 such that

    (i) 2nd row, 2nd column element of U_1 is $\cos\theta_0$.

    (ii) 2nd row, 3rd column element of U_1 is $-\sin\theta_0$.

    (iii) 3rd row, 2nd column element of U_1 is $\sin\theta_0$.

    (iv) 3rd row, 3rd column element of U_1 is $\cos\theta_0$.

    (v) The remaining principal diagonal element of U_1 are 1.

    (vi) The remaining off–diagonal elements are zero.

Step 6: Find U_1^{-1}. Note that $U_1^{-1} = U_1^{\mathrm{T}}$ as U_1 is an orthogonal matrix.

$$\therefore \qquad U_1^{-1} = U_1^{T}$$

Step 7: Compute $T_1 = U_1^{-1} A U_1$
If T_1 is in a diagonal form, then go to Step 9, else go to Step 8.

Step 8: Repeat the Steps 2 to 7 replacing A by T,
We get

$$T_2 = U_2^{-1} T_1 U_2$$

i.e., $$T_2 = U_2^{-1} U_1^{-1} A U_1 U_2$$

Suppose that $T_r = U_r^{-1} T_{r-1} U_r$
(r is a positive integer) is a diagonal matrix D (λ_1, λ_2, ... λ_n) whose off-diagonal matrix elements are either zero or nearly zero [Note that the value of r depend the elements of A and its order].

Step 9: The eigen values are λ_1, λ_2, ... λ_n.

Step 10: Compute $U = U_1 U_2 .... U_r$
The column vectors of U are the eigen vectors corresponding to the eigen values of symmetric matrix A.

SOLVED PROBLEMS

Problem 1: Find the eigen values and the eigen vector of the symmetric matrix $\begin{bmatrix} 1 & \sqrt{2} & \sqrt{2} \\ \sqrt{2} & 3 & \sqrt{2} \\ 2 & \sqrt{2} & 1 \end{bmatrix}$ By the Jacobi's method.

Solution: Let $A = \begin{bmatrix} 1 & \sqrt{2} & \sqrt{2} \\ \sqrt{2} & 3 & \sqrt{2} \\ 2 & \sqrt{2} & 1 \end{bmatrix}$. Now the off-diagonal elements are $a_{12} = \sqrt{2}$, $a_{13} = 2$, $a_{21} = \sqrt{2}$, $a_{23} = \sqrt{2}$, $a_{31} = 2$ and $a_{32} = 2$. Hence, the numerically, largest off-diagonal element is $a_{13} = 2$ and $a_{31} = 2$. The four elements of 2×2 orthogonal matrix are $a_{11} = 1$, $a_{13} = 2$, $a_{31} = 2$ and $a_{33} = 1$. By Eq. (11. 9)

$$\theta = \frac{1}{2}\tan^{-1}\frac{2a_{13}}{a_{11}-a_{33}}$$

$\therefore$
$$\theta = \frac{1}{2}\tan^{-1}\left(\frac{2\times 2}{0}\right) = \frac{\pi}{4}$$

Hence,
$$U_1 = \begin{bmatrix} \cos\dfrac{\pi}{4} & 0 & -\sin\dfrac{\pi}{4} \\ 0 & 1 & 0 \\ \sin\dfrac{\pi}{4} & 0 & \cos\dfrac{\pi}{4} \end{bmatrix} = \begin{bmatrix} \dfrac{1}{\sqrt{2}} & 0 & \dfrac{-1}{\sqrt{2}} \\ 0 & 1 & 0 \\ \dfrac{1}{\sqrt{2}} & 0 & \dfrac{1}{\sqrt{2}} \end{bmatrix}$$

Therefore,

$$U_1^{-1} = \begin{bmatrix} \dfrac{1}{\sqrt{2}} & 0 & \dfrac{1}{\sqrt{2}} \\ 0 & 1 & 0 \\ \dfrac{-1}{\sqrt{2}} & 0 & \dfrac{1}{\sqrt{2}} \end{bmatrix}$$

Now, the first rotation on A gives
$$T_1 = U_1^{-1}\, A\, U_1$$

i.e.,
$$T_1 = \begin{bmatrix} \dfrac{1}{\sqrt{2}} & 0 & \dfrac{1}{\sqrt{2}} \\ 0 & 1 & 0 \\ \dfrac{-1}{\sqrt{2}} & 0 & \dfrac{1}{\sqrt{2}} \end{bmatrix}\begin{bmatrix} 1 & \sqrt{2} & 2 \\ \sqrt{2} & 3 & \sqrt{2} \\ 2 & \sqrt{2} & 1 \end{bmatrix}\begin{bmatrix} \dfrac{1}{\sqrt{2}} & 0 & \dfrac{-1}{\sqrt{2}} \\ 0 & 1 & 0 \\ \dfrac{1}{\sqrt{2}} & 0 & \dfrac{1}{\sqrt{2}} \end{bmatrix}$$

$$= \begin{bmatrix} 3 & 2 & 0 \\ 2 & 3 & 0 \\ 0 & 0 & -1 \end{bmatrix}$$

Observe that the first row, third column element and third row first column element of T_1 are zero.

Now we repeat the above process replacing A by T_1.

The numerically largest off-diagonal elements of T_1 are $a_{12} = 2$ and $a_{21} = 2$ and the four elements of 2×2 matrix are $a_{11} = 3$, $a_{12} = 2$, $a_{21} = 2$ and $a_{22} = 3$. Thus,

$$\theta = \frac{1}{2} \tan^{-1} \left(\frac{2a_{12}}{a_{11} - a_{22}} \right)$$

i.e.,

$$\theta = \frac{1}{2} \tan^{-1} \left(\frac{4}{0} \right) = \frac{\pi}{4}$$

Therefore,

$$U_2 = \begin{bmatrix} \cos\dfrac{\pi}{4} & -\sin\dfrac{\pi}{4} & 0 \\ \sin\dfrac{\pi}{4} & \cos\dfrac{\pi}{4} & 0 \\ 0 & 0 & 0 \end{bmatrix} = \begin{bmatrix} \dfrac{1}{\sqrt{2}} & \dfrac{-1}{\sqrt{2}} & 0 \\ \dfrac{1}{\sqrt{2}} & \dfrac{1}{\sqrt{2}} & 0 \\ 0 & 0 & 1 \end{bmatrix}$$

and hence,

$$U_2^{-1} = \begin{bmatrix} \dfrac{1}{\sqrt{2}} & \dfrac{1}{\sqrt{2}} & 0 \\ \dfrac{-1}{\sqrt{2}} & \dfrac{1}{\sqrt{2}} & 0 \\ 0 & 0 & 1 \end{bmatrix}$$

$$\therefore \qquad T_2 = U_2^{-1} \, T_1 \, U_2$$

i.e.,

$$T_2 = \begin{bmatrix} \dfrac{1}{\sqrt{2}} & \dfrac{1}{\sqrt{2}} & 0 \\ \dfrac{-1}{\sqrt{2}} & \dfrac{1}{\sqrt{2}} & 0 \\ 0 & 0 & 1 \end{bmatrix} \begin{bmatrix} 1 & \sqrt{2} & 2 \\ \sqrt{2} & 3 & \sqrt{2} \\ 2 & \sqrt{2} & 1 \end{bmatrix} \begin{bmatrix} \dfrac{1}{\sqrt{2}} & \dfrac{-1}{\sqrt{2}} & 0 \\ \dfrac{1}{\sqrt{2}} & \dfrac{1}{\sqrt{2}} & 0 \\ 0 & 0 & 1 \end{bmatrix}$$

i.e.,

$$T_2 = \begin{bmatrix} 5 & 0 & 0 \\ 0 & 1 & 0 \\ 0 & 0 & -1 \end{bmatrix} = D(5, 1, -1)$$

which is in diagonal form. Thus, the eigen values of A are 5, 1 and -1.

To find the eigen vectors:

Recall $T_2 = U_2^{-1} T_1 U_2$. Since $T_1 = U_1^{-1} A U_1$, we have $T_2 = U_2^{-1} U_1^{-1} A U_1 U_2$. Thus, $T_2 = U^{-1} A U$ where $U = U_1 U_2$. Now we compute $U_1 U_2$,

$$U = U_1 U_2 = \begin{bmatrix} \dfrac{1}{\sqrt{2}} & 0 & \dfrac{-1}{\sqrt{2}} \\[2mm] 0 & 1 & 0 \\[2mm] \dfrac{1}{\sqrt{2}} & 0 & \dfrac{1}{\sqrt{2}} \end{bmatrix} \begin{bmatrix} \dfrac{1}{\sqrt{2}} & \dfrac{-1}{\sqrt{2}} & 0 \\[2mm] \dfrac{1}{\sqrt{2}} & \dfrac{1}{\sqrt{2}} & 0 \\[2mm] 0 & 0 & 1 \end{bmatrix}$$

$$= \begin{bmatrix} \dfrac{1}{2} & \dfrac{-1}{2} & \dfrac{-1}{\sqrt{2}} \\[2mm] \dfrac{1}{\sqrt{2}} & \dfrac{1}{\sqrt{2}} & 0 \\[2mm] \dfrac{1}{2} & \dfrac{-1}{2} & \dfrac{1}{\sqrt{2}} \end{bmatrix}$$

Therefore, the eigen vectors corresponding to the eigen values are

$$\begin{bmatrix} \dfrac{1}{2} \\[2mm] 0 \\[2mm] \dfrac{1}{2} \end{bmatrix}, \begin{bmatrix} \dfrac{-1}{2} \\[2mm] \dfrac{1}{\sqrt{2}} \\[2mm] \dfrac{-1}{\sqrt{2}} \end{bmatrix} \text{ and } \begin{bmatrix} \dfrac{-1}{\sqrt{2}} \\[2mm] 0 \\[2mm] \dfrac{1}{\sqrt{2}} \end{bmatrix}.$$

Problem 2: Using Jacobi's method, find the eigen values and eigen vector of $A = \begin{bmatrix} 4 & 0 & 1 \\ 0 & -2 & 0 \\ 1 & 0 & 4 \end{bmatrix}$.

Solution: The largest off-diagonal element of A is $a_{13} = a_{31} = 1$.

The four elements of 2×2 matrix are $a_{11} = 4$, $a_{13} = 1$, $a_{31} = 1$ and $a_{33} = 4$. Therefore,

$$\theta = \frac{1}{2} \tan^{-1} \left(\frac{2a_{13}}{a_{11} - a_{33}} \right) = \frac{1}{2} \tan^{-1} \left(\frac{2}{0} \right) = \frac{\pi}{4}$$

Hence,
$$U_1 = \begin{bmatrix} \dfrac{1}{\sqrt{2}} & 0 & \dfrac{-1}{\sqrt{2}} \\[3mm] 0 & 1 & 0 \\[3mm] \dfrac{1}{\sqrt{2}} & 0 & \dfrac{1}{\sqrt{2}} \end{bmatrix}$$

and
$$U^{-1}{}_1 = \begin{bmatrix} \dfrac{1}{\sqrt{2}} & 0 & \dfrac{1}{\sqrt{2}} \\[3mm] 0 & 1 & 0 \\[3mm] \dfrac{-1}{\sqrt{2}} & 0 & \dfrac{1}{\sqrt{2}} \end{bmatrix}$$

$\therefore \qquad T_1 = U_1^{-1} A \, U_1$

$$= \begin{bmatrix} \dfrac{1}{\sqrt{2}} & 0 & \dfrac{1}{\sqrt{2}} \\[3mm] 0 & 1 & 0 \\[3mm] \dfrac{-1}{\sqrt{2}} & 0 & \dfrac{1}{\sqrt{2}} \end{bmatrix} \begin{bmatrix} 4 & 0 & 1 \\ 0 & -2 & 0 \\ 1 & 0 & 4 \end{bmatrix} \begin{bmatrix} \dfrac{1}{\sqrt{2}} & 0 & \dfrac{-1}{\sqrt{2}} \\[3mm] 0 & 1 & 0 \\[3mm] \dfrac{1}{\sqrt{2}} & 0 & \dfrac{1}{\sqrt{2}} \end{bmatrix}$$

$$= \begin{bmatrix} \dfrac{1}{\sqrt{2}} & 0 & \dfrac{1}{\sqrt{2}} \\[3mm] 0 & 1 & 0 \\[3mm] \dfrac{-1}{\sqrt{2}} & 0 & \dfrac{1}{\sqrt{2}} \end{bmatrix} \begin{bmatrix} \dfrac{5}{\sqrt{2}} & 0 & \dfrac{-3}{\sqrt{2}} \\[3mm] 0 & -2 & 0 \\[3mm] \dfrac{5}{\sqrt{2}} & 0 & \dfrac{3}{\sqrt{2}} \end{bmatrix}$$

$$= \begin{bmatrix} 5 & 0 & 0 \\ 0 & -2 & 0 \\ 0 & 0 & 3 \end{bmatrix},$$ which is in diagonal form. Therefore, the eigen

values of given symmetric matrix A are 5, -2 and 3. Further, since $T_1 = U_1^{-1} A \, U_1 = D\,(5, -2, 3)$, the eigen vectors corresponding to its eigen values are

$$\begin{bmatrix} \dfrac{1}{\sqrt{2}} \\ 0 \\ \dfrac{1}{\sqrt{2}} \end{bmatrix}, \begin{bmatrix} 0 \\ 1 \\ 0 \end{bmatrix} \text{ and } \begin{bmatrix} \dfrac{-1}{\sqrt{2}} \\ 0 \\ \dfrac{1}{\sqrt{2}} \end{bmatrix}.$$

Problem 3: Find the eigen values of the symmetric matrix

$$A = \begin{bmatrix} 2 & 3 & 1 \\ 3 & 2 & 2 \\ 1 & 2 & 1 \end{bmatrix} \text{ by the Jacobi's method.}$$

Solution: The numerically greatest off-diagonal element of A is $a_{12} = a_{21} = 3$. Further, $a_{11} = 2$ and $a_{33} = 1$.

$$\therefore \qquad \theta = \frac{1}{2}\tan^{-1}\frac{2a_{12}}{a_{11} - a_{33}} = \frac{1}{2}\tan^{-1}\frac{6}{1} = 0.7854$$

Hence, the orthogonal matrix

$$U_1 = \begin{bmatrix} 0.7071 & -0.7071 & 0 \\ 0.7071 & 0.7071 & 0 \\ 0 & 0 & 1 \end{bmatrix}$$

and

$$U^{-1}_{1} = \begin{bmatrix} 0.7071 & 0.7071 & 0 \\ -0.7071 & 0.7071 & 0 \\ 0 & 0 & 1 \end{bmatrix}$$

Now the first rotation on A is given by

$$T_1 = U^{-1}_{1}AU_1 = \begin{bmatrix} 5 & 0 & 2.1213 \\ 0 & -1 & 0.7071 \\ 2.1213 & 0.7071 & 1 \end{bmatrix}$$

Now, the largest off-diagonal matrix of T_1 is $a_{13} = a_{31} = 2.1213$. Note that $a_{11} = 5$ and $a_{33} = 1$

$$\therefore \qquad \theta = \frac{1}{2}\tan^{-1}\frac{2a_{13}}{a_{11}-a_{33}} = \frac{1}{2}\tan^{-1}\left\{\frac{2\times 2.1213}{(5-1)}\right\} = 0.4074$$

Hence, the orthogonal matrix for the second rotation is

$$U_2 = \begin{bmatrix} 0.9181 & 0 & -0.3962 \\ 0 & 1 & 0 \\ 0.3962 & 0 & 0.9181 \end{bmatrix}$$

and

$$U^{-1}{}_2 = \begin{bmatrix} 0.9181 & 0 & 0.3962 \\ 0 & 1 & 0 \\ -0.3962 & 0 & 0.9181 \end{bmatrix}$$

Hence, second rotation on A gives

$$T_2 = U_2^{-1}T_1 U_2 = \begin{bmatrix} 5.9155 & 0.2802 & 0 \\ 0.2802 & -1 & 0.6492 \\ 0 & 0.6492 & 0.0845 \end{bmatrix}$$

The largest off-diagonal element numerically is T_2 is $a_{23}=a_{32}=0.6492$. The values of $a_{22}=-1$ and $a_{33}=0.0845$. Therefore,

$$\theta = \frac{1}{2}\tan^{-1}\left(\frac{2a_{23}}{a_{22}-a_{33}}\right) = \frac{1}{2}\tan^{-1}\frac{2\times 0.6492}{-1-0.0845}$$

$$= -0.4375.$$

The orthogonal matrix for the rotation is

$$U_3 = \begin{bmatrix} 1 & 0 & 0 \\ 0 & 0.9058 & -0.4236 \\ 0 & 0.4236 & 0.9058 \end{bmatrix}$$

and $\qquad U_3^{-1} = U_3^{T}.$

Now the third rotation on A gives

$$T_3 = U^{-1}{}_3 T_2 U_3 = \begin{bmatrix} 5.9155 & 0.2538 & 0.1187 \\ 0.2538 & -1.3036 & 0 \\ 0.1187 & 0 & 0.3881 \end{bmatrix}$$

Again, the numerically the largest off-diagonal element of T_3 is $a_{12} = a_{21} = 0.2538$. The values of $a_{11} = 5.9155$ and $a_{33} = 0.3881$

$$\therefore \qquad \theta = \frac{1}{2}\tan^{-1}\left(\frac{2a_{12}}{a_{11} - a_{22}}\right) = \frac{1}{2}\tan^{-1}\frac{0.5076}{7.2191} = 0.0351$$

Hence, orthogonal matrix for fourth rotation is

$$U_4 = \begin{bmatrix} 0.9994 & -0.0351 & 0 \\ 0.0351 & 0.9994 & 0 \\ 0 & 0 & 1 \end{bmatrix}$$

and

$$U_4^{-1} = U_4^{T}.$$

The fourth rotation on A gives

$$T_4 = U^{-1}{}_4 T_3 U_4 = \begin{bmatrix} 5.9244 & 0 & 0.1186 \\ 0 & -1.3125 & -0.0042 \\ 0.1186 & -0.0042 & 0.3811 \end{bmatrix}$$

The largest off-diagonal element of T_4 numerically is $a_{13} = 0.1186$. The values of $a_{11} = 5.9244$ and $a_{33} = 0.3881$. Hence,

$$\theta = \frac{1}{2}\tan^{-1}\left(\frac{2a_{13}}{a_{11} - a_{33}}\right)$$

$$= \frac{1}{2}\tan^{-1}\left(\frac{2 \times 0.1186}{5.9244 - 0.3881}\right)$$

$$= 0.0214$$

Thus, the orthogonal matrix for fifth rotation is

$$U_5 = \begin{bmatrix} 0.9998 & 0 & 0.0214 \\ 0 & 1 & 0 \\ 0.0214 & 0 & 0.9998 \end{bmatrix}$$

and $U_5^{-1} = U_5^{T}$.

Fifth rotation on A gives,

$$T_5 = U_5^{-1} T_4 U_5 = \begin{bmatrix} 5.9269 & -0.0001 & 0 \\ -0.0001 & -1.3125 & -0.0042 \\ 0 & -0.0042 & 0.3856 \end{bmatrix}$$

The largest off-diagonal element numerical is T_5 is $a_{23} = -0.0042$. Note that $a_{22} = -1.3125$ and $a_{33} = 0.3856$. Therefore,

$$\theta = \frac{1}{2} \tan^{-1} \left\{ \frac{2a_{23}}{(a_{22} - a_{33})} \right\}$$

$$= \frac{1}{2} \tan^{-1} \left\{ \frac{2(-0.0042)}{(-1.3125 - 0.3856)} \right\}$$

$$= 0.0025.$$

The orthogonal matrix for sixth rotation is

$$U_6 = \begin{bmatrix} 1 & 0 & 0 \\ 0 & 1 & -0.0025 \\ 0 & 0.0025 & 1.000 \end{bmatrix}$$

and

$$U_6^{-1} = U_6^T \text{ Thus, } T_6 = U_6^{-1} T_5 U_6 = \begin{bmatrix} 5.9269 & -0.0001 & 0 \\ -0.0001 & -1.3125 & 0 \\ 0 & 0 & 0.3856 \end{bmatrix}$$

Hence, the approximate eigen values are 5.9269, -1.3125 and 0.3856.

Problem 4: Find the eigen values of $A = \begin{bmatrix} 5 & 0 & 1 \\ 0 & -2 & 0 \\ 1 & 0 & 5 \end{bmatrix}$ using the Jacobi's method.

Solution: The numerically largest off-diagonal element of A is $a_{13} = a_{31} = 1$. Note that $a_{11} = 5$ and $a_{33} = 5$. Therefore,

$$\theta = \frac{1}{2} \tan^{-1} \frac{2a_{13}}{a_{11} - a_{33}}$$

$$= \frac{1}{2} \tan^{-1}\left(\frac{2}{0}\right) = 0.7854$$

The orthogonal matrix of first rotation on A is

$$U_1 = \begin{bmatrix} 0.7071 & 0 & -0.7071 \\ 0 & 1 & 0 \\ 0.7071 & 0 & 0.7071 \end{bmatrix}$$

and

$$U_1^{-1} = \begin{bmatrix} 0.7071 & 0 & 0.7071 \\ 0 & 1 & 0 \\ -0.7071 & 0 & 0.7071 \end{bmatrix}$$

Hence, the first rotation A gives

$$T_1 = U_1^{-1} A U_1 = \begin{bmatrix} 0.7071 & 0 & -0.7071 \\ 0 & 1 & 0 \\ 0.7071 & 0 & 0.7071 \end{bmatrix} \begin{bmatrix} 5 & 0 & 1 \\ 0 & -2 & 0 \\ 1 & 0 & 5 \end{bmatrix} \begin{bmatrix} 0.7071 & 0 & 0.7071 \\ 0 & 1 & 0 \\ -0.7071 & 0 & 0.7071 \end{bmatrix}$$

$$= \begin{bmatrix} 6 & 0 & 0 \\ 0 & -2 & 0 \\ 0 & 0 & 4 \end{bmatrix}$$

Hence, the eigen values of given matrix A are 6, –2 and 4.

Problem 5: Find the eigen values of the matrix $A = \begin{bmatrix} 1 & 2 & -1 \\ 2 & 1 & 2 \\ -1 & 2 & 1 \end{bmatrix}$ by Jacobi's method.

Solution: The numerically largest off-diagonal element of A is $a_{12} = a_{21} = 2$. Note that $a_{11} = 1$ and $a_{33} = 1$. Therefore,

$$\theta = \frac{1}{2} \tan^{-1} \frac{2a_{12}}{a_{11} - a_{33}} = \frac{1}{2} \tan^{-1}\left(\frac{2 \times 2}{1 - 1}\right) = 0.7854$$

The orthogonal matrix for the first rotation on A is

$$U_1 = \begin{bmatrix} 0.7071 & -0.7071 & 0 \\ 0.7071 & 0.7071 & 0 \\ 0 & 0 & 1 \end{bmatrix}$$

and $U_1^{-1} = U_1^T$. The first rotation on A gives

$$T_1 = U_1^{-1}AU_1 = \begin{bmatrix} 3 & 0 & 0.7071 \\ 0 & -1 & 2.1213 \\ 0.7071 & 2.1213 & 1.0000 \end{bmatrix}$$

Now, numerically the largest off-diagonal element of T_1 is $a_{23} = a_{32} = 2.1213$. Note that $a_{22} = -1$ and $a_{33} = 1$. Therefore,

$$\theta = \frac{1}{2}\tan^{-1}\frac{2a_{23}}{a_{22} - a_{33}} = \frac{1}{2}\tan^{-1}\left(\frac{2 \times 2.1213}{-1-1}\right)$$

$$= -0.5651.$$

Hence, the orthogonal matrix for second rotation on A is given by

$$U_2 = \begin{bmatrix} 1 & 0 & 0 \\ 0 & 0.8445 & -0.5355 \\ 0 & 0.5355 & 0.8445 \end{bmatrix}$$

and $U_2^{-1} = U_2^T$. Thus,

$$T_2 = U_2^{-1}T_1U_2 = \begin{bmatrix} 3 & -0.3787 & 0.5972 \\ -0.3787 & -2.3452 & 0 \\ 0.5972 & 0 & 2.3452 \end{bmatrix}$$

The numerically largest off-diagonal element in T_2 is $a_{13} = a_{31} = 0.592$. Note that $a_{11} = 3$ and $a_{33} = 2.3452$

$$\therefore \qquad \theta = \frac{1}{2}\tan^{-1}\frac{2a_{13}}{a_{11} - a_{33}} = 0.5347$$

The orthogonal matrix for third rotation is

$$U_3 = \begin{bmatrix} 0.8604 & 0 & -0.5095 \\ 0 & 1 & 0 \\ 0.5095 & 0 & 0.8604 \end{bmatrix}$$

and $U_3^{-1} = U_3^T$

The third rotation on A given

$$T_3 = U_3^{-1} A_2 U_3 = \begin{bmatrix} 3.3537 & -0.3258 & 0 \\ -0.3258 & -2.3452 & 0.1930 \\ 0 & 0.1930 & 1.9915 \end{bmatrix}$$

Again the numerically largest off-diagonal element is T_3 is $a_{12} = a_{21} = 0.3258$. Note that $a_{11} = 3.3537$ and $a_{33} = 1.9915$.
Therefore,

$$\theta = \frac{1}{2} \tan^{-1} \frac{2a_{12}}{a_{11} - a_{22}}$$

$$= \frac{1}{2} \tan^{-1} \left\{ \frac{2 \times (-0.3258)}{3.3537 - 1.9915} \right\}$$

$$= -0.0569$$

The orthogonal matrix for fourth rotation is

$$U_4 = \begin{bmatrix} 0.9984 & 0.0569 & 0 \\ -0.0569 & 0.9984 & 0 \\ 0 & 0 & 1 \end{bmatrix}$$

and $U_4^{-1} = U_4^T$

The fourth rotation on A gives

$$T_4 = U_4^{-1} T_3 U_4 = \begin{bmatrix} 3.3723 & 0 & -0.0110 \\ 0 & -2.3638 & 0.1927 \\ -0.0010 & 0.1927 & 1.9915 \end{bmatrix}$$

Similarly, we obtain

$$T_5 = U_5^{-1} T_4 U_5 = \begin{bmatrix} 3.3723 & 0.0005 & -0.0110 \\ 0.0005 & -2.2723 & 0 \\ -0.0010 & 0 & 2.00 \end{bmatrix},$$

$$T_6 = U_6^{-1} T_5 U_6 = \begin{bmatrix} 3.3724 & 0.0005 & 0 \\ 0.0005 & -2.3723 & 0 \\ 0 & 0 & 1.9999 \end{bmatrix}$$

$$\text{and} \qquad T_7 = U_7^{-1} T_6 U_7 = \begin{bmatrix} 3.3724 & 0 & 0 \\ 0 & -2.2723 & 0 \\ 0 & 0 & 1.9999 \end{bmatrix}$$

$$\text{where} \qquad U_5 = \begin{bmatrix} 1 & 0 & 0 \\ 0 & 0.9990 & 0.0441 \\ 0 & -0.441 & 0.9990 \end{bmatrix},$$

$$U_6 = \begin{bmatrix} 1 & 0 & 0.0080 \\ 0 & 1 & 0 \\ -0.0080 & 0 & 1 \end{bmatrix}$$

and

$$U_7 = \begin{bmatrix} 1 & -0.0001 & 0 \\ 0.0001 & 1.000 & 0 \\ 0 & 0 & 1 \end{bmatrix}.$$

Thus, the eigen values of given matrix A are 3.3724, −2.2723 and 1.9999.

11.11 GIVEN'S METHOD TO FIND EIGEN VALUES OF REAL SYMMETRIC MATRIX

In this method, the real symmetric matrix A is reduced into a tri-diagonal matrix using orthogonal transformation and the eigen values are determined by constructing Sturm sequence followed by some calculations, and once we found eigen values, we can find the eigen vectors using the procedure described in Solved problems of section 11.5. The drawback of Jacobi's method is that the zero obtained by a particular orthogonal transformation may not remain zero during the subsequent transformation. Further to reduce a symmetric matrix into tri-diagonal form, it requires $\dfrac{n(n-1)}{2}$ rotations by Jacobi's method, where it requires $\dfrac{(n-1)(n-2)}{2}$ rotation is the Given's method. Now, we proceed to describe Given's method.

Let $A = [a_{ij}]$ be real symmetric matrix of order n. Choose the orthogonal matrix.

$$U_1 = \begin{bmatrix} 1 & 0 & 0 & -- & -- & 0 \\ 0 & \cos\theta & -\sin\theta & -- & -- & 0 \\ 0 & \sin\theta & \cos\theta & -- & -- & 0 \\ 0 & 0 & 0 & 1 & -- & 0 \\ -- & -- & -- & -- & -- & - \\ 0 & 0 & -- & -- & -- & 1 \end{bmatrix} \qquad (11.13)$$

Not that it is obtained by

    (i) Replacing 2nd row, 2nd column element of I_n by $\cos\theta$.

    (ii) Replacing 2nd row, 3rd column of I_n by $-\sin\theta$.

    (iii) Replacing 3rd row, 2nd column of I_n by $\sin\theta$.

    (iv) Replacing 3rd row, 3rd column of I_n by $\cos\theta$.

The inverse of U_1 is

$$U_1^{-1} = \begin{bmatrix} 1 & 0 & 0 & 0 & -- & 0 \\ 0 & \cos\theta & \sin\theta & 0 & -- & 0 \\ 0 & -\sin\theta & \cos\theta & 1 & -- & 0 \\ -- & -- & -- & -- & -- & - \\ -- & -- & -- & -- & -- & - \\ 0 & 0 & 0 & 0 & -- & 1 \end{bmatrix} \qquad (11.14)$$

Now, suppose

$$T_1 = U_1^{-1} A U_1 = [a_{jj}^1] \qquad (11.15)$$

The matrix T_1 is obtained through the orthogonal transformation.

    The angle θ is to be chosen such that
$a_{13}^1 = a_{31}^1 = 0$. From Eq. (11.15), we have

$$a_{13}^1 = -a_{12}\sin\theta + a_{13}\cos\theta \qquad (11.16)$$

and $\qquad\qquad a_{31}^1 = -a_{21}\sin\theta + a_{31}\cos\theta \qquad (11.17)$

Since, $a_{12} = a_{21}$ and $a_{13} = a_{31}$, we have $a_{13}^1 = a_{31}^1$.

    Therefore, $a_{13}^1 = 0$ gives

$$\tan\theta = \frac{a_{13}}{a_{12}}$$

$$\therefore \qquad \theta = \tan^{-1}\left(\frac{a_{13}}{a_{12}}\right) \qquad (11.18)$$

With the value of θ given by (11. 18) and performing the plane rotation, the elements $a_{13}{}^1$ and $a_{31}{}^1$ become zero. We call this rotation as rotation in (2, 3) plane.

Similarly, performing the rotations is (2, 4) plane, we can reduce the element of first row, 4th column and 4th row, 1st column of transformed matrix, continuing the rotations in (2, 5), (2, 6), ... , (2, n) planes the elements of positions in

   (i) first row, 5th column and 5th row, 1st column
   (ii) first row, 6th column, and 6th row 1st column
   ____

   (iii) first row n^{th} column, and n^{th} row 1st column
        are zero respectively of transformed matrices.

Note that to reduce first row and first column elements of the matrix A to 1st row and 1st column of tridiagonal matrix, it requires $(n - 2)$ plane rotations.

The same process may be continued for 3rd row, 4th row, . . . and n^{th} row to reduce the given matrix into tri-diagonal matrix. The total number of rotations required $(n - 2) + (n - 3) + (n - 4) + \ldots . . 1$

$= \dfrac{(n-1)(n-2)}{2}$ in order to reduce the given matrix A is the tri-diagonal

matrix. Note that the tri-diagonal matrix is obtained through the rotation or orthogonal transformation on given matrix.

After performing the above-said plane rotation, we have a matrix of the form

$$T = \begin{bmatrix} \beta_1 & \gamma_1 & 0 & 0 & 0 & -- & 0 & 0 & 0 \\ \gamma_1 & \beta_2 & \gamma_2 & 0 & 0 & -- & 0 & 0 & 0 \\ 0 & \gamma_2 & \beta_3 & \gamma_3 & 0 & -- & 0 & 0 & 0 \\ - & - & - & - & - & - & - & - & - \\ 0 & 0 & 0 & 0 & 0 & -- & \gamma_n & \beta_{n-1} & \gamma_{n-1} \\ 0 & 0 & 0 & 0 & 0 & -- & 0 & \gamma_{n-1} & \beta_n \end{bmatrix}$$

which is a tridiagonal matrix.

Now, we construct Sturm sequence to get the eigen values of the matrix A.

Define

$$f_n(\lambda) = |\lambda I - T|$$

where n is an order of the matrix A (equivalently the order of T). Then

$$f_n(\lambda) = \begin{vmatrix} \lambda - \beta & -\gamma & 0 & 0 & ..... & ..... & 0 & 0 & 0 \\ -\gamma_1 & \lambda - \beta_2 & -\gamma_2 & 0 & ..... & ..... & 0 & 0 & 0 \\ 0 & -\gamma_2 & \lambda - \beta_3 & 0 & ..... & ..... & 0 & 0 & 0 \\ ..... & ..... & ..... & ..... & ..... & ..... & ..... & ..... & - \\ 0 & 0 & 0 & 0 & ..... & ..... & -\gamma_{n-1} & \lambda - \beta_{n-1} & -\gamma_{n-1} \\ 0 & 0 & 0 & 0 & 0 & ..... & 0 & -\gamma_{n-1} & \lambda - \beta_n \end{vmatrix} = 0$$

$$(11.19)$$

The roots of (11.9) are the eigen values of the matrix A as (11.9) is the characteristic equation of T and, A and T are similar matrices. The sequence $\{f_n\}$ satisfies the following recurrence relation:

$$f_k(\lambda) = (\lambda - \beta_k) f_{k-1} - \gamma^2_{k-1} f_{k-2}, \ 2 \le k \le n \ \text{with} \ f_1 = \lambda - \beta_1 \ \text{and} \ f_0(\lambda) = 1.$$

Let all $\gamma_i \ne 0$ $(i = 1, 2, ..., n-1)$, Then $\{f_n\}$ is Sturm sequence. Using the method described in chapter 0, section (0.11), we can find the roots of $f_n(\lambda) = 0$ and the roots of this equation are the eigen values of given matrix A.

Note: Suppose that after performing 2 plane rotations on A the tri-diagonal matrix is $\begin{bmatrix} \beta_1 & \gamma_1 & 0 \\ \gamma_1 & \beta_2 & \gamma_2 \\ 0 & \gamma_2 & \beta_3 \end{bmatrix}$. Then, the sturm sequence is

$$f_0(\lambda) = 1.$$
$$f_1(\lambda) = \lambda - \beta_1$$

$$f_2(\lambda) = (\lambda - \beta_2) f_1(\lambda) - \gamma_1^2 f_0(\lambda)$$

$$f_3(\lambda) = (\lambda - \beta_3) f_2(\lambda) - \gamma_2^2 f_1(\lambda).$$

11.12 PROCEDURE TO FIND THE EIGEN VALUES OF REAL SYMMETRIC MATRIX OF ORDER 3 BY GIVEN'S METHOD

Step 1: Let the given 3×3 real symmetric matrix be A.

Step 2: Note down the values of a_{12} and a_{13}.

Step 3: Compute $\theta = \tan^{-1} \dfrac{a_{13}}{a_{12}}$. $\left(\dfrac{-\pi}{4} \le \theta \le \dfrac{\pi}{4} \right)$ Let this angel be θ_1.

Step 4: Write $U_1 = \begin{bmatrix} 1 & 0 & 0 \\ 0 & \cos\theta_1 & -\sin\theta_1 \\ 0 & \sin\theta_1 & \cos\theta_1 \end{bmatrix}$.

Step 5: Compute U_1^{-1}.

Step 6: Compute $T_1 = U_1^{-1} A U_1$ Then T_1 will be in tri-diagonal form.

Step 7: Compare the corresponding elements of reduced tri-diagonal

matrix with $\begin{bmatrix} \beta_1 & \gamma_1 & 0 \\ \gamma_2 & \beta_2 & 0 \\ 0 & \gamma_2 & \beta_3 \end{bmatrix}$ note down the values of $\beta_1, \beta_2, \beta_3,$ γ_1 and γ_2.

Step 8: Construct the Sturm sequence

$$f_0(\lambda) = 1.$$

$$f_1(\lambda) = \lambda - \beta_1$$

$$f_2(\lambda) = (\lambda - \beta_2) f_1(\lambda) - \gamma_1^2 f_0(\lambda)$$

$$f_3(\lambda) = (\lambda - \beta_3) f_2(\lambda) - \gamma_2^2 f_1(\lambda).$$

Step 9: The roots of $f_3(\lambda) = 0$ are the eigen values.

If we fail to find the factors of $f_3(\lambda)$, then we follow the procedure described in Section (0.11) to get the root of $f_3(\lambda) = 0$.

Step 10: We can find the eigen vectors following the procedure given in solved problems (section 11.5).

SOLVED PROBLEMS

Problem 1: Using the Given's method transform the matrix

$$A = \begin{bmatrix} 1 & 2 & 3 \\ 2 & 1 & -1 \\ 3 & -1 & 1 \end{bmatrix} \text{ to tri-diagonal matrix.}$$

Solution: Here, $a_{12} = 2$ and $a_{13} = 3$, and therefore, $\tan \theta = \dfrac{a_{13}}{a_{12}} = \dfrac{3}{2}$,

hence, $\cos \theta = \dfrac{2}{\sqrt{13}}$ and $\sin \theta = \dfrac{3}{\sqrt{13}}$. Now orthogonal matrix

$$U_1 = \begin{bmatrix} 1 & 0 & 0 \\ 0 & \dfrac{2}{\sqrt{13}} & \dfrac{-3}{\sqrt{13}} \\ 0 & \dfrac{3}{\sqrt{13}} & \dfrac{2}{\sqrt{13}} \end{bmatrix}$$

and

$$U_1^{-1} = \begin{bmatrix} 1 & 0 & 0 \\ 0 & \dfrac{2}{\sqrt{13}} & \dfrac{3}{\sqrt{13}} \\ 0 & \dfrac{-3}{\sqrt{13}} & \dfrac{2}{\sqrt{13}} \end{bmatrix}$$

Thus, the transformed matrix T_1,

$$T_1 = U_1^{-1}\, A\, U_1$$

i.e.,

$$T_1 = \begin{bmatrix} 1 & 0 & 0 \\ 0 & \dfrac{2}{\sqrt{13}} & \dfrac{3}{\sqrt{13}} \\ 0 & \dfrac{-3}{\sqrt{13}} & \dfrac{2}{\sqrt{13}} \end{bmatrix} \begin{bmatrix} 1 & 2 & 3 \\ 2 & 1 & -1 \\ 3 & -1 & 1 \end{bmatrix} \begin{bmatrix} 1 & 0 & 0 \\ 0 & \dfrac{2}{\sqrt{13}} & \dfrac{-3}{\sqrt{13}} \\ 0 & \dfrac{3}{\sqrt{13}} & \dfrac{2}{\sqrt{13}} \end{bmatrix}$$

$$= \begin{bmatrix} 1 & \sqrt{13} & 0 \\ \sqrt{13} & \dfrac{1}{13} & \dfrac{5}{13} \\ 0 & \dfrac{5}{13} & \dfrac{25}{13} \end{bmatrix},$$

which is the required tri-diagonal matrix.

Problem 2: Using the Given's method transform $A = \begin{bmatrix} 2 & 1 & 1 \\ 1 & 1 & 0 \\ 1 & 0 & 1 \end{bmatrix}$ to tri-diagonal matrix.

Solution: Here, $a_{12} = 1$ and $a_{13} = 1$, and therefore,

$$\tan \theta = \frac{a_{13}}{a_{12}} = \frac{1}{1} = 1$$

$$\therefore \qquad \theta = \frac{\pi}{4}$$

Hence, the orthogonal matrix for (2, 3) plane rotation is

$$U_1 = \begin{bmatrix} 1 & 0 & 0 \\ 0 & \dfrac{1}{\sqrt{2}} & \dfrac{-1}{\sqrt{2}} \\ 0 & \dfrac{1}{\sqrt{2}} & \dfrac{1}{\sqrt{2}} \end{bmatrix}$$

and

$$U_1^{-1} = \begin{bmatrix} 1 & 0 & 0 \\ 0 & \dfrac{1}{\sqrt{2}} & \dfrac{1}{\sqrt{2}} \\ 0 & \dfrac{-1}{\sqrt{2}} & \dfrac{1}{\sqrt{2}} \end{bmatrix}.$$

Hence, transformed matrix

$$T_1 = U_1^{-1} A U_1 = \begin{bmatrix} 1 & 0 & 0 \\ 0 & \dfrac{1}{\sqrt{2}} & \dfrac{1}{\sqrt{2}} \\ 0 & \dfrac{-1}{\sqrt{2}} & \dfrac{1}{\sqrt{2}} \end{bmatrix} \begin{bmatrix} 2 & 1 & 1 \\ 1 & 1 & 0 \\ 1 & 0 & 1 \end{bmatrix} \begin{bmatrix} 1 & 0 & 0 \\ 0 & \dfrac{1}{\sqrt{2}} & \dfrac{-1}{\sqrt{2}} \\ 0 & \dfrac{1}{\sqrt{2}} & \dfrac{1}{\sqrt{2}} \end{bmatrix}$$

$$= \begin{bmatrix} 1 & 0 & 0 \\ 0 & \dfrac{1}{\sqrt{2}} & \dfrac{1}{\sqrt{2}} \\ 0 & \dfrac{-1}{\sqrt{2}} & \dfrac{1}{\sqrt{2}} \end{bmatrix} \begin{bmatrix} 2 & \dfrac{2}{\sqrt{2}} & 0 \\ 1 & \dfrac{1}{\sqrt{2}} & \dfrac{-1}{\sqrt{2}} \\ 1 & \dfrac{1}{\sqrt{2}} & \dfrac{1}{\sqrt{2}} \end{bmatrix}$$

$$= \begin{bmatrix} 2 & \dfrac{2}{\sqrt{2}} & 0 \\ \dfrac{2}{\sqrt{2}} & 1 & 0 \\ 0 & 0 & 1 \end{bmatrix} = \begin{bmatrix} 2 & \sqrt{2} & 0 \\ \sqrt{2} & 1 & 0 \\ 0 & 0 & 1 \end{bmatrix},$$

which is the required tri-diagonal matrix.

Problem 3: Find the eigen values of $\begin{bmatrix} 1 & 2 & 2 \\ 2 & 1 & 2 \\ 2 & 2 & 1 \end{bmatrix}$ by Given's method.

Solution: Let $A = \begin{bmatrix} 1 & 2 & 2 \\ 2 & 1 & 2 \\ 2 & 2 & 1 \end{bmatrix}$ since A is of order 3, we require

$\dfrac{(n-1)(n-2)}{2} = \dfrac{(3-1)(3-2)}{2} = 1$ rotation to reduce A into tri-diagonal

matrix using the orthogonal transformation.

Here, $a_{12} = 2$ and $a_{13} = 2$. Therefore,

$$\tan \theta = \frac{a_{13}}{a_{12}} = \frac{2}{2} = 1$$

Hence, $\theta = \dfrac{\pi}{4}$ and the orthogonal matrix is

$$U_1 = \begin{bmatrix} 1 & 0 & 0 \\ 0 & \cos\dfrac{\pi}{4} & -\sin\dfrac{\pi}{4} \\ 0 & \sin\dfrac{\pi}{4} & \cos\dfrac{\pi}{4} \end{bmatrix} = \begin{bmatrix} 1 & 0 & 0 \\ 0 & \dfrac{1}{\sqrt{2}} & \dfrac{-1}{\sqrt{2}} \\ 0 & \dfrac{1}{\sqrt{2}} & \dfrac{1}{\sqrt{2}} \end{bmatrix}$$

and

$$U_1^{-1} = \begin{bmatrix} 1 & 0 & 0 \\ 0 & \dfrac{1}{\sqrt{2}} & \dfrac{1}{\sqrt{2}} \\ 0 & \dfrac{-1}{\sqrt{2}} & \dfrac{1}{\sqrt{2}} \end{bmatrix}$$

Now, the first rotation on A gives

$$T_1 = U_1^{-1} A U_1 = \begin{bmatrix} 1 & 0 & 0 \\ 0 & \dfrac{1}{\sqrt{2}} & \dfrac{1}{\sqrt{2}} \\ 0 & \dfrac{-1}{\sqrt{2}} & \dfrac{1}{\sqrt{2}} \end{bmatrix} \begin{bmatrix} 1 & 2 & 2 \\ 2 & 1 & 2 \\ 2 & 2 & 1 \end{bmatrix} \begin{bmatrix} 1 & 0 & 0 \\ 0 & \dfrac{1}{\sqrt{2}} & \dfrac{-1}{\sqrt{2}} \\ 0 & \dfrac{1}{\sqrt{2}} & \dfrac{1}{\sqrt{2}} \end{bmatrix}$$

$$= \begin{bmatrix} 1 & 2\sqrt{2} & 0 \\ 2\sqrt{2} & 3 & 0 \\ 0 & 0 & -1 \end{bmatrix}$$

Comparing this matrix with $\begin{bmatrix} \beta_1 & \gamma_1 & 0 \\ \gamma_1 & \beta_2 & \gamma_2 \\ 0 & \gamma_2 & \beta_3 \end{bmatrix}$,

We have $\beta_1 = 1$, $\gamma_1 = 2\sqrt{2}$, $\beta_2 = 3$, $\gamma_2 = 0$ and $\beta_3 = -1$. Now we form the sturm sequence

$$f_0 = 1$$
$$f_1 = \lambda - \beta_1 = \lambda - 1$$
$$f_2 = (\lambda - \beta_2)f_1 - \gamma_1^2 f_0$$

$$= (\lambda - 3)(\lambda - 1) - (2\sqrt{2})^2 \tag{1}$$

$$= \lambda^2 - 4\lambda - 5$$

$$f_3 = (f - \beta_3)f_2 - \gamma_2^2 f_1$$

$$= (\lambda + 1)\{\lambda^2 - 4\lambda - 5\} - (0)^2.(\lambda - 1)$$

$$= (\lambda + 1)(\lambda - 5)(\lambda + 1)$$

Thus, the roots $f_3(\lambda) = 0$ are $-1, -1$ and 5 which are the eigen values of the given matrix A.

Problem 4: Transform $A = \begin{bmatrix} 1 & \sqrt{2} & \sqrt{2} & 2 \\ \sqrt{2} & -\sqrt{2} & -1 & \sqrt{2} \\ \sqrt{2} & -1 & \sqrt{2} & \sqrt{2} \\ 2 & \sqrt{2} & \sqrt{2} & -3 \end{bmatrix}$ to tri-diagonal form using Given's method.

Solution: Note that order of given matrix is 4, hence, the number of orthogonal transformations required are $\dfrac{(n-1)(n-2)}{2} = \dfrac{(4-1)(4-2)}{2} = 3$ to reduce A into tri-diagonal form. Now, we proceed perform first orthogonal transformation on A. Here $a_{12} = \sqrt{2}$ and $a_{13} = \sqrt{2}$. To perform the orthogonal transformation about $(2, 3)$ plane, the angle θ is given by

$$\tan\theta = \frac{a_{13}}{a_{12}} = \frac{\sqrt{2}}{\sqrt{2}} = 1$$

$$\therefore \qquad \theta = \frac{\pi}{4}.$$

The orthogonal matrix for the first rotation is

$$U_1 = \begin{bmatrix} 1 & 0 & 0 & 0 \\ 0 & \dfrac{1}{\sqrt{2}} & \dfrac{-1}{\sqrt{2}} & 0 \\ 0 & \dfrac{1}{\sqrt{2}} & \dfrac{1}{\sqrt{2}} & 0 \\ 0 & 0 & 0 & 1 \end{bmatrix}$$

and $U^{-1}_1 = U^{T}_1$,

Hence, the first rotation on A gives

$$T_1 = U^{-1}_1 A U_1 = \begin{bmatrix} 1 & 2 & 0 & 2 \\ 2 & -1 & \sqrt{2} & 2 \\ 0 & \sqrt{2} & 1 & 0 \\ 2 & 2 & 0 & -3 \end{bmatrix}$$

For the second orthogonal transformation $\tan \theta = \dfrac{a^1_{14}}{a^1_{12}} = \dfrac{2}{2} = 1$

where a^1_{14} and a^1_{12} are elements of $T_1 = [a^1_{ii}]$, and therefore, the orthogonal matrix for the second transformation is

$$U_2 = \begin{bmatrix} 1 & 0 & 0 & 0 \\ 0 & \dfrac{1}{\sqrt{2}} & 0 & \dfrac{-1}{\sqrt{2}} \\ 0 & 0 & 1 & 0 \\ 0 & \dfrac{1}{\sqrt{2}} & 0 & \dfrac{1}{\sqrt{2}} \end{bmatrix}$$

and $U^{-1}_2 = U^T_2$.

Hence, the second rotation on A (equivalently first rotation on T_1) gives

$$T_2 = U^{-1}_2 T_1 U_2 = \begin{bmatrix} 1 & 2\sqrt{2} & 0 & 0 \\ 2\sqrt{2} & 0 & 1 & -1 \\ 0 & 1 & 1 & -1 \\ 0 & -1 & -1 & -4 \end{bmatrix}$$

For, third orthogonal transformation, θ is given by

$$\tan \theta = \dfrac{a^{11}_{24}}{a^{11}_{23}} = \dfrac{-1}{1} = -1$$

i.e.,
$$\theta = \dfrac{-\pi}{4}$$

where $T_2 = [a^{11}_{ij}]$

The orthogonal matrix for third rotation is given by

$$U_3 = \begin{bmatrix} 1 & 0 & 0 & 0 \\ 0 & 1 & 0 & 0 \\ 0 & 0 & \dfrac{1}{\sqrt{2}} & \dfrac{1}{\sqrt{2}} \\ 0 & 0 & \dfrac{-1}{\sqrt{2}} & \dfrac{1}{\sqrt{2}} \end{bmatrix}$$

and $U^{-1}{}_3 = U^{T}{}_3$.

The third rotation of matrix on A is given by

$$T_3 = U_3^{-1} T_2 U_3 = \begin{bmatrix} 1 & 2\sqrt{2} & 0 & 0 \\ 2\sqrt{2} & 0 & \sqrt{2} & 0 \\ 0 & \sqrt{2} & \dfrac{-1}{2} & \dfrac{5}{2} \\ 0 & 0 & \dfrac{5}{2} & \dfrac{-5}{2} \end{bmatrix}$$

which is the required tri-diagonal matrix.

11.13   HOUSEHOLDER'S METHOD TO FIND THE EIGEN VALUES OF A SYMMETRIC MATRIX

In this method, a symmetric matrix A of order n is reduced to tri-diagonal form by orthogonal transformations, which represent reflections. Householder suggested the orthogonal transformations are of the form

$$R = I - 2ww^T$$

where $W^T = [x_1 \ x_2 \,........\, x_n]$ such that $x_1^2 + x_2^2 + \,........\, + x_n^2 = 1.$
Note that R is symmetric, since

$$R^T = (I - 2ww^T)^T$$

$$= I^T - 2(w^T)^T \, w^T$$

$$= I - 2ww^T$$

$$= R$$

Also, R is orthogonal since

$$R^T R = (I - 2ww^T)(I - 2ww^T) \ (\because R^T = R)$$

$$= I - 2ww^T - 2ww^T - 2ww^T + 4ww^T \, ww^T$$

$$= I - 4ww^T + 4w \, (w^T w) \, w^T$$

$$= I - 4ww^T + 4w \, (1) \, w^T$$

$$= I$$

Hence, $R^T = R^{-1}$. Thus, R is symmetric and orthogonal

The vectors w_r are constructed with first $(r-1)$ component as zero i.e.,

$$w_r{}^T = [0 \ 0 \ .... \ 0. \ x_r \ x_{r+1} \ .... \ x_n]$$

Since $w_r{}^T w_r = 1$, we obtain $x^2{}_r + x^2{}_{r-1} + ... + x^2{}_n = 1$

Put $A_1 = A$.

With $w_2 = [0, x_2{}^{(2)}, x_3{}^{(2)} \ .... \ x_n{}^{(2)}]$, $w_3 = [0, 0, x_3{}^{(3)} \ .... \ x_n{}^{(3)}] \ .... \ w_n$
$= [0, 0, ... \ 0 \ x_n{}^{(n)}]$ we from the matrices
$R_r = I - 2w_r w_r{}^T \ (r = 2, 3, ..., n)$ and have similar transformations

$$A_2 = R_2 A_1 R_2$$

$$A_3 = R_3 A_2 R_3$$

$$.....................$$

$$A_{n-1} = R_{n-1} A_{n-2} R_{n-1}$$

Using the transformation $A_2 = R_2 A R_2$, we can compute the components $x_2{}^{(2)}, x_3{}^{(2)}, ... \ x_n{}^{(2)}$ of w_2 such that the elements of (first row, 3rd column), (first row, fourth column), ..., (first row, nth column) and the elements of corresponding positions in the first column, i.e., (3rd row, first column), (4th row, first column) .... (nth row, 1st column)) are zero. This rotation gives $(n-2)$ zeros in the first row and first column. Similarly, after performing the transformations $A_3 = R_3 A_2 R_3, ..., A_{n-1} = R_{n-1} A_{n-2} R_{n-1}$, the matrix A will be reduce to tri-diagonal matrix. To find the eigen values of it, we construct the Sturm sequence as in the case of Given's method and solve the characteristic equation $f_n(\lambda) = 0$. The roots of $f_n(\lambda)$ are the eigen values.

11.14 PROCEDURE TO REDUCE 3×3 SYMMETRIC MATRIX *A* TO TRI-DIAGONAL MATRIX USING HOUSEHOLDER'S METHOD

Step 1: Let $A = \begin{bmatrix} a_{11} & a_{12} & a_{13} \\ a_{12} & a_{22} & a_{23} \\ a_{13} & a_{23} & a_{33} \end{bmatrix}$ and note down a_{12} and a_{13} values.

Step 2: Put $A_1 = A$.

Step 3: Suppose $w_2{}^T = [0 \ \ x_2 \ \ x_3]$ such that

$$w_2 w_2{}^T = 1, \text{ i.e., } x_2^2 + x_3^2 = 1. \tag{1}$$

Step 4: Compute $\qquad R_2 = I - 2\, w_2 w_2{}^T$

$$\therefore \qquad R_2 = \begin{bmatrix} 1 & 0 & 0 \\ 0 & 1 - 2x_2^2 & -2x_2 x_3 \\ 0 & -2x_2 x_3 & 1 - 2x_3^2 \end{bmatrix}. \tag{2}$$

Step 5: Find x_2 and x_3 using the following relations.

$$s = \sqrt{a_{12}^2 + a_{13}^2}$$

$$x_2 = \sqrt{\frac{1}{2}\left[1 + \frac{a_{12}\, sign(a_{12})}{s}\right]}$$

$x_3 = \sqrt{1 - x_2^2}$ or $x_2 = \dfrac{a_{13}\, sign(a_{12})}{2sx_2}$, where sign (a_{12}) is the sign of a_{12}. Note that if $x_2 = \sqrt{4}$ then either we can take 2 or –2. The sign of x_2 is immaterial.

Step 6: Substitute value of x_2 and x_3 in (2) to get R_2.

Step 7: Compute $A_2 = R_2\, A\, R_2$, which is in tri-diagonal form.

SOLVED PROBLEMS

Problem 1: Using the Householder's method, transform

$A = \begin{bmatrix} 1 & 3 & 4 \\ 3 & 2 & -1 \\ 4 & -1 & 1 \end{bmatrix}$ to a tri-diagonal matrix.

Solution: Let $w_2 = [0\ x_2\ x_3]^T$ such that $w_2 w_2{}^T = 1$. i.e., $x_2^2 + x_3^2 = 1$. Suppose $A_1 = A$. The values of the elements a_{12} and a_{13} of A are $a_{12} = 3$ and $a_{13} = 4$. The orthogonal matrix R_2 is given by

$$R_2 = I - 2w_2 w_2{}^T = \begin{bmatrix} 1 & 0 & 0 \\ 0 & 1-2x_2^2 & -2x_2 x_3 \\ 0 & -2x_2 x_3 & 1-2x_3^2 \end{bmatrix} \tag{1}$$

$\because$
$$s = \sqrt{a_{12}^2 + a_{13}^2}$$

we have
$$s = \sqrt{3^2 + 4^2} = 5$$

$\therefore$
$$x_2 = \sqrt{\frac{1}{2}\left[1 + \frac{a_{12}\,sign(a_{12})}{s}\right]}$$

$$= \sqrt{\frac{1}{2}\left[1 + \frac{3(+)}{5}\right]}$$

$$= \sqrt{\frac{1}{2} \times \frac{8}{5}} = \frac{2}{\sqrt{5}}$$

$\therefore$
$$x_3 = \sqrt{1 - x_2^2} = \sqrt{1 - \frac{4}{5}} = \frac{1}{\sqrt{5}}$$

Substituting the values of x_2 and x_3 in (1), we get

$$R_2 = \begin{bmatrix} 1 & 0 & 0 \\ 0 & \dfrac{-3}{5} & \dfrac{-4}{5} \\ 0 & \dfrac{-4}{5} & \dfrac{3}{5} \end{bmatrix}$$

Thus, the transformation matrix

$$A_2 = R_2 A_1 R_2$$

$$= \begin{bmatrix} 1 & 0 & 0 \\ 0 & \dfrac{-3}{5} & \dfrac{-4}{5} \\ 0 & \dfrac{-4}{5} & \dfrac{3}{5} \end{bmatrix} \begin{bmatrix} 1 & 3 & 4 \\ 3 & 2 & -1 \\ 4 & -1 & 1 \end{bmatrix} \begin{bmatrix} 1 & 0 & 0 \\ 0 & \dfrac{-3}{5} & \dfrac{-4}{5} \\ 0 & \dfrac{-4}{5} & \dfrac{3}{5} \end{bmatrix}$$

$$= \begin{bmatrix} 1 & -5 & 0 \\[4pt] -5 & \dfrac{2}{5} & \dfrac{1}{5} \\[8pt] 0 & \dfrac{1}{5} & \dfrac{3}{5} \end{bmatrix}$$

which is the required tri-diagonal matrix.

Problem 2: Reduce $A = \begin{bmatrix} 1 & 2 & 2 \\ 2 & 1 & 2 \\ 2 & 2 & 1 \end{bmatrix}$ to tri-diagonal matrix using Householder's method.

Solution: Note that $a_{12} = 2$ and $a_{13} = 2$. Let $w_2 = [0 \ x_2 \ x_3]^T$ such that $w_2 w_2^{\ T} = 1$, i.e., $x_2^{\ 2} + x_3^{\ 2} = 1$. Let $A_1 = A$. The orthogonal matrix

$$R_2 = I - 2w_2 w_2^{\ T} = \begin{bmatrix} 1 & 0 & 0 \\ 0 & 1-2x_2^2 & -2x_2 x_3 \\ 0 & -2x_2 x_3 & 1-2x_3^2 \end{bmatrix} \tag{1}$$

Since $s = \sqrt{a_{12}^2 + a_{13}^2}$, we have $\quad s = \sqrt{2^2 + 2^2} = 2\sqrt{2}$

Hence,

$$x_2 = \sqrt{\frac{1}{2}\left[1 + \frac{a_{12}\,sign(a_{12})}{s}\right]}$$

$$= \sqrt{\frac{1}{2}\left[1 + \frac{2}{2\sqrt{2}}\right]} = \sqrt{\frac{1}{2}\frac{(\sqrt{2}+1)}{\sqrt{2}}}$$

and

$$x_3 = \sqrt{1 - x_2^{\ 2}} = \sqrt{1 - \left\{\frac{2}{2\sqrt{2}}(\sqrt{2}+1)\right\}^2}$$

$$= \sqrt{\frac{(\sqrt{2}-1)}{2\sqrt{2}}}$$

Before substituting the values of x_2 and x_3 in R_2 (i.e., equation (1)), we compute

$$1 - 2x_2^{\ 2},\ x_2\,x_3 \text{ and } 1 - 2x_3^{\ 2},$$

$$1 - 2x_2^2 = 1 - 2 \times \frac{\left(\sqrt{2}+1\right)}{2\sqrt{2}} = \frac{-1}{\sqrt{2}}$$

$$x_2 x_3 = \sqrt{\frac{1}{2\sqrt{2}}\left(\sqrt{2}+1\right) \cdot \frac{1}{2\sqrt{2}}\left(\sqrt{2}-1\right)} = \frac{1}{2\sqrt{2}} \cdot$$

and

$$1 - 2x_3^2 = 1 - 2\left\{\frac{1}{2\sqrt{2}}\left(\sqrt{2}-1\right)\right\} = \frac{1}{\sqrt{2}}$$

Thus,

$$R_2 = \begin{bmatrix} 1 & 0 & 0 \\ 0 & \dfrac{-1}{\sqrt{2}} & \dfrac{-1}{\sqrt{2}} \\ 0 & \dfrac{-1}{\sqrt{2}} & \dfrac{1}{\sqrt{2}} \end{bmatrix}$$

The rotation matrix on A is given by

$$A_2 = R_2 \, A_1 \, R_2$$

i.e.,

$$A_2 = \begin{bmatrix} 1 & 0 & 0 \\ 0 & \dfrac{-1}{\sqrt{2}} & \dfrac{-1}{\sqrt{2}} \\ 0 & \dfrac{-1}{\sqrt{2}} & \dfrac{1}{\sqrt{2}} \end{bmatrix} \begin{bmatrix} 1 & 2 & 2 \\ 2 & 1 & 2 \\ 2 & 2 & 1 \end{bmatrix} \begin{bmatrix} 1 & 0 & 0 \\ 0 & \dfrac{-1}{\sqrt{2}} & \dfrac{-1}{\sqrt{2}} \\ 0 & \dfrac{-1}{\sqrt{2}} & \dfrac{1}{\sqrt{2}} \end{bmatrix}$$

$$= \begin{bmatrix} 1 & -2\sqrt{2} & 0 \\ -2\sqrt{2} & 3 & 0 \\ 0 & 0 & -1 \end{bmatrix},$$

which is the required tri-diagonal firm of transformed matrix of A.

Problem 3: Reduce $\begin{bmatrix} 2 & 1 & 1 \\ 1 & 1 & 0 \\ 1 & 0 & 1 \end{bmatrix}$ to tri-diagonal form using the House-holder's method.

Solution: Let A be the given matrix. Then $a_{12} = 1$ and $a_{13} = 1$. Let $w_2^T = [\, 0 \; x_2 \; x_3 \,]$ such that $w^T w = 1$. i.e., $x_2^2 + x_3^2 = 1$. Suppose $A_1 = A$. Now the orthogonal matrix R_2 is given by

$$R_2 = I - 2ww^T = \begin{bmatrix} 1 & 0 & 0 \\ 0 & 1-2x_2^2 & -2x_2x_3 \\ 0 & -2x_2x_3 & 1-2x_3^2 \end{bmatrix} \qquad (1)$$

By Householder's method, we have

$$S = \sqrt{a_{12}^2 + a_{13}^2} = \sqrt{1+1} = \sqrt{2},$$

$$x_2 = \sqrt{\frac{1}{2}\left[1 + \frac{a_{12}\,sign(a_{12})}{S}\right]}$$

$$= \sqrt{\frac{1}{2}\left[1 + \frac{1}{\sqrt{2}}\right]} = \sqrt{\frac{(\sqrt{2}+1)}{2\sqrt{2}}}$$

and $$x_3 = \sqrt{1 - x_2^2} = \sqrt{\frac{(\sqrt{2}-1)}{2\sqrt{2}}}$$

The values of $1 - 2x_2^2 = \dfrac{-1}{\sqrt{2}}$, $x_2x_3 = \dfrac{1}{2\sqrt{2}}$ and $1 - 2x_3^2 = \dfrac{1}{\sqrt{2}}$

$$R_2 = \begin{bmatrix} 1 & 0 & 0 \\ 0 & \dfrac{-1}{\sqrt{2}} & \dfrac{-1}{\sqrt{2}} \\ 0 & \dfrac{-1}{\sqrt{2}} & \dfrac{1}{\sqrt{2}} \end{bmatrix}$$

Therefore, the orthogonal transformation matrix is given by

$$A_2 = R_2 A R_1 = \begin{bmatrix} 1 & -\sqrt{2} & 0 \\ -\sqrt{2} & 1 & 0 \\ 0 & 0 & 1 \end{bmatrix},$$ which is the required tri-diagonal

matrix of transformed matrix of A.

EXERCISE 11

1. Find the eigen values and eigen vectors of $\begin{bmatrix} 1 & 4 \\ 3 & 2 \end{bmatrix}$.

2. Find the eigen values and eigen vectors of the matrix $A = \begin{bmatrix} 3 & 1 & 4 \\ 0 & 2 & 6 \\ 0 & 0 & 5 \end{bmatrix}$.

3. Find the eigen values of the matrix $A = \begin{bmatrix} 2 & -1 & 1 \\ -1 & 2 & -1 \\ 1 & -1 & 2 \end{bmatrix}$ and also find the

   eigen vector corresponding to the largest eigen vector.

4. If one of the eigen values of the matrix $A = \begin{bmatrix} 1 & 1 & 3 \\ 1 & 5 & 1 \\ 3 & 1 & 1 \end{bmatrix}$ is 3, then find

   the other eigen values of A.

5. If $A = \begin{bmatrix} 4 & 1 \\ 3 & 2 \end{bmatrix}$ find the eigen values of A^2.

6. Obtain the largest eigen value and corresponding eigen vector for the following matrices using Power method

   (i) $\begin{bmatrix} 1 & -3 & 2 \\ 4 & 4 & -1 \\ 6 & 3 & 5 \end{bmatrix}$　　(ii) $\begin{bmatrix} 7 & 4 & -4 \\ 4 & -8 & -1 \\ 4 & -1 & -8 \end{bmatrix}$

   (iii) $\begin{bmatrix} 8 & 1 & 2 \\ 0 & 10 & -1 \\ 6 & 2 & 15 \end{bmatrix}$　　(iv) $\begin{bmatrix} 5 & 0 & 1 \\ 0 & -2 & 0 \\ 1 & 0 & 5 \end{bmatrix}$

   (v) $\begin{bmatrix} 1 & 3 & -1 \\ 3 & 2 & 4 \\ -1 & 4 & 10 \end{bmatrix}$　(vi) $\begin{bmatrix} -4 & -5 \\ 1 & 2 \end{bmatrix}$　(vii) $\begin{bmatrix} 25 & 1 & 2 \\ 1 & 3 & 0 \\ 2 & 0 & -4 \end{bmatrix}$.

7. Obtain the smallest eigen value of the matrix $\begin{bmatrix} 1 & 2 \\ 3 & 4 \end{bmatrix}$ by implementing Power method on A^{-1}.

8. Using Jacobi's method, find the eigen values and eigen vectors corresponding to the eigen values of the matrix $A = \begin{bmatrix} 3 & 2 & 1 \\ 2 & 3 & 2 \\ 1 & 2 & 3 \end{bmatrix}$.

9. Apply Given's method to find the eigen values and eigen vectors of the matrix $A = \begin{bmatrix} 1 & \dfrac{1}{2} & \dfrac{1}{3} \\ \dfrac{1}{2} & \dfrac{1}{3} & \dfrac{1}{4} \\ \dfrac{1}{3} & \dfrac{1}{4} & \dfrac{1}{5} \end{bmatrix}$.

10. Find the eigen values and eigen vectors of $A = \begin{bmatrix} 2 & 2 \\ 2 & 2 \end{bmatrix}$ by Jacobi's method.

11. Find the eigen values and eigen vectors of the matrix $A = \begin{bmatrix} -2 & -2 & 6 \\ -2 & -3 & 4 \\ 6 & 4 & -1 \end{bmatrix}$ by Jacobi's method.

12. Find the eigen values and eigen vectors of the matrix $\begin{bmatrix} 3 & 2 & 1 \\ 2 & 3 & 2 \\ 1 & 2 & 3 \end{bmatrix}$.

13. Find the eigen values and eigen vector of the matrix $A = \begin{bmatrix} 2 & \sqrt{2} & 4 \\ \sqrt{2} & 6 & \sqrt{2} \\ 4 & \sqrt{2} & 2 \end{bmatrix}$ using Jacobi's method.

14. Using Given's method transform $\begin{bmatrix} 3 & 2 & 2 \\ 2 & 5 & 2 \\ 2 & 2 & 3 \end{bmatrix}$ into tri-diagonal matrix.

15. Determine the largest and smallest eigen values of the matrix

$$A = \begin{bmatrix} 9 & 10 & 8 \\ 10 & 5 & -1 \\ 8 & -1 & 3 \end{bmatrix}$$ using power method.

16. Determine the largest eigen value and corresponding eigen vector of

the matrix $\begin{bmatrix} 3 & 2 & 4 \\ -1 & 4 & 10 \\ 1 & 3 & -1 \end{bmatrix}$ using power method.

17. Using Householder's method, reduce the matrix $A = \begin{bmatrix} 6 & 1 & 1 \\ 1 & 4 & 2 \\ 1 & 2 & 3 \end{bmatrix}$ into

tri-diagonal form.

18. Find the eigen values and eigen vector of the matrix $\begin{bmatrix} 1 & -2 & 4 \\ -2 & 5 & -2 \\ 4 & -2 & 1 \end{bmatrix}$

using Jacobi's method.

19. Find the eigen values and eigen vectors of the matrix $\begin{bmatrix} 1 & \sqrt{3} & 4 \\ \sqrt{3} & 5 & \sqrt{3} \\ 4 & \sqrt{3} & 1 \end{bmatrix}$

using Householder's method.

20. Find the eigen values and eigen vector of the matrix $A = \begin{bmatrix} 3 & 2 & 2 \\ 2 & 5 & 2 \\ 2 & 2 & 3 \end{bmatrix}$ by

Householder's method.

ANSWERS

1. $-2, 5;\ \begin{pmatrix} -0.8 \\ 0.6 \end{pmatrix}, \begin{pmatrix} -0.7071 \\ -0.7071 \end{pmatrix}$
   2. $3, 2, 5\ \begin{pmatrix} 1 \\ 0 \\ 0 \end{pmatrix}, \begin{pmatrix} -0.7071 \\ 0.7071 \\ 0 \end{pmatrix}, \begin{pmatrix} 0.8018 \\ 0.5345 \\ 0.2673 \end{pmatrix}$

3. $1, 1, 4;\ \begin{pmatrix} -0.5774 \\ 0.5774 \\ 0.5774 \end{pmatrix}$ 4. $-2, 6$ 5. $25, 1$

6. (i) $7, \begin{pmatrix} 0.2868 \\ 0.0637 \\ 0.9559 \end{pmatrix}$ (ii) $-9, \begin{pmatrix} 0 \\ -0.7071 \\ -0.7071 \end{pmatrix}$ (iii) $16.0380, \begin{pmatrix} -0.2196 \\ 0.1594 \\ -0.9625 \end{pmatrix}$

 (iv) $6, \begin{pmatrix} 0.7071 \\ 0 \\ 0.7071 \end{pmatrix}$ (v) $11.662, \begin{pmatrix} 0.0229 \\ 0.3885 \\ 0.9212 \end{pmatrix}$ (vi) $-3, \begin{pmatrix} -0.9866 \\ 0.1961 \end{pmatrix}$

 (vii) $25.1822, \begin{pmatrix} -0.9967 \\ -0.0449 \\ -0.0683 \end{pmatrix}$

7. $-0.3723, \begin{pmatrix} -0.8246 \\ 0.5658 \end{pmatrix}$

8. $0.6277, 2, 6.3723 \begin{pmatrix} -0.4544 \\ 0.7662 \\ 0.4544 \end{pmatrix} \begin{pmatrix} -0.7071 \\ 0 \\ 0.7071 \end{pmatrix} \begin{pmatrix} 0.5418 \\ 0.6426 \\ 0.5418 \end{pmatrix}$

9. $0.0027, 0.1223, 1.4083 \begin{pmatrix} -0.1277 \\ 0.7137 \\ -0.6887 \end{pmatrix} \begin{pmatrix} 0.5474 \\ -0.5283 \\ -0.6490 \end{pmatrix} \begin{pmatrix} 0.8270 \\ 0.4599 \\ 0.3233 \end{pmatrix}$

10. $0, 4, \begin{pmatrix} -0.7071 \\ 0.7071 \end{pmatrix} \begin{pmatrix} 0.7071 \\ 0.7071 \end{pmatrix}$

11. $-9, 3, 6 \begin{pmatrix} -0.6667 \\ -0.3333 \\ 0.6667 \end{pmatrix} \begin{pmatrix} -0.6667 \\ 0.6667 \\ 0.3333 \end{pmatrix} \begin{pmatrix} 0.3333 \\ 0.6667 \\ 0.6667 \end{pmatrix}$

12. $0, 6227, 2, 6.3723 \begin{pmatrix} -0.4544 \\ 0.7662 \\ -0.4544 \end{pmatrix} \begin{pmatrix} -0.7071 \\ 0 \\ 0.7071 \end{pmatrix} \begin{pmatrix} 0.5418 \\ 0.6426 \\ 0.5418 \end{pmatrix}$

13. $-2, 4, 8$ $\begin{pmatrix} -0.7071 \\ 0 \\ 0.7071 \end{pmatrix} \begin{pmatrix} 0.5 \\ 0.7071 \\ 0.5 \end{pmatrix} \begin{pmatrix} 0.5 \\ 0.7071 \\ 0.5 \end{pmatrix}$

14. $\begin{pmatrix} 3 & 2\sqrt{2} & 0 \\ 2\sqrt{2} & 6 & -1 \\ 0 & -1 & 2 \end{pmatrix}$

15. $19.2861, 4.7914$ $\begin{pmatrix} -0.7786 \\ -0.5205 \\ -0.3505 \end{pmatrix} \begin{pmatrix} -0.0894 \\ 0.6449 \\ -0.7591 \end{pmatrix}$

16. $7.6142,$ $\begin{pmatrix} 0.5987 \\ 0.7323 \\ 0.3245 \end{pmatrix}$

17. $\begin{pmatrix} 6 & \sqrt{2} & 0 \\ \sqrt{2} & 5.5 & -0.5 \\ 0 & -0.5 & 1.5 \end{pmatrix}$

18. $-3, 2.1716, 7.8284$ $\begin{pmatrix} 0.7071 \\ 0 \\ -0.7071 \end{pmatrix} \begin{pmatrix} 0.5 \\ -0.7071 \\ 0.5 \end{pmatrix} \begin{pmatrix} 0.5 \\ 0.7071 \\ 0.5 \end{pmatrix}$

19. $-3, 2.5505, 7.4495$ $\begin{pmatrix} -0.7071 \\ 0 \\ 0.7071 \end{pmatrix} \begin{pmatrix} 0.5 \\ -0.7071 \\ 0.5 \end{pmatrix} \begin{pmatrix} 0.5 \\ 0.7071 \\ 0.5 \end{pmatrix}$

20. $1, 2.1716, 7.8284$ $\begin{pmatrix} 0.7071 \\ 0 \\ -0.7071 \end{pmatrix} \begin{pmatrix} 0.5 \\ -0.7071 \\ 0.5 \end{pmatrix} \begin{pmatrix} 0.5 \\ 0.7071 \\ 0.5 \end{pmatrix}.$

12

Difference Equations

In this chapter, we discuss the methods of solving only ordinary difference equations. We know that the solution of $\dfrac{dy}{dx} - y = 0$ is $y = e^x + c$, where c is an arbitrary constant. Suppose, we were given $\Delta y_n - y_n = 0$. We want to know for what function $y_n = f(n)$, the equation $\Delta y_n - y_n = 0$. Assuming the interval of differencing h as unity, the function y_n is $c2^n$, where c is an arbitrary constant. The equation $\Delta y_n - y_n = 0$ is regarded an ordinary difference equation. There is a close similarity between the methods of solving difference equations and differential equations and also with regard to solutions like complementary function and particular solutions of them. In this chapter, we assume that the values of $y = y(x)$ are defined at equidistant points. Further, we also assume that the variable has been transformed in such a way that the length of interval is 1. i.e., the interval of differencing $h = 1$. In view of this, we shall use y_n for $y(n)$ and $\Delta y_n = y_{n+1} - y_n$. The values of $y = x^2$ at 2, 4, 6 and 8 are same as $y = 4n^2$ at $n = 1, 2, 3$ and 4, respectively. Later function is obtained by the transformation $n = x/2$.

12.1   ORDINARY DIFFERENCE EQUATION

An ordinary difference equation is an equation that

- (i) may or may not contain independent variable;
- (ii) may or may not contain dependent variable; and
- (iii) must contain one or several differences like Δy, $\Delta^2 y$, $\Delta^3 y$, ..., $\Delta^n y$,

where $y = y(x)$ is a function of one independent variable.

A partial difference equation contains several independent variables.

The equations $\Delta^2 y - \Delta y = 0$, $x^2 \Delta^2 y - 3x \Delta y = 0$ and $x^2 \Delta y - y = 0$ are the examples of ordinary difference equation. Hereafter, a difference equation means an ordinary difference equation.

Note: A difference equation can be reduced to $\phi(x, y(x), y(x + 1), ..., y(x + n)) = 0$ using the relation of the operators Δ and E, i.e., $\Delta = E - 1$. For example, $\Delta y(x) - y(x) = x^2$ can be reduced to $(E - 1)y(x) - y(x) = x^2$, i.e., $y(x + 1) - 2y(x) = x^2$. It is called a recurrence relation.

12.2   ORDER, DEGREE AND SOLUTION OF DIFFERENCE EQUATION

The order of a difference equation is the difference between the largest and smallest arguments occurring in it, when the equation expressed is free from the operator Δ. The degree of a difference equation is the highest power of $y(x + rh)$ for some integer r.

For example, the order of $y(x + 2h) - y(x + h) = 0$ is one and the order of $y(x + 3h) - 4y(x) = 0$ is three. The degree of each of these equations is one. The degree of $y(x + 3) - 3y^4(x + 2) + y^2(x + 1) = 0$ is four and the order of this equation is two.

A solution of a difference equation is an expression for $y(x)$, which satisfies that difference equation.

A solution in which the number of arbitrary constants is equal to that of the order of the difference equation is called the general solution.

A solution obtained from the general solution by assigning particular values to the arbitrary constants is called a particular solution.

For the difference equation $\Delta y_n - y_n = 0$, $y_n = c.2^n$ (where c is an arbitrary constant) is the general solution of difference equation where as $y_n = 6.2^n$ is a particular solution of it.

Note: Finding the solution general difference equation is highly complicated. Therefore, we consider only, the solutions of the difference equations of the form.

$$P_0 y_{n+k} + P_1 y_{n+k-1} + P_2 y_{n+k-2} + ... + P_k y_n = f(n), \qquad (12.1)$$

where $P_0, P_1, ..., P_k$ are constants, $f(n)$ is function of n and the interval differencing is unity. The equations of the form (12.1) are called linear difference equations. The difference equations of the type $q_0 y(x + n + k) + q_1 y(x + n + k - 1) + ... + q_k q(x + n) = g(x)$, where

$q_0, q_1, \ldots, q_k$ are constants and $g(x)$ is a function of x can be reduced to (12.1) using suitable transformation, which moves the origin to x.

SOLVED PROBLEMS

Problem 1: Find the order and degree of each of the following difference equations.

(i) $y_{n+2} - 8y_{n+1} + 15y_n = 5^n$

(ii) $y_{n+4}^2 - 6y_{n+1} + y_n^6 = 0$

(iii) $y_{n+2} - 2y_{n+1}\, y_n + 2y_{n-1} = 0.$

Solution:

(i) The highest argument in the given difference equation is $n + 2$ and the lowest argument in it is n. Therefore, the order of difference equation is $(n + 2) - n = 2$. The degree of the difference equation is 1 as the powers of y_{n+2}, y_{n+1}, and y_n is one.

(ii) The highest argument in the given difference equation is $(n + 4)$ and the lowest argument in it is n. Therefore, the order of difference equation is $(n + 4 - n) = 4$. The degree of the difference equation is 2 as the power of y_{n+4} is 2.

(iii) The highest argument in the given difference equation is $(n + 2)$ and the lowest argument in it is $(n - 1)$. Therefore, the order of the difference equation is $\{(n + 2) - (n - 1)\} = 3$. Clearly, the degree of the given difference equation is one.

Problem 2: Express the difference equation $\Delta^3 y_n + 2\Delta^2 y_n + \Delta y_n + y_n = 0$ in subscript notation of y.

Solution: The given difference equation can be expressed as

$$(E - 1)^3 y_n + 2\,(E - 1)^2 y_n + (E - 1) y_n + y_n = 0$$

i.e., $\quad (E^3 - 3E^2 + 3E - 1)y_n + 2\,(E^2 - 2E + 1)y_n + (E - 1)y_n + y_n = 0$

i.e., $\quad (E^3 - 3E^2 + 3E - 1 + 2E^2 - 4E + 2 + E - 1 + 1)y_n = 0$

i.e., $\quad (E^3 - E^2 + 2)y_n = 0$

i.e., $\quad y_{n+3} - y_{n+2} + 2y_n = 0,$

which is the required equation.

Problem 3: Find the order of the difference equation $\Delta^3 y_n - 6\Delta^2 y_n + 2\Delta y_n + y_n = 0$.

Solution: First, we shall express the given difference equation in subscript notation of y. The given difference equation can be expressed as

$$(E - 1)^3 y_n - 6 \ (E - 1)^2 y_n + 2 \ (E - 1) y_n + y_n = 0$$

i.e., $$(E^3 - 9E^2 + 17E - 6) y_n = 0$$

i.e., $$y_{n+3} - 9y_{n+2} + 17y_{n+1} - 6y_n = 0$$

Therefore, the order of the given difference equation is $(n + 3) - n = 3$.

Problem 4: Show that $y_n = c_1 3^n + c_2 2^n$ is a solution of the difference equation $y_{n+2} - 5y_{n+1} + 6y_n = 0$, where c_1 and c_2 are constants.

Solution: Since, $y_n = c_1 3^n + c_2 2^n$, we have $y_{n+1} = c_1 3^{n+1} + c_2 2^{n+1}$ and $y_{n+2} = c_1 3^{n+2} + c_2 2^{n+2}$.

Consider

$$y_{n+2} - 5y_{n+1} + 6y_n$$
$$= (c_1 3^{n+2} + c_2 2^{n+2}) - 5(c_1 3^{n+1} + c_2 2^{n+1}) + 6(c_1 3^n + c_2 2^n)$$
$$= c_1 3^n \ \{3^2 - 5(3) + 6\} + c_2 2^n \ \{2^2 - 5(2) + 6\}$$
$$= c_1 3^n \ (0) + c_2 2^n \ (0)$$
$$= 0.$$

$\therefore y_n = c_1 3^n + c_2 2^n$ is the solution of given difference equation.

Problem 5: Show that $y_n = (c_1 + c_2 n) \ 2^n$ is a solution of the difference equation $y_{n+2} - 4y_{n+1} + 4y_n = 0$.

Solution: Given that $y_n = (c_1 + c_2 n) \ 2^n$; therefore, $y_{n+1} = \{c_1 + c_2 \ (n + 1)\} \ 2^{n+1}$ and $y_{n+2} = \{c_1 + c_2 \ (n + 2)\} \ 2^{n+2}$.

Now,

$$y_{n+2} - 4y_{n+1} + 4y_n$$
$$= \{c_1 + c_2 \ (n + 2)\} \ 2^{n+2} - 4\{c_1 + c_2 \ (n + 1)\} \ 2^{n+1} + 4\{c_1 + c_2 n\} \ 2^n$$
$$= c_1 2^n \ \{2^2 - 4(2) + 4\} + c_2 2^n \ \{4 \ (n + 2) - 8 \ (n + 1) + 4n\}$$
$$= 0.$$

Thus, $y_n = (c_1 + c_2 n) \ 2^n$ is the solution of given difference equation.

12.3   FORMATION OF DIFFERENCE EQUATION

For which differential equation, $y = ce^x$ (c is arbitrary constant) will be its general solution? Clearly $\dfrac{dy}{dx} - y = 0$ is the desired differential equation. Note that the general solution contains one arbitrary constant and the order of differential equation is one. Similar to the case of forming a differential equation when its general solution is given, a difference equation can be formed when its general solution is known. If the general solution of difference equation contain k (positive integer) arbitrary constants the order of difference equation for which the given solution is a general solution is k. To eliminate k arbitrary constants contained in the given general solution of a difference equation we need $(k+1)$ equations involving the k arbitrary constants. Among the $(k+1)$ equations, the given general solution is one of them. So, we need k more equations. These k equations can be obtained by operating $E, E^2, \ldots. E^k$ successively on the given general solution. The eliminant of these k arbitrary equation yields a difference equation for which the given solution is a general solution.

SOLVED PROBLEMS

Problem 1: Form a difference equation for which $y_n = c_1 e^n$ is a solution.

Solution: Given general solution contain one arbitrary constant, i.e., c_1. Therefore, we need one more equation containing c_1. Operating E on both sides of $y_n = c_1 e^n$, we get $y_{n+1} = c_1 e^{n+1}$. Thus, we have two equations $y_n = c_1 e^n$ and $y_{n+1} = c_1 e^{n+1}$ containing c_1. Expressing $y_{n+1} = c_1.e.e^n$ and using $y_n = c_1 e^n$. We have $y_{n+1} = ey_n$, i.e., $y_{n+1} - ey_n = 0$, which is the required difference equation.

Problem 2: For a difference equation by eliminating the arbitrary constants c_1 and c_2 from $y = c_1 3^n + c_2 4^n$.

Solution: Since,

$$y_n = c_1 3^n + c_2 4^n \tag{1}$$

containing two arbitrary constants, the required difference equation must be of order 2. Operating E and E^2 respectively on (1), we get

$$y_{n+1} = c_1 3^{n+1} + c_2 4^{n+1} \tag{2}$$

and
$$y_{n+2} = c_1 3^{n+2} + c_2 4^{n+2} \tag{3}$$

The equations (1), (2), and (3) can be expressed as

$$c_1 3^n + c_2 4^n - y_n = 0$$
$$3c_1 3^n + 4c_2 4^n - y_{n+1} = 0$$
$$9c_1 3^n + 16c_2 4^n - y_{n+2} = 0$$

Eliminating $c_1 3^n$, $c_2 4^n$, -1 from the above equations, we obtain

$$\begin{vmatrix} 1 & 1 & y_n \\ 3 & 4 & y_{n+1} \\ 9 & 16 & y_{n+2} \end{vmatrix} = 0$$

i.e., $y_{n+2} - 7y_{n+1} - 12y_n = 0$, which is the required difference equation.

Problem 3: Form a difference equation by eliminating the arbitrary constants c_1 and c_2 from $y_n = (c_1 n + c_2)\, 4^n$.

Solution: Given that

$$y_n = (c_1 n + c_2)4^n$$

i.e.,
$$c_1 n 4^n + c_2 4^n - y_n = 0 \tag{1}$$

Operating E and E^2 separately on (1), we obtain

$$c_1 (n+1)\, 4^{n+1} + c_2 4^{n+1} - y_{n+1} = 0 \tag{2}$$
$$c_1 (n+2)\, 4^{n+2} + c_2 4^{n+2} - y_{n+2} = 0 \tag{3}$$

Now, eliminating $c_1 4^n$, $c_2 4^n$ and (-1) from the equations (1) to (3), we have

$$\begin{vmatrix} n & 1 & y_n \\ 4(n+1) & 4 & y_{n+1} \\ 16(n+2) & 16 & y_{n+2} \end{vmatrix} = 0$$

i.e.,
$$n\{4y_{n+2} - 16y_{n+1}\} - 1\{4\,(n+1)y_{n+2} - 16\,(n+2)y_{n+1}\}$$
$$+ y_n\{4(n+1)\,(16) - 16\,(n+2)\,(4)\} = 0$$

i.e., $y_{n+2} - 8y_{n+1} + 16y_n = 0$, which is the required difference equation.

EXERCISE 12 (a)

1. Find the order and degree of each of the following difference equation.

   (i) $y_{n+2} + 3y_{n+1} + 2y_n = \cos 2n.$

   (ii) $y_{n+3}^2 - 2y_n y_{n+1} + 16 y_n^2 y_{n+1}^4 = n^2.$

   (iii) $y_{n+1} - y_n = 6^n.$

   (iv) $y_{n+2} - 5y_{n+1} + 6y_n = 0.$

   (v) $y_{n+2} + y_{n+1} y_n^4 + 7y_{n+1}^3 = n^2.$

   (vi) $y_{n+2} + y_{n+1}^2 \, y_n - 6y_n = 0.$

   (vii) $y_{n+4} - 4y_{n+1} + 6y_n = n^2 . 3^n.$

   (viii) $\Delta^3 y_n - 4\Delta y_n + 7y_n = \sin 4n.$

   (ix) $E^2 y_n + 3Ey_{n-1} + y_n = \cos n.$

   (x) $\Delta^2 y_n + 8\Delta y_n + 3y_n = \cos \pi n.$

2. Show that $y_n = c_1 3^n + c_2 4^n$ is a solution of $y_{n+2} - 7 y_{n+1} + 12y_n = 0.$

3. Show that $y_n = (c_1 + c_2 n) 3^n$ is a solution of $y_{n+2} - 6y_{n+1} + 9y_n = 0.$

4. Show that $y_n = c_1 (1)^n + c_2 (2)^n + c_3 3^n$ is a solution of

   $y_{n+3} - 6y_{n+2} + 8y_{n+1} - 6y_n = 0.$

5. Show that $y_n = \left\{ c_1 \cos \dfrac{2n\pi}{3} + c_2 \sin \dfrac{2n\pi}{3} \right\} (1)^n$ is a solution of

   $y_{n+2} + y_{n+1} + y_n = 0.$

6. Show that $y_n = c_1 (1)^n + c_2 (2)^n + \dfrac{5^n}{12} + n . 2^{n-1}$ is a solution of

   $y_{n+2} - 3y_{n+1} + 2y_n = 5^n + 2^n.$

7. Form a difference equation by eliminating c from each of the following equation:

   (i) $y_n = c \, 6^n$ (ii) $y_n = c (1/2)^n$ (iii). $y_n = c. n$

8. Form a difference equation by eliminating c_1 and c_2 from each of the following equation:

   (i) $y_n = c_1 . 6^n + c_2 . 4^n.$

   (ii) $y_n = c_1 (-2)^n + c_2 3^n.$

(iii) $y_n = c_1(-3)^n + c_2(-4)^n$.

(iv) $y_n = (c_1 n + c_2)4^n$.

(v) $y_n = (\sqrt{2})^n \left[c_1 \cos\dfrac{n\pi}{4} + c_2 \sin\dfrac{n\pi}{4} \right]$.

ANSWERS

1. (i) 2,1   (ii) 3,4   (iii) 1,1   (iv) 2,1   (v) 2,4

  (vi) 2,2   (vii) 4,1   (viii) 3,1   (ix) 3,1   (x) 2,1

7. (i) $y_{n+1} - 6y_n = 0$ (ii) $2y_{n+1} - y_n = 0$ (iii) $ny_{n+1} - (n+1)y_n = 0$

8. (i) $y_{n+2} - 10y_{n+1} + 24y_n = 0$ (ii) $y_{n+2} - y_{n+1} - 6y_n = 0$

  (iii) $y_{n+2} + 7y_{n+1} + 12y_n = 0$ (iv) $y_{n+2} - 8y_{n+1} + 16y_n = 0$

  (v) $y_{n+2} - 2y_{n+1} + 2y_n = 0$.

12.4 LINEAR DIFFERENCE EQUATIONS

The difference equation of the form

$$P_0\, y_{n+k} + P_1\, y_{n+k-1} + P_2\, y_{n+k-2} + \ldots\ldots + P_k\, y_n = f(n), \qquad (12.2)$$

where $P_0, P_1, \ldots, P_k$ and $f(n)$ are functions of n, is called Linear difference equation (LDE).

If $f(n) = 0$, then the Eq. (12.2) can be reduced to

$$P_0\, y_{n+k} + P_1\, y_{n+k-1} + \ldots\ldots + P_k\, y_n = 0 \qquad (12.3)$$

which is called homogeneous linear difference equation (HLDE). Hereafter, we use HLDE for homogeneous linear difference equation, otherwise i.e., $f(n) \neq 0$, it is called non-homogeneous linear difference equation. Further if, $P_0, P_1, \ldots, P_k$ are constants, then (12.2) is called LDE with constant coefficients.

The linear difference equation with constant coefficients can be reduced to the following form:

$$y_{n+k} + a_1\, y_{n+k-1} + \ldots.. + a_k\, y_n = f(n),$$

where $a_1, a_2, \ldots\ldots, a_k$ are constants and $f(n)$ is function of n.

Theorem:

If $\phi(n)$ is general solution of linear homogeneous difference equation with constant coefficients,

$$y_{n+k} + a_1 \, y_{n+k-1} + a_2 y_{n+k-2} + \,..... + a_k y_n = 0, \tag{12.4}$$

and $\psi(n)$ is a solution of

$$y_{n+k} + a_1 \, y_{n+k-1} + a_2 \, y_{n+k-2} + \,...... + a_k y_n = f(n), \tag{12.5}$$

where $a_1, a_2, ... a_n$ are constants and $f(n)$ is a function of n, then $\phi(n) + \psi(n)$ is a solution of (12.5)

Proof: Since $\phi(n)$ is a general solution of (12.4), we have

$$\phi_{n+k} + a_1\phi_{n+k-1} + a_2\phi_{n+k-2} + \,...... + a_k\phi_n = 0, \tag{12.6}$$

and $\psi(n)$ is solution (12.5), we have

$$\psi_{n+k} + a_1\psi_{n+k-1} + a_2\psi_{n+k-2} + \,...... + a_k\psi_n = f(n). \tag{12.7}$$

Adding (12.6) and (12.7), we get

$$(\phi_{n+k} + \psi_{n+k}) + a_1 \, (\phi_{n+k-1} + \psi_{n+k-1}) + a_2 \, (\phi_{n+k-2} + \psi_{n+k-2}) + \,......$$
$$+ a_k \, (\phi_n + \psi(n)) = f(n),$$

which implies that $\phi(n) + \psi(n)$ is a solution of (12.5). This solution $\phi(n) + \psi(n)$ is called complete solution of (12.5). The solution $\phi(n)$ is called complementary function and $\psi(n)$ is called particular integral of (12.5).

Thus, $$\text{C.S} = \text{C. F} + \text{P. I}$$

Note that C. F of non-homogeneous linear difference Eq. (12.5) is the general solution of Eq. (12.4).

12.5   GENERAL SOLUTION OF HOMOGENEOUS LINEAR DIFFERENCE EQUATION WITH CONSTANT COEFFICIENTS

Let

$$y_{n+k} + a_1 y_{n+k-1} + a_2 y_{n+k-2} + \,.......... + a_k y_n = 0$$

be a homogeneous linear differential equation with constant coefficients.

Let $y_n = \lambda^n (\lambda \neq 0)$ be solution of (12.4). Then

$$\lambda^{n+k} + a_1\lambda^{n+k-1} + a_2\lambda^{n+k-2} + \,............ + a_k\lambda^n = 0$$

$$\lambda^n(\lambda^k + a_1\lambda^{k-1} + a_2\lambda^{k-2} + \,....... + a_k) = 0$$

Hence,

$$\lambda^k + a_1 \lambda^{k-1} + a_2 \lambda^{k-2} + \ldots\ldots + a_k = 0 \tag{12.8}$$

The Eq. (12.8) is called auxiliary Eq. of (12.4). Eq. (12.8) is a polynomial in λ of kth degree, and hence it will have k roots.

Case (i): When all the roots of auxiliary equation are distinct and real.

Let a_1, a_2, ..., a_k be the distinct and real roots of (12.8).

Then,

$$(E - a_1)(E - a_2) \ldots\ldots (E - a_k)\, y_n = 0$$

$\therefore$
$$(E - a_i)\, y_n = 0.\ (i = 1, 2 \ldots k)$$

Now, consider

$$(E - a_1)\, y_n = 0$$

i.e.,
$$y_{n+1} - a\, y_n = 0.$$

The solution of this difference equation is $y_n = c_1 a_1{}^n$, where c_1 is an arbitrary constant or c_1 is a periodic function with period 1. Hereafter, we use the phrases "arbitrary constants" and "arbitrary periodic function with period one" interchangeably in the rest of the chapter. Similarly, it can be shown that

$y_n = c_2 a_2{}^n, y_n = c_3 a_3{}^n, \ldots y_n = c_k a_k{}^n$ are solutions of $(E - a_i)\, y_n = 0$ for $i = 2, 3,\ldots\ldots k$. Thus,

$$y_n = c_1\, a_1{}^n + c_2\, a_2{}^n + {}_{\ldots} + c_k\, a_k{}^n.$$

Case (ii): The Auxiliary equation has repeated real roots.

Sub-case (i): The Auxiliary equation has a double root. Let a be a double root of auxiliary Eq. (12.8). Then

$$(E - a)^2\, y_n = 0$$

i.e.,
$$(E - a)\, (E - a)\, y_n = 0 \tag{12.9}$$

Let
$$(y - a)\, y_n = z_n \tag{12.10}$$

Eq. (12.9) becomes

$$(E - a)\, E_n = 0.$$

Clearly,
$$z_n = c_2{}^1 a^n. \tag{12.11}$$

In view of (12.10), Eq. (12.11) becomes

$$(E - a)\, y_n = c_2^{\,1} a^n$$

i.e.,
$$y_{n+1} - a y_n = c_2^{\,1} a^n.$$

Multiplying by a^{-n-1}, we get

$$a^{-n-1}\, y_{n+1} - a^{-n}\, y_n = c_2^{\,1} a^{-1}$$

$$\Delta\{a^{-n} y_n\} = c_2^{\,1} a^{-1}$$

$$\therefore \qquad a^{-n}.\, y_n = \frac{1}{\Delta}\, (c_2^{\,1} a^{-1})$$

i.e.,
$$a^{-n} y_n = c_2^{\,1} a^{-1}\, n + c_1$$

i.e.,
$$y_n = (c_1^{\,1}\, a^{-1} n + c_1)\, a^n$$

or
$$y_n = (c_2\, n + c_1)\, a^n, \text{ where } c_2^{\,1}\, a^{-1} = c_2.$$

Sub-case (ii): If a root a of auxiliary equation repeated m times.

Using the induction, it can be proved that if a root a is repeated m times of an auxiliary equation then the solution corresponding to this root is of

$$(c_1 + c_2 n + c_3 n^2 + \ldots + c_m n^{m-1})\, a^n,$$

which is the part of the complementary function or it may be a complementary function.

Case (iii): The auxiliary equation has non-repeated complex roots.

Suppose the auxiliary Eq. (12.8) has two complex roots. As the complex roots of a polynomial equation with real coefficients occurs in pairs, and they are such that one is conjugate to the other, we suppose them as

$$a_1 = a + i\beta \text{ and } a_2 = a - i\beta$$

The part of the complementary function of (12.4) corresponding to these two roots are

$$c_1\, (a + i\beta)^n + c_2\, (a - i\beta)^n \qquad (12.12)$$

Expressing $a + i\beta$ in the polar form, we have

$$a + i\beta = r(\cos\theta + i\sin\theta), \qquad (12.13)$$

where

$$r = \sqrt{a^2 + \beta^2},\ \theta = \tan^{-1}\left(\frac{\beta}{a}\right) \qquad (12.14)$$

and

$$a - i\beta = r(\cos\theta - i\sin\theta) \tag{12.15}$$

In view of (12.13) and (12.15), the expression (12.12) reduces to

$$c_1(a + i\beta)^n + c_2(a - i\beta)^n$$

$$= c_1 r^n(\cos n\theta + i\sin n\theta) + c_2 r^n(\cos n\theta - i\sin n\theta)$$

$$= r^n\{(c_1 + c_2)\cos n\theta + (c_1 - c_2)\sin n\theta\}$$

$$= r^n(A\cos n\theta + B\sin n\theta),$$

where $\qquad A = c_1 + c_2$ and $B = (c_1 - c_2)i$

Thus, if $a + i\beta$ and $a - i\beta$ are roots of Eq. (12.8), then the solution corresponding to these roots is

$$r^n(A\cos n\theta + B\sin n\theta),$$

where r and θ are given by (12.14).

Case (iv): The auxiliary equation has repeated complex roots.

Suppose the roots of auxiliary Eq. (12.8) be $a + i\beta$, $a - i\beta$ and each is repeated twice. Then, the roots are $a + i\beta$, $a - i\beta$, $a + i\beta$, $a - i\beta$. Using the argument given in case (ii) and case (iii), it can be shown that the complementary function corresponding to these four roots is

$$[\{(c_1 + c_2 n)\cos n\theta + (c_3 + c_4 n)\sin n\theta\}] r^n,$$

where r and θ are given by (12.14).

Note: The summary of (12.5) is given below, which may be useful in finding the complementary function of 12.8.

(i) The part of the C. F., which is a solution of difference equation corresponding to each non-repeated root a of auxiliary equation is ca^n.

(ii) The part of C. F., which is a solution of difference equation corresponding to

    (a) double root a (a is real) of A.E. is

$$(c_1 + c_2 n)\, a^n$$

    (b) triple root a (a is real) of A.E is

$$(c_1 + c_2 n + c_3 n^2)\, a^n$$

   (c) quadruple root a (a is real) of A.E is

$$(c_1 + c_2 n + c_3 n^2 + c_4 n^3)\, a^{\,n}$$

   (d) a real root a repeated m (m is positive integer ≥ 2) times is

$$(c_1 + c_2 n + c_3 n^2 + \ldots\ldots + c_m n^{m-1})\, a^n.$$

(iii) The part of C. F., which is a solution of difference equation, corresponding to two complex roots $a + i\beta$, $a - i\beta$ of its auxiliary equation is

$$(c_1 \cos n\theta + c_2 \sin n\theta)\, r^n,$$

where r and θ are given by (12.14).

(iv) The part of C. F., which is a solution of difference equation corresponding to the roots $a + i\beta$, $a + i\beta$, $a - i\beta$, $a - i\beta$ of its auxiliary equation is

$$[(c_1 + c_2\, n) \cos n\theta + (c_3 + c_4\, n) \sin n\theta]\, r^n,$$

where r and θ are given by (12.14).

EXAMPLE

The complementary functions of HLDEs are given in the following table for the given roots of their auxiliary equations as illustrations.

	Roots of A.E.	C. F.
1	$2, 3, 7$	$c_1 2^n + c_2 3^n + c_3 7^n$
2	$1, -2, 1/2$	$c_1\, (1)^n + c_2\, (-2)^n + c_3\, (1/2)^n$
3	$3, 3$	$(c_1 + c_2 n)\, 3^n$
4	$4, 4, 4$	$(c_1 + c_2 n + c_3 n^2)\, 4^n$
5	$6, 6, 6, 6$	$(c_1 + c_2 n + c_3 n^2 + c_4\, n^3)\, 6^n$
6	$2 + 3i, 2 - 3i$	$r^n (c_1 \cos n\theta + c_2 \sin n\theta)$, where $r = \sqrt{2^2 + 3^2} = \sqrt{13}$ and $\theta = \tan^{-1}\left(\dfrac{3}{2}\right)$
7	$0, 2, 3$	$c_1\, (0)^n + c_2\, (2)^n + c_3\, (3)^n$

8	$2, 2, 3 + i, 3 - i$	$(c_1 + c_2 n)\, 2^n + r^n\, (c_3 \cos n\theta + c_4 \sin n\theta)$, where $r = \sqrt{3^2 + (-1)^2} = \sqrt{10}$ and $\theta = \tan^{-1}\left(\dfrac{1}{3}\right)$
9	$3 + 2i, 3 - 2i,$ $3 + 2i, 3 - 2i$	$[(c_1 + c_2 n) \cos n\theta + (c_3 + c_4 n)]\, r$, where $r = \sqrt{3^2 + 2^2} = \sqrt{13}$ and $\theta = \tan^{-1}\left(\dfrac{2}{3}\right)$
10	$\dfrac{1}{2}, \dfrac{2}{3}, -4, 6$	$c_1 \left(\dfrac{1}{2}\right)^n + c_2 \left(\dfrac{2}{3}\right)^n + c_3 (-4)^n + c_4 (6)^n$

12.6 WORKING RULE TO FIND THE COMPLEMENTARY FUNCTION OF HLDE

Suppose $y_{n+k} + a'_1 y_{n+k-1} + \ldots + a'_k y_n = 0$ be a HLDE with constant coefficients.

Step 1: Express the given HLDE in the form

$$\phi(E)\, y_n = 0,$$

where $\phi(E) = E^{n+k} + a_1 E^{n+k-1} + \ldots + a_k.$

Step 2: Write the auxiliary equation $\phi(\lambda) = 0$

($\phi(\lambda)$ is obtained by replace E by λ in $\phi(E)$).

Step 3: Find the roots of $\phi(\lambda) = 0$.

Step 4: Separate the roots of $\phi(\lambda) = 0$ Such that

   (i)   They are real and distinct;

   (ii)   A real root and its repeated;

   (iii)   Pair of complex roots that are non-repeated and one will be conjugate to other;

   (iv)   A pair of two complex roots (one is conjugate to other) repeated (for example; $\alpha + i\beta, \alpha - i\beta, \alpha + i\beta, \alpha - i\beta$).

Step 5: Following the rules stated in Note of section 12.5 form the solution of each category specified in Step 4 and take their sum, which will be the required complementary function.

SOLVED PROBLEMS

Problem 1: Solve $y_{n+2} - 4y_{n+1} + 3y_n = 0$.

Solution: The given difference equation can be written as $(E^2 - 4E + 3)$ $y_n = 0$. Hence its auxiliary equation is

$$\lambda^2 - 4\lambda + 3 = 0$$

The roots of it are 1 and 3. These roots are real and distinct.

$$\therefore \qquad \text{C. F.} = c_1 (1)^n + c_2 (3)^n$$

Thus, $\qquad\qquad\qquad y_n = c_1 (1)^n + c_2 (2)^n.$

Problem 2: Solve the difference equation $y_{n+2} - 4y_{n+1} + 4y_n = 0$.

Solution: The given difference equation can be written as $(E^2 + 4E + 4)$ $y_n = 0$. Hence, its auxiliary equation is $\lambda^2 + 4\lambda + 4 = 0$. i.e., $(\lambda + 2)^2 = 0$. Hence, its roots are -2 and -2. i.e., -2 is a double root. Therefore,

$$\text{C. F.} = (c_1 + c_2 n) \, (-2)^n$$

Thus,

$$y_n = (c_1 + c_2 \, n) \, (-2)^n.$$

Problem 3: Solve $y_{n+2} - 4y_{n+1} + 13y_n = 0$.

Solution: The given difference equation can be expressed as $(E^2 - 4E + 13) \, y_n = 0$. Hence, its auxiliary equation is $\lambda^2 - 4\lambda + 13 = 0$. Therefore, its roots are $\dfrac{\pm 4 \pm \sqrt{16 - 4 \times 1 \times 13}}{2} = \dfrac{4 \pm 6i}{2} = 2 \pm 3i$. Thus, the roots of auxiliary equation are $2 + 3i$ and $2 - 3i$, which are complex.

$$\therefore \text{C. F.} = (c_1 \cos n\theta + c_2 \sin n\theta) \, r^n,$$

where $r = \sqrt{(2)^2 + (3)^2} = \sqrt{13}$ and $\theta = \tan^{-1}\left(\dfrac{3}{2}\right)$. Thus,

$$y_n = (c_1 \cos n\theta + c_2 \sin n\theta) \, r^n.$$

Problem 4: Solve $y_{n+3} + 4y_{n+2} + 4y_{n+1} = 0$.

Solution: The given difference equation can be expressed as $(E^3 + 4E^2 + 4E) \, y_n = 0$. Hence, its auxiliary equation is $\lambda^3 + 4\lambda^2 + 4\lambda = 0$. i.e.,

$\lambda(\lambda^2 + 4\lambda + 4) = 0$, i.e., $\lambda(\lambda + 2)^2 = 0$. Therefore, the roots of auxiliary equation are 0, –2, –2. Note that the real root –2 is a double root.

Therefore,

$$\text{C. F} = c_1 0^n + (c_2 + c_3 n)\,(-2)^n$$

Thus,

$$y_n = c_1 0^n + (c_2 + c_3 n)\,(-2)^n.$$

Problem 5: Solve $(y_{n+3} - 3y_{n+2} + 3y_{n+1} - y_n) = 0$.

Solution: The given difference equation can be written as $(E^3 - 3E^2 + 3E - 1)\,y_n = 0$. Hence, its auxiliary equation is $\lambda^3 - 3\lambda^2 + 3\lambda - 1 = 0$, i.e., $(\lambda - 1)^3 = 0$. Therefore, the roots of auxiliary equation are 1, 1, 1. The real root 1 is repeated thrice (3 times).

$\therefore$
$$\text{C. F.} = (c_1 + c_2 n + c_3 n^2)\,(1)^n$$

Thus,

$$y_n = (c_1 + c_2 n + c_3 n^2)\,(1)^n.$$

Problem 6: Solve $y_{n+3} + 3y_{n+2} - 4y_n = 0$.

Solution: The given difference equation can be written in the form $(E^3 + 3E^2 - 4)\,y_n = 0$. Hence, its auxiliary equation is

$$\lambda^3 + 3\lambda^2 - 4 = 0 \tag{1}$$

By trial-and-error method, we can observe that $\lambda = 1$ is a real root of Eq. (1) $[\because 1^3 + 3(1)^2 - 4 = 0]$.

Therefore, Eq. (1) becomes

$$(\lambda - 1)\,(\lambda^2 + 4\lambda + 4) = 0$$

$$(\lambda - 1)\,(\lambda + 2)^2 = 0$$

Therefore, the roots of auxiliary equation are 1, –2, and –2. Note that –2 is a double root.

$\therefore$
$$\text{C. F.} = c_1\,(1)^n + (c_2 + c_3 n)\,(-2)^n$$

Thus,

$$y_n = c_1\,(1)^n + (c_2 + c_3 n)\,(-2)^n.$$

Problem 7: Solve $(6y_{n+2} - 5y_{n+1} + y_n) = 0$.

Solution: The given difference equation can be written in the form $(6E^2 - 5E + 1)\,y_n = 0$. Hence, its auxiliary equation is $6\lambda^2 - 5\lambda + 1 = 0$, i.e., $(3\lambda - 1)(2\lambda - 1) = 0$. Hence, the roots of auxiliary equation are 1/3, 1/2.

$$\therefore \qquad C.\,F. = c_1\,(1/3)^n + c_2\,(1/2)^n.$$

Thus,

$$y_n = c_1\left(\frac{1}{3}\right)^n + c_2\left(\frac{1}{2}\right)^n.$$

Problem 8: Solve $y_{n+3} - 2y_{n+2} + y_{n+1} - 2y_n = 0$.

Solution: The given difference equation can be written as $(E^3 - 2E^2 + E - 2)\,y_n = 0$. Hence, its auxiliary equation is $\lambda^3 - 2\lambda^2 + \lambda - 2 = 0$, i.e., $\lambda^2(\lambda - 2) + (\lambda - 2) = 0$, i.e., $(\lambda - 2)(\lambda^2 + 1) = 0$. Therefore, the roots of auxiliary equation are $2, i, -i$.

$$\therefore \qquad C.\,F. = c_1 2^n + (c_2\,\cos n\theta + c_3\,\sin n\theta)\,r^n,$$

where $r = \sqrt{0^2 + 1^2} = 1$ and $\theta = \tan^{-1}\left(\frac{1}{0}\right) = \frac{\pi}{2}$

Thus, $\qquad y_n = c_1 2^n + (c_2\,\cos n\pi/2 + c_3\,\sin n\pi/2)\,(1)^n.$

Problem 9: Solve $y_{n+4} - 8y_{n+3} + 24y_{n+2} - 32y_{n+1} + 16y_n = 0$.

Solution: The given difference equation can be written as $(E^4 - 8E^3 + 24E^2 - 32E + 16)\,y_n = 0$. Hence, its auxiliary equation is $\lambda^4 - 8\lambda^3 + 24\lambda^2 - 32\lambda + 16 = 0$, i.e., $(\lambda - 2)^4 = 0$. Therefore, the roots of auxiliary equation are $2, 2, 2, 2$, i.e., the real root 2 is repeated 4 times.

$$\therefore \qquad C.\,F. = (c_1 + c_2 n + c_3 n^2 + c_4 n^3)\,(2)^n$$

Thus,

$$y_n = (c_1 + c_2 n + c_3 n^2 + c_4 n^3)\,(2)^n.$$

Problem 10: Solve $y_{n+2} - 6y_{n+1} + 8y_n = 0$, given that $y_0 = 3$, and $y_1 = 10$.

Solution: The given difference equation can be written in the form $(E^2 - 6E + 8)\,y_n = 0$. Hence, its auxiliary equation is $\lambda^2 - 6\lambda + 8 = 0$, i.e., $(\lambda - 2)(\lambda - 4) = 0$. Therefore, the roots of auxiliary equation are 2 and 4, which are real and distinct.

$$\therefore \qquad C.\,F. = c_1\,(2)^n + c_2\,(4)^n$$

Thus,

$$y_n = c_1\,(2)^n + c_2\,(4)^n. \tag{1}$$

Now, for $n = 0$, $y_0 = c_1 + c_2$ and for $n = 1$, we have $y_1 = 2c_1 + 4c_2$. Given that $y_0 = 3$ and $y_1 = 10$; Therefore,

$$c_1 + c_2 = 3 \tag{2}$$

$$2c_1 + 4c_2 = 10 \tag{3}$$

Solving (2) and (3), we get $c_1 = 1$, $c_2 = 2$.

$$\therefore \qquad y_n = 1.(2)^n + 2.(4)^n.$$

Problem 11: Given that $2 + 3i$ is a root of $t^4 - 8t^3 + 36t^2 - 104t + 169 = 0$. Solve the difference equation $y_{n+4} - 8y_{n+3} + 36y_{n+2} - 104y_{n+1} + 169y_n = 0$.

Solution: The given difference equation can be written as $(E^4 - 8E^3 + 36E^2 - 104E + 169)\, y_n = 0$. Hence, its auxiliary equation is

$$\lambda^4 - 8\lambda^3 + 36\lambda^2 - 104\lambda + 169 = 0. \tag{1}$$

From the hypothesis, $2 + 3i$ is a root of (1). As the complex root of a polynomial equation with real coefficients occurs in pairs such that one will be conjugate to other, the equation (1) will have $(2 - 3i)$ as a root. Hence, $\{\lambda - (2 + 3i)\}\,\{\lambda - (2 - 3i)\} = \lambda^2 - 4\lambda + 13$ is a factor of L.H.S of (1). Dividing L.H.S of (1) by $\lambda^2 - 4\lambda + 13$, we will have quotient $\lambda^2 - 4\lambda + 13$ and remainder is zero. Hence, Eq. (1) reduces to

$$(\lambda^2 - 4\lambda + 13)\,(\lambda^2 - 4\lambda + 13) = 0$$

Thus, the roots of auxiliary equation are $2 + 3i$, $2 - 3i$, and $2 + 3i$, $2 - 3i$. Hence,

$$\text{C. F.} = [(c_1 + c_2 n)\,\cos n\theta + (c_3 + c_4 n)\,\sin n\theta],$$

where

$$r = \sqrt{2^2 + 3^2} = \sqrt{13}, \quad \theta = \tan^{-1}\left(\frac{3}{2}\right) \tag{2}$$

Thus,

$$y_n = [(c_1 + c_2 n)\,\cos n\theta + (c_3 + c_4 n)\,\sin n\theta]\,r^n,$$

where r and θ are given (2).

Problem 12: If $F_1 = 0$ and $F_2 = 1$, solve the recurrence relation of Fibonacci sequence of numbers $F_{n+2} = F_n + F_{n+1}$ $(n = 1, 2...)$.

Solution: The given recurrence relation is $F_{n+2} - F_{n+1} - F_n = 0$. It can be written as $(E^2 - E - 1)\,F_n = 0$. Hence, its auxiliary equation is $\lambda^2 - \lambda - 1 = 0$. The roots of it are

$$\frac{1 \pm \sqrt{1+4}}{2} = \frac{1 \pm \sqrt{5}}{2}$$

$$\text{C. F.} = c_1 \left(\frac{1+\sqrt{5}}{2} \right)^n + c_2 \left(\frac{1-\sqrt{5}}{2} \right)^n$$

Thus,

$$F_n = c_1 \left(\frac{1+\sqrt{5}}{2} \right)^n + c_2 \left(\frac{1-\sqrt{5}}{2} \right)^n \tag{1}$$

Given that $F_1 = 0$,

$$\therefore \qquad 0 = c_1 \left(\frac{1+\sqrt{5}}{2} \right)^1 + c_2 \left(\frac{1-\sqrt{5}}{2} \right)^1 \tag{2}$$

and also given that $F_2 = 1$,

$$\therefore \qquad 1 = c_1 \left(\frac{1+\sqrt{5}}{2} \right)^2 + c_2 \left(\frac{1-\sqrt{5}}{2} \right)^2 \tag{3}$$

multiplying (2) by $\left(\dfrac{1+\sqrt{5}}{2} \right)$ and subtracting from (3), we get

$$c_2 \left[\left(\frac{1-\sqrt{5}}{2} \right)^2 - \left(\frac{1-\sqrt{5}}{2} \right)\left(\frac{1+\sqrt{5}}{2} \right) \right] = 1$$

$$c_2 \left[\frac{(1-2\sqrt{5}+5)-(1-5)}{4} \right] = 1$$

$$\frac{c_2}{4} \times (10 - 2\sqrt{5}) = 1$$

$$\therefore \qquad c_2 = \frac{5+\sqrt{5}}{10}$$

Similarly, it can be shown that $c_1 = \dfrac{5-\sqrt{5}}{10}$. Thus,

$$F_n = \left(\frac{5-\sqrt{5}}{10} \right)\left(\frac{1+\sqrt{5}}{2} \right)^n + \left(\frac{5+\sqrt{5}}{10} \right)\left(\frac{1-\sqrt{5}}{2} \right)^n .$$

12.7   PARTICULAR INTEGRAL OF NHLDE

Let $\phi(E) y_n = f(n)$ be a non-homogeneous linear difference equation, where

$$\phi(E) = E^k + a_1 E^{k-1} + a_2 E^{k-2} \ldots + a_k.$$

and $a_1, a_2, \ldots a_k$ are constants

Then, $\dfrac{1}{\phi(E)} f(n)$ is defined as a particular integral of $\phi(E) y_n = f(n)$.

Note that, if $\dfrac{1}{\phi(E)} f(n) = g(n)$ then $\phi(E) g(n) = f(n)$, and $g(n)$ is the particular integral of $\phi(E) y_n = f(n)$.

Theorem:

If $\phi(E) = E^k + a_1 E^{k-1} + a_2 E^{k-2} \ldots + a_k$ and $f(n)$ and $g(n)$ are functions of n, then

(i) $\dfrac{1}{\phi(E)} a = \dfrac{a}{\phi(1)}$ if $\phi(1) \neq 0$, where a is a constant and

$$\frac{1}{(E-1)^m} a = \frac{a n^{(m)}}{\underline{|m}}$$

(ii) $\dfrac{1}{\phi(E)} a f(n) = a \dfrac{1}{\phi(E)} f(n)$, where a is any constant

(iii) $\dfrac{1}{\phi(E)} \{f(n) + g(n)\} = \dfrac{1}{\phi(E)} f(n) + \dfrac{1}{\phi(E)} g(n)$

(iv) $\dfrac{1}{\phi(E)} \{a f(n) + \beta g(n)\} = a \dfrac{1}{\phi(E)} f(n) + \beta \dfrac{1}{\phi(E)} g(n),$

where a, β are constants.

Proof (i): Consider $\phi(E) a = (E^k + a_1 E^{k-1} + \ldots + a_k) a$

$$= a + a_1 a + \ldots + a_k a \ (\because E^m a = a)$$

$$= \phi(1) a$$

$$\frac{1}{\phi(1)} \phi(E) a = a \ \text{if} \ \phi(1) \neq 0.$$

$$\phi(E)\left\{\frac{a}{\phi(1)}\right\} = a$$

$$\Rightarrow \frac{1}{\phi(E)}a = \frac{a}{\phi(1)} \tag{12.16}$$

If $\phi(1) = 0$, then $E = 1$ is a root of $\phi(E) = 0$

Let
$$\frac{1}{E-1}a = g(n) \tag{12.17}$$

$$a = (E-1)\,g(n)$$

i.e.,
$$g(n+1) - g(n) = a$$

$$\Delta g(n) = a$$

$$g(n) = \frac{1}{\Delta}a$$

$$g(n) = an^{(1)}$$

i.e.,
$$\frac{1}{E-1}a = an^{(1)} \tag{By 12.17}$$

Hence,
$$\frac{1}{E-1}a = an^{(1)}$$

Similarly, it can be shown that

$$\frac{1}{(E-1)^m}a = \frac{a.n^{(m)}}{\lfloor m}$$

(ii) Let
$$\frac{1}{\phi(E)}f(n) = g(n) \tag{12.18}$$

Then,
$$\phi(E)\,g(n) = f(n)$$

$$\therefore a\phi(E)\,g(n) = af(n)$$

i.e.,
$$\phi(E)\left\{ag(n)\right\} = af(n)$$

$$\therefore \frac{1}{\phi(E)}a\,.\,f(n) = a.g(n)$$

i.e.,
$$\frac{1}{\phi(E)}a\,.\,f(n) = a\,.\,\frac{1}{\phi(E)}f(n) \tag{By 12.18}$$

(iii) Let
$$\frac{1}{\phi(E)} f(n) = f_p(n) \tag{12.19}$$

$$\frac{1}{\phi(E)} g(n) = g_p(n); \tag{12.20}$$

Then,
$$\phi(E) f_p(n) = f(n)$$

$$\phi(E) g_p(n) = g(n).$$

Adding these two equations, we get

$$\phi(E) \{f_p(n) + g_p(n)\} = f(n) + g(n)$$

Evalution of $\dfrac{1}{f(t)} f(n)$ when $f(N) = A^N$

$$\frac{1}{\phi(E)} \{f(n) + g(n)\} = f_p(n) + g_p(n)$$

$$\frac{1}{\phi(E)} \{f(n) + g(n)\} = \frac{1}{\phi(E)} f(n) + \frac{1}{\phi(E)} g(n). \text{ (By 12.19 and 12.20)}$$

(iv)   The result (iv) can be proved using (ii) and (iii).

12.8   EVALUATION OF $\dfrac{1}{\phi(E)} f(n)$ WHEN $f(n) = a^n$, WHERE a IS A CONSTANT

Let
$$\phi(E) = E^k + a_1 E^{k-1} + \ldots + a_k.$$

Then
$$\phi(E) a^n = a^{n+k} + a_1 a^{n+k-1} + a_2 a^{n+k-2} + \ldots + a_k a^n$$

$$= a^n (a^k + a_1 a^{k-1} + a_2 a^{k-2} + \ldots + a_k)$$

$$= a^n \phi(a)$$

Suppose, $\phi(a) \neq 0$, then

$$\frac{1}{\phi(a)} \phi(E).a^n = a^n$$

$$\phi(E) \left\{ \frac{1}{\phi(a)}.a^n \right\} = a^n$$

Operating, $\dfrac{1}{\phi(E)}$ on both sides, we have

$$\frac{1}{\phi(E)}\, a^n = \frac{1}{\phi(a)}.a^n \text{ if } \phi(a) \neq 0 \qquad (12.21)$$

Suppose $\phi(a) = 0$

Case (i): Let a be non-repeated root of $\phi(E)$.

Then, $\phi(E) = (E-a)\psi(E)$ and $\psi(a) \neq 0$

Now
$$\frac{1}{\phi(E)}\, a^n = \frac{1}{(E-a)\psi(E)}.a^n$$

i.e.,
$$\frac{1}{\phi(E)}\, a^n = \frac{1}{(E-a)\psi(a)}.a^n$$

i.e.,
$$\frac{1}{\phi(E)}\, a^n = \frac{1}{\psi(a)}\frac{1}{(E-a)}.a^n \qquad (12.22)$$

Let
$$\frac{1}{(E-a)}.a^n = g(n); \qquad (12.23)$$

Then
$$(E-a)\, g(n) = a^n$$

$$g(n+1) - ag(n) = a^n$$

Dividing by a^{n+1}

$$\frac{g(n+1)}{a^{n+1}} - \frac{g(n)}{a^n} = \frac{1}{a}$$

$$\Delta\left\{\frac{g(n)}{a^n}\right\} = \frac{1}{a}$$

$$\frac{g(n)}{a^n} = \frac{1}{a}\frac{1}{\Delta}1$$

$$= \frac{1}{a}n^{(1)}$$

$\therefore$
$$g(n) = n.\, a^{n-1}$$

Thus,
$$\frac{1}{(E-a)}.a^n = n.a^{n-1} \qquad \text{(By 12.23)}$$

$$\therefore \frac{1}{\phi(E)}\, a^n = \frac{1}{\psi(a)}n.a^{n-1}$$

Case (ii): Let a be a repeated root of $\phi(E)$ and it repeats m times. $(m \le k)$. Then

$$\phi(E) = (E-a)^m \, \psi(n) \text{ and } \psi(a) \ne 0.$$

Now,

$$\frac{1}{\phi(E)} a^n = \frac{1}{(E-a)^m \, \psi(E)} . a^n$$

$$= \frac{1}{(E-a)^m \, \psi(a)} . a^n = \frac{1}{\psi(a)} \frac{1}{(E-a)^m} . a^n$$

$$= \frac{1}{\psi(a)} \frac{1}{(E-a)^{m-1}} \left\{ \frac{1}{E-a} a^n \right\}$$

$$= \frac{1}{\psi(a)} \frac{1}{(E-a)^{m-1}} \left(n^{(1).} . a^{n-1} \right)$$

$$= \frac{1}{\psi(a)} \frac{1}{(E-a)^{m-2}} \frac{1}{E-1} n^{(1)} . a^{n-1}$$

$$= \frac{1}{\psi(a)} = \frac{1}{(E-a)^{m-2}} \times \left\{ \frac{n^{(2)} a^{n-2}}{\lfloor 2} \right\}$$

$$= \frac{1}{\psi(a)} . \frac{n^{(m)} a^{n-m}}{\lfloor m}$$

Remark: $\dfrac{1}{(E-a)^m} . a^n = \dfrac{n^{(m)} a^{n-m}}{\lfloor m}$, $\qquad$ (12.24)

where m is a positive integer.

12.9   EVALUATION OF $\dfrac{1}{\phi(E)} f(n)$ WHEN $f(n) = $ Sin mn OR $f(n) = $ Cos mn

$$\frac{1}{\phi(E)} \cos mn = \text{Real part of } \frac{1}{\phi(E)} . e^{imn}$$

$$= \text{Real part of } \frac{1}{\phi(E)} \left\{ e^{im} \right\}^n$$

Now, $\dfrac{1}{\phi(E)}\{e^{im}\}^{n}$ can be evaluated using the method described in section 12.8. Note that $\dfrac{1}{\phi(E)}\sin(mn) = $ Imaginary part of $\dfrac{1}{\phi(E)}e^{imn}$.

12.10 EVALUATION OF $\dfrac{1}{\phi(E)}f(n)$, WHERE $f(n)$ IS A POLYNOMIAL IN n

Let $f(n)$ be a polynomial of mth degree and $f(n)$ in factorial notation be

$$f(n) = a_0\, n^{(m)} + a_1\, n^{(m-1)} + a_2\, n^{(m-2)} + \dots + a_m \qquad (12.23)$$

Now,

$$\frac{1}{\phi(E)}\cdot f(n) = \frac{1}{\phi(1+\Delta)}\{a_0 n^{(m)} a_1 n^{(m-1)} + a_2 n^{(m-2)} + \dots + a_m\} \text{ (By 12.23)}$$

We express $\dfrac{1}{\phi(1+\Delta)}$ in ascending powers of Δ, and operating these terms on $\{a_0\, n^{(m)} + a_1\, n^{(m-1)} + a_2\, n^{(m-2)} + \dots + a_m\}$, we get the particular integral of $\dfrac{1}{\phi(E)}f(n)$ when $f(n)$ is given by (12.23).

If $\phi(1+\Delta) = \Delta^t\{1 + h(\Delta)\}$ (h (Δ) is a function of Δ) for some positive integer t, then

$$\frac{1}{\phi(1+\Delta)}f(n) = \frac{1}{\Delta^t}\times\frac{1}{\{1+h(\Delta)\}}f(n)$$

$\left\{\dfrac{1}{1+h(\Delta)}\right\}f(n)$ is evaluated using the process described above. Let this result be $g(n)$;

Now

$$\frac{1}{\phi(1+\Delta)}f(n) = \frac{1}{\Delta^t}\cdot\frac{1}{1+h(\Delta)}f(n)$$

$$= \frac{1}{\Delta^t}\cdot g(n)$$

Applying $\dfrac{1}{\Delta}$ operator t times on $g(n)$ we get $\dfrac{1}{\phi(E)} f(n)$. The following formula is useful in such cases.

$$\frac{1}{\Delta} n^{(k)} = \frac{n^{(k+1)}}{k+1}, \text{ where } k \text{ is a non-negative integer with } n^{(0)} = 1.$$

12.11   EVALUATE $\dfrac{1}{\phi(E)} f(n)$, WHEN $f(n) = a^n g(n)$

Consider $\phi(E)\{a^n g(n)\}$

$$\phi(E)\{a^n g(n)\}$$

$$= (E^k + a_1 E^{k-1} + a_2 E^{k-2} \dots + a_k) a^n g(n)$$

$$= a^{n+k} g(n+k) + a_1 a^{n+k-1} g(n+k-1) + a_2 a^{n+k-2} g(n+k-2) + \dots + a_k a^n g(n)$$

$$= a^n[a^k g(n+k) + a_1 a^{k-1} g(n+k-1) + a_2 a^{k-2} g(n+k-2) + \dots + a_k g(n)]$$

$$= a^n[a^k E^k g(n) + a_1 a^{k-1} E^{k-1} g(n) + a_2 a^{k-2} E^{k-2} g(n) + \dots + a_k g(n)]$$

$$= a^n[(a E)^k + a_1 (a E)^{k-1} + a_2 (a E)^{k-2} + \dots + a_n]g(n)$$

$$= a^n \phi(a E) g(n)$$

Thus,

$$\phi(E)\left\{a^n g(n)\right\} = a^n \phi(aE) g(n) \tag{12.24}$$

Now, suppose

$$\frac{1}{\phi(E)} a^n g(n) = h(n) \tag{12.25}$$

Then,

$$\phi(E) h(n) = a^n g(n) \tag{12.26}$$

It is possible only when $h(n)$ is product of a^n and some function on n. Therefore, let

$$h(n) = a^n \Psi(n) \tag{12.27}$$

In view of (12.27), Eq. (12.26) reduces to

$$\phi(E)\{a^n \Psi(n)\} = a^n g(n) \tag{12.28}$$

$$\therefore \qquad a^n \phi(aE) \Psi(n) = a^n g(n) \tag{12.24}$$

i.e., $\qquad \phi(aE) \Psi(n) = g(n)$

$$\therefore \qquad \psi(n) = \frac{1}{\phi(aE)} g(n) \tag{12.29}$$

From (12.28), we have

$$\frac{1}{\phi(E)} a^n g(n) = a^n \psi(n) \qquad (12.30)$$

From (12.29) and (12.30), we get

$$\frac{1}{\phi(E)} a^n \cdot g(n) = a^n \cdot \frac{1}{\phi(aE)} g(n) \qquad (12.31)$$

Note: (i) The particular integral of HLDE is zero, i.e., $\dfrac{1}{\phi(E)} 0 = 0$.

(ii) The particular integral of a difference equation should be free from the arbitrary constants.

(iii) The complete solution of a non-homogeneous linear difference equation have two parts, namely complementary function and particular integral. We express it as
$$C.S = C.\ F + P.\ I$$

(iv) The complementary function of $\phi\ (E)\ y_n = f(n)$ is the complete solution of $\phi\ (E)\ y_n = 0$.

12.12   WORKING RULE TO FIND THE COMPLETE SOLUTION OF NON-HOMOGENEOUS LINEAR DIFFERENCE EQUATION WITH CONSTANT COEFFICIENTS

Step 1: Express the given NHLDE in the form $\phi\ (E)\ y_n = f(n)$.

Step 2: Write the auxiliary equation $\phi\ (\lambda) = 0$.

Step 3: Find the roots of $\phi\ (\lambda) = 0$.

Step 4: Write the C. F. corresponding to the nature of the roots λ using the rules specified in note of section 12.5.

Step 5: Compute the P. I of difference equation $\dfrac{1}{\phi(E)} f(n)$.

Step 6: Write the $y_n = $ (C. F. Obtained in step 4) + (P. I obtained in step 5).

SOLVED PROBLEMS

Problem 1: Solve $y_{n+2} - 7y_{n+1} + 12y_n = 2^n$.

Solution: The given difference equation is a NHLDE with constant coefficients. Therefore, it has C. F and P. I. Further, its complete solution is the sum of C. F and P. I. Now, we proceed to find them.

The given difference equation can be expressed as

$$(E^2 - 7E + 12)\, y_n = 2^n \tag{1}$$

Hence, the auxiliary equation of (1) is

$$\lambda^2 - 7\lambda + 12 = 0$$

i.e., $$(\lambda - 3)(\lambda - 4) = 0$$

Therefore, the roots of auxiliary equation are 2 and 3.

Hence,

$$\text{C. F.} = c_1 3^n + c_2\, 4^n \tag{2}$$

Now,

$$\text{P. I} = \frac{1}{E^2 - 7E + 12}\,.2^n$$

$$= \frac{2^n}{2^2 - 7\times 2 + 12} \qquad \left(\because \frac{1}{\phi(E)}a^n = \frac{a^n}{\phi(a)}\text{ if }\phi(a) \neq 0\right)$$

$$= \frac{2^n}{2} = 2^{n-1}$$

Thus,

$$\text{C.S} = \text{C. F} + \text{P. I}$$

$$y_n = c_1 3^n + c_2 4^n + 2^{n-1}.$$

Problem 2: Solve $y_{n+2} - 5y_{n+1} + 6y_n = 3^n$.

Solution: The given difference equation can be expressed as

$$(E^2 - 5E + 6)\, y_n = 3^n. \tag{1}$$

The auxiliary equation of (1) is $\lambda^2 - 5\lambda + 6 = 0$ i.e., $(\lambda - 2)(\lambda - 3) = 0$.
Therefore, 2 and 3 are roots of auxiliary equation.

$$\therefore \qquad \text{C. F.} = c_1 2^n + c_2 3^n \tag{2}$$

Now,

$$\text{P. I} = \frac{1}{E^2 - 5E + 6}\,.3^n \tag{3}$$

Observe that $\phi\,(3) = 0$, where $\phi\,(E) = E^2 - 5E + 6$. Therefore,

$$\text{P. I} = \frac{1}{(E-3)(E-2)}\,.3^n$$

$$= \frac{1}{(E-3)(3-2)} 3^n$$

$$= \frac{1}{E-3} 3^n \qquad \left[\because \frac{1}{\phi(E)} a^n = \frac{a^n}{\phi(a)} \text{ if } \phi(a) \neq 0 \right]$$

$$= \frac{n^{(1)} . 3^{n-1}}{\underline{|1}} \qquad\qquad\qquad [\text{By } (12.22)]$$

$\therefore \qquad$ C.S = C. F + P. I.

i.e., $\qquad y_n = c_1 2^n + c_2 3^n + n.3^{n-1}.$

Problem 3: Solve $y_{n+3} - 3y_{n+1} + 2y_n = 2^n.$

Solution: The given difference equation can be written as $(E^3 - 3E + 2)$ $y_n = 2^n$. Hence, its auxiliary equation is $\lambda^3 - 3\lambda + 2 = 0$, i.e., $(\lambda - 1)(\lambda - 1)$ $(\lambda + 2) = 0$. Thus, the roots of auxiliary equation are $1, 1, -2$. Therefore,

$$\text{C. F.} = (c_1 + c_2 n)(1)^n + c_3 (-2)^n$$

Now,

$$\text{P. I} = \frac{1}{E^3 - 3E + 2} .2^n$$

$$= \frac{1}{2^3 - 3 \times 2 + 2} 2^n$$

$$= 2^{n-2}$$

$\therefore \qquad\qquad y_n = [(c_1 + c_2 n)(1)^n + c_3 (-2)^n] + 2^{n-2}.$

Problem 4: Solve $(y_{n+3} - 6y_{n+2} + 12y_{n+1} - 8)y_n = 2^n.$

Solution: The given difference equation can be written as $(E^3 - 6E^2 + 2E - 8) y_n = 2^n$. Hence, its auxiliary equation is $\lambda^3 - 6\lambda^2 + 12\lambda - 8 = 0$, i.e., $(\lambda - 2)^3 = 0$. Hence, its roots are $2, 2, 2$.

$$\therefore \text{C. F.} = (c_1 + c_2 n + c_3 n^2) 2^n$$

Now,

$$\text{P. I} = \frac{1}{E^3 - 6E^2 + 12E - 8} .2^n$$

$$= \frac{1}{(E-2)^3} .2^n$$

$$= \frac{n^{(3)}.2^{n-3}}{\lfloor 3}$$

[By (12.22)]

$$= \frac{n(n-1)(n-2)}{6}2^{n-3}$$

$\therefore$ $\text{C.S} = \text{C. F} + \text{P. I.}$

i.e., $y_n = \left(c_1 + c_2 n + c_3 n^2\right)2^n + \dfrac{n(n-1)(n-2)}{6}2^{n-3}.$

Problem 5: Find $\dfrac{1}{(E^2 - 6E + 9)}.3^n.$

Solution: $\dfrac{1}{(E^2 - 6E + 9)}.3^n = \dfrac{1}{(E-3)^2}.3^n$

$$= \frac{n^{(2)}.3^{n-2}}{\lfloor 2}$$

$$= \frac{n(n-1)}{2}.3^{n-2}.$$

Problem 6: Find $\dfrac{1}{(E-2)(E-3)(E-4)}\left\{2^n + 3^n + 4^n\right\}.$

Solution: Required expression

$$= \frac{1}{(E-2)(E-3)(E-4)}2^n + \frac{1}{(E-3)(E-2)(E-4)}3^n$$

$$+ \frac{1}{(E-4)(E-2)(E-3)}4^n$$

$$= \frac{1}{(E-2)(2-3)(2-4)}.2^n + \frac{1}{(E-3)(3-2)(3-4)}.3^n$$

$$+ \frac{1}{(E-4)(4-2)(4-3)}.4^n$$

$$= \frac{1}{2}\frac{1}{E-2}.2^n + \frac{1}{(-1)}\frac{1}{(E-3)}.3^n + \frac{1}{2}\frac{1}{(E-4)}.4^n$$

$$= \frac{1}{2} \cdot \frac{n^{(1)} \cdot 2^{n-1}}{\lfloor 1} + \frac{1}{(-1)} \cdot \frac{n^{(1)} \cdot 3^{n-1}}{\lfloor 1} + \frac{1}{2} \cdot \frac{n^{(1)} \cdot 4^{n-1}}{\lfloor 1}$$

$$= \frac{1}{2} n \cdot 2^{n-1} - n \cdot 3^{n-1} + \frac{1}{2} \cdot n \cdot 4^{n-1} \quad (\because n^{(1)} = n).$$

Problem 7: Solve $y_{n+2} - 4y_n = 9n^2$.

Solution: The given difference equation can be expressed as $(E^2 - 4)$ $y_n = 9n^2$. Hence, its auxiliary equation is $\lambda^2 - 4 = 0$. Hence, its roots are 2 and -2.

$$\text{C. F.} = c_1 (-2)^n + c_2 2^n.$$

Now,

$$\text{P. I} = \frac{1}{(E^2 - 4)} 9n^2 \tag{1}$$

First, we express n^2 in factorial notation. Let $n^2 = A \, n^{(2)} + B \, n^{(1)} + C$, where A, B and C are constants. Then

$$n^2 = A \, n(n - 1) + Bn + C$$

i.e., $$n^2 = A \, n^2 + (-A + B) \, n + C \tag{2}$$

Comparing the coefficients of like powers of n in (2), we get

$$A = 1 \tag{3}$$
$$-A + B = 0 \tag{4}$$
$$C = 0 \tag{5}$$

From (3), (4) and (5), we have $A = 1$, $B = 1$ and $C = 0$.

$\therefore$ $$n^2 = 1. \, n^{(2)} + 1. \, n^{(1)}$$

i.e., $$n^2 = n^{(2)} + n^{(1)} \tag{6}$$

In view of (6), Eq. (1) becomes

$$\text{P. I} = 9. \frac{1}{(E^2 - 4)} \cdot \left\{ n^{(2)} + n^{(1)} \right\}$$

$$= 9. \frac{1}{(1 + \Delta)^2 - 4} \cdot \left\{ n^{(2)} + n^{(1)} \right\}$$

$$= 9 \cdot \frac{1}{\left(1 + 2\Delta + \Delta^2 - 4\right)} \cdot \left\{ n^{(2)} + n^{(1)} \right\}$$

$$= 9 \cdot \frac{1}{-3 + 2\Delta + \Delta^2} \cdot \left\{ n^{(2)} + n^{(1)} \right\}$$

$$= 9 \cdot \frac{1}{-3\left[1 - \dfrac{\Delta^2 + 2\Delta}{3} \right]} \cdot \left\{ n^{(2)} + n^{(1)} \right\}$$

$$= -3 \cdot \left[1 + \frac{\Delta^2 + 2\Delta}{3} + \left(\frac{\Delta^2 + 2\Delta}{3} \right)^2 + \ \ldots \right] \cdot \left\{ n^{(2)} + n^{(1)} \right\}$$

$$= -3 \left[1 + \frac{2\Delta}{3} + \frac{\Delta^2}{3} + \frac{4}{9}\Delta^2 \right] \left[n^{(2)} + n^{(1)} \right]$$

$$\left(\because \Delta^t \left\{ n^{(2)} + n^{(1)} \right\} = 0 \ \text{ for } t \geq 3 \right)$$

$$= -3 \left[1 + \frac{2\Delta}{3} + \frac{7}{9}\Delta^2 \right] \left\{ n^{(2)} + n^{(1)} \right\}$$

$$= -3 \left[\left\{ n^{(2)} + n^{(1)} \right\} + \frac{2}{3} \left\{ 2n^{(1)} + n^{(0)} \right\} + \frac{7}{9} \left\{ 2 \right\} \right]$$

$$= -3 \left[n^{(2)} + n^{(1)} + \frac{4}{3} n^{(1)} + \frac{2}{3} + \frac{14}{9} \right]$$

$$= -3 \left[n^{(2)} + \frac{7}{3} n^{(1)} + \frac{20}{9} \right]$$

$$= -3 \left[n(n-1) + \frac{7}{3} n + \frac{20}{9} \right]$$

$$= -3 \left(n^2 + \frac{4n}{3} + \frac{20}{9} \right)$$

Thus,

$$y_n = c_1 (-2)^n + c_2 (2)^n - 3 \left(n^2 + \frac{4n}{3} + \frac{20}{9} \right).$$

Problem 8: Solve $y_{n+2} + y_{n+1} + y_n = n^2 + n + 2$.

Solution: The given difference equation can be expressed as $(E^2 + E + 1)\, y_n = n^2 + n + 2$. Its auxiliary equation is $\lambda^2 + \lambda + 1 = 0$. The roots of it are $\dfrac{-1}{2} + \dfrac{\sqrt{3}}{2}\,i$ and $\dfrac{-1}{2} + \dfrac{\sqrt{3}}{2}\,i$. As the roots are complex,

$$\text{C. F.} = r^n\,(c_1\,\cos n\theta + c_2\,\sin n\theta),$$

where $\qquad r = \sqrt{\left(\dfrac{-1}{2}\right)^2 + \left(\dfrac{\sqrt{3}}{2}\right)^2} = 1$ and $\theta = \tan^{-1}\left(-\sqrt{3}\right) = \dfrac{2\pi}{3}$

$$\therefore \qquad \text{C. F.} = \left(c_1\,\cos\frac{2n\pi}{3} + c_2\,\sin\frac{2n\pi}{3}\right)(1)^n \qquad\qquad (1)$$

Now,

$$\text{P. I.} = \frac{1}{(E^2 + E + 1)}\cdot\left(n^2 + n + 2\right) \qquad\qquad (2)$$

We express $n^2 + n + 2$ in factorial notation so that their differences are computed easily, using $\Delta n^{(r)} = r n^{(r-1)}$. Let

$$n^2 + n + 2 = A\, n^{(2)} + B\, n^{(1)} + C,$$

where A, B and C are constants. Then

$$n^2 + n + 2 = An\,(n-1) + Bn + C$$

i.e., $\qquad n^2 + n + 2 = An^2 + (-A + B)\,n + C \qquad\qquad (3)$

Comparing the coefficients of like-powers of n in Eq. (2), we get

$$A = 1 \qquad\qquad (4)$$
$$-A + B = 1 \qquad\qquad (5)$$
$$C = 2 \qquad\qquad (6)$$

Solving (4), (5) and (6) for A, B, and C, we get $A = 1$, $B = 2$ and $C = 2$. Thus,

$$n^2 + n + 1 = n^{(2)} + 2n^{(1)} + 2. \qquad\qquad (7)$$

In view of (7), Eq. (2) becomes

$$\text{P. I.} = \frac{1}{(E^2 + E + 1)}\cdot\left(n^{(2)} + 2n^{(1)} + 2\right)$$

$$= \frac{1}{\left\{(1+\Delta)^2 + (1+\Delta) + 1\right\}}\left(n^{(2)} + 2n^{(1)} + 2\right)$$

$$= \frac{1}{3 + 3\Delta + \Delta^2} \cdot \left\{ n^{(2)} + 2n^{(1)} + 2 \right\}$$

$$= \frac{1}{3\left(1 + \dfrac{3\Delta + \Delta^2}{3}\right)} \cdot \left\{ n^{(2)} + 2n^{(1)} + 2 \right\}$$

$$= \frac{1}{3}\left(1 + \frac{3\Delta + \Delta^2}{3}\right)^{-1} \cdot \left\{ n^{(2)} + 2n^{(1)} + 2 \right\}$$

$$= \frac{1}{3}\left[1 - \Delta - \frac{\Delta^2}{3} + \Delta^2 \right]\left\{ n^{(2)} + 2n^{(1)} + 2 \right\}$$

$$\left(\because \Delta^t \left\{ n^{(2)} + n^{(1)} + 2 \right\} = 0 \text{ for } t \geq 3 \right)$$

$$= \frac{1}{3}\left[1 - \Delta + \frac{2}{3}\Delta^2 \right]\left\{ n^{(2)} + 2n^{(1)} + 2 \right\}$$

$$= \frac{1}{3}\left[\left\{ n^{(2)} + 2n^{(1)} + 2 \right\} - \Delta\left\{ n^{(2)} + 2n^{(1)} + 2 \right\} + \frac{2}{3}\Delta^2 \left\{ n^{(2)} + 2n^{(1)} + 2 \right\} \right]$$

$$= \frac{1}{3}\left[\left\{ n^{(2)} + 2n^{(1)} + 2 \right\} - \left\{ 2n^{(1)} + 2 \right\} + \frac{2}{3}(2) \right]$$

$$= \frac{1}{3}\left[n^{(2)} + \frac{4}{3} \right]$$

$$= \frac{1}{3}\left(n^2 - n + \frac{4}{3} \right) \tag{8}$$

$\therefore y_n = \text{C. F} + \text{P. I}$, where C. F. is given by R.H.S. of Eq. (1) and P. I is given by R.H.S. of Eq. (8).

Problem 9: Solve $y_{n+2} - y_n = n^2$.

Solution: The given difference equation can be expressed as $(E^2 - 1)$ $y_n = n^2$. Hence, its auxiliary equation is $\lambda^2 - 1 = 0$. As the roots are -1 and 1.

$$\text{C. F.} = c_1 (-1)^n + c_2 (2)^n \tag{1}$$

Now,

$$\text{P. I.} = \frac{1}{E^2 - 1} n^2$$

$$= \frac{1}{E^2 - 1}\left\{n^2 + n^{(1)}\right\}$$

$$= \frac{1}{(1+\Delta)^2 - 1}\left\{n^2 + n^{(1)}\right\}$$

$$= \frac{1}{\Delta(1+\Delta)}\left\{n^{(2)} + n^{(1)}\right\}$$

$$= \frac{1}{\Delta}\left\{\left(1 - \Delta + \Delta^2\right)\right\}\left\{n^{(2)} + n^{(1)}\right\}$$

$$= \frac{1}{\Delta}\left[\left\{n^{(2)} + n^{(1)}\right\} - \Delta\left\{n^{(2)} + n^{(1)}\right\}\right.$$

$$\left. + \Delta^2\left\{n^{(2)} + n^{(1)}\right\}\right]$$

$$= \frac{1}{\Delta}\left[\left\{n^{(2)} + n^{(1)}\right\} - \left\{2n^{(1)} + 1\right\} + \left\{2\right\}\right]$$

$$= \frac{1}{\Delta}\left[n^{(2)} - n^{(1)} + 1\right]$$

$$= \frac{n^{(3)}}{3} - \frac{n^{(2)}}{2} + \frac{n^{(1)}}{1} \tag{2}$$

Thus, y_n = C. F + P. I, where C. F and P. I are given by R. H. S. of (1) and (2) respectively.

Problem 10: Solve $y_{n+2} - 7y_{n+1} + 12y_n = \cos n$.

Solution: The given difference equation can be written as $(E^2 - 7E + 12)$ $y_n = \cos n$. Hence, its auxiliary equation is $\lambda^2 - 7\lambda + 12 = 0$. Therefore, its roots are 3 and 4.

$$\text{C. F.} = c_1 3^n + c_2 4^n \tag{1}$$

Now,

$$\text{P. I} = \frac{1}{E^2 - 7E + 12}\cos n$$

$$= \text{Real part of } \frac{1}{E^2 - 7E + 12}e^{in}$$

$$= \text{Real part of } \frac{1}{E^2 - 7E + 12}(e^i)^n$$

$$= \text{Real part of } \frac{1}{(e^i)^2 - 7(e^i) + 12}(e^i)^n \left(\because \frac{1}{\phi(E)}a^n = \frac{1}{\phi(a)}a^n \right)$$

$$= \text{Real part of } \frac{1}{e^{2i} - 7e^i + 12} \cdot e^{in}$$

$$= \text{Real part of } \frac{1}{(\cos 2 + i\sin 2) - 7(\cos 1 + i\sin 1) + 12} \cdot e^{in}$$

$$= \text{Real part of } \frac{(\cos n + i\sin n)}{(\cos 2 - 7\cos 1 + 12) + i(\sin 2 - 7\sin 1)}$$

$$= \text{Real part of}$$

$$\frac{\left[(\cos 2 - 7\cos 1 + 12) - i(\sin 2 - 7\sin 1)\right](\cos n + i\sin n)}{(\cos 2 - 7\cos 1 + 12)^2 + (\sin 2 - 7\sin 1)^2}$$

$$= \frac{\cos(n)(\cos 2 - 7\cos 1 + 12) + \sin(n)\{\sin 2 - 7\sin 1\}}{(\cos 2 - 7\cos 1 + 12)^2 + (\sin 2 - 7\sin 1)^2} \tag{2}$$

$\therefore y_n = $ C. F + P. I., where C. F and P. I are given by right-hand sides of Eqs. (1) and (2) respectively.

Problem 11: Solve $y_{n+2} - y_n = \sin(2n)$.

Solution: Given difference equation is $(E^2 - 1)y_n = \sin(2n)$. Its auxiliary equation is $\lambda^2 - 1 = 0$ and its roots are -1 and 1.

$$\therefore \qquad\qquad \text{C. F} = c_1 (1)^n + c_2 (-1)^n \tag{1}$$

Now,

$$\text{P. I} = \frac{1}{E^2 - 1}\sin(2n)$$

$$= \text{Imaginary part of } \frac{e^{2in}}{E^2 - 1}$$

$$= \text{Imaginary part of } \frac{(e^{2i})^n}{(e^{2i})^2 - 1}$$

$$= \text{Imaginary part of } \frac{e^{2in}}{(e^{4i} - 1)}$$

$$= \text{Imaginary part of } \frac{(\cos 2n + i\sin 2n)}{\{(\cos 4 - 1) + i\sin 4\}}$$

$$= \text{Imaginary part of } \frac{(\cos 2n + i\sin 2n)\{(\cos 4 - 1) - i\sin 4\}}{(\cos 4 - 1)^2 + \sin^2 4}$$

$$= \text{Imaginary part of } \frac{(\cos 2n + i\sin 2n)\{(\cos 4 - 1) - i\sin 4\}}{2(1 - \cos 4)}$$

$$= \frac{\sin(2n)(\cos 4 - 1) - \sin 4 . \cos 2n}{2(1 - \cos 4)} \tag{2}$$

Thus, $y_n = $ C. F + P. I, where C. F and P. I. are respectively given by the right-hand side of (1) and (2).

Problem 12: Solve the difference equation $y_{n+2} - 4y_{n+1} + 3y_n = 2^n(n^2 - n)$.

Solution: The given difference equation is $(E^2 - 4E + 3)\, y_n = 2^n (n^2 - n)$. Its auxiliary equation is $\lambda^2 - 4\lambda + 3 = 0$. i.e., $(\lambda - 3)(\lambda - 1) = 0$. Therefore, its roots are 1 and 3. Hence,

$$\text{C. F} = c_1 (1)^n + c_2 (3)^n \tag{1}$$

$$\text{P. I.} = \frac{1}{E^2 - 4E + 3}.2^n\left(n^2 - n\right)$$

$$= 2^n \frac{1}{(2E)^2 - 4(2E) + 3}.\left(n^2 - n\right) \qquad (\text{by}(12.31))$$

$$= 2^n . \frac{1}{4E^2 - 8E + 3} . n^{(2)} \qquad (\because n^{(2)} = n(n-1) = n^2 - n)$$

$$= 2^n \frac{1}{4(1 + \Delta)^2 - 8(1 + \Delta) + 3} . n^{(2)}$$

$$= 2^n . \frac{1}{4\Delta^2 - 1} . n^{(2)}$$

$$= 2^n \frac{1}{(-1)}\left(1 - 4\Delta^2\right)^{-1} . n^{(2)}$$

$$= -2^n \left[1 + 4\Delta^2\right] n^{(2)} \quad \left(\because \Delta^t n^{(2)} = 0 \text{ for } t \ge 3\ t \text{ is an integer}\right)$$

$$= -2^n \left[n^{(2)} + 4\Delta^2 n^{(2)}\right]$$

$$= -2^n \left[n^{(2)} + 8 \right]$$

$$= -2^n \left(n^2 - n + 8 \right)$$

$$\therefore \qquad y_n = \text{C. F} + \text{P. I}$$

i.e., $\qquad y_n = c_1 (1)^n + c_2 (3)^n - 2^n (n^2 - n + 8).$

Problem 13: Solve $y_{n+2} + y_n = n\, 3^n$, give that $y_0 = \dfrac{91}{50}$ and $y_1 = \dfrac{119}{25}$.

Solution: Given difference equation can be expressed as $(E^2 + 1)\, y_n = n\, 3^n$. Its auxiliary equation is $\lambda^2 + 1 = 0$ and its roots are i and $-i$. Hence,

$$\therefore \qquad \text{C. F} = \left(c_1 \cos\frac{n\pi}{2} + c_2 \sin\frac{n\pi}{2} \right)(1)^n$$

Now,

$$\text{P. I} = \frac{1}{\left(E^2 + 1\right)} . 3^n . n$$

$$= 3^n . \frac{1}{\left(3E\right)^2 + 1} . n \qquad\qquad (\text{by}(12.31))$$

$$= 3^n . \frac{1}{9E^2 + 1} . n$$

$$= 3^n . \frac{1}{9(1 + \Delta)^2 + 1} . n$$

$$= 3^n . \frac{1}{10 + 18\Delta + 9\Delta^2} . n$$

$$= \frac{3^n}{10} . \left[1 + \frac{18\Delta + 9\Delta^2}{10} \right]^{-1} n$$

$$= \frac{3^n}{10} . \left[1 - \frac{18\Delta + 9\Delta^2}{10} + ... \right] n$$

$$= \frac{3^n}{10} . \left[1 - \frac{18\Delta}{10} \right] n \qquad\qquad \left(\because \Delta^t n = 0 \text{ for } t \geq 2 \right)$$

$$= \frac{3^n}{10} . \left[n - \frac{9}{5} \right]$$

$$= \frac{3^n}{50} \cdot (5n - 9)$$

Thus,

$$y_n = \left(c_1 \cos \frac{n\pi}{2} + c_2 \sin \frac{n\pi}{2} \right)(1)^n + \frac{3^n}{50}(5n - 9) \tag{1}$$

Given that $y_0 = 91/50$ and $y_1 = 119/25$, therefore, substituting $n = 0$ and 1 in Eq. (1), we get

$$\frac{91}{50} = c_1 + \left(\frac{-9}{50} \right) \tag{2}$$

and

$$\frac{119}{25} = c_2 + \frac{3}{50}(-4) \tag{3}$$

From (2) we get $c_1 = 2$ and from Eq. (3) we get $c_2 = 5$.

$$\therefore \qquad y_n = \left(2 \cos \frac{n\pi}{2} + 5 \sin \frac{n\pi}{2} \right)(1)^n + \frac{3^n}{50}(5n - 9),$$

which is the required solution.

Problem 14: Solve $y_{n+2} - 4y_{n+1} + 3y_n = 4n^2 + 5^n$.

Solution: The given difference equation is $(E^2 - 4E + 3)y_n = 4n^2 + 5^n$. Its auxiliary equation is $\lambda^2 - 4\lambda + 3 = 0$ and the roots of it are 1 and 3.

$$\text{C. F} = c_1 (1)^n + c_2 (3)^n \tag{1}$$

Now,

$$\text{P. I} = \frac{1}{\left(E^2 - 4E + 3\right)}\left(4n^2 + 5^n\right)$$

$$= 4 \cdot \frac{1}{E^2 - 4E + 3} n^2 + \frac{1}{E^2 - 4E + 3} 5^n$$

$$= 4 \cdot \frac{1}{\left(1 + \Delta\right)^2 - 4\left(1 + \Delta\right) + 3} n^2 + \frac{1}{5^2 - 4 \times 5 + 3} 5^n$$

$$= 4 \cdot \frac{1}{\left(-2\Delta + \Delta^2\right)}\left\{ n^{(2)} + n^{(1)} \right\} + \frac{5^n}{8}$$

$$= 4.\dfrac{1}{-2\Delta}\dfrac{1}{\left(1-\dfrac{\Delta}{2}\right)}\left\{n^{(2)}+n^{(1)}\right\}+\dfrac{5^n}{8}$$

$$= -2.\dfrac{1}{\Delta}\left(1+\dfrac{\Delta}{2}+\dfrac{\Delta^2}{4}\right)\left\{n^{(2)}+n^{(1)}\right\}+\dfrac{5^n}{8}$$

$$= -2.\dfrac{1}{\Delta}\left[\left\{n^{(2)}+n^{(1)}\right\}+\dfrac{\Delta}{2}\left\{n^{(2)}+n^{(1)}\right\}+\dfrac{\Delta^2}{4}\left\{n^{(2)}+n^{(1)}\right\}\right]+\dfrac{5^n}{8}$$

$$= -2.\dfrac{1}{\Delta}\left[n^{(2)}+n^{(1)}+\dfrac{1}{2}\left\{2n^{(1)}+1\right\}+\dfrac{1}{4}\left\{2\right\}\right]+\dfrac{5^n}{8}$$

$$= -2.\dfrac{1}{\Delta}\left[n^{(2)}+2n^{(1)}+1\right]+\dfrac{5^n}{8}$$

$$= -2.\left[\dfrac{n^{(3)}}{3}+n^{(2)}+n^{(1)}\right]+\dfrac{5^n}{8}\qquad\left(\because\ \dfrac{1}{\Delta}n^{(k)}=\dfrac{n^{(k+1)}}{k+1}\right)$$

$$\therefore\qquad y_n = c_1(1)^n+c_2(3)^n-2\left[\dfrac{n^{(3)}}{3}+n^{(2)}+n^{(1)}\right]+\dfrac{5^n}{8},$$

which is the required solution.

Problem 15: Solve $y_{n+1}-4y_n=6\sin 3n$.

Solution: The given difference equation is $(E-4)\,y_n=6\sin 3n$. Hence, its auxiliary equation is $\lambda-4=0$ and 4 is a root of the auxiliary equation.

$$\therefore\ \text{C. F}=c_1(4)^n$$

Now,

$$\text{P. I}=\dfrac{1}{E-4}6.\sin(3n)$$

$$= 6\dfrac{1}{(E-4)}\sin(3n)$$

$$= 6.\ \text{Imaginary part of }\dfrac{e^{3in}}{(E-4)}$$

$$= 6.\ \text{Imaginary part of }\dfrac{e^{3in}}{e^{3i}-4}$$

$$= 6.\ \text{Imaginary part of }\dfrac{(\cos 3n+i\sin 3n)}{(\cos 3-4)+i\sin 3}$$

$$= 6. \text{ Imaginary part of } \frac{(\cos 3n + i\sin 3n)\{(\cos 3 - 4) - i\sin 3\}}{(\cos 3 - 4)^2 + \sin^2 3}$$

$$= 6. \text{ Imaginary part of } \frac{(\cos 3n + i\sin 3n)\{(\cos 3 - 4) - i\sin 3\}}{(17 - 8\cos 3)}$$

$$= 6. \frac{\sin 3n(\cos 3 - 4) - \sin 3\cos 3n}{17 - 8\cos 3}$$

$$= 6. \left\{ \frac{\sin 3(n-1) - 4\sin 3n}{17 - 8\cos 3} \right\}$$

$$\therefore y_n = c_1 (4)^n + 6. \left\{ \frac{\sin 3(n-1) - 4\sin 3n}{17 - 8\cos 3} \right\},$$

which is the required solution.

Problem 16: Solve $y_{n+1} = \sqrt{y_n}$.

Solution: Given difference equation is

$$y_{n+1} = \sqrt{y_n} \tag{1}$$

Taking logarithm on both sides of (1), we get

$$\log y_{n+1} = \frac{1}{2}\log y_n$$

or
$$2\log y_{n+1} - \log y_n = 0 \tag{2}$$

Let $v_n = \log y_n$ $\qquad\qquad\qquad (3)$

Then, Eq. (3) reduces to

$$2v_{n+1} - v_n = 0 \tag{4}$$

The difference Eq. 4 is a HLDE with constant coefficients. Hence, its auxiliary equation is $2\lambda - 1 = 0$, since the Eq. (4) is equivalent to $(2E - 1) v_n = 0$. The root of auxiliary equation is ½.

Therefore,

$$v_n = c_1 (1/2)^n \tag{5}$$

In view of (3), the Eq. (5) becomes

$$\log y_n = c_1 (2^{-n}).$$

$$\therefore \qquad y_n = e^{c_1 . 2^{-n}}$$

and it is the required solution of given difference Eq. (1).

EXERCISE 12 (b)

1. Solve the following difference equations:
   (i) $9y_{n+2} - 6y_{n+1} + y_n = 0.$

   (ii) $y_{n+2} - 6y_{n+1} + 8y_n = 0.$

   (iii) $y_{n+3} - 3y_{n+2} + 4y_n = 0.$

   (iv) $y_{n+2} - 11y_{n+1} + 28y_n = 0.$

   (v) $12y_{n+2} + 7y_{n+1} + y_n = 0.$

   (vi) $y_{n+3} - y_{n+2} - y_{n+1} - 2y_n = 0.$

   (vii) $y_{n+3} - 5y_{n+2} + 17y_{n+1} - 13y_n = 0.$

   (viii) $y_{n+3} - 15y_{n+2} + 75y_{n+1} - 125y_n = 0.$

   (ix) $y_{n+3} - 7y_{n+2} + 14y_{n+1} - 8y_n = 0.$

   (x) $6y_{n+2} - 5y_{n+1} + y_n = 0.$

   (xi) $\left(E^2 - 6E + 9\right) y_n = 0.$

   (xii) $\Delta^2 y_n - 7\Delta y_n + 12y_n = 0.$

   (xiii) $\left(E^2 - 4E + 1\right) y_n = 0.$

   (xiv) $\Delta^2 y_n + 3\Delta y_n + y_n = 0.$

   (xv) $\Delta y_{n+1} = 6y_n.$

2. Solve the following difference equations:
   (i) $y_{n+2} - 10y_{n+1} + 16y_n = 0,$ when $y_0 = 1$ and $y_1 = 8.$

   (ii) $y_{n+2} - 4y_{n+1} + 4y_n = 0,$ when $y_0 = 0$ and $y_1 = 2.$

   (iii) $y_{n+2} - 6y_{n+1} + 9y_n = 0,$ when $y_0 = 1$ and $y_1 = 0.$

   (iv) $y_{n+3} - 3y_{n+1} + 2y_n = 0$ when $y_0 = 6$ and $y_1 = 9.$

   (v) $y_{n+2} + 3y_{n+1} + y_n = 0$ when $y_0 = 2$ and $y_1 = 16.$

3. Solve the following difference equations:
   (i) $y_{n+2} - 4y_{n+1} + 4y_n = 3^n.$

   (ii) $y_{n+2} - 4y_{n+1} + 4y_n = 2^n.$

   (iii) $y_{n+2} - 7y_n + 12y_n = e^{3n}.$

   (iv) $y_{n+1} - y_n = \cos 3n.$

   (v) $y_{n+3} - 5y_{n+2} + 3y_{n+1} + 9y_n = 2^n + 4^n.$

(vi) $y_{n+2} - 2y_{n+1} + y_n = n^2$.

(vii) $y_{n+2} - 4y_n = 6n^2 + 4n + 2$.

(viii) $y_{n+2} - 2y_{n+1} + y_n = \sin(2n)$.

(ix) $y_{n+2} - 4y_n = n - 3$.

(x) $y_{n+1} - y_n = n^2 + 3n + 6$.

4. Solve the following difference equations:

(i) $y_{n+1} - 3y_n = n^2 + \cos n + 4^n$.

(ii) $y_{n+1} - 2y_n = 2^n + 3^n + 6 . 5^n$.

(iii) $y_{n+1} - 4y_n = \cos n + \sin n + n^2 + 4^n$.

(iv) $y_{n+1} - y_n = \cos 5n$.

5. Solve the following difference equations:

(i) $y_{n+2} - 7y_{n+1} - 8y_n = (n^2 - n). \, 3^n$.

(ii) $y_{n+2} + y_{n+1} + y_n = 6n.2^n$.

(iii) $y_{n+2} - 4y_{n+1} + 4y_n = 2^n (n^2 + 1)$.

(iv) $y_{n+2} + 6y_{n+1} + 8y_n = n.2^n + 3^n + n^2$.

6. Solve the following difference equations:

(i) $y_{n+2} - 4y_{n+1} + 3y_n = 2^n (n^2 - n)$ when $y_0 = 4$ and $y_1 = 10$.

(ii) $y_{n+1} - y_n = 6n \, 3^n$, when $y_0 = 6$.

ANSWERS

1. (i) $y_n = (c_1 + c_2)\left(\dfrac{1}{3}\right)^n$

(ii) $y_n = c_1 \, 2^n + c_2 \, 4^n$

(iii) $y_n = (c_1 \, n + c_2) \, 2^n + c_3 \, (-1)^n$

(iv) $y_n = c_1 \, 7^n + c_2 \, 4^n$

(v) $y_n = c_1 \left(-\dfrac{1}{4}\right)^n + c_2 \left(-\dfrac{1}{3}\right)^n$

(vi) $y_n = c_1 \, 2^n + c_2 \, \cos(2\pi n/3) + c_3 \, \sin(2\pi n/3)$

(vii) $y_n = c_1 \, 1^n + c_2 \, (2 + \sqrt{3})^n + c_3 \, (2 - \sqrt{3})^n$

(viii) $y_n = (c_1 n^2 + c_2 n + c_3)(5)^n$

(ix) $y_n = c_1 1^n + c_2 2^n + c_3 4^n$

(x) $y_n = c_1 \left(\dfrac{1}{2}\right)^n + c_2 \left(\dfrac{1}{3}\right)^n$

(xi) $y_n = (c_1 n + c_2) 3^n$

(xii) $y_n = c_1 4^n + c_2 5^n$

(xiii) $y_n = c_1 (2 + \sqrt{3})^n + c_2 (2 - \sqrt{3})^n$

(xiv) $y_n = c_1 \left(-\dfrac{1}{2} + \dfrac{\sqrt{5}}{2}\right)^n + c_2 \left(-\dfrac{1}{2} - \dfrac{\sqrt{5}}{2}\right)^n$

(xv) $y_n = c_1 3^n + c_2 (-2)^n$

2. (i) $y_n = 8^n$
    (ii) $y_n = n\, 2^n$

(iii) $y_n = (-n + 1)\, 3^n$
(iv) $y_n = 5\, 1^n + 1\, 2^n$

(v) $y_n = \dfrac{(19 + \sqrt{5})}{\sqrt{5}}\left(\dfrac{-3}{2} + \dfrac{\sqrt{5}}{2}\right)^n + \dfrac{(-19 + \sqrt{5})}{\sqrt{5}}\left(-\dfrac{3}{2} - \dfrac{\sqrt{5}}{2}\right)^n$

3. (i) $y_n = (c_1 n + c_2)\, 2^n + (3)^n$

(ii) $y_n = (c_1 n + c_2)\, 2^n + n(n-1)\, 2^{(n-3)}$

(iii) $y_n = c_1 4^n + c_2 3^n + \dfrac{1}{\left(e^6 - 7e^3 + 12\right)} e^{3n}$

(iv) $y_n = c_1 1^n + \dfrac{1}{\left(2\sin\left(\dfrac{3}{2}\right)\right)} \sin\left(\dfrac{3}{2}(2n - 1)\right)$

(v) $y_n = c_1 (-1)^n + (c_2 n + c_3)\, 3^n + \dfrac{1}{3} 2^n + \dfrac{1}{45} 4^n$

(vi) $y_n = (c_1 n + c_2)\, 1 + \dfrac{n^{(4)}}{12} + \dfrac{n^{(3)}}{6}$

(vii) $y_n = c_1 2^n + c_2 (-2)^n - \left(2n^{(2)} + 6n^{(1)} + 6\right)$

(viii) $y_n = (c_1 n + c_2)\, (1)^n + \dfrac{\left[(2\sin 2 - \sin 4)\cos 2n + (\sin 2n)(1 - 2\cos 2 + \cos 4)\right]}{4(1 - \cos 2)^2}$

(ix) $y_n = c_1\, 2^n + c_2\, (-2)^n - \dfrac{1}{9}(3n-7)$

(x) $y_n = c_1\, (1)^n + \dfrac{1}{3}(n^{(3)} + 6n^{(2)} + 18)$

4. (i) $y_n = c_1\, 3^n - \dfrac{1}{2}n^{(2)} - n^{(1)} - \dfrac{1}{2} + \dfrac{\cos(n-1) - 3\cos n}{(10 - 6\cos 1)} + 4^n$

(ii) $y_n = c_1\, 2^n + n\, 2^{(n-1)} + 3^n + 2\,(5)^n$

(iii) $y_n = c_1\, 4^n +$
$$\dfrac{\cos(n-1) - 4\cos n}{(17 - 4\cos 1)} + \dfrac{\sin(n-1) - 4\sin n}{17 - 4\cos 1} - \dfrac{n^2}{3} - \dfrac{2n}{9} - \dfrac{1}{3} + n4^{(n-1)}$$

(iv) $y_n = c_1\, 1^n + \dfrac{\cos 5(n-1) - \cos 5n}{2(1 - \cos 5)}$

5. (i) $y_n = c_1\,(8)^n + c_2(-1)^n - \dfrac{3}{20}\left(n^{(2)} - \dfrac{6}{20}n^{(1)} - \dfrac{189}{2100}\right)$

(ii) $y_n = c_1\, \cos(2\pi n/3) + c_2\, \sin(2\pi n/3) + \dfrac{6(2)^n}{7}\left(n - \dfrac{10}{7}\right)$

(iii) $y_n = (c_1\, n + c_2)\,(2)^n + \dfrac{2^n}{48}(n)(n-1)^2(n-2) + \dfrac{n(n-1)}{2}2^n$

(iv) $y_n = c_1\,(-4)^n + c_2(-2)^n + \dfrac{2^n}{24}\left(n - \dfrac{5}{6}\right) + \dfrac{3^n}{35} + \dfrac{1}{15}\left(n^{(2)} - \dfrac{1}{15}n^{(1)} - \dfrac{22}{225}\right)$

6. (i) $y_n = 7\,(3)^n + 5\,(1)^n - (2)^n\,(n^2 - n + 8)$

(ii) $y_n = \dfrac{21}{2}1^n + \dfrac{3^{n+1}}{2}(2n - 3).$

13

Boundary Value Problems

Some problems in applied mathematics like hydrodynamics and elasticity, electromagnetic theory, quantum mechanics and engineering, when formulated mathematically, yield partial differential equations. For example, wave motion in elastic solids is given by $\dfrac{\partial^2 u}{\partial x^2} = \dfrac{1}{c^2}\dfrac{\partial^2 u}{\partial t^2}$.

To interpret the physical behavior of the problem, it is necessary to know the solution of the partial differential equation. Only a few of the partial differential equations can be solved by the analytical methods like method of seperation of variables. The importance of numerical methods is that the solutions of a partial differential equations with some specified conditions can be obtained numerically although it is not possible to obtain the solution of them by analytical methods. Moreover, to implement these numerical methods on a computer to obtain the numerical solution of a partial differential equation with specified condition, one can write a program in C or C^{++} or FORTRAN etc. and obtain its solution by electronic computer.

In this chapter, we restrict our study to find the numerical solution of some linear partial differential equations of second order with some specified conditions.

13.1 DEFINITIONS

Boundary conditions of a partial differential equation are the conditions on the solution to be satisfied at two or more points. The partial differential equation and the boundary conditions constitute the boundary value problem. For example,

$$\frac{\partial^2 u}{\partial x^2} = \frac{1}{c^2}\frac{\partial^2 u}{\partial t^2}, \; 0 \le x \le l, \, t \ge 0,$$

subject to the conditions (i) $u(0,\,t) = 0$, $t > 0$ (ii) $u(l,\,t) = 0$, $t > 0$

(iii) $u(x,\,0) = 10\sin\left(\dfrac{\pi x}{l}\right)$, $0 \le x \le l$

(iv) $\dfrac{\partial u}{\partial x}(x,\,0) = 0$, $0 \le x \le l$, is a boundary value problem. The conditions (i), (ii), (iii) and (iv) are the boundary conditions.

13.2 CLASSIFICATIONS OF LINEAR PARTIAL DIFFERENTIAL EQUATIONS OF SECOND ORDER

A general second-order linear partial differential equation in two independent variables is of the from

$$Au_{xx} + Bu_{xy} + Cu_{yy} + Du_x + Eu_y + Fu = G \qquad (13.1)$$

where $u_{xx} = \dfrac{\partial^2 u}{\partial x^2}, \; u_{xy} = \dfrac{\partial^2 u}{\partial x \partial y}, \; u_{yy} = \dfrac{\partial^2 u}{\partial y^2}, \; u_x = \dfrac{\partial u}{\partial x}, u_y = \dfrac{\partial u}{\partial y}, \; u = u(x, y)$

and A, B, C, D, E, F and G are functions of x and y. For example, $xu_{xx} + u_{yy} = 0$ is a linear partial differential equation of second order.

A partial differential equation given by Eq. (13.1) is defined as

  (i) elliptic if $B^2 - 4AC < 0$;

  (ii) parabolic if $B^2 - 4AC = 0$;

  (iii) hyperbolic if $B^2 - 4AC > 0$.

If $B^2 - 4AC < 0$ for a set of points S, which is a subset of the domain of the function $u(x, y)$, then we say that Eq. (13.1) is elliptic on S. Similar interpretation can extended to parabolic and hyperbolic partial differential equations. The partial differential equation $u_{xx} + u_{yy} = 0$ is elliptic, whereas $u_{tt} = c^2 u_{xx}$ is hyperbolic and $u_t = c^2 u_{xx}$ is parabolic.

SOLVED PROBLEMS

Problem 1: Classify the partial differential equation
$$\frac{1}{4}u_{xx} + 3u_{xy} + 9u_{yy} = 0.$$

Solution: Comparing the given partial differential equation with Eq. (13.1), we have $A = \dfrac{1}{4}, B = 3$ and $C = 9$.

$$\therefore \qquad B^2 - 4AC = 9 - 4.\frac{1}{4}.9 = 0$$

Therefore, the given partial differential equation is parabolic on the domain of the function $u(x, y)$.

Problem 2: Classify the partial differential equation $x^2 u_{xx} + (1 - y^2) u_{yy} = 0$.

Solution: Here $A = x^2$, $B = 0$ and $C = (1 - y^2)$

$$\therefore \quad B^2 - 4AC = 0 - 4(x^2)(1 - y^2) = 4x^2(y^2 - 1)$$

Clearly, $x^2 > 0 \ \forall x \in \mathbb{R} - \{0\}$ and $y^2 < 1 \ \forall y \in (-1, 1)$

Hence, the equation is elliptic in the region, $x \neq 0$ and $-1 < y < 1$. Now, $x^2 > 0 \forall x \in \mathbb{R} - \{0\}$, $y^2 > 1$ if $y \in (-\infty, -1) \cup (1, \infty)$. Therefore, the given partial differential equation is hyperbolic since $B^2 - 4AC > 0 \ \forall x \in \mathbb{R} - \{0\}$ and $y \in (-\infty, -1) \cup (1, \infty)$.

Further, $B^2 - 4AC = 0$ for $x = 0$ or $y^2 = 1$.

Hence, the given partial differential equation is parabolic of $x = 0$ or $y = 1$ or $y = -1$.

Problem 3: Classify the partial differential equation $y^2 u_{xx} - 2xy u_{xy} + x^2 u_{yy} = \dfrac{y^2}{x} u_x + \dfrac{x^2}{y} u_y$

Solution: The given partial differential equation can be expressed as

$$y^2 u_{xx} - 2xy u_{xy} + x^2 u_{yy} - \frac{y^2}{x} u_x - \frac{x^2}{y} u_y = 0 \cdot$$

Therefore, $A = y^2$, $B = -2xy$, $C = x^2$, $D = \dfrac{-y^2}{x}$, and $E = \dfrac{-x^2}{y}$. Now, $B^2 - 4AC = 4x^2 y^2 - 4x^2 y^2 = 0$. Hence, the given partial differential equation is parabolic $\forall \ x, y \in \mathbb{R}$.

Problem 4: Classify the partial differential equation

$$(2+x^2)u_{xx} + (5+2x^2)u_{xy} + (3+x^2)u_{yy} = 0.$$

Solution: $A = 2+x^2$, $B = 5+2x^2$ and $c = (3+x^2)$. Hence, $B^2 - 4AC$
$= (5+2x^2)^2 - 4(2+x^2)(3+x^2) = 1 > 0$. Therefore, the given partial
differential equation is hyperbolic $\forall\ x, y \in \mathbb{R}$.

13.3   FINITE DIFFERENCE APPROXIMATIONS OF PARTIAL DERIVATIVES

Consider a rectangular region $R = [0,a] \times [0,b]$ in the *xy–plane* where
a, b are real numbers. Divide the region R into a of rectangular net-
work of sides h and k. The points of intersection of the family of lines
$x = 0, h, 2h, 3h,.... nh$ and $y = 0, k, 2k, 3k,....., mk$ are called mesh points
or grid points or nodal points or lattice points, (see Fig. 13.1). Here m,
n are positive integers such that $nh = a$ and $mk = b$.

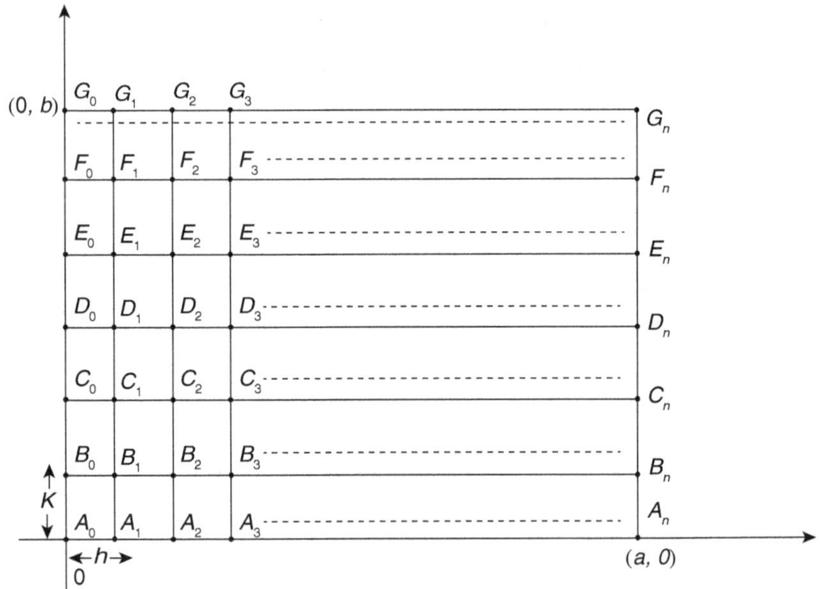

Figure 13.1

The points $A_0, A_1, A_2 ..... B_0, B_1, ...$ and $G_0, G_1, G_2, G_3 .... G_n$ are
mesh points and h and k are known as mesh lengths.

For example, if $R = [0, 8] \times [0, 4]$, $h = 2$ and $k = 1$, then the mesh points of this rectangular region R are shown in Fig. 13.2.

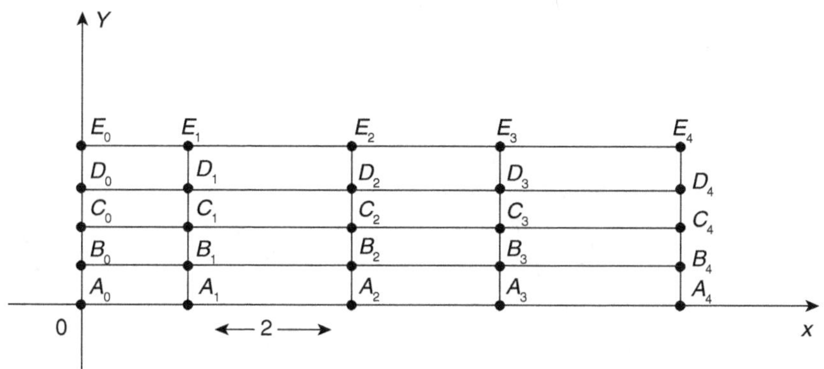

Figure 13.2

The points A_0, A_1, A_2, A_3, A_4, B_0, B_1 ... B_4, E_0, E_1, E_2, E_3, E_4 are the mesh points of the region $R = [0, 8] \times [0, 4]$ with mesh lengths $h = 2$ and $k = 1$.

Let $u(x, y)$ be a function of two independent variables x and y. From the definition of partial derivative, we have

$$\frac{\partial u}{\partial x} = \lim_{h \to 0} \frac{u(x + h, y) - u(x, y)}{h} \qquad (13.2)$$

The Taylor's expansion of $u(x + h, y)$ is

$$u(x + h, y) = u(x, y) + \frac{1}{\lfloor 1} \left[h \frac{\partial u}{\partial x} + 0 \frac{\partial u}{\partial y} \right]$$

$$+ \frac{1}{\lfloor 2} \left[h^2 \frac{\partial^2 u}{\partial x^2} + 2h0 \frac{\partial^2 u}{\partial x \partial y} + 0^2 \frac{\partial^2 u}{\partial y^2} \right] + \dots$$

$$u(x + h, y) - u(x, y) = h \frac{\partial u}{\partial x} + \frac{h^2}{\lfloor 2} \frac{\partial^2 u}{\partial x^2} + \dots\dots$$

i.e.,
$$\frac{u(x + h, y) - u(x, y)}{h} = \frac{\partial u}{\partial x} + \frac{h}{\lfloor 2} \frac{\partial^2 u}{\partial x^2} + \dots\dots \qquad (13.3)$$

Let $0(h^m)$ denote the sum of the terms involving $h^m, h^{m+1}, h^{m+2}, \dots$ where m is a positive integer.

Now, the Eq. (13.3), reduces to

$$\frac{\partial u}{\partial x} = \frac{u(x+h,y)-u(x,y)}{h} + 0(h) \qquad (13.4)$$

Similarly, we have

$$\frac{\partial u}{\partial y} = \frac{u(x,y+k)-u(x,y)}{k} + 0(k) \qquad (13.5)$$

Further, it can be shown that

$$\frac{\partial u}{\partial x} = \frac{u(x+h,y)-u(x-h,y)}{2h} + 0(h^2) \qquad (13.6)$$

$$\frac{\partial u}{\partial y} = \frac{u(x,y+k)-u(x,y-k)}{2k} + 0(k^2) \qquad (13.7)$$

$$\frac{\partial^2 u}{\partial x^2} = \frac{u(x-h,y)-2u(x,y)+u(x+h,y)}{h^2} + 0(h^2) \qquad (13.8)$$

$$\frac{\partial^2 u}{\partial y^2} = \frac{u(x,y-k)-2u(x,y)+u(x,y+k)}{k^2} + 0(k^2) \qquad (13.9)$$

Neglecting $0(h)$, $0(h^2)$, $0(k)$, $0(k^2)$ in the above equations, the approximations of partial derivatives are

$$\frac{\partial u}{\partial x} \approx \frac{u(x+h,y)-u(x,y)}{h} \qquad (13.10)$$

$$\frac{\partial u}{\partial y} \approx \frac{u(x,y+k)-u(x,y)}{k} \qquad (13.11)$$

$$\frac{\partial u}{\partial x} \approx \frac{u(x+h,y)-u(x-h,y)}{2h} \qquad (13.12)$$

$$\frac{\partial u}{\partial y} \approx \frac{u(x,y+k)-u(x,y-k)}{2k} \qquad (13.13)$$

$$\frac{\partial^2 u}{\partial x^2} \approx \frac{u(x-h,y)-2u(x,y)+u(x+h,y)}{h^2} \qquad (13.14)$$

and

$$\frac{\partial^2 u}{\partial y^2} \approx \frac{u(x,y-k)-2u(x,y)+u(x,y+k)}{k^2} \qquad (13.15)$$

Note that the approximation of $\dfrac{\partial u}{\partial x}$ obtained by Eq. (13.12) is a better approximation compared to its approximation obtained by

Eq. (13.10). Similarly the approximation of $\dfrac{\partial u}{\partial y}$ obtained by (13.13) is a better approximation than its approximation obtained by (13.11).

Now, we introduce the notation $u\,(ih, jk) = u_{i,j}$ where i, j are positive integers and h, k are mesh lengths.

Let $u\,(x, y) = u(ih, jk) = u_{i,j}$, then Eqs. (13.10) to (13.15) reduce to

$$\frac{\partial u}{\partial x} \simeq \frac{u_{i+1,j} - u_{i,j}}{h} \tag{13.16}$$

$$\frac{\partial u}{\partial y} \simeq \frac{u_{i,j+1} - u_{i,j}}{k} \tag{13.17}$$

$$\frac{\partial u}{\partial x} \simeq \frac{u_{i+1,j} - u_{i-1,j}}{2h} \tag{13.18}$$

$$\frac{\partial u}{\partial y} \simeq \frac{u_{i,j+1} - u_{i,j-1}}{2k} \tag{13.19}$$

$$\frac{\partial^2 u}{\partial x^2} \simeq \frac{u_{i-1,j} - 2u_{i,j} + u_{i+1,j}}{h^2} \tag{13.20}$$

and

$$\frac{\partial^2 u}{\partial y^2} \simeq \frac{u_{i,j-1} - 2u_{i,j} + u_{i,j+1}}{k^2} \tag{13.21}$$

Now, replacing the partial derivatives appeared in the given partial differential equations by their corresponding difference approximations, we get finite difference equation analogous to the given partial differential equation. The values of $u_{i,j}$ at the mesh points lying on the boundary are called boundary values.

SOLVED PROBLEMS

Problem 1: Obtain the finite difference equation analogous to $\dfrac{\partial^2 u}{\partial x^2} + 2\dfrac{\partial u}{\partial x} = 0$.

Solution: The finite difference approximations of $\dfrac{\partial u}{\partial x}$ and $\dfrac{\partial^2 u}{\partial x^2}$ are given by Eqs. (13.16) and (13.20). Hence, the required equation is

$$\frac{u_{i-1,j} - 2u_{i,j} + u_{i+1,j}}{h^2} + 2\left\{\frac{u_{i-1,j} - u_{i,j}}{h}\right\} = 0$$

i.e. $\qquad u_{i-1,j} - 2u_{i,j} + u_{i+1,j} - 2h(u_{i+1,j} - u_{i,j}) = 0,$

where $u_{i,j} = u(ih, jk)(i = 1, 2, \ldots m; j = 1, 2 \ldots, n)$ and h, k are mesh lengths.

Problem 2: Obtain the finite difference equation analogous to $\frac{\partial^2 u}{\partial x^2} - 5\frac{\partial u}{\partial t} = 0$.

Solution: The finite difference approximation of $\frac{\partial^2 u}{\partial x^2}$ is given by Eq. (13.20), i.e.,

$$\frac{\partial^2 u}{\partial x^2} = \frac{u_{i-1,j} - 2u_{i,j} + u_{i+1,j}}{h^2} \qquad (1)$$

Further, the finite difference approximation of $\frac{\partial u}{\partial t}$ can be obtained by (3.17) replacing the independent variable y by t and it is given by $\frac{\partial u}{\partial t} = \frac{u_{i,j+1} - u_{i,j}}{k}.$

Hence, the required equation is

$$\frac{u_{i-1,j} - 2u_{i,j} + u_{i+1,j}}{h^2} + \frac{1}{k}(u_{i,j+1} - u_{i,j}) = 0,$$

i.e., $\qquad k\left\{u_{i-1,j} - 2u_{i,j} + u_{i+1,j}\right\} + h^2(u_{i,j+1} - u_{i,j}) = 0.$

Problem 3: Obtain the finite difference equation analogous to $\frac{\partial^2 u}{\partial x^2} + 5\frac{\partial^2 u}{\partial y^2} = 0$.

Solution: The finite difference approximations of $\frac{\partial^2 u}{\partial x^2}$ and $\frac{\partial^2 u}{\partial y^2}$ are given by Eqs. (13.20) and (13.21).

Hence, the required difference equation analogous to $\frac{\partial^2 u}{\partial x^2} + 5\frac{\partial^2 u}{\partial y^2} = 0$ is

$$\frac{u_{i-1,j} - 2u_{i,j} + u_{i+1,j}}{h^2} + \frac{5}{k^2}\{u_{i,j-1} - 2u_{i,j} + u_{i,j+1}\} = 0,$$

i.e., $k^2\{u_{i-1,j} - 2u_{i,j} + u_{i+1,j}\} + 5h^2\{u_{i,j-1} - 2u_{i,j} + u_{i,j+1}\} = 0$

13.4   ELLIPTIC EQUATION–LAPLACE'S EQUATION

In this section, we describe the numerical method to obtain the solution of two-dimensional Laplace equation.

$$u_{xx} + u_{yy} = 0, \tag{13.22}$$

which is an elliptic equation. These types of equations arise in potential and steady-state flow problems.

Let R be a rectangular region. We suppose that the domain of $u(x, y)$ is R and $u(x, y)$ is known at the boundary of R. We divide R into rectangles of length h and breadth k. Then finite difference analogous of Eqn. (13.22) is

$$\frac{1}{h^2}(u_{i+1,j} - 2u_{i,j} + u_{i-1,j})$$

$$+ \frac{1}{k^2}(u_{i,j+1} - 2u_{i,j} + u_{i,j-1}) = 0 \tag{13.23}$$

If $h = k$, Eq. (13.23) reduces to

$$u_{i,j} = \frac{1}{4}[u_{i+1,j} + u_{i-1,j} + u_{i,j+1} + u_{i,j-1}] \tag{13.24}$$

This shows that the value of u at any interior mesh point is the mean of the four neighboring values of it (Fig. 13.3).

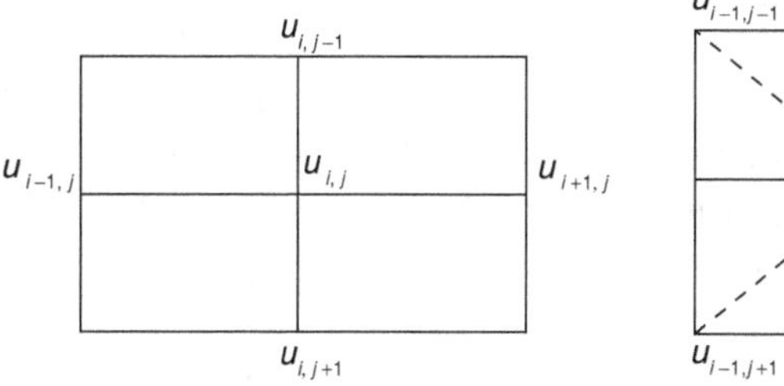

Figure 13.3   **Figure 13.4**

Note that the four neighboring values of $u_{i,j}$ are immediately left and right of $u_{i,j}$ and immediate top and bottom values of $u_{i,j}$.

Eq. (13.24) is called standard five point formula (SFPF). On some occasions, we can also use

$$u_{i,j} = \frac{1}{4}[u_{i-1,j+1} + u_{i+1,j-1} + u_{i+1,j+1} + u_{i-1,j-1}], \qquad (13.25)$$

which is known as diagonal five point formula (DFPF), (see Fig. 13.4), which can be derived from Eq. (13.23) expanding the terms on the right-hand side of it in Taylor series.

Thus, using the boundary conditions, Eqs. (13.24) and (13.25), we can find all the values of $u_{i,j}$ at mesh points for given boundary value problem.

The approximations $u_{i,j}$ of $u\,(x,\,y)$ at $x = ih$ and $y = jk$ obtained by the above process can be improved either by Jacobi method or Gauss–Seidel method. The iterative formula for computation of $u_{i,j}^{(n+1)}$ of Jacobi's method is

$$u_{i,j}^{(n+1)} = \frac{1}{4}\left[u_{i-1,j}^{(n)} + u_{i+1,j}^{(n)} + u_{i,j-1}^{(n)} + u_{i,j+1}^{(n)} \right] \qquad (13.26)$$

and that of Gauss–Seidel method is

$$u_{i,j}^{(n+1)} = \frac{1}{4}\left[u_{i-1,j}^{(n+1)} + u_{i+1,j}^{(n)} + u_{i,j-1}^{(n+1)} + u_{i,j+1}^{(n)} \right] \qquad (13.27)$$

Obtaining the numerical solution of Laplace equation with boundary condition using Eqs. (13.24) and (13.25) and improving these approximations by Gauss–Seidel method is known as Liebmann's iteration process.

Note: In case, if it is not possible to find $u_{i,j}$ either by SFPF or by DFPF because of the one unknown on the right-hand side of them then we approximate the unknown value suitably observing the boundary conditions or their variations, i.e., the values of $u_{i,j}$ at mesh points lying on the boundary or variation of $u\,(x,\,y)$ along a row or column.

13.5   PICTORIAL REPRESENTATION OF THE VALUES OF $u\,(x,\,y)$ AT THE MESH POINTS OF A SQUARE

We illustrate this for a square of $[0,\,2h] \times [0,\,2h]$. Note that $k = h$. Then mesh points are $(0,\,0)$, $(h,\,0)$, $(2h,\,0)$, $(0,\,h)$, $(h,\,h)$, $(2h,\,h)$,

$(0, 2h)$, $(h, 2h)$ and $(2h, 2h)$. The values of $u(x, y)$ at these points are $u_{0,0}$, $u_{1,0}$, $u_{2,0}$, $u_{0,1}$, $u_{1,1}$, $u_{2,1}$, $u_{0,2}$, $u_{1,2}$ and u_{22}. These values are shown pictorially in Fig. 13.5.

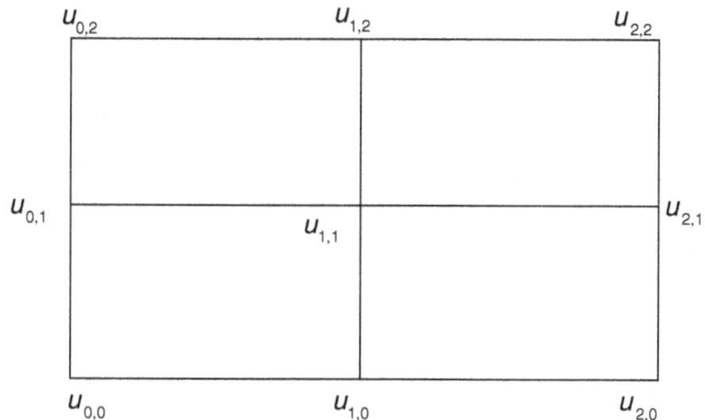

Figure 13.5

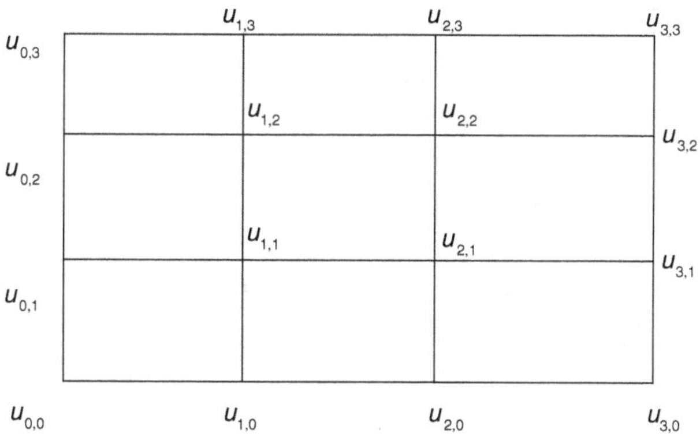

Figure 13.6

If the square is $[0, 3h] \times [0, 3h]$, then $u(i, j)$ values at the mesh points are as shown in Fig. 13.6. The same representation can be extended to rectangular region $[0, mh] \times [0, nk]$ where m and n are positive integers and h, k are mesh lengths.

Usually, $u(x, y)$ values at the interior points are supposed as u_1, u_2, u_3, .., u_k where k is a positive integer. Further we refer $u_{0,1}, u_{1,1}, u_{2,1}, \ldots\ldots$ etc. as first row entries of $u(x, y)$ and $u_{0,2}, u_{1,2}, u_{2,2} \ldots$ etc are referred as second row entries of $u(x, y)$ and so on when the values shown in the format stated as in Fig 13.6. Similarly, $u_{1,0}$, $u_{1,1}$, $u_{1,2}$, $u_{1,3}$ are reformed as column entries and so on.

Note: If the region R is shown pictorially and assigned certain values at the mesh points lying on the boundary, then these are the values of $u(x, y)$ at the boundary mesh points and they are called boundary values, i.e., they represent boundary values. For example, the following Fig. 13.7 indicates $u_{0,0} = 0$, $u_{1,0} = 100$, $u_{2,0} = 200$, $u_{3,0} = 300$, $u_{0,1} = 10$, $u_{3,1} = 40$, $u_{0,2} = 20$, $u_{3,2} = 50$, $u_{0,3} = 80$ $u_{1,3} = 0$, $u_{2,3} = 0$ and $u_{3,3} = 60$

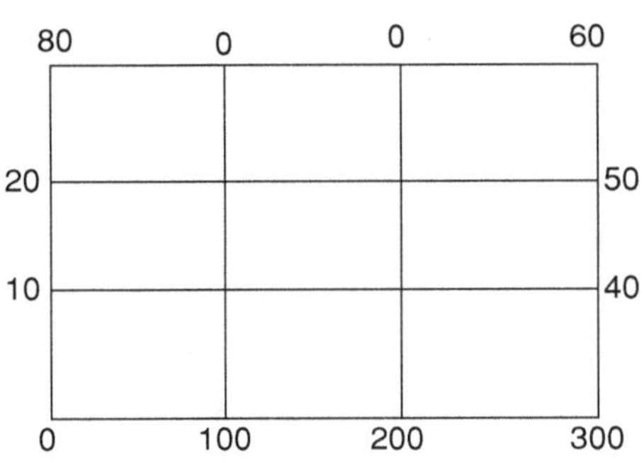

Figure 13.7

SOLVED PROBLEMS

Problem 1: Solve the Laplace equation (13.22) for the square region shown in Fig. 13.8 by (i) Gauss–Jacobi's method and (ii) Gauss–Seidel method.

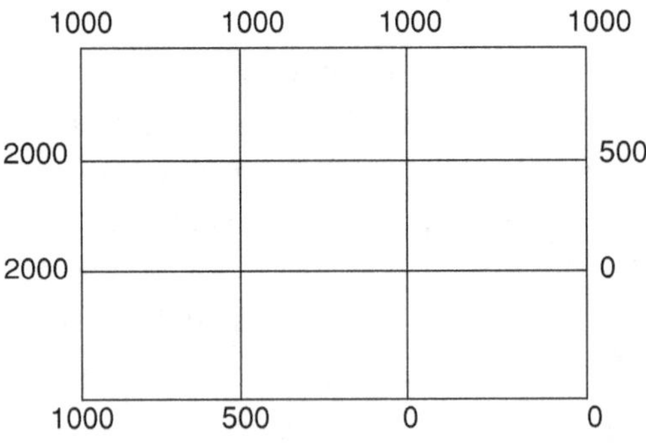

Figure 13.8

Solution: Let the mesh points be $A_1, A_2, A_3, A_4; B_1, B_2, B_3, B_4; C_1, C_2, C_3, C_4$ and $D_1, D_2, D_3, D_4;$ (See Fig. 13.9)

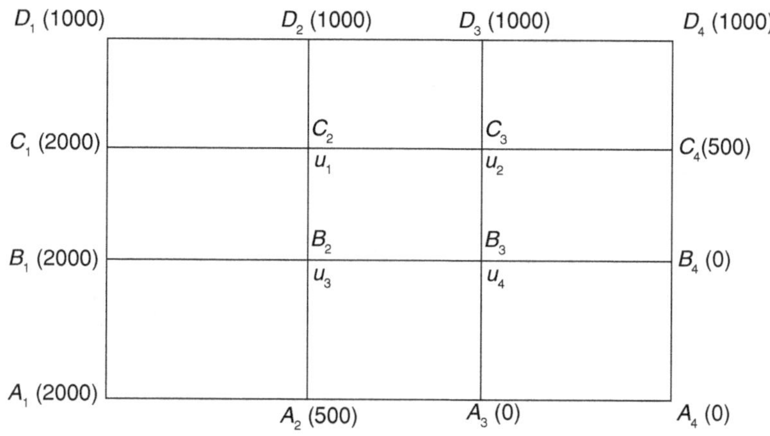

Figure 13.9

Note that the numbers shown in brackets indicate the values of $u(x, y)$ at the corresponding mesh points. Of course, they are boundary values. Further, the mesh points C_2, C_3 and B_2, B_3 are the interior mesh points.

Let u_1, u_2, u_3 and u_4 be the values of $u(x, y)$ at the mesh points C_2, C_3, B_2 and B_3, respectively.

We need an approximation of one of u_1, u_2, u_3 and u_4 to get the initial approximations of u_1, u_2, u_3 and u_4 either by SFPF or DFPF.

We assume $u_4 = 0$ (as mesh point B_3 is close to B_4 and its value is 0).

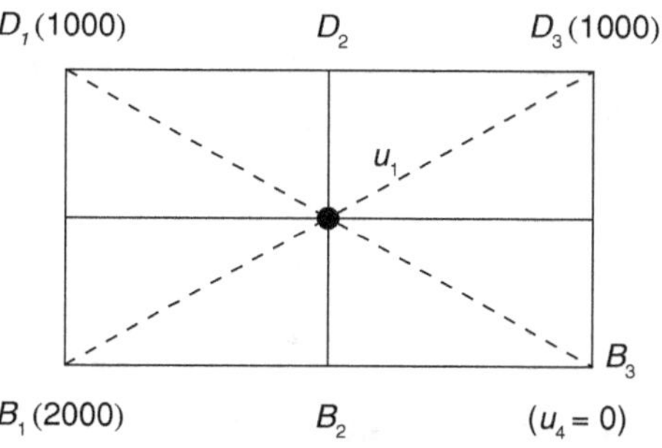

Figure 13.10

Therefore, applying DFPF for B_1, B_3, D_3, D_1 (Fig. 13.10)

$$u_1 = \frac{1}{4}[(1000 + 0) + (2000 + 1000)] = 1000$$

Thus, $u_1^{(0)} = 1000$. Now, consider the square $B_2\, B_3\, D_4\, D_2$. (See Fig. 13.11). By SFPF, we get

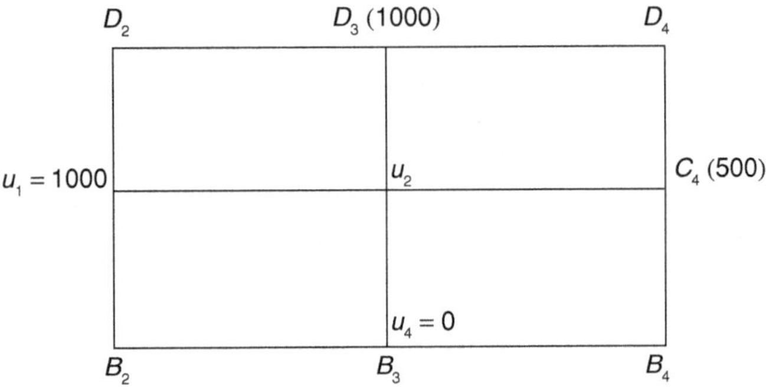

Figure 13.11

$$u_2 = \frac{1}{4}[(1000 + 0) + (1000 + 500)] = 625$$

To obtain u_3, consider the square $A_1\, A_3\, C_3\, C_1$. (See Fig. 13.12). Applying SFPF, we have

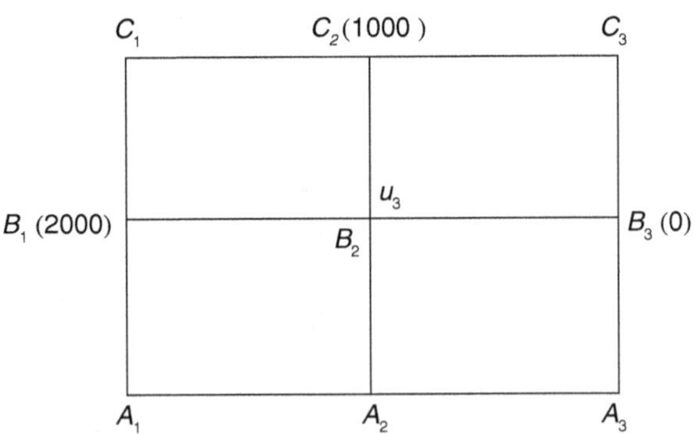

Figure 13.12

$$u_3 = \frac{[(2000 + 0) + (1000 + 500)]}{4} = 875$$

Finally, consider the square $A_2\,A_4\,C_4\,C_2$. (See Fig. 13.13). Again applying SFPF for this square,

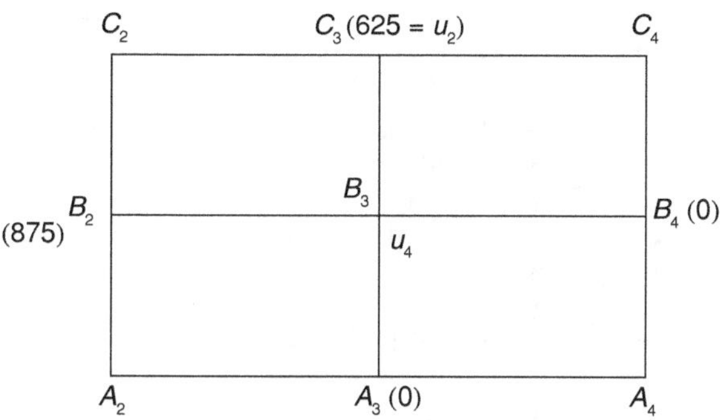

Figure 13.13

We obtain,

$$u_4 = \frac{1}{4}[(625 + 0) + (875 + 0)] = 375 .$$

Thus, the values of $u(x, y)$ at the interior mesh points and its boundary values are shown in Fig. 13.14:

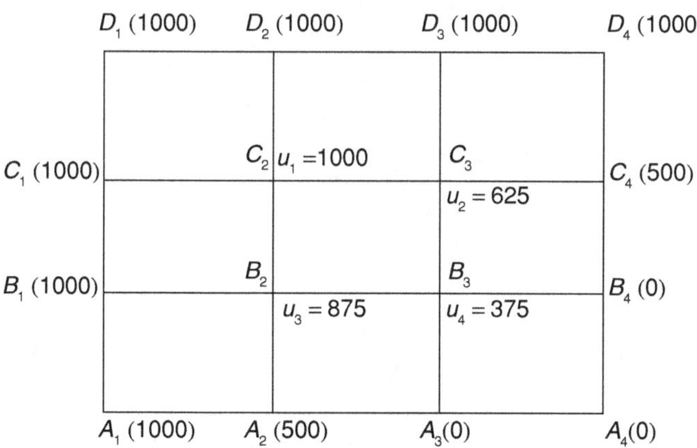

Figure 13.14

We suppose $u_1^{(0)} = 1000,\ u_2^{(0)} = 625,\ u_3^{(0)} = 875$ and $u_4^{(0)} = 375$

(a) Jacobi's method

Now, we improve the approximations of u_1, u_2, u_3, u_4 by Jacobi's method.

By SFPF, we have

$$u_1^{(n+1)} = \frac{1}{4}[2000 + u_2^{(n)} + 1000 + u_3^{(n)}] \tag{1}$$

$$u_2^{(n+1)} = \frac{1}{4}[u_1^{(n)} + 500 + 1000 + u_4^{(n)}] \tag{2}$$

$$u_3^{(n+1)} = \frac{1}{4}[2000 + u_4^{(n)} + u_1^{(n)} + 500] \tag{3}$$

$$u_4^{(n+1)} = \frac{1}{4}[u_3^{(n)} + 0 + u_2^{(n)} + 0] \tag{4}$$

where n is a non-negative integer and $u_i^{(k)}$ represent k^{th} approximation of $u_i (i = 1, 2, 3, 4)$, where k is a non-negative integer.

Putting $n = 0$ in Eqs. (1) to (4), we get

$$u_1^{(1)} = \frac{1}{4}(2000 + u_2^{(0)} + 1000 + u_3^{(0)}) \tag{5}$$

$$u_2^{(1)} = \frac{1}{4}(u_1^{(0)} + 500 + 1000 + u_4^{(0)}) \tag{6}$$

$$u_3^{(1)} = \frac{1}{4}(2000 + u_4^{(0)} + u_1^{(0)} + 500) \tag{7}$$

$$u_4^{(1)} = \frac{1}{4}(u_1^{(0)} + 0 + u_2^{(0)} + 0) \tag{8}$$

Iteration 1

Substituting the values of $u_1^{(0)}, u_2^{(0)}, u_3^{(0)}$ and $u_4^{(0)}$ in the Eqs. (5) to (8), we get,

$$u_1^{(1)} = (2000 + 625 + 1000 + 875)/4 = 1125$$

$$u_2^{(1)} = (1125 + 500 + 1000 + 375)/4 = 718.75$$

$$u_3^{(1)} = (2000 + 375 + 1125 + 500)/4 = 968.75$$

$$u_4^{(1)} = (875 + 0 + 625 + 0)/4 = 375$$

Similarly substituting $n = 1, 2, 3 \dots\dots 7$ in the Eqs. (1) to (4) and repeating the above process, we get successive approximations of $u_i^{(n)} (i = 1, 2, 3, 4 : n = 2, 3,\dots,)$ these values are as given below.

Iteration 2

$$u_1^{(2)} = (2000.000000 + 718.750000 + 1000.000000 + 968.750000)/4$$
$$= 1171.875000$$

$$u_2^{(2)} = (1125.000000 + 500.000000 + 1000.000000 + 375.000000)/4$$
$$= 750.000000$$

$$u_3^{(2)} = (2000.000000 + 375.000000 + 1125.000000 + 500.000000)/4$$
$$= 1000.000000$$

$$u_4^{(2)} = (968.750000 + 0.000000 + 718.750000 + 0.000000)/4$$
$$= 421.875000$$

Iteration 3

$$u_1^{(3)} = (2000.000000 + 750.000000 + 1000.000000 + 1000.000000)/4$$
$$= 1187.500000$$

$$u_2^{(3)} = (1171.875000 + 500.000000 + 1000.000000 + 421.8750000)/4$$
$$= 773.437500$$

$$u_3^{(3)} = (2000.000000 + 421.875000 + 1171.875000 + 500.000000)/4$$
$$= 1023.437500$$

$$u_4^{(3)} = (1000.000000 + 0.000000 + 750.000000 + 0.000000)/4$$
$$= 437.500000$$

Iteration 4

$$u_1^{(4)} = (2000.000000 + 773.437500 + 1000.000000 + 1023.437500)/4$$
$$= 1199.218750$$

$$u_2^{(4)} = (1187.500000 + 500.000000 + 1000.000000 + 437.500000)/4$$
$$= 781.250000$$

$$u_3^{(4)} = (2000.000000 + 437.500000 + 1187.500000 + 500.000000)/4$$
$$= 1031.250000$$

$$u_4^{(4)} = (1023.437500 + 0.000000 + 773.437500 + 0.000000)/4$$
$$= 449.218750$$

Iteration 5

$$u_1^{(5)} = (2000.000000 + 781.250000 + 1000.000000 + 1031.250000)/4$$
$$= 1203.125000$$

$$u_2^{(5)} = (1199.218750 + 500.000000 + 1000.000000 + 449.218750)/4$$
$$= 787.109375$$

$$u_3^{(5)} = (2000.000000 + 449.218750 + 1199.218750 + 500.000000)/4$$
$$= 1037.109375$$

$$u_4^{(5)} = (1031.250000 + 0.000000 + 781.250000 + 0.000000)/4$$
$$= 453.125000$$

Iteration 6

$$u_1^{(6)} = (2000.000000 + 787.109375 + 1000.000000 + 1037.109375)/4$$
$$= 1206.054688$$

$$u_2^{(6)} = (1203.125000 + 500.000000 + 1000.000000 + 453.125000)/4$$
$$= 789.062500$$

$$u_3^{(6)} = (2000.000000 + 453.125000 + 1203.125000 + 500.000000)/4$$
$$= 1039.062500$$

$$u_4^{(6)} = (1037.109375 + 0.000000 + 787.109375 + 0.000000)/4$$
$$= 456.054688$$

Iteration 7

$$u_1^{(7)} = (2000.000000 + 789.062500 + 1000.000000 + 1039.062500)/4$$
$$= 1207.031250$$

$$u_2^{(7)} = (1206.054688 + 500.000000 + 1000.000000 + 456.054688)/4$$
$$= 790.527344$$

$$u_3^{(7)} = (2000.000000 + 456.054688 + 1206.054688 + 500.000000)/4$$
$$= 1040.527344$$

$$u_4^{(7)} = (1039.062500 + 0.000000 + 789.062500 + 0.000000)/4$$
$$= 457.031250$$

Iteration 8

$$u_1^{(8)} = (2000.000000 + 790.527344 + 1000.000000 + 1040.527344)/4$$
$$= 1207.763672$$

$$u_2^{(8)} = (1207.031250 + 500.000000 + 1000.000000 + 457.031250)/4$$
$$= 791.015625$$

$$u_3^{(8)} = (2000.000000 + 457.031250 + 1207.031250 + 500.000000)/4$$
$$= 1041.015625$$

$$u_4^{(8)} = (1040.527344 + 0.000000 + 790.527344 + 0.000000)/4$$
$$= 457.763672$$

By Jacobi's method, up to 8 iterations, the values of u_1, u_2, u_3 and u_4 are approximately $u_1 = 1208$, $u_2 = 791$, $u_3 = 1041$, and $u_4 = 458$.

(b) Gauss–Seidel method

Here also, we suppose $u_1^{(0)} = 1000$, $u_2^{(0)} = 625$ $u_3^{(0)} = 875$ and $u_4^{(0)} = 375$.

By the SFPF, the improved approximations of u_1, u_2, u_3 and u_4 as per Gauss–Seidel method are given by

$$u_1^{(n+1)} = \frac{1}{4}\left[2000 + u_2^{(n)} + 1000 + u_3^{(n)}\right] \tag{9}$$

$$u_2^{(n+1)} = \frac{1}{4}\left[u_1^{(n+1)} + 500 + 1000 + u_4^{(n)}\right] \qquad (10)$$

$$u_3^{(n+1)} = \frac{1}{4}\left[2000 + u_4^{(n)} + u_1^{(n+1)} + 500\right] \qquad (11)$$

$$u_4^{(n+1)} = \frac{1}{4}\left[u_3^{(n+1)} + 0 + u_2^{(n+1)} + 0\right] \qquad (12)$$

Iteration 1

Substituting $n = 0$ in Eqs. (9) to (12), we get

$$u_1^{(1)} = \frac{1}{4}(2000 + u_2^{(0)} + 1000 + u_3^{(0)}) \qquad (13)$$

$$u_2^{(1)} = \frac{1}{4}(u_1^{(1)} + 500 + 1000 + u_4^{(0)}) \qquad (14)$$

$$u_3^{(1)} = \frac{1}{4}(2000 + u_4^{(0)} + u_1^{(1)} + 500) \qquad (15)$$

$$u_4^{(1)} = \frac{1}{4}(u_1^{(1)} + 0 + u_2^{(1)} + 0) \qquad (16)$$

Substituting $u_2^{(0)}$ and $u_3^{(0)}$ values in (13) we have

$$u_1^{(1)} = \frac{1}{4}(2000 + 625 + 1000 + 875) = 1125$$

Now substituting $u_1^{(1)}$ and $u_4^{(0)}$ values in Eq. (14), we get

$$u_2^{(1)} = \frac{1}{4}(1125 + 800 + 1000 + 375) = 750$$

Similarly, substituting $u_4^{(0)}$ and $u_1^{(1)}$ values in the Eq. (15), we have

$$u_3^{(1)} = \frac{1}{4}(2000 + 375 + 1125 + 500) = 1000$$

Finally, substituting $u_1^{(1)}$ and $u_2^{(1)}$ values in Eq. (16), we obtain

$$u_4^{(1)} = \frac{1}{4}(1000 + 0 + 750 + 0) = 437.5$$

Similarly, substituting $n = 1, 2... 7$ in Eqs. (9) to (12) and repeating above process we get the following successive approximations of u_1, u_2, u_3 and u_4

Iteration 2

$$u_1^{(2)} = (2000.000000 + 750.000000 + 1000.000000 + 1000.000000)/4$$
$$= 1187.500000$$

$$u_2^{(2)} = (1187.500000 + 500.000000 + 1000.000000 + 437.5000000)/4$$
$$= 781.250000$$

$$u_3^{(2)} = (2000.000000 + 437.500000 + 1187.500000 + 500.000000)/4$$
$$= 1031.250000$$

$$u_4^{(2)} = (1031.250000 + 0.000000 + 781.250000 + 0.000000)/4$$
$$= 453.125000$$

Iteration 3

$$u_1^{(3)} = (2000.000000 + 781.250000 + 1000.000000 + 1031.250000)/4$$
$$= 1203.125000$$

$$u_2^{(3)} = (1203.125000 + 500.000000 + 1000.000000 + 453.125000)/4$$
$$= 789.062500$$

$$u_3^{(3)} = (2000.000000 + 453.125000 + 1203.125000 + 500.000000)/4$$
$$= 1039.06250$$

$$u_4^{(3)} = (1039.062500 + 0.000000 + 789.062500 + 0.000000)/4$$
$$= 457.031250$$

Iteration 4

$$u_1^{(4)} = (2000.000000 + 789.062500 + 1000.000000 + 1039.062500)/4$$
$$= 1207.031250$$

$$u_2^{(4)} = (1207.031250 + 500.000000 + 1000.000000 + 457.031250)/4$$
$$= 791.015625$$

$$u_3^{(4)} = (2000.000000 + 457.031250 + 1207.031250 + 500.000000)/4$$
$$= 1041.015625$$

$$u_4^{(4)} = (1041.015625 + 0.000000 + 791.015625 + 0.000000)/4$$
$$= 458.007813$$

Iteration 5

$$u_1^{(5)} = (2000.000000 + 791.015625 + 1000.000000 + 1041.015625)/4$$
$$= 1208.007813$$

$$u_2^{(5)} = (1208.007813 + 500.000000 + 1000.000000 + 458.007813)/4$$
$$= 791.503906$$

$$u_3^{(5)} = (2000.000000 + 458.007813 + 1208.007813 + 500.000000)/4$$
$$= 1041.503906$$

$$u_4^{(5)} = (1041.503906 + 0.000000 + 791.503906 + 0.000000)/4$$
$$= 458.251953$$

Iteration 6

$$u_1^{(6)} = (2000.000000 + 791.503906 + 1000.000000 + 1041.503906)/4$$
$$= 1208.251953$$

$$u_2^{(6)} = (1208.251953 + 500.000000 + 1000.000000 + 458.251953)/4$$
$$= 791.625977$$

$$u_3^{(6)} = (2000.000000 + 458.251953 + 1208.251953 + 500.000000)/4$$
$$= 1041.625977$$

$$u_4^{(6)} = (1041.625977 + 0.000000 + 791.625977 + 0.000000)/4$$
$$= 458.312988$$

Iteration 7

$$u_1^{(7)} = (2000.000000 + 791.625977 + 1000.000000 + 1041.625977)/4$$
$$= 1208.312988$$

$$u_2^{(7)} = (1208.312988 + 500.000000 + 1000.000000 + 458.312988)/4$$
$$= 791.656494$$

$$u_3^{(7)} = (2000.000000 + 458.312988 + 1208.312988 + 500.000000)/4$$
$$= 1041.656494$$

$$u_4^{(7)} = (1041.656494 + 0.000000 + 791.656494 + 0.000000)/4$$
$$= 458.328247$$

Iteration 8

$$u_1^{(8)} = (2000.000000 + 791.656494 + 1000.000000 + 1041.656494)/4$$
$$= 1208.328247$$

$$u_2^{(8)} = (1208.328247 + 500.000000 + 1000.000000 + 458.328247)/4$$
$$= 791.664124$$

$$u_3^{(8)} = (2000.000000 + 458.328247 + 1208.328247 + 500.000000)/4$$
$$= 1041.664124$$

$$u_4^{(8)} = (1041.664124 + 0.000000 + 791.664124 + 0.000000)/4$$
$$= 458.332062$$

Upto 8 iterations, the values of u_1, u_2, u_3 and u_4 are 1208,792,1042 and 458 respectively.

Problem 2: Given the Values of $u(x, y)$ on the boundary of the square shown in Fig 13.15 solve $u_{xx} + u_{yy} = 0$ at the interior mesh points of the square.

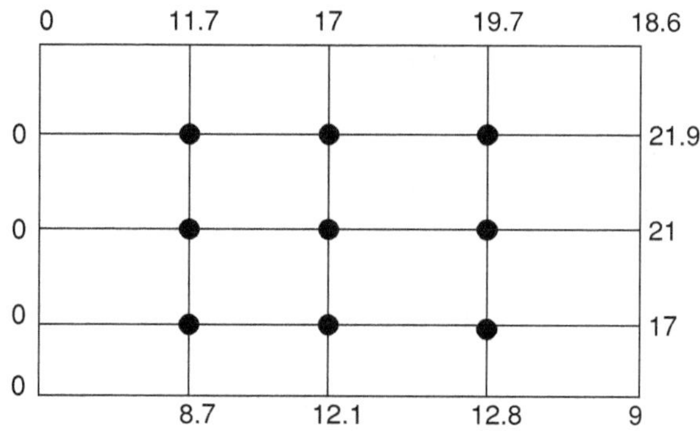

Figure 13. 15

Solution: Let $u_1, u_2, u_3, \ldots\ldots$ and u_9 be the values of $u\,(x, y)$ at the interior mesh points $A, B, C, D, E, E\,G, H$ and I respectively (see Fig. 13.16)

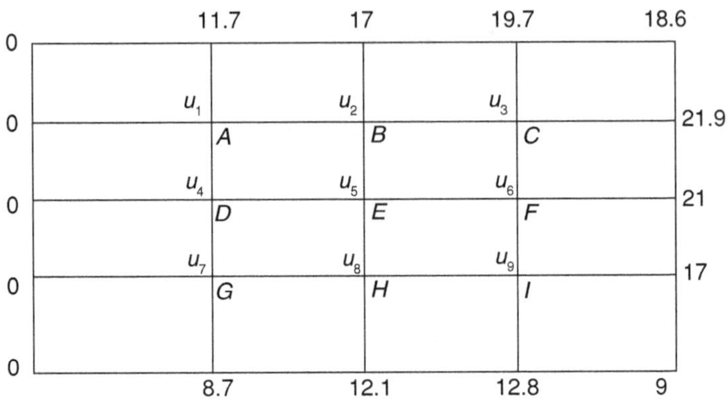

Figure 13.16

By the SFPF, we have

$$u_5 = \frac{1}{4}\left[(u_2 + u_8) + (u_4 + u_6)\right]$$

We assume $u_2 = 17$, $u_8 = 12.1$, $u_4 = 0$ and $u_6 = 21$ as the values of u at mesh points B, H, D and F are close to the Boundary Values 17, 12.1, 0 and 21 respectively. Hence

$$u_5 = \frac{1}{4}[17 + 12.1 + 0 + 21] = 12.525$$

By the DFPF (see Fig. 13.17),

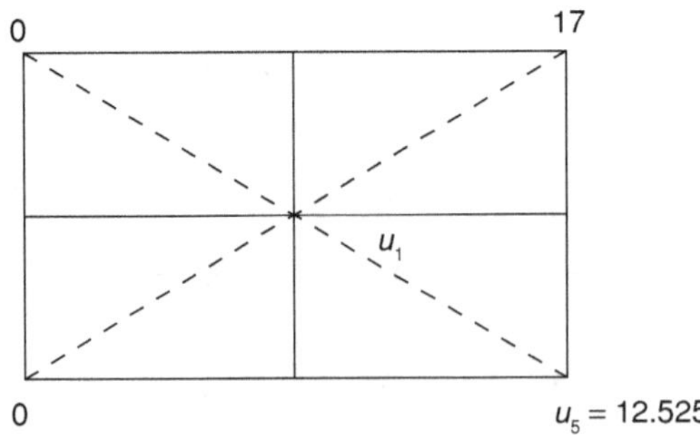

Figure 13.17

we obtain

$$u_1 = \frac{1}{4}[0 + 17 + 0 + 12.525] = 7.38125$$

Again by the DFPF (see Fig. 13.18)

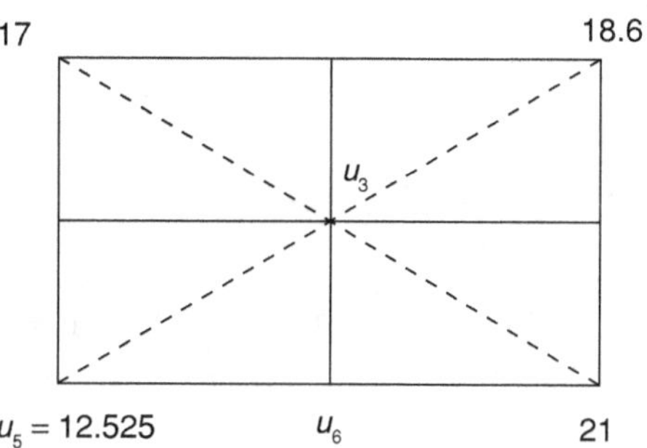

Figure 13.18

we obtain

$$u_3 = \frac{1}{4}[18.6 + 12.525 + 17 + 21] = 17.28125$$

From the Fig. 13.19 and applying DFPF,

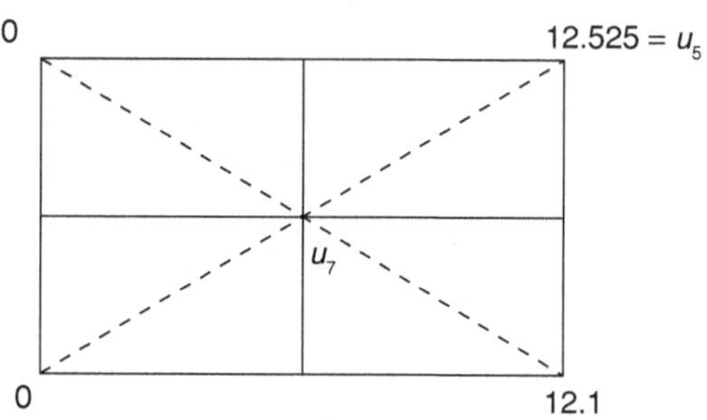

Figure 13.19

We get

$$u_9 = \frac{1}{4}[12.525 + 0 + 0 + 12.1] = 6.15625$$

Now, from the Fig. 13.20 and applying DFPF, we get

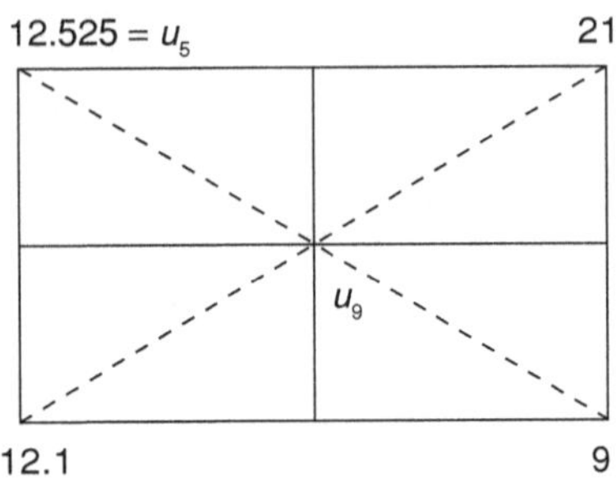

Figure 13.20

$$u_9 = \frac{1}{4}[21 + 12.1 + 12.52 + 9]$$

$$= 13.65625.$$

We evaluate u_2, u_4, u_6 and u_8 by using SFPF repeatedly. From the Fig. 13.21,

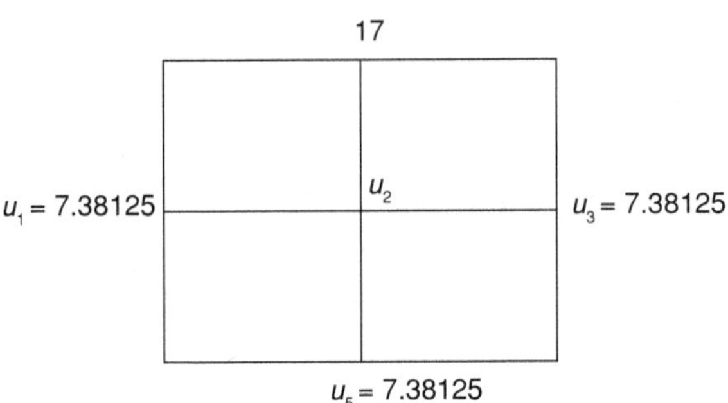

Figure 13.21

we have

$$u_2 = \frac{1}{4}[17 + u_5 + u_1 + u_3]$$

Substituting u_5, u_1 and u_3 values in the above equations, we get

$$u_2 = \frac{1}{4}[17.12.525 + 7.38125 + 17.28125]$$

i.e., $u_2 = 13.546875$

From the Figs 13.22, 13.23 and 13.24 and using SFPF, we obtain

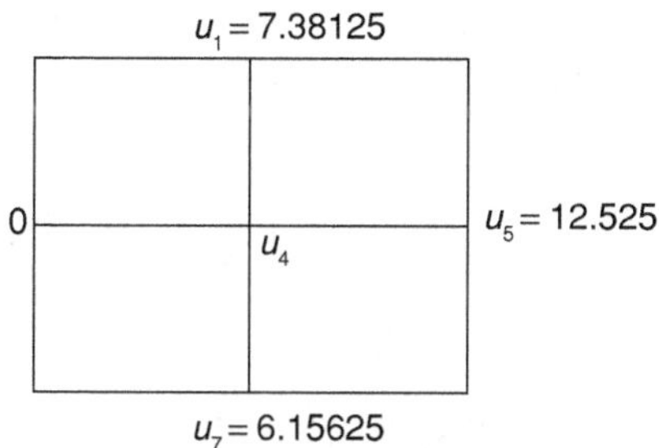

Figure 13.22

$$u_4 = \frac{1}{4}[7.38125 + 6.15625 + 12.525 + 0]$$

i.e., $\qquad u_4 = 6.515625$

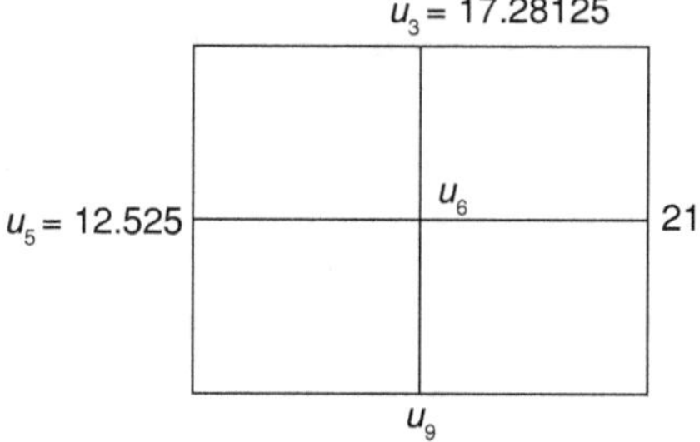

Figure 13.23

$$u_6 = \frac{1}{4}[17.28125 + 13.65625 + 12.525 + 21]$$

i.e., $\qquad u_6 = 16.115625$

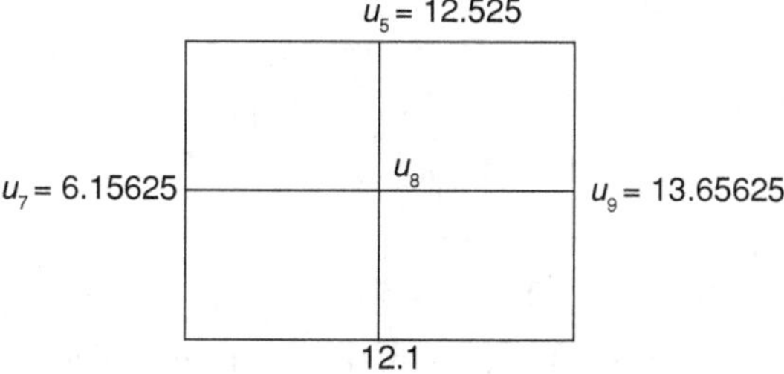

Figure 13.24

$$u_8 = \frac{1}{4}[u_5 + 12.1 + u_7 + u_9]$$

$$= \frac{1}{4}[12.525 + 12.1 + 6.15625 + 13.65625)$$

i. e., $$u_8 = 11.109375$$

The values of u_1 to u_9 are improved by Liebmann's iteration process. The iterative formulas are as follows:

$$u_1^{(n+1)} = \frac{1}{4}[11.7 + u_4^{(n)} + 0 + u_2^{(n)}] \tag{1}$$

$$u_2^{(n+1)} = \frac{1}{4}[17 + u_5 + u_1^{(n+1)} + u_3^{(n)}] \tag{2}$$

$$u_3^{(n+1)} = \frac{1}{4}[19.7 + u_6^{(n)} + u_2^{(n+1)} + 21.9] \tag{3}$$

$$u_4^{(n+1)} = \frac{1}{4}[0 + u_5^{(n)} + u_1^{(n+1)} + u_7^{(n)}] \tag{4}$$

$$u_5^{(n+1)} = \frac{1}{4}[u_2^{(n+1)} + u_8^{(n)} + u_4^{(n+1)} + u_6^{(n)}] \tag{5}$$

$$u_6^{(n+1)} = \frac{1}{4}[u_3^{(n+1)} + u_9^{(n)} + u_5^{(n+1)} + 21] \tag{6}$$

$$u_7^{(n+1)} = \frac{1}{4}[u_4^{(n+1)} + 8.7 + 0 + u_8^{(n)}] \tag{7}$$

$$u_8^{(n+1)} = \frac{1}{4}[u_5^{(n+1)} + 12.1 + u_7^{(n+1)} + u_9^{(n)}] \tag{8}$$

$$u_9^{(n+1)} = \frac{1}{4}[u_8^{(n+1)} + 17.8 + u_8^{(n+1)} + 17] \tag{9}$$

Suppose, $u_1^{(0)} = 7.38125$, $u_2^{(0)} = 13.546875$, $u_3^{(0)} = 17.28125$, $u_4^{(0)} = 6.515625$, $u_5^{(0)} = 12.525$, $u_6^{(0)} = 16.115625$, $u_7^{(0)} = 6.15625$, $u_8^{(0)} = 11.109375$, $u_9^{(0)} = 13.65625$;

Substituting $n = 0$ in Eqs. (1) to (9) and substituting the relevant values in these equations, we get the first approximations of $u_i^{(1)}$ $(i = 1, 2, ... 9)$ and they are as follows:

Iteration 1

$$u_1^{(1)} = (0.000000 + 13.546875 + 11.700000 + 6.515625)/4$$
$$= 7.940625$$

$$u_2^{(1)} = (7.940625 + 17.281250 + 17.000000 + 12.525000)/4$$
$$= 13.686719$$

$$u_3^{(1)} = (13.686719 + 21.900000 + 19.700000 + 16.115625)/4$$
$$= 17.850586$$

$$u_4^{(1)} = (0.000000 + 12.525000 + 7.940625 + 6.156250)/4$$
$$= 6.655469$$

$$u_5^{(1)} = (6.655469 + 16.115625 + 13.688719 + 11.109375)/4$$
$$= 11.891797$$

$$u_6^{(1)} = (11.891797 + 21.000000 + 17.850586 + 13.656250)/4$$
$$= 16.099658$$

$$u_7^{(1)} = (0.000000 + 11.109375 + 6.655469 + 8.700000)/4$$
$$= 6.616211$$

$$u_8^{(1)} = (6.6616211 + 13.656250 + 11.898797 + 12.100000)/4$$
$$= 11.066064$$

$$u_9^{(1)} = (11.066064 + 17.000000 + 16.099658 + 12.800000)/4$$
$$= 14.241431$$

Similarly, substituting $n = 1, 2\ldots\ldots14$ in Eqs. (1) to (9), we get the approximations of
$u_i^{(n)}$ $(i = 1, 2, \ldots\ldots 9: n = 2, 3, 4 \ldots\ldots 15)$. These values are shown below:

Iteration 2

$$u_1^{(2)} = (0.000000 + 13.686719 + 11.700000 + 6.655469)/4$$
$$= 8.010547$$

$$u_2^{(2)} = (8.010547 + 17.850586 + 17.000000 + 11.891797)/4$$
$$= 13.688232$$

$$u_3^{(2)} = (13.688232 + 21.900000 + 19.700000 + 16.099658)/4$$
$$= 17.846973$$
$$u_4^{(2)} = (0.000000 + 11.891797 + 8.010547 + 6.616211)/4$$
$$= 6.629639$$
$$u_5^{(2)} = (6.629639 + 16.099658 + 13.688232 + 11.066064)/4$$
$$= 11.870898$$
$$u_6^{(2)} = (11.870898 + 21.000000 + 17.846973 + 14.241431)/4$$
$$= 16.239825$$
$$u_7^{(2)} = (0.000000 + 11.066064 + 6.629639 + 8.700000)/4$$
$$= 6.598926$$
$$u_8^{(2)} = (6.598926 + 14.241431 + 11.870898 + 12.100000)/4$$
$$= 11.202814$$
$$u_9^{(2)} = (11.202814 + 17.000000 + 16.239825 + 12.800000)/4$$
$$= 14.310660$$

Iteration 3

$$u_1^{(3)} = (0.000000 + 13.688232 + 11.700000 + 6.629639)/4$$
$$= 8.004468$$
$$u_2^{(3)} = (8.004468 + 17.846973 + 17.000000 + 11.870898)/4$$
$$= 13.680585$$
$$u_3^{(3)} = (13.680585 + 21.900000 + 19.700000 + 16.239825)/4$$
$$= 17.880103$$
$$u_4^{(3)} = (0.000000 + 11.870898 + 8.004468 + 6.598926)/4$$
$$= 6.618573$$
$$u_5^{(3)} = (6.618573 + 16.239825 + 13.680585 + 11.202814)/4$$
$$= 11.935449$$
$$u_6^{(3)} = (11.935449 + 21.000000 + 17.880103 + 14.310660)/4$$
$$= 16.281553$$

$$u_7^{(3)} = (0.000000 + 11.202814 + 6.618573 + 8.700000)/4$$
$$= 6.630347$$

$$u_8^{(3)} = (6.630347 + 14.310660 + 11.935449 + 12.100000)/4$$
$$= 11.244114$$

$$u_9^{(3)} = (11.244114 + 17.000000 + 16.281553 + 12.800000)/4$$
$$= 14.331417$$

Iteration 4

$$u_1^{(4)} = (0.000000 + 13.680585 + 11.700000 + 6.618573)/4$$
$$= 7.999789$$

$$u_2^{(4)} = (7.999789 + 17.880103 + 17.000000 + 11.935449)/4$$
$$= 13.703835$$

$$u_3^{(4)} = (13.703835 + 21.900000 + 19.700000 + 16.281553)/4$$
$$= 17.896347$$

$$u_4^{(4)} = (0.000000 + 11.935449 + 7.999789 + 6.630347)/4$$
$$= 6.641396$$

$$u_5^{(4)} = (6.641396 + 16.281553 + 13.703835 + 11.244114)/4$$
$$= 11.967725$$

$$u_6^{(4)} = (11.967725 + 21.000000 + 17.896347 + 14.331417)/4$$
$$= 16.298872$$

$$u_7^{(4)} = (0.000000 + 11.244114 + 6.641896 + 8.700000)/4$$
$$= 6.646378$$

$$u_8^{(4)} = (6.646378 + 14.331417 + 11.967725 + 12.100000)/4$$
$$= 11.261380$$

$$u_9^{(4)} = (11.261380 + 17.000000 + 16.298872 + 12.800000)/4$$
$$= 14.340063$$

Iteration 5

$$u_1^{(5)} = (0.000000 + 13.703835 + 11.700000 + 6.641396)/4$$
$$= 8.011308$$

$$u_2^{(5)} = (8.011308 + 17.896347 + 17.000000 + 11.967725)/4$$
$$= 13.718845$$

$$u_3^{(5)} = (13.718845 + 21.900000 + 19.700000 + 16.298872)/4$$
$$= 17.904429$$

$$u_4^{(5)} = (0.000000 + 11.967725 + 8.011308 + 6.646378)/4$$
$$= 6.656353$$

$$u_5^{(5)} = (6.656353 + 16.298872 + 13.718845 + 11.261380)/4$$
$$= 11.983862$$

$$u_6^{(5)} = (11.983862 + 21.000000 + 17.904429 + 14.340063)/4$$
$$= 16.307089$$

$$u_7^{(5)} = (0.000000 + 11.261380 + 6.656353 + 8.700000)/4$$
$$= 6.654433$$

$$u_8^{(5)} = (6.654433 + 14.340063 + 11.983862 + 12.100000)/4$$
$$= 11.269590$$

$$u_9^{(5)} = (11.269590 + 17.000000 + 16.307089 + 12.800000)/4$$
$$= 14.344170$$

Iteration 6

$$u_1^{(6)} = (0.000000 + 13.718845 + 11.700000 + 6.656353)/4$$
$$= 8.018799$$

$$u_2^{(6)} = (8.018799 + 17.904429 + 17.000000 + 11.983862)/4$$
$$= 13.726773$$

$$u_3^{(6)} = (13.726773 + 21.900000 + 19.700000 + 16.307089)/4$$
$$= 17.908465$$

$$u_4^{(6)} = (0.000000 + 11.983862 + 8.018799 + 6.654433)/4$$
$$= 6.664274$$

$$u_5^{(6)} = (6.664274 + 16.307089 + 13.726773 + 11.269590)/4$$
$$= 11.991931$$

$$u_6^{(6)} = (11.991931 + 21.000000 + 17.908465 + 14.344170)/4$$
$$= 16.311142$$

$$u_7^{(6)} = (0.000000 + 11.269590 + 6.664274 + 8.700000)/4$$
$$= 6.658466$$

$$u_8^{(6)} = (6.658466 + 14344170 + 11.991931 + 12.100000)/4$$
$$= 11.273642$$

$$u_9^{(6)} = (11.273642 + 17.000000 + 16.311142 + 12.800000)/4$$
$$= 14.346196$$

Iteration 7

$$u_1^{(7)} = (0.000000 + 13.726773 + 11.700000 + 6.664274)/4$$
$$= 8.022762$$

$$u_2^{(7)} = (8.022762 + 17.908465 + 17.000000 + 11.991931)/4$$
$$= 13.730790$$

$$u_3^{(7)} = (13.730790 + 21.900000 + 19.700000 + 16.311142)/4$$
$$= 17.910483$$

$$u_4^{(7)} = (0.000000 + 11.991931 + 8.022762 + 6.658466)/4$$
$$= 6.668290$$

$$u_5^{(7)} = (6.668290 + 16.311142 + 13.730790 + 11.273642)/4$$
$$= 11.995966$$

$$u_6^{(7)} = (11.995966 + 21.000000 + 17.910483 + 14.346196)/4$$
$$= 16.313161$$

$$u_7^{(7)} = (0.000000 + 11.273642 + 6.668290 + 8.700000)/4$$
$$= 6.660483$$

$$u_8^{(7)} = (6.660483 + 14.346196 + 11.995966 + 12.100000)/4$$
$$= 11.275661$$

$$u_9^{(7)} = (11.275661 + 17.000000 + 16.313161 + 12.800000)/4$$
$$= 14.347206$$

Iteration 8

$$u_1^{(8)} = (0.000000 + 13.730790 + 11.700000 + 6.668290)/4$$
$$= 8.024770$$

$$u_2^{(8)} = (8.024770 + 17.910483 + 17.000000 + 11.995966)/4$$
$$= 13.732805$$

$$u_3^{(8)} = (13.732805 + 21.900000 + 19.700000 + 16.313161)/4$$
$$= 17.911491$$

$$u_4^{(8)} = (0.000000 + 11.995966 + 8.024770 + 6.660483)/4$$
$$= 6.670305$$

$$u_5^{(8)} = (6.670305 + 16.313161 + 13.732805 + 11.275661)/4$$
$$= 11.997983$$

$$u_6^{(8)} = (11.997983 + 21.000000 + 17.911491 + 14.347206)/4$$
$$= 16.314170$$

$$u_7^{(8)} = (0.000000 + 11.275661 + 6.670305 + 8.700000)/4$$
$$= 6.661491$$

$$u_8^{(8)} = (6.661491 + 14.347206 + 11.997983 + 12.100000)/4$$
$$= 11.276670$$

$$u_9^{(8)} = (11.276670 + 17.000000 + 16.314170 + 12.800000)/4$$
$$= 14.347710$$

Iteration 9

$$u_1^{(9)} = (0.000000 + 13.732805 + 11.700000 + 6.670305)/4$$
$$= 8.025777$$

$$u_2^{(9)} = (8.025777 + 17.911491 + 17.000000 + 11.997983)/4$$
$$= 13.733813$$

$$u_3^{(9)} = (13.733813 + 21.900000 + 19.700000 + 16.314170)/4$$
$$= 17.911996$$

$$u_4^{(9)} = (0.000000 + 11.997983 + 8.025777 + 6.661491)/4$$
$$= 6.671313$$

$$u_5^{(9)} = (6.671313 + 16.314170 + 13.733813 + 11.276670)/4$$
$$= 11.998991$$

$$u_6^{(9)} = (11.998991 + 21.000000 + 17.911996 + 14.347710)/4$$
$$= 16.314674$$

$$u_7^{(9)} = (0.000000 + 11.276670 + 6.671313 + 8.700000)/4$$
$$= 6.661996$$

$$u_8^{(9)} = (6.661996 + 14.347710 + 11.998991 + 12.100000)/4$$
$$= 11.277174$$

$$u_9^{(9)} = (11.277174 + 17.000000 + 16.314674 + 12.800000)/4$$
$$= 14.347962$$

Iteration 10

$$u_1^{(10)} = (0.000000 + 10.733813 + 11.700000 + 6.671313)/4$$
$$= 8.026281$$

$$u_2^{(10)} = (8.026281 + 17.911996 + 17.000000 + 11.998991)/4$$
$$= 13.734317$$

$$u_3^{(10)} = (13.734317 + 21.900000 + 19.700000 + 16.314674)/4$$
$$= 17.912248$$

$$u_4^{(10)} = (0.000000 + 11.998991 + 8.026281 + 6.661996)/4$$
$$= 6.671817$$

$$u_5^{(10)} = (6.671817 + 16.314674 + 13.734317 + 11.277174)/4$$
$$= 11.999496$$

$$u_6^{(10)} = (11.999496 + 21.000000 + 17.912248 + 14.347962)/4$$
$$= 16.314926$$

$$u_7^{(10)} = (0.000000 + 11.277174 + 6.671817 + 8.700000)/4$$
$$= 6.662248$$

$$u_8^{(10)} = (6.662248 + 14.347962 + 11.999496 + 12.100000)/4$$
$$= 11.277426$$

$$u_9^{(10)} = (11.277426 + 17.000000 + 16.314926 + 12.800000)/4$$
$$= 14.348088$$

Iteration 11

$$u_1^{(11)} = (0.000000 + 13.734317 + 11.700000 + 6.671817)/4$$
$$= 8.026534$$

$$u_2^{(11)} = (8.026534 + 17.912248 + 17.000000 + 11.999496)/4$$
$$= 13.734569$$

$$u_3^{(11)} = (13.734569 + 21.900000 + 19.700000 + 16.314926)/4$$
$$= 17.912374$$

$$u_4^{(11)} = (0.000000 + 11.999496 + 8.026534 + 6.662248)/4$$
$$= 6.672069$$

$$u_5^{(11)} = (6.672069 + 16.314926 + 13.734569 + 11.277426)/4$$
$$= 11.999748$$

$$u_6^{(11)} = (11.999748 + 21.000000 + 17.912374 + 14.348088)/4$$
$$= 16.315052$$

$$u_7^{(11)} = (0.000000 + 11.277426 + 6.672069 + 8.700000)/4$$
$$= 6.662374$$

$$u_8^{(11)} = (6.662374 + 14.348088 + 11.999748 + 12.100000)/4$$
$$= 11.277552$$

$$u_9^{(11)} = (11.277552 + 17.000000 + 16.315052 + 12.800000)/4$$
$$= 14.348151$$

Iteration 12

$$u_1^{(12)} = (0.000000 + 13.734569 + 11.700000 + 6.672069)/4$$
$$= 8.026660$$

$$u_2^{(12)} = (8.026660 + 17.912374 + 17.000000 + 11.999748)/4$$
$$= 13.734695$$

$$u_3^{(12)} = (13.734695 + 21.900000 + 19.700000 + 16.315052)/4$$
$$= 17.912437$$

$$u_4^{(12)} = (0.000000 + 11.999748 + 8.026660 + 6.662374)/4$$
$$= 6.672195$$

$$u_5^{(12)} = (6.672195 + 16.315052 + 13.734695 + 11.277552)/4$$
$$= 11.999874$$

$$u_6^{(12)} = (11.999874 + 21.000000 + 17.912437 + 14.348151)/4$$
$$= 16.315116$$

$$u_7^{(12)} = (0.000000 + 11.277552 + 6.672195 + 8.700000)/4$$
$$= 6.662437$$

$$u_8^{(12)} = (6.662437 + 14.348151 + 11.999874 + 12.100000)/4$$
$$= 11.277616$$

$$u_9^{(12)} = (11.277616 + 17.000000 + 16.315116 + 12.800000)/4$$
$$= 14.348183$$

Iteration 13

$$u_1^{(13)} = (0.000000 + 13.734695 + 11.700000 + 6.672195)/4$$
$$= 8.026723$$

$$u_2^{(13)} = (8.026723 + 17.912437 + 17.000000 + 11.999874)/4$$
$$= 13.734758$$

$$u_3^{(13)} = (13.734758 + 21.900000 + 19.700000 + 16.315116)/4$$
$$= 17.912468$$

$$u_4^{(13)} = (0.000000 + 11.999874 + 8.026723 + 6.662437)/4$$
$$= 6.672258$$

$$u_5^{(13)} = (6.672258 + 16.315116 + 13.734758 + 11.277616)/4$$
$$= 11.999937$$

$$u_6^{(13)} = (11.999937 + 21.000000 + 17.912468 + 14.348183)/4$$
$$= 16.315147$$

$$u_7^{(13)} = (0.000000 + 11.277616 + 6.672258 + 8.700000)/4$$
$$= 6.662468$$

$$u_8^{(13)} = (6.662468 + 14.348183 + 11.999937 + 12.100000)/4$$
$$= 11.277647$$

$$u_9^{(13)} = (11.277647 + 17.000000 + 16.315147 + 12.800000)/4$$
$$= 14.348199$$

Iteration 14

$$u_1^{(14)} = (0.000000 + 13.734758 + 11.700000 + 6.672258)/4$$
$$= 8.026754$$

$$u_2^{(14)} = (8.026754 + 17.912468 + 17.000000 + 11.999937)/4$$
$$= 13.734790$$

$$u_3^{(14)} = (13.734790 + 21.900000 + 19.700000 + 16.315147)/4$$
$$= 17.912484$$

$$u_4^{(14)} = (0.000000 + 11.999937 + 8.026754 + 6.662468)/4$$
$$= 6.672290$$

$$u_5^{(14)} = (6.672290 + 16.315147 + 13.734790 + 11.277647)/4$$
$$= 11.999968$$

$$u_6^{(14)} = (11.999968 + 21.000000 + 17.912484 + 14.348199)/4$$
$$= 16.315163$$

$$u_7^{(14)} = (0.000000 + 11.277647 + 6.672290 + 8.700000)/4$$
$$= 6.662484$$

$$u_8^{(14)} = (6.662484 + 14.348199 + 11.999968 + 12.100000)/4$$
$$= 11.277663$$

$$u_9^{(14)} = (11.277663 + 17.000000 + 16.315163 + 12.800000)/4$$
$$= 14.348206$$

Iteration 15

$$u_1^{(15)} = (0.000000 + 13.734790 + 11.700000 + 6.672290)/4$$
$$= 8.026770$$

$$u_2^{(15)} = (8.026770 + 17.912484 + 17.000000 + 11.999968)/4$$
$$= 13.734806$$

$$u_3^{(15)} = (13.734806 + 21.900000 + 19.700000 + 16.315163)/4$$
$$= 17.912492$$

$$u_4^{(15)} = (0.000000 + 11.999968 + 8.026770 + 6.662484)/4$$
$$= 6.672306$$

$$u_5^{(15)} = (6.672306 + 16.315163 + 13.734806 + 11.277663)/4$$
$$= 11.999984$$

$$u_6^{(15)} = (11.999984 + 21.000000 + 17.912492 + 14.348206)/4$$
$$= 16.315171$$

$$u_7^{(15)} = (0.000000 + 11.277663 + 6.672306 + 8.700000)/4$$
$$= 6.662492$$

$$u_8^{(15)} = (6.662492 + 14.348206 + 11.999984 + 12.100000)/4$$
$$= 11.277671$$

$$u_9^{(15)} = (11.277671 + 17.000000 + 16.315171 + 12.800000)/4$$
$$= 14.348210$$

Problem 3: Solve $u_{xx} + u_{yy} = 0$ at the interior mesh points of the following square with boundary values as shown in Fig. 13.25 using Leibmann's iteration process.

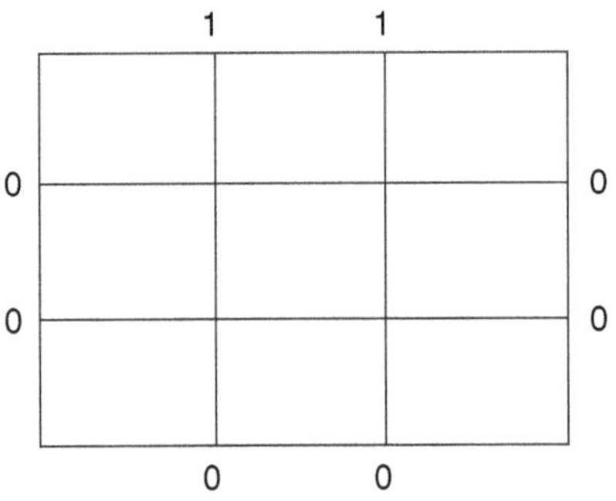

Figure 13.25

Solution: Let u_1, u_2, u_3 and u_4 be the values of $u(x, y)$ at the interior mesh points A, B, C and D respectively (Fig. 13.26).

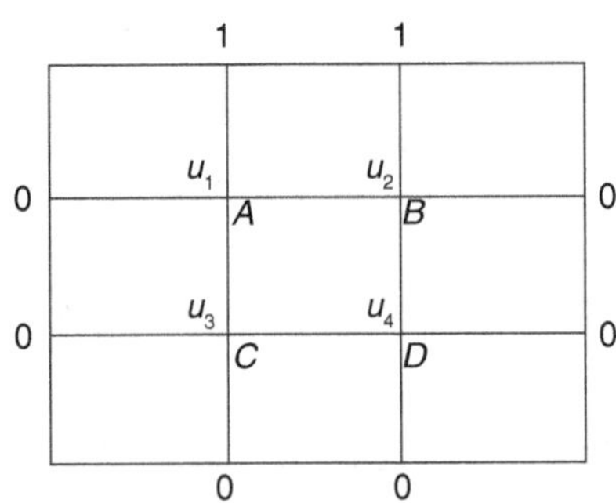

Figure 13.26

By SFPF, we have

$$u_3 = \frac{1}{4}(0 + u_4 + u_1 + 0)$$

We assume $u_4 = 0$ and $u_1 = 1$; then

$$u_3 = \frac{1}{4}(0 + 0 + 1 + 0) = 0.25$$

Similarly,

$$u_4 = \frac{1}{4}(u_3 + 0 + u_2 + 0)$$

We assume $u_2 = 1$; then

$$u_4 = \frac{1}{4}(0.25 + 0 + 1 + 0) = 0.3125$$

$$u_1 = \frac{1}{4}(0 + u_2 + 1 + u_3)$$

$$= \frac{1}{4}(0 + 1 + 1 + 0.25) = 0.5625$$

$$u_2 = \frac{1}{4}(u_1 + 0 + 1 + u_4)$$

$$= \frac{1}{4}(0.5625 + 0 + 1 + 0.3125) = 0.46875$$

We suppose $u_1^{(0)} = 0.5625$; $u_2^{(0)} = 0.46875$ $u_3^{(0)} = 0.25$ and $u_4^{(0)} = 0.3125$

The iterative formulas to improve the approximations of u_1, u_2, u_3 and u_4 by the Leibmann's iteration process are

$$u_1^{(n+1)} = \frac{1}{4}(1 + u_4^{(n)} + 0 + u_3^{(n)}) \tag{1}$$

$$u_2^{(n+1)} = \frac{1}{4}(1 + u_4^{(n)} + u_1^{(n+1)} + 0) \tag{2}$$

$$u_3^{(n+1)} = \frac{1}{4}(u_1^{(n+1)} + 0 + 0 + u_4^{(n)}) \tag{3}$$

$$u_4^{(n+1)} = \frac{1}{4}(u_2^{(n)} + 0 + u_3^{(n)} + 0) \tag{4}$$

Substituting $n = 0$ in the Eqs. (1) to (4) and putting relevant values in these equations, we obtain the first approximations of $u_i^{(1)}$ $(i = 1, 2, 3, 4)$ and they are given below.

Iteration 1

$$u_1^{(1)} = (0.000000 + 0.468750 + 1.000000 + 0.250000)/4$$
$$= 0.429688$$

$$u_2^{(1)} = (0.429688 + 0.000000 + 1.000000 + 0.312500)/4$$
$$= 0.435547$$

$$u_3^{(1)} = (0.000000 + 0.312500 + 0.429688 + 0.000000)/4$$
$$= 0.185547$$

$$u_4^{(1)} = (0.185547 + 0.000000 + 0.435547 + 0.000000)/4$$
$$= 0.155273$$

Similarly, substituting $n = 1, 2, ..., 9$ in Eqs. (1) to (4) and repeating the above process, we obtain $x_i^{(n)}$ $(i = 1, 2, 3, 4; n = 2, 3, 4, 5 .., 10)$

Iteration 2

$$u_1^{(2)} = (0.000000 + 0.435547 + 1.000000 + 0.185547)/4$$
$$= 0.405273$$

$$u_2^{(2)} = (0.405273 + 0.000000 + 1.000000 + 0.155273)/4$$
$$= 0.390137$$

$$u_3^{(2)} = (0.000000 + 0.155273 + 0.405273 + 0.000000)/4$$
$$= 0.140137$$

$$u_4^{(2)} = (0.140137 + 0.000000 + 0.390137 + 0.000000)/4$$
$$= 0.132568$$

Iteration 3

$$u_1^{(3)} = (0.000000 + 0.390137 + 1.000000 + 0.140137)/4$$
$$= 0.382568$$

$$u_2^{(3)} = (0.382568 + 0.000000 + 1.000000 + 0.132568)/4$$
$$= 0.378784$$

$$u_3^{(3)} = (0.000000 + 0.132568 + 0.382568 + 0.000000)/4$$
$$= 0.128784$$

$$u_4^{(3)} = (0.128784 + 0.000000 + 0.378784 + 0.000000)/4$$
$$= 0.126892$$

Iteration 4

$$u_1^{(4)} = (0.000000 + 0.378784 + 1.000000 + 0.128784)/4$$
$$= 0.376892$$

$$u_2^{(4)} = (0.376892 + 0.000000 + 1.000000 + 0.126892)/4$$
$$= 0.375946$$

$$u_3^{(4)} = (0.000000 + 0.126892 + 0.376892 + 0.000000)/4$$
$$= 0.125946$$

$$u_4^{(4)} = (0.125946 + 0.000000 + 0.375946 + 0.000000)/4$$
$$= 0.125473$$

Iteration 5

$$u_1^{(5)} = (0.000000 + 0.375946 + 1.000000 + 0.125946)/4$$
$$= 0.375473$$

$$u_2^{(5)} = (0.375473 + 0.000000 + 1.000000 + 0.125473)/4$$
$$= 0.375237$$

$$u_3^{(5)} = (0.000000 + 0.125473 + 0.375473 + 0.000000)/4$$
$$= 0.125237$$

$$u_4^{(5)} = (0.125237 + 0.000000 + 0.375237 + 0.000000)/4$$
$$= 0.125118$$

Iteration 6

$$u_1^{(6)} = (0.000000 + 0.375237 + 1.000000 + 0.125237)/4$$
$$= 0.375118$$

$$u_2^{(6)} = (0.375118 + 0.000000 + 1.000000 + 0.128118)/4$$
$$= 0.375059$$

$$u_3^{(6)} = (0.000000 + 0.125118 + 0.375118 + 0.000000)/4$$
$$= 0.125059$$

$$u_4^{(6)} = (0.125059 + 0.000000 + 0.375059 + 0.000000)/4$$
$$= 0.125030$$

Iteration 7

$$u_1^{(7)} = (0.000000 + 0.375059 + 1.000000 + 0.125059)/4$$
$$= 0.375030$$

$$u_2^{(7)} = (0.375030 + 0.000000 + 1.000000 + 0.125030)/4$$
$$= 0.375015$$

$$u_3^{(7)} = (0.000000 + 0.125030 + 0.375030 + 0.000000)/4$$
$$= 0.125015$$

$$u_4^{(7)} = (0.125015 + 0.000000 + 0.375015 + 0.000000)/4$$
$$= 0.125007$$

Iteration 8

$$u_1^{(8)} = (0.000000 + 0.375015 + 1.000000 + 0.125015)/4$$
$$= 0.375007$$

$$u_2^{(8)} = (0.375007 + 0.000000 + 1.000000 + 0.125007)/4$$
$$= 0.375004$$

$$u_3{}^{(8)} = (0.000000 + 0.125007 + 0.375007 + 0.000000)/4$$
$$= 0.125004$$

$$u_4{}^{(8)} = (0.125004 + 0.000000 + 0.375004 + 0.000000)/4$$
$$= 0.125002$$

Iteration 9

$$u_1{}^{(9)} = (0.000000 + 0.375004 + 1.000000 + 0.125004)/4$$
$$= 0.375002$$

$$u_2{}^{(9)} = (0.375002 + 0.000000 + 1.000000 + 0.125002)/4$$
$$= 0.375001$$

$$u_3{}^{(9)} = (0.000000 + 0.125002 + 0.375002 + 0.000000)/4$$
$$= 0.125001$$

$$u_4{}^{(9)} = (0.125001 + 0.000000 + 0.375001 + 0.000000)/4$$
$$= 0.125000$$

Iteration 10

$$u_1{}^{(10)} = (0.000000 + 0.375001 + 1.000000 + 0.125001)/4$$
$$= 0.375000$$

$$u_2{}^{(10)} = (0.375000 + 0.000000 + 1.000000 + 0.125000)/4$$
$$= 0.375000$$

$$u_3{}^{(10)} = (0.000000 + 0.125000 + 0.375000 + 0.000000)/4$$
$$= 0.125000$$

$$u_4{}^{(10)} = (0.125000 + 0.000000 + 0.375000 + 0.000000)/4$$
$$= 0.125000$$

Thus, $u_1 = 0.375$, $u_2 = 0.375$, $u_3 = 0.125$ and $u_4 = 0.125$ upto 10 iterations or of three-decimal accuracy.

Problem 4: Solve $u_{xx} + u_{yy} = 0$ for the following square mesh with boundary as indicated in Fig. 13.27 using Liebmann's iteration process.

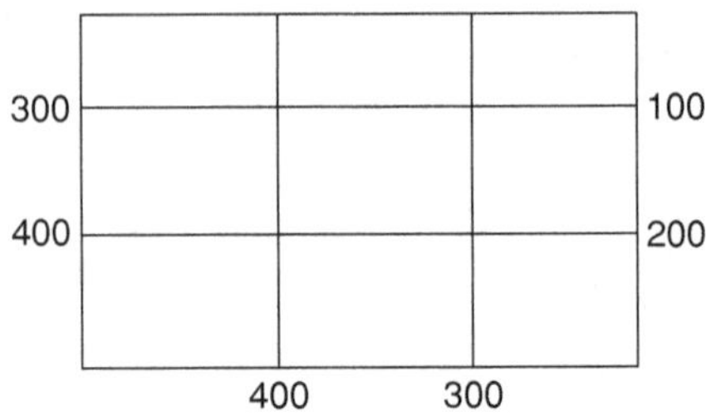

Figure 13.27

Solution: Let u_1, u_2, u_3 and u_4 be the values of $u(x, y)$ at the interior mesh points A, B, C and D of the given square mesh $PQRS$ (see Figure 13.28).

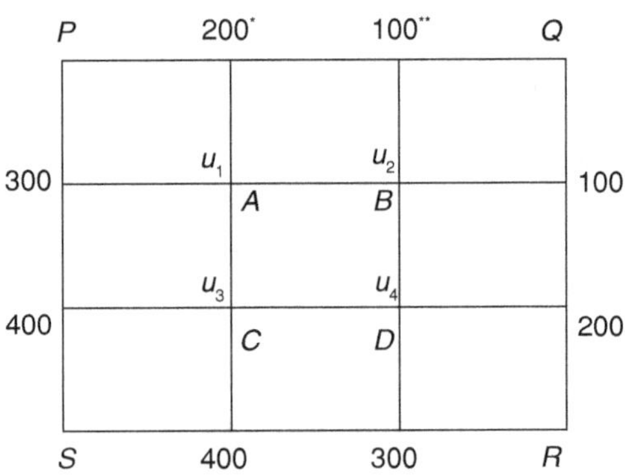

Figure 13.28

We can observe that the boundary values are symmetric about the diagonal SQ of the square $PQRS$. Hence, the boundary values at the mesh points on the top of A and B are 200^* and 100^{**} respectively. The asterisk and double asterisk marked values indicate that the values are obtained by symmetry. Further, we also have $u_1 = u_4$ because of symmetry.

The variations of $u(x, y)$ from the boundary PS to QR of the square along AB is 200. Therefore, we assume the value of of u_1 as

$$u_1 = 300 - \frac{200}{3} = \frac{700}{3} \approx 233$$

Since, $u_1 = u_4$, we have $u_4 = 233$. Now

$$\therefore \qquad u_2 = \frac{1}{4}[u_1 + 100 + 100 + u_4]$$

$$\therefore \qquad u_2 = \frac{1}{4}[233 + 100 + 100 + 233] = 167$$

Similarly, $\qquad u_3 = \frac{1}{4}[400 + u_4 + u_1 + 400]$

$$= \frac{1}{4}[400 + 2300 + 233 + 400] = 317.$$

Now, we suppose that $u_1^{(0)} = 233$, $u_2^{(0)} = 167$, $u_3^{(0)} = 317$ and $u_4^{(0)} = u_1^{(0)} = 233$.

The iterative formulas to improve the approximations of u_1, u_2, u_3 and u_4 by Liebmann's iteration process are

$$u_1^{(n+1)} = \frac{1}{4}(300 + u_2^{(n)} + 200 + u_3^{(n)})$$

$$u_2^{(n+1)} = \frac{1}{4}(u_1^{(n+1)} + 100 + 100 + u_4^{(n)})$$

$$u_3^{(n+1)} = \frac{1}{4}(400 + u_4^{(n)} + u_1^{(n)} + 400)$$

$$u_4^{(n+1)} = u_1^{(n+1)}$$

Proceeding as in Problem (3), we get the successive approximations of $u_i\,(i = 1, 2, 3, 4)$ and they are as follows:

Iteration 1

$$u_1^{(1)} = (300.000000 + 167.000000 + 200.000000 + 317.000000)/4$$
$$= 246.000000$$

$$u_2^{(1)} = (246.000000 + 100.000000 + 100.000000 + 233.000000)/4$$
$$= 169.750000$$

$$u_3^{(1)} = (400.000000 + 233.000000 + 246.000000 + 400.000000)/4$$
$$= 319.75000$$

$$u_4^{(1)} = (319.750000 + 200.000000 + 169.750000 + 300.000000)/4$$
$$= 247.375000$$

Iteration 2

$$u_1^{(2)} = (300.000000 + 169.750000 + 200.000000 + 319.750000)/4$$
$$= 247.375000$$
$$u_2^{(2)} = (247.375000 + 100.000000 + 100.000000 + 247.375000)/4$$
$$= 173.6875$$
$$u_3^{(2)} = (400.000000 + 247.375000 + 247.375000 + 400.000000)/4$$
$$= 323.68750$$
$$u_4^{(2)} = (323.687500 + 200.000000 + 173.687500 + 300.000000)/4$$
$$= 249.343750$$

Iteration 3

$$u_1^{(3)} = (300.000000 + 173.687500 + 200.000000 + 323.687500)/4$$
$$= 249.343750$$
$$u_2^{(3)} = (249.343750 + 100.000000 + 100.000000 + 249.343750)/4$$
$$= 174.671875$$
$$u_3^{(3)} = (400.000000 + 249.343750 + 249.343750 + 400.000000)/4$$
$$= 324.671875$$
$$u_4^{(3)} = (324.671875 + 200.000000 + 174.671875 + 300.000000)/4$$
$$= 249.835938$$

Iteration 4

$$u_1^{(4)} = (324.999980 + 200.000000 + 174.999980 + 300.000000)/4$$
$$= 249.999990$$
$$u_2^{(4)} = (249.835938 + 100.000000 + 100.000000 + 249.835938)/4$$
$$= 174.917969$$
$$u_3^{(4)} = (400.000000 + 249.835938 + 249.835938 + 400.000000)/4$$
$$= 324.917969$$
$$u_4^{(4)} = (324.917969 + 200.000000 + 174.917969 + 300.000000)/4$$
$$= 249.958984$$

Iteration 5

$$u_1^{(5)} = (300.000000 + 174.917969 + 200.000000 + 324.917969)/4$$
$$= 249.958984$$

$$u_2^{(5)} = (249.958984 + 100.000000 + 100.000000 + 249.958984)/4$$
$$= 174.979492$$

$$u_3^{(5)} = (400.000000 + 249.958984 + 249.958984 + 400.000000)/4$$
$$= 324.979492$$

$$u_4^{(5)} = (324.979492 + 200.000000 + 174.979492 + 300.000000)/4$$
$$= 249.989746$$

Iteration 6

$$u_1^{(6)} = (300.000000 + 174.979492 + 200.000000 + 324.979492)/4$$
$$= 249.989746$$

$$u_2^{(6)} = (249.989746 + 100.000000 + 100.000000 + 249.989746)/4$$
$$= 174.994873$$

$$u_3^{(6)} = (400.000000 + 249.989746 + 249.989746 + 400.000000)/4$$
$$= 324.994873$$

$$u_4^{(6)} = (324.994873 + 200.000000 + 174.994873 + 300.000000)/4$$
$$= 249.997437$$

Iteration 7

$$u_1^{(7)} = (300.000000 + 174.994873 + 200.000000 + 324.994873)/4$$
$$= 249.997437$$

$$u_2^{(7)} = (249.997437 + 100.000000 + 100.000000 + 249.997437)/4$$
$$= 174.998718$$

$$u_3^{(7)} = (400.000000 + 249.997437 + 249.997437 + 400.000000)/4$$
$$= 324.998718$$

$$u_4^{(7)} = (324.998718 + 200.000000 + 174.998718 + 300.000000)/4$$
$$= 249.999359$$

Iteration 8

$$u_1^{(8)} = (300.000000 + 174.998718 + 200.000000 + 324.998718)/4$$
$$= 249.999359$$
$$u_2^{(8)} = (249.999359 + 100.000000 + 100.000000 + 249.999359)/4$$
$$= 174.999680$$
$$u_3^{(8)} = (400.000000 + 249.999359 + 249.999359 + 400.000000)/4$$
$$= 324.999680$$
$$u_4^{(8)} = (324.99980 + 200.000000 + 174.999680 + 300.000000)/4$$
$$= 249.999840$$

Iteration 9

$$u_1^{(9)} = (300.000000 + 174.999680 + 200.000000 + .324.999680)/4$$
$$= 249.999840$$
$$u_2^{(9)} = (249.999840 + 100.000000 + 100.000000 + 249.999840)/4$$
$$= 174.999920$$
$$u_3^{(9)} = (400.000000 + 249.999840 + 249.999840 + 400.000000)/4$$
$$= 324.999920$$
$$u_4^{(9)} = (324.9999920 + 200.000000 + 174.999980 + 300.0000000)/4$$
$$= 249.999960$$

Iteration 10

$$u_1^{(10)} = (300.000000 + 174.999920 + 200.000000 + 324.999920)/4$$
$$= 249.999960$$
$$u_2^{(10)} = (249.999960 + 100.000000 + 100.000000 + 249.999960)/4$$
$$= 174.99980$$
$$u_3^{(10)} = (400.000000 + 249.999960 + 249.999960 + 400.000000)/4$$
$$= 324.999980$$
$$u_4^{(10)} = (324.999980 + 200.000000 + 174.999980 + 300.000000)/4$$
$$= 249.999990$$

Thus, the approximate values of u_1, u_2, u_3 and u_4 are

$$u_1 = 250, \ u_2 = 175, \ u_3 = 325 \ \text{and} \ u_4 = 250.$$

Problem 5: Solve $u_{xx} + u_{yy} = 0$ over the region $[0, 4] \times [0, 4]$ subject to the boundary conditions:

  (i) $u(x, 0) = 3x$ for $0 \leq x \leq 4$

  (ii) $u(x, 4) = x^2$ for $0 \leq x \leq 4$

  (iii) $u(0, y) = 0$ for $0 \leq y \leq 4$

  (iv) $u(4, y) = 12 + y$ for $0 \leq y \leq 4$

Solution: We divide the square $[0, 4] \times [0, 4]$ into 16 square meshes having the side 1, hence $h = 1$ and $k = 1$. Thus, dividing $[0, 4] \times [0, 4]$ into 16 square meshes, we have 25 mesh points and out of them, 9 are the interior mesh points. Thus, we have to find 9 values of $u(x, y)$ at these 9 interior mesh points. We suppose 25 mesh points as $A_1, A_2, A_3, A_4, A_5, B_1, B_2, B_3, B_4, B_5, \ldots E_1, E_2, E_3, E_4,$ and E_5. Further, let $u_1, u_2, u_3, u_4, u_5, u_6, u_7, u_8$ and u_9 be the values of $u(x, y)$ at the interior mesh points are $D_2, D_3, D_4, C_2, C_3, C_4, B_2, B_3$ and B_4 respectively (Fig. 13.29).

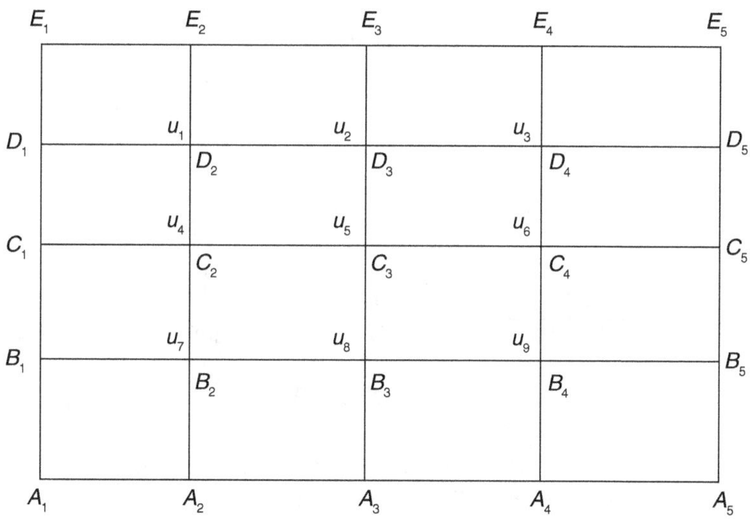

Figure 13.29

Now, we evaluate the values of $u(x, y)$ at the mesh points lying on the boundary, using the given boundary conditions.

Given that $u(x, 0) = 3x$ for $0 \leq x \leq 4$. Therefore, $u(0, 0) = 0$, $u(1, 0) = 3$, $u(2, 0) = 6$, $u(3, 0) = 9$ and $u(4, 0) = 12$. Hence, the values of $u(x, y)$ at the boundary points A_1, A_2, A_3, A_4 and A_5 are 0, 3, 6, 9 and 12, respectively.

From the boundary conditions, ii), we have $u(x, 4) = x^2, 0 \le x \le 4$. Therefore, $u(0,4)=0, u(1, 4) = 1, u(2,4) = 4, u(3, 4) = 9,$ and $u(4,4) = 16$. Thus, the values of $u(x, y)$ at the boundary points E_1, E_2, E_3, E_4 and E_5 are 0, 1, 4, 9 and 16, respectively. Since $u(0, y) = 0$ for $0 \le y \le 4$, we have $u(0,0) = 0,\ u(0, 1) = 0, u(0, 2) = 0, u(0, 3) = 0$ and $u(0,4) = 0$. Therefore, the boundary values of $u(x, y)$ at A_1, B_1, C_1, D_1 and E_1 are all zero. Finally given that $u(x, y) = 12 + y, 0 \le y \le 4$. Therefore, $u(4, 0) = 12, u(4, 1) = 13, u(4, 2) = 14,\ u(4, 3) = 15$ and $u(4, 4) = 16$. Hence, the values of $u(x, y)$ at A_5, B_5, C_5, D_5 and E_5 are 12, 13, 14, 15 and 16, respectively.

The mesh square with obtained boundary values is shown in Fig. 13.30.

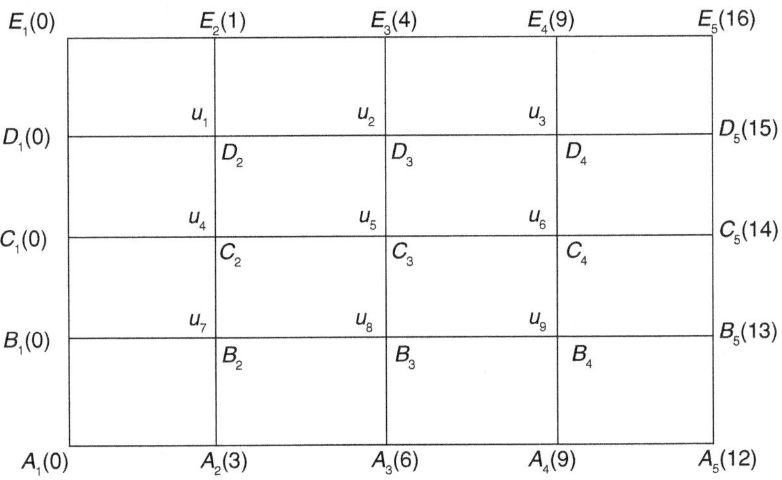

Figure 13.30

The values shown in brackets at the mesh point indicate the values of $u(x, y)$ at the corresponding mesh points.

Now, we proceed to find approximate values of $u_1, u_2, \ldots$ and u_9 using SFPF and DFPF. By SFPF.

$$u_5 = \frac{1}{4}(u_4 + u_6 + u_2 + u_8)$$

We assume $u_4 = 0$, $u_6 = 14$, $u_2 = 4$ and $u_8 = 6$. (Since u_4 is close to $C_1(0)$, u_6 is close to $C_5(14)$, u_2 is close to $E_3(4)$ and u_8 is close to $A_3(6)$).

$$\therefore u_5 = \frac{1}{4}[0 + 14 + 4 + 6] = 6.$$

By applying DFPF, we have

$$u_1 = \frac{1}{4}[0 + 6 + 4 + 0] = 2.5$$

$$u_3 = \frac{1}{4}[16 + 6 + 4 + 14] = 10$$

$$u_7 = \frac{1}{4}[6 + 0 + 0 + 6] = 3$$

$$u_9 = \frac{1}{4}[14 + 6 + 6 + 12] = 9.5$$

and applying SFPF we obtain

$$u_2 = \frac{1}{4}[4 + 6 + 2.5 + 10] = 5.625$$

$$u_4 = \frac{1}{4}[2.5 + 3 + 0 + 6] = 2.875$$

$$u_6 = \frac{1}{4}[10 + 9.5 + 6 + 14] = 9.875$$

$$u_8 = \frac{1}{4}[6 + 6 + 3 + 9.5] = 6.125.$$

We suppose $u_1^{(0)} = 2.5$, $u_2^{(0)} = 5.625$, $u_3^{(0)} = 10$, $u_4^{(0)} = 2.875$, $u_5^{(0)} = 4$, $u_6^{(0)} = 9.875$, $u_7^{(0)} = 3$, $u_8^{(0)} = 6.125$, $u_9^{(0)} = 9.5$.

The iterative formulas to improve the approximations of $u_i (i = 1, 2, ..9)$ by Leibmann's iteration process are as follows:

$$u_1^{(n+1)} = \frac{1}{4}(1 + u_4^{(n)} + 0 + u_2^{(n)}) \tag{1}$$

$$u_2^{(n+1)} = \frac{1}{4}(u_1^{(n+1)} + u_3^{(n)} + 4 + u_5^{(n)}) \tag{2}$$

$$u_3^{(n+1)} = \frac{1}{4}(u_2^{(n+1)} + 15 + 9 + u_6^{(n)}) \tag{3}$$

$$u_4^{(n+1)} = \frac{1}{4}(0 + u_5^{(n)} + u_1^{(n+1)} + u_7^{(n)}) \tag{4}$$

$$u_5^{(n+1)} = \frac{1}{4}(u_4^{(n+1)} + u_6^{(n)} + u_2^{(n+1)} + u_8^{(n)}) \tag{5}$$

$$u_6^{(n+1)} = \frac{1}{4}(u_5^{(n+1)} + 14 + u_3^{(n+1)} + u_9^{(n)}) \tag{6}$$

$$u_7^{(n+1)} = \frac{1}{4}(0 + u_8^{(n)} + u_4^{(n+1)} + 3) \tag{7}$$

$$u_8^{(n+1)} = \frac{1}{4}(u_7^{(n+1)} + u_9^{(n)} + u_5^{(n+1)} + 6) \tag{8}$$

$$u_9^{(n+1)} = \frac{1}{4}(u_8^{(n+1)} + 13 + u_6^{(n+1)} + 9) \tag{9}$$

Substituting $n = 0$ in the Eqs. (1) to (9) and substituting the relevant values in those equations we get the following approximations of $u_i^{(1)} (i = 1, 2....9)$.

Iteration 1

$u_1^{(1)} = (0.000000 + 5.625000 + 1.000000 + 2.875000)/4 = 2.875000$

$u_2^{(1)} = (2.375000 + 10.000000 + 4.000000 + 6.000000)/4 = 5.593750$

$u_3^{(1)} = (5.593750 + 15.000000 + 9.000000 + 9.875000)/4 = 9.867188$

$u_4^{(1)} = (0.000000 + 6.000000 + 2.375000 + 3.000000)/4 = 2.843750$

$u_5^{(1)} = (2.843750 + 9.875000 + 5.593750 + 6.125000)/4 = 6.109375$

$u_6^{(1)} = (6.109375 + 14.000000 + 9.867188 + 9.500000)/4 = 9.869141$

$u_7^{(1)} = (0.000000 + 6.125000 + 2.843750 + 3.000000)/4 = 2.992188$

$u_8^{(1)} = (2.992188 + 9.500000 + 6.109375 + 6.000000)/4 = 6.150391$

$u_9^{(1)} = (6.150391 + 13.000000 + 9.869141 + 9.000000)/4 = 9.504883$

Similarly, substituting $n = 1, 2, ..., 14$ in the Eqs. (1) to (9), we obtain the approximations $u_i^{(n)}$ $(i = 1, 2, ... 9: n = 2, 3, ..., 15)$ and they are given below:

Iteration 2

$$u_1^{(2)} = (0.000000 + 5.593750 + 1.000000 + 2.843750)/4 = 2.359375$$

$$u_2^{(2)} = (2.359375 + 9.867188 + 4.000000 + 6.109375)/4 = 5.583984$$

$$u_3^{(2)} = (5.583984 + 15.000000 + 9.000000 + 9.869141)/4 = 9.863281$$

$$u_4^{(2)} = (0.000000 + 6.109375 + 2.359375 + 2.992188)/4 = 2.865234$$

$$u_5^{(2)} = (2.865234 + 9.869141 + 5.583984 + 6.150391)/4 = 6.117188$$

$$u_6^{(2)} = (6.117188 + 14.000000 + 9.863281 + 9.504883)/4 = 9.871338$$

$$u_7^{(2)} = (0.000000 + 6.150391 + 2.865234 + 3.000000)/4 = 3.003906$$

$$u_8^{(2)} = (3.003906 + 9.504883 + 6.117188 + 6.000000)/4 = 6.156494$$

$$u_9^{(2)} = (6.156494 + 13.000000 + 9.871338 + 9.000000)/4 = 9.506958$$

Iteration 3

$$u_1^{(3)} = (0.000000 + 5.583984 + 1.000000 + 2.865234)/4 = 2.362305$$

$$u_2^{(3)} = (2.362305 + 9.863281 + 4.000000 + 6.117188)/4 = 5.585693$$

$$u_3^{(3)} = (5.585693 + 15.000000 + 9.000000 + 9.871338)/4 = 9.864258$$

$$u_4^{(3)} = (0.000000 + 6.117188 + 2.362305 + 3.003906)/4 = 2.870850$$

$$u_5^{(3)} = (2.870850 + 9.871338 + 5.585693 + 6.156494)/4 = 6.121094$$

$$u_6^{(3)} = (6.121094 + 14.000000 + 9.864258 + 9.506958)/4 = 9.873077$$

$$u_7^{(3)} = (0.000000 + 6.156494 + 2.870850 + 3.000000)/4 = 3.006836$$

$$u_8^{(3)} = (3.006836 + 9.506958 + 6.121094 + 6.000000)/4 = 6.158722$$

$$u_9^{(3)} = (6.158722 + 13.000000 + 9.873077 + 9.000000)/4 = 9.507950$$

Iteration 4

$$u_1^{(4)} = (0.000000 + 5.585693 + 1.000000 + 2.870850)/4 = 2.364136$$

$$u_2^{(4)} = (2.364136 + 9.864258 + 4.000000 + 6.121094)/4 = 5.587372$$

$$u_3^{(4)} = (5.587372 + 15.000000 + 9.000000 + 9.873077)/4 = 9.865112$$

$$u_4^{(4)} = (0.000000 + 6.121094 + 2.364136 + 3.006836)/4 = 2.873016$$

$$u_5^{(4)} = (2.873016 + 9.873077 + 5.587372 + 6.158722)/4 = 6.123047$$

$$u_6^{(4)} = (6.123047 + 14.000000 + 9.865112 + 9.507950)/4 = 9.874027$$

$$u_7^{(4)} = (0.000000 + 6.158722 + 2.873016 + 3.000000)/4 = 3.007935$$

$$u_8^{(4)} = (3.007935 + 9.507950 + 6.123047 + 6.000000)/4 = 6.159733$$

$$u_9^{(4)} = (6.159733 + 13.000000 + 9.874027 + 9.000000)/4 = 9.508440$$

Iteration 5

$$u_1^{(5)} = (0.000000 + 5.587372 + 1.000000 + 2.873016)/4 = 2.365097$$

$$u_2^{(5)} = (2.365097 + 9.865112 + 4.000000 + 6.123047)/4 = 5.588314$$

$$u_3^{(5)} = (5.588314 + 15.000000 + 9.000000 + 9.874027)/4 = 9.865585$$

$$u_4^{(5)} = (0.000000 + 6.123047 + 2.365097 + 3.007935)/4 = 2.874020$$

$$u_5^{(5)} = (2.874020 + 9.874027 + 5.588314 + 6.159733)/4 = 6.124023$$

$$u_6^{(5)} = (6.124023 + 14.000000 + 9.865585 + 9.508440)/4 = 9.874512$$

$$u_7^{(5)} = (0.000000 + 6.159733 + 2.874020 + 3.000000)/4 = 3.008438$$

$$u_8^{(5)} = (3.008438 + 9.508440 + 6.124023 + 6.000000)/4 = 6.160225$$

$$u_9^{(5)} = (6.160225 + 13.000000 + 9.874512 + 9.000000)/4 = 9.508684$$

Iteration 6

$$u_1^{(6)} = (0.000000 + 5.588314 + 1.000000 + 2.874020)/4 = 2.365583$$

$$u_2^{(6)} = (2.365583 + 9.865585 + 4.000000 + 6.124023)/4 = 5.588798$$

$$u_3^{(6)} = (5.588798 + 15.000000 + 9.000000 + 9.874512)/4 = 9.865828$$

$$u_4^{(6)} = (0.000000 + 6.124023 + 2.365583 + 3.008438)/4 = 2.874511$$

$$u_5^{(6)} = (2.874511 + 9.874512 + 5.588798 + 6.160225)/4 = 6.124512$$

$$u_6^{(6)} = (6.124512 + 14.000000 + 9.865828 + 9.508684)/4 = 9.874756$$

$$u_7^{(6)} = (0.000000 + 6.160225 + 2.874511 + 3.000000)/4 = 3.008684$$

$$u_8^{(6)} = (3.008684 + 9.508684 + 6.124512 + 6.000000)/4 = 6.160470$$

$$u_9^{(6)} = (6.160470 + 13.000000 + 9.874756 + 9.000000)/4 = 9.508806$$

Iteration 7

$$u_1^{(7)} = (0.000000 + 5.588798 + 1.000000 + 2.874511)/4 = 2.365827$$

$$u_2^{(7)} = (2.365827 + 9.865828 + 4.000000 + 6.124512)/4 = 5.589042$$

$$u_3^{(7)} = (5.589042 + 15.000000 + 9.000000 + 9.874756)/4 = 9.865949$$

$$u_4^{(7)} = (0.000000 + 6.124512 + 2.365827 + 3.008684)/4 = 2.874756$$

$$u_5^{(7)} = (2.874756 + 9.874756 + 5.589042 + 6.160470)/4 = 6.124756$$

$$u_6^{(7)} = (6.124756 + 14.000000 + 9.865949 + 9.508806)/4 = 9.874878$$

$$u_7^{(7)} = (0.000000 + 6.160470 + 2.874756 + 3.000000)/4 = 3.008806$$

$$u_8^{(7)} = (3.008806 + 9.508806 + 6.124756 + 6.000000)/4 = 6.160592$$

$$u_9^{(7)} = (6.160592 + 13.000000 + 9.874878 + 9.000000)/4 = 9.508868$$

Iteration 8

$$u_1^{(8)} = (0.000000 + 5.589042 + 1.000000 + 2.874756)/4 = 2.365949$$

$$u_2^{(8)} = (2.365949 + 9.865949 + 4.000000 + 6.124756)/4 = 5.589164$$

$$u_3^{(8)} = (5.589164 + 15.000000 + 9.000000 + 9.874878)/4 = 9.866010$$

$$u_4^{(8)} = (0.000000 + 6.124756 + 2.365949 + 3.008806)/4 = 2.874878$$

$$u_5^{(8)} = (2.874878 + 9.874878 + 5.589164 + 6.160592)/4 = 6.124878$$

$$u_6^{(8)} = (6.124878 + 14.000000 + 9.866010 + 9.508868)/4 = 9.874939$$

$$u_7^{(8)} = (0.000000 + 6.160592 + 2.874878 + 3.000000)/4 = 3.008868$$

$$u_8^{(8)} = (3.008868 + 9.508868 + 6.124878 + 6.000000)/4 = 6.160653$$

$$u_9^{(8)} = (6.160653 + 13.000000 + 9.874939 + 9.000000)/4 = 9.508898$$

Iteration 9

$$u_1^{(9)} = (0.000000 + 5.589164 + 1.000000 + 2.874878)/4 = 2.366010$$

$$u_2^{(9)} = (2.366010 + 9.866010 + 4.000000 + 6.124878)/4 = 5.589225$$

$$u_3^{(9)} = (5.589225 + 15.000000 + 9.000000 + 9.874939)/4 = 9.866041$$

$$u_4^{(9)} = (0.000000 + 6.124878 + 2.366010 + 3.008868)/4 = 2.874939$$

$$u_5^{(9)} = (2.874939 + 9.874939 + 5.589225 + 6.160653)/4 = 6.124939$$

$$u_6^{(9)} = (6.124939 + 14.000000 + 9.866041 + 9.508898)/4 = 9.874969$$

$$u_7^{(9)} = (0.000000 + 6.160653 + 2.874939 + 3.000000)/4 = 3.008898$$

$$u_8^{(9)} = (3.008898 + 9.508898 + 6.124939 + 6.000000)/4 = 6.160684$$

$$u_9^{(9)} = (6.160684 + 13.000000 + 9.874969 + 9.000000)/4 = 9.508913$$

Iteration 10

$$u_1^{(10)} = (0.000000 + 5.589225 + 1.000000 + 2.874939)/4 = 2.366041$$

$$u_2^{(10)} = (2.366041 + 9.866041 + 4.000000 + 6.124939)/4 = 5.589255$$

$$u_3^{(10)} = (5.589255 + 15.000000 + 9.000000 + 9.874969)/4 = 9.866056$$

$$u_4^{(10)} = (0.000000 + 6.124939 + 2.366041 + 3.008898)/4 = 2.874969$$

$$u_5^{(10)} = (2.874969 + 9.874969 + 5.589255 + 6.160684)/4 = 6.124969$$

$$u_6^{(10)} = (6.124969 + 14.000000 + 9.866056 + 9.508913)/4 = 9.874985$$

$$u_7^{(10)} = (0.000000 + 6.160684 + 2.874969 + 3.000000)/4 = 3.008913$$

$$u_8^{(10)} = (3.008913 + 9.508913 + 6.124969 + 6.000000)/4 = 6.160699$$

$$u_9^{(10)} = (6.160699 + 13.000000 + 9.874985 + 9.000000)/4 = 9.508921$$

Iteration 11

$$u_1^{(11)} = (0.000000 + 5.589255 + 1.000000 + 2.874969)/4 = 2.366056$$

$$u_2^{(11)} = (2.366056 + 9.866056 + 4.000000 + 6.124969)/4 = 5.589270$$

$$u_3^{(11)} = (5.589270 + 15.000000 + 9.000000 + 9.874985)/4 = 9.866064$$

$$u_4^{(11)} = (0.000000 + 6.124969 + 2.366056 + 3.008913)/4 = 2.874985$$

$$u_5^{(11)} = (2.874985 + 9.874985 + 5.589270 + 6.160699)/4 = 6.124985$$

$$u_6^{(11)} = (6.124985 + 14.000000 + 9.866064 + 9.508921)/4 = 9.874992$$

$$u_7^{(11)} = (0.000000 + 6.160699 + 2.874985 + 3.000000)/4 = 3.008921$$

$$u_8^{(11)} = (3.008921 + 9.508921 + 6.124985 + 6.000000)/4 = 6.160707$$

$$u_9^{(11)} = (6.160707 + 13.000000 + 9.874992 + 9.000000)/4 = 9.508925$$

Iteration 12

$$u_1^{(12)} = (0.000000 + 5.589270 + 1.000000 + 2.874985)/4 = 2.366064$$

$$u_2^{(12)} = (2.366064 + 9.866064 + 4.000000 + 6.124985)/4 = 5.589278$$

$$u_3^{(12)} = (5.589278 + 15.000000 + 9.000000 + 9.874992)/4 = 9.866068$$

$$u_4^{(12)} = (0.000000 + 6.124985 + 2.366064 + 3.008921)/4 = 2.874992$$

$$u_5^{(12)} = (2.874992 + 9.874992 + 5.589278 + 6.160707)/4 = 6.124992$$

$$u_6^{(12)} = (6.124992 + 14.000000 + 9.866068 + 9.508925)/4 = 9.874996$$

$$u_7^{(12)} = (0.000000 + 6.160707 + 2.874992 + 3.000000)/4 = 3.008925$$

$$u_8^{(12)} = (3.008925 + 9.508925 + 6.124992 + 6.000000)/4 = 6.160710$$

$$u_9^{(12)} = (6.160710 + 13.000000 + 9.874996 + 9.000000)/4 = 9.508927$$

Iteration 13

$$u_1^{(13)} = (0.000000 + 5.589278 + 1.000000 + 2.874992)/4 = 2.366068$$

$$u_2^{(13)} = (2.366068 + 9.866068 + 4.000000 + 6.124992)/4 = 5.589282$$

$$u_3^{(13)} = (5.589282 + 15.000000 + 9.000000 + 9.874996)/4 = 9.866070$$

$$u_4^{(13)} = (0.000000 + 6.124992 + 2.366068 + 3.008925)/4 = 2.874996$$

$$u_5^{(13)} = (2.874996 + 9.874996 + 5.589282 + 6.160710)/4 = 6.124996$$

$$u_6^{(13)} = (6.124996 + 14.000000 + 9.866070 + 9.508927)/4 = 9.874998$$

$$u_7^{(13)} = (0.000000 + 6.160710 + 2.874996 + 3.000000)/4 = 3.008927$$

$$u_8^{(13)} = (3.008927 + 9.508927 + 6.124996 + 6.000000)/4 = 6.160712$$

$$u_9^{(13)} = (6.160712 + 13.000000 + 9.874998 + 9.000000)/4 = 9.508928$$

Iteration 14

$$u_1^{(14)} = (0.000000 + 5.589282 + 1.000000 + 2.874996)/4 = 2.366070$$

$$u_2^{(14)} = (2.366070 + 9.866070 + 4.000000 + 6.124996)/4 = 5.589284$$

$$u_3^{(14)} = (5.589284 + 15.000000 + 9.000000 + 9.874998)/4 = 9.866070$$

$$u_4^{(14)} = (0.000000 + 6.124996 + 2.366070 + 3.008927)/4 = 2.874998$$

$$u_5^{(14)} = (2.874998 + 9.874998 + 5.589284 + 3.160712)/4 = 6.124998$$

$$u_6^{(14)} = (6.124998 + 14.000000 + 9.866070 + 9.508928)/4 = 9.874999$$

$$u_7^{(14)} = (0.000000 + 6.160712 + 2.874998 + 3.000000)/4 = 3.008928$$

$$u_8^{(14)} = (3.008928 + 9.508928 + 6.124998 + 6.000000)/4 = 6.160713$$

$$u_9^{(14)} = (6.160713 + 13.000000 + 9.874999 + 9.000000)/4 = 9.508928$$

Iteration 15

$$u_1^{(15)} = (0.000000 + 5.589284 + 1.000000 + 2.874998)/4 = 2.366070$$

$$u_2^{(15)} = (2.366070 + 9.866070 + 4.000000 + 6.124998)/4 = 5.589285$$

$$u_3^{(15)} = (5.589285 + 15.000000 + 9.000000 + 9.874999)/4 = 9.866071$$

$$u_4^{(15)} = (0.000000 + 6.124998 + 2.366070 + 3.008928)/4 = 2.874999$$

$$u_5^{(15)} = (2.874999 + 9.874999 + 5.589285 + 6.160713)/4 = 6.124999$$

$$u_6^{(15)} = (6.124999 + 14.000000 + 9.866071 + 9.508928)/47 = 9.875000$$

$$u_7^{(15)} = (0.000000 + 6.160713 + 2.874999 + 3.000000)/4 = 3.008928$$

$$u_8^{(15)} = (3.008928 + 9.508928 + 6.124999 + 6.000000)/4 = 6.160714$$

$$u_9^{(15)} = (6.160714 + 13.000000 + 9.875000 + 9.000000)/4 = 9.508928$$

Thus, the approximate values of u_1, to u_9 are $u_1 = 2.3660, u_2 = 5.5892, u_3 = 9.8660, u_4 = 2.8749, u_5 = 6.1249, u_6 = 9.8750, u_7 = 3.0089, u_8 = 6.1607,$ and $u_9 = 9.5089.$

13.6   ELLIPTIC EQUATION–POISSON EQUATION

The Poisson equation in two independent variables is given by

$$\frac{\partial^2 u}{\partial x^2} + \frac{\partial^2 u}{\partial y^2} = f(x, y), \tag{13.28}$$

where $f(x, y)$ is a function of x and y. This type of partial differential equations occur in fluid mechanics, elasticity and in some fields of physics and engineering.

The SFPF for Eq. 13.28 is

$$u_{i-1,j} + u_{i+1,j} + u_{i,j+1} + u_{i,j-1} - u_{i,j} = h^2 f(ih, jh) \tag{13.29}$$

Substituting the values of interior points (i.e., the values of i and j) in Eq. (13.29), we get a system of linear equations, and these equations can be solved by one of the following methods: (i) Gauss elimination method; (ii) Cramer's rule; (iii) Gauss–Jordan method; (iv) Jacobi's method; and (v) Gauss–Seidel method.

SOLVED PROBLEMS

Problem 1: Solve the Poisson equation $u_{xx} + u_{yy} = x^2 y^2$ in the square mesh shown in Fig. 13.31 with side 1, given that $u(x, y) = 0$, on the boundary AB, BC, CD and DA of the square $ABCD$.

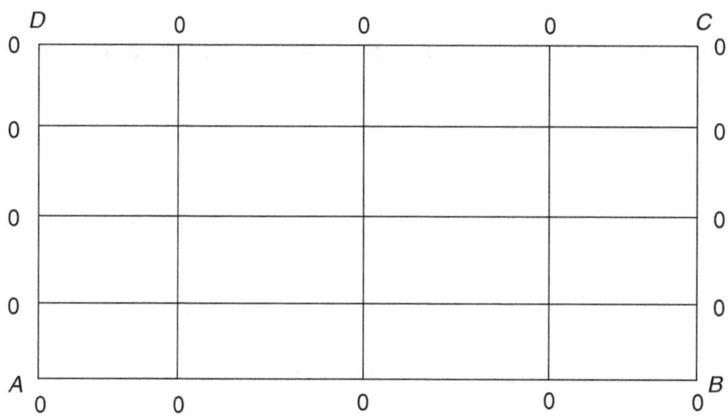

Figure 13.31

Solution: Let $u_1, u_2, ..., u_9$ be the values of $u\,(x, y)$ at the interior mesh points. Choose the Coordinate system having origin at u_5, the horizontal line through u_5 as x-axis and vertical line through it as y-axis (Fig. 13.32).

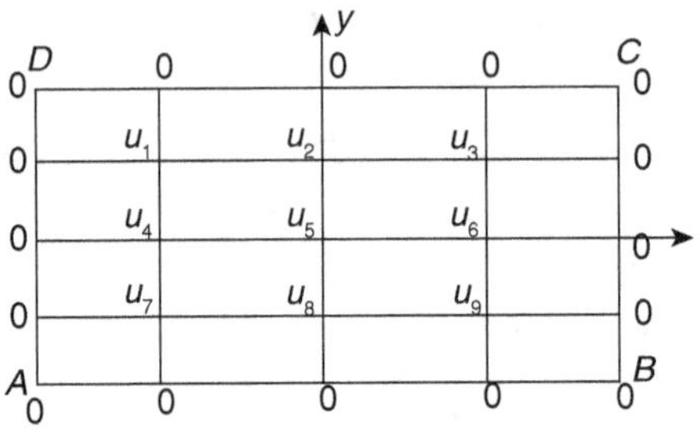

Figure 13.32

Note that at u_5, the value of $i = 0, j = 0$ (i.e., $u_5 = u(0, 0)$), at u_4 the value of $i = -1, j = 0$ (i.e., $u_4 = u(-1, 0)$) at u_7 the value of $i = -1, j = -1$, etc. The given partial differential equation is symmetric about x-axis, y-axis and the line $y = x$, Therefore,

$$u_2 = u_4 = u_6 = u_8 \tag{1}$$

$$u_1 = u_3 = u_7 = u_9 \tag{2}$$

Now, suppose

$$u_2 = u_4 = u_6 = u_8 = z_1 \tag{3}$$

and

$$u_1 = u_3 = u_7 = u_9 = z_2 \tag{4}$$

$$u_5 = z_3 \tag{5}$$

For the given partial differential equation, Eq. 13.29 reduces to

$$u_{i-1,j} + u_{i+1,j} + u_{i,j-1} + 4_{i,j+1} - 4u_{i,j} = i^2 j^2 \tag{6}$$

Putting $i = -1, j = -1$ in Eq. (6), we get

$$u_{-2,-1} + u_{0,-1} + u_{-1,-2} + u_{-1,0} - 4u_{-1,-1} = (-1)^2(-1)^2 \tag{7}$$

Since
$$u_{-2,-1} = u(-2,-1) = 0, u_{0,-1} = u(0,-1) = u_8$$

$$u_{-1,-2} = u(-1,-2) = 0, \ u_{-1,0} = u(-1,0) = u_4 \text{ and}$$

$$u_{-1,-1} = u(-1,-1) = 7$$

Eq. (7) reduce to $\quad u_8 + u_4 - 4u_7 = 1 \tag{8}$

Now, substituting $i = 0, j = 1$, in (6), we obtain

$$u_{-1,1} + u_{1,1} + u_{0,0} + u_{0,2} - 4u_{0,1} = 0$$

$$u_1 + u_3 + u_5 + 0 - 4u_2 = 0$$

$$u_1 + u_3 + u_5 - 4u_2 = 0 \tag{9}$$

Similarly, by substituting $i = 0, j = 0$ in Eq. (6), we get

$$u_{-1,0} + u_{1,0} + u_{0,-1} + 4_{0,1} - 4u_{0,0} = 0$$

$$u_4 + u_6 + u_8 + u_2 - 4u_5 = 0 \tag{10}$$

In view of Eqs. (3), (4) and (5), the Eqs. (8), (9) and (10) reduce to

$$z_1 + z_7 - 4z_2 = 1 \quad \text{i.e, } 2z_1 - 4z_2 = 1 \tag{11}$$

$$2z_2 + z_3 - 4z_1 = 0 \text{ i.e., } -4z_1 + 2z_4 + z_3 = 0 \tag{12}$$

$$4z_1 - 4z_3 = 0 \text{ i.e., } z_2 - z_3 = 0 \tag{13}$$

Solving (11), (12) and (13), we get $z_1 = 0.42$, $z_2 = -0.4$ and $z_3 = -0.4$. Thus, the values of $u(x, y)$ at the interior mesh points are

$$u_1 = u_3 = u_7 = u_9 = -0.4, \ u_5 = -0.4$$

and $$u_2 = u_4 = u_6 = u_8 = 0.42$$

Problem 2: Solve $u_{xx} + u_{yy} = -10 \ (x^2 + y^2 + 10)$ over the square $[0, 3] \times [0, 3]$ having 1 as length of mesh square with $u = 0$ on the boundary of the square.

Solution: Since $h = 1$ is the length of mesh square, the region $[0, 3] \times [0, 3]$ has 9 mesh squares, which is shown in Fig. 13.33. The values shown along

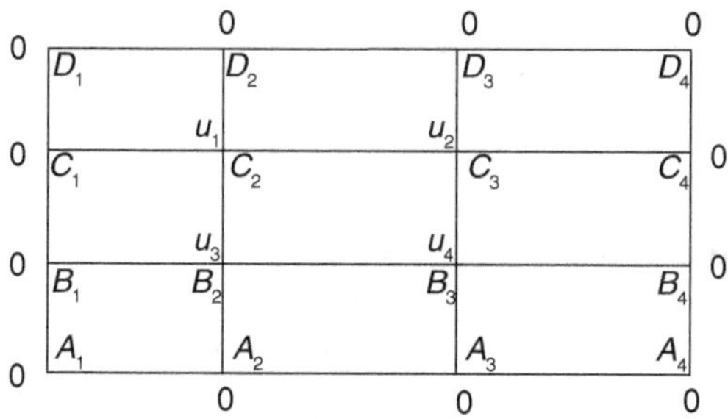

Figure 13.33

the boundary indicate the boundary values. Let, u_1, u_2, u_3 and u_4 be the values of $u(x, y)$ at the interiors points C_2, C_3, B_2 and B_3 respectively.

The given partial differential equation, (13.29) reduces to

$$u_{i-1,j} + u_{i+1,j} + u_{i,j+1} + u_{i,j-1} - 4u_{i,j} = -10(i^2 + j^2 + 10). \quad (1)$$

Substituting $i = 1, j = 2$ in the eq(1), we get

$$u_{0,2} + u_{2,2} + u_{1,1} + u_{1,3} - 4u_{1,2} = -10 (1^2 + 2^2 + 10)$$

$$0 + u_2 + u_3 + 0 - 4u_1 = -150$$

i.e., $\qquad\qquad\qquad u_2 + u_3 - 4u_1 = -150 \qquad\qquad\qquad (2)$

Similarly, substituting $(i = 2, j = 2)$, $(i = 1, j = 1)$ and $(i = 2, j = 1)$ in Eq. (1), we obtain

$$u_1 + u_4 - 4u_2 = -180 \qquad\qquad (3)$$

and $\qquad\qquad u_1 + u_4 - 4u_3 = -120 \qquad\qquad (4)$

$$u_2 + u_3 - 4u_4 = -150, \qquad\qquad (5)$$

respectively.

Clearly from Eqs. (2) and (5), we get $u_1 = u_4$. Thus, we need to find only u_1, u_2 and u_3. Hence, we need to solve Eqs. (2), (3) and (4) only. i. e.,

$$u_2 + u_3 - 4u_1 = -150$$

$$u_1 + u_4 - 4u_2 = -180$$

$$u_1 + u_4 - 4u_3 = -120,$$

which can be written as

$$u_1 = \frac{1}{4}(u_2 + u_3 + 150) \qquad\qquad (6)$$

$$u_2 = \frac{1}{4}(u_2 + u_4 + 180) \qquad\qquad (7)$$

$$u_1 = \frac{1}{4}(u_1 + u_4 + 120) \qquad\qquad (8)$$

Since $u_1 = u_4$, the Eqs. (6) to (8) reduce to

$$u_1 = \frac{1}{4}(u_2 + u_3 + 150) \qquad\qquad (9)$$

$$u_2 = \frac{1}{2}(u_1 + 90) \qquad\qquad (10)$$

$$u_3 = \frac{1}{2}(u_1 + 60) \qquad\qquad (11)$$

The iterative formulas to implement Gauss–Seidel method are

$$u_1^{(n+1)} = \frac{1}{4}(u_2^{(n)} + u_3^{(n)} + 150) \tag{12}$$

$$u_2^{(n+1)} = \frac{1}{2}(u_1^{(n+1)} + 90) \tag{13}$$

$$u_3^{(n+1)} = \frac{1}{2}(u_1^{(n+1)} + 60). \tag{14}$$

Suppose $u_2^{(0)} = u_3^{(0)}$. Substituting $n = 0, 1, 2, 3...$ in Eqs. (12) to (14), we get the values $(u_1^{(1)}, u_2^{(1)}, u_3^{(1)})$, $(u_1^{(2)}, u_2^{(2)}, u_3^{(3)})$,.. and so on, upon substituting relevant values and simplifying. The value obtained at different iterations are shown below:

Iteration no.	u_1	u_2	u_3	u_4
1	37.500000	63.750000	48.750000	37.500000
2	65.625000	77.812500	62.812500	65.625000
3	72.656250	81.328125	66.328125	72.656250
4	74.414063	82.207031	67.207031	74.414063
5	74.853516	82.426758	67.426758	74.853516
6	74.963379	82.481689	67.481689	74.963379
7	74.990845	82.495422	67.495422	74.990845
8	74.997711	82.498856	67.498856	74.997711
9	74.999428	82.499714	67.499714	74.999428
10	74.999857	82.499928	67.499928	74.999857
11	74.999964	82.499982	67.499982	74.999964
12	74.999991	82.499996	67.499996	74.999991

13.7 PARABOLIC EQUATIONS

In this section, we describe two different methods of finding the solutions of one-dimensional heat equation $\dfrac{\partial^2 u}{\partial x^2} = a\dfrac{\partial u}{\partial t}$, which is a parabolic equation subject to the following boundary conditions:

(i) $u(0, t) = T_0 \ \forall \ t \in [0, b]$

(ii) $u(l, t) = T_1 \ \forall \ t \in [0, b]$

   (iii) $u(x,0) = f(x) \ \forall \ x \in [0, l]$,

where the domain of $u(x, t)$ is $[0, a] \times [0, b]$.

   (i) **Solution of one dimensional heat equation by Bender–Schmidt method.**

Let the heat conduction equation is one dimensional be

$$\frac{\partial^2 u}{\partial x^2} - a \frac{\partial u}{\partial t} = 0 \tag{13.30}$$

where a is a constant.

The approximations of $\dfrac{\partial u}{\partial t}$ and $\dfrac{\partial^2 u}{\partial x^2}$ by the finite differences used in this method are

$$\frac{\partial u}{\partial t} = \frac{u_{i,j+1} - u_{i,j}}{k} \tag{13.31}$$

$$\frac{\partial^2 u}{\partial x^2} = \frac{1}{h^2}[u_{i+1,j} - 2u_{i,j} + u_{i-1,j}] \tag{13.32}$$

Substituting Eqs. (13.31) and (13.32) in Eq. (13.30), we get

$$\frac{1}{h^2}[u_{i+1,j} - 2u_{i,j} + u_{i-1,j}] - \frac{a}{k}[u_{i,j+1} - u_{i,j}] = 0,$$

which can be expressed as

$$u_{i,j+1} = u_{i,j} + \lambda\,[u_{i-1,j} - 2u_{i,j} + u_{i+1,j}] \tag{13.33}$$

where

$$\lambda = \frac{k}{ah^2} \tag{13.34}$$

Note that λ is a known constant as h and k are the lengths of mesh rectangular and a is the coefficient of $\dfrac{\partial u}{\partial t}$. Therefore, it is possible to choose k so that we can have a particular value of λ. For example, if $h = 1$, and $a = 4$, we can choose k so that $\lambda = \dfrac{1}{2}$. The value of k in this case is $(k = ah^2\lambda)\,k = 4 \times 1^2 \times 1/2 = 2.$

The formula (13.33) is valid only for $0 < \lambda \le 1/2$.

If $\lambda = \dfrac{1}{2}$, Eq (13.33) reduces to

$$u_{i,j+1} = \frac{1}{2}(u_{i-1,j} + u_{i+1,j}) \tag{13.35}$$

which is known as Bender–Schmidt's explicit formula . The formula (13.35) can be used to solve one-dimensional heat equation only when $\lambda = 1/2$ and Eq. 13.33 can be used to solve one- dimensional heat equation for $0 < \lambda \le 1/2$.

   (ii) **Crank–Nicolson method to solve one-dimensional heat equation.**

Let the one-dimensional heat equation be given by Eq. (13.30) i. e..,

$$\frac{\partial^2 u}{\partial x^2} - a\frac{\partial u}{\partial t} = 0$$

The following finite difference approximations are used in this method.

$$\frac{\partial u}{\partial t} = \frac{u_{i,j+1} - u_{i,k}}{k} \tag{13.36}$$

and

$$\frac{\partial^2 u}{\partial x^2} = \frac{1}{2}\left[\left\{\frac{u_{i+1,j} - 2u_{i,j} + u_{i-1,j}}{h^2}\right\} + \left\{\frac{u_{i+1,j+1} - 2u_{i,j+1} + u_{i-1,j+1}}{h^2}\right\}\right] \tag{13.37}$$

Observe that u_{xx} given by Eq. (13.37) is the average of its finite difference approximations on j th and $j + 1$th rows.

   In view of Eqs. (13.36) and (13.37), Eq. (13.30) is reduced to

$$\frac{1}{2h^2}\left[(u_{i+1,j} - 2u_{i,j} + u_{i-1,j}) + (u_{i+1,j+1} - 2u_{i,j+1} + u_{i-1,j+1})\right]$$

$$-a\left[\frac{u_{i,j+1} - u_{i,j}}{k}\right] = 0$$

which can be expressed as

$$\lambda(u_{i+1,j+1} + u_{i-1,j+1}) - 2(1+\lambda)u_{i,j+1} = 2(\lambda-1)u_{i,j} - \lambda(u_{i+1,j} + u_{i-1,j}) \tag{13.38}$$

where

$$\lambda = \frac{k}{ah^2} \tag{13.39}$$

   Eq. (13.38) is called Crank–Nicholson difference method. When $\lambda = 1$, the Crank–Nicholson method becomes

$$u_{i,j+1} = \frac{1}{4}[u_{i-1,j+1} + u_{i+1,j+1} + u_{i+1,j} + u_{i-1,j}] \tag{13.40}$$

The Crank–Nicholson method given by Eq. (13.38) is convergent for all finite values of λ. The Crank–Nicolson formula gives n linear equations in n unknown values of $u(x, y)$ at the interior mesh points. Solving these equations, we can obtain the numerical solution of Eq. (13.30).

Note: For convenience, the values of $u_{i,j}$ are shown in the following tabular form:

j \ i	0	1	2	3	-----------	n
0	$u_{0,0}$	$u_{1,0}$	$u_{2,0}$	$u_{3,0}$	-----------	$u_{n,0}$
1	$u_{0,1}$	$u_{1,1}$	$u_{2,1}$	$u_{3,1}$	-----------	$u_{n,1}$
2	$u_{0,2}$	$u_{1,2}$	$u_{2,2}$	$u_{3,2}$	-----------	$u_{n,2}$
----	----	----	----	----	-----------	----
n	$u_{0,n}$	$u_{1,n}$	$u_{2,n}$	$u_{3,n}$	-----------	$u_{n,n}$

We refer $u_{0,0}$, $u_{1,0}$ .... $u_{n,0}$ as first row elements, $u_{0,1}$ $u_{1,1}$ ...., $u_{n,1}$ as second row element, and so on in the table. Note that it is not a mesh square of rectangle.

13.8 V-SHAPED CALCULATIONS

To avoid laborious calculations by Bender–Schmidt's process, we define V-shaped calculations that can be used to determine $u(x, y)$ values at interior mesh points as per Bender–Schmidt's formula.

The entry $u(x, y)$ value at an interior mesh points A is the average of top entries of two adjacent entries of A. We call such computations as V-shaped calculations. For example, a_1, b_1, c_1, d_1, e_1 ..... are entries of a row. The next row entries can be computed by V-shaped calculations, which are shown in the following Fig. 13.34.

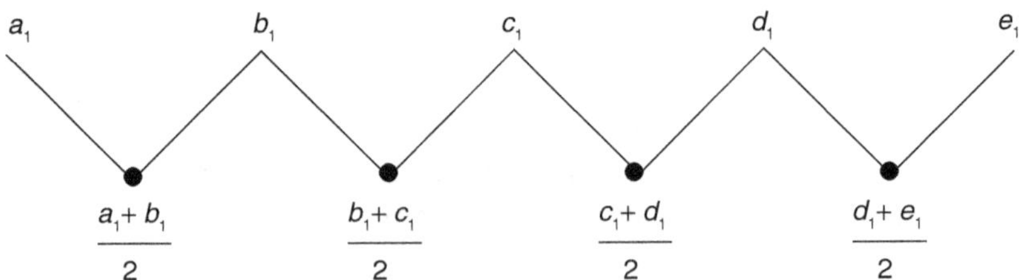

Figure 13.34

Note that V-shaped calculations are introduced to obtain the $u\ (x, y)$ values at interior points in a short-cut method.

13.9 PROCEDURE TO SOLVE $\dfrac{\partial^2 u}{\partial x^2} - a\dfrac{\partial u}{\partial t} = 0$
BY BLENDER–SCHMIDT METHOD

Step 1: Express the given one-dimensional heat conduction equation in the form $\dfrac{\partial^2 u}{\partial x^2} - a\dfrac{\partial u}{\partial t} = 0$ and note down the value of a.

Step 2: Select h value.

Step 3: Choose k such that $\lambda = 1/2$

i. e., $$k = \frac{1}{2}ah^2$$

(for Blender–Schmidt method $\lambda = \dfrac{1}{2}$)

Step 4: Using the boundary conditions

$$u(0, x) = T_0$$
$$u(l, t) = T_1$$
$$u(x, 0) = f(x)$$

Find the values of $u(x, y)$ at the points, lying on the boundary. At this stage, we know the following values:

(i) $u_{0,0}, u_{1,0}, u_{2,0}, u_{3,0} \ldots\ldots\ldots\ldots u_{n,0}$

(ii) $u_{0,1}, u_{0,2}, u_{0,3}, u_{0,4} \ldots\ldots\ldots u_{0,n}$

(iii) $u_{n,1}, u_{n,2}, u_{n,3}, u_{n,4} \ldots\ldots\ldots u_{n,n}$

Step 5: Construct the following table and enter the values found in Step 4

i \ j	0	1	2	3	....	n
0	$u_{0,0}$	$u_{1,0}$	$u_{2,0}$	$u_{3,0}$	....	$u_{n,0}$
1	$u_{0,1}$					$u_{n,1}$
2	$u_{0,2}$					$u_{n,2}$
3	$u_{0,3}$					$u_{n,3}$
$\vdots$	$\vdots$					$\vdots$
n	$u_{0,n}$					$u_{n,n}$

Step 6: The second row entries can be computed by using V-shaped calculations. After the second row is filled, we can compute the third row entries using V-shaped computations. Similarly, we can find entries of other rows using the same V-shaped calculations. The table so obtained is the numerical solution of the given boundary value problem.

SOLVED PROBLEMS

Problem 1: Using Bender–Schmidt method, find the values of $u(x, y)$ satisfying the parabolic equation $u_{xx} - 2u_t = 0$ and the boundary conditions (i) $u(0, t) = 0$ (ii) $u(5, t) = 0$ and (iii) $u(x, 0) = x(5 - x)$ at the points $x = i(i = 0, 1, 2, 3, 4)$ and t such that $0 < t \leq 6$.

Solution: Comparing the given partial differential equation with $u_{xx} - au_t = 0$, we have $a = 2$. To implement Bender–Schmidt formula, we should have $\lambda = 1/2$. Hence, $k = \lambda\, ah^2 = \dfrac{1}{2} \times 2 \times 1 = 1$. Thus, we have to find $u(x,t)$ values at $x = 0, 1, 2, 3, 4, 5$ and $t = 0, 1, 2, 3, 4, 5, 6$.

Now, from the boundary condition (i), we have $u(0, 0) = 0, u(0, 1) = 0$, $u(0, 2) = 0$, $u(0, 3) = 0$, $u(0, 4) = 0$, $u(0, 5) = 0$ and $u(0, 6) = 0$. These are the entries of first column of the table as stated in Step 5 of Section 13.9.

From the boundary condition (ii) i. e., $u(5, t) = 0$, we have $u(5, 0) = u(5, 2) = u(5, 3) = u(5, 4)\ u(5, 5) = u(5, 6) = 0$. These are the last column entries of the table as stated in Step 5 of Section 13.9.

From the boundary condition (iii), we have

$$u(0, 0) = 0,\ u(1, 0) = 1(5 - 1) = 4$$

$$u(2, 0) = 2(5 - 2) = 6,\ u(3, 0) = 3(5 - 3) = 6$$

$$u(4, 0) = 4(5 - 4) = 4 \text{ and } (5, 0) = 0$$

These are the entries of first of first row of the table:

$j\backslash i$	0	1	2	3	4	5
0	0	4	6	6	4	0
1	0	3	5	5	3	0

2	0	2.5	4	4	2.5	0
3	0	2	3.25	3.25	2	0
4	0	1.625	2.625	2.625	1.625	0
5	0	1.3125	2.125	2.125	1.3125	0
6	0	1.0625	1.71875	1.71875	1.0625	0

The Blender–Schmidt's formula

$$u_{i,j+1} = \frac{1}{2}(u_{i-1,j} + u_{i+1,j}) \tag{1}$$

The second-row entries can be computed by (1) in the following manner, using V-shaped calculations (see Fig. 13.35)

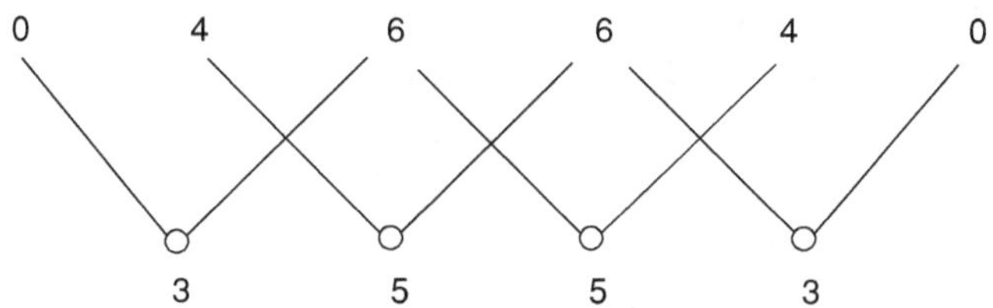

Figure 13.15

Similarly 3rd row entries can be computed as follows using V-shaped calculations.

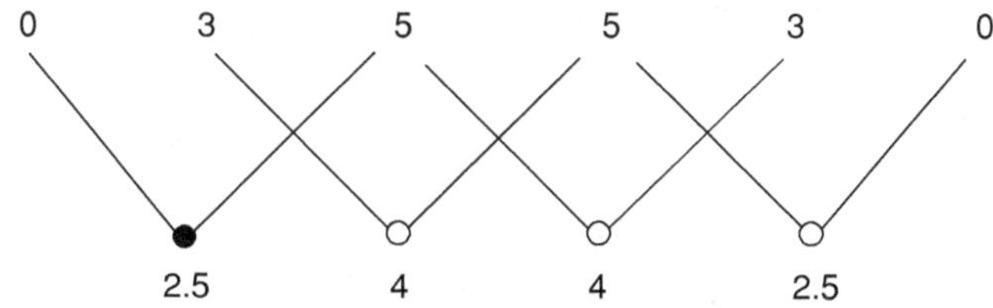

Repeating this process, we can find the entries of the remaining entries of each row. The entries are shown in the table.

Problem 2: Solve by Bender–Schmidt method $u_{xx} - \frac{1}{5}u_t = 0$ with the boundary conditions: (i) $u\,(0, t) = 0$ (ii) $u\,(5, t) = 60$

(iii) $u(x,0) = \begin{cases} 20x & \text{for} \quad 0 \le x \le 3 \\ 60 & \text{for} \quad 3 \le x \le 5 \end{cases}$

for 5 appropriate values of t with $h = 1$.

Solution: Given that $h = 1$. The value of $a = \dfrac{1}{5}$

To implement Bender–Schmidt method, we should have $\lambda = \dfrac{1}{2}$.

Hence, $k = a\dfrac{h^2}{2} = \dfrac{1}{5} \times \dfrac{1}{2} = \dfrac{1}{10} = 0.1$. Thus, we have to find $u\,(x,\,t)$ at $x = 0, 1, 2, 3, 4, 5$ and $t = 0, 0.1, 0.2, 0.3, 0.4, 0.5$.

From the boundary conditions (i), (ii) and (iii), we have

$u_{0,0} = 0,\ u_{0,1} = 0,\ u_{0,2} = 0,\ u_{0,3} = 0,\ u_{0,4} = 0,\ u_{0,5} = 0,\ u_{5,0} = 60,\ u_{5,1} = 60,$
$u_{5,2} = 60,\ u_{5,3} = 60,\ u_{5,4} = 60,\ u_{5,5} = 60,\ u_{0,0} = 0,\ u_{1,0} = 20,\ u_{2,0} = 40,$
$u_{3,0} = 60,\ u_{4,0} = 60,\ u_{5,0} = 60.$

Now, we fill the values found in the table and compute the remaining entries by using V-shaped calculations.

i \ j	0	1	2	3	4	5
0 ($t = 0$)	0	20	40	60	60	60
1 ($t = 0.1$)	0	20	40	50	60	60
2 ($t = 0.2$)	0	20	35	50	55	60
3 ($t = 0.3$)	0	17.5	35	45	55	60
4 ($t = 0.4$)	0	17.5	31.25	45	52.5	60
5 ($t = 0.5$)	0	15.625	31.25	41.875	52.5	60

V-shaped calculations are of a short-cut method for obtaining entries that are equal to the values to be obtained from Bender–Schmidt's formula $u_{i,\,j+1} = \dfrac{1}{2}\left(u_{i+1,j} + u_{i-1,j}\right).$

Problem 3: Using Crank–Nicolson method, solve $\dfrac{\partial^2 u}{\partial x^2} - \dfrac{\partial u}{\partial t} = 0$ with the boundary conditions (i) $u(x, 0) = 0$, (ii) $u(0, t) = 0$ and (iii) $u(1, t) = t$. Compute u for 2 steps in the variable t and $h = \dfrac{1}{4}$.

Solution: Given that $h = \dfrac{1}{4}$, comparing the given partial differential equation with $u_{xx} - au_t = 0$, we have $a = 1$. We choose k such that $\lambda = 1$.

The relation between λ, a, h and k is $\lambda = \dfrac{k}{ah^2}$

$$\therefore k = \lambda\, ah^2 = 1.1\left(\frac{1}{4}\right)^2 = \frac{1}{16}$$

For $\lambda = 1$, the Crank–Nicholson formula is

$$u_{i,j+1} = \frac{1}{4}(u_{i+1,j+1} + u_{i-1,j+1} + u_{i+1,j} + u_{i-1,j}) \tag{1}$$

From the boundary conditions, we have

(i) $u_{0,0} = 0, u_{0,1} = 0, u_{0,2} = 0$

(ii) $u_{1,0} = 0, u_{2,0} = 0, u_{3,0} = 0, u_{4,0} = 0$

(iii) $u_{4,1} = \dfrac{1}{16}, u_{4,2} = \dfrac{2}{16}$

i j	0 $(x = 0)$	1 $\left(x = \dfrac{1}{4}\right)$	2 $\left(x = \dfrac{1}{2}\right)$	3 $\left(x = \dfrac{3}{4}\right)$	4 $(x = 1)$
0 $t = 0$	0	0	0	0	0
1 $t = \dfrac{1}{16}$	0	u_1	u_2	u_3	$\dfrac{1}{16}$
2 $t = \dfrac{2}{16}$	0	u_4	u_5	u_6	$\dfrac{2}{16}$

Let the values of $u(x, t)$ at the interior mesh points be u_1, u_2, u_3, u_4, u_5 and u_6 (see the table). From (1), we have

$$u_1 = \frac{1}{4}u_2 \qquad (2)$$

$$u_2 = \frac{1}{4}(u_1 + u_3) \qquad (3)$$

$$u_3 = \frac{1}{4}\left(u_2 + \frac{1}{16}\right) \qquad (4)$$

Note that u_1 is the average of its immediate adjacent entries and their top entries (Fig 13.36). The same is applicable to u_2, u_3, u_4, u_5, and u_6.

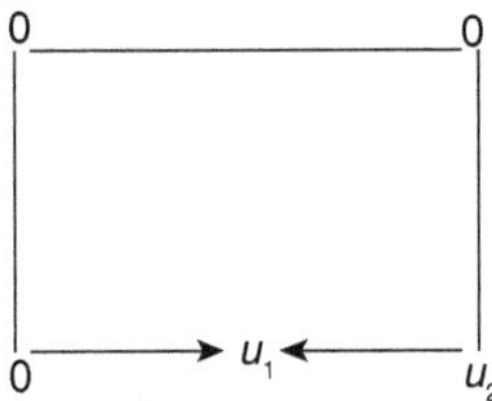

Figure 13.36

Eqs. (2), (3), and (4) can be written as

$$4u_1 - u_2 = 0 \qquad (5)$$

$$u_1 - 4u_2 + u_3 = 0 \qquad (6)$$

$$-u_2 + 4u_3 = \frac{1}{16} \qquad (7)$$

Eqs. (5), (6), and (7) can be expressed as

$$AX = B,$$

where
$$A = \begin{pmatrix} 4 & -1 & 0 \\ 1 & -4 & 1 \\ 0 & -1 & 4 \end{pmatrix}, \ X = \begin{pmatrix} u_1 \\ u_2 \\ u_3 \end{pmatrix} \text{ and } B = \begin{pmatrix} 0 \\ 1 \\ \dfrac{1}{16} \end{pmatrix}$$

Then,
$$A^{-1} = \begin{pmatrix} 0.2679 & -0.0714 & 0.0179 \\ 0.0714 & -0.2857 & 0.0714 \\ 0.0179 & -0.0714 & 0.2679 \end{pmatrix}$$

and
$$X = A^{-1}\,B = \begin{pmatrix} 0.0011 \\ 0.0045 \\ 0.0167 \end{pmatrix}$$

Thus, $\qquad u_1 = 0.0011,\ u_2 = 0.0045$ and $u_3 = 0.0167$

with $u_1,\ u_2,\ u_3$ values the last 2 rows are

$$\begin{array}{ccccc} 0 & 0.0011 & 0.0045 & 0.0167 & 0 \\ 0 & u_4 & u_5 & u_6 & 0 \end{array}$$

By the formula (1), we have

$$u_4 = \frac{1}{4}[0 + 0 + 0.0045 + u_5]$$

$$u_5 = \frac{1}{4}[0.0011 + u_4 + 0.0167 + u_6]$$

$$u_6 = \frac{1}{4}[0.0045 + u_5 + 0 + 0]$$

which can be written as

$$4u_4 - u_5 = 0.0045 \tag{8}$$

$$u_4 - 4u_5 + u_6 = -0.0178 \tag{9}$$

$$-u_5 + 4u_6 = 0.192 \tag{10}$$

Solving Eqs. (8), (9) and (10), we get $u_4 = 0.0059$, $u_5 = 0.191$ and $u_6 = 0.0528$.

Problem 4: Using Crank–Nicolson method, solve $u_{xx} - u_t = 0$ with the boundary (i) $u(0, t) = 0$ (ii) $u(1, t) = 0$ and (iii) $u(x, 0) = \dfrac{\pi}{2}$ for $0 < x < 1$ and $h = \dfrac{1}{3}$, $k = \dfrac{1}{36}$ compute u for 2 steps in the variable t.

Solution: Here, $a = 1$, $h = \dfrac{1}{3}$ and $k = \dfrac{1}{36}$. Hence, $\lambda = \dfrac{k}{ah^2} = \dfrac{1/36}{1 \times \dfrac{1}{(9)}} = \dfrac{1}{4}$,

From the boundary condition $u(0, t) = 0$, we have $u_{0,0} = 0$, $u_{0,1} = 0$, $u_{0,2} = 0$ and from $u(1, t) = 0$, we have $u_{3,0} = 0$, $u_{3,1} = 0$ and $u_{3,2} = 0$.

Since $u(x, 0) = \sin \pi x$, we have $u_{0,0} = 0$, $u_{1,0} = \dfrac{\sqrt{3}}{2}$, $u_{2,0} = \dfrac{\sqrt{3}}{2}$ and $u_{3,0} = 0$.

Let $u_{1,1} = u_1$, $u_{2,1} = u_2$ and $u_{1,2} = u_3$ $u_{2,2} = 4$. These values are shown is the following table:

$j\backslash i$	0 $(x = 0)$	1 $\left(x = \dfrac{1}{3}\right)$	2 $\left(x = \dfrac{2}{3}\right)$	3 $\left(x = \dfrac{3}{3}\right)$
0 $(t = 0)$	$u_{0,0} = 0$	$u_{1,0} = \sqrt{3}/2$	$u_{2,0} = \sqrt{3}/2$	$u_{3,0} = 0$
1 $\left(t = \dfrac{1}{16}\right)$	$u_{0,1} = 0$	$u_{1,1} = u_1$	$u_{2,1} = u_2$	$u_{3,1} = 0$
2 $\left(t = \dfrac{2}{16}\right)$	$u_{0,2} = 0$	$u_{1,2} = u_3$	$u_{2,2} = u_4$	$u_{3,2} = 0$

Note that, since $u_{i,j} = u(ih, jk)$, we have $u_{0,1} = u\left(0, \dfrac{1}{16}\right)$,

$u_{0,2} = u\left(0, \dfrac{2}{16}\right)$, $u_{3,1} = u\left(1, \dfrac{1}{16}\right)$, $u_{3,2} = u\left(1, \dfrac{2}{16}\right)$, etc.

For $\lambda = \dfrac{1}{4}$ the equation (13.38) reduces to

$$\frac{1}{4}(u_{i+1,j+1} + u_{i-1,j+1} - \frac{5}{2}u_{i,j+1})$$

$$= -\frac{3}{2}u_{i,j} - \frac{1}{4}(u_{i+1,j} + u_{i-1,j}) \tag{1}$$

Substituting $i = 1, j = 0$ in (1), we obtain

$$\frac{1}{4}(u_{2,1} + u_{0,1} - \frac{5}{2}u_{1,1}) = -\frac{3}{2}u_{1,0} - \frac{1}{4}(u_{2,0} + u_{0,0})$$

$$\frac{1}{4}(u_2 + 0 - \frac{5}{2}u_1) = -\frac{3}{2}\left(\frac{\sqrt{3}}{2}\right) - \frac{1}{4}\left(\frac{\sqrt{3}}{2} + 0\right)$$

$$u_2 - 10u_1 = -\frac{7}{2}\sqrt{3} \tag{2}$$

Similarly, substituting $i = 2, j = 0$ in (1), we get

$$u_1 - 10u_2 = \frac{7}{2}\sqrt{3} \tag{3}$$

Next substituting $i = 1, j = 1$ in (1), we obtain

$$u_4 - 10u_3 = -6u_1 - u_2 \tag{4}$$

Solving Eqs. (2) and (3), we obtain $u_1 = u_2 = 0.6735$.

Substituting u_1, u_2 values in Eqs. (4) and (5) and solving for u_3 and u_4, we get

$$u_3 = u_4 = 0.5238.$$

Thus, the values of $u(x, y)$ at the interior mesh points are $u_{1,1} = 0.6735$,

$u_{2,1} = 0.6735$, $u_{1,2} = 0.5238$ and $u_{2,2} = 0.5238$ i.e., $u\left(\frac{1}{3}, \frac{1}{16}\right) = 0.6735$,

$u\left(\frac{2}{3}, \frac{1}{16}\right) = 0.6735$, $u\left(\frac{1}{3}, \frac{2}{16}\right) = 0.5238$, and $u\left(\frac{2}{3}, \frac{2}{16}\right) = 0.5238$.

13.10   HYPERBOLIC EQUATION

The one-dimensional wave equation is given by

$$\frac{\partial^2 u}{\partial x^2} = \frac{1}{c^2}\frac{\partial^2 u}{\partial t^2} \tag{13.41}$$

which is a hyperbolic equation. It represents the wave motion of a string. Eq. 13.41 can be written in the form.

$$c^2 u_{xx} - u_{tt} = 0 \tag{13.42}$$

In this section, we describe the method of obtaining the numerical solution of Eq. (13.42) subject to the following conditions.

$$u(0, t) = 0 \tag{13.43}$$
$$u(l, t) = 0 \tag{13.44}$$
$$u(x, 0) = f(x) \tag{13.45}$$
$$u(x, 0) = 0 \tag{13.46}$$

where l is the length of a string and $f(x)$ is a real valued function. The conditions given by Eqs. (13.43) and (13.44) are called boundary conditions whereas the conditions given by Eqs. (13.45) and 13.46 are called initial conditions.

The analogous finite difference equation of Eq. (13.42) is

$$c^2 \frac{(u_{i+1,j} - 2u_{i,j} + u_{i-1,j})}{h^2} - \frac{1}{k^2}\left(u_{i,j+1} - 2u_{i,j} + u_{i,j-1}\right) = 0$$

i. e., $u_{i+1,j} - 2u_{i,j} + u_{i-1,j} = \dfrac{h^2}{k^2 c^2}(u_{i,j+1} - 2u_{i,j} + 2u_{i,j-1})$ (13.47)

Introducing

$$\lambda = \frac{k}{h} \tag{13.48}$$

The Eq. (13.47) reduces to

$$\lambda^2 c^2 (u_{i+1,j} - 2u_{i,j} + u_{i-1,j}) = u_{i,j+1} - 2u_{i,j} + u_{i,j-1}$$

which can be expressed as

$$u_{i,j+1} = 2(1 - \lambda^2 c^2)u_{i,j} + \lambda^2 c^2 (u_{i+1,j} + u_{i-1,j}) - u_{i,j-1} \tag{13.49}$$

Eq. (13.49) is called explicit scheme to obtain the solution of Eq. (13.42). To obtain Eq. (13.49) in a simplest form, we choose λ such that $1 - \lambda^2 c^2 = 0$ i.e., $1 - \dfrac{k^2}{h^2} c^2 = 0$,

i. e., $$k = \frac{h}{c}$$

(It means that the length of mesh along the time variable is h/c.)

Hence, Eq. (13.49) reduces to $u_{i,j+1} = u_{i+1,j} + u_{i-1,j} - u_{i,j-1}$ (13.50)

Note that this formula is applicable to solve the wave equation only when $k = \dfrac{h}{c}$.

The values of $u_{0,0}$, $u_{0,1}$, $u_{0,2}$, $u_{0,3}$ ... etc can be obtained from Eq. (13.43), and the values of $u(x, y)$ at the mesh points along the line $x = l$ can be obtained from 13.44 and the values of $u_{0,0}$, $u_{1,0}$, $u_{2,0}$, $u_{3,0}$... etc. can be obtained from Eq. (13.45). Further, $u_{0,1}$, $u_{1,1}$, $u_{2,1}$, $u_{3,1}$... etc. can be obtained from

$$u_{i,1} = \frac{1}{2}(u_{i-1,0} + u_{i+1,0}) \qquad (13.51)$$

which is the finite difference analogue of Eq. (13.46).

The values of $u(x, y)$ at the interior mesh points can be obtained from Eq. (13.49). If $k = \dfrac{h}{c}$, then these values can be obtained from Eq. (13.50).

Note: In this section, the numerical solution of the boundary value problem of the type (13.41) is shown in the following tabular form:

i / j	0	1	2	3	....	n
0	$u_{0,0}$	$u_{1,0}$	$u_{2,0}$	$u_{3,0}$	....	$u_{n,0}$
1	$u_{0,1}$	$u_{1,1}$	$u_{2,1}$	$u_{3,1}$	....	$u_{n,1}$
2	$u_{0,2}$	$u_{1,2}$	$u_{2,2}$	$u_{3,2}$	....	$u_{n,2}$
–	....	....	....	....	....	....
m	$u_{0,m}$	$u_{1,m}$	$u_{2,m}$	$u_{3,m}$	....	$u_{n,m}$

13.11 SHORT-CUT METHOD TO FIND $u(x, y)$ AT THE INTERIOR POINTS ON AND AFTER THIRD ROW WHEN $k = h/c$

Form a diamond-shaped figure to determine the $u(x, y)$ at the mesh point A in the following way:

(i) Join the mesh point A to the mesh point that is immediate top of its immediate following mesh point. Let this point be B.

(ii) Join the mesh point A to the mesh point that is immediate top of its immediate preceeding mesh point. Let this point be C.

(iii) Join B and C to a mesh point that lies in the preceeding row of lying B and C, and the column in which A lies. Let this point be D.

The value of $u(x, y)$ at the mesh point A = entry at B + entry at C − entry at D.

For convenience, we call this type of calculation as diamond-shaped computations.

Note that it is equivalent to the computations of $u\ (x,\ y)$ by the formula (13.50).

The first row entries of the table shown in Note of 13.10 are obtained from the boundary conditions and its second row entries at interior mesh points can be obtained from V-shaped calculations.

Now, we illustrate diamond-shaped computations. Suppose the two rows' entries are as follows:

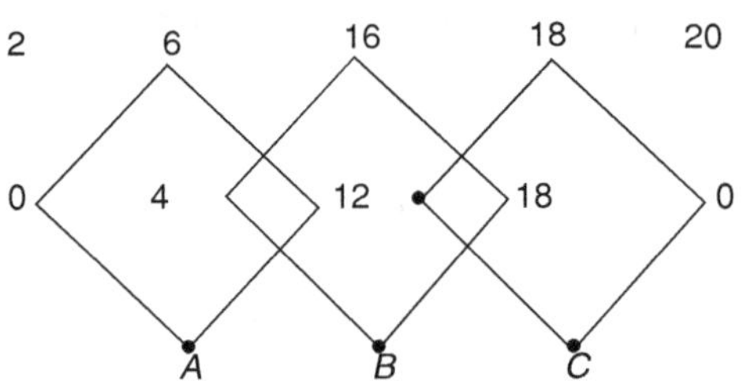

Fig 13.16

Suppose A, B and C are to be determined by diamond-shaped computations. Form the diamond-shaped figures as shown this Fig. 13.16. Now, the value of $u\ (x,\ y)$ at $A = 0 + 12 - 6$, at $B = 4 + 18 - 16$, and at $C = 12 + 0 - 18$. Note that the preceeding to A and succeeding to C are the boundary values. They are known values, repeating this process we can compute the entries of the other rows.

SOLVED PROBLEMS

Problem 1: Solve $u_{xx} = u_{tt}$ subject to the conditions (i) $u(0,\ t) = 0$ (ii) $u(4,\ t) = 0$ (iii) $u(x,\ 0) = \dfrac{x(4-x)}{2}$ and (iv) $u_t(x,0) = 0$ and take $h = 1$ and obtain the solution up to five steps in t-directions.

Solution: Comparing the given equation with $c^2 u_{xx} - u_{tt} = 0$, we have

$c = 1$ given that $h = 1$. We choose k such that $k = \dfrac{h}{c} = \dfrac{1}{1} = 1$.

From the boundary condition (i) i.e., $u(0,\ t) = 0$, we have

$$u_{0,0} = 0,\ u_{0,1} = 0,\ u_{0,2} = 0,\ u_{0,3} = 0,\ u_{0,4} = 0 \text{ and } u_{0,5} = 0.$$

Similarly, from the boundary condition (ii) we have,

$$u_{4,0} = 0 \,, u_{4,1} = 0 \,, u_{4,2} = 0 \,, u_{4,3} = 0 \,, u_{4,4} = 0 \text{ and } u_{4,5} = 0$$

Now from the boundary condition (iii) i.e., $u(x,0) = x\dfrac{(4-n)}{2}$, we have

$$u_{0,0} = 0,$$

$$u_{1,0} = 1\frac{(4-1)}{2} = 1.5$$

$$u_{2,0} = 2\frac{(4-2)}{2} = 2$$

$$u_{3,0} = 3\frac{(4-3)}{2} = 1.5$$

$$u_{4,0} = \frac{4(4-4)}{2} = 0$$

The boundary condition (iv) is equivalent to $u_{i,j} = \dfrac{1}{2}(u_{i-1,0} + u_{i+1,0})$

This can be used to find the second-row entries. The remaining row entries can be found from the relation.

$$u_{i,j+1} = u_{i+1,j} + u_{i-1,j} - u_{i,j-1}$$

$$j = 2, 3, 4, 5 \text{ and } i = 1, 2, 3$$

The second-row entries can be found using V-shaped calculations and the remaining rows by diamond-shaped computations.

i / j	0	1	2	3	4
0	0	1.5	2	1.5	0
1	0	1	1.5	1	0
2	0	0	0	0	0
3	0	−1	−1.5	−1	0
4	0	−1.5	−2	−1.5	0
5	0	−1	−1.5	−1	0

Entries of second row are obtained as follows:

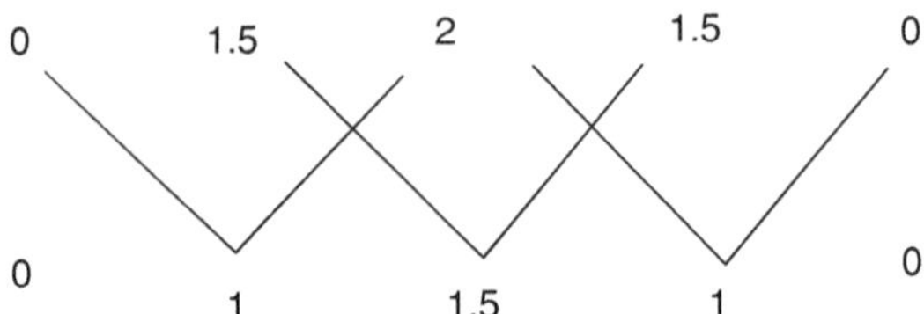

Entries of third row are obtained as follows using diamond-shaped computations:

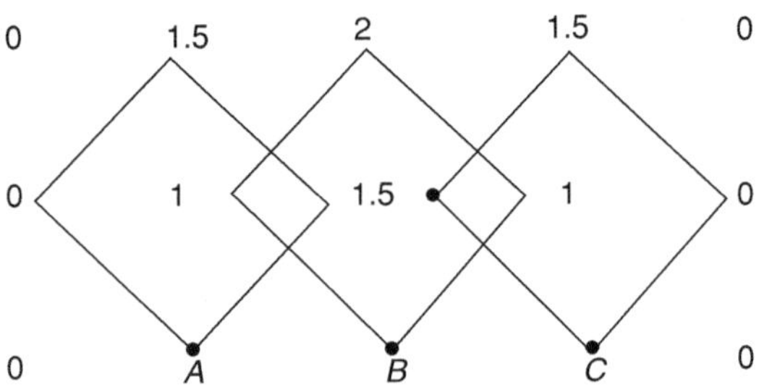

The value at mesh point $A = 0 + 1.5 - 1.5 = 0$
The value at mesh point $B = 1 + 1 - 1 = 0$
The value at mesh point $C = 1.5 + 0 - 1.5 = 0$
The entries of fourth are obtained as follows:

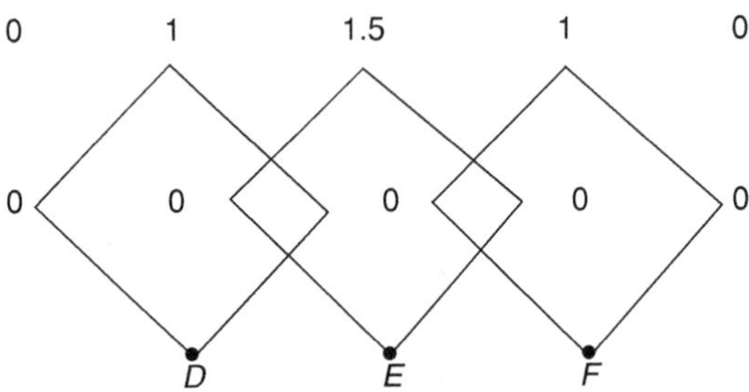

The value of $u(x, y)$ at $D = 0 + 0 - 1 = -1$
The value of $u(x, y)$ at $E = 0 + 0 - 1.5 = -1.5$
The value of $u(x, y)$ at $F = 0 + 0 - 1 = -1$
Repeating the above process, we can compute the entries of the other rows. These values are shown in the table.

Problem 2: Solve $16\,u_{xx} - u_{tt} = 0$ subject to the conditions (i) $u(0,t)=0$, (ii) $u(5, t) = 0$ (iii) $u(x, 0) = x^2(x - 5)$ (iv) $u_t(x,0) = 0$ with $h = 1$ obtain the solution in five steps in t-direction.

Solution: Here, $c = 4$ and $h = 1$ and we choose k such that

$$k = \frac{h}{c} = \frac{1}{4} = 0.25.$$

Thus, we have to find $u(x, t)$ at $x = 0, 1, 2, 3, 4, 5$ and $t = 0,$

$$\frac{1}{4}, \frac{2}{4}, \frac{3}{4}, \frac{4}{4}, \frac{5}{4}.$$

From the boundary condition $u(0, t) = 0$, we have

$$u_{0,0} = 0,\; u_{0,1} = 0,\; u_{0,2} = 0,\; u_{0,3} = 0,\; u_{0,4} = 0,\; u_{0,5} = 0$$

and from the boundary condition

$u(5, t) = 0$ gives $u(5, 0) = u(5, 1) = u(5, 2) = u(5, 3) = u(5, 4) = u(5, 5) = 0$

From the boundary condition (iii), we have

$$u_{0,0} = 0,\; u_{1,0} = -4,\; u_{2,0} = -12,\; u_{3,0} = -18,\; u_{4,0} = -16 \text{ and } u_{5,0} = 0.$$

The boundary condition (iv) is equivalent to

$$u_{i,1} = \frac{1}{2}(u_{i-1,0} + u_{i+1,0})$$

Using this relation, we can find the entries of second row. However, these can be obtained using V-shaped computations (short-cut method). The entries of other rows can be found using diamond-shaped computations (short-cut method) of course, these can be found from

$$u_{i,j+1} = u_{i+1,j} + u_{i-1,j} - u_{i,j-1}$$

The numerical solution of the given boundary value problem is shown below:

i / j	$0(x = 0)$	$1(x = 1)$	$2(x = 2)$	$3(x = 3)$	$4(x = 4)$	$5(x = 5)$
0 $t = 0$	0	−4	−12	−18	−16	0
1 $t = \dfrac{1}{4}$	0	−6	−11	−14	−9	0
2 $t = \dfrac{2}{4}$	0	−7	−8	−2	2	0

3 $t = \dfrac{3}{4}$	0	-2	2	8	7	0
4 $t = \dfrac{4}{4}$	0	9	14	11	6	0
5 $t = \dfrac{5}{4}$	0	16	18	12	4	0

The entries of second row are computed in the following way:

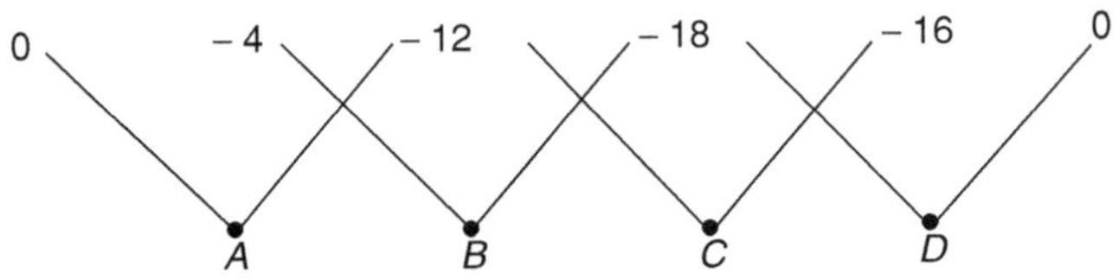

The value of $u\,(x,\ t)$ at $A = \dfrac{0 + (-12)}{2} = -6$

The value of $u\,(x,\ t)$ at $B = \dfrac{-4 - 18}{2} = -11$

The value of $u\,(x,\ t)$ at $C = \dfrac{-12 - 16}{2} = -14$

The value of $u\,(x,\ t)$ at $D = \dfrac{-18 - 0}{2} = -9$

The third-row entries are computed as follows:

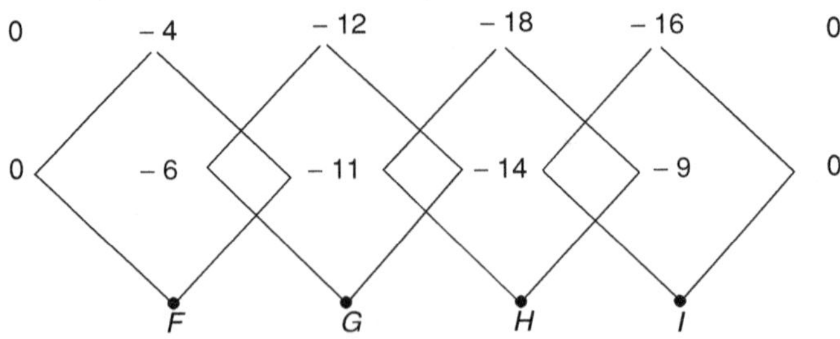

The value of u (x, t) at $F = 0 - 11 - (-4) = -7$
The value of u (x, t) at $G = -6 - 14 - (-12) = -8$
The value of u (x, t) at $H = -11 - 9 - (-18) = -2$
The value of u (x, t) at $I = -14 + 0 - (-16) = 2$
Repeating the above process, we can compute the entries of the other rows.

Problem 3: The Function u (x, t) satisfies the equation $u_{xx} - u_{tt} = 0$ and the conditions (i) u $(0, t) = 0$ (ii) u $(1,t) = 0$ (iii) u $(x, 0) = \sin^3 \pi x$ (iv) $u_t(x,0) = 0$; use the explicit scheme to calculate u (x, t) for $x = 0(0.25)$ and $t = (0.2)0.4$.

Solution: Comparing the given partial differential equation with $c^2 u_{xx} - u_{tt} = 0$, we have $c = 1$, given that $h = 0.25 = \dfrac{1}{4}$ and $k = 0.2$.

Hence, $\lambda = \dfrac{k}{h} = \dfrac{2/10}{1/4} = 4/5$.

We have to find $u_{i,j}$ $(i = 0, 1, 2, 3, 4$ and $j = 0, 1, 2)$, i. e., we have to find

u (x, t) at $n = 0$, $\dfrac{1}{4}, \dfrac{2}{4}, \dfrac{3}{4}, \dfrac{4}{4}$ and $t = 0, 0.2$ and $t = 0.4$.

Note that $\qquad\qquad u_{i,j} = u(ih, jk)$

From the condition (i), $u(0, t) = 0$, we have

$$u_{0,0} = 0, \; u_{0,1} = 0 \text{ and } u_{0,2} = 0$$

From the condition (ii), u $(1, t) = 0$, we have

$$u_{4,0} = 0, u_{4,1} = 0 \text{ and } u_{4,2} = 0 \text{ since } u \text{ } (1, t) = u \text{ } (4, t)$$

From the condition (iii), we have

$$u_{i,0} = \sin^3 ((\pi/x) = \sin(\pi/4) \; (i = 0, 1, 2, 3, 4);$$

Therefore,

$$u_{0,0} = 0$$

$$u_{1,0} = \left(\dfrac{1}{\sqrt{2}} \right) 3 = 0.3535$$

$$u_{2,0} = 1$$

$$u_{3,0} = 0.3535$$

The condition $u_{t(x,0)} = 0$ gives

$$\frac{u_{i,j+1} - u_{i,j-1}}{2k} = 0$$

$$u_{i,j+1} = u_{i,j-1} \tag{1}$$

In view of Eq. (1), Eq. 13.49 reduces to

$$u_{i,j+1} = (1 - \lambda^2 c^2)u_{i,j} + \frac{\lambda^2 c^2}{2}(u_{i+1,j} + u_{i-1,j}) \tag{2}$$

Substituting the values of $\lambda = 4/5$ and $c = 1$ in Eq. (2) we get

$$u_{i,j+1} = 0.36 u_{i,j} + 0.32(u_{i+1,j} + u_{i-1,j}) \tag{3}$$

Substituting $j = 0$ in Eq. (3), we have

$$u_{i,1} = 0.36\, u_{i,0} + 0.32(u_{i+1,0} + u_{i-1,0}) \tag{4}$$

Substituting $i = 1$ in Eq. (4), we get

$$u_{1,1} = 0.36\, u_{1,0} + 0.32(u_{2,0} + u_{0,0})$$
$$= 0.36(0.3535) + 0.32(1 + 0) = 0.44726.$$

Similarly, substituting $i = 2$ in Eq. (4), we have

$$u_{2,1} = 0.36 u_{2,0} + 0.32\,(u_{3,0} + u_{1,0})$$
$$= 0.36(1) + 0.32(0.3535 + 0.3535) = 0.58624.$$

Next, substituting $i = 3$ in Eq. (4) we have

$$u_{3,1} = 0.36 u_{3,0} + 0.32(u_{4,0} + u_{2,0})$$

$$= 0.36(0.3535) + 0.32(0 + 1) = 0.44726.$$

Thus, $u_{1,1} = 0.44726$, $u_{2,1} = 0.58624$ and $u_{3,1} = 0.44726$.

To find $u_{1,2}$, $u_{2,2}$ and $u_{3,2}$, substituting $S = 1$ is Eq. (3), we obtain

$$u_{i,2} = 0.36 u_{i,1} + 0.32(u_{i+1,1} + u_{i-1,1})$$

Putting $i = 1$ in Eq. (5), we obtain

$$u_{i,2} = 0.36 u_{i,1} + 0.32(u_{2,1} + u_{0,1})$$

$$= 0.36(0.44726) + 0.32(0.58624 + 0) = 0.3486104.$$

Putting $i = 2$ i Eq. (5), we get

$$u_{2,2} = 0.36 u_{2,1} + 0.32(u_{3,1} + u_{1,1})$$

$$= 0.36(0.58624) + 0.32(0.44726 + 0344726) = 0.4972928.$$

Finally, substituting $i = 3$ in Eq. (5), we have

$$u_{3,2} = 0.36u_{3,1} + 0.32(u_{4,1} + u_{2,1})$$

$$= 0.36(0.44726) + 0.32(0 + 0.58624) = 0.3486104.$$

The numerical solution of the given boundary value problem is shown in the following table.

i \ j	0 $(x=0)$	1 $\left(x=\dfrac{1}{4}\right)$	2 $\left(x=\dfrac{1}{4}\right)$	3 $\left(x=\dfrac{3}{4}\right)$	4 $\left(x=\dfrac{4}{4}\right)$
0 $(t=0)$	0	0.3535	0	0.3535	0
1 $(t=0.2)$	0	0.44726	0.58624	0.44726	0
2 $(t=0.4)$	0	0.3486104	0.4972928	0.3486104	0

EXERCISE 13

1. Classify the following differential equation

   (i) $u_{xx} + 3\,u_{xy} + 2\,u_{yy} = 0$

   (ii) $u_{xx} - 2\,u_{xy} + u_{yy} = 0$

   (iii) $u_{xx} + 4\,u_{yy} + 6\,u_{yy} - u_x + 6u_y = 0$

   (iv) $4\,u_{xx} + 2\,u_{xy} + 6\,u_{yy} + 7\,u_x - 8\,u_y + 6u = 0$

   (v) $u_{xx} - 2\,u_{xy} - 8\,u_{yy} = 0.$

2. Determine the region where the partial differential equation

   $(x+1)\,u_{xx} - 2(x+2)\,u_{xy} + (x+3)\,u_{yy} = 0$ is hyperbolic.

3. Determine the region where the partial differential equation $x\,u_{xx} + 4\,u_{yy} = 0\ x > 0,\ y > 0$ is elliptic.

4. Determine the regions where the partial differential equation

   $u_{xx} + 4\,u_{xy} + (x^2 + y^2)\,u_{yy} = \sin(x+y)$ is parabolic, hyperbolic and elliptic.

5. Obtain the finite-difference analogues of each of the following partial

differential equations:

(i) $u_{xx} - 5u_{xy} + 3u_{yy} = 0$

(ii) $u_{xx} + 4u_{yy} = 0$

(iii) $u_{xx} + 6u_{xy} + 9u_{yy} = 0$

(iv) $u_{xx} + 2u_{xy} + u_{yy} - u_x + u_y = 0$

(v) $u_{xx} + 4u_t = 0$

(vi) $u_{xx} - 16u_t = 0$

6. Solve $u_{xx} + u_{yy} = 0$ subject to the conditions $u\,(0, y) = 0$, $u\,(4, y) =$ $12 + y$ for $0 \le y \le 4$, $u\,(x, 0) = 3x$ and $u\,(x, 4) = x^2$ for $0 \le x \le 4$ take $h = 1$ and $k = 1$

7. Solve $u_{xx} + u_{yy} = 0$ for the following square mesh with boundary values as shown in the following figure:

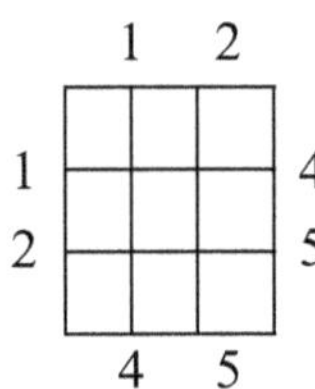

8. Solve $u_{xx} + u_{yy} = 0$ taking $h = 1$, $k = 1$ with boundary values shown in the following figure:

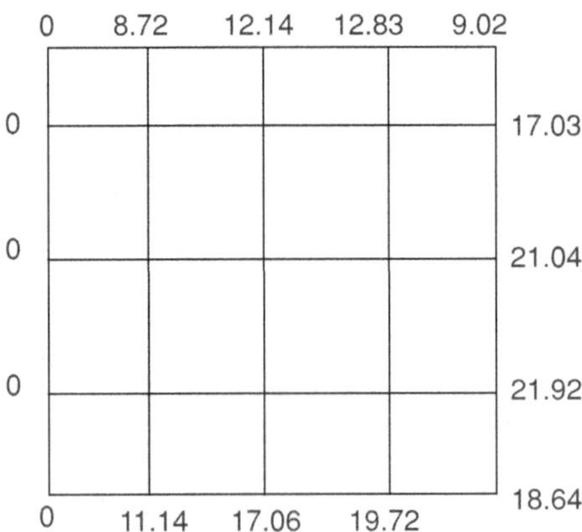

9. Solve $u_{xx} + u_{yy} = 0$ for the following square mesh with the boundary values as indicated in the following figure:

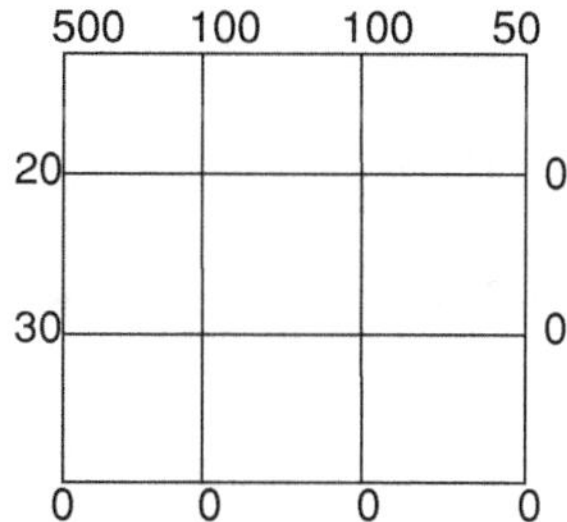

10. Solve the Poisson equation $u_{xx} + u_{yy} = 12x^2 y^2$ for the following square mesh with the boundary values as indicated at the mesh points lying on the boundary.

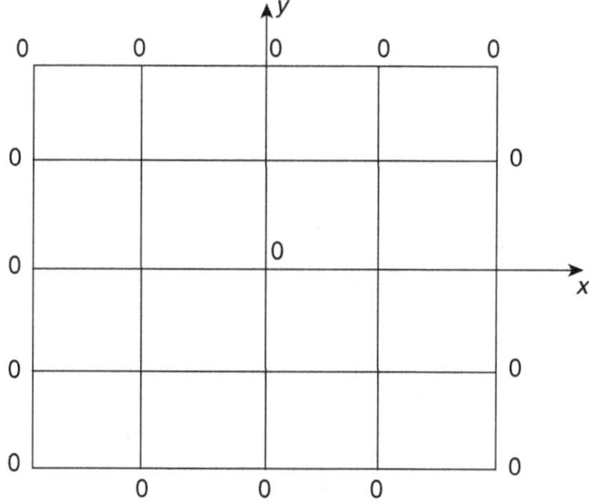

11. Solve the Poisson equation $u_{xx} + u_{yy} = 2(x^2 + y^2)$ in square region $[0, 4] \times [0, 4]$ given that $u(x, y) = 4$ on the boundaries of the region, taking $h = 1$ and $k = 1$.

12. Solve $u_{xx} = u_t$ subject to the conditions $u\,(0, t) = 0$, $u\,(5, t) = 0$ and $u(x, 0) = 25x^2 - x^4$, taking $h = 1$ and $k = \dfrac{1}{2}$.

13. Compute u for four time-steps with $h = \dfrac{1}{4}, k = \dfrac{1}{32}$ from $u_t = u_{xx}$ with the conditions $u\,(0, t) = u\,(1, t) = 0$ and $u\,(x, 0) = 100\,(x - x^2)$

14. Find the solution of $u_{xx} - 2u_t = 0 \quad 0 \le x \le 4$, given that $u\,(0, t) = 0$ $u(4, t) = 0$ and $u(x, 0) = x(4 - x)$ taking $h = 1$ using: Blender–Schmidt process $h = 1$.

15. Solve $u_{xx} - u_t = 0$ subject to the conditions $u(0, t) = 0$, $u(5, t) = 0$ and $u(x, 0) = x^2 (25 - x^2)$ with $h = 1$ and $k = \dfrac{1}{2}$ for four time steps.

16. Solve by Schmidth's method $u_{xx} - u_t = 0$ subject to the condition $u(0, t) = 0$, $u(1, t) = 0$, $u(x, 0) = \operatorname{Sin} \pi x$ take $h = \dfrac{1}{3}$, $k = \dfrac{1}{36}$. Carry out computations for two time-levels.

17. Solve $4 \dfrac{\partial^2 u}{\partial t^2} - \dfrac{\partial u}{\partial t} = 0$, $0 \le x \le 8$ subject to the boundary conditions $u(0, t) = 0$, $u(8, t) = 0$ and $u(x, 0) = 4x - \dfrac{x^2}{2}$ and carry out computations for time levels by taking $h = 1$ and $k = \dfrac{1}{8}$.

18. Solve $u_{xx} - u_{tt} = 0$ subject to the conditions u $(0, t) = 0$, $u(x, 0) = \dfrac{x(1-x)}{2}$ $(0 < x < 1)$ and $u_t(x, 0) = 0$, taking $h = 0.25$ and $k = 0.2$. Carry out computations for three time-levels.

19. Solve $4u_{xx} - u_{tt} = 0$ subject to the conditions u $(0, t) = 0$, $u(4, t) = 0$ $u_t(x, 0) = 0$ and $u(x, 0) = x (4 - x)$. Take $h = 1$ and carry out computations for five time-levels.

20. Solve $16u_{xx} - u_{tt} = 0$ subject to the conditions $u(0, t) = 0$, $u(5, t) = 0$, $u(x, 0) = x^2(5 - x)$ and $u_t(x, 0) = 0$. Take $h = 1$ and carry out computations for four steps in t-direction.

ANSWERS

1. (i) Hyperbolic (ii) Parabolic (iii) Elliptic (iv) Elliptic (v) Hyperbolic
2. $R \times R$, where R is the set of all real numbers
3. $\left\{ (x, y) \varepsilon R^+ \times R^+ / (y^2 - 4x < 0 \right\}$, where R^+ is set of all positive real numbers

4. $\left\{ (x, y) \varepsilon R \times R / \dfrac{x^2}{4} + \dfrac{y^2}{1} = 1 \right\}$, $\left\{ (x, y) \varepsilon R \times R / \dfrac{x^2}{4} + \dfrac{y^2}{1} < 1 \right\}$,

$\left\{ (x, y) \varepsilon R \times R / \dfrac{x^2}{4} + \dfrac{y^2}{1} > 1 \right\}$

5. (i) $\dfrac{u_{i-1,j} - 5u_{i,j} + u_{i+1,j}}{h^2} - 5\left[\dfrac{u_{i+1,j+1} - u_{i+1,j} - u_{i,j+1} + u_{i,j}}{hk}\right] + 3$

$\left[\dfrac{u_{i,j-1} - 2u_{i,j} + u_{i,j+1}}{k^2}\right] = 0$

(ii) $\left[\dfrac{u_{i-1,j} - 2u_{i,j} + u_{i+1,j}}{h^2}\right] + 4\left[\dfrac{u_{i,j-1} - 4u_{i,j} + u_{i,j+1}}{k^2}\right] = 0$

(iii) $\dfrac{u_{i-1,j} - 5u_{i,j} + u_{i+1,j}}{h^2} + 6\left[\dfrac{u_{i+1,j+1} - u_{i+1,j} - u_{i,j+1} + u_{i,j}}{hk}\right] + 9$

$\left[\dfrac{u_{i,j-1} - 2u_{i,j} + u_{i,j+1}}{k^2}\right] = 0$

(iv) $\dfrac{u_{i-1,j} - 5u_{i,j} + u_{i+1,j}}{h^2} + 2\left[\dfrac{u_{i+1,j+1} - u_{i+1,j} - u_{i,j+1} + u_{i,j}}{hk}\right] +$

$\left[\dfrac{u_{i,j-1} - 2u_{i,j} + u_{i,j+1}}{k^2}\right] - \dfrac{u_{i+1,j} - u_{i,j}}{h} + \dfrac{u_{i,j+1} - u_{i,j}}{k} = 0$

(v) $\left[\dfrac{u_{i-1,j} - 2u_{i,j} + u_{i+1,j}}{h^2}\right] + 4\left[\dfrac{u_{i,j-1} - 4u_{i,j} + u_{i,j+1}}{k^2}\right] = 0$

(vi) $\left[\dfrac{u_{i-1,j} - 2u_{i,j} + u_{i+1,j}}{h^2}\right] - 16\left[\dfrac{u_{i,j+1} - u_{i,j}}{k}\right] = 0$.

6. $u_1 = 2.37$, $u_2 = 5.59$, $u_3 = 9.87$, $u_4 = 2.88$, $u_5 = 6.13$, $u_6 = 9.88$, $u_7 = 3.01$, $u_8 = 6.16$, $u_9 = 9.51$

7. 3, 4, 2, 3

8. 6.658615, 11.288614, 14.369307, 6.628614, 11.988613, 16.329307, 7.869307, 13.709307, 17.919653

9. 37.5, 37.5 12.5, 12.5

10.

−4.5	−3	−4.5
−3	−3	−3
−4.5	−3	−4.5

11.

4	4	4	4	4	
4	2	2	2		4
4	2	2	2		4
4	2	2	2		4
4					4
	4	4	4	4	

12.

i \\ j	0 $x=0$	1 $x=1$	2 $x=2$	3 $x=3$	4 $x=4$	5 $x=5$
0 $t=0.0$	0	24	84	144	149	0
1 $t=0.5$	0	42	84	114	72	0
2 $t=1.0$	0	42	78	78	57	0
3 $t=1.5$	0	39	60	67.5	39	0
4 $t=2.0$	0	30	53.25	49.5	33.75	0
5 $t=2.5$	0	26.625	39.75	42.5	24.75	0
6 $t=3.0$	0	19.875	35.0625	32.25	21.75	0
7 $t=3.5$	0	17.531	26.0625	28.406	16.125	0
8 $t=4.0$	0	13.031	22.968	21.0938	14.2031	0
9 $t=4.5$	0	11.484	17.062	18.5859	10.5489	0
10 $t=5.0$	0	8.531	15.035	13.8047	9.2929	0

13.

	4.7	6.2	4.7		
0	6.2	9.4	6.2		0
0	9.4	12.5	9.4		0
0	12.5	18.75	12.5		0
	18.75	25	18.75		

14.

j \ i	0 $x=0$	1 $x=1$	2 $x=2$	3 $x=3$	4 $x=4$
0 $t=0$	0	0	0	0	0
1 $t=1$	0	2	3	2	0
2 $t=2$	0	1.5	2	1.5	0
3 $t=3$	0	1	1.5	1	0
4 $t=4$	0	0.75	1	0.75	0
5 $t=5$	0	0.5	.75	.5	0
6 $t=6$	0	.375	.5	.375	0
7 $t=7$	0	.25	.375	.25	0
8 $t=8$	0	.1875	.25	0.1875	0
9 $t=9$	0	.125	.1875	.125	0
10 $t=10$	0	0.094	0.125	0.094	0

15.

i / j	0 $x = 0$	1 $x = 1$	2 $x = 2$	3 $x = 3$	4 $x = 4$	5 $x = 5$
0 $t = 0$	0	24	84	144	144	0
1 $t = 0.5$	0	42	84	114	72	0
2 $t = 1$	0	42	78	78	57	0
3 $t = 1.5$	0	39	60	67.5	39	0
4 $t = 2.0$	0	30	53.25	49.5	33.75	0

16.

i / j	0 $x = 0$	1 $x = \dfrac{1}{3}$	2 $x = \dfrac{2}{3}$	3 $x = \dfrac{3}{3}$
0 $t = 0$	0	$\dfrac{\sqrt{3}}{2}$	$\dfrac{\sqrt{3}}{2}$	0
1 $t = \dfrac{1}{36}$	0	0.6495	0.6495	0
2 $t = \dfrac{2}{36}$	0	0.4871	0.4871	0

17.

j	0 $x = 0$	1 $x = 1$	2 $x = 2$	3 $x = 3$	4 $x = 4$	5 $x = 5$	6 $x = 6$	7 $x = 7$	8 $x = 8$
0 $t = 0$	0	3.5	6	7.5	8	7.5	6	3.5	0

		0	1	2	3	4	5	6	7	8
1 $t = 0.5$		0	3	5.5	7	7.5	7	5.5	3	0
2 $t = 1$		0	2.75	5	6.5	7	6.5	5	2.75	0
3 $t = 1.5$		0	2.5	4.625	6	6.5	6	4.625	2.5	0
4 $t = 2.0$		0	2.3125	4.25	5.6635	6	5.5625	4.25	2.3125	0
5 $t = 2.5$		0	2.125	3.9375	5.125	5.5625	5.125	3.9375	2.125	0

18.

j \ i	0 $x = 0$	1 $x = 1$	2 $x = 2$	3 $x = 3$	4 $x = 4$
0 $t = 0$	0	0.09375	0.125	0.09375	0
1 $t = \dfrac{1}{4}$	0	0.0625	0.09375	0.0625	0
2 $t = \dfrac{2}{4}$	0	0	0	0	0
3 $t = \dfrac{3}{4}$	0	−0.0625	−0.09375	−0.0625	0

19.

j \ i	0 $x = 0$	1 $x = 1$	2 $x = 2$	3 $x = 3$	4 $x = 4$
0 $t = 0$	0	−3	4	3	0

1 $t = 0.5$	0	2	3	2	0
2 $t = 1$	0	0	0	0	0
3 $t = 1.5$	0	−2	−3	−2	0
4 $t = 2.0$	0	−3	−4	−3	0
5 $t = 2.5$	0	−2	−3	−2	0

20.

i j	0 $x = 0$	1 $x = 1$	2 $x = 2$	3 $x = 3$	4 $x = 4$	5 $x = 5$
0 $t = 0$	0	−4	−12	−18	−16	0
1 $t = 0.5$	0	−6	−11	−14	−9	0
2 $t = 1$	0	−7	−8	−2	2	0
3 $t = 1.5$	0	−2	2	8	7	0
4 $t = 2.0$	0	9	14	11	6	0
5 $t = 2.5$	0	16	18	12	4	0

Bibliography

1. Curtis F. Gerald, Patrick O. Wheatley, Applied Numerical Analysis, Pearson Education, (Singapore) Private Limited, Delhi, India 2002.

2. Carl-Erick Froberg, Introduction to Numerical Analysis, Addison-Wesley, Publishing Company, INC., 1981.

3. Ralph G. Stanton, Numerical Methods for Science and Engineering, Prentice-Hall, Englewood Cliffs, New Jersey 1961.

4. James B. Scarborough, Numerical Mathematical Analysis, Oxford and IBH Publishing Company Private Limited, New Delhi.

5. Jain M.K., Iyengar S.R.K., Jain R.K., Numerical Methods for Scientific and Engineering Computations, New Age International Private Limited, Delhi, India 2003.

6. Francis Scheid, Numerical Analysis, Tata McGraw-Hill Publishing Company Limited, New Delhi, India 2004.

7. Sastry S.S., Introductory Methods of Numerical Analysis, PHI Learning Private Limited, New Delhi, India 2009.

8. Sankara Rao K., Numerical Methods for Scientists and Engineers, PHI Learning Private Limited, New Delhi, India 2009.

9. Saxena H.C., Finite Differences and Numerical Analysis, S. Chand and Company Limited, New Delhi, India 2000.

10. Arumugam S., Thangapandi Issac A., Somasundaram A., Numerical Methods, Scitech Publications (India) Private Limited, Chennai, India 2009.